Applied Fluid Mechanics

Merle C. Potter • David C. Wiggert

Applied Fluid Mechanics

Sixth Edition

Springer

Merle C. Potter
Department of Mechanical Engineering
Michigan State University
East Lansing, MI, USA

David C. Wiggert
Department of Civil Engineering
Michigan State University
East Lansing, MI, USA

ISBN 978-3-031-65275-2 ISBN 978-3-031-65276-9 (eBook)
https://doi.org/10.1007/978-3-031-65276-9

This Springer imprint is published by the registered company Springer Nature Switzerland AG
The registered company address is: Gewerbestrasse 11, 6330 Cham, Switzerland

Contents

CHAPTER 9 FLOWS IN PIPING SYSTEMS 480

CHAPTER 10 TURBOMACHINERY 530

Preface

This is a fluids textbook written for a second course in Fluid Mechanics in hopes of presenting an improved view of fluid mechanics so that the student can better understand the physical concepts and follow the mathematics. This is not an easy task: Fluid mechanics is a subject that contains many difficult-to-understand phenomena. For example, how would you explain the hole scooped out in the sand by river water on the upstream side of an abutment or the snow scooped out by the wind in front of a tree? Or the unexpected strong wind around the corner of a tall building in Chicago? Or the vibration and subsequent collapse of a large cable-suspension bridge due to the wind? After this second course the student should better understand and be able to analyze many of the important phenomena encountered by the engineer.

Here are several additional subtleties that are often either not included in most fluids texts or are presented incompletely:

1. **Laminar and turbulent flows.** The definitions of laminar and turbulent flows must be done carefully. The measurement of a laminar flow of honey next to an erratically moving piston would appear to be turbulent.
2. **Entrance flow.** Both laminar and turbulent entrance flows are invariably presented incorrectly. To indicate that a parabolic profile exists when the viscous wall layers merge in a laminar pipe flow is simply incorrect.
3. **The wake and separated regions.** A separated region is contained inside the wake (see Figure 6.3), a fact that most textbooks do not make clear.
5. **Boundary layers.** Texts typically do not display an actual boundary layer. It is import-tant to realize that boundary layers are extremely thin. See Figure 6.24.
6. **Inviscid (potential) flow.** The role that an inviscid flow plays in an external flow, accounting for viscous effects, is often not clear. See Figure 6.21.
7. **Flow through a contraction.** The flow would surprisingly resemble that presented in Figure 1.19.
8. **Losses.** Why are losses through elbows and bends so high even though the area does not change?

A first course in fluid mechanics may not have presented fluid flows as fields. That is, they have presented primarily those flows that can be approximated as one-dimensional flows and have treated other flows using experimental data. We must recognize that when a fluid flows around an object, such as a building or an abutment, its velocity possesses all three components which depend on all three space variables and often, time. If we present the equations that describe such a general flow, the equations are referred to as field equations, and velocity and pressure fields become of interest. This is quite analogous to electrical and magnetic fields in electrical engineering. For the difficult problems of the future, such as large-scale environmental pollution, to be analyzed by engineers, it is imperative that we understand fluid fields. Thus, in Chapter 5 we introduce the field equations and discuss several solutions for some relatively simple geometries. The more conventional manner of treating the flows individually may be of particular interest in cuvil engineering so there is sufficient time to include Chapters 8, 9, and 10.

Sections of the material included in Chapters 1 through 8 can be used to present a 2nd course in fluid mechanics for mechanical engineering students, and sections in Chapters 1 through 6 and 8 through 10 can be used for civil engineering students. The instructor can fit the material to a selected course outline.

Answers to selected home problems are presented just prior to the Index. We have also included design-type problems in several of the chapters. Of course, there are numerous classes of problems that are extremely difficult to solve; even for an experienced engineer to solve these more difficult problems, the engineer must often gain more information from other sources than is included in this text. There are, however, many problems that can be solved successfully using the material and concepts presented herein.

We have emphasized SI units; however, all properties and dimensional constants are given in English units as well. Approximately one-fifth of the examples and problems are presented using English units, but, there are sufficient examples and problems to use only SI units.

The authors are very much indebted to both their former professors, to their present colleagues, and to comments from students.

Merle C. Potter and David C. Wiggert

Nomenclature

For quick reference

A - area

A_2, A_3 - profile type

a - acceleration, speed of a pressure wave

a - acceleration vector

a_x, a_y, a_z - acceleration components

B - bulk modulus of elasticity, free surface width

b - channel bottom width

C - centroid, Chezy coefficient

C_1, C_3 - profile type

C_D - drag coefficient

C_d - discharge coefficient

C_f - skin-friction coefficient

C_H - head coefficient

C_L - lift coefficient

C_p - pressure recovery factor, pressure coefficient

C_{NPSH} - net positive suction head coefficient

C_Q - flow-rate coefficient

C_{thrust} - thrust coefficient

C_{torque} - torque coefficient

C_V - velocity coefficient

C_W - power coefficient

c - specific heat, speed of sound, chord length, celerity

c_f - local skin-friction coefficient

c_p - constant pressure specific heat

c_v - constant volume specific heat

c.s. - control surface

c.v. - control volume

D - diameter

$\frac{D}{Dt}$ - substantial derivative

d - diameter

dx - differential distance

$d\theta$ - differential angle

E - energy, specific energy, coefficient

E_c - critical energy

EGL - energy grade line

Eu - Euler number

e - the exponential, specific energy, wall roughness height, pipe wall thickness

exp - the exponential e

$\mathbf{F}$ - force vector

F - force

F_B - buoyant force

F_D - drag force

F_L - lift force

F_H - horizontal force component

F_V - vertical force component

F_W - body force equal to the weight

f - friction factor, frequency

G - center of gravity

$\overline{GM}$ - metacentric height

$\mathbf{g}$ - gravity vector

g - gravity

H - enthalpy, height, total energy

H_2, H_3 - profile type

H_D - design head

H_P - pump head

H_T - turbine head

HGL - hydraulic grade line

h - distance, height, specific enthalpy

h_j - head loss across a hydraulic jump

I - second moment of an area

$\overline{I}$ - second moment about the centroidal axis

I_{xy} - product of inertia

$\hat{\mathbf{i}}$ - unit vector in the x-direction

$\hat{\mathbf{j}}$ - unit vector in the y-direction

$\hat{\mathbf{k}}$ - unit vector in the z-direction

K - thermal conductivity, flow coefficient

K_c - contraction coefficient

K_e - expansion coefficient

K_{uv} - correlation coefficient

k - ratio of specific heats

L- length

L_E - entrance length

L_e - equivalent length

ℓ - length

ℓ_m - mixing length

M - molar mass, momentum function

M - Mach number

M_1, M_2, M_3, - profile type

m - mass, side-wall slope

$\dot{m}$ - mass flux

$\dot{m}_r$ - relative mass flux

m_a - added mass

m_1, m_2, - side-wall slopes

$\dot{mom}$ - momentum flux

N - general extensive property, number of jets

$NPSH$ - net positive suction head

n - normal direction, number of moles, Manning number

$\hat{\mathbf{n}}$ - unit normal vector

P - power, force, wetted perimeter

p - pressure

Q - flow rate (discharge), heat transfer

Q_D - design discharge

$\dot{Q}$ - rate of heat transfer

q - source strength, specific discharge, heat flux

R - radius, gas constant, hydraulic radius, radius of curvature

Re - Reynolds number

Re_{crit} - critical Reynolds number

R_u - universal gas constant

R_x, R_y - force components

r - radius, coordinate variable

$\mathbf{r}$ - position vector

S - specific gravity, entropy, distance, slope of channel, slope of EGL

S_1, S_2, S_3; - profile type

S_c - critical slope

St - Strouhal number

$\mathbf{S}$ - position vector

S_0 - slope of channel bottom

s - specific entropy, streamline coordinate

$\hat{\mathbf{s}}$ - unit vector tangent to streamline

sys - system

T - temperature, torque, tension

t - time, tangential direction

U - average velocity

U_{∞} - free-stream velocity away from a body

u - x-component velocity, circumferential blade speed

u' - velocity perturbation

$\tilde{u}$ - Specific internal energy

$\overline{u}$ - time average velocity

u_{τ} - shear velocity

V - velocity

V_c - critical velocity

V_{ss} - steady-state velocity

$\mathbf{V}$ - velocity vector

$\overline{V}$ - spatial average velocity

$\forall$ - volume

V_B - blade velocity

V_n - normal component of velocity

V_r - relative speed

V_t - tangential velocity

v - velocity, y-component velocity

v' - velocity perturbation

v_r, v_z, v_{θ}, v_{ϕ} - velocity components

W - work, weight, change in hydraulic grade line

$\dot{W}$ - work rate (power)

$\dot{W}_f$ - actual power

We - Weber number

$\dot{W}_s$ - shaft work (power)

w - width

w - z-component velocity, velocity of a hydraulic bore

X_T - distance where transition begins

x - coordinate variable

x_m - origin of moving reference frame

$\tilde{x}$ - distance relative to a moving reference frame

$\overline{x}$ - x-coordinate of centroid

Y - upstream water height above top of weir

y - coordinate variable, flow energy head

y_p - distance to center of pressure

$\overline{y}$ - y-coordinate of centroid

y_c - critical depth

z - coordinate variable

α - angle, angle of attack, lapse rate, thermal diffusivity, kinetic-energy correction factor, blade angle

β - angle, momentum correction factor, fixed jet angle, blade angle

Δ - a small increment

∇ - gradient operator

∇^2 - Laplacian

δ - boundary layer thickness

$\delta(x)$ - Dirac-delta function

δ_d - displacement thickness

δ_v - viscous wall layer thickness

ε - a small volume

$\boldsymbol{\varepsilon}_{xx}$, $\boldsymbol{\varepsilon}_{xy}$, $\boldsymbol{\varepsilon}_{xx}$ - three rate-of-strain components

ϕ - angle, coordinate variable, velocity potential function, speed factor

Γ - circulation, vortex strength

λ - mean free path, a constant, wave length

γ - specific weight

η - a general intensive property, eddy viscosity, efficiency, a position variable

η_P - pump efficiency

η_T - turbine efficiency

μ - viscosity, doublet strength

υ - kinematic viscosity

π - a pi term

θ - angle, momentum thickness

ρ - density

Ω - angular velocity

Ω_P - specific speed of a pump

Ω_T - specific speed of a turbine

$\boldsymbol{\Omega}$ - angular velocity vector

ψ - stream function

σ - surface tension, cavitation number, circumferential stress

σ_{xx}, σ_{yy}, σ_{zz} - normal stress components

τ - stress vector

$\overline{\tau}$ - time average stress

τ_{xy}, τ_{xz}, τ_{yz}, - shear stress components

ω - angular velocity, vorticity

$\boldsymbol{\omega}$ - vorticity vector

$\frac{\partial}{\partial x}$ - partial derivative

1 Introduction to Fluids in Motion

Outline

Chapter Objectives

The objectives of this chapter are to:

- Mathematically describe the motion of a fluid.
- Express the acceleration and vorticity of a fluid particle given the velocity components.
- Describe the deformation of a fluid particle.
- Classify various fluid flows. Is a flow viscous, is it turbulent, is it incompressible, or is it a uniform flow?
- Derive the Bernoulli equation and identify its restrictions.
- Present several examples and numerous problems that demonstrate how fluid flows are described, how flows are classified, and how Bernoulli's equation is used to estimate flow variables.

1.1 INTRODUCTION

This chapter serves as an introduction to all the following chapters. Fluid motions manifest themselves in many different ways. Some can be described very easily, while others require a thorough understanding of physical laws. In engineering applications, it is important to describe the fluid motions as simply as can be justified. This usually depends on the required accuracy. Often, accuracies of ±10% are acceptable, although in some applications higher accuracies have to be achieved. The general equations of motion are very difficult to solve; consequently, it is the engineer's responsibility to know which simplifying assumptions can be made. This, of course, requires experience and, more importantly, an understanding of the physics involved.

M. C. Potter and D. C. Wiggert, *Applied Fluid Mechanics*,
https://doi.org/10.1007/978-3-031-65276-9_1

Some common assumptions used to simplify a flow situation are related to fluid properties. For example, under certain conditions, the viscosity can affect the flow significantly; in others, viscous effects can be neglected, greatly simplifying the equations without significantly altering the predictions. It is well known that the compressibility of a gas in motion should be taken into account if the velocities are very high. But compressibility effects do not have to be taken into account to predict wind forces on buildings or to predict any other physical quantity that is a direct effect of wind. Wind speeds are simply not high enough, even in a hurricane. Numerous examples could be cited. After our study of fluid motions, the appropriate assumptions used should become more obvious.

This chapter has three sections. In the first section we introduce the reader to some important general approaches used to analyze fluid mechanics problems. In the second section we give a brief overview of different types of flow, such as compressible and incompressible flows, and viscous and inviscid flows. Detailed discussions of each of these flow types follow in later chapters. The third section introduces the reader to the commonly used Bernoulli equation, an equation that establishes how pressures and velocities vary in a flow field. The use of this equation, however, requires many simplifying assumptions, and its application is, therefore, limited.

1.2 DESCRIPTION OF FLUID MOTION

The analysis of complex fluid flow problems is often aided by the visualization of flow patterns, which permit the development of a better intuitive understanding and help in formulating the mathematical problem. The flow in a washing machine is a good example. An easier, yet difficult, problem is the flow in the vicinity of where a wing attaches to a fuselage, or where a bridge support interacts with the water at the bottom of a river. In Section 1.2.1 we discuss the description of physical quantities as a function of space and time coordinates. The second topic in this section introduces the different flow lines that are useful in our objective of describing a fluid flow. Finally, the mathematical description of motion is presented.

1.2.1 Lagrangian and Eulerian Descriptions of Motion

In the description of a flow field, it is convenient to think of individual particles, each of which is considered to be a small mass of fluid consisting of a large number of molecules, that occupies a small volume $\Delta V\!\!\!\!-$ that moves with the flow. If the fluid is incompressible, the volume does not change in magnitude but may deform. If the fluid is compressible, as the volume deforms, it also changes its magnitude. In both cases the particles are considered to move through a flow field as an entity.

In the study of particle mechanics, where attention is focused on individual particles, motion is observed as a function of time. The position, velocity, and acceleration of each particle are listed as $\mathbf{s}(x_0, y_0, z_0, t)$, $\mathbf{V}(x_0, y_0, z_0, t)$, and $\mathbf{a}(x_0, y_0, z_0, t)$, and quantities of interest can be calculated. (A bold letter represents a vector.) The point (x_0, y_0, z_0) locates the starting point—the name—of each particle. This is the **Lagrangian description**, named after Joseph L. Lagrange (1736–1813), of motion that is used in a course on dynamics. In the Lagrangian description many particles can be followed and their influence on one another noted. This becomes, however, a difficult task as the number of particles becomes extremely large in even the simplest fluid flow.

An alternative to following each fluid particle separately is to identify points in space and then observe the velocity of particles passing each point; we can observe the rate of change of velocity as the particles pass each point, that is, $\partial \mathbf{V}/\partial x$, $\partial \mathbf{V}/\partial y$, and $\partial \mathbf{V}/\partial z$, and we can observe if the velocity is changing with time at each particular point, that is, $\partial \mathbf{V}/\partial t$. In this **Eulerian description** of motion, named after Leonhard Euler (1707–1783), the flow properties, such as velocity, are functions of both space and time. In Cartesian coordinates the velocity is expressed as $\mathbf{V} = \mathbf{V}(x, y, z, t)$. The region of flow that is being considered is called a **flow field**.

An example may clarify these two ways of describing motion. An engineering firm is hired to make recommendations that would improve the traffic flow in a large city. The engineering firm has two alternatives: Hire college students to travel in automobiles throughout the city recording appropriate observations (the Lagrangian approach), or hire college students to stand at the intersections and record the required information (the Eulerian approach). A correct interpretation of each set of data would lead to the same set of recommendations, that is, the same solution. In this example it may not be obvious which approach would be preferred; in an introductory course in fluids, however, the Eulerian description is used exclusively since the physical laws using the Eulerian description are easier to apply to actual situations. Yet, there are examples where a Lagrangian description is needed, such as drifting buoys used to study ocean currents.

If the quantities of interest do not depend on time, that is, $\mathbf{V} = \mathbf{V}(x, y, z)$, the flow is said to be a steady flow. Most of the flows of interest are steady flows. For a steady flow, all flow quantities at a particular point are independent of time, that is,

$$\frac{\partial \mathbf{V}}{\partial t} = 0 \qquad \frac{\partial p}{\partial t} = 0 \qquad \frac{\partial \rho}{\partial t} = 0 \tag{1.2.1}$$

to list a few. It is implied by the partial derivative that x, y, and z are held fixed in the above. Note that the properties of a fluid particle do, in general, vary with time; the velocity and pressure vary with time as a particular fluid particle progresses along its path in a flow, even in a steady flow. In a steady flow, however, properties do not vary with time at a fixed point.

1.2.2 Pathlines, Streaklines, and Streamlines

Three different lines help us describe a flow field. A **pathline** is the locus of points traversed by a given particle as it travels in a field of flow; the pathline provides us with a "history" of the particle's locations. A photograph of a pathline would require a time exposure of an illuminated particle. A photograph showing pathlines of particles below a water surface with waves is given in Figure 1.1.

A **streakline** is defined as an instantaneous line whose points are occupied by all particles originating from some specified point in the flow field. Streaklines tell us where the particles are "right now."A photograph of a streakline would be a snapshot of the set of illuminated

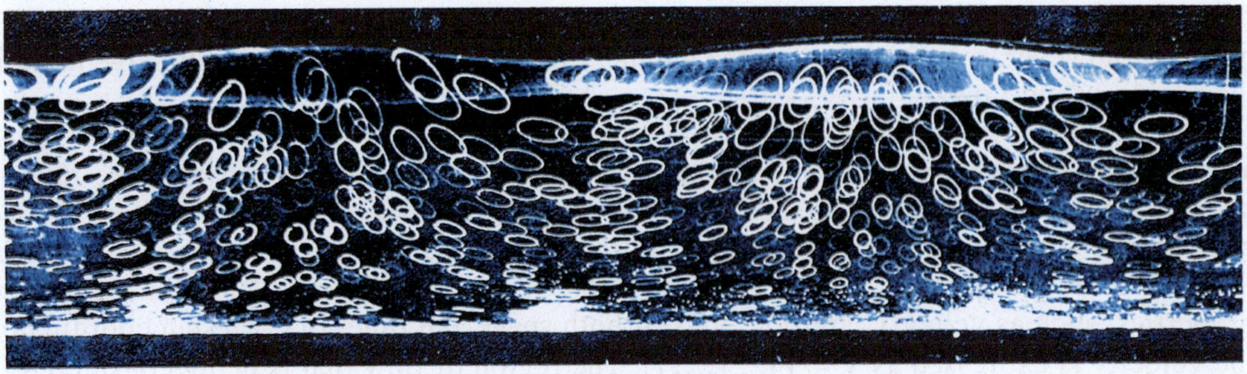

Figure 1.1 Pathlines underneath a wave in a tank of water.

particles that passed a certain point. Figure 1.2 shows streaklines produced by the continuous release of a small-diameter stream of smoke as it moves around a cylinder.

A **streamline** is a line in the flow possessing the following property: the velocity vector of each particle occupying a point on the streamline is tangent to the streamline. This is shown graphically in Figure 1.3. An equation that expresses that the velocity vector is tangent to a streamline is

$$\mathbf{V} \times d\mathbf{r} = 0 \tag{1.2.2}$$

since $\mathbf{V}$ and $d\mathbf{r}$ are in the same direction, as shown in the figure; recall that the cross product of two vectors in the same direction is zero. This equation will be used in future chapters as the mathematical expression of a streamline. A photograph of a streamline cannot be made directly. For a general unsteady flow the streamlines can be inferred from photographs of short pathlines of a large number of particles.

A **streamtube** is a tube whose walls are streamlines. Since the velocity is tangent to a streamline, no fluid can cross the walls of a streamtube. The streamtube is of particular interest in fluid mechanics. A pipe is a streamtube since its walls are streamlines; an open channel is a streamtube since no fluid crosses the walls of the channel. We often sketch a streamtube with a small cross section in the interior of a flow for demonstration purposes.

Figure 1.2 Streaklines in the unsteady flow around a cylinder.

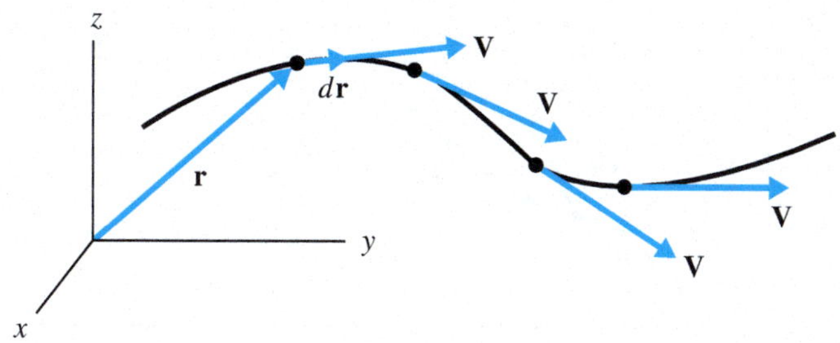

Figure 1.3 Streamline in a flow field.

In a steady flow, pathlines, streaklines, and streamlines are all coincident. All particles passing a given point will continue to trace out the same path since the velocity in our Eulerian system does not change with time; hence the pathlines and streaklines coincide. In addition, the velocity vector of a particle at a given point will be tangent to the line that the particle is moving along; thus the line is also a streamline. Since the flows that we observe in laboratories are invariably steady flows, we call the lines that we observe streamlines even though they may actually be streaklines, or for the case of time exposures, pathlines.

1.2.3 Acceleration

The acceleration of a fluid particle is found by considering a particular particle shown in Figure 1.4. Its velocity changes from $\mathbf{V}(t)$ at time t to $\mathbf{V}(t + dt)$ at time $t + dt$. The acceleration is, by definition,

$$\mathrm{a} = \frac{d\mathbf{V}}{dt} \tag{1.2.3}$$

where $d\mathbf{V}$ is shown in Figure 1.4. The velocity vector $\mathbf{V}$ is given in component form as

$$\mathbf{V} = u\hat{\mathbf{i}} + v\hat{\mathbf{j}} + w\hat{\mathbf{K}} \tag{1.2.4}$$

where (u, v, w) are the velocity components in the x-, y-, and z-directions, respectively, and $\hat{\mathbf{i}}$, $\hat{\mathbf{j}}$ and $\hat{\mathbf{k}}$ are unit vectors. The quantity $d\mathbf{V}$ is, using the chain rule from calculus with $\mathbf{V} = \mathbf{V}(x, y, z, t)$,

$$d\mathbf{V} = \frac{\partial \mathbf{V}}{\partial x}dx + \frac{\partial \mathbf{V}}{\partial y}dy + \frac{\partial \mathbf{V}}{\partial z}dz + \frac{\partial \mathbf{V}}{\partial t}dt \tag{1.2.5}$$

This gives the acceleration using Eq. 1.2.3 as

$$\mathbf{a} = \frac{\partial \mathbf{V}}{\partial x}\frac{dx}{dt} + \frac{\partial \mathbf{V}}{\partial y}\frac{dy}{dt} + \frac{\partial \mathbf{V}}{\partial z}\frac{dz}{dt} + \frac{\partial \mathbf{V}}{\partial t} \tag{1.2.6}$$

Since we have followed a particular particle, as in Figure 1.4, we recognize that

$$\frac{dx}{dt} = u \qquad \frac{dy}{dt} = v \qquad \frac{dz}{dt} = w \tag{1.2.7}$$

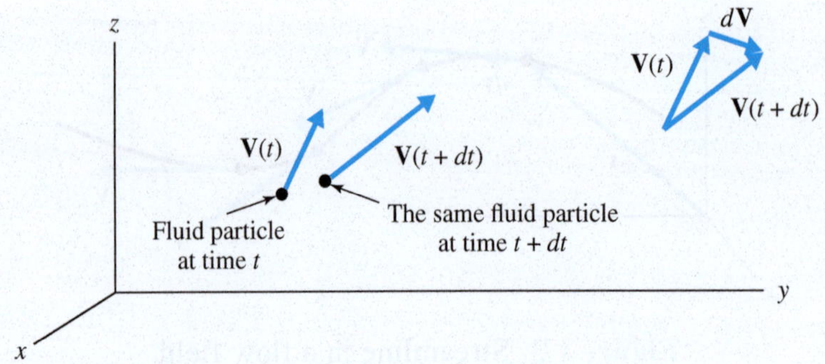

Figure 1.4 Velocity of a fluid particle.

The acceleration is then expressed as

$$\mathbf{a} = u\frac{\partial \mathbf{V}}{\partial x} + v\frac{\partial \mathbf{V}}{\partial y} + w\frac{\partial \mathbf{V}}{\partial z} + \frac{\partial \mathbf{V}}{\partial t} \tag{1.2.8}$$

The scalar component equations of the above vector equation using rectangular[1] (also called Cartesian) coordinates are written as

$$\begin{aligned} a_x &= \frac{\partial u}{\partial t} + u\frac{\partial u}{\partial x} + v\frac{\partial u}{\partial y} + w\frac{\partial u}{\partial z} \\ a_y &= \frac{\partial v}{\partial t} + u\frac{\partial v}{\partial x} + v\frac{\partial v}{\partial y} + w\frac{\partial v}{\partial z} \\ a_z &= \frac{\partial w}{\partial t} + u\frac{\partial w}{\partial x} + v\frac{\partial w}{\partial y} + w\frac{\partial w}{\partial z} \end{aligned} \tag{1.2.9}$$

We often return to Eq. 1.2.3 and write Eq. 1.2.8 in a simplified form as

$$\mathbf{a} = \frac{D\mathbf{V}}{Dt} \tag{1.2.10}$$

where, in rectangular coordinates,

$$\frac{D}{Dt} = u\frac{\partial}{\partial x} + v\frac{\partial}{\partial y} + w\frac{\partial}{\partial z} + \frac{\partial}{\partial t} \tag{1.2.11}$$

This derivative is called the **substantial derivative**, or **material derivative**. It is given a special name and special symbol (D/Dt instead of d/dt) because we followed a particular fluid particle, that is, we followed the substance (or material). It represents the relationship between a Lagrangian derivative in which a quantity depends on time t and an Eulerian derivative in which a quantity depends on position (x, y, z) and time t. The substantial derivative can be used with other dependent variables; for example, DT/Dt would represent the rate of change of the temperature of a fluid particle as we followed the particle along.

The substantial derivative and acceleration components in cylindrical and spherical coordinates are presented in Table 1.1 and in Appendix A.4.

---

[1]Rectangular coordinates are often referred to as Cartesian coordinates. Cartesian coordinates have the same unit length in each coordinate direction, an unnecessary condition in rectangular coordinates.

Table 1.1 The Substantial Derivative, Acceleration, and Vorticity in Rectangular, Cylindrical, and Spherical Coordinates

Substantial Derivative

Rectangular

$$\frac{D}{Dt} = u\frac{\partial}{\partial x} + v\frac{\partial}{\partial y} + w\frac{\partial}{\partial z} + \frac{\partial}{\partial t}$$

Cylindrical

$$\frac{D}{Dt} = v_r\frac{\partial}{\partial r} + \frac{v_\theta}{r}\frac{\partial}{\partial \theta} + v_z\frac{\partial}{\partial z} + \frac{\partial}{\partial t}$$

Spherical

$$\frac{D}{Dt} = v_r\frac{\partial}{\partial r} + \frac{v_\theta}{r}\frac{\partial}{\partial \theta} + \frac{v_\phi}{t\sin\theta}\frac{\partial}{\partial \phi} + \frac{\partial}{\partial t}$$

Acceleration

Rectangular

$$a_x = \frac{\partial u}{\partial t} + u\frac{\partial u}{\partial x} + v\frac{\partial u}{\partial y} + w\frac{\partial u}{\partial z}$$

$$a_y = \frac{\partial v}{\partial t} + u\frac{\partial v}{\partial x} + v\frac{\partial v}{\partial y} + w\frac{\partial v}{\partial z}$$

$$a_z = \frac{\partial w}{\partial t} + u\frac{\partial w}{\partial x} + v\frac{\partial w}{\partial y} + w\frac{\partial w}{\partial z}$$

Cylindrical

$$a_r = \frac{\partial v_r}{\partial t} + v_r\frac{\partial v_r}{\partial r} + \frac{v_\theta}{r}\frac{\partial v_r}{\partial \theta} + v_z\frac{\partial v_r}{\partial z} - \frac{v_\theta^2}{r}$$

$$a_\theta = \frac{\partial v_\theta}{\partial t} + v_r\frac{\partial v_\theta}{\partial r} + \frac{v_\theta}{r}\frac{\partial v_\theta}{\partial z} + v_z\frac{\partial v_\theta}{\partial z} + \frac{v_r v_\theta}{r}$$

$$a_z = \frac{\partial v_z}{\partial t} + v_r\frac{\partial v_z}{\partial r} + \frac{v_\theta}{r}\frac{\partial v_z}{\partial \theta} + v_z\frac{\partial v_z}{\partial z}$$

Spherical

$$a_r = \frac{\partial v_r}{\partial t} + v_r\frac{\partial v_r}{\partial r} + \frac{v_\theta}{r}\frac{\partial v_r}{\partial \theta} + \frac{v_\phi}{r\sin\theta}\frac{\partial v_r}{\partial \phi} - \frac{v_\phi^2 + v_\theta^2}{r}$$

$$a_\theta = \frac{\partial v_\theta}{\partial t} + v_r\frac{\partial v_\theta}{\partial t} + \frac{v_\theta}{r}\frac{\partial v_\theta}{\partial \theta} + \frac{v_\phi}{r\sin\phi}\frac{\partial v_\theta}{\partial \phi} + \frac{v_r v_\theta - v_\phi^2\cot\theta}{r}$$

$$a_\phi = \frac{\partial v_\phi}{\partial t} + v_r\frac{\partial v_\phi}{\partial r} + \frac{v_\theta}{r}\frac{\partial v_\phi}{\partial \theta} + \frac{v_\phi}{r\sin\theta}\frac{\partial v_\phi}{\partial \phi} + \frac{v_r v_\phi + v_\theta v_\phi\cot\theta}{r}$$

Vorticity

Rectangular

$$\omega_x = \frac{\partial w}{\partial y} - \frac{\partial v}{\partial z}$$

$$\omega_y = \frac{\partial u}{\partial z} - \frac{\partial w}{\partial x}$$

$$\omega_z = \frac{\partial v}{\partial x} - \frac{\partial u}{\partial y}$$

Cylindrical

$$\omega_r = \frac{1}{r}\left(\frac{\partial v_z}{\partial \theta}\right) - \frac{\partial v_\theta}{\partial z}$$

$$\omega_\theta = \frac{\partial v_r}{\partial z} - \frac{\partial v_z}{\partial r}$$

$$\omega_z = \frac{1}{r}\left(\frac{\partial (r v_\theta)}{\partial r} - \frac{\partial v_r}{\partial \theta}\right)$$

Spherical

$$\omega_r = \frac{1}{r\sin\theta}\left[\frac{\partial}{\partial \theta}(v_\theta \sin\theta) - \frac{\partial v_\theta}{\partial \phi}\right]$$

$$\omega_\phi = \frac{1}{r}\left[\frac{\partial}{\partial r}(r v_\theta) - \frac{\partial v_r}{\partial \theta}\right]$$

$$\omega_\theta = \frac{1}{r}\left[\frac{1}{\sin\theta}\frac{\partial v_r}{\partial \phi} - \frac{\partial}{\partial r}(r v_\phi)\right]$$

The time-derivative term on the right side of Eqs. 1.2.8 and 1.2.9 for the acceleration is called the **local acceleration** and the remaining terms on the right side in each equation form the **convective acceleration**. Hence the acceleration of a fluid particle is the sum of the local acceleration and convective acceleration. In a pipe, local acceleration results if, for example, a valve is being opened or closed; and convective acceleration occurs in the vicinity of a change in the pipe geometry, such as a pipe contraction or an elbow. In both cases fluid particles change speed, but for very different reasons.

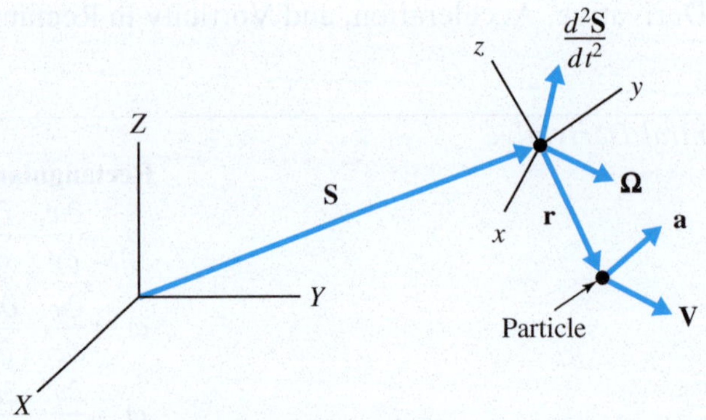

Figure 1.5 Motion relative to a noninertial xyz-reference frame.

We must note that the foregoing expressions for acceleration give the acceleration relative to an observer in the observer's reference frame only. In certain situations the observer's reference frame may be accelerating; then the acceleration of a particle relative to a fixed reference frame may be needed. It is given by

$$\mathbf{A} = \mathbf{a} + \underbrace{d^2\mathbf{S}/\partial t^2}_{\text{acceleration of reference frame}} + \underbrace{2\boldsymbol{\Omega}\times\mathbf{V}}_{\text{Coriolis acceleration}} + \underbrace{\boldsymbol{\Omega}\times(\boldsymbol{\Omega}\times\mathbf{r})}_{\text{normal acceleration}} + \underbrace{d\boldsymbol{\Omega}/\partial t\times\mathbf{r}}_{\text{angular acceleration}} \tag{1.2.12}$$

where **a** is given by Eq. 1.2.8, $d^2\mathbf{S}/\partial t^2$ is the acceleration of the observer's reference frame, **V** and **r** are the velocity and position vectors of the particle, respectively, in the observer's reference frame, and Ω is the angular velocity of the observer's reference frame (see Figure 1.5). Note that all vectors are written using the unit vectors of the XYZ-reference frame. For most engineering applications, reference frames attached to the earth yield $\mathbf{A} = \mathbf{a}$, since the other terms in Eq. 1.2.12 are often negligible with respect to **a**. We may decide, however, to attach the xyz-reference frame to an accelerating device (a rocket), or to a rotating device (a sprinkler arm); then certain terms of Eq. 1.2.12 must be included along with **a** of Eq. 1.2.8.

If the acceleration of all fluid particles is given by $\mathbf{A} = \mathbf{a}$ in a selected reference frame, it is an **inertial** reference frame. If $\mathbf{A} \neq \mathbf{a}$, it is a **noninertial** reference frame. A reference frame that moves with constant velocity without rotating is an inertial reference frame. When analyzing flow about, for example, an airfoil moving at a constant speed, we attach the reference frame to the airfoil so that a steady flow is observed in that reference frame.

1.2.4 Angular Velocity and Vorticity

A fluid flow may be thought of as the motion of a collection of fluid particles. As a particle travels along it may rotate or deform. The rotation and deformation of the fluid particles are of particular interest in our study of fluid mechanics. There are certain flows, or regions of a flow, in which the fluid particles do not rotate; such flows are of special importance, particularly in flows around objects, and are referred to as **irrotational flows**. Flow outside a thin boundary layer on airfoils, outside the separated flow region around autos and other moving vehicles, the flow around submerged objects, and many other flows are examples of irrotational flows. Irrotational flows are extremely important.

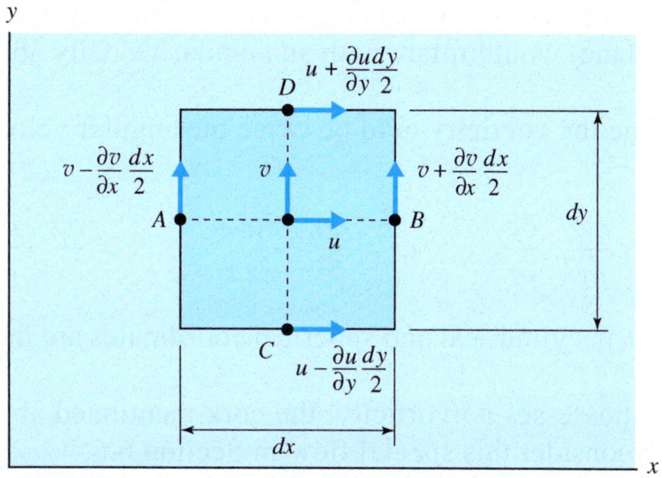

Figure 1.6 Fluid particle occupying an infinitesimal parallelepiped at a particular instant.

Let us consider a small fluid particle that occupies an infinitesimal volume that has the *xy*-face as shown in Figure 1.6. The **angular velocity** Ω_z about the *z*-axis is the average of the angular velocity of line segment *AB* and line segment *CD*. The angular velocities of the two segments, counterclockwise being positive, are

$$
\begin{aligned}
\Omega_{AB} &= \frac{v_B - v_A}{dx} \\
&= \left[v + \frac{\partial v}{\partial x}\frac{dx}{2} - \left(v - \frac{\partial v}{\partial x}\frac{dx}{2} \right) \right] \Big/ dx = \frac{\partial v}{\partial x}
\end{aligned}
\tag{1.2.13}
$$

$$
\begin{aligned}
\Omega_{CD} &= -\frac{u_D - u_C}{dy} \\
&= -\left[u + \frac{\partial u}{\partial y}\frac{dy}{2} - \left(u - \frac{\partial u}{\partial y}\frac{dy}{2} \right) \right] \Big/ dy = -\frac{\partial u}{\partial y}
\end{aligned}
\tag{1.2.14}
$$

Consequently, the angular velocity Ω_z of the fluid particle is

$$
\begin{aligned}
\Omega_z &= \frac{1}{2}(\Omega_{AB} + \Omega_{CD}) \\
&= \frac{1}{2}\left(\frac{\partial v}{\partial x} - \frac{\partial u}{\partial y} \right)
\end{aligned}
\tag{1.2.15}
$$

If we had considered the *xz*-face, we would have found the angular velocity about the *y*-axis to be

$$
\Omega_y = \frac{1}{2}\left(\frac{\partial u}{\partial z} - \frac{\partial w}{\partial x} \right) \tag{1.2.16}
$$

and the *yz*-face would provide us with the angular velocity about the *x*-axis:

$$
\Omega_x = \frac{1}{2}\left(\frac{\partial w}{\partial y} - \frac{\partial v}{\partial z} \right) \tag{1.2.17}
$$

These are the three components of the angular velocity vector. A cork placed in a water flow in a wide channel (the xy-plane) would rotate with an angular velocity about the z-axis, given by Eq. 1.2.15.

It is common to define the **vorticity** $\boldsymbol{\omega}$ to be twice the angular velocity; its three components are then

$$\omega_x = \frac{\partial w}{\partial y} - \frac{\partial v}{\partial z} \qquad \omega_y = \frac{\partial u}{\partial z} - \frac{\partial w}{\partial x} \qquad \omega_z = \frac{\partial v}{\partial x} - \frac{\partial u}{\partial y} \tag{1.2.18}$$

The vorticity components in cylindrical and spherical coordinates are included in Table 1.1 and in Appendix A.4.

An irrotational flow possesses no vorticity; the cork mentioned above would not rotate in an irrotational flow. We consider this special flow in Section 6.5.

The deformation of the particle of Figure 1.6 is the rate of change of the angle that line segment AB makes with line segment CD. If AB is rotating with an angular velocity different from that of CD, the particle is deforming. The deformation is represented by the **rate-of-strain tensor**; its component ϵ_{xy} in the xy-plane is given by

$$\begin{aligned}\epsilon_{xy} &= \frac{1}{2}(\Omega_{AB} - \Omega_{CD}) \\ &= \frac{1}{2}\left(\frac{\partial v}{\partial x} + \frac{\partial u}{\partial y}\right)\end{aligned} \tag{1.2.19}$$

For the xz-plane and the yz-plane we have

$$\epsilon_{xz} = \frac{1}{2}\left(\frac{\partial w}{\partial x} + \frac{\partial u}{\partial z}\right) \qquad \epsilon_{yz} = \frac{1}{2}\left(\frac{\partial w}{\partial y} + \frac{\partial v}{\partial z}\right) \tag{1.2.20}$$

Observe that $\epsilon_{xy} = \epsilon_{yx}$, $\epsilon_{xz} = \epsilon_{zx}$, and $\epsilon_{yz} = \epsilon_{zy}$. By observation, we see that the rate-of-strain tensor is symmetric.

The fluid particle could also deform by being stretched or compressed in a particular direction. For example, if point B of Figure 1.6 is moving faster than point A, the particle would be stretching in the x-direction. This normal rate of strain is measured by

$$\begin{aligned}\epsilon_{xx} &= \frac{u_B - u_A}{dx} \\ &= \left[u + \frac{\partial u}{\partial x}\frac{dx}{2} - \left(u - \frac{\partial u}{\partial x}\frac{dx}{2}\right)\right] \Big/ dx = \frac{\partial u}{\partial x}\end{aligned} \tag{1.2.21}$$

Similarly, in the y- and z-directions we would find that

$$\epsilon_{yy} = \frac{\partial v}{\partial y} \qquad \epsilon_{zz} = \frac{\partial w}{\partial z} \tag{1.2.22}$$

The symmetric rate-of-strain tensor can be displayed as

$$\epsilon_{ij} = \begin{pmatrix} \epsilon_{xx} & \epsilon_{xy} & \epsilon_{xz} \\ \epsilon_{xy} & \epsilon_{yy} & \epsilon_{yz} \\ \epsilon_{xz} & \epsilon_{yz} & \epsilon_{zz} \end{pmatrix} \quad (1.2.23)$$

where the subscripts i and j take on numerical values 1, 2, or 3. Then ϵ_{12} represents ϵ_{xy} in row 1 column 2.

We will see in Chapter 3 that the normal and shear stress components in a flow are related to the foregoing rate-of-strain components. In fact, in a one-dimensional flow, the shear stress was related to $\partial u/\partial y$ with $\tau = \mu\ \partial u/\partial y$; note that $\partial u/\partial y$ is twice the rate-of-strain component given by Eq. 1.2.19 with $v = 0$.

Example 1.1

The velocity field is given by $\mathbf{V} = 2x\hat{\mathbf{i}} - yt\hat{\mathbf{j}}$ m/s, where x and y are in meters and t is in seconds. Find the equation of the streamline passing through (2, −1) and a unit vector normal to the streamline at (2, −1) at t = 4 s.

Solution

The velocity vector is tangent to a streamline so that $\mathbf{V} \times d\mathbf{r} = 0$ (the cross product of two parallel vectors is zero). For the given velocity vector we have, at t = 4 s,

$$\mathbf{V} \times \mathbf{dr} = (2x\hat{\mathbf{i}} - 4y\hat{\mathbf{j}}) \times (dx\hat{\mathbf{i}} + dy\hat{\mathbf{j}}) = (2x\,dy + 4y\,dx)\hat{\mathbf{k}} = 0$$

where we have used $\hat{\mathbf{i}} \times \hat{\mathbf{j}} = \mathbf{k}$, $\hat{\mathbf{j}} \times \hat{\mathbf{i}} = -\mathbf{k}$, and $\hat{\mathbf{i}} \times \hat{\mathbf{i}} = 0$. Consequently,

$$2x\,dy = -4y\,dx \qquad \text{or} \qquad \frac{dy}{y} = -2\frac{dx}{x}$$

Integrate both sides:

$$\ln y = -2\ln x + \ln C$$

where we used ln C for convenience. This is written as

$$\ln y = \ln x^{-2} + \ln C = \ln(Cx^{-2})$$

Hence

$$x^2 y = C$$

At (2, −1) we find $C = -4$, so that the streamline passing through (2, −1) has the equation

$$x^2 y = -4$$

A normal vector is perpendicular to the streamline, hence the velocity vector; using $\hat{\mathbf{n}} = n_x\hat{\mathbf{i}} + n_y\hat{\mathbf{j}}$ we have at (2, −1) and t = 4 s

$$\mathbf{V} \cdot \hat{\mathbf{n}} = (4\hat{\mathbf{i}} + 4\hat{\mathbf{j}}) \cdot (n_x\hat{\mathbf{i}} + n_y\hat{\mathbf{j}}) = 0$$

Using $\hat{\mathbf{i}} \cdot \hat{\mathbf{i}} = 1$ and $\hat{\mathbf{i}} \cdot \hat{\mathbf{j}} = 0$, this becomes

$$4n_x + 4n_y = 0 \qquad \therefore n_x - n_y$$

Then, because $\hat{\mathbf{n}}$ is a unit vector, $n_x^2 + n_y^2 = 1$ and we find that

$$n_x^2 = 1 - n_x^2 \qquad \therefore n_n = \frac{\sqrt{2}}{2} \qquad \text{and} \qquad n_y = -\frac{\sqrt{2}}{2}$$

The unit vector normal to the streamline is written as

$$\hat{\mathbf{n}} = \frac{\sqrt{2}}{2}(\hat{\mathbf{i}} - \hat{\mathbf{j}})$$

Example 1.2

A velocity field in a particular flow is given by $\mathbf{V} = 20y^2\hat{\mathbf{i}} - 20xy\hat{\mathbf{j}}$ m/s. Calculate the acceleration, the angular velocity, the vorticity vector, and any nonzero rate-of-strain components at the point (1, −1, 2).

Solution

We could use Eq. 1.2.9 and find each component of the acceleration, or we could use Eq. 1.2.8 and find a vector expression. Using Eq. 1.2.8, we have

$$\begin{aligned}
\mathbf{a} &= u\frac{\partial \mathbf{V}}{\partial x} + v\frac{\partial \mathbf{V}}{\partial y} + w\cancelto{0}{\frac{\partial \mathbf{V}}{\partial z}} + \cancelto{0}{\frac{\partial \mathbf{V}}{\partial t}} \\
&= 20y^2(-20y\hat{\mathbf{j}}) - 20xy(40y\hat{\mathbf{i}} - 20x\hat{\mathbf{j}}) \\
&= -800xy^2\hat{\mathbf{i}} - 400(y^3 - x^2y)\hat{\mathbf{j}}
\end{aligned}$$

where we have used $u = 20\,y^2$ and $v = -20xy$, as given by the velocity vector. All particles passing through the point (1, −1, 2) have the acceleration

$$\underline{\mathbf{a} = -800\hat{\mathbf{i}}\ \text{m/s}^2}$$

The angular velocity has two zero components:

$$\Omega_x = \frac{1}{2}\left(\cancelto{0}{\frac{\partial w}{\partial y}} - \cancelto{0}{\frac{\partial v}{\partial z}}\right) = 0, \quad \Omega_y = \frac{1}{2}\left(\cancelto{0}{\frac{\partial u}{\partial z}} - \cancelto{0}{\frac{\partial w}{\partial x}}\right) = 0$$

The non-zero z-component, at the point (1, −1, 2) is

$$\Omega_z = \frac{1}{2}\left(\frac{\partial v}{\partial x} - \frac{\partial u}{\partial y}\right)$$

$$= \frac{1}{2}(-20y - 40y) = \underline{30 \text{ rad/s}}$$

The vorticity vector is twice the angular velocity vector:

$$\boldsymbol{\omega} = 2\Omega_z \hat{\mathbf{k}} = \underline{60\hat{\mathbf{k}} \text{ rad/s}}$$

The nonzero rate-of-strain components are

$$\epsilon_{xy} = \frac{1}{2}\left(\frac{\partial v}{\partial x} + \frac{\partial u}{\partial y}\right)$$

$$= \frac{1}{2}(-20y + 40y) = \underline{-10 \text{ rad/s}}$$

$$\epsilon_{yy} = \frac{\partial v}{\partial y}$$

$$= -20x = \underline{-20 \text{ rad/s}}$$

The rate-of-strain tensor would be written as

$$\epsilon_{ij} = \begin{bmatrix} 0 & -10 & 0 \\ -10 & -20 & 0 \\ 0 & 0 & 0 \end{bmatrix} \text{rad/s}$$

1.3 CLASSIFICATION OF FLUID FLOWS

In this section we provide an overview of some of the aspects of fluid mechanics that are considered in more depth in subsequent sections and chapters. Although most of the notions presented here are redefined and discussed in detail later, it will be helpful at this point to introduce the general classification of fluid flows.

1.3.1 One-, Two-, and Three-Dimensional Flows

In the Eulerian description of motion the velocity vector, in general, depends on three space variables and time, that is, $\mathbf{V} = \mathbf{V}(x, y, z, t)$. Such a flow is a **three-dimensional flow**, because the velocity vector depends on three space coordinates. The solutions to problems in such a flow are very difficult and are beyond the scope of an introductory course. Even if the flow could be assumed to be steady [i.e., $\mathbf{V} = \mathbf{V}(x, y, z)$], it would remain a three-dimensional flow. A particular flow is shown in Figure 1.7. This flow is normal to a plane surface; the fluid decelerates and comes to rest at the **stagnation point**. The velocity components, u, v, and w depend on x, y and z; that is, $u = u(x, y, z)$, $v = v(x, y, z)$ and $w = w(x, y, z)$.

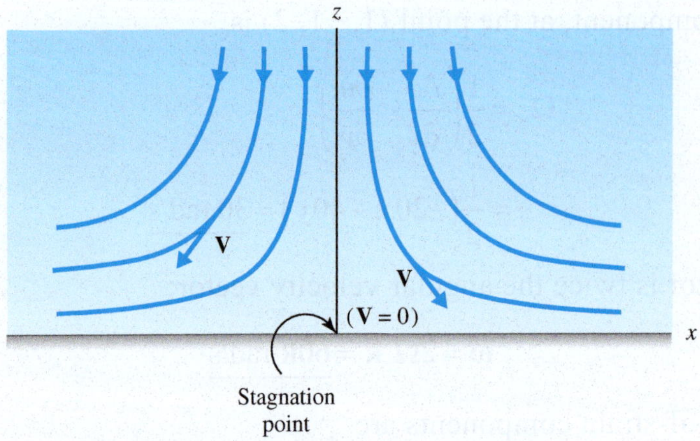

Figure 1.7 A stagnation point flow.

Often a three-dimensional flow can be approximated as a two-dimensional flow. For example, the flow over a wide dam is three-dimensional because of the end conditions, but the flow in the central portion away from the ends can be treated as two-dimensional. In general, a **two-dimensional flow** is a flow in which the velocity vector depends on only two space variables. An example is a **plane flow**, in which the velocity vector depends on two spatial coordinates, x and y, but not z, $\mathbf{V} = \mathbf{V}(x, y)$. In an axisymmetric flow, the velocity vector would depend on r and z, i.e., $\mathbf{V} = \mathbf{V}(r, z)$; the flow in Figure 1.7 would be considered an axisymmetric, two-dimensional flow if described in a cylindrical coordinate system.

A **one-dimensional flow** is a flow in which the velocity vector depends on only one space variable. Such flows occur away from geometry changes in long, straight pipes or between parallel plates, as shown in Figure 1.8. The velocity in the pipe varies only with r, that is $u = u(r)$. The velocity between parallel plates varies only with the coordinate y, i.e., $u = u(y)$. Even if the flow is unsteady so that $u = u(y, t)$, as would be the situation during startup, the flow is one-dimensional, except near the inlet.

The flows shown in Figure 1.8 may also be referred to as **developed flows**; that is, the velocity profile does not vary with respect to the space coordinate in the direction of flow. This demands that the region of interest be a substantial distance from an entrance or a sudden change in geometry.

There are many engineering problems in fluid mechanics in which a flow field is simplified to a **uniform flow**: the velocity, and other fluid properties, are constant over the area, as in Figure 1.9. This simplification is made when the velocity is essentially constant over the area, a rather common occurrence. Examples of such flows are relatively high-speed flow in a pipe section, and flow in a stream. The average velocity may change from one section to another; the flow conditions depend only on the space variable in the flow direction. The pressure is assumed to be constant normal to the streamlines but varies in the flow direction if the pipe geometry changes. For large conduits, however,

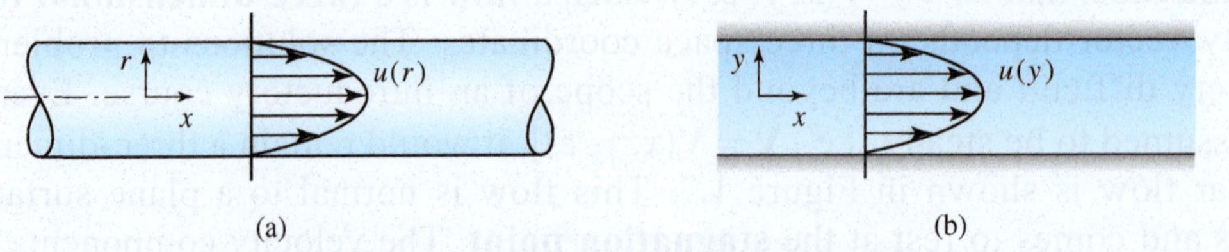

Figure 1.8 One-dimensional flow: (a) flow in a pipe; (b) flow between parallel plates.

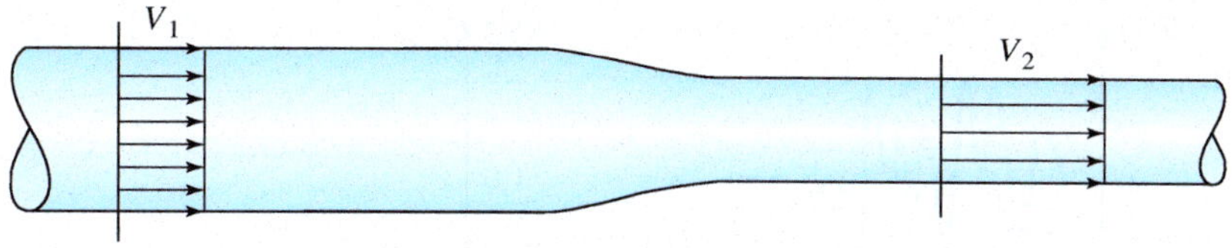

Figure 1.9 Uniform velocity profiles.

it may be necessary to consider hydrostatic variation in the pressure normal to the streamlines. Much of Chapter 5 will be devoted to studying flow in pipes, a very important flow.

1.3.2 Viscous and Inviscid Flows

A fluid flow may be broadly classified as either a viscous flow or an inviscid flow. An **inviscid flow** is one in which viscous effects do not significantly influence the flow and are thus neglected. In a **viscous flow** the effects of viscosity are important and cannot be ignored.

To model an inviscid flow analytically, we can simply let the viscosity be zero; this will obviously make all viscous effects zero. It is more difficult to create an inviscid flow experimentally, because all fluids of interest (such as water and air) have viscosity. The question then becomes: Are there flows of interest in which the viscous effects are negligibly small? The answer is "yes, if the shear stresses in the flow are small and act over such small areas that they do not significantly affect the flow field." This statement is very general, of course, and it will take considerable analysis to justify the inviscid flow assumption.

Based on experience, it has been found that the primary flows, which can be modeled as inviscid flows, are **external flows**, that is, flows which exist exterior to bodies. Inviscid flows are of primary importance in flows around streamlined bodies, such as flow around an airfoil or a hydrofoil. Any viscous effects that may exist are confined to a thin layer, called a **boundary layer**, which is attached to the boundary, such as that shown in Figure 1.10; the velocity in a boundary layer is always zero at a fixed wall, a result of viscosity. On a re-entry vehicle the zero wall velocity causes very large gradients so that the shear stresses can actually burn up the vehicle. For many flow situations, boundary layers are so thin that they can simply be ignored when studying the gross features of a flow around a streamlined body. For example, the inviscid flow (often referred to as the free-stream flow) solution provides an excellent prediction to the flow around the airfoil, except inside the boundary layer and possibly near the trailing edge. Inviscid flow is also encountered

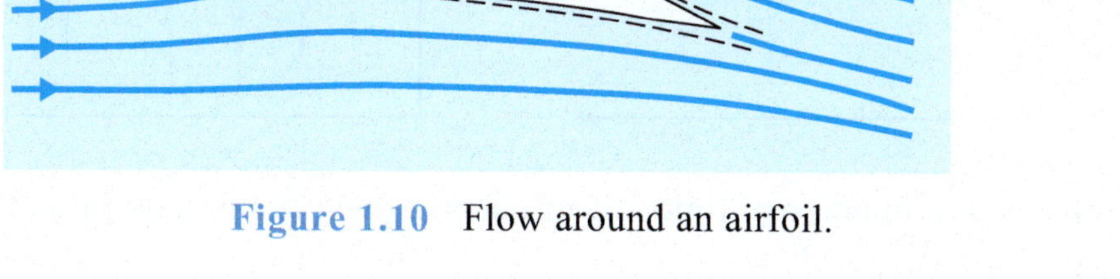

Figure 1.10 Flow around an airfoil.

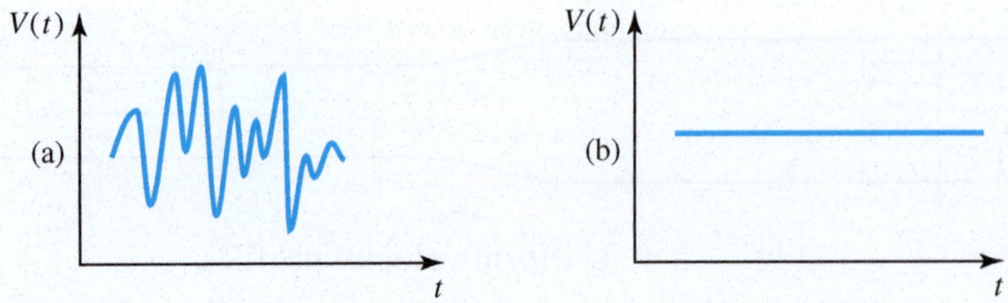

Figure 1.11 Velocity as a function of time in a laminar flow: (a) unsteady flow; (b) steady flow.

in contractions inside piping systems and in short regions of internal flows where viscous effects can be ignored.

Viscous flows include the broad class of internal flows, such as flows in pipes and conduits and in open channels. In such flows viscous effects cause substantial "losses" and account for the huge amounts of energy that must be used to transport oil and gas in pipelines. The no-slip condition resulting in zero velocity at the wall and the resulting shear stresses lead directly to these losses.

1.3.3 Laminar and Turbulent Flows

A viscous flow can be classified as either a laminar flow or a turbulent flow. In a **laminar flow** the fluid flows with no significant mixing of neighboring fluid particles. If dye were injected into the flow, it would not mix with the neighboring fluid except by molecular activity; it would retain its identity for a relatively long period of time. Viscous shear stresses always influence a laminar flow. The flow may be highly time dependent, due to the erratic motion of a piston in the flow of honey as shown by the output of a velocity probe in Figure 1.11a, or it may be steady, as shown in Figure 1.11b.

In a **turbulent flow** fluid motions vary irregularly so that quantities such as velocity and pressure show a random variation with time and space coordinates. The physical quantities are often described by statistical averages. In this sense we can define a "steady" turbulent flow as a flow in which the time-average physical quantities do not change in time. Figure 1.12 shows instantaneous velocity measurements in an unsteady and a steady turbulent flow. A dye injected into a turbulent flow would mix immediately by the action of the randomly moving fluid particles; it would quickly lose its identity in this diffusion process.

It is important to note, however, that erratic fluctuations of the velocity components are necessary but not sufficient for a flow to be turbulent; a correlation of velocity components is also required. The free-stream flow around the airfoil in Figure 1.10 contains random fluctuations,

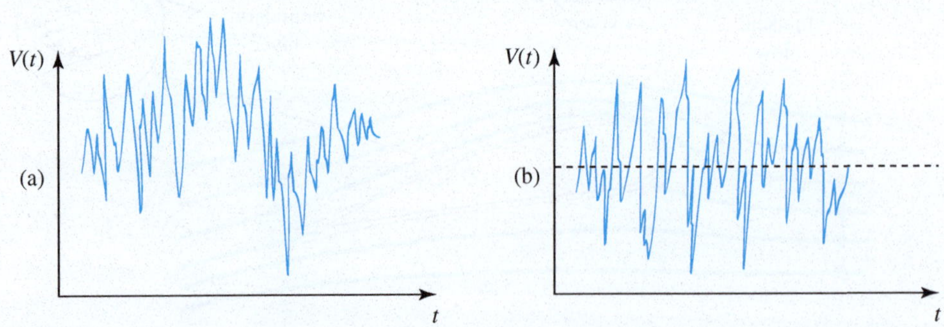

Figure 1.12 Velocity as a function of time in a turbulent flow: (a) unsteady flow; (b) "steady" flow.

referred to as free-stream fluctuations or free-stream "turbulence," although it is not a turbulent flow. Section 5.6 presents the details of a turbulent flow.

A laminar flow and a turbulent flow can be observed by performing a simple experiment with a water faucet. Turn the faucet on so the water flows out very slowly as a silent stream. This is laminar flow. Open the faucet slowly and observe the flow becoming turbulent. Note that a turbulent flow develops with a relatively small flow rate.

The reason why a flow can be laminar or turbulent has to do with what happens to a small flow disturbance, a perturbation to the velocity components. A flow disturbance can either increase or decrease in size. If a flow disturbance in a laminar flow increases (i.e., the flow is unstable), the flow may become turbulent if the disturbance decreases, the flow remains laminar. In certain situations the flow may develop into a different laminar flow, as is the case between concentrically rotating cylinders at low rotational speed the flow would be in simple circles. But at a sufficiently high speed the flow becomes unstable and vortices suddenly appear; it is a much more complex laminar flow called Taylor–Conette flow.

Whether a flow is laminar or turbulent depends on three physical parameters describing the flow conditions. The first parameter is a length scale of the flow field, such as the thickness of a boundary layer or the diameter of a pipe. If this length scale is sufficiently large, a flow may be turbulent. The second parameter is a velocity scale such as a spatial average of the velocity; for a large enough velocity the flow may be turbulent. The third parameter is the kinematic viscosity; for a small enough viscosity the flow may be turbulent.

The three parameters can be combined into a single parameter that can serve as a tool to predict the flow regime. This parameter is the **Reynolds number**, named after Osborne Reynolds (1842–1912), a dimensionless parameter, defined as

$$\mathrm{Re} = \frac{VL}{\nu} \tag{1.3.1}$$

where L and V are a characteristic length and velocity, respectively, and ν is the kinematic viscosity; for example, in a pipe flow L would be the pipe diameter and V would be the average velocity. If the Reynolds number is relatively small, the flow is laminar; if it is large, the flow is turbulent. This is more precisely stated by defining a **critical Reynolds number**, $\mathrm{Re}_{\mathrm{crit}}$, so that the flow is laminar if $\mathrm{Re} < \mathrm{Re}_{\mathrm{crit}}$. For example, in a flow inside a rough-walled pipe it is found that $\mathrm{Re}_{\mathrm{crit}} \approx 2000$. This is the minimum critical Reynolds number and is used for most engineering applications. If the pipe wall is extremely smooth and free of vibration, the critical Reynolds number can be increased as the fluctuation level in the flow is decreased; values in excess of 40 000 have been measured. The critical Reynolds number is different for every geometry, for example, it is 1500 for flow between parallel plates using the average velocity and the distance between the plates.

The flow can also be intermittently turbulent and laminar; this is called an **intermittent flow**. This phenomenon can occur when the Reynolds number is close to $\mathrm{Re}_{\mathrm{crit}}$. Figure 1.13 shows the output of a velocity probe for such a flow.

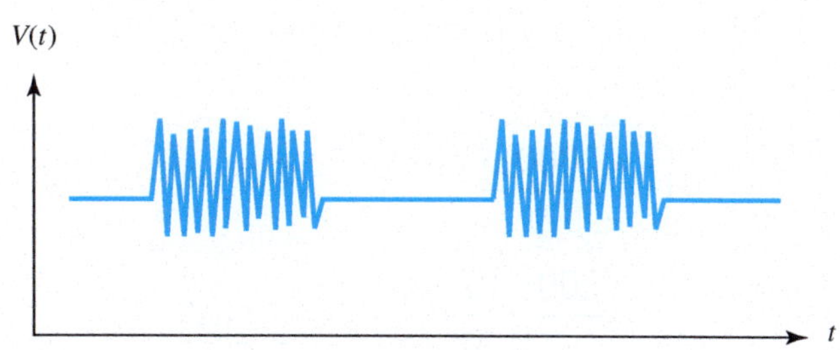

Figure 1.13 Velocity versus time signal from a velocity probe in an intermittent flow.

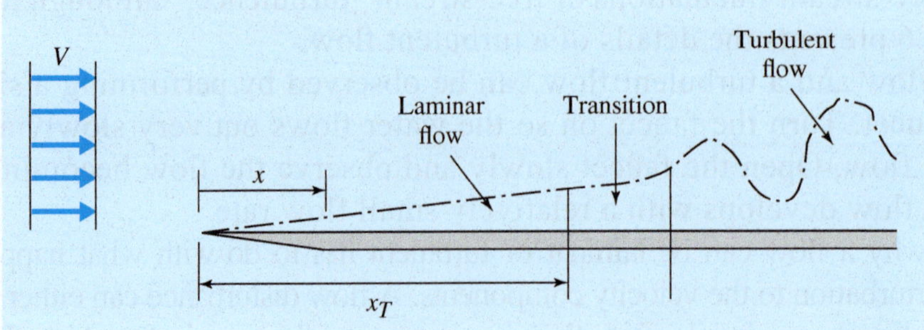

Figure 1.14 Boundary layer flow on a flat plate.

In a boundary layer that exists on a flat plate, due to a constant-velocity fluid stream, as shown in Figure 1.14, the length scale changes with distance from the upstream edge. A Reynolds number is calculated using the length x as the characteristic length. For a certain x_T, Re becomes $\mathrm{Re}_{\mathrm{crit}}$ and the flow undergoes transition from laminar to turbulent. For a smooth rigid plate in a uniform flow with a low free-stream fluctuation level, values as high as $\mathrm{Re}_{\mathrm{crit}} = 10^6$ have been observed. In most engineering applications we assume a rough wall, or high free-stream fluctuation level, with an associated critical Reynolds number of approximately 3×10^5.

It is not appropriate to refer to an inviscid flow as laminar or as turbulent. The inviscid flow of Figure 1.10 is often called the **free stream**. In fact, a sensor in the free stream of Figure 1.10 could appear as in Figure 1.12b but the flow would not be turbulent. An additional requirement, to be presented in Chapter 5, is necessary for a flow to be considered turbulent. The free stream can be irrotational or it can possess vorticity; most often it is irrotational.

Example 1.3

The 2-cm-diameter pipe of Figure E1.3 is used to transport water at 20°C. What is the maximum average velocity that may exist in the pipe for which laminar flow is guaranteed?

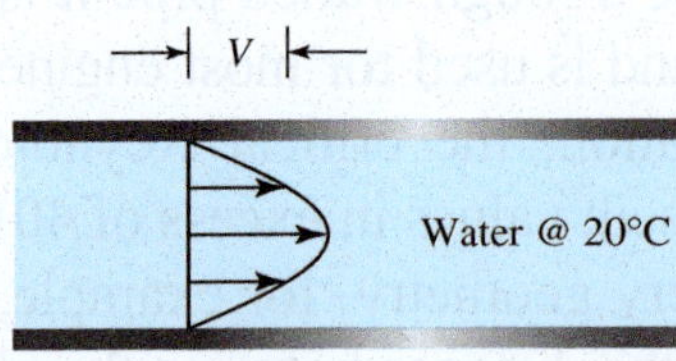

Figure E1.3

Solution

The kinematic viscosity is found in Appendix B to be $\nu = 10^{-6}$ m²/s, the value used if the temperature is not given. Using a Reynolds number of 2000 so that a laminar flow is guaranteed, we find that

$$V = \frac{2000\,\nu}{D}$$

$$= \frac{200 \times 10^{-6}\ \mathrm{m^2/s}}{0.02\ \mathrm{m}} = \underline{0.1\,\mathrm{m/s}}$$

This average velocity is quite small. Velocities this small are not usually encountered in actual situations; hence, laminar flow is seldom of engineering interest except for specialized topics such as lubrication. Most internal flows are turbulent flows, and thus the study of turbulence gains much attention.

1.3.4 Incompressible and Compressible Flows

The last major classification of fluid flows to be considered in this chapter separates flows into incompressible and compressible flows. An **incompressible flow** exists if the density of each fluid particle remains relatively constant as it moves through the flow field, that is,

$$\frac{D\rho}{Dt} = 0 \tag{1.3.2}$$

This does not demand that the density is constant everywhere. If the density is constant, then obviously, the flow is incompressible, but that would be a more restrictive condition. Atmospheric and oceanographic flows, in which $\rho = \rho(z)$, where z is vertical, and flows that involve adjacent layers of fresh and salt water, as happens when rivers enter the ocean, are examples of incompressible flows in which the density varies.

In addition to liquid flows, low-speed gas flows, such as the atmospheric flow referred to above, are also considered to be incompressible flows. The **Mach number**, named after Ernst Mach (1838–1916), is defined as

$$\mathrm{M} = \frac{V}{c} \tag{1.3.3}$$

where V is the gas speed and the wave speed $c = \sqrt{kRT}$. Equation 1.3.3 is useful in deciding whether a particular gas flow can be studied as an incompressible flow. If $\mathrm{M} < 0.3$, density variations are at most 3% and the flow is assumed to be incompressible; for standard air this corresponds to a velocity below about 100 m/s or 300 ft/s. If $\mathrm{M} > 0.3$, the density variations influence the flow and compressibility effects should be accounted for; such flows are **compressible flows** and are considered in Chapter 7.

Incompressible gas flows include atmospheric flows, the aerodynamics of landing and takeoff of commercial aircraft, heating and air-conditioning airflows, flow around automobiles and through radiators, and the flow of air around buildings, to name a few. Compressible flows include the aerodynamics of high-speed aircraft, airflow through jet engines, steam flow through the turbine in a power plant, airflow in a compressor, and the flow of the air-gas mixture in an internal combustion engine.

1.4 THE BERNOULLI EQUATION

In this section we present an equation that is probably used more often in fluid flow applications than any other equation. It is also often misused; it is thus important to understand its limitations. Its limitations are a result of several assumptions made in the derivation. One of the assumptions is that viscous effects are neglected. In other words, shear stresses introduced by velocity gradients are not taken into consideration. These stresses are often very small

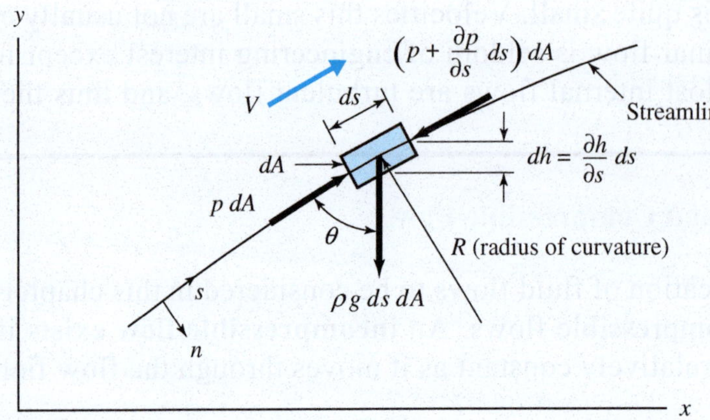

Figure 1.15 Particle moving along a streamline.

compared with pressure differences in the flow field. Locally, these stresses have little effect on the flow field and the assumption is justified. However, over long distances or in regions of high-velocity gradients, these stresses may affect the flow conditions so that viscous effects must be included.

The derivation of this important equation, the Bernoulli equation, starts with the application of Newton's second law to a fluid particle. Let us use an infinitesimal cylindrical particle positioned as shown in Figure 1.15, with length *ds* and cross-sectional area *dA*. The forces acting on the particle are the pressure forces and the weight, as shown. Summing forces in the direction of motion, the *s*-direction, there results

$$p\,dA - \left(p + \frac{\partial p}{\partial s} ds\right) dA - \rho g\, ds\, dA \cos\theta = \rho\, ds\, dA\, a_s \tag{1.4.1}$$

where a_s is the acceleration of the particle in the s-direction. It is given by[2]

$$a_s = V\frac{\partial V}{\partial s} + \frac{\partial V}{\partial t} \tag{1.4.2}$$

where $\partial V/\partial t = 0$ since we will assume steady flow. Also, we see that

$$dh = ds \cos\theta = \frac{\partial h}{\partial s} ds \tag{1.4.3}$$

so that

$$\cos\theta = \frac{\partial h}{\partial s} \tag{1.4.4}$$

[2]This can be verified by considering Eq. 1.2.9a, assuming that $v = w = 0$. Think of the x-direction being tangent to the streamline at the instant shown, so that $u = V$.

Then, after dividing by $ds\, dA$, and using the above equations for a_s and $\cos\theta$, Eq. 1.4.1 takes the form

$$-\frac{\partial p}{\partial s}-\rho g\frac{\partial h}{\partial s}=\rho V\frac{\partial V}{\partial s} \tag{1.4.5}$$

Now, divide by the constant density ρ and note that $V\,\partial V/\partial s=\partial(V^2/2)/\partial s$; then we can write Eq. 1.4.5 as

$$\frac{\partial}{\partial s}\left(\frac{V^2}{2}+\frac{p}{\rho}+gh\right)=0 \tag{1.4.6}$$

This is satisfied if, along the streamline,

$$\frac{V^2}{2}+\frac{p}{\rho}+gh=\text{const} \tag{1.4.7}$$

where the constant may have a different value on a different streamline. Between two points on the same streamline,

$$\frac{V_1^2}{2}+\frac{p_1}{\rho}+gh_1=\frac{V_2^2}{2}+\frac{p_2}{\rho}+gh_2 \tag{1.4.8}$$

This is the well-known *Bernoulli equation*, named after Daniel Bernoulli (1700–1782). Note the five assumptions:

–Inviscid flow (no shear stresses)
–Steady flow ($\partial V/\partial t=0$)
–Along a streamline ($a_s=V\,\partial V/\partial s$)
–Constant density ($\partial\rho/\partial s=0$)
–Inertial reference frame ($\mathbf{A}=\mathbf{a}$ as in Eq. 1.2.12)

If Eq. 1.4.8 is divided by g, the dimension of each term is length and Bernoulli's equation takes the alternate form

$$\frac{V_1^2}{2g}+\frac{p_1}{\gamma}+h_1=\frac{V_2^2}{2g}+\frac{p_2}{\gamma}+h_2 \tag{1.4.9}$$

The sum of the two terms ($p/\gamma+h$) is called the *piezometric head* and the sum of the three terms the *total head.* The pressure p is often referred to as **static pressure**, and the sum of the two terms

$$p+\rho\frac{V^2}{2}=p_T \tag{1.4.10}$$

is called the *total pressure* p_T or **stagnation pressure**, the pressure at a stagnation point (see Figure 1.7) in the flow.

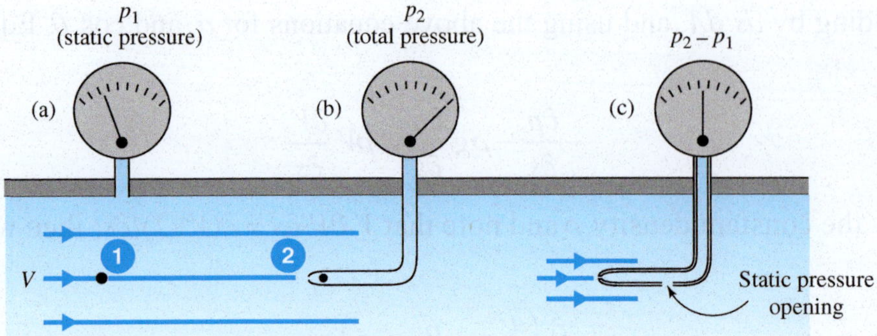

Figure 1.16 Pressure probes: (a) piezometer; (b) pitot probe; (c) pitot-static probe.

The static pressure in a pipe can be measured simply by installing a so-called **piezometer**, shown[3] in Figure 1.16a. A device, known as a **pitot probe**, sketched in Figure 1.16b, is used to measure the total pressure in a fluid flow. Point 2 just inside the pitot tube is a stagnation point; the velocity there is zero. The difference between the readouts can be used to determine the velocity at point 1. A **pitot-static probe** is also used to measure the difference between total and static pressure with one probe (Figure 1.16c). The velocity at point 1 (using the readings of the piezometer and pitot probes, or the reading from the pitot-static probe) can be determined by applying the Bernoulli equation between points 1 and 2:

$$\frac{V_1^2}{2g} + \frac{p_1}{\gamma} = \frac{p_2}{\gamma} \tag{1.4.11}$$

where we have assumed point 2 to be a stagnation point so that $V_2 = 0$. This gives

$$V_1 = \sqrt{\frac{2}{\rho}(p_2 - p_1)} \tag{1.4.12}$$

We will find many uses for Bernoulli's equation in our study of fluids. We must be careful, however, never to use it in an unsteady flow or if viscous effects are significant (the primary reasons for making Bernoulli's equation inapplicable). We must also never confuse Bernoulli's equation with the energy equation; they are independent equations as Example 1.6 illustrates.

The Bernoulli equation can be used to determine how high the water from a fireman's hose will reach, to find the pressure on the surface of a low-speed airfoil,[4] and to find the wind force on the window in a house. These examples are all external flows, flows around objects submerged in the fluid.

Another class of problems where inviscid flow can be assumed and where the Bernoulli equation finds frequent application involves internal flows over relatively short distances, for example, flow through a contraction, as shown in Figure 1.17a, or flow from a plenum, as shown in Figure 1.17b. For a given velocity profile entering the short contraction, the pressure drop $(p_1 - p_2)$ and the velocity profile at section 2 can be approximated assuming an inviscid flow. Viscous effects are typically very small and require substantial distances and areas over which to

[3] When drilling the hole in the wall necessary for the piezometer, burrs are often formed on the inner surface. It is important that such burrs be removed since they may cause errors as high as 30% in the pressure reading.

[4] To consider the flow around an aircraft as a steady flow, we simply make the aircraft stationary and move the air, as is done in model studies using a wind tunnel. The pressures and forces remain unchanged.

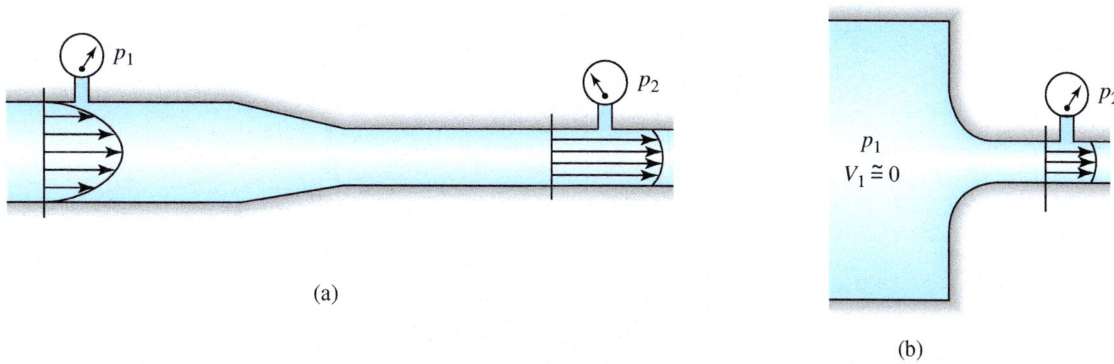

Figure 1.17 Internal inviscid flows: (a) flow through a contraction; (b) flow from a plenum.

operate in order to become significant; so in situations such as those shown in Figure 1.17, viscous effects can often be neglected.

Inviscid flow does not always give a good approximation to the actual flow that exists around a body. Consider the inviscid flow at relatively large Reynolds numbers (e.g., Re = 800) around the sphere (or cylinder), shown in Figure 1.18. A stagnation point where $V = 0$ exists at both the front and the back of the sphere. Bernoulli's equation predicts a maximum pressure at the stagnation points A and C because the velocity is zero at such points. A maximum velocity, and thus a minimum pressure, would exist at point B. In the inviscid flow of part (a), the fluid flowing from B to C must flow from the low-pressure region near B to the high-pressure region near C. In an actual flow for a relatively large Reynolds number, there exists a thin boundary layer in which the velocity goes to zero at the surface of the sphere. This slow-moving fluid near the boundary does not have sufficient momentum to move into the higher-pressure region near C; the result is that the fluid *separates* from the boundary—the boundary streamline leaves the boundary—creating a **separated region**, a region of recirculating flow, as shown in the actual flow sketched in part (b). The pressure does not increase but remains relatively low over the rear part of the sphere. The high pressure that exists near the front stagnation point is never recovered on the rear of the sphere, resulting in a relatively large drag force in the direction of flow. A similar situation occurs in the flow around an automobile.

The flow on the front of the sphere is well approximated by an inviscid flow, but it is obvious that the flow over the rear of the sphere of Figure 1.18 deviates radically from an

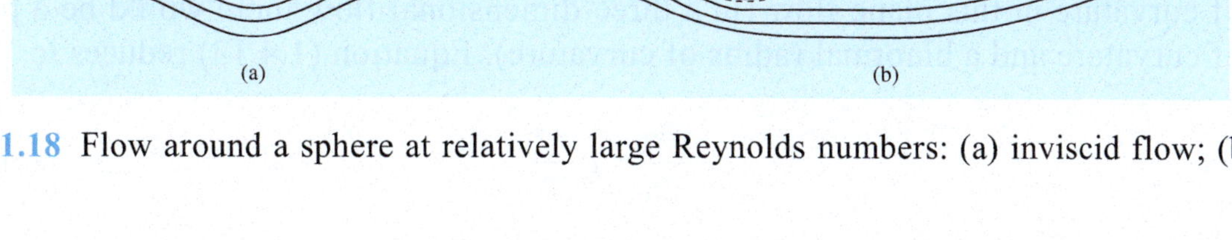

Figure 1.18 Flow around a sphere at relatively large Reynolds numbers: (a) inviscid flow; (b) actual flow.

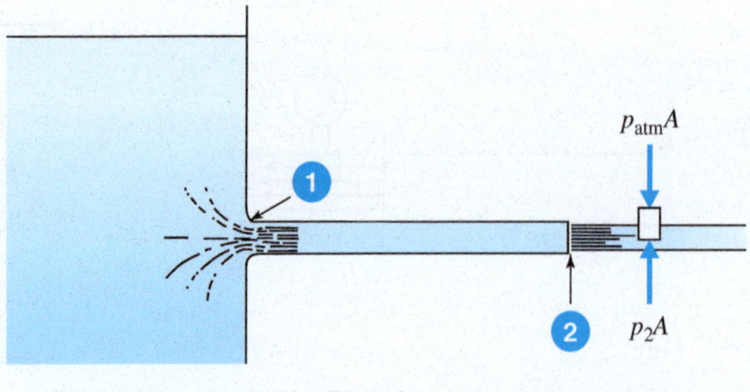

Figure 1.19 Exit flow into the atmosphere.

Inviscid flow. Viscous effects in the boundary layer have led to a separated flow, a phenomenon that is often undesirable. For example, separated flow on an airfoil is called **stall** and must never occur, except on the wings of special stunt planes. On the blades of a turbine separated flows lead to substantially reduced efficiency. The air deflector on the roof of the cab of a semitruck reduces the separated region, thereby reducing drag and fuel consumption.

If viscous effects are negligible in a steady liquid flow, we can use Bernoulli's equation to locate points of possible **cavitation**. This condition occurs when the local pressure becomes equal to the vapor pressure of the liquid. It is to be avoided, if at all possible, because of damage to solid surfaces or because the vaporized liquid may cause devices to not operate effectively. At the point where cavitation occurs, small vapor bubbles are generated, and these bubbles collapse when they enter a higher-pressure region. The collapse is accompanied by very large local pressures that last for only a small fraction of a second. These pressure spikes may reach a wall, where they can, after repeated applications, result in significant damage.

An important observation regarding pressure changes in a fluid needs to be made regarding entrances and exits to a pipe or conduit. Consider a flow from a reservoir through a pipe, as sketched in Figure 1.19. At the entrance the streamlines are curved and the pressure is not constant across section 1, so the pressure at section 1 cannot be assumed to be uniform. At the exit, however, the streamlines are straight, so that no acceleration exists normal to the streamlines; hence, the pressure forces acting on the ends of the small cylindrical control volume must be equal. We write this as $p_2 = p_{\text{atm}}$ or $p_2 = 0$ gage pressure.

In Figure 1.15 we summed forces on the fluid element along the streamline and derived Bernoulli's equation. We can gain additional insight into the pressure field if we sum forces normal to the streamline. Let us consider the fluid particle to be a parallelepiped with thickness dn in the n-direction shown and area dA_s with length ds on the side. Applying Newton's second law in the n-direction provides

$$pdA_s - \left(p + \frac{\partial p}{\partial n} dn \right) dA_s = \rho \, dA_s dn \frac{V^2}{R} \tag{1.4.13}$$

where we have neglected the weight since we are not intending to integrate over significant distances. We have assumed the acceleration in the normal direction to be V^2/R, where R is the radius of curvature in this plane flow (in a three-dimensional flow there would be a principal radius of curvature and a binormal radius of curvature). Equation (1.4.13) reduces to

$$-\frac{\partial p}{\partial n} = \rho \frac{V^2}{R} \tag{1.4.14}$$

From this equation we can qualitatively describe how the pressure changes normal to a streamline (Bernoulli's equation predicts the pressure changes along a streamline). If we replace $\partial p/\partial n$ with $\Delta p/\Delta n$, the incremental pressure change Δp over the short distance Δn normal to the streamline is given by

$$\Delta p = -\rho \frac{V^2}{R} \Delta n \tag{1.4.15}$$

This says that the pressure decreases in the n-direction; this decrease is directly proportional to ρ and V^2 and inversely proportional to R. Consequently, a tornado, with $p = 0$ outside the tornado, will have a very low pressure at its center where R is relatively small and V is quite large. This low pressure may lift a roof from a house.

In Figure 1.19 the pressure would be relatively low at the corner of section 1 and relatively high at the center of section 1. Such qualitative descriptions can be quite helpful in understanding fluid flow behavior.

Example 1.4

The wind reaches a speed of 90 mph in a storm. Calculate the force acting on the 3 ft × 6 ft window of Figure E1.4 facing the storm. The window is in a high-rise building, so the wind speed is not reduced due to ground effects. Use $\rho = 0.0024$ slug/ft^3.

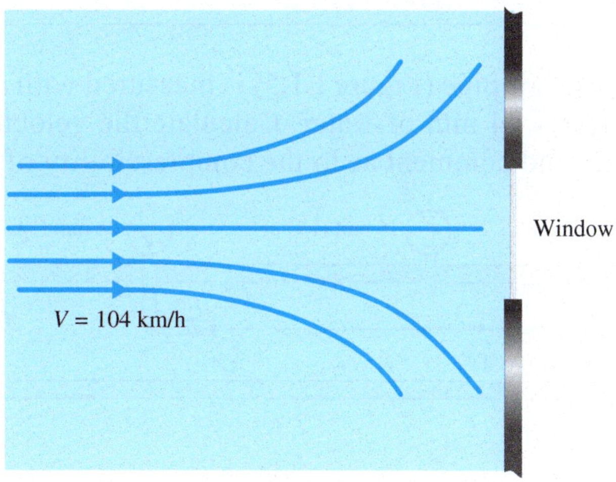

Figure E1.4

Solution

The window facing the storm will be in a stagnation region where the wind speed is brought to zero. Working with gage pressures, the pressure p upstream in the wind is zero. The velocity V must have units of ft/s. It is

$$V = 90 \frac{\text{mi}}{\text{hr}} \times \frac{1\ \text{hr}}{3600\ \text{s}} \times \frac{5280\ \text{ft}}{1\ \text{mi}} = 132\ \text{ft/s}$$

Bernoulli's equation can be used in this situation since we can neglect viscous effects, and steady flow occurs along a streamline at constant density (air is incompressible at speeds below about 220 mph). We calculate the pressure on the window selecting state 1 in the free stream and state 2 on the window, as follows:

$$\frac{V_1^2}{2g} + \cancelto{0}{\frac{p_1}{\gamma}} + h_1 = \cancelto{0}{\frac{V_2^2}{2g}} + \frac{p_2}{\gamma} + h_2$$

$$\therefore\ p_2 = \frac{\rho V_1^2}{2}$$

$$= \frac{0.0024\ \text{slug/ft}^3 \times 13.2^2\ \text{ft}^2/\text{s}^2}{2} = 20.9\ \text{lb/ft}^2$$

where we have used $\gamma = \rho g$, $h_2 = h_1$, $p_1 = 0$, and $V_2 = 0$. Multiply by the area and find the force to be

$$F = pA = 20.9 \times 3 \times 6 = \underline{376\ \text{lb}}$$

We recommend that you verify the units of lb/ft^2 on the pressure calculation above. To do this, use $F = ma$ which provides slug = lb-s^2/ft. When using English units, always use mass in slugs, length in feet, force in pounds, and time in seconds.

Example 1.5

The static pressure head in an air pipe (Figure E1.5) is measured with a piezometer as 16 mm of water. A pitot probe indicates 24 mm of water. Calculate the velocity of the 20°C air. Also, calculate the Mach number and comment as to the compressibility of the flow.

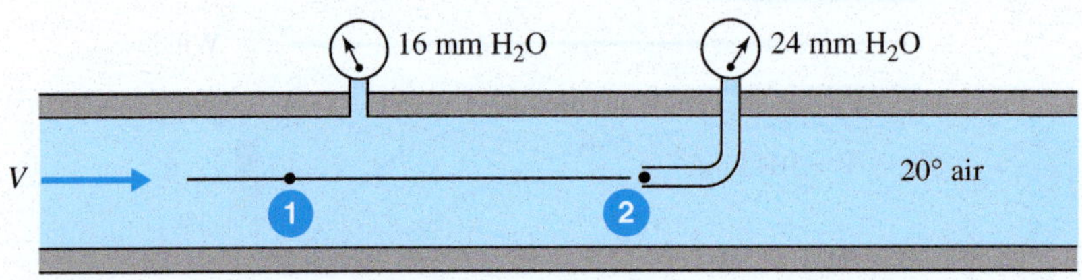

Figure E1.5

Solution

Bernoulli's equation is applied between two points on the streamline that terminates at the stagnation point of the pitot probe. Point 1 is upstream and p_2 is the total pressure at point 2; then, with no elevation change,

$$\frac{V_1^2}{2g} + \frac{p_1}{\gamma} = \frac{p_T}{\gamma}$$

The pressure measured with the piezometer is $p_1 = \gamma h = 9810 \times 0.016 = 157$ Pa. We use the ideal gas law to calculate the density:

$$\rho = \frac{p}{RT}$$
$$= \frac{(157 + 101000)\,\text{Pa}}{287\ \text{kJ/kg} \cdot \text{K} \times (273 + 20)\,\text{K}} = 1.203\ \text{kg/m}^3$$

where standard atmospheric pressure, which is 101 000 Pa (if no elevation is given, assume standard conditions), is added since absolute pressure is needed in the ideal gas law. The velocity is then

$$V_1 = \sqrt{\frac{2}{\rho}(p_T - p_1)}$$
$$= \sqrt{\frac{2(0.024 \times 9810 - 157)\text{Pa}}{1.203\ \text{kg/m}^3}} = \underline{11.42\ \text{m/s}}$$

To find the Mach number, we must calculate the speed of sound. It is

$$c = \sqrt{kRT}$$
$$= \sqrt{1.4 \times 287\ \text{kJ / kg} \cdot \text{K} \times 293\ \text{K}} = 343\ \text{m/s}$$

The Mach number is then

$$\text{M} = \frac{V}{c} = \frac{11.44}{343} = \underline{0.0334}$$

Obviously, the flow can be assumed to be incompressible since $\text{M} < 0.3$. The velocity would have to be much higher before compressibility would be significant.

Example 1.6

Bernoulli's equation, in the form of Eq. 1.4.8, looks very much like the energy equation developed in thermodynamics for a control volume. Discuss the differences between the two equations.

Solution

From thermodynamics we recall that the steady-flow energy equation for a control volume with one inlet and one outlet takes the form

$$\dot{Q} - \dot{W}_s = \dot{m}\left(\frac{V_2^2}{2} + \frac{p_2}{\rho_2} + \tilde{u}_2 + gz_2\right) - \dot{m}\left(\frac{V_1^2}{2} + \frac{p_1}{\rho_1} + \tilde{u}_1 + gz_1\right)$$

This becomes, after dividing through by g,

$$\frac{V_2^2}{2g} + \frac{p_2}{\gamma} + z_2 = \frac{V_1^2}{2g} + \frac{p_1}{\gamma} + z_1$$

where we have made the following assumptions:

No heat transfer $(\dot{Q} = 0)$
No shaft work $(\dot{W}_s = 0)$
No temperature change ($\tilde{u}_2 = \tilde{u}_1$, i.e., no losses due to shear stresses)
Uniform velocity profiles at the two sections
Steady flow
Constant density $(\gamma_2 = \gamma_1)$

Even though several of these assumptions are the same as those made in the derivation of the Bernoulli equation (steady flow, constant density, and no shear stress), we must not confuse the two equations; the Bernoulli equation is derived from Newton's second law and is valid along a streamline, whereas the energy equation is derived from the first law of thermodynamics and is valid between two sections in a fluid flow. The energy equation can be used across a pump to determine the horsepower required to provide a particular pressure rise; the Bernoulli equation can be used along a stagnation streamline to determine the pressure at a stagnation point, a point where the velocity is zero. The equations are quite different, and just because the energy equation degenerates to the Bernoulli equation for particular situations, the two should not be used out of context.

Example 1.7

Explain why a burr on the upstream side of the piezometer opening of Figure 1.18a will result in a low reading of the pressure.

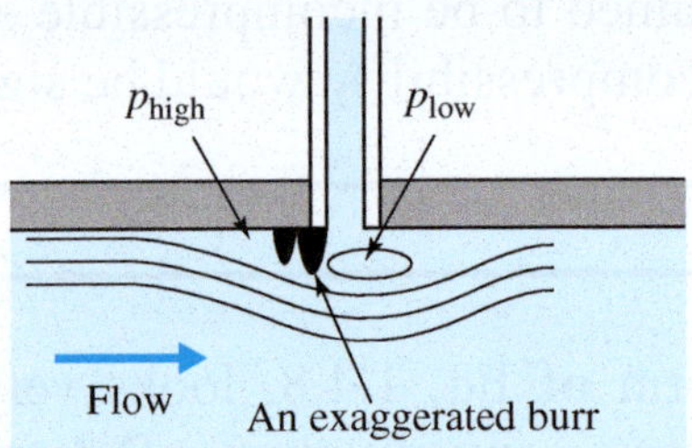

Figure E1.7

Solution

A burr on the upstream side of the piezometer opening would result in a flow in the vicinity of the burr somewhat like that shown in Figure E1.7. A streamline pattern would develop so that a relatively high pressure would occur on the upstream side of the burr and a relatively low pressure on the downstream side at the opening of the piezometer tube. Consequently, since the center of the curvature of the streamline is in the vicinity of the opening, a lower reading of the pressure would be recorded. If the burr were on the downstream side of the opening, a higher pressure reading would be recorded.

1.5 SUMMARY

The Eulerian description of motion was used to express the acceleration as

$$\mathbf{a} = u\frac{\partial \mathbf{V}}{\partial x} + v\frac{\partial \mathbf{V}}{\partial y} + w\frac{\partial \mathbf{V}}{\partial z} + \frac{\partial \mathbf{V}}{\partial t} \tag{1.5.1}$$

Fluid motion may result in fluid particles being rotated and/or deformed. For a flow in the xy-plane a particle would rotate with angular velocity

$$\Omega_z = \frac{1}{2}\left(\frac{\partial v}{\partial x} - \frac{\partial u}{\partial y}\right) \tag{1.5.2}$$

and deform with

$$\epsilon_{xx} = \frac{\partial u}{\partial x} \qquad \epsilon_{yy} = \frac{\partial v}{\partial y} \qquad \epsilon_{xy} = \frac{1}{2}\left(\frac{\partial v}{\partial x} + \frac{\partial u}{\partial y}\right) \tag{1.5.3}$$

Fluid flows are classified as steady or unsteady; viscous or inviscid; laminar, turbulent, or free stream; and incompressible or compressible. Any of these can be uniform, one-, two-, or three-dimensional flows. Experience and practice are needed to appropriately classify a particular flow of interest. It is only for the simpler flows (e.g., a steady, laminar, incompressible, one-dimensional flow) that we hope to obtain a relatively simple solution.

Finally, the famous Bernoulli equation

$$\frac{V_1^2}{2g} + \frac{p_1}{\gamma} + h_1 = \frac{V_2^2}{2g} + \frac{p_2}{\gamma} + h_2 \tag{1.5.4}$$

was presented for a steady, inviscid, constant-density flow along a streamline in an inertial reference frame. The estimate of the pressure change normal to a streamline,

$$\Delta p = -\rho\frac{V^2}{R}\Delta n \tag{1.5.5}$$

was also obtained.

PROBLEMS

Flow Fields

1.1 A fire is started and the smoke from the chimney goes straight up; no wind is present. After a few minutes a wind arises but the smoke continues to rise slowly. Sketch the streakline of the smoke, the pathline of the first particles leaving the chimney, and a few streamlines when that the wind blows parallel to the ground in a constant direction.

1.2 A researcher has large number of small flotation devices, each equipped with a battery and light bulb. Explain how the pathlines and streaklines would be determined near the surface of a stream with some unknown currents that vary with time.

1.3 A little boy chases his dad around the yard with the water hose of Figure P1.3. Sketch a pathline and a streakline if the boy is running perpendicular to the water jet.

Figure P1.3

1.4 The hot-air balloon of Figure P1.4 travels with the wind. The wind velocity vector is $\mathbf{V} = 6\hat{\mathbf{i}} + 10\hat{\mathbf{j}}$ ft/s for the first hour and then it is $10\hat{\mathbf{i}} + 5\hat{\mathbf{j}}$ ft/s for 2 hr. On *xy*-coordinates, sketch the pathline of the balloon and streamlines at $t = 2$ hrs. If a number of hot-air balloons started from the same location, sketch the streakline formed by the balloons at $t = 3$ hr. The balloons start at the origin.

Figure P1.4

1.5 A velocity field is given by $\mathbf{V} = (2t + 2)\hat{\mathbf{i}} + 2t\hat{\mathbf{j}}$ m/s. Sketch the pathlines of two particles up to $t = 5$ s, one that originates from the origin at $t = 0$, and another that originates from the origin at $t = 2$ s. Also, sketch the streamlines at $t = 5$ s.

1.6 Using rectangular coordinates, express the *z*-component of Eq. 1.2.2.

1.7 The traffic situation on Mackinac Island, Michigan, where no automobiles are allowed (bikes are permitted), is to be studied. Comment on how such a study could be performed using a Lagrangian approach and an Eulerian approach.

1.8 Determine the speed of a fluid particle at the origin and at the point (1, –2, 0) for each of the following velocity fields when $t = 2$ s. All distances are in feet and *t* is in seconds.

(a) $\mathbf{V} = (x + 2)\hat{\mathbf{i}} + xt\hat{\mathbf{j}} - z\hat{\mathbf{k}}$ ft/s

(b) $\mathbf{V} = xy\hat{\mathbf{i}} - 2y^2\hat{\mathbf{j}} + tyz\hat{\mathbf{k}}$ ft/s

(c) $\mathbf{V} = x^2t\hat{\mathbf{i}} - (xz + 2t)\hat{\mathbf{j}} + xyt\hat{\mathbf{k}}$ ft/s

1.9 Calculate the angle that the velocity vector makes with the *x*-axis and a unit vector normal to the streamline at (1, –2) for the following velocity fields when $t = 2$ s. All distances are in meters and t is in seconds.

(a) $\mathbf{V} = (x+2)\hat{\mathbf{i}} + xt\hat{\mathbf{j}}$ m/s

(b) $\mathbf{V} = xy\hat{\mathbf{i}} - 2y^2\hat{\mathbf{j}}$ m/s

(c) $\mathbf{V} = (x^2+4)\hat{\mathbf{i}} - y^2t\hat{\mathbf{j}}$ m/s

1.10 Find the equation of the streamline that passes through (1, –2) at $t = 2$ s for the flow of:
(a) Problem 1.9a
(b) Problem 1.9b
(c) Problem 1.9c

1.11 Find the acceleration vector field for a fluid flow that possesses the following velocity field where x, y, z are in meters. Evaluate the acceleration at (2, –1, 3) at $t = 2$ s.

(a) $\mathbf{V} = 20(1-y^2)\hat{\mathbf{i}}$ m/s

(b) $\mathbf{V} = 2x\hat{\mathbf{i}} + 2y\hat{\mathbf{j}}$ m/s

(c) $\mathbf{V} = x^2t\hat{\mathbf{i}} + 2xyt\hat{\mathbf{j}} + 2yzt\hat{\mathbf{k}}$ m/s

(d) $\mathbf{V} = x\hat{\mathbf{i}} - 2xyz\hat{\mathbf{j}} + tz\hat{\mathbf{k}}$ m/s

1.12 Find the angular velocity vector for the following flow fields. Evaluate the angular velocity at (2, –1, 3) at $t = 2$ s.
(a) Problem 1.11a
(b) Problem 1.11b
(c) Problem 1.11c
(d) Problem 1.11d

1.13 Find the vorticity vector for the following flow fields. Evaluate the vorticity at (2, –1, 3) at $t = 2$ s.
(a) Problem 1.11a
(b) Problem 1.11b
(c) Problem 1.11c
(d) Problem 1.11d

1.14 Determine the components of the rate-of-strain tensor for the velocity field of the following at (2, –1, 3) at $t = 2$ s.
(a) Problem 1.11a
(b) Problem 1.11b
(c) Problem 1.11c
(d) Problem 1.11d

1.15 The velocity components, in m/s, in cylindrical coordinates are given by

$$v_r = \left(10 - \frac{40}{r^2}\right)\cos\theta, \quad v_\theta = -\left(10 + \frac{40}{r^2}\right)\sin\theta$$

(a) Calculate the acceleration of a fluid particle occupying the point (4 m, 180°).
(b) Calculate the vorticity component at (4 m, 180°).

1.16 The velocity components, in m/s, in spherical coordinates are given by

$$v_r = \left(10 - \frac{80}{r^3}\right)\cos\theta, \quad v_\theta = -\left(10 + \frac{80}{r^3}\right)\sin\theta$$

(a) Calculate the acceleration of a fluid particle occupying the point (4 m, 180°).
(b) Calculate the vorticity component at (4 m, 180°).

1.17 An unsteady flow occurs between parallel plates such that $u = u(y, t)$, $v = 0$, and $w = 0$. Write an expression for the acceleration. What is the acceleration if the flow is steady, that is, $u = u(y, t)$, $v = 0$, and $w = 0$?

1.18 Consider a symmetrical steady flow in a pipe with the axial and radial velocity components designated $u(r, x)$ and $v(r, x)$, respectively. Write the equations for the two acceleration components a_r and a_x. Use equations from Table 1.1. See Figure P1.18 for the coordinates.

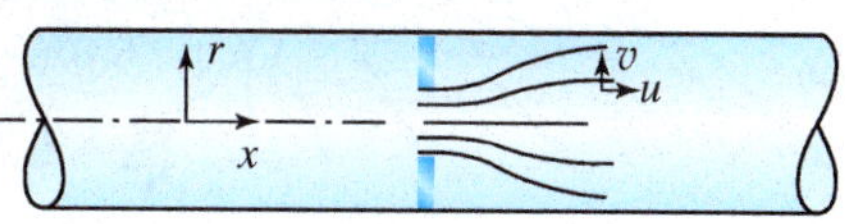

(a)

Figure P1.18

1.19 The velocity in the 2-cm-diameter pipe of Figure P1.19 has only one nonzero velocity component given by $u(r,t) = 2(1 - r^2/r_0^2)$ $(1 - e^{-t/10})$ m/s, where r_0 is the radius of the pipe and t is in seconds. Calculate the maximum velocity and the maximum acceleration:

(a) Along the centerline of the pipe
(b) Along a streamline at $r = 0.5$ cm
(c) Along a streamline just next to the pipe wall [*Hint:* Let $v_z = u(r, t)$, $v_r = 0$, and

$v_\theta = 0$ in the appropriate equations in Table 1.1.]

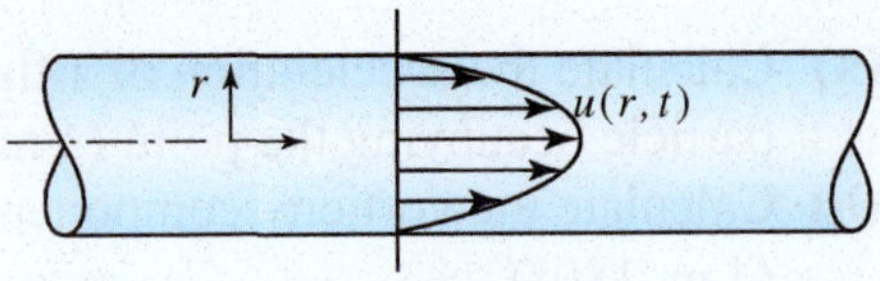

(a)

Figure P1.19

1.20 The temperature changes periodically in a flow according to $T(y, t) = 20(1 - y^2)\cos \pi t/100$°C. If the velocity is given by $u = 2(1 - y^2)$ m/s², determine the rate of change of the temperature of a fluid particle located at $y = 0$ if $t = 20$ s.

1.21 The density of air in the atmosphere varies according to $\rho(z) = 1.23e^{-10^{-4}z}$ kg/m³. Air flowing over the mountain of Figure P1.21 has the velocity vector $\mathbf{V} = 20\hat{\mathbf{i}} + 10\hat{\mathbf{k}}$ m/s at a location of interest where $z = 3000$ m. Find the rate at which a particle's density is changing at that location.

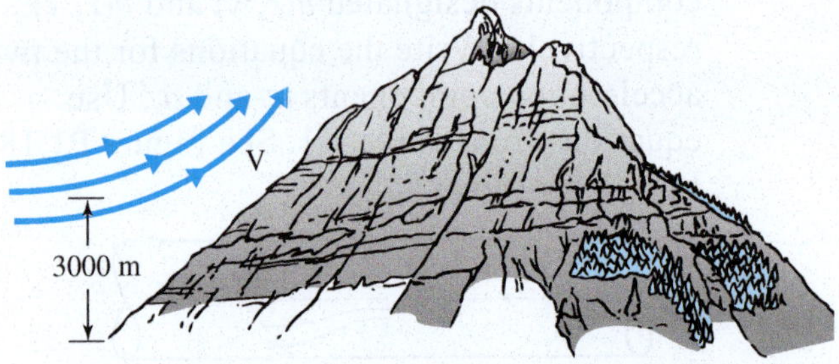

Figure P1.21

1.22 The density variation with elevation is given by $\rho(z) = 1000(1 - z/4)$ kg/m³. At a location where $\mathbf{V} = 10\hat{\mathbf{i}} + 10\hat{\mathbf{k}}$ m/s, find $D\rho/Dt$.

1.23 Salt is slowly added to water in a pipe so that $\partial\rho/\partial x = 0.01$ kg/m⁴. Determine $D\rho/Dt$ if the velocity is uniform at 4 m/s.

1.24 Express the substantial derivative in terms of the gradient ∇ and the velocity vector $\mathbf{V}$. Recall from calculus that in rectangular coordinates

$$\nabla = \frac{\partial}{\partial x}\hat{\mathbf{i}} + \frac{\partial}{\partial y}\hat{\mathbf{j}} + \frac{\partial}{\partial z}\hat{\mathbf{k}}$$

1.25 We can write Eqs. 1.2.9 in a simplified vector form. The gradient is a vector operator expressed in rectangular coordinates as $\nabla = \frac{\partial}{\partial x}\hat{\mathbf{i}} + \frac{\partial}{\partial y}\hat{\mathbf{j}} + \frac{\partial}{\partial z}\hat{\mathbf{k}}$.

Write the dot product of the velocity vector $\mathbf{V}$ and the gradient ∇ and then write Eqs. 1.2.9 as one vector equation expressing the acceleration **a** as the sum of the local acceleration and the convective acceleration.

1.26 For the flow shown in Figure P1.26, relative to a fixed reference frame, find the acceleration of a fluid particle at:

(a) Point A
(b) Point B

The water at B makes an angle of 30° with respect to the ground and the sprinkler arm is horizontal.

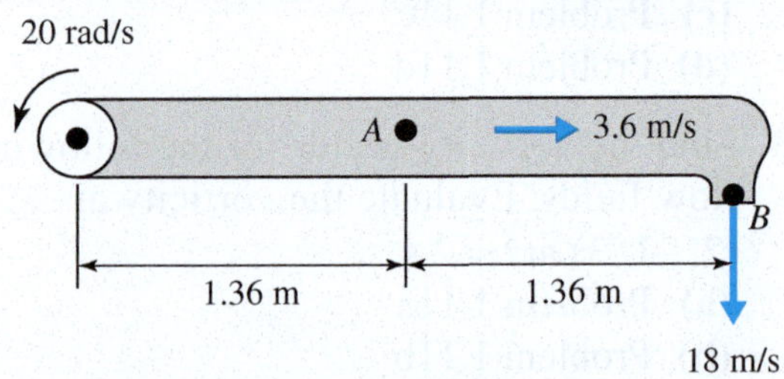

Figure P1.26

1.27 A river is flowing due south at 5 m/s at a latitude of 45°. Calculate the acceleration of a particle floating with the river relative to a fixed reference frame. The radius of the Earth is 6400 km.

Classification of Fluid

1.28 Consider each of the following flows and state whether it could be approximated as a one-, two-, or three-dimensional flow or as a uniform flow:

- **(a)** Flow from a vertical pipe striking a horizontal wall
- **(b)** Flow in the waves of the ocean near a beach
- **(c)** Flow near the entrance of a pipe
- **(d)** Flow around a rocket with a blunt nose
- **(e)** Flow around an automobile
- **(f)** Flow in an irrigation channel
- **(g)** Flow through an artery
- **(h)** Flow through a vein

1.29 Which of the flows in Problem 1.28 could be assumed to be a steady flow? Which must be modeled as an unsteady flow?

1.30 Which flow in Problem 1.28 could best be modeled as a plane flow?

1.31 Select the flows in Problem 1.28 that would possess a stagnation point. Sketch each of the selected flows indicating the location of the stagnation point.

1.32 Which of the flows in Problem 1.28 could be modeled as a developed flow?

1.33 State whether each of the flows in Problem 1.28 could be considered as primarily an inviscid flow or a viscous flow.

1.34 Select the flows in Problem 1.28 that are external flows. Which of the external flows possess a stagnation point?

1.35 Sketch the flow around a razor blade positioned parallel to the flow showing the boundary layers.

1.36 The 32°C water exiting the 1.5-cm-diameter faucet of Figure P1.36 has an average velocity of 2 m/s. Would you expect the flow to be laminar or turbulent?

Figure P1.36

1.37 The Red Cedar River flows placidly through Michigan State University's campus. In a certain section the depth is 2.5 ft and the average velocity was estimated from a floating leaf to be 0.6 fps. A PhD student stated that the flow was laminar. Was he correct?

1.38 Air at 40°C flows in a rectangular 30 cm × 6 cm heating duct at an average velocity of 4 m/s. Is the flow laminar or turbulent?

1.39 The sphere of diameter D of Figure P1.39 is traveling with a velocity V of 1.2 m/s in 20°C atmospheric air. If $\mathrm{Re} = VD/\nu$ is less than 4×10^4, the boundary layer around the front of the sphere is completely laminar. Determine if the boundary layer is completely laminar on a sphere of diameter:

- **(a)** 1 cm
- **(b)** 1 m

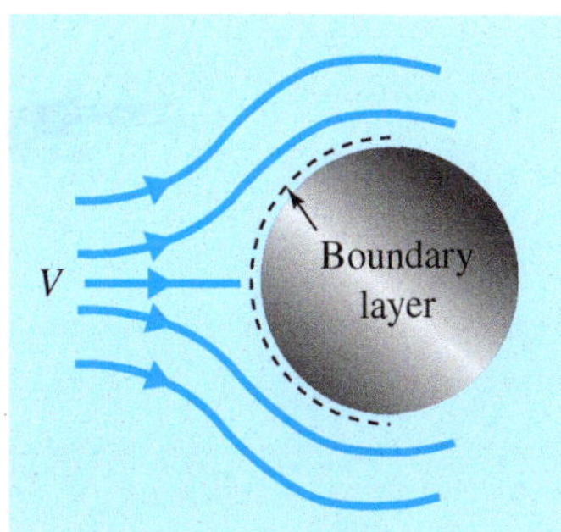

Figure P1.39

1.40 The airfoil on a commercial airliner is approximated as a flat plate sketched in Figure P1.40. How long would you expect the laminar portion of the boundary layer to be if it is flying:

(a) At an altitude of 10 000 m and a speed of 900 km/h?

(b) At an altitude of 30 000 ft and a speed of 600 mph?

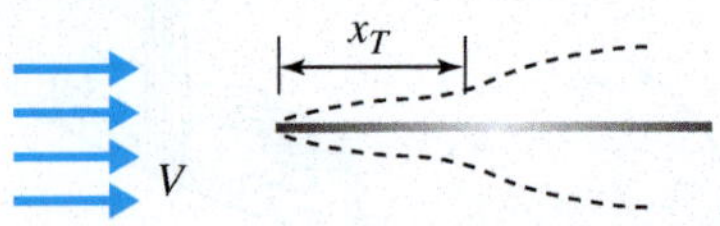

Figure P1.40

1.41 A leaf keeps cool by transpiration, a process in which water moves from the leaf to the atmosphere. An experimenter wonders if the boundary layer on a leaf influences the transpiration, so an "experimental" leaf is set up in the laboratory and air is blown parallel to it at 6 m/s. Comment as to whether the boundary layer is expected to be laminar or turbulent.

1.42 For the following situations state whether a compressible flow is required or if the flow can be approximated with an incompressible flow:

(a) An aircraft flying at 100 m/s at an elevation of 8000 m

(b) A golf ball traveling at 240 ft/s

(c) Flow around an object being studied in a high-temperature wind tunnel if the temperature is 100°C and the air velocity is 100 m/s

1.43 Write out Eq. 1.3.2 using Eq. 1.2.11. For a steady, plane flow what relationship must exist for an incompressible flow in which the density is allowed to vary?

1.44 If $\rho = \rho_0(1 + cz)$ models the density variation in a channel (there is heavy salt water at the bottom and fresh water at the top) in which $u(y, z)$ is the only velocity component, is the flow incompressible?

Bernoulli's Equation

1.45 A pitot tube is used to measure the velocity of a small aircraft flying at 3000 ft. Calculate its velocity if the pitot tube measures:
(a) 0.3 psi
(b) 0.9 psi
(c) 0.09 psi

1.46 Approximate the force acting on the 15-cm-diameter headlight shown in Figure P1.46 of an automobile traveling at 120 kph.

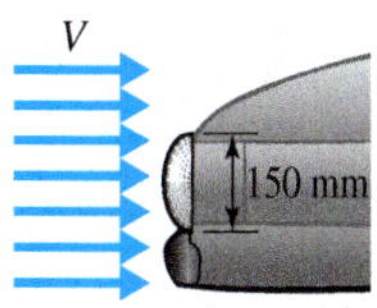

Figure P1.46

1.47 A vacuum cleaner is capable of creating a vacuum of 2 kPa just inside the hose of Figure P1.47. What maximum average velocity would be expected in the hose?

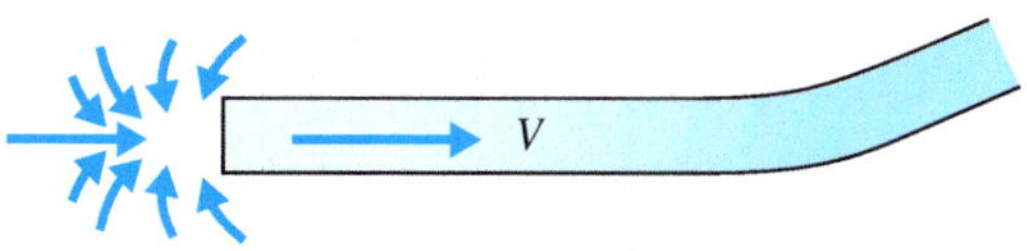

Figure P1.47

1.48 The inviscid, incompressible flow in the vicinity of a stagnation point (Figure P1.48) is approximated by $u = -10x$, $v = 10y$. If the pressure at the origin is p_0, find an expression for the pressure neglecting gravity effects:
(a) Along the negative x-axis
(b) Along the positive y-axis

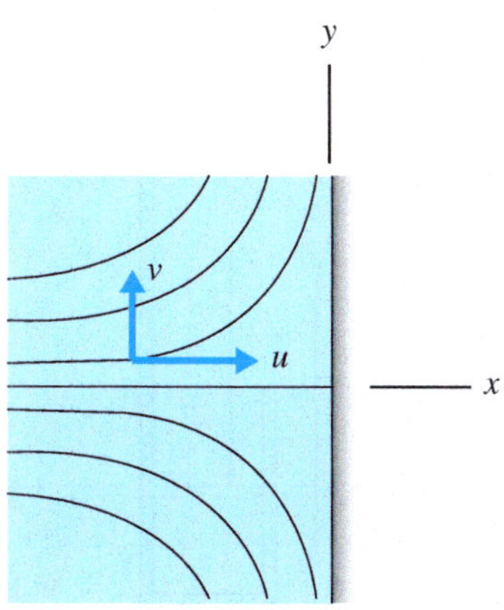

Figure P1.48

1.49 The inviscid, incompressible flow field exterior to the cylinder shown in Figure P1.49 is given by

$$v_r = U_\infty\left(1-\frac{r_c^2}{r^2}\right)\cos\theta, \quad v_\theta = U_\infty\left(1+\frac{r_c^2}{r^2}\right)\sin\theta$$

If the pressure at $r = \infty$ is zero (i.e., $p_\infty = 0$), find an expression for the pressure neglecting gravity effects:
(a) Along the negative x-axis
(b) At the stagnation point
(c) On the surface of the cylinder
(d) On the surface of the cylinder at $\theta = 90°$

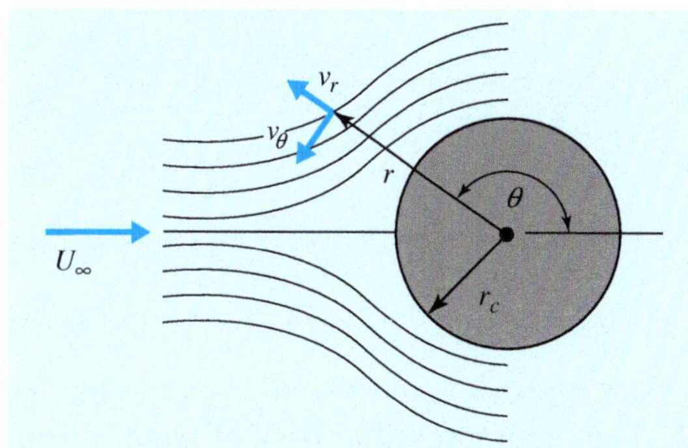

Figure P1.49

1.50 The inviscid, incompressible flow field exterior to a sphere (see Figure P1.49) is given by

$$v_r = U_\infty\left(1-\frac{r_c^3}{r^3}\right)\cos\theta, \quad v_\theta = U_\infty\left(1+\frac{r_c^3}{r^3}\right)\sin\theta$$

If the pressure at $r = \infty$ is zero, find an expression for the pressure neglecting gravity effects:
(a) Along the negative x-axis
(b) At the stagnation point
(c) On the surface of the sphere
(d) On the surface of the sphere at $\theta = 90°$

1.51 The velocity along the negative x-axis in the inviscid, incompressible flow field exterior to the body shown in Figure P1.51 is given by $u(x) = U_\infty + q/2\pi x$. If the pressure at $x = -\infty$ is zero, find an expression for the pressure neglecting gravity effects:

(a) Along the negative x-axis if $U_\infty = 10$ m/s and $q = 20\pi$ m^2/s
(b) At the stagnation point if $U_\infty = 10$ m/s and $q = 20\pi$ m^2/s
(c) Along the negative x-axis if $U_\infty = 30$ ft/s and $q = 60\pi$ ft^2/s
(d) At the stagnation point if $U_\infty = 30$ ft/s and $q = 60\pi$ ft^2/s

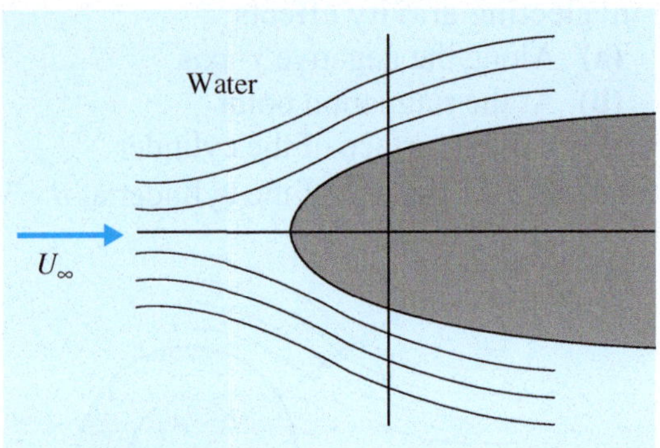

Figure P1.51

1.52 The incompressible flow of water through the short contraction of Figure 1.19a is assumed to be inviscid. If a pressure drop of 20 kPa is measured, estimate the velocity at the wall at section 2 just downstream of the contraction. (Actually, a boundary layer would develop, and the velocity calculated at the wall would be the velocity at the edge of the boundary layer; see the inset of Figure 1.10.)

1.53 Air flows from a relatively large plenum in a furnace out a relatively small rectangular duct. If the pressure in the plenum measures 60 Pa and in the duct 10.2 Pa, estimate the velocity of the 40°C air in the duct.

1.54 A short contraction is followed by an expansion, as shown in Figure P1.54. The manometer is used to determine the velocity of the fluid providing the viscous effects are negligible. If water is flowing steadily, determine the velocity V_1 if $H = 12$ cm. Losses through the contraction are negligible.

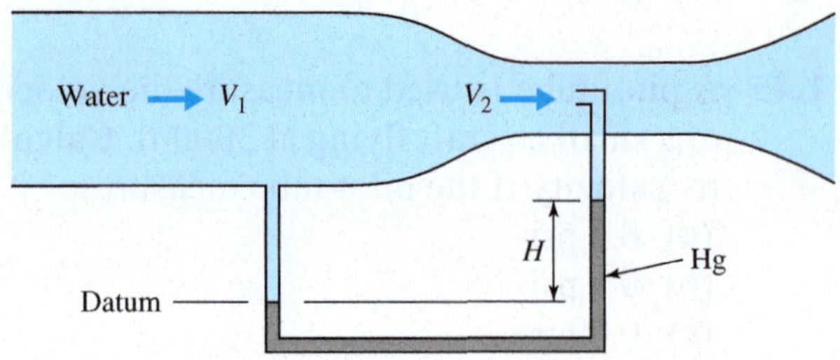

Figure P1.54

1.55 What is the velocity of the water in the pipe if the manometer shown in Figure P1.55 reads:
(a) 4 cm?
(b) 10 cm?
(c) 2 in.?
(d) 4 in.?

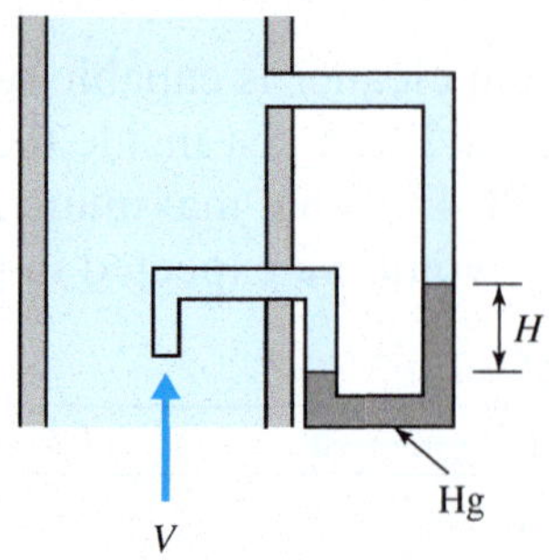

Figure P1.55

1.56 Air at 120 kPa absolute and 30°C flows vertically upward in a pipe, as shown in Figure P1.56. If the water manometer deflection $H = 5$ cm, determine the velocity in the smaller pipe. Assume the air to be incompressible.

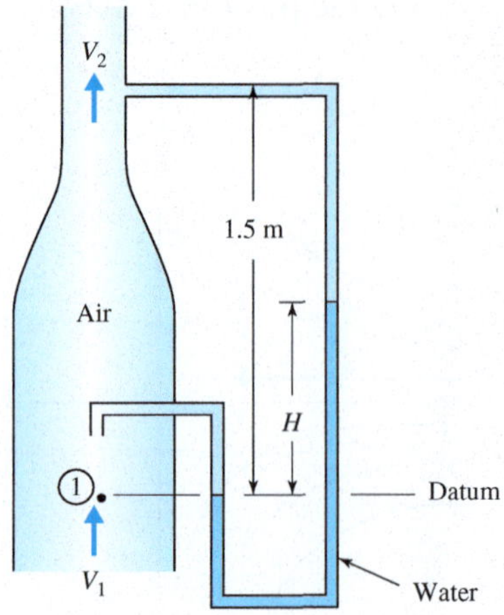

Figure P1.56

1.57 A manometer, positioned inside a cylinder as shown in Figure P1.57, reads 4 cm of water. Estimate U_∞ assuming an inviscid flow. Refer to the velocity field in Problem 1.49.

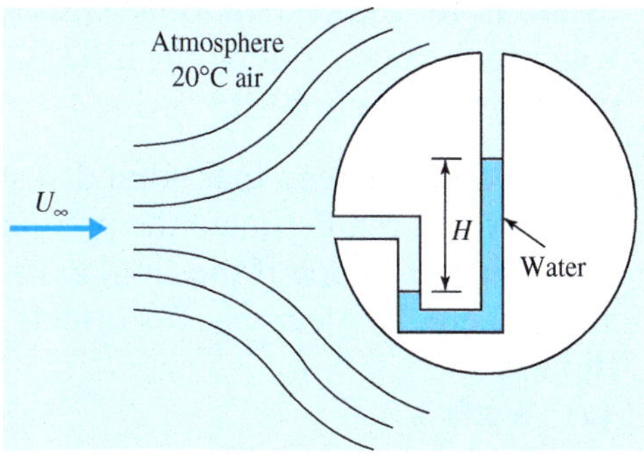

Figure P1.57

1.58 A manometer, positioned as shown in Figure P1.57 inside a sphere, reads 4 cm of water. Estimate U_∞ assuming an inviscid flow. Refer to the velocity field in Problem 1.50.

1.59 For the flow shown in Figure P1.59, estimate the pressure p_1 and velocity V_1 if $V_2 = 20$ m/s and:

(a) $H = 1$ cm
(b) $H = 5$ cm
(c) $H = 10$ cm

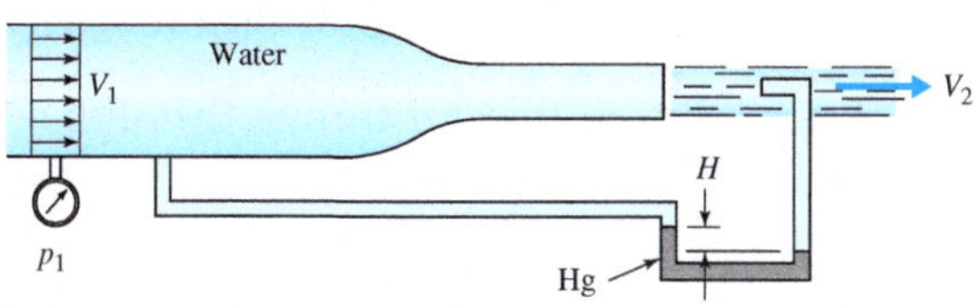

Figure P1.59

1.60 Water at 15°C flows steadily through the contraction shown in Figure P1.60 such that $V_2 = 4V_1$ If the gage reading is maintained at 120 kPa, determine the maximum velocity V_1 possible before cavitation occurs.

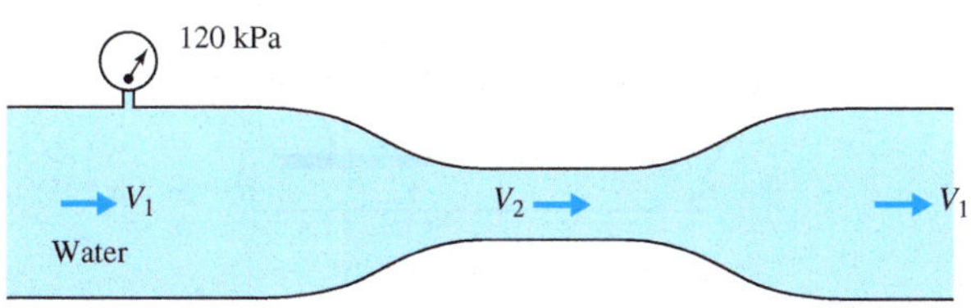

Figure P1.60

1.61 In the pipe contraction shown in Figure P1.61, water flows steadily with a velocity of $V_1 = 0.5$ m/s and $V_2 = 1.125$ m/s. Two piezometer tubes are attached to the pipe at sections 1 and 2. Determine the height H. Neglect any losses through the contraction.

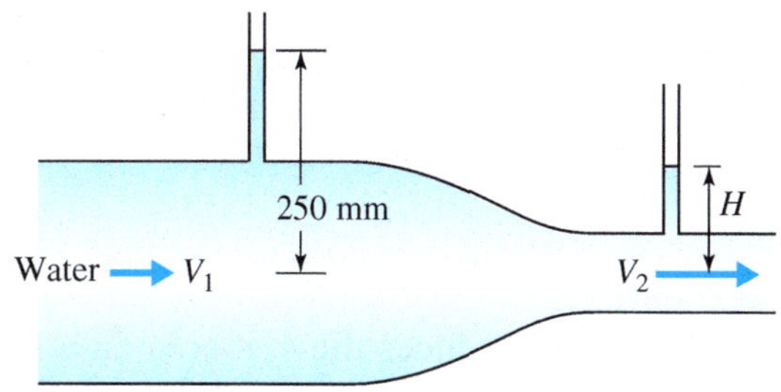

Figure P1.61

1.62 Air at 20°C is drawn into the hose of a vacuum cleaner through a head that is relatively free of obstructions (the flow can be assumed to be inviscid). Estimate the velocity in the hose if the vacuum in the hose measures:

(a) 2 cm of water
(b) 8 cm of water
(c) 1 in. of water
(d) 4 in. of water

1.63 A wind tunnel is designed to draw in air from the atmosphere and produce a velocity of 100 m/s in the test section. The fan is located downstream of the test section. What pressure is to be expected in the test section if the atmospheric temperature and pressure are:

(a) –20°C, 90 kPa?
(b) 0°C, 95 kPa?
(c) 20°C, 92 kPa?
(d) 40°C, 100 kPa?

1.64 The pump shown in Figure P1.64 creates a flow such that $V = 14$ m/s. Predict the pressure at the gage shown assuming an inviscid flow in the entrance and a uniform flow at the gage. Use a streamline starting at:

(a) Point A
(b) Point B

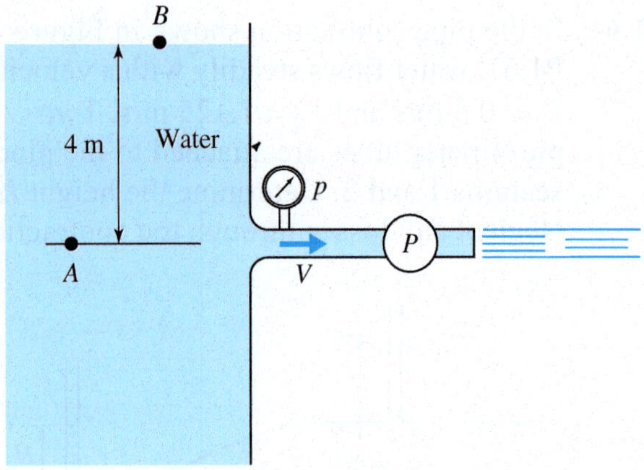

Figure P1.64

1.65 A fireman reduces the exit area on a nozzle so that the velocity inside the hose is quite small relative to the exiting velocity. What is the maximum exiting velocity and what is the maximum height the water can reach if the pressure inside the hose is:
(a) 700 kPa?
(b) 1400 kPa?
(c) 100 psi?
(d) 200 psi?

1.66 The velocity downstream of a sluice gate is assumed to be uniform (Figure P1.66). Express V in terms of H and h for this inviscid flow. Use a streamline:
(a) Along the top downstream
(b) Along the bottom downstream

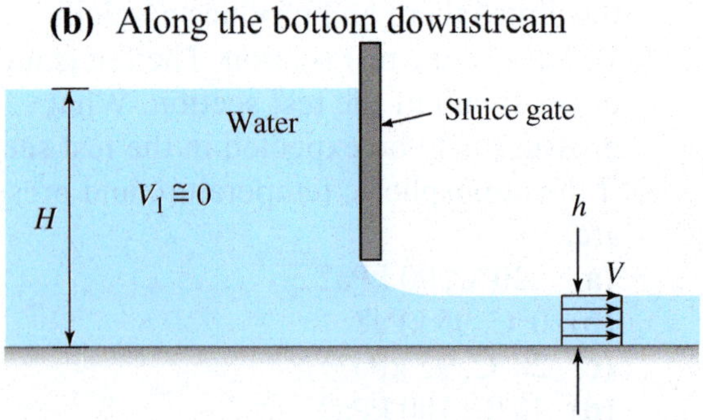

Figure P1.66

1.67 To what maximum velocity can water be accelerated before it reaches the turbine blades of a hydroturbine if it enters with relatively low velocity at:
(a) 600 kPa?
(b) 300 kPa?
(c) 80 psi?
(d) 40 psi?

1.68 Water exists in a city's water system at a pressure of 500 kPa at a particular location. The water pipe must traverse over a hill. How high could the hill be, above that location, for the system to possibly supply water to the other side of the hill? A negative pressure is not allowed.

1.69 Fluid flows between the radial disks shown in Figure P1.69. Estimate the pressure in the 2-cm-diameter pipe if the fluid exits to the atmosphere. Neglect viscous effects. The fluid is:
(a) Water
(b) Benzene
(c) Gasoline
(d) Air

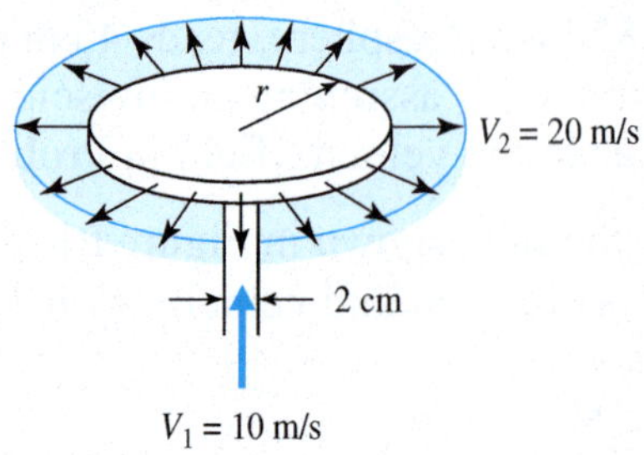

Figure P1.69

1.70 Estimate the pressure at $r = 10$ cm if the velocity there is 30 m/s in Problem 1.69d.

1.71 It is proposed that air blown through a tube attached to a metal disk be used to pick up envelopes, as shown in Figure P1.71. Would such a setup actually pick up an envelope? Explain. Assume an inviscid flow with the air slowing down as it moves radially outward.

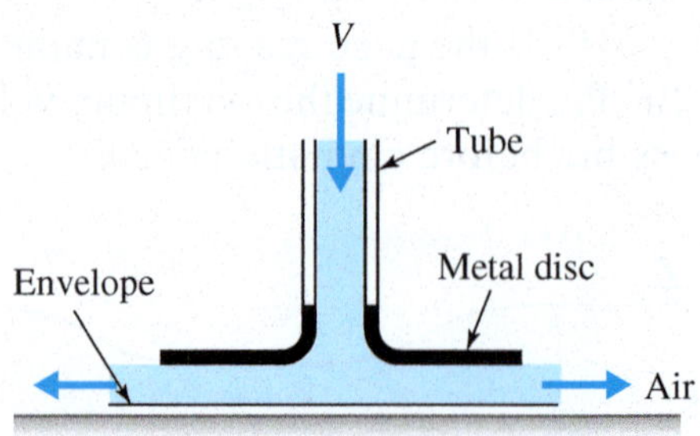

Figure P1.71

1.72 From which of the following objects would you expect the flow to separate and form a substantial separated region?

(a) A golf ball
(b) A telephone wire
(c) A wind machine blade
(d) A 2-mm-diameter wire in a low-speed wind tunnel
(e) An automobile
(f) An aircraft

(**Note:** Separation occurs whenever the Reynolds number exceeds a value around 20 over a blunt object.)

1.73 Explain, with the use of a sketch, why a burr on the downstream side of a piezometer opening in a pipe (see Figure 1.16a) will result in a reading of the pressure that is too high.

1.74 An incompressible, inviscid flow of water enters a bend with uniform velocity of V_1 = 10 m/s (Figure P1.74). Estimate the pressure difference between points A and B if the average radius of curvature at the bend is 5 cm. Sketch an anticipated velocity profile along AB. Assume that $p_A < p_1$ and $p_B < p_1$.

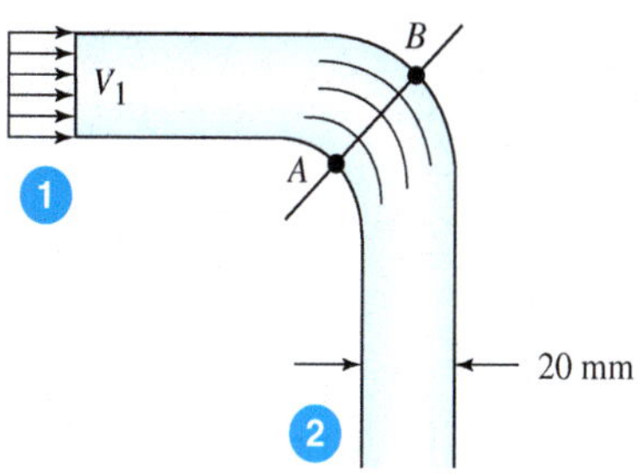

Figure P1.74

1.75 Viscosity causes a fluid to stick to a surface. If the fluid in Problem 1.74 sticks to the surface, explain why, in a viscous flow, a secondary flow is caused by the bend. Sketch such a flow on a circular cross section at section 2.

1.76 In Figure P1.76, assuming an inviscid flow, insert one of these signs between the pressures: >, <, ≅. Neglect changes due to elevation.

$p_A \quad p_B$

$p_C \quad p_D$

$p_B \quad p_D$

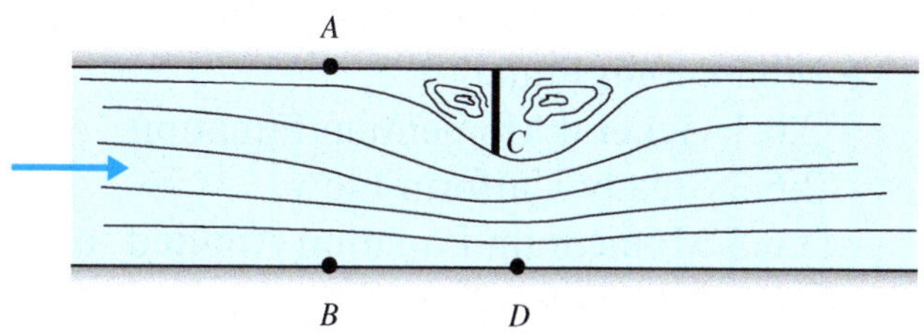

Figure P1.76

2 The Integral Forms of the Fundamental Laws

Outline

Chapter Objectives

The objectives of this chapter are to:

- Derive an equation that will allow us to convert the three basic laws that are stated for a system to a form that is applicable to a control volume.
- Apply the conservation of mass to control volumes of interest.
- Analyze the work-rate term of the energy equation.
- Apply the energy equation to numerous engineering situations.
- Apply Newton's second law to control volumes of interest.
- Apply the moment-of-momentum equation to rotating devices.
- Present numerous examples of the basic laws applied to control volumes so that students will be able to correctly solve fluid flow problems involving many of the control volumes of interest to engineers.
- Express the basic laws in their most general control-volume form so that complex problems encountered in engineering applications can be properly analyzed and hopefully solved.

M. C. Potter and D. C. Wiggert, *Applied Fluid Mechanics*,
https://doi.org/10.1007/978-3-031-65276-9_2

2.1 INTRODUCTION

Quantities of interest to engineers can often be expressed in terms of integrals. For example, volume flow rate is the integral of the velocity over an area; heat transfer is the integral of the heat flux over an area; force is the integral of a stress over an area; mass is the integral of the density over a volume; and kinetic energy is the integral of $V^2/2$ over each mass element in a volume. There are, of course, many other integral quantities. To determine an integral quantity, the integrand must be known, or information must be available so that a good approximation to the integrand can be made. If the integrand is not known or cannot be approximated with any degree of certainty, appropriate differential equations (see Chapter 3) must be solved yielding the needed integrand; the integration is then performed giving the engineer the desired integral quantity.

In this chapter we present the integral quantities of interest, develop equations that relate the integral quantities, and provide a number of examples for which the integrands are given or can be approximated. This includes a surprisingly large variety of problems. There are, however, many integral quantities that cannot be determined since the integrands are unknown. These would include the lift and drag on an airfoil, the torque on the blades of a wind machine, and the kinetic energy in the wake of a submarine. To determine such integrands, it would be necessary to solve the appropriate differential equations, a task that is often quite difficult; some relatively simple flow geometries for which solutions to the differential equation can be determined are considered in subsequent chapters.

Also, there are many quantities of interest that are not integral in nature. Included would be the point of separation of the flow around a body, the concentration of a pollutant in a stream at a certain location, the pressure distribution on the side of a building, and the wave-shore interaction along a lake. To study subjects such as these, it is necessary to consider the differential equations that describe the flow. Most of the topics mentioned are relegated to specialized graduate courses; however, some topics that require the solution of the more easily solvable differential equations are included in this book.

2.2 THE THREE BASIC LAWS

The integral quantities of primary interest in fluid mechanics are contained in three basic laws: conservation of mass, first law of thermodynamics, and Newton's second law. These basic laws are expressed using a Lagrangian description in terms of a **system**, a fixed collection of material particles. For example, if we consider flow through a pipe, we could identify a fixed quantity of fluid at time t as the system (Figure 2.1); this system would then move due to velocity to a downstream location at timer $t + \Delta t$. Any of the three basic laws could be applied to this system. This is, however, not an easy task. First let us state the basic laws in their general form.

Conservation of Mass: The law stating that mass must be conserved is:

The mass of a system remains constant

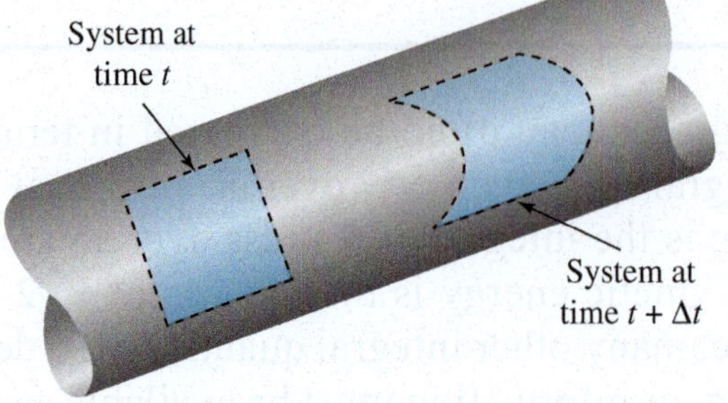

Figure 2.1 Example of a system in fluid mechanics.

The mass of a fluid particle is $\rho\, d\mathcal{V}$, where $d\mathcal{V}$ is the volume occupied by the particle and ρ is its density. Knowing that the density can change from point to point in the system, the conservation of mass can be expressed in integral form as

$$\frac{D}{Dt}\int_{\text{sys}} \rho d\mathcal{V} = 0 \tag{2.2.1}$$

where D/Dt is used since we are following a specified group of material particles, a system.

First Law of Thermodynamics: The law that relates heat transfer, work, and energy change is the first law of thermodynamics; it states:

> The rate of heat transfer to a system minus the rate at which the system does work equals the rate at which the energy of the system is changing

Recognizing that both density and specific energy may change from point to point in the system, it may be expressed as

$$\dot{Q} - \dot{W} = \frac{D}{Dt}\int_{\text{sys}} e\rho d\mathcal{V} \tag{2.2.2}$$

where the **specific energy** e accounts for kinetic energy $V^2/2$ per unit mass, potential energy gz per unit mass, and internal energy $\tilde{u}$ per unit mass. Other forms of energy—chemical, electrical, nuclear—are not included in a course in fluid mechanics. Equation (2.2.2) is often referred to as the **energy equation**.

In its basic form stated here, the first law of thermodynamics applies only to a system, a collection of fluid particles; therefore, D/Dt is used. We study $\dot{Q}$ and $\dot{W}$ in Section 2.5 where we consider the energy equation in detail.

Newton's Second Law: Newton's second law, also called the momentum equation, states:

> The resultant force acting on a system equals the rate
> at which the momentum of the system is changing

The momentum of fluid particle of mass is a vector quantity given by $\mathbf{V}\rho\, d\forall$; consequently, Newton's second law may be expressed in an inertial reference frame as

$$\Sigma \mathbf{F} = \frac{D}{Dt}\int_{\text{sys}} \mathbf{V}\rho d\forall \tag{2.2.3}$$

recognizing that both density and velocity may change from point to point in the system. This equation reduces to $\Sigma\, \mathbf{F} = m\mathbf{a}$ if $\mathbf{V}$ and ρ are constant throughout the entire system; ρ is often a constant, but in fluid mechanics the velocity vector invariably changes from point to point. Again, D/Dt is used to provide the rate-of-change since Newton's second law is applied to a system.

Moment-of-Momentum Equation: The moment-of-momentum equation results from Newton's second law; it states:

> The resultant moment acting on a system equals the rate
> of change of the angular momentum of the system

In equation form this becomes, relative to an inertial reference frame,

$$\Sigma \mathbf{M} = \frac{D}{Dt}\int_{\text{sys}} \mathbf{r} \times \mathbf{V}\rho d\forall \tag{2.2.4}$$

where $\mathbf{r} \times \mathbf{V}\rho d\forall$ represents the angular momentum of a fluid particle with mass $\rho d\forall$. The vector $\mathbf{r}$ locates the volume element $d\forall$ and is measured from the origin of the coordinate axes, the point relative to which the resultant moment is measured.

Note that in each of the basic laws the integral quantity is an extensive property (it depends on the mass) of the system. We will use the symbol N_{sys} to denote this extensive property; for example, N_{sys} could represent the mass, the momentum, or the energy of the system. The left-hand side of Eq. 2.2.1 and the right-hand sides of Eqs. 2.2.2, 2.2.3, and 2.2.4 may all be expressed as

$$\frac{DN_{\text{sys}}}{Dt} \tag{2.2.5}$$

where N_{sys} represents an integral quantity, either a scalar quantity or a vector quantity.

It is also useful to introduce the variable η for the intensive property (it is independent of the mass), the property of the system per unit mass. The relation between N_{sys} and η is given by

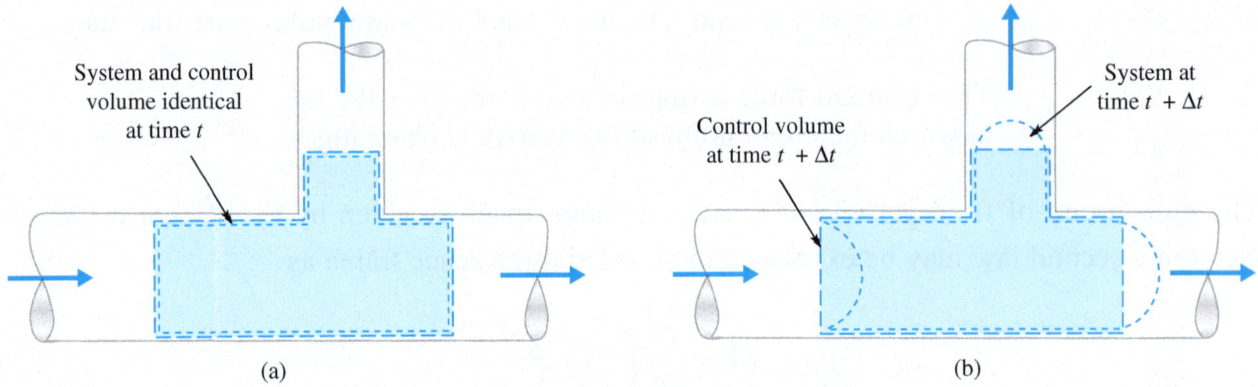

Figure 2.2 Example of a fixed control volume and a system: (a) time t; (b) time $t+\Delta t$.

$$N_{sys} = \int_{sys} \eta \rho d\mathcal{V} \tag{2.2.6}$$

As an example, the extensive property of Newton's second law is the momentum

$$\textbf{momentum}_{system} = \int_{sys} \mathbf{V}\rho \, d\mathcal{V} \tag{2.2.7}$$

which is a vector quantity. A corresponding intensive property would be the velocity vector **V**. Note that the density and velocity, which may vary from point to point within the system, may also be functions of time, as in an unsteady flow.

Our interest is most often focused on a device, or a region of space, into which fluid enters and/or from which fluid leaves; we identify this region as a **control volume**. An example of a fixed control volume is shown in Figure 2.2a; it's the tee in the pipe. A control volume need not be fixed; it could deform as in a piston-cylinder during exhaust or in a balloon as it deflates. We will, however, consider only fixed control volumes in this book. This will not limit us in most problem situations.

The difference between a control volume and a system is illustrated in Figure 2.2b. The figure indicates that the system occupies the control volume at time t and has partially moved out of it at time $t+\Delta t$. Since it is often more convenient to focus on a control volume (e.g., a pump) rather than on a system, the first order of business is to find a transformation that will allow us to express the substantial derivative of a system (a Lagrangian description) in terms of quantities associated with a control volume (an Eulerian description) so that the basic laws can be applied directly to a control volume. This will be done in general and then applied to the specific laws.

2.3 SYSTEM-TO-CONTROL-VOLUME TRANSFORMATION

We are interested in the time rate of change of the extensive property N_{sys} as we follow a system along, that is, DN_{sys}/Dt, and we would like to express this in terms of quantities that pertain to a control volume. In this section we present the derivation of such a transformation.

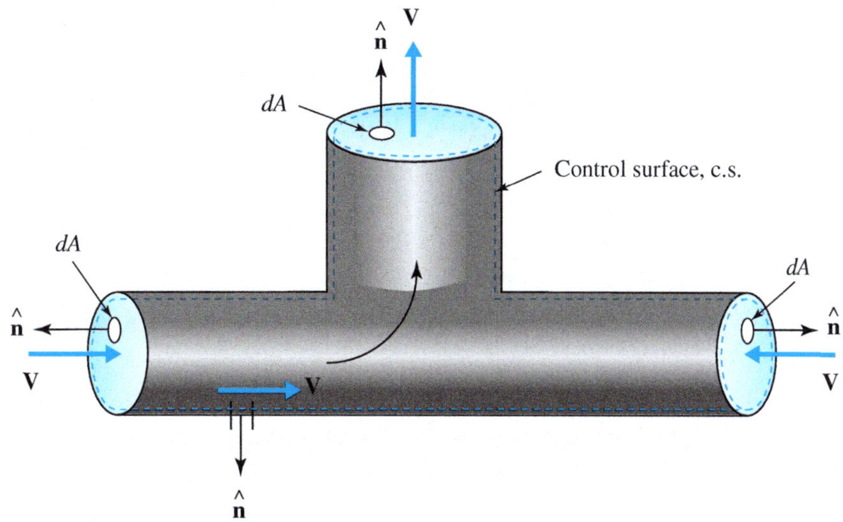

Figure 2.3 Illustration showing the flux of an extensive property.

The derivation involves fluxes of the extensive property in and out of the control volume. A flux is a measure of the rate at which an extensive property crosses an area; for example, a mass flux is the rate at which mass crosses an area. It is useful to introduce vector notation to describe these fluxes. Consider an area element dA of the **control surface**, the surface area that completely encloses a control volume. The property flux across an elemental area dA (see Figure 2.3) may be expressed by

$$\text{flux across } dA = \eta\rho\hat{\mathbf{n}}\cdot\mathbf{V}\ dA \tag{2.3.1}$$

where $\hat{\mathbf{n}}$, a unit vector normal to the area element dA, always points out of the control volume, and η represents the intensive property associated with N_{sys} (see Eq. 2.2.6). Only the normal component $\hat{\mathbf{n}}\cdot\mathbf{V}$ of the velocity vector contributes to this flux. If there is no normal component of velocity on a particular area, such as the wall of a pipe, no flux occurs across that area. A positive $\hat{\mathbf{n}}\cdot\mathbf{V}$ indicates a flux out of the volume; a negative $\hat{\mathbf{n}}\cdot\mathbf{V}$, that is, $\mathbf{V}$ has a component in the opposite direction of $\hat{\mathbf{n}}$, indicates a flux into the volume. We must always use $\hat{\mathbf{n}}$ pointing out of the volume. The velocity vector $\mathbf{V}$ may be at some angle to the unit vector $\hat{\mathbf{n}}$; the dot product $\hat{\mathbf{n}}\cdot\mathbf{V}$ accounts for the appropriate component of $\mathbf{V}$ that produces a flux through the area.

The net property flux out of the control surface is then obtained by integrating over the entire control surface:

$$\text{net flux of property} = \int_{\text{c.s.}} \eta\rho\hat{\mathbf{n}}\cdot\mathbf{V}dA \tag{2.3.2}$$

If the net flux is positive, the flux out is larger than the flux in.

Let us return now to the derivative DN_{sys}/Dt. The definition of a derivative from calculus allows us to write

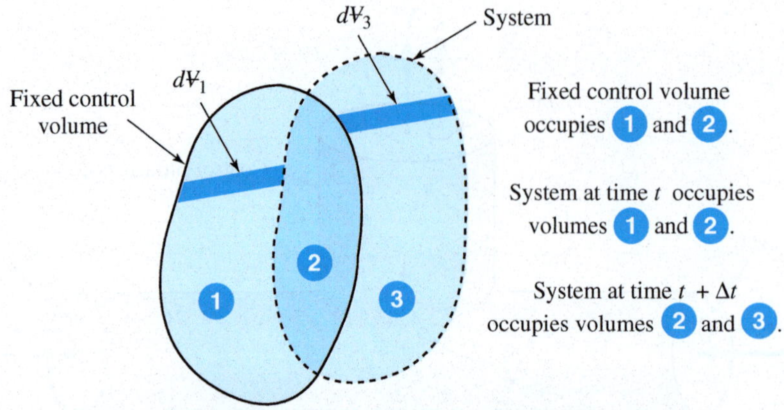

Figure 2.4 The system occupies the fixed control volume at time t.

$$\frac{DN_{\text{sys}}}{Dt} = \lim_{\Delta t \to 0} \frac{N_{\text{sys}}(t+\Delta t) - n_{\text{sys}}(t)}{\Delta t} \tag{2.3.3}$$

The system is shown in Figure 2.4 at times t and $t + \Delta t$. Assume that the system occupies the full control volume at time t; if we were considering a device, such as a pump, the particles of the system would just fill the device at time t. Since the device, the control volume shown in Figure 2.4, is assumed to be fixed in space, the system will move through the device. Equation 2.3.3 can then be written

$$\frac{DN_{\text{sys}}}{Dt} = \lim_{\Delta t \to 0} \frac{N_3(t+\Delta t) + N_2(t+\Delta t) - N_2(t) - N_1(t)}{\Delta t} \tag{2.3.4a}$$

$$= \lim_{\Delta t \to 0} \frac{N_2(t+\Delta t) + N_1(t+\Delta t) - N_2(t) - N_1(t)}{\Delta t} + \lim_{\Delta t \to 0} \frac{N_3(t+\Delta t) - N_1(t+\Delta t)}{\Delta t} \tag{2.3.4b}$$

where, in this second expression, we have simply added and subtracted N_1 $(t + \Delta t)$ in the numerator. In the equations above, the numerical subscript denotes the region; for example, $N_2(t)$ signifies the extensive property in region 2 at time t. Now, we observe that the first limit on the right-hand side of Eq. 2.3.4b refers to the control volume, so we can write

$$\frac{DN_{\text{sys}}}{Dt} = \lim_{\Delta t \to 0} \frac{N_{\text{c.v.}}(t+\Delta t) - N_{\text{c.v.}}(t)}{\Delta t} + \lim_{\Delta t \to 0} \frac{N_3(t+\Delta t) - N_1(t+\Delta t)}{\Delta t} \tag{2.3.5}$$

The first ratio on the right-hand side is $dN_{\text{c.v.}}/dt$, where we use an ordinary derivative since we are not following specific fluid particles. Thus there results

$$\frac{DN_{\text{sys}}}{Dt} = \frac{dN_{\text{c.v.}}}{dt} + \lim_{\Delta t \to 0} \frac{N_3(t+\Delta t) - N_1(t+\Delta t)}{\Delta t} \tag{2.3.6}$$

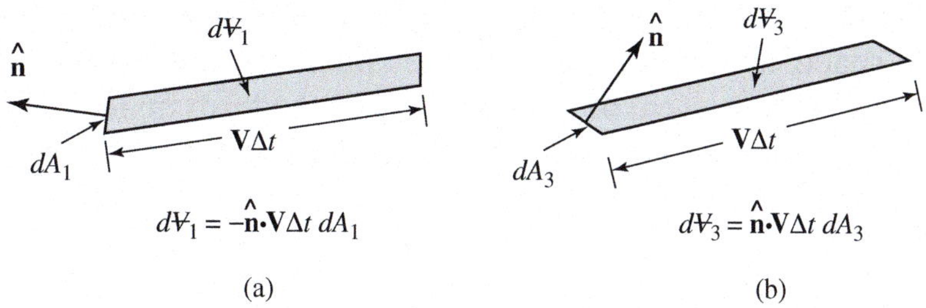

Figure 2.5 Differential volume elements.

Now, we must find expressions for the extensive quantities $N_3\,(t + \Delta t)$ and $N_1\,(t + \Delta t)$. They, of course, depend on the mass contained in the volume elements $d\Psi_1$ and $d\Psi_3$ shown in Figure 2.4 and enlarged in Figure 2.5. Note that the unit vector $\hat{\mathbf{n}}$ always points out of the volume, and hence to obtain a positive differential volume a negative sign is required for region 1. Also, note that the cosine of the angle between the velocity vector and the normal vector is required,[1] thus the presence of the dot product $(\hat{\mathbf{n}} \cdot \mathbf{V} = V\cos\theta)$. Referring to Figures 2.4 and 2.5 we have

$$
\begin{aligned}
N_3(t+\Delta t) &= \int_{A_3} \eta\rho\hat{\mathbf{n}}\cdot\mathbf{V}\Delta t dA_3 \\
N_1(t+\Delta t) &= -\int_{A_1} \eta\rho\hat{\mathbf{n}}\cdot\mathbf{V}\Delta t dA_1
\end{aligned}
\qquad (2.3.7)
$$

Recognizing that A_3 plus A_1 completely surrounds the control volume, we combine the two integrals into one integral. That is,

$$
N_3(t+\Delta t) - N_1(t+\Delta t) = \int_{\text{c.s.}} \eta\rho\hat{\mathbf{n}}\cdot\mathbf{V}\Delta t dA \qquad (2.3.8)
$$

where the control surface, denoted by c.s., is the area that completely surrounds the control volume. Substituting Eq. 2.3.8 back into Eq. 2.3.6 yields the desired result, the system-to-control-volume transformation, or equivalently, the **Reynolds transport theorem**:

$$
\frac{DN_{\text{sys}}}{Dt} = \frac{d}{dt}\int_{\text{c.v.}} \eta\rho\, d\Psi + \int_{\text{c.s.}} \eta\rho\hat{\mathbf{n}}\cdot\mathbf{V}dA \qquad (2.3.9)
$$

This is a Lagrangian-to-Eulerian transformation of the rate of change of an extensive integral quantity.

The first integral represents the rate of change of the extensive property in the control volume. The second integral represents the flux of the extensive property across the control surface; it is nonzero only where fluid crosses the control surface. We study this flux term in

[1]To obtain the volume of a box, we multiply the height by the area of the base, provided that the box is upright. If it is completely collapsed, its volume is zero. Hence for some intermediate position, the volume is the height times the area of the base times the cosine of the appropriate angle.

considerable detail in the following sections. Thus we can now express the basic laws in terms of a fixed volume in space. We will do this in subsequent sections for each of the basic laws.

We can move the time derivative of the control volume term inside the integral since, for a fixed control volume, the limits on the volume integral are independent of time; we then write an equivalent form of the Reynolds transport theorem as

$$\frac{DN_{\text{sys}}}{Dt} = \int_{\text{c.v.}} \frac{\partial}{\partial t}(\rho\eta)d\mathcal{V} + \int_{\text{c.s.}} \eta\rho\hat{\mathbf{n}}\cdot\mathbf{V}dA \tag{2.3.10}$$

In this form we have used $\partial/\partial t$ since ρ and η are, in general, dependent on the position variables, as well as the time t.

2.3.1 Simplifications of the Reynolds Transport Theorem

Many flows of interest are steady flows, so that $\partial(\eta\rho)\partial t = 0$. Our system-to-control-volume transformation then takes the form

$$\frac{DN_{\text{sys}}}{Dt} = \int_{\text{c.v.}} \eta\rho\hat{\mathbf{n}}\cdot\mathbf{V}dA \tag{2.3.11}$$

Furthermore, there is often only one area A_1 across which fluid enters the control volume and one area A_2 across which fluid leaves the control volume; assuming that the velocity vector is normal to the area (Figure 2.6), we can write $\hat{\mathbf{n}}\cdot\mathbf{V}_1 = -V_1$ over area A_1 and $\hat{\mathbf{n}}\cdot\mathbf{V}_2 = V_2$ over area A_2. Then Eq. 2.3.11 becomes

$$\frac{DN_{\text{sys}}}{Dt} = \int_{A_2} \eta_2\rho_2V_2dA - \int_{A_1} \eta_1\rho_1V_1dA \tag{2.3.12}$$

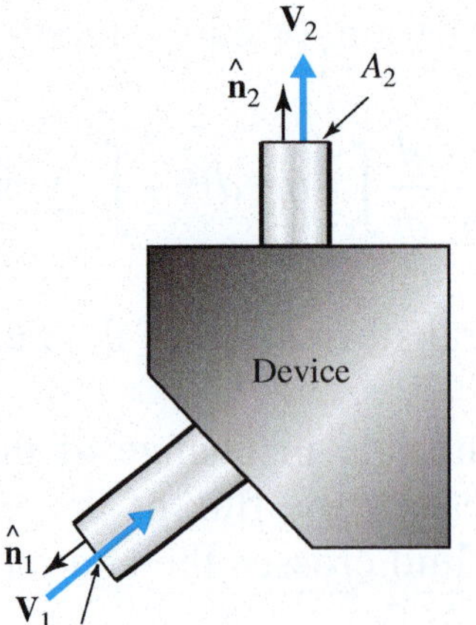

Figure 2.6 Flow into and from a device.

Finally, there are many situations that are modeled acceptably by assuming uniform properties over each plane area (see Figure 1.9); then the equation simplifies to

$$\frac{DN_{\text{sys}}}{Dt} = \eta_2 \rho_2 V_2 A_2 - \eta_1 \rho_1 V_1 A_1 \tag{2.3.13}$$

We will find that the system-to-control-volume transformation in this simplified form is most often used in the application of the basic laws to problems of interest. Some applications will, however, be included that will illustrate nonuniform distributions and unsteady flows.

If we generalize Eq. 2.3.13 to include several areas across which the fluid flows, we could write

$$\frac{DN_{\text{sys}}}{Dt} = \sum_{i=1}^{N} \eta_i \rho_i \hat{\mathbf{n}}_i \cdot \mathbf{V}_i A_i \tag{2.3.14}$$

where N is the number of areas. The dot product $\hat{\mathbf{n}} \cdot \mathbf{V}$ would provide us with the appropriate sign at each area; for an inlet area, $\hat{\mathbf{n}} \cdot \mathbf{V}$ introduces a negative sign, and for an exit area, $\hat{\mathbf{n}} \cdot \mathbf{V}$ retains the positive sign.

For an unsteady flow in which flow properties are assumed to be uniform throughout the control volume, the system-to-control-volume equation takes the form

$$\frac{DN_{\text{sys}}}{Dt} = \forall_{\text{c.v.}} \frac{d(\eta\rho)}{dt} + \eta_2 \rho_2 V_2 A_2 - \eta_1 \rho_1 V_1 A_1 \tag{2.3.15}$$

for one inlet and one outlet with uniform properties.

2.4 CONSERVATION OF MASS

A system is a given collection of fluid particles; hence its mass remains fixed. This is expressed as

$$\frac{Dm_{\text{sys}}}{Dt} = \frac{D}{Dt} \int_{\text{sys}} \rho \, d\forall = 0 \tag{2.4.1}$$

In Eq. 2.2.6, N_{sys} represents the mass of the system, so we simply let $\eta = 1$. Thus the conservation of mass, referring to Eq. 2.3.9, becomes

$$0 = \frac{d}{dt} \int_{\text{c.v.}} \rho \, d\forall + \int_{\text{c.s.}} \rho \hat{\mathbf{n}} \cdot \mathbf{V} dA \tag{2.4.2}$$

or, if we prefer, it takes the equivalent form for a fixed control volume,

$$0 = \int_{\text{c.v.}} \frac{\partial \rho}{\partial t} d\mathcal{V} + \int_{\text{c.s.}} \rho \hat{\mathbf{n}} \cdot \mathbf{V} dA \tag{2.4.3}$$

If the flow is steady, there results (see Eq. 2.3.11)

$$\int_{\text{c.s.}} \rho \hat{\mathbf{n}} \cdot \mathbf{V} dA = 0 \tag{2.4.4}$$

which, for a uniform flow with one entrance and one exit, takes the form (see Eq. 2.3.13)

$$\rho_2 A_2 V_2 = \rho_1 A_1 V_1 \tag{2.4.5}$$

where for an inlet we have used $\hat{\mathbf{n}}_1 \cdot \mathbf{V}_1 = -V_1$ and for an exit $\hat{\mathbf{n}} \cdot \mathbf{V}_2 = V_2$. Recall $\hat{\mathbf{n}}$ always points out of the control volume.

If the density is constant in the control volume, the derivative $\partial \rho / \partial t = 0$ even if the flow is unsteady; the continuity equation (2.4.3) then reduces to

$$A_1 V_1 = A_2 V_2 \tag{2.4.6}$$

This form of the continuity equation is used quite often, particularly with liquids and low-speed gas flows.

At this point we wish to discuss again the use of uniform velocity profiles (see also Section 1.3.1). Suppose that the velocity profiles at the entrance and the exit are not uniform, such as sketched in Figure 2.7. Furthermore, suppose that the density is uniform over each area. Then the continuity equation takes the form

$$\rho_1 \int_{A_1} V_1 dA = \rho_2 \int_{A_2} V_2 dA \tag{2.4.7}$$

or, letting an overbar denote an average, we can write

$$\rho_1 \overline{V}_1 A_1 = \rho_2 \overline{V}_2 A_2 \tag{2.4.8}$$

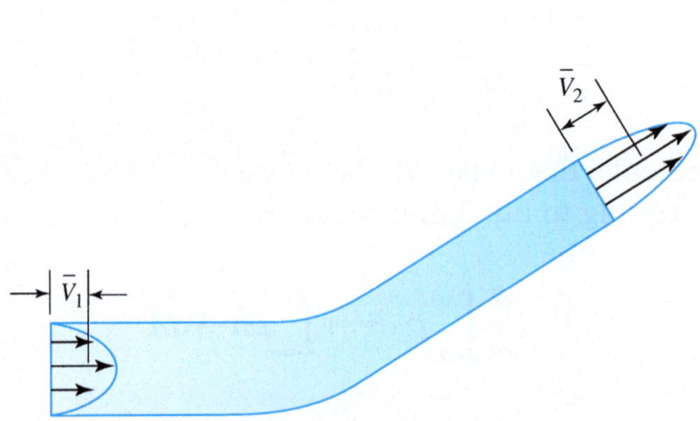

Figure 2.7 Nonuniform velocity profiles.

where $\overline{V}_1$ and $\overline{V}_2$ are the **average velocities** over the areas at sections 1 and 2, respectively. In examples and problems the overbar is often omitted. It should be kept in mind, however, that actual velocity profiles are usually not uniform; Eqs. 2.4.5 and 2.4.6 are used with the velocities representing average velocities.

Any one of the above equations (2.4.2) through (2.4.8) is referred to as the **continuity equation**.

Before presenting some examples applying the continuity equation, two fluxes are defined that will be useful in specifying the quantity of flow. The **mass flux** $\dot{m}$, or the mass rate of flow, is

$$\dot{m} = \int_A \rho V_n dA \tag{2.4.9}$$

and has units of kg/s (slug/s); V_n is the normal component of velocity. The **flow rate** Q, or the volume rate of flow, is

$$Q = \int_A V_n dA \tag{2.4.10}$$

and has units of m^3/s (ft^3/s) or sometimes L/s. The mass flux is usually used in specifying the quantity of flow for a compressible flow and the flow rate for an incompressible flow. We may also refer to the flow rate as **discharge**.

In terms of average velocity, we have

$$Q = A\overline{V} \tag{2.4.11}$$

$$\dot{m} = \rho A\overline{V} \tag{2.4.12}$$

where for the mass flux we assume a uniform density profile; we also assume that the velocity is normal to the area.

The following examples are solved by first selecting a control volume. If you study the examples carefully, you will notice that often there is only one proper choice for the control volume. We must position the inlet and exit areas at locations where the integrands are either known or where they can be approximated; also, the quantity being sought is often included at an inlet or exit area. In a few cases there may be more freedom in the selection of the control volume (Example 2.5).

This first example represents the primary use of the continuity equation. It allows us to calculate the velocity at one section if it is known at another section.

Example 2.1

Water flows at a uniform velocity of 3 m/s into a nozzle that reduces the diameter from 10 cm to 2 cm (Figure E2.1). Calculate the water's velocity leaving the nozzle and the flow rate.

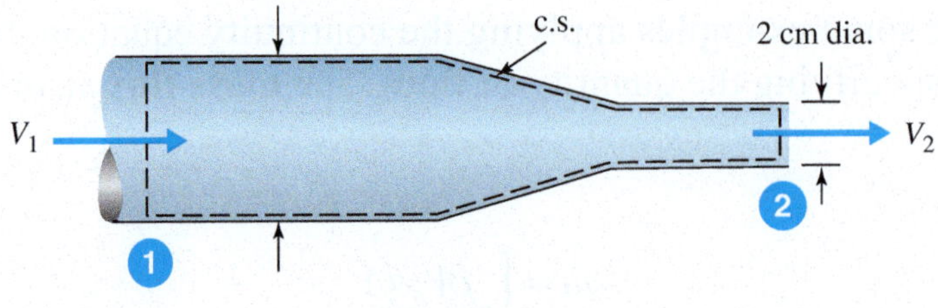

Figure E2.1

Solution

The control volume is selected to be the inside of the nozzle as shown. Flow enters the control volume at section 1 and leaves at section 2. The simplified continuity equation (2.4.6) is used since the density of water is assumed constant and the velocity profiles are uniform:

$$A_1V_1 = A_2V_2$$

$$\therefore V_2 = V_1\frac{A_1}{A_2} = 3\frac{\pi \times 0.1^2/4}{\pi \times 0.2^2/4} = \underline{75 \text{ m/s}}$$

The flow rate, or discharge, is found to be

$$Q = V_1A_1$$
$$= 3\times\pi\times 0.1^2/4 = \underline{0.0236 \text{ m}^3\text{/s}}$$

Example 2.2

Water flows in and out of a device as shown in Figure E2.2a. Calculate the rate of change of the mass of water (dm/dt) in the device.

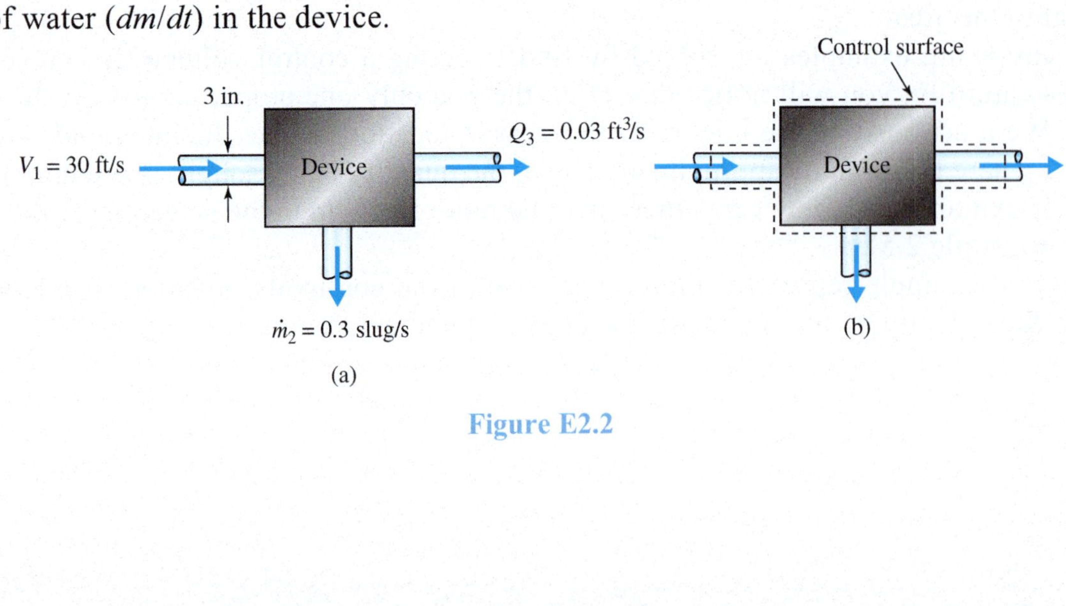

Figure E2.2

Solution

The control surface of the control volume selected is shown in Figure E2.2b. The continuity equation (2.4.2), with three surfaces across which water flows, takes the following form:

$$0 = \frac{d}{dt}\int_{\text{c.v.}} \rho d\forall + \int_{\text{c.s.}} \rho \hat{\mathbf{n}} \cdot \mathbf{V} dA$$
$$= \frac{dm}{dt} - \rho_1 A_1 V_1 + \rho_2 A_2 V_2 + \rho_3 A_3 V_3$$

where we have assumed the density to be constant over the volume and we have used $\mathbf{V}_1 \cdot \hat{\mathbf{n}} = -V_1$, since $\hat{\mathbf{n}}_1$ points out of the volume, opposite to the direction of $\mathbf{V}_1$. The last three terms come from the area integral. In terms of the quantities given, the above can be expressed as

$$0 = \frac{dm}{dt} - \rho_1 A_1 V_1 + \dot{m}_2 + \rho_3 Q_3$$
$$= \frac{dm}{dt} - 1.94 \text{ slug/ft}^3 \times \left(\pi \times \frac{1.5^2}{144} \right) \text{ft}^2 \times 30 \text{ ft/s} + 0.3 \text{ slug/s} + 1.94 \text{ slug/ft}^3 \times 0.3 \text{ ft}^3\text{/s}$$

This is solved to yield

$$\frac{dm}{dt} = \underline{1.975 \text{ slug/s}}$$

Hence, the mass is increasing at the rate of 1.975 slug/s. To accomplish this, the device could contain a spongelike material that absorbs water.

Example 2.3

A uniform flow of air approaches a cylinder as shown in Figure E2.3a. The symmetrical velocity distribution at the location shown downstream in the wake of the cylinder is approximated by

$$u(y) = 1.25 + \frac{y^2}{4} \qquad -1 < y < 1$$

where $u(y)$ is in m/s and y is in meters. Determine the mass flux across the surface AB per meter of depth (into the paper). Use $\rho = 1.23$ kg/m^3.

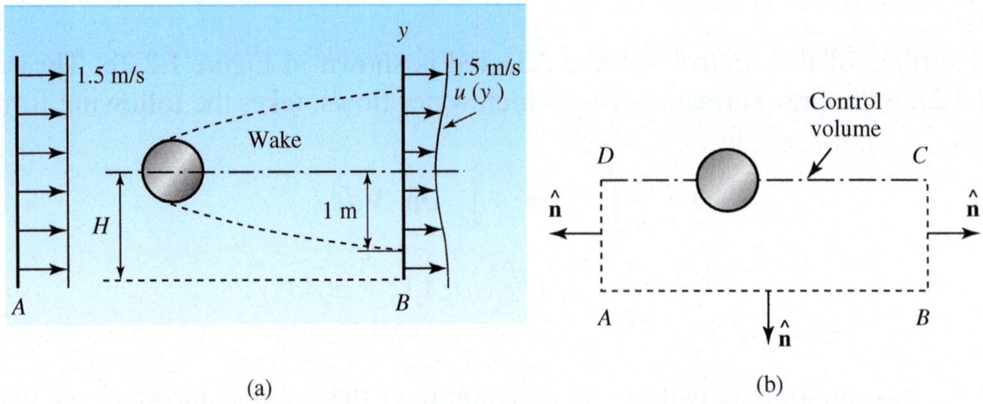

Figure E2.3

Solution

Select *ABCD* as the control volume (Figure E2.3b). Outside the wake (a region of retarded flow) the velocity is constant at 1.5 m/s. Hence the velocity normal to plane *AD* is 1.5 m/s. No mass flux crosses the surface *CD* because of symmetry. Assuming a steady flow, the continuity equation (2.4.3) becomes

$$0 = \int_{\text{c.s.}} \rho \mathbf{V} \cdot \hat{\mathbf{n}} dA$$

Mass flux occurs across three surfaces: *AB*, *BC*, and *AD*. Thus the equation above takes the form, assuming 1 m depth into the paper,

$$\begin{aligned} 0 &= \int_{A_{AD}} \rho \mathbf{V} \cdot \hat{\mathbf{n}} dA + \int_{A_{BC}} \rho \mathbf{V} \cdot \hat{\mathbf{n}} dA + \int_{A_{AD}} \rho \mathbf{V} \cdot \hat{\mathbf{n}} dA \\ &= \dot{m}_{AB} + \int_0^H \rho u(y)\, 1 \times dy - \rho \text{ kg/m}^3 \times 1.5 \text{ m/s} \times H \text{ m} \times 1 \text{ m} \end{aligned}$$

where the negative sign for surface *AD* results from the fact that the unit vector points out of the volume to the left while the velocity vector points to the right. Recall that a negative sign in the steady-flow continuity equation is always associated with an influx and a positive sign with an outflux. Now, we integrate out to 1 m instead of *H*, since the mass that enters on the left beyond 1 m simply leaves on the right with no net gain or loss. So, letting $H = 1$ m, we have

$$0 = \dot{m}_{AB} + \int_0^1 1.23\left(1.25 + \frac{y^2}{4}\right) dy - (1.23 \times 1 \times 1.5) \text{ kg/s}$$

Perform the integration and there results

$$\dot{m}_{AB} = \underline{0.205 \text{ kg/s per meter}}$$

Example 2.4

A balloon is being inflated with a water supply of 0.6 m^3/s (Figure E2.4a). Find the rate of increase of the radius at the instant when $R = 0.5$ m.

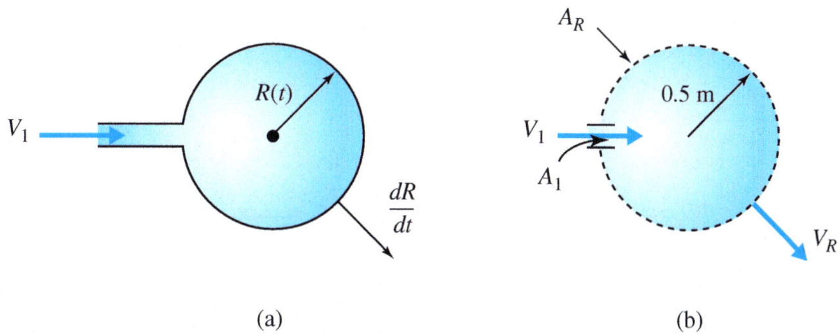

Figure E2.4

Solution

The objective is to find dR/dt when the radius $R = 0.5$ m. This growth rate $V_R = dR/dt$ is the same as the water velocity normal to the wall of the balloon. Therefore, we select as our fixed control volume a sphere with a constant radius of 0.5 m (see Figure E2.4b) so that we can calculate the velocity of the water at the surface at the instant shown moving radially out at $R = 0.5$ m. The continuity equation is written as

$$0 = \int_{\text{c.v.}} \underbrace{\frac{\partial \rho}{\partial t}}_{0} d\forall + \int_{\text{c.s.}} \rho \mathbf{V} \cdot \hat{\mathbf{n}}\, dA$$

The first term is zero because the density of water inside the control volume does not change in time. Further, the water crosses two areas: the inlet area A_1 with a velocity V_1 and the remainder of the sphere surface A_R with a velocity V_R. We will assume that $A_1 << A_R$. The continuity equation then takes the form

$$0 = -\rho A_1 V_1 + \rho A_R V_R$$

Since the flow rate into the volume is $A_1 V_1 = 0.6$ m^3/s and $A_R \simeq 4\pi R^2$ assuming that A_1 is quite small, we can solve for V_R. At $R = 0.5$ m

$$V_R = \frac{A_1 V_1}{4\pi R^2} = \frac{0.6 \text{ m}^3/\text{s}}{4\pi \times 0.5^2 \text{ m}^2} = 0.191 \text{ m/s}$$

$$\therefore \frac{dR}{dt} = \underline{0.191 \text{ m/s}}$$

We have used a fixed control volume and allowed the moving surface of the balloon to pass through it at the instant considered. With this approach it is possible to model situations in which surfaces, such as a piston, are allowed to move.

Example 2.5

This example shows that there may be more than one good choice for a control volume. We want to determine the rate at which the water level rises in an open container if the water coming in through a 0.10-m^2 pipe has a velocity of 0.5 m/s and the flow rate going out is 0.2 m^3/s (Figure E2.5a). The container has a circular cross section with a diameter of 0.5 m.

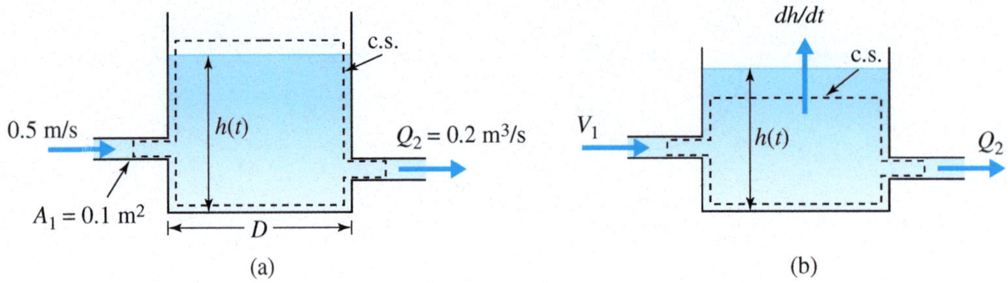

Figure E2.5

Solution

First we select a control volume that extends above the water surface as shown in Figure E2.5a. Apply the continuity equation (Eq. 2.4.2):

$$\frac{d}{dt}\int_{\text{c.v.}} \rho d\forall + \rho(-V_1)A_1 + \rho V_2 A_2 = 0$$

in which the first term describes the rate of change of mass in the control volume. Hence, neglecting the airmass above the water, we have

$$\frac{d(\rho h \pi D^2/4)}{dt} - \rho V_1 A_1 + \rho Q_2 = 0$$

Divide by the constant ρ,

$$\frac{\pi D^2}{4}\frac{dh}{dt} - V_1 A_1 + Q_2 = 0$$

The rate at which the water level is rising is then

$$\frac{dh}{dt} = \frac{V_1 A_1 - Q_2}{\pi D^2/4}$$

Thus,

$$\frac{dh}{dt} = \frac{(0.5 \times 0.1 - 0.2)\ \text{m}^3/\text{s}}{(\pi \times 0.5^2/4)\ \text{m}^2} = \underline{-0.764\ \text{m/s}}$$

The negative sign indicates that the water level is actually decreasing.

Let's solve this problem again but with another choice for the control volume, one with its top surface below the water level (Figure E2.5b). The velocity at the top surface is then equal to the rate at which the surface rises, i.e., *dh*/*dt*. The flow condition inside the control volume is steady. Hence we can apply Eq. 2.4.4. There are three areas across which fluid flows. On the third area, the velocity is *dh*/*dt*; hence the continuity equation takes the form

$$\rho(-V_1)A_1 + \rho Q_2 + \rho\frac{dh}{dt}\frac{\pi}{4}D^2 = 0$$

so that

$$\frac{dh}{dt} = \frac{V_1 A_1 - Q_2}{\pi D^2/4}$$

This is the same result as given above.

2.5 ENERGY EQUATION

Many problems involving fluid motion demand that the first law of thermodynamics, often referred to as the *energy equation*, be used to relate quantities of interest. If the heat transferred to a device (a boiler or compressor), or the work done by a device (a pump or turbine), is desired, the energy equation is obviously needed. It is also used to relate pressures and velocities when Bernoulli's equation is not applicable; this is the case whenever viscous effects cannot be neglected, such as flow through a piping system in an industrial plant or a golf course, or in an open channel bringing water to Los Angeles.

Let us express the energy equation in control volume form. For a system it is

$$\dot{Q} - \dot{W} = \frac{D}{Dt}\int_{\text{sys}} e\rho\, d\mathcal{V} \tag{2.5.1}$$

where the specific energy e includes specific kinetic energy $V^2/2$, specific potential energy gz, and specific internal energy $\tilde{u}$; that is,

$$e = \frac{V^2}{2} + gz + \tilde{u} \tag{2.5.2}$$

We will not include other forms of energy, such as energy due to magnetic or electric field-flow field interactions or those due to chemical reactions. In terms of a control volume, Eqs 2.5.1 becomes

$$\dot{Q} - \dot{W} = \frac{d}{dt}\int_{\text{c.v.}} e\rho \, d\forall + \int_{\text{c.s.}} \rho e \mathbf{V} \cdot \hat{\mathbf{n}} \, dA \tag{2.5.3}$$

This can be put in simplified forms for certain restricted flows, but first let us discuss the rate-of-heat transfer term $\dot{Q}$ and the work-rate term $\dot{W}$.

The term $\dot{Q}$ represents the rate-of-energy transfer across the control surface due to a temperature difference. (Do not confuse this term with the flow rate Q.) The rate-of-heat transfer term is either given or results from using Eq. 2.5.3. The calculation of $\dot{Q}$ from given temperatures is the objective of a course in heat transfer and is quite difficult, in general, to calculate. An entire course in heat transfer is often devoted to the determination of $\dot{Q}$ from given temperatures. The work-rate term is discussed in detail in the following section.

2.5.1 Work-Rate Term

The work-rate term results from work being done by the system. Or, since we consider the instant that the system occupies the control volume, we can also state that the work-rate term results from work being done by the control volume. Work of interest in fluid mechanics is due to a force moving through a distance while it acts on the control volume. The rate of doing work $\dot{W}$, or power, is given by the dot product of a force **F** with its velocity:

$$\dot{W} = -\mathbf{F} \cdot \mathbf{V}_I \tag{2.5.4}$$

where $\mathbf{V}_I$ is the velocity measured with respect to a fixed inertial reference frame. The negative sign results because we have selected the convention that work done on the control volume is negative.

If the force results from a variable stress acting over the control surface, we must integrate,

$$\dot{W} = -\int_{\text{c.s.}} \boldsymbol{\tau} \cdot \mathbf{V}_I \, dA \tag{2.5.5}$$

where $\boldsymbol{\tau}$ is the stress vector acting on the elemental area dA, the differential force being represented by $d\mathbf{F} = \boldsymbol{\tau} \, dA$, as shown in Figure 2.8.

For a moving control volume, such as a car, we have to evaluate the velocity with respect to a fixed reference frame. For example, let us consider an automobile traveling at constant speed (see Example 2.10). If we want to apply the energy equation, we could make the car the control volume. In that case, the velocity in Eq. 2.5.4 would be measured relative to a fixed reference and not relative to the car. If the velocity relative to the car were used, the drag force would have a zero velocity which would result in no work done; but we know that at high speed, energy from the gasoline goes primarily to overcome the drag. Thus the stationary reference frame is needed.

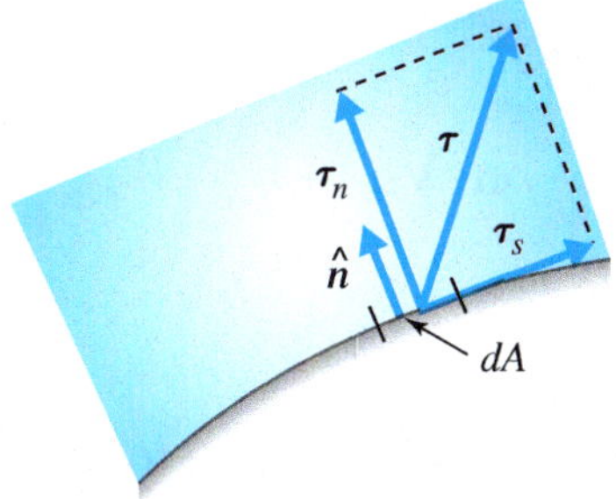

Figure 2.8 Stress vector acting on the control surface.

In general, for moving control volumes the velocity vector $\mathbf{V}_I$ is related to a relative velocity $\mathbf{V}$, observed in a reference frame attached to the control volume by

$$\mathbf{V}_I = \mathbf{V} + \dot{\mathbf{S}} + \Omega \times \mathbf{r} \quad (2.5.6)$$

where $\dot{\mathbf{S}}$ is the velocity of the control volume (see Figure 1.5). We can now write the work rate of Eq. 2.5.5 as

$$\dot{W} = -\int \boldsymbol{\tau} \cdot \mathbf{V} dA + \dot{W}_1 \quad (2.5.7)$$

where the "inertial work-rate" term is given by

$$\dot{W}_I = -\int_{\text{c.s.}} \boldsymbol{\tau} \cdot (\dot{\mathbf{S}} + \Omega \times \mathbf{r}) dA \quad (2.5.8)$$

Next, express the stress vector as the sum of a normal component and a shear component, as shown in Figure 2.8, that is,

$$\boldsymbol{\tau} = -p\hat{\mathbf{n}} + \boldsymbol{\tau}_s \quad (2.5.9)$$

where the normal stress component is assumed to be negative the pressure p. Then

$$\dot{W} = \int_{\text{c.s.}} p\hat{\mathbf{n}} \cdot \mathbf{V} dA - \int_{\text{c.s.}} \boldsymbol{\tau}_s \cdot \mathbf{V} dA + \dot{W}_1 \quad (2.5.10)$$

We will allow the shear stress term to consist of two parts. One part accounts for work called **shaft work** $\dot{W}_S$ transmitted by a rotating shaft that is cut by the control surface; this term is important when we deal with flows in pumps and turbines. The other part will be denoted **shear work** $\dot{W}_{\text{shear}}$ and results from moving boundaries; this term is required if the control surface itself moves relative to the control volume, as occurs with a moving belt.

Hence, the work-rate term becomes

$$\dot{W} = \int_{\text{c.s.}} p\hat{\mathbf{n}} \cdot \mathbf{V} dA + \dot{W}_S + \dot{W}_{\text{shear}} + \dot{W}_I \tag{2.5.11}$$

The terms are summarized as follows:

$\int p\hat{\mathbf{n}} \cdot \mathbf{V}\, dA$ Work rate resulting from the force due to pressure moving at the control surface. It is often referred to as **flow work**.

$\dot{W}_S$ Work rate resulting from rotating shafts such as that of a pump or turbine, or the equivalent electric power.

$\dot{W}_{\text{shear}}$ Work rate due to the shear acting on a moving boundary, such as a moving belt.

$\dot{W}_I$ Work rate that occurs when the control volume moves relative to a fixed reference frame.

We should note that the work-rate terms $\dot{W}_{\text{shear}}$ and $\dot{W}_I$ are seldom encountered in problems in textbooks. They are included here for completeness.

2.5.2 General Energy Equation

When the work-rate term of Eq. 2.5.11 is substituted into Eq. 2.5.3, we obtain the energy equation in the form

$$\dot{Q} - \dot{W}_S - \dot{W}_{\text{shear}} - \dot{W}_I = \frac{d}{dt}\int_{\text{c.v.}} e\rho\, d\mathcal{V} + \int_{\text{c.s.}} \left(e + \frac{p}{\rho}\right) \rho\hat{\mathbf{n}} \cdot \mathbf{V} dA \tag{2.5.12}$$

The work-rate term needed to move the pressure force has been moved to the right-hand side, as is common, and is treated like an energy flux term.

Substitution of Eq. 2.5.2 into Eq. 2.5.12 results in

$$\dot{Q} - \dot{W}_S - \dot{W}_{\text{shear}} - \dot{W}_I = \frac{d}{dt}\int_{\text{c.v.}} \left(\frac{V_I^2}{2} + gz + \tilde{u}\right) \rho\, d\mathcal{V}$$
$$+ \int_{\text{c.s.}} \left(\frac{V_I^2}{2} + gz + \tilde{u} + \frac{p}{\rho}\right) \rho \mathbf{V} \cdot \hat{\mathbf{n}} dA \tag{2.5.13}$$

This general form of the energy equation is useful in analyzing fluid flow problems that may include time-dependent effects and nonuniform profiles. Before we simplify the equation for steady flow and uniform profiles, let us introduce the losses.

In many fluid flows, useful forms of energy (kinetic energy and potential energy) and flow work are converted into unusable energy forms (internal energy or heat transfer). If we assume that the temperature of the control volume remains unchanged, the internal energy does not change and the losses are balanced by heat transfer across the control surface. This heat transfer can be the result of convection, radiation, or conduction at the control surfaces. The theory of heat transfer presents a detailed description of these effects. However, in a fluid mechanics course the sum of these effects is lumped together and denoted $\dot{Q}$. Thus, we define **losses** as the sum of all the terms representing unusable forms of energy:

$$\text{losses} = -\dot{Q} + \frac{d}{dt}\int_{\text{c.v.}} \tilde{u}\rho dV + \int_{\text{c.s.}} \tilde{u}\rho \mathbf{V}\cdot\hat{\mathbf{n}}\, dA \qquad (2.5.14)$$

We can now rewrite the energy equation as

$$-\dot{W}_S - \dot{W}_{\text{shear}} - \dot{W}_I = \frac{d}{dt}\int_{\text{c.v.}} \left(\frac{V_I^2}{2} + gz\right)\rho d\forall$$
$$+ \int_{\text{c.s.}} \left(\frac{V_I^2}{2} + gz + \frac{p}{\rho}\right)\rho\mathbf{V}\cdot\hat{\mathbf{n}}\, dA + \text{losses} \qquad (2.5.15)$$

Losses are due to two primary effects:

1. Viscosity causes internal friction that results in increased internal energy (temperature increase) or heat transfer.
2. Changes in geometry result in separated flows that require useful energy to maintain the resulting secondary motions in which viscous dissipation occurs.

In a conduit, the losses due to viscous effects are distributed over the entire length, whereas the loss due to a geometry change (a valve, an elbow, an enlargement) is concentrated in the vicinity of the geometry change.

The analytical calculation of losses is rather difficult, particularly when the flow is turbulent. In general, the prediction of losses is based on empirical formulas. Such formulas will be given in subsequent chapters. In this chapter we discuss losses qualitatively, and, in examples and problems, losses will be given. For a pump or a turbine the losses are expressed in terms of the efficiency. For example, if the efficiency of a pump is 80%, the losses would be 20% of the energy input to the pump.

It may be that the objective in a particular fluid flow is to change the internal energy of the fluid, such as in the steam generator (boiler) of a power plant, by the transfer of heat; then the definition of losses above must be altered so that the loss term includes only the dissipative effects caused by the viscosity of the fluid. Generally, for problems of interest in fluid mechanics, Eq. 2.5.15 is acceptable.

2.5.3 Steady Uniform Flow

Consider a steady-flow situation in which there is one entrance and one exit across which uniform profiles can be assumed. Also, assume that $\dot{W}_{\text{shear}} = \dot{W}_I = 0$ with $V_I = V$. For such a flow the term $(V^2/2 + gz + p/\rho)$ in Eq. 2.5.15 is constant across the cross section because V is constant (we assume a uniform velocity profile) and the sum of $p/\rho + gz$ is constant if the streamlines at each section are parallel. The energy equation (Eq. 2.5.15) then simplifies to

$$-\dot{W}_S = \rho_2 V_2 A_2 \left(\frac{V_2^2}{2} + \frac{p_2}{\rho_2} + gz_2 \right) - \rho_1 V_1 A_1 \left(\frac{V_1^2}{2} + \frac{p_1}{\rho_1} + gz_1 \right) + \text{losses} \tag{2.5.16}$$

where the subscripts 1 and 2 refer to the entrance and exit, respectively. The mass flux is given by $\dot{m} = \rho_1 A_1 V_1 = \rho_2 A_2 V_2$. After dividing by $\dot{m}g$ we have

$$-\frac{\dot{W}_S}{\dot{m}g} = \frac{V_2^2 - V_1^2}{2g} + \frac{p_2}{\gamma_2} - \frac{p_1}{\gamma_1} + z_2 - z_1 + h_L \tag{2.5.17}$$

where we have introduced the **head loss** h_L, defined to be

$$h_L = \frac{\tilde{u}_2 - \tilde{u}_1}{g} - \frac{\dot{Q}}{\dot{m}g} \tag{2.5.18}$$

It is often written in terms of a **loss coefficient** K as

$$h_L = K\frac{V^2}{2g} \tag{2.5.19}$$

where V is most often V_1 or V_2; if it is not obvious, it will be specified. Loss coefficients are discussed in some detail in Chapter 5 and are tabulated in Table 5.2.

The head loss is referred to as a "head" since it has dimensions of length. We may also refer to $V^2/2g$ as the **velocity head** and p/γ as the **pressure head** since those terms also have dimensions of length. Also recall from Chapter 1 that $p/\gamma + z$ is called the **piezometric head**. Further, the sum of the piezometric head and the velocity head is called the **total head**.

The energy equation, in the form of Eq. 2.5.17, is useful in many applications and is, perhaps, the most often used form of the energy equation. If the losses are negligible and if there is no shaft work, we note that the energy equation takes the form

$$\frac{V_2^2}{2g} + \frac{p_2}{\gamma_2} + z_2 = \frac{V_1^2}{2g} + \frac{p_1}{\gamma_1} + z_1 \tag{2.5.20}$$

Observe that the energy equation has been reduced to a form identical with Bernoulli's equation when $\gamma_2 = \gamma_1$ (a constant density flow). We must remember, however, that Bernoulli's equation is a momentum equation applicable along a streamline and the equation above is an energy equation applied between two sections of a flow. It is not surprising that both should predict identical results from the conditions stated because the velocity head is constant over a cross section and the sum of pressure head and elevation remains constant over a cross section.

The energy equation (2.5.17) may be applied to any steady, uniform flow with one entrance and one exit. The control volume is usually selected such that the entrance and exit sections have a uniform total head. For example, it may be applied to water flow through a long pipeline;

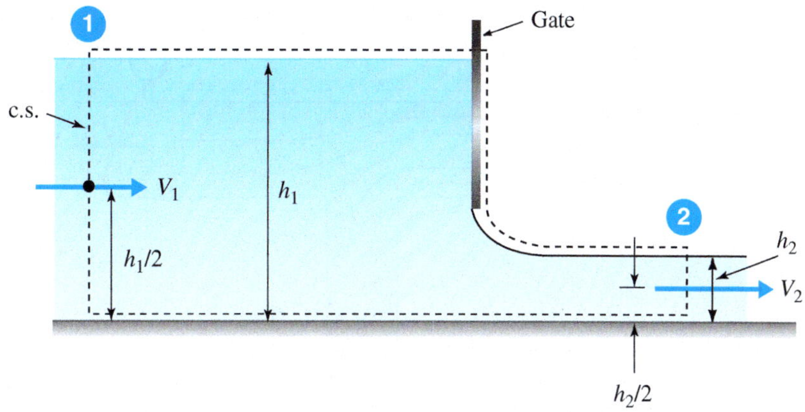

Figure 2.9 Application of the energy equation to a gate in an open channel.

the total head at the entrance and exit may then be evaluated conveniently at the center of the pipe entrance and exit. The energy equation may be applied to the flow passing a gate (Figure 2.9). An appropriate control volume is shown. The total head at the entrance and exit can be evaluated at any point at the entrance and exit, respectively. However, a convenient choice would be the points at the water surface. Thus the energy equation becomes

$$\frac{V_1^2}{2g} + \frac{p_1}{\gamma} + \cancelto{0}{h_1} = \frac{V_2^2}{2g} + \frac{p_2}{\gamma} + \cancelto{0}{h_2} + h_L \tag{2.5.21}$$

where the pressures at the surfaces have been set to zero. If we had picked the centroids of the entrance and exit, as shown in Figure 2.9, we would have obtained

$$\frac{V_1^2}{2g} + \frac{p_1}{\gamma} + \frac{h_1}{2} = \frac{V_2^2}{2g} + \frac{p_2}{\gamma} + \frac{h_2}{2} + h_L \tag{2.5.22}$$

We see that this result is the same as in Eq. 2.5.21 if we substitute $p_1 = \gamma h_1/2$ and $p_2 = \gamma h_2/2$.

For completeness, it is noted that the losses between 1 and 2 in Figure 2.9 could be neglected because internal viscous effects occur only over a relatively short distance and no significant secondary flows are generated.

The energy equation (2.5.15) can be applied to any control volume. For example, consider steady, uniform incompressible flow through a T-section in a pipe (Figure 2.10) in which there is one entrance and two exits. The energy equation could be applied to each of two control volumes, one for the mass flux that exits section 2 and the other for the mass flux that exits section 3:

$$\begin{aligned} \frac{V_1^2}{2g} + \frac{p_1}{\gamma} + z_1 &= \frac{V_2^2}{2g} + \frac{p_2}{\gamma} + z_2 + h_{L_{1-2}} \\ \frac{V_1^2}{2g} + \frac{p_1}{\gamma} + z_1 &= \frac{V_3^2}{2g} + \frac{p_3}{\gamma} + z_3 + h_{L_{1-3}} \end{aligned} \tag{2.5.23}$$

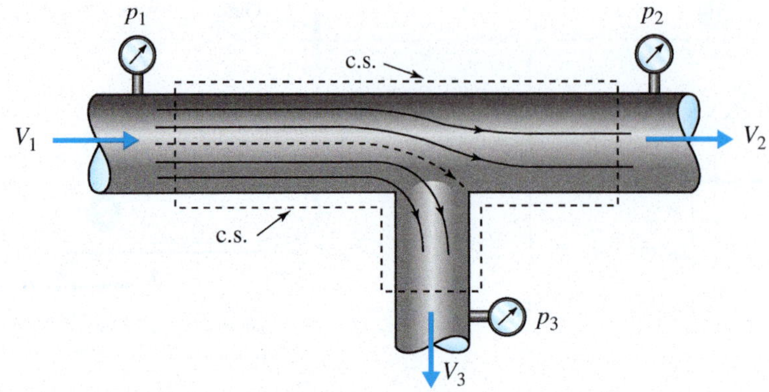

Figure 2.10 Application of the energy equation to a T-section.

The loss terms in Eq. 2.5.23 include the losses between the inlet and the respective exits. If the losses are negligible, the energy equation reduces to a form similar to the Bernoulli equation being applied along a streamline going from 1 to 2 or a streamline going from 1 to 3.

A final note to this section regards nomenclature for pumps and turbines in a flow system. It is often conventional to call the energy term $(\dot{W}_S/\dot{m}g)$ associated with a pump the **pump head** H_P, and the term $(\dot{W}_S/\dot{m}g)$ associated with a turbine the **turbine head** H_T. Then the energy equation (2.5.17), for an incompressible flow, takes the form

$$H_P + \frac{V_1^2}{2g} + \frac{p_1}{\gamma} + z_1 = H_T + \frac{V_2^2}{2g} + \frac{p_2}{\gamma} + z_2 + h_L \tag{2.5.24}$$

In this form we have equated the energy at the inlet plus added energy to the energy at the exit plus extracted energy (energy per unit weight, of course). If any of the quantities is zero (e.g., there is no pump), the appropriate term is simply omitted. The terms H_P and H_T above represent the energy that is transferred to and from the fluid, respectively. If the energy delivered by the turbine or required by the pump is desired, the efficiency of each device must be used.

The power generated by the turbine with an efficiency of η_T is simply

$$\dot{W}_T = \dot{m}gH_T\eta_T = \gamma Q H_T \eta_T \tag{2.5.25}$$

where we have used $\dot{m} = \rho AV$. The power requirement by a pump with an efficiency of η_P would be

$$\dot{W}_P = \frac{\dot{m}gH_P}{\eta_P} = \frac{\gamma Q H_P}{\eta_P} \tag{2.5.26}$$

We will calculate power in watts, ft-lb/s, or horsepower. Recall that one horse-power is equivalent to 746 W or 550 ft-lb/s.

2.5.4 Steady Nonuniform Flow

If the assumption of uniform velocity profiles is not acceptable for a problem of interest, as is sometimes the situation, we have to consider the control-surface integral in Eq. 2.5.15 with the proper expression for the velocity distribution. In practice, a velocity distribution can be accounted for by introducing the **kinetic-energy correction factor** α, defined by

$$\alpha = \frac{\int V^3 dA}{\bar{V}^3 A} \tag{2.5.27}$$

where $\bar{V}$ is the average velocity over the area A, given by Eq. 2.4.11. Then the term that accounts for the flux of kinetic energy in Eq. 2.5.15 is

$$\frac{1}{2}\rho \int_A V^3 dA = \frac{1}{2}\alpha\rho\bar{V}^3 A \tag{2.5.28}$$

where we have used $\mathbf{V}\cdot\hat{\mathbf{n}} = V$ and $V_I = V$. Using this factor, we can account for nonuniform velocity distributions by modifying Eq. 2.5.24 to read

$$H_P + \alpha_1 \frac{\bar{V}_1^2}{2g} + \frac{p_1}{\gamma} + z_1 = H_T + \alpha_2 \frac{\bar{V}_2^2}{2g} + \frac{p_2}{\gamma} + z_2 + h_L \tag{2.5.29}$$

where $\bar{V}_1$ and $\bar{V}_2$ are the average velocities at sections 1 and 2, respectively. For a flow with a parabolic profile in a pipe we can calculate α = 2.0 (see Example 2.9). For most internal turbulent flows, however, the profile is nearly uniform with $\alpha \simeq 1.05$. Hence we simply let $\alpha =$ 1 since it is so close to unity; this will always be done unless otherwise stated, since most of the internal flows that we encounter are, in fact, turbulent flows.

Example 2.6

The pump of Figure E2.6 is to increase the pressure of 0.2 m^3/s of water from 200 kPa to 600 kPa. If the pump is 85% efficient, how much electrical power will the pump require? The exit area is 20 cm above the inlet area. Assume inlet and exit areas are equal.

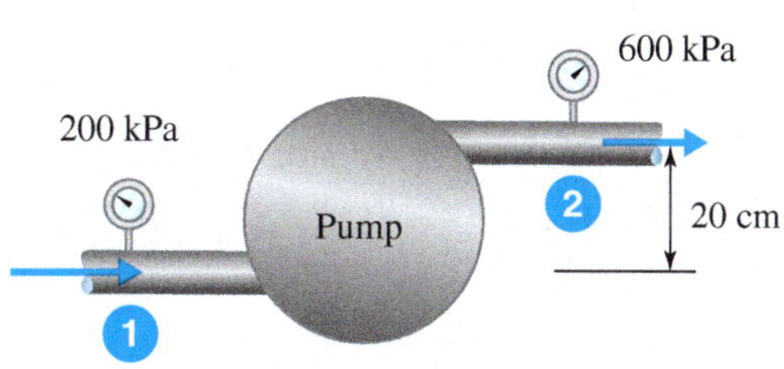

Figure E2.6

Solution

The energy equation (2.5.24) across the pump provides the energy delivered to the water as a pump head:

$$H_P = \frac{p_2 - p_1}{\gamma} + z_2 - z_1$$
$$= \frac{(600\,000 - 200\,000)\ \text{N/m}^2}{9810\ \text{N/m}^3} + 0.2\ \text{m} = 41.0\ \text{m}$$

where $V_2 = V_1$ since the inlet and exit areas are equal, and any losses are accounted for with the efficiency of Eq. 2.5.26. That equation provides the power required by the pump:

$$\dot{W}_P = \frac{\gamma Q H_P}{\eta_P}$$
$$= \frac{9810\,\text{N/m}^3 \times 0.2\,\text{m}^3/\text{s} \times 41.0\,\text{m}}{0.85} = 94600\ \text{J/s} \quad \text{or} \quad \underline{94.6\ \text{kW}}$$

Example 2.7

Water flows from a reservoir through a 2.5-ft-diameter pipeline to a turbine-generator unit and exits to a river that is 100 ft below the reservoir surface. If the flow rate is 90 ft^3/s and the turbine-generator efficiency is 88%, calculate the power output. Assume the loss coefficient in the pipeline (including the exit) to be $K = 2$.

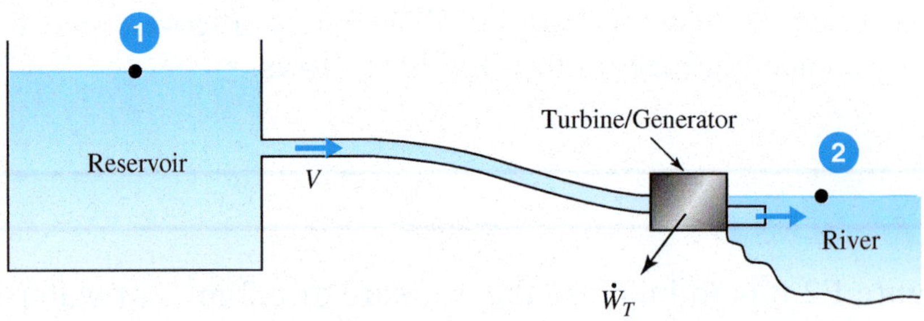

Figure E2.7

Solution

Referring to Figure E2.7, we select the control volume to extend from section 1 to section 2 on the reservoir and river surfaces, where we know the velocities, pressures, and elevations; we consider the water surface of the left reservoir to be the entrance and the water surface of the river to be the exit. The velocity in the pipe is

$$V = \frac{Q}{A} = \frac{90}{\pi \times 2.5^2/4} = 18.3\ \text{ft/s}$$

Now, consider the energy equation. We will use gage pressures so that $p_1 = p_2 = 0$; the datum is placed through the lower section 2 so that $z_2 = 0$; the velocities V_1 and V_2 on the reservoir surfaces are negligibly small; K is assumed to be based on the 2.5-ft-diameter pipe velocity. The energy equation (2.5.24) then becomes

$$\overset{0}{\cancel{H_P}} + \overset{0}{\cancel{\frac{V_1^2}{2g}}} + \overset{0}{\cancel{\frac{p_1}{\gamma}}} + z_1 = H_T + \overset{0}{\cancel{\frac{V_2^2}{2g}}} + \overset{0}{\cancel{\frac{p_2}{\gamma}}} + \overset{0}{\cancel{z_2}} + K\frac{V^2}{2g}$$

$$100 = H_T + 2\frac{18.3^2\,\text{ft}^2/s^2}{2\times 32.2\,\text{ft}/s^2}$$

$$\therefore H_T = 89.6\,\text{ft}$$

From this the power output is found using Eq. 2.5.25 to be

$$\dot{W}_T = Q\gamma H_T \eta_T$$
$$= 90\,\text{ft}^3/\text{s}\times 62.4\,\text{lb/ft}^3 \times 89.6\,\text{ft}\times 0.88 = 443{,}000\ \text{ft-lb/s} \quad \text{or} \quad \underline{805\ \text{hp}}$$

In this example we have used gage pressure; the potential-energy datum was assumed to be placed through section 2, V_1 and V_2 were assumed to be insignificantly small, and K was assumed to be based on the 2.5-ft-diameter pipe velocity.

Example 2.8

The venturi meter shown reduces the pipe diameter from 10 cm to a minimum of 5 cm (Figure E2.8). Calculate the flow rate and the mass flux assuming ideal conditions.

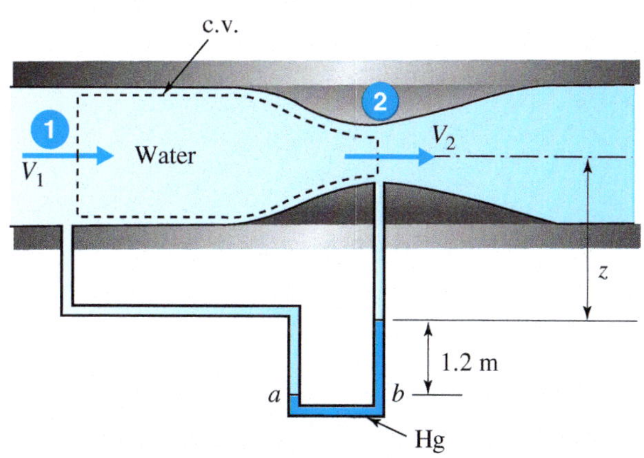

Figure E2.8

Solution

The control volume is selected as shown such that the entrance and exit correspond to the sections where the pressure information of the manometer can be applied. The manometer's reading is interpreted as follows:

$$p_a = p_b$$
$$p_1 + \gamma(z + 1.2) = p_2 + \gamma z + 13.6\gamma \times 1.2$$

where z is the distance from the pipe centerline to the top of the mercury column. The manometer then gives

$$\frac{p_1 - p_2}{\gamma} = (13.6 - 1) \times 1.2 = 15.12\text{m}$$

Continuity (2.4.6) allows us to relate V_2 to V_1 by

$$V_1 A_1 = V_2 A_2$$
$$\therefore V_2 = \frac{A_1}{A_2} V_1 = \frac{\pi \times 10^3/4}{\pi \times 5^2/4} V_1 = 4V_1$$

The energy equation (2.5.17) assuming ideal conditions (no losses and uniform flow) with $h_L = \dot{W}_S = 0$ takes the form

$$0 = \frac{V_2^2 - V_1^2}{2g} + \frac{p_2 - p_1}{\gamma} + \cancelto{0}{(z_2 - z_1)}$$
$$= \frac{16V_1^2 - V_1^2}{2g} - 15.12$$
$$\therefore V_1 = 4.45\text{m/s}$$

The flow rate is

$$Q = A_1 V_1 = (\pi \times 0.05^2) \times 4.45 = \underline{0.0350\ \text{m}^3/\text{s}}$$

The mass flux is

$$\dot{m} = \rho Q = 1000 \times 0.035 = \underline{35.0\ \text{kg/s}}$$

Example 2.9

The velocity distribution for a certain flow in a pipe is $V(r) = V_{max}(1 - r^2/r_0^2)$, where r_0 is the pipe radius (Figure E2.9). Determine the kinetic-energy correction factor.

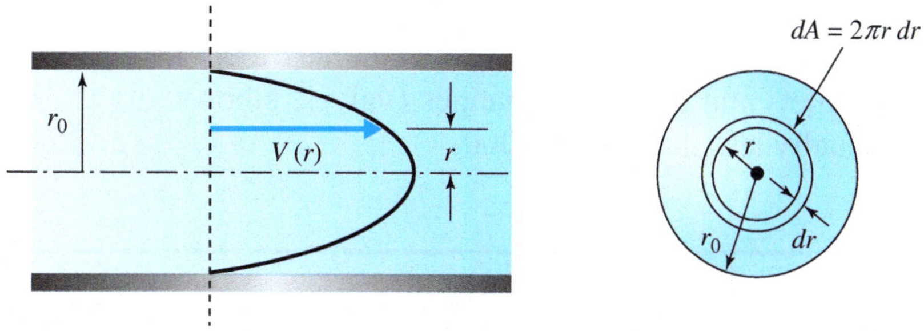

Figure E2.9

Solution

To find the kinetic-energy correction factor α, we must know the average velocity. It is (combine Eqs. 2.4.10 and 2.4.11)

$$\begin{aligned}\bar{V} &= \frac{\int V dA}{A} \\ &= \frac{1}{\pi r_0^2}\int_0^{r_0} V_{max}\left(1-\frac{r^2}{r_0^2}\right)2\pi r dr = \frac{2\pi V_{max}}{\pi r_0^2}\int_0^{r_0}\left(r-\frac{r^3}{r_0^2}\right)dr \\ &= \frac{2V_{max}}{r_0^2}\left(\frac{r_0^2}{2}-\frac{r_0^4}{4r_0^2}\right) = \frac{1}{2}V_{max}\end{aligned}$$

Using Eq. 2.5.27, there results

$$\begin{aligned}\alpha &= \frac{\int V^3 dA}{\bar{V}^3 A} \\ &= \frac{\int_0^{r_0} V_{max}(1-r^2/r_0^2)^3 2\pi r\, dr}{\left(\frac{1}{2}V_{max}\right)^3 \pi r_0^2} = \frac{16}{r_0^2}\int_0^{r_0}\left(1-\frac{3r^2}{r_0^2}+\frac{3r^4}{r_0^4}-\frac{r^6}{r_0^6}\right)rdr \\ &= \frac{16}{r_0^2}\left(\frac{r_0^2}{2}-\frac{3r_0^2}{4}+\frac{3r_0^2}{6}-\frac{r_0^2}{8}\right) = 2\end{aligned}$$

Consequently, the kinetic energy flux associated with a parabolic velocity distribution across a circular area is given by

$$\int \rho \mathbf{V} \cdot \hat{\mathbf{n}} \frac{V^2}{2} dA = 2 \times \frac{\dot{m}\bar{V}^2}{2} = \dot{m}\bar{V}^2$$

Parabolic velocity distributions are encountered in laminar flows in pipes and between parallel plates, downstream of inlets and geometry changes (valves, elbows, etc.). The Reynolds number must be quite small, usually less than about 2000.

Example 2.10

The drag force on an automobile (Figure E2.10) is approximated by the expression $0.15\rho V_\infty^2 A$, where A is the projected cross-sectional area and V_∞ is the automobile's speed. If $A = 1.2\ \text{m}^2$, calculate the efficiency η of the engine if the rate of fuel consumption f (the gas mileage) is 15 km/L and the automobile travels at 90 km/h. Assume that the fuel releases 44 000 kJ/kg during combustion. Neglect the energy lost due to the exhaust gases and coolant and assume that the only resistance to motion is the drag force. Use $\rho_{\text{air}} = 1.12\ \text{kg/m}^3$ and $\rho_{\text{fuel}} = 0.68$ kg/L.

Figure E2.10

Solution

If the car is taken as the moving control volume (note that the control volume is fixed), as shown, we can simplify the energy equation (Eq. 2.5.3 in combination with 2.5.11) to

$$\dot{Q} - \dot{W}_I = 0$$

since all other terms are negligible; there is no velocity crossing the control volume, so $\mathbf{V} \cdot \hat{\mathbf{n}} = 0$ (neglect the energy of the exhaust gases); there is no shear or shaft work; the energy of the c.v. remains constant. The energy input $\dot{Q}$ which accomplishes useful work is η times the energy released during combustion; that is,

$$\dot{Q} = \dot{m}_f \times 44000\eta \text{ kJ/s}$$

where $\dot{m}_f$ is the mass flux of the fuel. The mass flux of fuel is determined knowing the rate of fuel consumption $\dot{f}$ and the density of fuel as 0.68 kg/L, as follows:

$$\dot{f} = \frac{\text{distance}}{\text{volume}} = \frac{V_\infty \times \text{time}}{Q \times \text{time}} = \frac{V_\infty}{\dot{m}_f / \rho_f} = \frac{\rho_f V_\infty}{\dot{m}_f}$$

with $V_\infty = 90\,000/3600 = 25$ m/s, we have, using $\dot{f} = 15 \times 1000$ m/L,

$$15 \times 1000 = \frac{0.68 \times 25}{\dot{m}_f}$$
$$\therefore \dot{m}_f = 0.001133 \text{ kg/s}$$

The inertial work-rate term is

$$\dot{W}_1 = V_\infty \times \text{drag}$$
$$= 0.15 \rho V_\infty^3 A = 0.15 \times 1.12 \times 25^3 \times 1.2 = 3150 \text{ J/s}$$

Equating $\dot{Q} = \dot{W}_1$, we have

$$44\,000 \eta \times 0.001133 = 3.15$$
$$\therefore \quad \eta = 0.0632 \quad \text{or} \quad \underline{6.32\%}$$

This is obviously a very low percentage, perhaps surprisingly low to the reader. Very little power (3.15 kJ/s = 4.22 hp) is actually needed to propel the automobile at 90 km/h. The relatively large engine, needed primarily for acceleration, is quite inefficient when simply propelling the automobile.

Note the importance of using a stationary reference frame. The reference frame attached to the automobile is an inertial reference frame since it is moving at constant velocity. Yet the energy equation demands a stationary reference frame allowing the energy required by the drag force to be properly included.

2.6 MOMENTUM EQUATION

2.6.1 General Momentum Equation

Newton's second law, often called the **momentum equation**, states that the resultant force acting on a system equals the rate of change of momentum of the system when measured in an inertial reference frame; that is,

$$\Sigma \mathbf{F} = \frac{D}{Dt}\int_{\text{sys}} \rho \mathbf{V}\, d\mathcal{V} \tag{2.6.1}$$

Using Eq. 2.3.9, with η replaced by $\mathbf{V}$, this is written for a control volume as

$$\Sigma \mathbf{F} = \frac{d}{dt}\int_{\text{c.v.}} \rho \mathbf{V}\, d\mathcal{V} + \int_{\text{c.s.}} \rho \mathbf{V}(\mathbf{V}\cdot\hat{\mathbf{n}})\, dA \tag{2.6.2}$$

where $\mathbf{V}\cdot\hat{\mathbf{n}}$ is simply a scalar for each differential area dA. The control surface integral on the right represents the net momentum flux across the control surface of the fluid entering and/or leaving the control volume.

When applying Newton's second law the quantity $\Sigma \mathbf{F}$ represents all forces acting on a control volume. The forces include the surface forces resulting from the surroundings acting on a control surface and body forces that result from gravity and magnetic fields. The momentum equation is often used to determine the forces induced by a flow. For example, the equation allows us to calculate the force on the support of an elbow in a pipeline or the force on a submerged body in a free-surface flow.

When we apply the momentum equation the surrounding fluid and sometimes the entire conduit or container is separated from the control volume. For example, in the horizontal nozzle of Figure 2.11a, the nozzle and the fluid in the nozzle are isolated. Thus, care must be taken to include the pressure forces shown and the force $\mathbf{F}_{\text{joint}}$. It is convenient to use gage pressures so that the pressure acting on the exterior of the pipe is then zero. Alternatively, we could have selected a control volume that includes only the fluid in the nozzle (Figure 2.11b). In that case we have to consider the pressure forces at the entrance and exit and the resultant pressure force $\mathbf{F}_{\text{nozzle}}$ of the interior wall of the nozzle on the fluid. A free body of the nozzle excluding the fluid shows that the force $\mathbf{F}_{\text{joint}}$ and $\mathbf{F}_{\text{nozzle}}$ are equal in magnitude. If the problem is to determine the forces exerted by the flow on the nozzle (Figure 2.11b), we have to reverse the direction of the calculated force $\mathbf{F}_{\text{nozzle}}$. Examples at the end of this section illustrate this.

2.6.2 Steady Uniform Flow

Equation 2.6.2 can be simplified considerably if a device has entrances and exits across which the flow may be assumed to be uniform and if the flow is steady. Then there results

$$\Sigma \mathbf{F} = \sum_{i=i}^{N} \rho_i A_i \mathbf{V}_i (\mathbf{V}_i \cdot \hat{\mathbf{n}}) \tag{2.6.3}$$

where N is the number of flow exit and entrance areas.

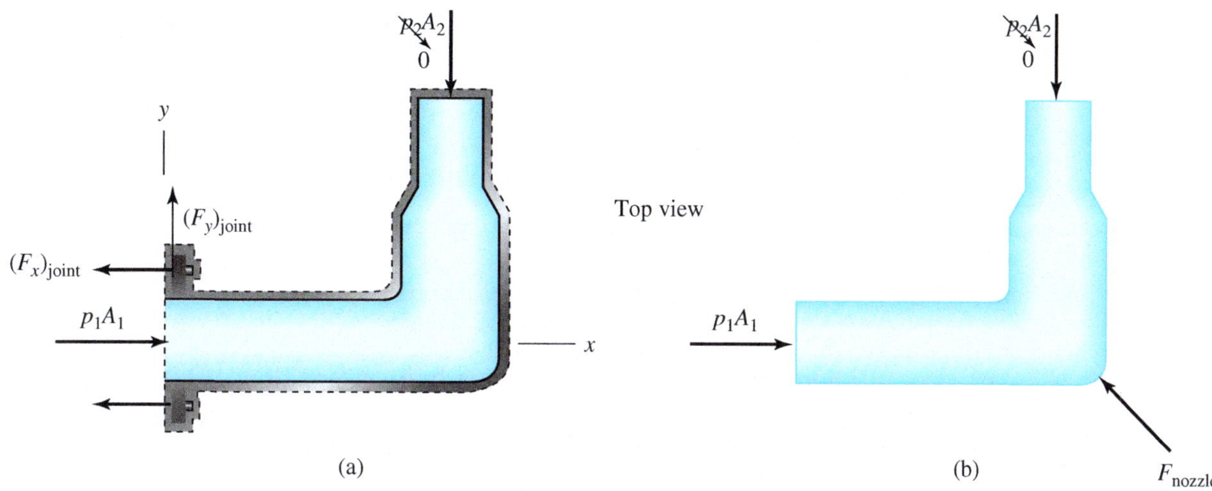

Figure 2.11 Forces acting on the control volume of a horizontal nozzle: (a) control volume includes nozzle and fluid in nozzle; (b) control volume includes fluid in nozzle only. We have neglected body forces.

At an entrance $\mathbf{V}\cdot\hat{\mathbf{n}} = -V$ since the unit vector points out of the volume, and at the exit $\mathbf{V}\cdot\hat{\mathbf{n}} = V$. If there is only one entrance and one exit, as in Figure 2.11, the momentum equation becomes

$$\Sigma\mathbf{F} = \rho_2 A_2 V_2 \mathbf{V}_2 - \rho_1 A_1 V_1 \mathbf{V}_1 \tag{2.6.4}$$

Using continuity,

$$\dot{m} = \rho_1 A_1 V_1 = \rho_2 A_2 V_2$$

the momentum equation takes the simplified form

$$\Sigma\mathbf{F} = \dot{m}(V_2 - V_1) \tag{2.6.5}$$

This is the form most often used in this text.

Note that the momentum equation is a vector equation which represents the three scalar equations

$$\begin{aligned}\Sigma F_x &= \dot{m}(V_{2x} - V_{1x})\\ \Sigma F_y &= \dot{m}(V_{2y} - V_{1y})\\ \Sigma F_z &= \dot{m}(V_{2z} - V_{1z})\end{aligned} \tag{2.6.6}$$

If we consider the nozzle of Figure 2.11a and we want to determine the x-component of the force of the joint on the nozzle, $(\mathbf{V}_1)_x = V_1$ and $(\mathbf{V}_2)_x = 0$ so that the momentum equation for the x-direction becomes

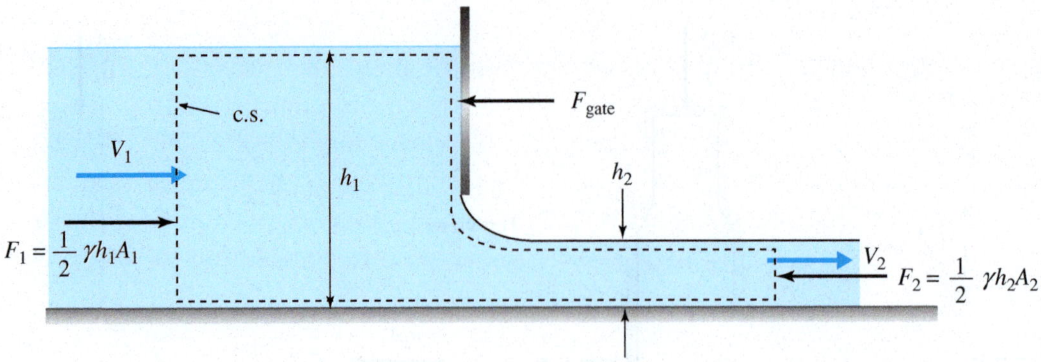

Figure 2.12 Force of the flow on a gate in a free-surface flow.

$$\Sigma F_x = -(F_x)_{\text{joint}} + p_1 A_1 = -\dot{m} V_1 \tag{2.6.7}$$

Similarly, we could write the y-component equation that would contain the term $(F_y)_{\text{joint}}$.

An example of a free-surface flow in a rectangular channel is shown in Figure 2.12. If the force of the gate on the flow is desired, the following expression can be derived from the momentum equation:

$$\Sigma F_x = -F_{\text{gate}} + F_1 - F_2 = \dot{m}(V_2 - V_1) \tag{2.6.8}$$

where F_1 and F_2 are pressure forces (see Figures 2.9 and 2.12).

Example 2.11

Water flows through a horizontal pipe bend and exits into the atmosphere (Figure E2.11a). The flow rate is 0.3 ft^3/s. Calculate the force in each of the rods holding the pipe bend in position. Neglect body forces and viscous effects and shear force in the rods.

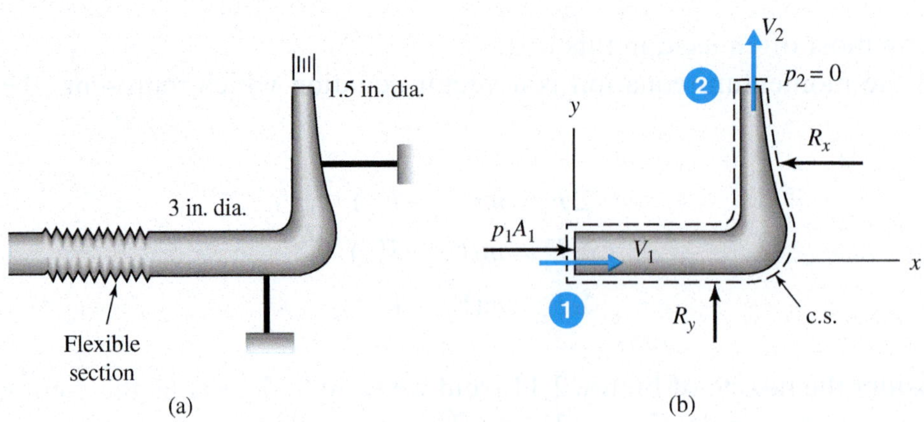

Figure E2.11

Solution

We have selected a control volume that surrounds the bend, as shown in Figure E2.11b. Since the rods have been cut, the forces that the rods exert on the control volume are included. The pressure force at the entrance of the control volume is also shown. The flexible section is capable of resisting the interior pressure, but it transmits no axial force or moment. The body force (weight of the control volume) does not act in the x- or y-direction but normal to it. Therefore, no other forces are shown. The average velocities are found to be

$$V_1 = \frac{\dot{Q}}{A_1} = \frac{0.3}{\pi \times (3/12)^2/4} = 6.11 \text{ ft/s}; \qquad V_2 = \frac{\dot{Q}}{A_2} = \frac{0.3}{\pi \times (1.5/12)^2/4} = 24.4 \text{ ft/s}$$

Before we can calculate the forces R_x and R_y we need to find the pressures p_1 and p_2. The pressure p_2 is zero because the flow exits into the atmosphere. The pressure at section 1 can be determined using the energy equation or the Bernoulli equation. Neglecting losses between sections 1 and 2, the energy equation gives

$$\frac{V_1^2}{2g} + \frac{p_1}{\gamma} = \frac{V_2^2}{2g} + \overset{0}{\cancel{\frac{p_2}{\gamma}}}$$

$$\therefore p_1 = \frac{\gamma}{2g}(V_2^2 - V_1^2) = \frac{62.4 \text{ lb/ft}^3}{2 \times 32.2 \text{ ft/s}^2}(24.4^2 - 6.11^2) \text{ ft}^2/\text{s}^2 = 541 \text{ psf}$$

Now we can apply the momentum equation (2.6.5) in the x-direction to find R_x and in the y-direction to find R_y:

x-direction:

$$p_1 A_1 - R_x = \dot{m}(\overset{0}{\cancel{V_{2x}}} - V_{1x})$$

$$541 \times \frac{\pi}{4}\left(\frac{3}{12}\right)^2 - R_x = 1.94 \text{ slug/ft}^3 \times 0.3 \text{ ft}^3/\text{s}\,(-6.11) \text{ ft/s}$$

$$R_x = \underline{30.1 \text{ lb}}$$

y-direction:

$$R_y = \dot{m}(V_{2y} - \overset{0}{\cancel{V_{xy}}})$$

$$= 1.94 \times 0.3 \times 24.4 = \underline{14.2 \text{ lb}}$$

Note that we have assumed uniform profiles and steady flow and used $\dot{m} = \rho\dot{Q}$. These are the usual assumptions if information is not given otherwise.

Example 2.12

When the velocity of a flow in an open rectangular channel of width w is relatively large, it is possible for the flow to "jump" from a depth y_1 to a depth y_2 over a relatively short distance, as shown in Figure E2.12a; this is referred to as a **hydraulic jump**. Express y_2 in terms of y_1 and V_1; assume a horizontal uniform flow.

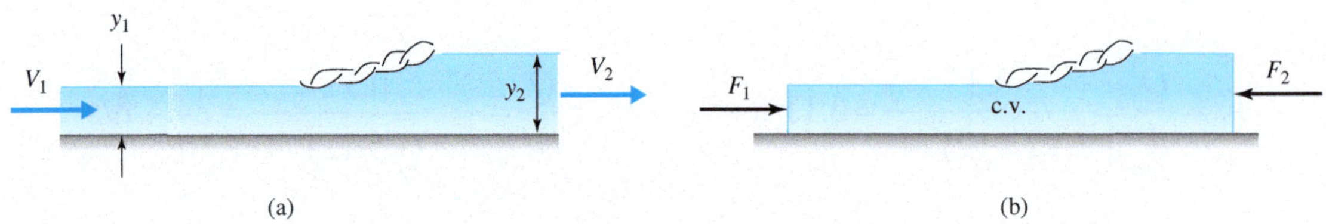

Figure E2.12

Solution

A control volume is selected as shown in Figure E2.12b with inlet and exit areas upstream and downstream of the "jump" sufficiently far that the streamlines are parallel to the wall with hydrostatic pressure distributions. Neglecting the drag that is present on the walls (if the distance between the sections is relatively small, the drag force should be negligible), the momentum equation can be manipulated as follows:

$$\Sigma F_x = \dot{m}(V_{2x} - V_{1x})$$
$$F_1 - F_2 = \rho A_1 V_1 (V_2 - V_1)$$
$$\gamma \frac{y_1}{2}(y_1 w) - \gamma \frac{y_2}{2}(y_2 w) = \rho y_1 w V_1 \left(V_1 \frac{y_1}{y_2} - V_1 \right)$$

where we have expressed F_1 and F_2 using the pressure as γh and continuity in the form of Eq. 2.4.6, so that

$$V_2 = \frac{y_1}{y_2} V_1$$

The above momentum equation can be simplified to

$$\frac{\gamma}{2}\left(y_1^2 - y_2^2\right) = \rho y_1 V_1^2 \frac{y_1 - y_2}{y_2}$$

or

$$\frac{g}{2}(y_1 - y_2)(y_1 + y_2) = \frac{y_1}{y_2}V_1^2(y_1 - y_2)$$

The factor $(y_1 - y_2)$ is divided out and y_2 is found assuming y_1 and V_1 are known as follows:

$$\frac{g}{2}(y_1 + y_2) = \frac{y_1}{y_2}V_1^2$$

$$y_2^2 + y_1 y_2 - \frac{2}{g}y_1 V_1^2 = 0$$

$$\therefore y_2 = \frac{1}{2}\left(-y_1 + \sqrt{y_1^2 + \frac{8}{g}y_1 V_1^2}\right)$$

where the quadratic formula has been used. The energy equation could now be used to provide an expression for the losses in the hydraulic jump.

Example 2.13

Consider the symmetrical flow of air around the cylinder. The control volume, excluding the cylinder, is shown in Figure E2.13. The velocity distribution downstream of the cylinder is approximated with the parabola, as shown. Determine the drag force per meter of length acting on the cylinder. Use $\rho = 1.23$ kg/m^3.

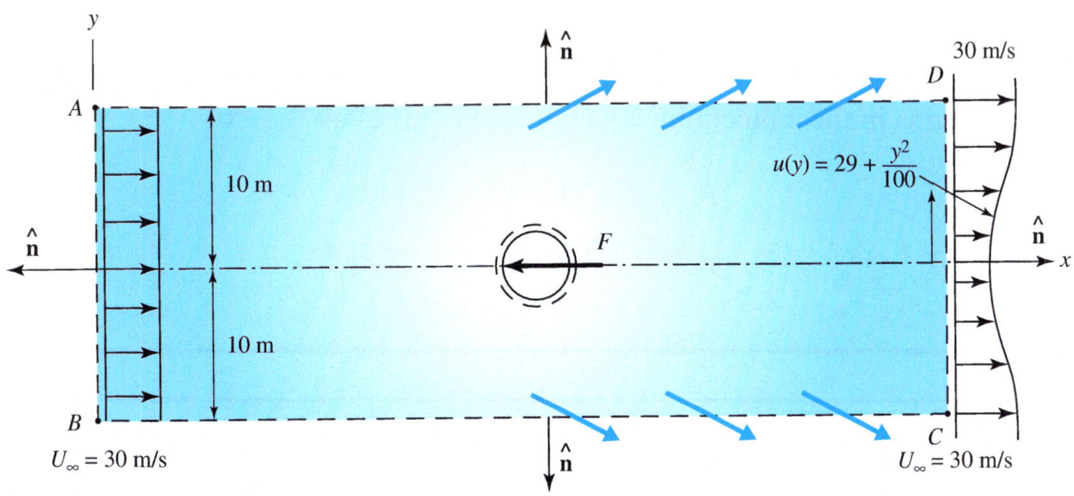

Figure E2.13

Solution

First, we must recognize that not all of the mass flux entering through AB exits through CD; consequently, some mass flux must exit AD and BC, as shown. The momentum equation (2.6.2) for the steady flow, applied to the control volume $ABCD$, takes the form

$$
\begin{aligned}
-F &= \int_{c.s} \rho V_x \mathbf{V}\cdot\hat{\mathbf{n}}dA = \int_{A_{CD}} \rho u\, \mathbf{V}\cdot\hat{\mathbf{n}}dA + \int_{A_{AD}} \rho u\, \mathbf{V}\cdot\hat{\mathbf{n}}dA + \int_{A_{BC}} \rho u\, \mathbf{V}\cdot\hat{\mathbf{n}}dA \\
&\quad + \int_{A_{AB}} \rho u\, \mathbf{V}\cdot\hat{\mathbf{n}}dA \\
&= \int_{A_{CD}} \rho u^2 dA + U_\infty \dot{m}_{AD} + U_\infty \dot{m}_{BC} - \int_{A_{AB}} \rho u^2 dA \\
&= 2\int_0^{10} 1.23\left(29 + \frac{y^2}{100}\right)^2 dy + 2\times 30\dot{m}_{AD} - 1.23\times 30^2\times 20
\end{aligned}
$$

where $\dot{m}_{BC} = \dot{m}_{AD}$ is the mass flux crossing BC and AD with the x-component velocity equal to 30 m/s. The limit of 10 m was used since at $y = 10$ m, the parabola gives $u(10) = 30$ m/s. We have used Eq. 2.4.9 for $\dot{m}_{AD}$ and $\dot{m}_{BC}$ recognizing that $\mathbf{V}\cdot\hat{\mathbf{n}} = V_n$, which would be the small y-component velocity. We now use continuity to find $\dot{m}_{AD}$:

$$
\begin{aligned}
0 = \int_{c.s} \rho\hat{\mathbf{n}}\cdot\mathbf{V}dA &= \int_{A_{AD}} \rho\hat{\mathbf{n}}\cdot\mathbf{V}dA + \int_{A_{BC}} \rho\hat{\mathbf{n}}\cdot\mathbf{V}dA + \int_{A_{CD}} \rho\hat{\mathbf{n}}\cdot\mathbf{V}dA + \int_{A_{AB}} \rho\hat{\mathbf{n}}\cdot\mathbf{V}dA \\
&= \dot{m}_{AD} + \dot{m}_{BC} + 2\int_0^{10} \rho u(y)\, dy - \rho\times 20\times 30 \\
&= 2\dot{m}_{AD} + 2\int_0^{10} 1.23\times\left(29 + \frac{y^2}{100}\right) dy - 1.23\times 20\times 30
\end{aligned}
$$

$$\therefore\quad \dot{m}_{AD} = 8.2 \text{ kg/s per meter of length}$$

Evaluating the terms in the momentum equation above gives us

$$
\begin{aligned}
F &= -21170 - 492 + 22\,140 \\
&= \underline{478 \text{ N/m}}
\end{aligned}
$$

Example 2.14

Find an expression for the head loss in a sudden expansion in a pipe in terms of V_1 and the area ratio (Figure E2.14a). Assume uniform velocity profiles and assume that the pressure at the sudden enlargement is p_1.

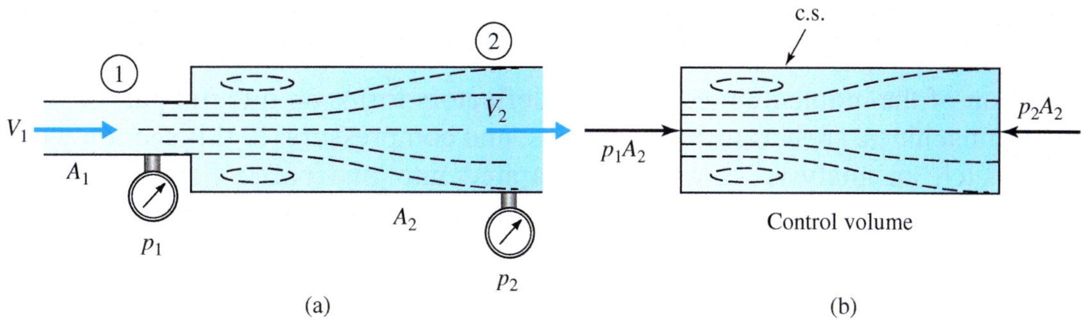

Figure E2.14

Solution

Figure E2.14a shows a sudden expansion with the diameter changing from d_1 to d_2. The pressure at the sudden enlargement is closest to p_1 since the streamlines are approximately parallel as shown (there is no pressure variation normal to parallel streamlines); they take some distance to again fill the pipe. Hence, the force acting on the left end of the control volume shown in Figure E2.14b is p_1A_2. Newton's second law applied to the control volume yields, assuming uniform profiles,

$$\Sigma F_x = \dot{m}(V_2 - V_1)$$
$$(p_1 - p_2)A_2 = \rho A_2 V_2 (V_2 - V_1)$$
$$\therefore \frac{p_1 - p_2}{\rho} = V_2(V_2 - V_1)$$

The energy equation (2.5.17) provides

$$0 = \frac{V_2^2 - V_1^2}{2g} + \frac{p_2 - p_1}{\gamma} + \cancelto{0}{z_2 - z_1} + h_L$$
$$\therefore h_L = \frac{p_1 - p_2}{\rho g} - \frac{V_2^2 - V_1^2}{2g}$$
$$= \frac{V_2(V_2 - V_1)}{g} - \frac{(V_2 + V_1)(V_2 - V_1)}{2g} = \frac{(V_1 - V_2)^2}{2g}$$

To express this in terms of only V_1, we can use continuity and relate

$$V_2 = \frac{A_1}{A_2}V_1$$

Then the expression above for the head loss becomes

$$h_L = \left(1 - \frac{A_1}{A_2}\right)^2 \frac{V_1^2}{2g}$$

2.6.3 Momentum Equation Applied to Deflectors

The application of the momentum equation to deflectors forms an integral part of the analysis of many turbomachines, such as turbines, pumps, and compressors. In this section we illustrate the steps in such an analysis. It will be separated into two parts: fluid jets deflected by stationary deflectors and fluid jets deflected by moving deflectors. For both problems we will assume the following:

- The pressure external to the fluid jets is everywhere constant so that the pressure in the fluid as it moves over a deflector remains constant.
- The frictional resistance due to the fluid-deflector interaction is negligible so that the relative speed between the deflector surface and the jet stream remains unchanged, a result of Bernoulli's equation.
- Lateral spreading of a plane jet is neglected.
- The body force, the weight of the control volume, is small and will be neglected.

Stationary Deflector: Let us first consider the stationary deflector, illustrated in Figure 2.13. Bernoulli's equation allows us to conclude that the magnitudes of the velocity vectors are equal (i.e., $V_2 = V_1$), since the pressure is assumed to be constant external to the fluid jet and elevation changes are negligible. Assuming steady, uniform flow, the momentum equation takes the form of Eq. 2.6.5, which for the x- and y-directions becomes

$$\begin{aligned} -R_x &= \dot{m}(V_2\cos\alpha - V_1) = \dot{m}V_1(\cos\alpha - 1) \\ R_y &= \dot{m}V_2\sin\alpha = \dot{m}V_1\sin\alpha \end{aligned} \tag{2.6.9}$$

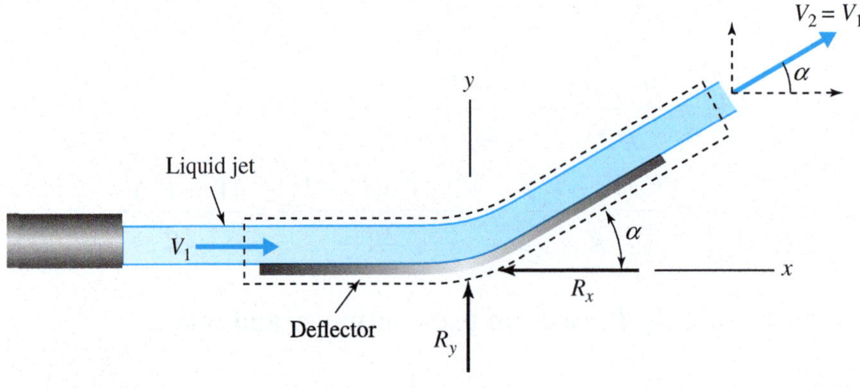

Figure 2.13 A stationary deflector. (Assume no lateral spreading of the jet.)

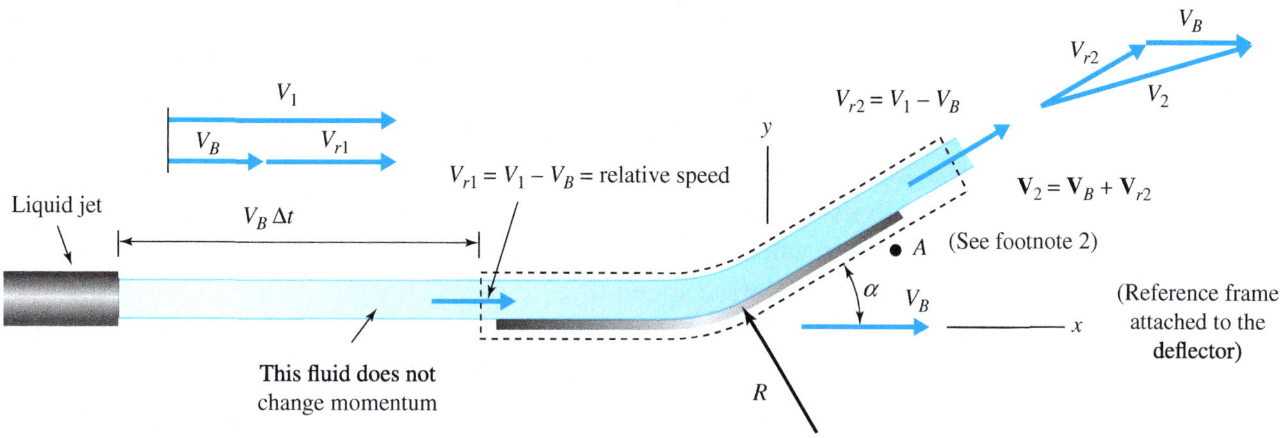

Figure 2.14 A moving deflector.

For given jet conditions the reactive force components can be calculated.

Moving Deflectors: The situation involving a moving deflector depends on whether a single deflector is moving (the blade on a snowplow or a water scoop used to slow a high-speed train) or whether a series of deflectors is moving (the vanes on a turbine). Let us first consider the single deflector shown in Figure 2.14 to be moving in the positive x-direction with the speed V_B. In a reference frame attached to the stationary nozzle, from which the fluid jet issues, the flow is unsteady[2]; that is, at a particular point in space, the flow situation varies with time. A steady flow is observed, however, from a reference frame attached to the deflector. From this inertial reference frame, moving with the constant velocity V_B, we observe the relative speed V_{r1} entering the control volume to be $V_1 - V_B$, as shown. It is this relative speed that remains constant as the fluid flows relative to the deflector; it does not change since the pressure does not change. Hence, from this moving frame, the momentum equation (2.6.5) takes the forms

$$\begin{aligned} -R_x &= \dot{m}_r (V_1 - V_B)(\cos\alpha - 1) \\ R_y &= \dot{m}_r (V_1 - V_B)\sin\alpha \end{aligned} \qquad (2.6.10)$$

where $\dot{m}_r$ represents only that part of the mass flux exiting the fixed jet that has its momentum changed. Since the deflector moves away from the fixed jet some of the fluid that exits the fixed jet never experiences a momentum change; this fluid is represented by the distance $V_B \Delta t$, shown in Figure 2.14. Hence

$$\dot{m}_r = \rho A (V_1 - V_B) \qquad (2.6.11)$$

where the relative speed $(V_1 - V_B)$ is used in the calculation; the mass flux $\rho A V_B$ is subtracted from the exiting mass flux $\rho A V_1$ to provide the mass flux $\dot{m}_r$ that experiences a momentum change.

[2]This is a rather subtle point. To determine whether a flow is steady, we observe the flow at a given point in space. If a flow property changes with time at that point, the flow is unsteady. In this situation, if we focus our attention on a particular point just before the blade, such as point A in Figure 2.14, first there is no flow, then the blade and the jet pass through the point, then there is again no flow. This is an unsteady flow.

For a series of vanes (a cascade), the jets may be oriented to the side, as shown in Figure 2.15. The actual force on a particular vane would be zero until the jet strikes the vane; then the force would increase to a maximum and decrease to zero as the vane leaves the jet. We will idealize the situation as follows: Assume that, on the average, the jet is deflected by the vanes as shown in Figures 2.15 and 2.16a as viewed from a stationary reference frame; the fluid jet enters the vanes with an angle β_1 and exits with a time-averaged angle β_2. What is desired, however, is that the relative velocity enter the vanes tangent to the leading edge of the vanes, that is, V_{r1} in Figure 2.16b is at the angle α_1. The relative speed then remains constant as the fluid travels over the vane with the exiting relative velocity V_{r2} leaving with the vane angle α_2. The relative and absolute velocities are related with the velocity equations which are displayed by the velocity polygons of Figures 2.16b and 2.16c.

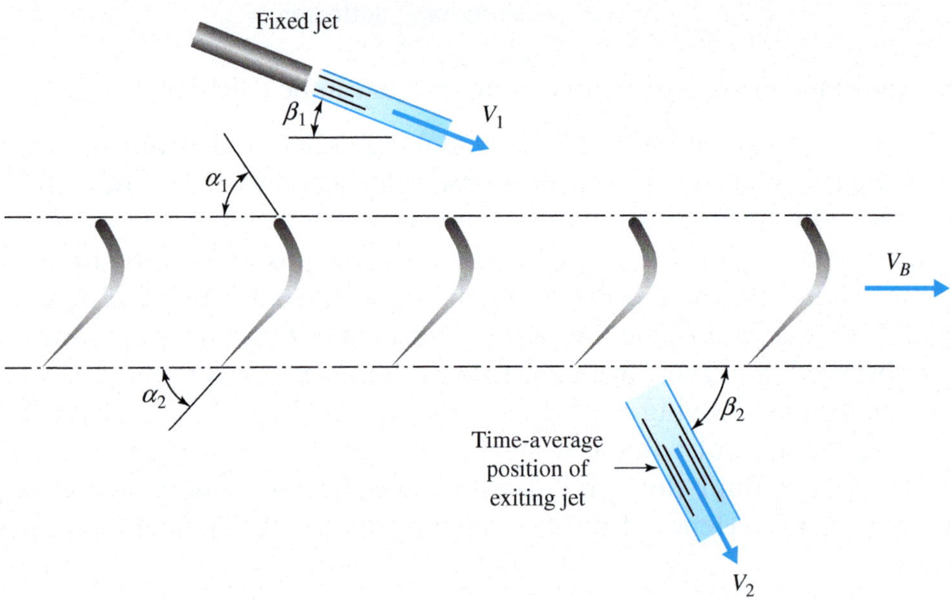

Figure 2.15 Fluid striking a series of vanes.

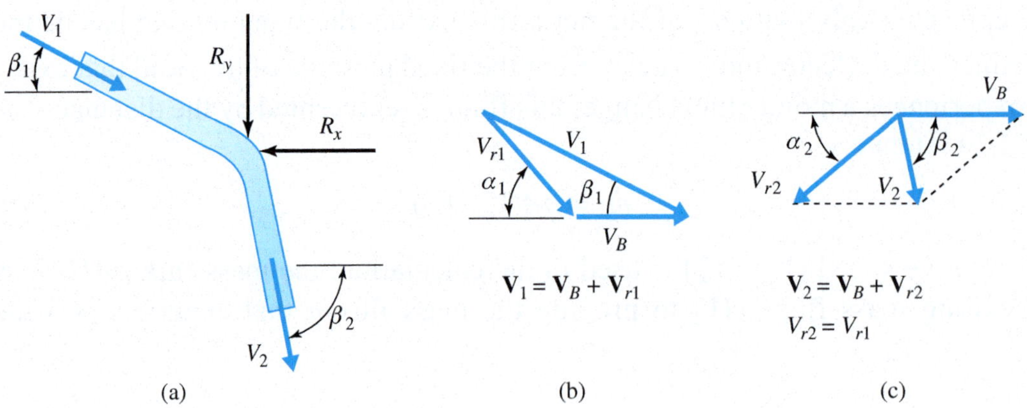

Figure 2.16 Detail of the flow situation involving a series of vanes: (a) average position of jet; (b) entrance velocity polygon; (c) exit velocity polygon.

Assuming that all of the mass exiting the fixed jet has its momentum changed, we can write the momentum equation as

$$-R_x = \dot{m}(V_{2x} - V_{1x}) \tag{2.6.12}$$

Example 2.17 will illustrate the details.

Interest is usually focused on the x-component of force since it is this component that is related to the power output (or requirement). The power would be found by multiplying the x-component force by the blade speed for each jet; this takes the form

$$\dot{W} = NR_x V_B \tag{2.6.13}$$

where N represents the number of jets. The y-component force does not move in the y-direction so it produces no power.

Example 2.15

A deflector turns a sheet of water through an angle of 30° as shown in Figure E2.15. What force components are necessary to hold the deflector in place if $\dot{m} = 32$ kg/s?

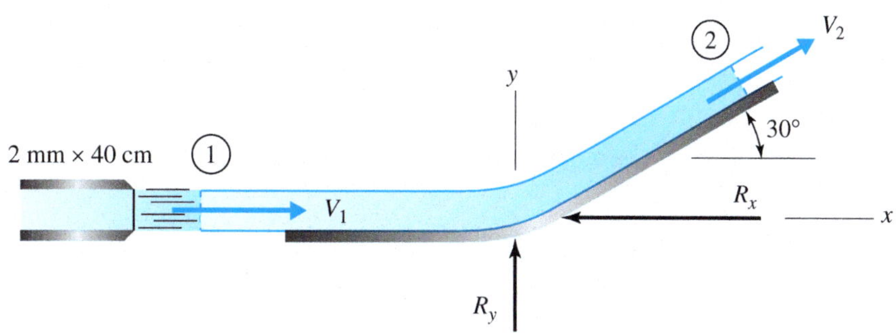

Figure E2.15

Solution

The control volume we have selected includes the deflector and the water adjacent to it. The only force that is acting on the control volume is due to a support needed to hold the deflector. This force has been decomposed into R_x and R_y.

The velocity V_1 is found to be

$$V_1 = \frac{\dot{m}}{\rho A_1}$$

$$= \frac{32 \text{ kg/s}}{1000 \text{ kg/m}^3 \times (0.002 \times 0.4) \text{ m}^2} = 40 \text{ m/s}$$

Bernoulli's equation (1.4.8) shows that if the pressure does not change, then the magnitude of the velocity does not change, provided that there is no significant change in elevation and that viscous effects are negligible; thus we can conclude that $V_2 = V_1$ since $p_2 = p_1$. Next, the momentum equation is applied in the x-direction to find R_x and then in the y-direction for R_y:

$$x\text{-direction:} \qquad -R_x = \dot{m}(V_{2x} - V_{1x})$$
$$= 32 \text{ kg/s } (40 \cos 30° - 40) \text{ m/s}$$
$$\therefore R_x = \underline{172 \text{ N}}$$

$$y\text{-direction:} \qquad R_y = \dot{m}(V_{2y} - \overset{0}{\cancel{V_{1y}}})$$
$$= 32\,(40 \sin 30°) = \underline{640 \text{ N}}$$

Example 2.16

The deflector shown in Figure E2.16 moves to the right at 30 m/s while the nozzle remains stationary. Determine (a) the force components needed to support the deflector, (b) $\mathbf{V}_2$ as observed from a fixed observer, and (c) the power generated by the vane. The jet velocity is 80 m/s.

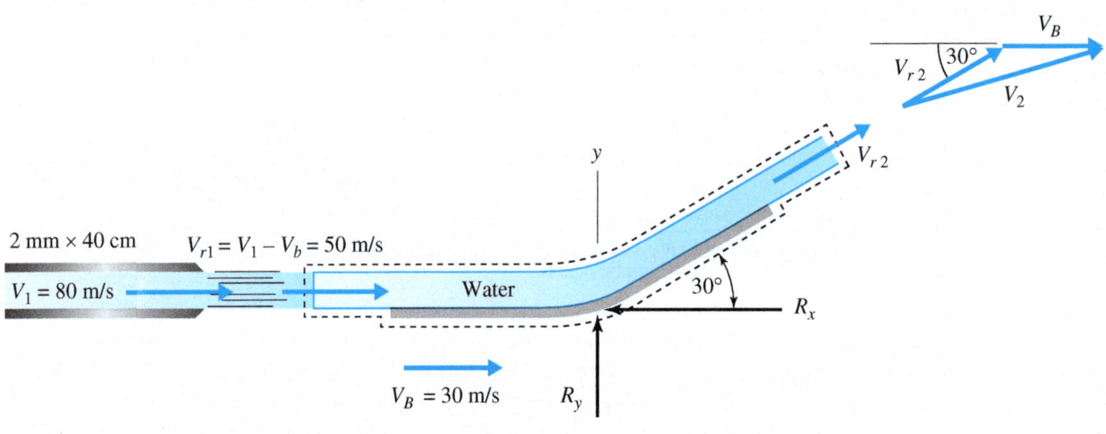

Figure E2.16

Solution

(a) To solve the problem of a moving deflector, we observe the flow from a reference frame attached to the deflector. In this moving reference frame the flow is steady and Bernoulli's equation with $p_1 = p_2$ can then be used to show that $V_{r1} = V_{r2} = 50$ m/s, the velocity of the sheet of water as observed from the deflector. Note that we cannot apply Bernoulli's equation in a fixed reference frame since the flow would not be steady. Applying the momentum equation to the moving control volume, which is indicated again by the dashed line, we obtain the following:

x-direction: $$-R_x = \dot{m}_r[(V_{r2})_x - (V_{r1})_x]$$
$$= 1000 \text{ kg/m}^3 \times 0.002 \text{ m} \times 0.4 \text{ m} \times 50 \text{ m/s } (50 \cos 30° - 50) \text{ m/s}$$
$$\therefore R_x = \underline{268 \text{ N}}$$

y-direction: $$R_y = \dot{m}_r[(V_{r2})_y - \overset{0}{\cancel{(V_{r1})_y}}]$$
$$= 1000 \times 0.002 \times 0.4 \times 50(50 \sin 30°) = \underline{1000 \text{ N}}$$

When calculating $\dot{m}_r$ we must use only that water which has its momentum changed; hence the velocity used is 50 m/s.

(b) Observed from a fixed observer the velocity $\mathbf{V}_2$ of the fluid after the deflection is $\mathbf{V}_2 = \mathbf{V}_{r2} + \mathbf{V}_B$, where $\mathbf{V}_{r2}$ is directed tangential to the deflector at the exit and has a magnitude equal to $\mathbf{V}_{r1}$ (see the velocity diagram above). Thus

$$(V_2)_x = V_{r2} \cos 30° + V_B$$
$$= 50 \times 0.866 + 30 = 73.3 \text{ m/s}$$
$$(V_2)_y = V_{r2} \sin 30°$$
$$= 50 \times 0.5 = 25 \text{ m/s}$$

Finally,

$$\mathbf{V}_2 = \underline{73.3\hat{\mathbf{i}} + 25\hat{\mathbf{j}} \text{ m/s}}$$

(c) The power generated by the moving vane is equal to the velocity of the vane times the force the vane exerts in the direction of the motion. Therefore,

$$\dot{W} = V_B \times R_x = 30 \text{ m/s} \times 260 \text{ N} = \underline{8040 \text{ W}}$$

Example 2.17

High-speed air jets strike the blades of a turbine rotor tangentially while the 1.5-m-diameter rotor rotates at 140 rad/s (Figure E2.17a). There are 10 such 4-cm-diameter jets. Calculate the maximum power output. The air density is 2.4 kg/m^3. Neglect any lateral spreading.

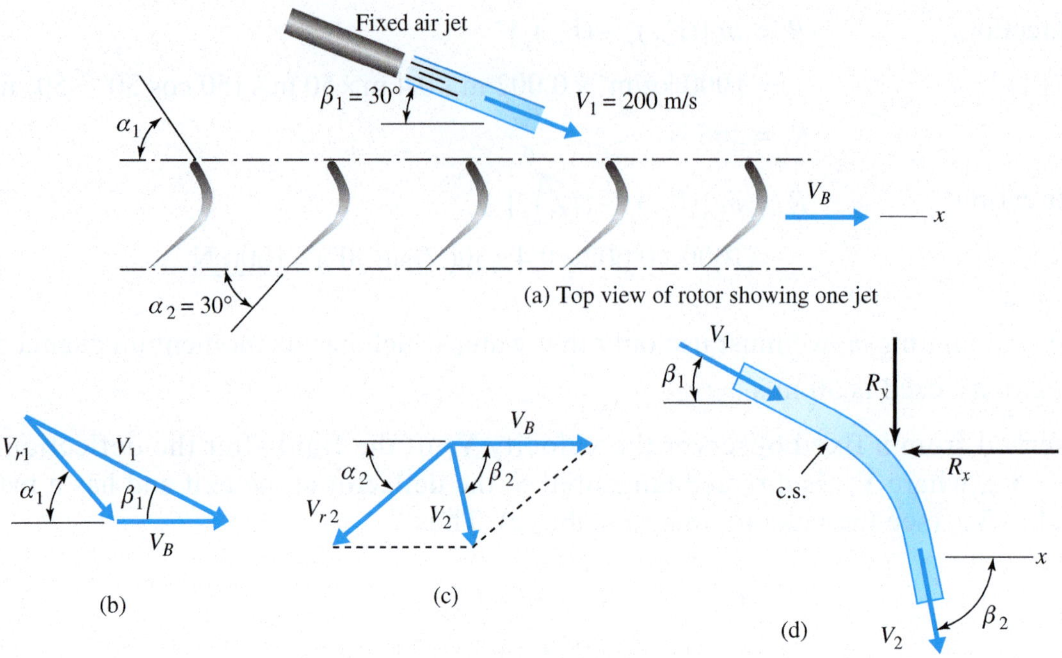

Figure E2.17

Solution

The blade angle α_1 is set by demanding that the air jet enter the blades tangentially, as observed from the moving blade; that is, the relative velocity vector $\mathbf{V}_r$ must make the angle α_1 with respect to the blade velocity $\mathbf{V}_B$. This is shown in Figure E2.17b. The relative entrance velocity is $\mathbf{V}_{r1}$ (Figure E2.17b), and the relative exit velocity is $\mathbf{V}_{r2}$ (Figure E2.17c). Both velocity polygons are presented by the vector equation

$$\mathbf{V} = \mathbf{V}_r + \mathbf{V}_B$$

which states that the absolute velocity equals the relative velocity plus the blade velocity. From the polygon at the entrance we have

$$\begin{aligned} V_1 \sin\beta_1 &= V_{r1} \sin\alpha_1 \\ V_1 \cos\beta_1 &= V_{r1} \cos\alpha_1 + V_B \\ \therefore \quad 200 \sin 30° &= V_{r1} \sin\alpha_1 \\ 200 \cos 30° &= V_{r1} \cos\alpha_1 + 0.75 \times 140 \end{aligned}$$

where $\mathbf{V}_B$ is the radius multiplied by the angular velocity. A simultaneous solution yields

$$V_{r1} = 121 \text{ m/s} \qquad \alpha_1 = 55.7°$$

The friction between the air and the blade is quite small and can be neglected when calculating the maximum output. This allows us to assume $V_{r2} = V_{r1}$. From the exiting velocity polygon we can write

$$V_B - V_{r2}\cos\alpha_2 = V_2\cos\beta_2$$
$$V_{r2}\sin\alpha_2 = V_2\sin\beta_2$$
$$\therefore \quad 0.75\times 140 - 121\cos 30° = V_2\cos\beta_2$$
$$121\sin 30° = V_2\sin\beta_2$$

A simultaneous solution results in

$$V_2 = 60.5 \text{ m/s} \qquad \beta_2 = 89.8°$$

The momentum equation applied to the control volume, shown in Figure E2.17d, gives

$$-R_x = \dot{m}(V_{2x} - V_{1x})$$
$$= 2.4 \text{ kg/m}^3 \times \pi \times 0.02^2 \text{ m}^2 \times 200 \text{ m/s}(60.5 \cos 89.8° - 200\cos 30°) \text{ m/s}$$
$$\therefore R_x = 104.3 \text{ N}$$

There are 10 jets, each producing the force above. The maximum power output is then

$$\text{power} = 10\times R_x \times V_B$$
$$= 10\times 104.3 \text{ N}\times(0.75\times 140) \text{ m/s} = 109600 \text{W} \quad \text{or} \quad \underline{109.6 \text{ kW}}$$

2.6.4 Momentum Equation Applied to Propellers

The application of the momentum equation to propellers is also of sufficient interest that this section will be devoted to illustrating the procedure. Consider the propeller of Figure 2.17 with the streamlines shown forming the surface of a control volume in which the fluid enters with a uniform velocity V_1 and exits with a uniform velocity V_2. The outer streamlines just touch the tips of the propeller. This flow situation can be seen to be identical to that of a propeller moving with velocity V_1 in a stagnant fluid by adding V_1 to the left in Figure 2.17. The momentum equation, applied to the control volume between sections 1 and 2 shown, gives

$$F = \dot{m}(V_2 - V_1) \tag{2.6.14}$$

This control volume is not sufficient, however, since the areas A_1 and A_2 are unknown. We know the flow area A of the propeller. So a control volume is drawn close to the propeller such that $V_3 \cong V_4$ and $A_3 \cong A_4 = A$. The momentum equation (2.6.5) in the x-direction takes the form

$$F + p_3 A - p_4 A = 0 \tag{2.6.15}$$

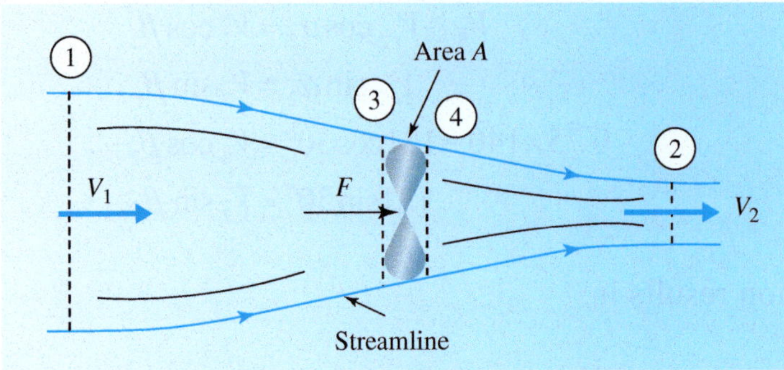

Figure 2.17 A propeller in a fluid flow.

or

$$F = (p_4 - p_3) = A \tag{2.6.16}$$

Now, since viscous effects would be quite small in this flow situation the energy equation up to the propeller and then downstream from the propeller is used to obtain

$$\frac{V_1^2 - V_3^2}{2} + \frac{p_1 - p_3}{\rho} = 0 \quad \text{and} \quad \frac{V_4^2 - V_2^2}{2} + \frac{p_4 - p_2}{\rho} = 0 \tag{2.6.17}$$

Adding these equations together, recognizing that $p_1 = p_2 = p_{\text{atm}}$, we have

$$\left(V_2^2 - V_1^2\right)\frac{\rho}{2} = p_4 - p_3 \tag{2.6.18}$$

Inserting this and Eq. 2.6.16 into Eq. 2.6.14 results in

$$V_3 = \frac{1}{2}(V_2 + V_1) \tag{2.6.19}$$

where we have used $\dot{m} = \rho A V_3$ since the propeller area is the only area known. This result shows that the velocity of the fluid moving through the propeller is the average of the upstream and downstream velocities.

The input power needed to produce this effect is found by applying the energy equation between sections 1 and 2, where the pressures are atmospheric; neglecting losses, Eq. 2.5.17 takes the form

$$\dot{W}_{\text{fluid}} = \frac{V_2^2 - V_1^2}{2}\dot{m} \tag{2.6.20}$$

where $\dot{W}_{\text{fluid}}$ is the energy input between the two sections. The moving propeller requires power given by

$$\begin{aligned}\dot{W}_{\text{prop}} &= F \times V_1 \\ &= \dot{m}(V_2 - V_1)V_1\end{aligned} \tag{2.6.21}$$

The theoretical propeller efficiency is then

$$\eta_P = \frac{\dot{W}_{\text{prop}}}{\dot{W}_{\text{fluid}}} = \frac{V_1}{V_3} \tag{2.6.22}$$

In contrast to the propeller, a wind turbine extracts energy from the airflow; the downstream velocity is reduced and the diameter is increased.

2.6.5 Control-Volume Analysis of Wind Turbines

Wind turbines, also referred to sometimes as wind machines or windmills, are devices that can convert wind power to mechanical power, and then to electrical power through the use of electric generators. Some of the earliest wind energy conversion devices were used to propel ships.

There are two primary types of wind turbines: vertical-axis and horizontal-axis. We will consider a horizontal-axis wind turbine, as sketched in Figure 2.18. These turbines have blades up to 80 m in length and hub heights up to 135 m.

The power available in a wind stream is equal to the rate of incoming kinetic energy of the wind within a streamtube, as shown in Figure 2.19. The power can be defined as

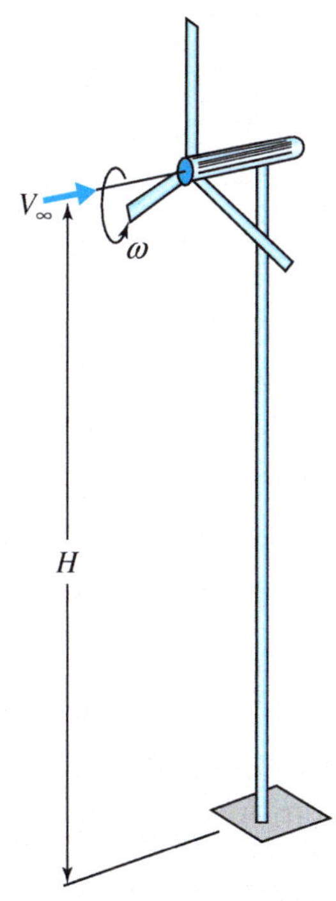

Figure 2.18 A typical horizontal-axis wind turbine.

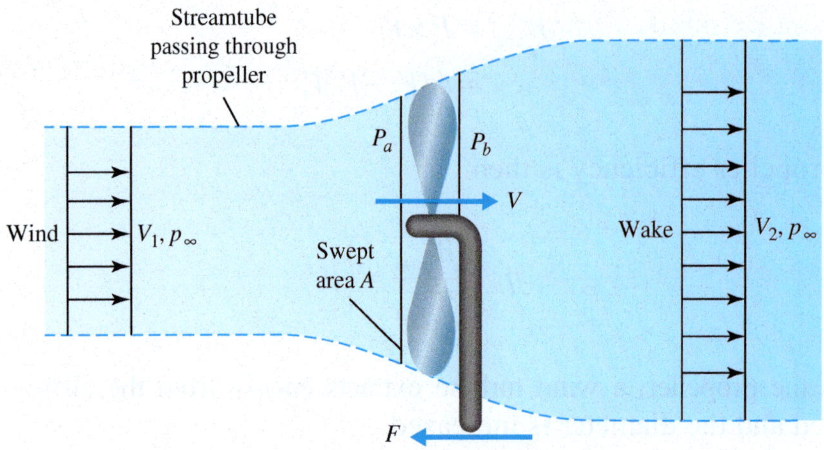

Figure 2.19 The streamtube containing the turbine.

$$\dot{W} = \frac{1}{2}\dot{m}V_1^2 \tag{2.6.23}$$

where V_1 is the upstream wind velocity and $\dot{m}$ is the mass flow rate through the streamtube. The mass flow rate is given by

$$\dot{m} = \rho A_1 V_1 \tag{2.6.24}$$

Hence, the available power $\dot{W}$ can be written as

$$\dot{W} = \frac{1}{2}\rho A_1 V_1^3 \tag{2.6.25}$$

This shows that the power of a wind stream is directly proportional to its density, area, and the cube of the velocity.

The total energy E that is available over a time period from 0 to t due to a varying wind speed can be found by recognizing that $\dot{W} = dE/dt$ (power is the rate of doing work) so that

$$E = \int_0^t \dot{W}dt = \frac{1}{2}\rho A \int_0^t V^3 dt \tag{2.6.26}$$

Now, consider the air flow between section 1 and section a, where section 1 is considered to be at some distance upstream of the wind turbine, and section a is just at the inlet to the wind turbine. Assuming the adiabatic flow is incompressible and that changes in potential energy are negligible, Bernoulli's equation between these two sections yields

$$p_1 + \frac{1}{2}\rho V_1^2 = p_a + \frac{1}{2}\rho V_a^2 \tag{2.6.27}$$

Similarly, applying Bernoulli's equation between section b and section 2 gives

$$p_b + \frac{1}{2}\rho V_b^2 = p_2 + \frac{1}{2}\rho V_2^2 \tag{2.6.28}$$

The wind velocity between sections a and b will decrease since the kinetic energy of the wind is being converted to mechanical work. The difference in the pressure across the turbine can be determined from the above two equations as follows:

$$p_a - p_b = (p_1 - p_2) + \frac{1}{2}\rho\left(V_1^2 - V_2^2\right) + \frac{1}{2}\rho\left(V_b^2 - V_a^2\right) \tag{2.6.29}$$

Since sections 1 and 2 are considered to be at some distance from the turbine where the pressure is equal to the ambient pressure, we assume

$$p_1 = p_2 = p_\infty \tag{2.6.30}$$

Moreover, since the thickness of a turbine blade is small compared to the length, the change in the wind velocity across the turbine is small, that is,

$$V_a \approx V_b \approx V_t \tag{2.6.31}$$

where V_t is the average velocity across the turbine. Equation (2.6.29) then simplifies to

$$p_a - p_b = \frac{1}{2}\rho\left(V_1^2 - V_2^2\right) \tag{2.6.32}$$

Note, that the velocity from section 1 to section a and from section b to section 2 will decrease gradually, as shown in Figure 2.20.

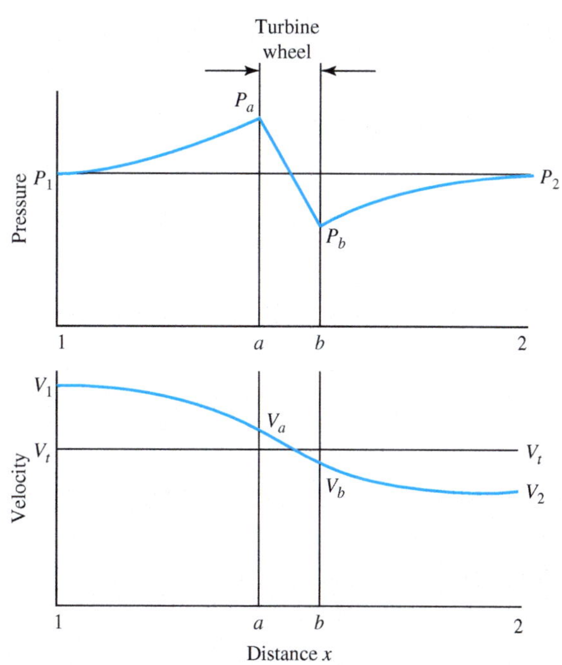

Figure 2.20 Representative pressure and velocity profiles across a horizontal-axis turbine.

The force F on the turbine support can be found by applying the momentum equation to the control volume between sections 1 and 2:

$$-F = \dot{m}(V_2 - V_1) \tag{2.6.33}$$

Similarly, the momentum equation applied between sections a and b yields

$$-F + (p_a - p_b)A = \dot{m}(V_b - V_a) \tag{2.6.34}$$

where A is the swept area of the turbine blades. Substituting Eq. 2.6.33 into Eq. 2.6.34, assuming $V_a \approx V_b$, we get

$$\dot{m}(V_1 - V_2) = A(p_a - p_b) \tag{2.6.35}$$

Combine Eqs. 2.6.32 and 2.6.35 and use $\dot{m} = \rho V_t A$ to find

$$pV_t A(V_1 - V_2) = \frac{1}{2}\rho\left(V_1^2 - V_2^2\right)A \tag{2.6.36}$$

which can be simplified to

$$V_t = \frac{1}{2}(V_1 + V_2) \tag{2.6.37}$$

Equation (2.6.23) for wind power can be rewritten as

$$\dot{W} = \frac{1}{2}\rho A V_t\left(V_1^2 - V_2^2\right) \tag{2.6.38}$$

or

$$\dot{W} = \frac{1}{4}\rho A(V_1 + V_2)\left(V_1^2 - V_2^2\right) \tag{2.6.39}$$

which reduces to Eq. 2.6.25 if $V_2 = 0$. (Use Eq. 2.6.37 and continuity to find $A = 2A_1$.) However, the fluid cannot come to a complete stop at the turbine exit. Hence, there is an optimum exit velocity that will result in maximum power. This optimum velocity can be obtained by differentiating Eq. 2.6.39 with respect to V_2 assuming a constant V_1 and equating the derivative to zero, that is $d\dot{W}/dV_2 = 0$. This gives

$$3V_2^2 + 2V_1V_2 - V_1^2 = 0 \tag{2.6.40}$$

Solving the equation for V_2 gives the optimum exit velocity as

$$V_2 = \frac{1}{3} V_1 \tag{2.6.41}$$

Using this exit velocity we determine the maximum power as

$$\dot{W}_{\max} = \frac{8}{27} \rho A V_1^3 \tag{2.6.42}$$

The maximum efficiency (also known as the **power coefficient**) of a wind turbine is defined as

$$\eta_{\max} = \frac{\dot{W}_{\max}}{\frac{1}{2}\rho A V_1^3} = \frac{\frac{8}{27}\rho A V_1^3}{\frac{1}{2}\rho A V_1^3} = 0.593 \quad \text{or} \quad 59.3\% \tag{2.6.43}$$

Example 2.18

A wind turbine with a rotor diameter of 50 m and is exposed to a 10 mph wind at standard atmospheric conditions (100 kPa, 25°C). Calculate the following:

(a) The available wind power
(b) The maximum power that can be obtained from the wind
(c) The maximum torque if the turbine rotates at 40 rpm

Solution

(a) The available power is $\dot{W} = \frac{1}{2}\dot{m}V_1^2$ or $\dot{W} = \frac{1}{4}\rho A V_1^3$. The density of the air is calculated as

$$\rho = \frac{p}{RT} = \frac{100}{0.287 \times 298} = 1.17 \text{ kg/m}^3$$

The area swept by the rotor is $A = \pi(25)^2 = 1963 \text{ m}^2$. The wind velocity is

$$V_1 = 10 \text{ mph} \times \frac{1.609 \text{ km/hr}}{\text{mph}} \times \frac{\text{m/s}}{3.6 \text{ km/hr}} = 4.47 \text{ m/s}$$

and the available power is

$$\dot{W} = \frac{1}{2}(1.17 \text{ kg/m}^3 \times 1963 \text{ m}^2)(4.47 \text{ m/s})^3 = 103000 \text{ W} \quad \text{or} \quad \underline{103 \text{ kW}}$$

(b) The maximum power that could be obtained is

$$\dot{W}_{\max} = \eta \dot{W}_{\text{available}} = 0.593 \times 103 = \underline{61.1\ \text{kW}}$$

(c) Torque is related to power by $\dot{W} = 2\pi n T$ if n is measured in rpm. Then

$$T = \frac{\dot{W}_{\max}}{2\pi n} = \frac{61.1 \times 10^3\ \text{N} \cdot \text{m/s}}{(2\pi\ \text{rad/rev}) \times 40\ \text{rev}/60\ \text{s}} = \underline{14600\ \text{N} \cdot \text{m}}$$

Example 2.19

The wind turbine considered in the previous example operates in a wind that has a velocity pattern during a day that could be approximated by half a sine wave as $V = V_m \sin(\pi t/24)$, where t is time in hours and $V_m = 8.5$ m/s. Calculate:

(a) The theoretical maximum wind turbine energy produced during a 24-hr period
(b) The wind mean energy velocity (the constant wind speed that will produce the same energy for a day that the variable wind speed produces)
(c) The electric energy generated during a 24-hr period if the overall efficiency of the wind turbine-generator is 0.3 (the overall efficiency includes efficiency of the generator and friction losses in the gear system used to couple the wind turbine to the generator)
(d) If the wind turbine has a cut-in velocity of 3.1 m/s and a cut-out velocity of 8 m/s, calculate the theoretical maximum wind turbine energy during a 24-hr period

Solution

(a) The total wind energy during a 24-hr period is

$$E = \frac{1}{2}\rho A \int_0^{24} V^3 dt = \frac{1}{2}\rho A \int_0^{24} [V_m \sin(\pi t/24)]^3 dt$$

Integrating we get

$$E = \frac{1}{2}\rho A V_m^3 \left[\frac{\cos^3(\pi t/24)}{\pi/8} - \frac{\cos(\pi t/24)}{\pi/24} \right]_0^{24} = \frac{1}{2}\rho A V_m^3 \left(\frac{32}{\pi} \right)$$

Using the density and area from the previous example and $t = 24$ hrs, we calculate the energy

$$E = \frac{1}{2}\left(1.17\ \text{kg/m}^3 \times 1963\ \text{m}^2\right)(8.5\ \text{m/s})^3 \left(\frac{32}{\pi}\ \text{hr} \right) = \underline{7180\ \text{kW-hr}}$$

The maximum wind energy is $E_{\max} = \eta_{\max} \times E = 0.593 \times 7180 = \underline{4260\ \text{kW-hr}}$

(b) The wind mean energy velocity can be calculated using

$$V_{\text{mean}} = \left(\frac{E_{\max}}{\frac{1}{2}\rho A \Delta t} \right)^{1/3} = \left(\frac{4260 \times 10^3}{\frac{1}{2}(1.17 \times 1963 \times 24)} \right)^{1/3} = \underline{5.37 \text{ m/s}}$$

(c) The total electric energy generated during the 24-hr period is

$$E_e = \eta_e \times E_{\max} = 0.3 \times 4260 = 1280 \text{ kW-hr}$$

(d) In this case, we have to determine the time during the day when the wind turbine is operational, that is the wind speed is between the cut-in and cut-out velocities. Using the given expression for V we calculate the time during the day when $V = 3.1$ m/s as

$$t_1 = \frac{24}{\pi}\sin^{-1}(3.1/8.5) = 2.85 \text{ hr}$$

where $\sin^{-1}(3.1/8.5) = 0.372$ rad. Similarly, we calculate t_2 when $V = 8$ m/s as

$$t_2 = \frac{24}{\pi}\sin^{-1}(8/8.5) = 9.4 \text{ hr}$$

To determine the total energy during the day we integrate between the above two times. Note that since the wind velocity pattern is symmetrical with respect to 12 h during the day, the cut-in and cut-out times occur twice during the day. Hence, we calculate the energy as follows:

$$E = \frac{1}{2}\rho A \left(2 \times \int_{t_1}^{t_2} V^3 dt \right) = \frac{1}{2}\rho A \left(2 \times \int_{t_1}^{t_2} [V_m \sin(\pi t/24)]^3 \, dt \right)$$

$$= \frac{1}{2}\rho A V_m^3 \left[\frac{\cos^3(\pi t/24)}{\pi/8} - \frac{\cos(\pi t/24)}{\pi/24} \right]_{t_1}^{t_2} = \frac{1}{2}\rho A V_m^3 (2 \times 2.602) = \underline{3702 \text{ kW-hr}}$$

The maximum wind energy is

$$E_{\max} = \eta_{\max} \times E = 0.593 \times 3702 = \underline{2190 \text{ kW-hr}}$$

The electric power generated is

$$E_e = \eta_e \times E_{\max} = 0.3 \times 2190 = \underline{658 \text{ kW-hr}}$$

In order to protect the wind turbine, generator, and gearing system from wear and high forces, the wind turbine is designed to shut down when the wind velocity is below the cut-in velocity and above the cut-out velocity.

2.6.6 Aerodynamic Analysis of a Rotating Wind-Turbine Blade

As discussed earlier in Section 2.6.5, wind turbines convert wind energy into mechanical energy. In the conservation of momentum used in the previous section, the geometry of the rotor blade is not considered. However, since the flow through a wind turbine depends on the blade geometry a more detailed approach is needed. Consider the flow over the blade (actually, an airfoil) profile at a cross-section of the wind turbine blade, as shown in Figure 2.21. The blade is considered to be rotating at an angular speed ω and is oriented at an angle of attack α with respect to the relative velocity of the wind. The pitch angle β is between the chord line and the plane of rotation of the wind turbine rotor. The drag and lift forces are defined with respect to the direction of the relative velocity. We also define a normal force component acting perpendicular to the chord line, and a tangential force component acting along the chord line. The thrust F_{thrust} acts in a direction perpendicular to the plane of rotation of the rotor, and the torque acts in the plane of rotation.

The lift and drag forces can be calculated using the relative velocity and the coefficients of lift and drag for the airfoil. The relative velocity vector is the velocity of the wind relative to the airfoil and is defined as

$$\mathbf{V}_r = \mathbf{V}_\infty - \mathbf{V}_B \tag{2.6.44}$$

where $V_B = \omega R$ is the blade velocity, ω is the angular velocity of the rotor blade, R is the radial distance between the center of rotation and the position on the airfoil where the forces are determined, and V_∞ is the wind free-stream velocity. Referring to Figure 2.21 and the velocity triangle, the relative velocity can be written as

$$V_r = \frac{V_\infty}{\sin(\alpha + \beta)} \tag{2.6.45}$$

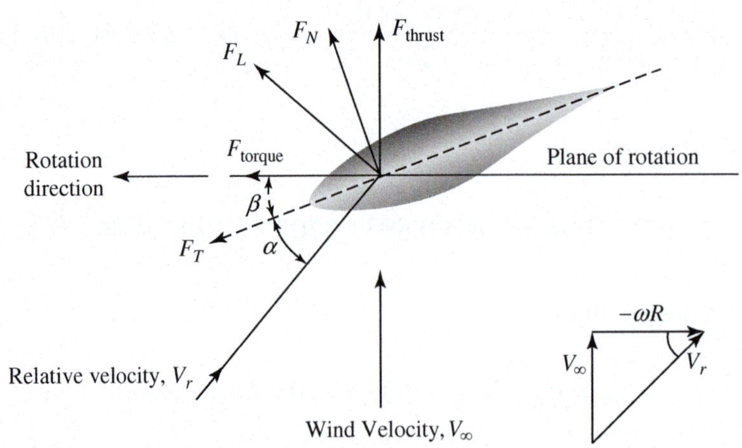

Figure 2.21 A rotating wind-turbine blade, and a velocity triangle.

Even though the flow over the rotor blade is three-dimensional, we use a two-dimensional approach to estimate the lift and drag forces per unit length as follows:

$$\begin{aligned} &\text{Lift force:} \quad F_L = \frac{1}{2} C_L \rho V_r^2 c \\ &\text{Drag force:} \quad F_D = \frac{1}{2} C_D \rho V_r^2 c \end{aligned} \tag{2.6.46}$$

where c is the chord length (the distance from the nose to the tail) and C_L and C_D are the lift and drag coefficients, respectively. The normal and tangential forces can be determined by projecting the lift and drag forces on to the normal and tangential directions of the rotor plane and can be written as

$$\begin{aligned} F_N &= F_L \cos\alpha + F_D \sin\alpha \\ F_T &= F_L \sin\alpha - F_D \cos\alpha \end{aligned} \tag{2.6.47}$$

The normal coefficient and tangential coefficient are defined and related as follows:

$$C_N = \frac{F_N}{\frac{1}{2}\rho V_r^2 c} \qquad C_T = \frac{F_T}{\frac{1}{2}\rho V_r^2 c} \tag{2.6.48}$$

$$\begin{aligned} C_N &= C_L \cos\alpha + C_D \sin\alpha \\ C_T &= C_L \sin\alpha - C_D \cos\alpha \end{aligned} \tag{2.6.49}$$

Similarly, the thrust and torque forces can be determined as

$$\begin{aligned} F_{\text{thrust}} &= F_L \cos(\alpha+\beta) + F_D \sin(\alpha+\beta) \\ F_{\text{torque}} &= F_L \sin(\alpha+\beta) - F_D \cos(\alpha+\beta) \end{aligned} \tag{2.6.50}$$

The thrust and torque coefficients are defined as follows:

$$C_{\text{thrust}} = \frac{F_{\text{thrust}}}{\frac{1}{2}\rho V_r^2 c} \qquad C_{\text{torque}} = \frac{F_{\text{torque}}}{\frac{1}{2}\rho V_r^2 c} \tag{2.6.51}$$

Alternatively, these coefficients can be determined in terms of the lift and drag coefficients using

$$\begin{aligned} C_{\text{thrust}} &= C_L \cos(\alpha+\beta) + C_D \sin(\alpha+\beta) \\ C_{\text{torque}} &= C_L \sin(\alpha+\beta) - C_D \cos(\alpha+\beta) \end{aligned} \tag{2.6.52}$$

The torque, per unit length, acting on a single blade is

$$T = F_{\text{torque}} \times R \tag{2.6.53}$$

where R is the radius at which F_{torque} is calculated. Since wind turbine rotors contain more than one blade, the total torque acting on the rotor can be obtained by multiplying the torque of a single blade by the number of blades.

The power generated by the rotor can be calculated using

$$P = N \times 2\pi n T \tag{2.6.54}$$

where N is the number of blades and n is the rotational speed in rev/s.

In general, wind turbine blades do not have a uniform pitch angle (sometimes referred to as twist) between the tip and the root where the blade attaches to the hub. Moreover, the blades do not have a constant chord length from root to tip, and hence a three-dimensional analysis of the aerodynamic forces is needed. Computational fluid dynamics can be used to analyze the flow and to calculate the aerodynamic forces more accurately than the approximations using a two-dimensional approach.

Example 2.20

A wind at 1 atm and 15°C is blowing at 20 m/s over a 10-m-diameter wind turbine that has a cutout velocity of 15 m/s. In this case, since the wind velocity is higher than the cutout velocity the rotor is "parked" (the rotor is locked so it is stationary) to prevent damage to the wind turbine. At a location midway between the root and tip of the blade, the chord length is 0.46 m, the angle of attack is 14°, and the lift and drag coefficients are 0.93 and 0.74, respectively. Calculate the normal and tangential forces acting at that location on this parked rotor.

Solution

Since the rotor is stationary the relative velocity is equal to the free-stream wind velocity, that is, $V_r = V_\infty = 20$ m/s. The air density is

$$\rho = \frac{p}{RT} = \frac{101}{0.287 \times 288} = 1.23\ \text{kg/m}^3$$

The lift and drag forces are

$$F_L = \frac{1}{2} C_L \rho V_r^2 c = \frac{0.93}{2} \times (1.23\ \text{kg/m}^3) \times (20\ \text{m/s})^2 \times 0.46\ \text{m} = 105\ \text{N/m}$$

$$F_D = \frac{1}{2} C_D \rho V_r^2 c = \frac{0.74}{2} \times (1.23\ \text{kg/m}^3) \times (20\ \text{m/s})^2 \times 0.46\ \text{m} = 83.7\ \text{N/m}$$

The normal and tangential forces are then calculated to be

$$F_N = F_L \cos\alpha + F_D \sin\alpha = 105\cos 14 + 83.7\sin 14 = \underline{122 \text{ N/m}}$$
$$F_T = F_L \sin\alpha - F_D \cos\alpha = 105\sin 14 - 83.7\cos 14 = \underline{-55.8 \text{ N/m}}$$

The minus sign indicates that the force is acting in a direction opposite to that shown in Figure 2.21. Note that in this case, $F_{\text{thrust}} = F_D$ and $F_{\text{torque}} = F_L$.

Example 2.21

A three-bladed rotor wind turbine has a diameter of 10 m, is operating in a 7 m/s wind at 1 atm and 25°C. The rotational speed is 72 rpm and the chord length is 0.46 m. At a location on the rotor blade where the radial distance is 50% of the blade length, the pitch angle is 9°, and the tangential and normal coefficients are 0.09 and 0.83, respectively. If the blade has a uniform chord length, airfoil profile, and pitch angle, calculate the torque and thrust force acting at that location. Then make an estimate of the torque and the power produced assuming that the average value acts over the entire length of the blade.

Solution

Since this is a rotating blade we calculate, using $\omega = 2\pi$ rad/rev × 1.2 rev/s = 7.54 rad/s and $R = 0.5 \times 5 = 2.5$ m, the relative velocity of the wind is

$$V_r = \sqrt{V_\infty^2 + (\omega R)^2} = \sqrt{49 + (7.54 \times 2.5)^2} = 20 \text{ m/s}$$

The local angle of attack is calculated using

$$\alpha = \sin^{-1}(V_\infty / V_r) - \beta = \sin^{-1}(7/20) - 9° = 11.5°$$

We can determine the lift and drag coefficients by manipulating Eqs. 2.6.49. The result is

$$C_L = C_N \cos\alpha + C_T \sin\alpha$$
$$C_D = C_N \sin\alpha - C_T \cos\alpha$$

The lift and drag coefficients can then be calculated to be

$$C_L = 0.83 \cos 11.5 + 0.09 \sin 11.5 = 0.831$$
$$C_D = 0.83 \sin 11.5 - 0.09 \cos 11.5 = 0.077$$

Next, we calculate the thrust and torque coefficients:

$$C_{\text{thrust}} = C_L \cos(\alpha+\beta) + C_D \sin(\alpha+\beta) = 0.831 \cos 20.5 + 0.077 \sin 20.5 = 0.850$$
$$C_{\text{torque}} = C_L \sin(\alpha+\beta) - C_D \cos(\alpha+\beta) = 0.831 \sin 20.5 - 0.077 \cos 20.5 = 0.219$$

The forces are calculated as follows

$$F_{\text{thrust}} = \frac{1}{2} C_{\text{thrust}} \rho V_r^2 c = \frac{0.805}{2} \times (1.23\ \text{kg/m}^3) \times (20\ \text{m/s})^2 \times 0.46\ \text{m} = 91\ \text{N/m}$$
$$F_{\text{torque}} = \frac{1}{2} C_{\text{torque}} \rho V_r^2 c = \frac{0.219}{2} \times (1.23\ \text{kg/m}^3) \times (20\ \text{m/s})^2 \times 0.46\ \text{m} = 25\ \text{N/m}$$

Assuming that these forces represent the average value over the entire blade length we multiply by the blade length to get the total thrust and torque forces acting on a single blade. The blade length is equal to the radius of the rotor, so the forces are

$$F_{\text{thrust}} = 91 \frac{\text{N}}{\text{m}} \times 5\ \text{m} = \underline{455\ \text{N}}$$
$$F_{\text{torque}} = 25 \frac{\text{N}}{\text{m}} \times 5\ \text{m} = \underline{125\ \text{N}}$$

The torque acting on a single blade can then be estimated to be

$$T = F_{\text{torque}} \times R = 125\ \text{N} \times 2.5\ \text{m} = \underline{312\ \text{N} \cdot \text{m}}$$

The power produced by the three blades on the rotor is approximately

$$P = N \times 2\pi n T = 3 \times 2\pi \left(72 \frac{rev}{60\ \text{s}} \right) \times 312\ \text{N} \cdot \text{m} = \underline{7070\ \text{W}}$$

This is only an approximation because of the two-dimensional model used.

2.6.7 Steady Nonuniform Flow

If we cannot assume uniform velocity profiles, we can let the momentum flux be expressed as

$$\int_A V^2 dA = \beta \overline{V}^2 A \tag{2.6.58}$$

where we have introduced the **momentum-correction factor** β, expressed explicitly as

$$\beta = \frac{\int V^2 dA}{\overline{V}^2 A} \tag{2.6.59}$$

The momentum equation (2.6.5), for a steady flow with one inlet and one exit, can then be written as

$$\Sigma \mathbf{F} = \dot{m}(\beta_2 V_2 - \beta_1 V_1) \tag{2.6.60}$$

For a laminar flow with a parabolic profile in a circular pipe, $\beta = 4/3$. If a profile is given, however, the integral is usually integrated and Eq. 2.6.2 is used.

Example 2.22

Calculate the momentum correction factor for a parabolic profile **(a)** between parallel plates of width w and **(b)** in a circular pipe of radius R. The parabolic profiles are shown in Figure E2.22.

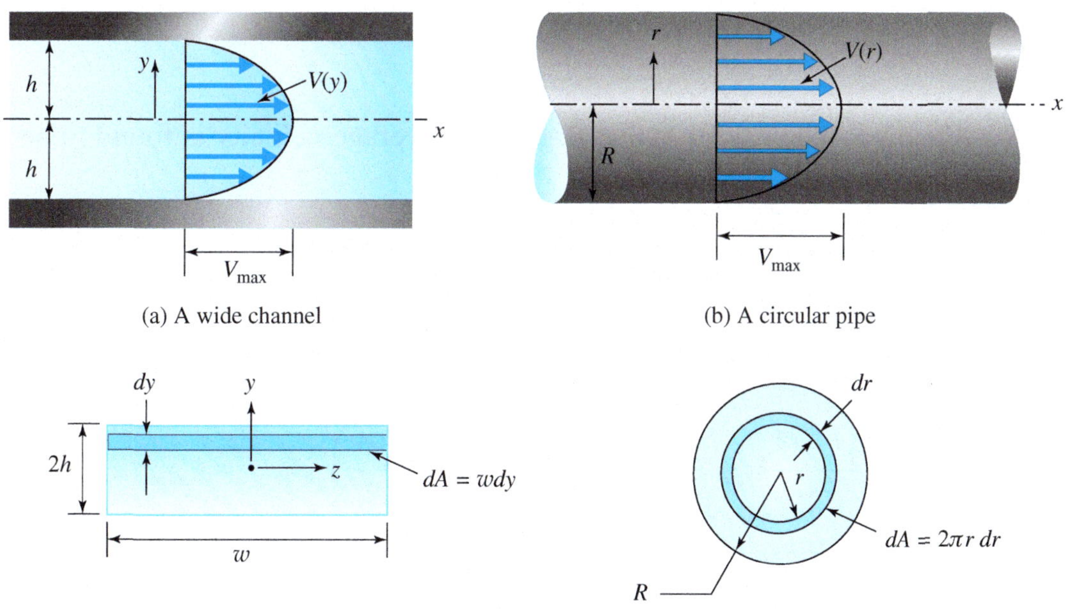

Figure E2.22

Solution

(a) A parabolic profile between parallel plates can be expressed as

$$V(y) = V_{\max}\left(1 - \frac{y^2}{h^2}\right)$$

where y is measured from the centerline, the velocity is zero at the walls where $y = \pm h$, and $V_{\max}$ is the centerline velocity at $y = 0$. First, let us find the average velocity. It is

$$\overline{V} = \frac{1}{A}\int V dA$$
$$= \frac{1}{hw}\int_0^h V_{\max}\left(1-\frac{y^2}{h^2}\right)w\,dy = \frac{V_{\max}}{h}\left(h-\frac{1}{3}h\right) = \frac{2}{3}V_{\max}$$

where we have integrated over the top half of the cross-section. Then

$$\beta = \frac{\int V^2 dA}{\overline{V}^2 A} = \frac{2}{\frac{4}{9}V_{\max}^2 \times 2hw}\int_0^h V_{\max}^2\left(1-\frac{y^2}{h^2}\right)^2 w\,dy = \underline{1.2}$$

where the factor "2" in the numerator accounts for the bottom half of the channel.

(b) For a circular pipe a parabolic profile can be written as

$$V(r) = V_{\max}\left(1-\frac{r^2}{R^2}\right)$$

where R is the pipe radius and $V = 0$ at $r = R$. The average velocity is found to be

$$\overline{V} = \frac{1}{A}\int V\,dA = \frac{1}{\pi R^2}\int_0^R V_{\max}\left(1-\frac{r^2}{R^2}\right)2\pi r\,dr = \frac{1}{2}V_{\max}$$

The momentum correction factor is then

$$\beta = \frac{\int V^2 dA}{\overline{V}^2 A} = \frac{1}{\frac{1}{4}V_{\max}^2 \pi R^2}\int_0^R V_{\max}^2\left(1.2.12-\frac{r^2}{R^2}\right)^2 2\pi r dr = \underline{1.33}$$

The correction factors above can be used to express the momentum flux across a cross-sectional area as $\beta\rho A\overline{V}^2$.

2.6.8 Noninertial Reference Frames

In certain situations it may be necessary to choose a noninertial reference frame in which the velocity is measured. This would be the case if we were to study the flow through a dishwasher arm, around a turbine blade, or from a rocket. Relative to a noninertial reference frame, Newton's second law takes the form (refer to Eq. 3.2.12)

$$\begin{aligned}\Sigma \mathbf{F} &= \frac{D}{Dt}\int_{\text{sys}} \rho \mathbf{V} d\mathcal{V} \\ &+ \int_{\text{sys}} \left[\frac{d^2\mathbf{S}}{dt^2} + 2\boldsymbol{\Omega} \times \mathbf{V} + \boldsymbol{\Omega} \times (\boldsymbol{\Omega} \times r) + \frac{d\boldsymbol{\Omega}}{dt} \times \mathbf{r} \right] \rho \, d\mathcal{V}\end{aligned} \tag{2.6.61}$$

where **V** is the velocity relative to the noninertial frame; the acceleration **a** of each particle in the system is already accounted for in the first term. Equation 2.6.61 is often written as

$$\begin{aligned}\Sigma \mathbf{F} - \mathbf{F}_I &= \frac{D}{Dt}\int_{\text{sys}} \rho \mathbf{V} d\mathcal{V} \\ &= \frac{d}{dt}\int_{\text{c.v.}} \rho \mathbf{V} d\mathcal{V} + \int_{\text{c.s.}} \rho \mathbf{V}(\mathbf{V} \cdot \hat{\mathbf{n}}) dA\end{aligned} \tag{2.6.62}$$

where $\mathbf{F}_I$ is called the "inertial body force," given by

$$\mathbf{F}_I = \int_{\text{sys}} \left[\frac{d^2\mathbf{S}}{dt^2} + 2\boldsymbol{\Omega} \times \mathbf{V} + \boldsymbol{\Omega} \times (\boldsymbol{\Omega} \times r) + \frac{d\boldsymbol{\Omega}}{dt} \times \mathbf{r} \right] \rho \, d\mathcal{V} \tag{2.6.63}$$

Since the system and control volume are identical at time t the system integration can be replaced with a control volume integration in the integral of Eq. 2.6.63. Example 2.23 will illustrate the use of a noninertial reference frame.

Example 2.23

The rocket shown in Figure E2.23, with an initial mass of 150 kg, burns fuel at the rate of 10 kg/s with a constant exhaust velocity of 700 m/s. What is the initial acceleration of the rocket and the velocity after 1 s? Neglect the drag on the rocket.

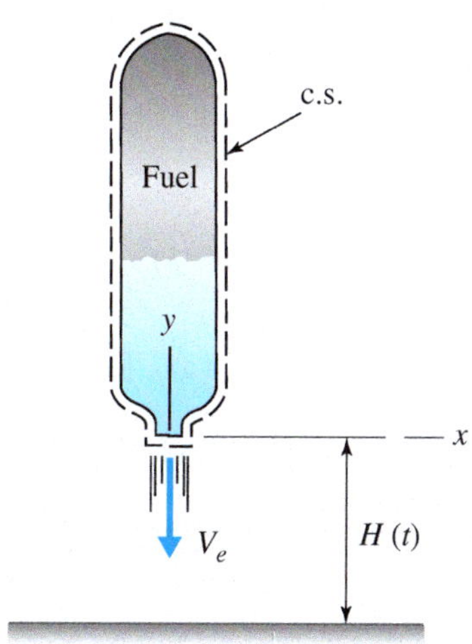

Figure E2.23

Solution

The control volume is sketched and includes the entire rocket. The reference frame attached to the rocket is accelerating upward at d^2H/dt^2. Newton's second law is written as, using z upward,

$$\Sigma F_z - (F_I)_z = \frac{d}{dt}\int_{\text{c.v.}} \rho V_z \, dV\!\!\!- + \int_{\text{c.s.}} \rho V_z \mathbf{V}\cdot\hat{\mathbf{n}}\, dA$$

$$\therefore\ -W - \frac{d^2H}{dt^2} m_{\text{c.v.}} = \rho_e(-V_e)V_e A_e$$

where

$$\frac{d}{dt}\int_{\text{c.v.}} \rho V_z \, dV\!\!\!- \approx 0$$

since V_z is the velocity of each mass element $\rho dV\!\!\!-$ relative to the reference frame attached to the control volume, the only vertical force is the weight W, and $m_{\text{c.v.}}$ is the mass of the control volume. From continuity we see that

$$m_{\text{c.v.}} = 150 - \dot{m}t = 150 - 10t$$

$$\therefore\ W = (150 - 10t)\times 9.81$$

The momentum equation becomes

$$-(150 - 10t)\times 9.81 - \frac{d^2H}{dt^2}(150 - 10t) = -\dot{m}_e V_e = -10\times 700 = -7000$$

This is written as

$$\frac{d^2H}{dt^2} = \frac{700}{15 - t} - 9.81$$

The initial acceleration is found by letting $t = 0$:

$$\left.\frac{d^2H}{dt^2}\right|_{t=0} = \frac{700}{15} - 9.81 = 36.9 \text{ m/s}^2$$

Integrate the expression for d^2H/dt^2 and obtain

$$\frac{dH}{dt} = -700 \ln(15 - t) - 9.81t + C$$

The constant $C = 700 \ln 15$ since $dH/dt = 0$ at $t = 0$. Thus at $t = 1$ s the velocity is

$$\frac{dH}{dt} = -700 \ln\frac{15}{14} - 9.81\times 1 = \underline{38.5 \text{ m/s}}$$

2.7 MOMENT-OF-MOMENTUM EQUATION

In the preceding section we determined the magnitude of force components in a variety of flow situations. To determine the line of action of a given force component, it is often necessary to apply the moment-of-momentum equation. Also, in analyzing the flow situation in devices that have rotating components the moment-of-momentum equation is needed to relate the rotational speed to the other flow parameters. Since it may be advisable to attach the reference frame to the rotating component, we will write the general equation with the inertial forces included. It is (see Eq. 2.2.4)

$$\Sigma \mathbf{M} - \mathbf{M}_I = \frac{D}{Dt}\int_{\text{sys}} \mathbf{r} \times \mathbf{V}\ \rho d\mathcal{V} \tag{2.7.1}$$

where

$$\mathbf{M}_I = \int \mathbf{r}\left[\frac{d^2\mathbf{S}}{dt^2} + 2\mathbf{\Omega} \times \mathbf{V} + \mathbf{\Omega} \times (\mathbf{\Omega} \times \mathbf{r}) + \frac{d\mathbf{\Omega}}{dt} \times \mathbf{r}\right] \rho\ d\mathcal{V} \tag{2.7.2}$$

This inertial moment $\mathbf{M}_I$ accounts for the fact that a noninertial reference frame attached to, for example a rotating arm was selected; it is simply the moment of $\mathbf{F}_I$ (see Eq. 2.6.63). Applying the system-to-control volume transformation, the moment-of-momentum equation for a control volume becomes

$$\Sigma \mathbf{M} - \mathbf{M}_I = \frac{d}{dt}\int_{\text{c.v.}} \mathbf{r} \times \mathbf{V}\ \rho d\mathcal{V} + \int_{\text{c.s.}} \mathbf{r} \times \mathbf{V}(\mathbf{V} \cdot \hat{\mathbf{n}})\rho\ dA \tag{2.7.3}$$

Examples will illustrate the application of this equation.

Example 2.24

A sprinkler has four 50-cm-long arms with nozzles at right angles with the arms and 45° with the ground (Figure E2.20). If the total flow rate is 0.01 m^3/s and a nozzle exit diameter is 12 mm, find the rotational speed of the sprinkler. Neglect friction.

Solution

The velocity exiting each nozzle shown is

$$\begin{aligned} V_e &= \frac{Q}{A} \\ &= \frac{0.01/4}{\pi \times 0.006^2} = 22.1 \text{ m/s} \end{aligned}$$

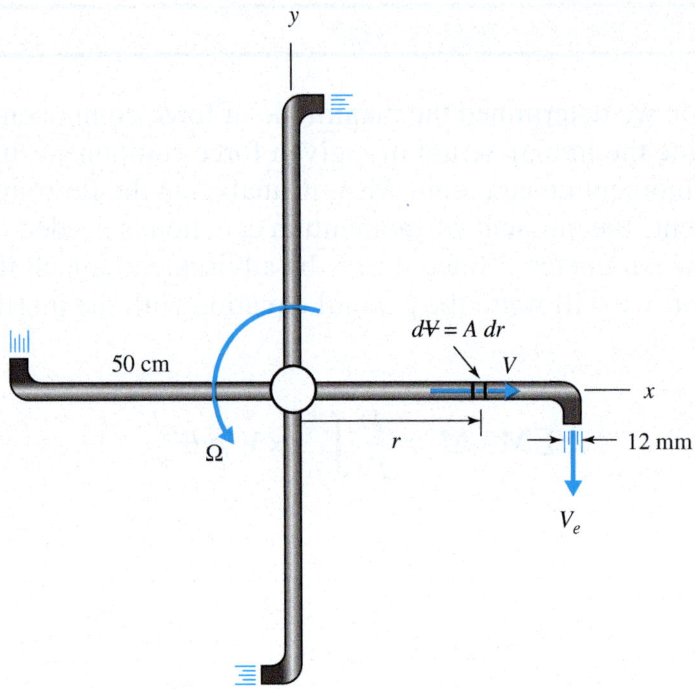

Figure E2.24

where the factor 4 accounts for the four exit areas. Attach the reference frame to the rotating arms as shown. Then, recognizing that $\mathbf{r} \times [\mathbf{\Omega} \times (\mathbf{\Omega} \times \mathbf{r})] = 0$ and assuming a stationary sprinkler so that $d^2\mathbf{S}/dt^2 = 0$ and constant angular velocity so that $d\mathbf{\Omega}/dt = 0$, we have

$$\begin{aligned}\mathrm{M}_I &= \int_{\text{c.v.}} \mathbf{r} \times (2\mathbf{\Omega} \times \mathbf{V})\rho d\mathcal{V} \\ &= 4\int_0^{0.5} r\hat{\mathbf{i}} \times (2\Omega\hat{\mathbf{k}} \times V\hat{\mathbf{i}})\rho A dr \\ &= 8\rho AV\Omega\hat{\mathbf{k}}\int_0^{0.5} r dr = \rho AV\Omega\hat{\mathbf{k}}\end{aligned}$$

where the small mass of water in the end nozzles is neglected compared to that in the long arms; the factor 4 again accounts for the four arms (each arm would provide the unit vector $\hat{\mathbf{k}}$). Since there are no external moments applied to the sprinkler about the vertical z-axis, $\Sigma M_z = 0$. For the steady flow Eq. 2.7.3 provides

$$\begin{aligned}\overset{0}{\cancel{\Sigma M_z}} - (\mathrm{M}_I)_z &= \int_{\text{c.s.}} (\mathbf{r} \times \mathbf{V})_z \mathbf{V} \cdot \hat{\mathbf{n}}\, \rho\, dA \\ -\rho AV\Omega\hat{\mathbf{k}} &= 4\int_{A_{\text{exit}}} \left[0.5\hat{\mathbf{i}} \times (0.707V_e\hat{\mathbf{k}} - 0.707V_e\hat{\mathbf{j}})\right]_z V_e \rho\, dA \\ -VA\Omega &= -4 \times 0.5 \times 0.707 V_e^2 A_e \\ \therefore \Omega &= 4 \times 0.5 \times 0.707 \times 22.1 = \underline{31.25 \text{ rad/s}}\end{aligned}$$

where we have used $AV = A_eV_e$ from continuity considerations.

Example 2.25

The nozzles of Example 2.24 make an angle of 0° with the ground and 90° with the arms. The water is suddenly turned on at $t = 0$ with the sprinkler motionless. Determine the resulting $\Omega(t)$ if the arm diameter is 24 mm. Neglect friction.

Solution

The reference frame is again attached to the rotating arms, as sketched in Example 2.24. Referring to the control volume integral of Eq. 2.7.3, we observe that $\mathbf{r} \times \mathbf{V} = 0$ since $\mathbf{r}$ is in the same direction as $\mathbf{V}$ along an arm. Thus Eq. 2.7.3, along with Eq. 2.7.2, takes the form

$$\overset{0}{\cancel{\Sigma \mathbf{M}}} - 4\int_0^{0.5} r\hat{\mathbf{i}} \times \left[2\Omega\hat{\mathbf{k}} \times V\hat{\mathbf{i}} + \Omega\hat{\mathbf{k}} \times (\Omega\hat{\mathbf{k}} \times r\hat{\mathbf{i}}) + \frac{d\Omega}{dt}\hat{\mathbf{k}} \times r\hat{\mathbf{i}} \right] \rho A dr$$
$$= \int_{0.5} \frac{d}{dt}\int_{0.5} r\hat{\mathbf{i}} \times V\hat{\mathbf{i}} \rho dA + 4\int_{A_{\text{exit}}} 0.5\hat{\mathbf{i}} \times V_e(-\hat{\mathbf{j}})V_e \rho dA$$

Perform the vector operations and divide by 4ρ,

$$-2AV\Omega\int_0^{0.5} r\,dr - \frac{d\Omega}{dt}A\int_0^{0.5} r^2\,dr = -0.5V_e^2 A_e$$

The required integration, using $AV = A_e V_e = 0.01\ \text{m}^3/\text{s}$ and $V_e = 2.21$ m/s gives

$$\frac{d\Omega}{dt} + 132.6\,\Omega = 5862$$

This linear, first-order differential equation is solved by adding the homogeneous solution (suppress the right-hand side) to the particular solution to obtain

$$\Omega(t) = Ce^{-132.6t} + 44.2$$

Using the initial condition $\Omega(0) = 0$, we find that $C = -44.2$. Then

$$\underline{\Omega(t) = 44.2(1 - e^{-132.6t})\ \text{rad/s}}$$

Observe that as time becomes large, the angular velocity is limited to 44.2 rad/s. If friction were included, this value would be reduced. If 44.2 is multiplied by 0.707 to account for the 45° angle, we obtain the value of Example 2.24.

2.8 SUMMARY

In this chapter we have presented the control-volume formulation of the fundamental laws. This formulation is useful when the integrands (the velocities and pressure) are known or can be approximated with an acceptable degree of accuracy. If this is not the case, the differential equations of Chapter 3 must be solved (numerically, or analytically as in Chapter 5), or experimental methods must be used to obtain the desired information; much of the remainder of this book is devoted to this task. After unknown velocities and pressures are determined, we often return to the control-volume formulation and calculate integral quantities of interest. Examples would include the lift and drag on an airfoil, the torque on a row of turbine blades, and the oscillating force on a suspension cable of a bridge.

As we have observed in the examples and problems of this chapter, the task of applying the control-volume equations is highly dependent on the appropriate selection of the boundaries of the control volume. Such boundaries are selected at locations where information is known or where the unknowns appear. Experience is often required in the selection of a control volume, as it is in the selection of a free-body diagram in statics, dynamics, and solid mechanics. The student has undoubtedly gained some of this experience while working through the sections of this chapter. Table 2.1 presents the various forms of the fundamental laws to aid the user in selecting an appropriate form for a particular problem.

Table 2.1 Integral Forms of the Fundamental Laws

Continuity	*Energy*	*Momentum*
General Form		
$0=\frac{d}{dt}\int_{\text{c.v.}}\rho d\forall+\int_{\text{c.s.}}\rho\mathbf{V}\cdot\hat{\mathbf{n}}dA$	$-\Sigma\dot{W}=\frac{d}{dt}\int_{\text{c.v.}}\left(\frac{V^2}{2}+gz\right)\rho\, d\forall$ $+\int_{\text{c.s.}}\left(\frac{V^2}{2}+\frac{p}{\rho}+gz\right)\rho\mathbf{V}\cdot\hat{\mathbf{n}}dA+\text{losses}$	$\Sigma\mathbf{F}=\frac{d}{dt}\int_{\text{c.v.}}\rho\mathbf{V}\, d\forall+\int_{\text{c.s.}}\rho\mathbf{V}(\mathbf{V}\cdot\hat{\mathbf{n}})dA$
Steady Flow		
$0=\int_{\text{c.s.}}\rho\mathbf{V}\cdot\hat{\mathbf{n}}\, dA$	$-\Sigma\dot{W}=\int_{\text{c.s.}}\left(\frac{V^2}{2}+\frac{p}{\rho}+gz\right)\rho\mathbf{V}\cdot\hat{\mathbf{n}}\, dA+\text{losses}$	$\Sigma\mathbf{F}=\int_{\text{c.s.}}\rho\mathbf{V}(\mathbf{V}\cdot\hat{\mathbf{n}})dA$
Steady Nonuniform Form[1]		
$\dot{m}=\rho_1A_1\overline{V}_1=\rho_2A_2\overline{V}_2$	$\frac{-\Sigma\dot{W}}{\dot{m}g}=\alpha_2\frac{\overline{V}_2^2}{2g}+\frac{p_2}{\gamma_2}+z_2-\alpha_1\frac{\overline{V}_1^2}{2g}-\frac{p_1}{\gamma_1}-z_1+h_L$	$\Sigma F_x=\dot{m}(\beta_2\overline{V}_{2x}-\beta_1\overline{V}_{1x})$ $\Sigma F_y=\dot{m}(\beta_2\overline{V}_{2y}-\beta_1\overline{V}_{1y})$
Steady Uniform Form[1]		
$\dot{m}=\rho_1A_1V_1=\rho_2A_2V_2$	$-\frac{\Sigma\dot{W}}{\dot{m}g}=\frac{V_2^2}{2g}+\frac{p^2}{\gamma^2}+z_2-\frac{V_1^2}{2g}-\frac{p_1}{\gamma}-z_1+h_L$	$\Sigma\mathbf{F}=\dot{m}(\mathbf{V}_2-\mathbf{V}_1)$
Steady Uniform Incompressible Flow[1]		
$Q=A_1V_1=A_2V_2$	$-\frac{\Sigma\dot{W}}{\dot{m}g}=\frac{V_2^2}{2g}+\frac{p^2}{\gamma}+z_2-\frac{V_1^2}{2g}-\frac{p_1}{\gamma}-z_1+h_L$ $H_p+\frac{V_1^2}{2g}+\frac{p_1}{\gamma}+z_1=H_T+\frac{V_2^2}{2g}+\frac{p_2}{\gamma}+z_2+h_L$	$\Sigma\mathbf{F}=\dot{m}(\mathbf{V}_2-\mathbf{V}_1)$
$\dot{m}$ = mass flux Q = flow rate $\overline{V}$ = average velocity $=\frac{\int VdA}{A}$	α = kinetic energy correction factor $=\frac{\int V^3dA}{\overline{V}^3A}$ β = momentum correction factor $=\frac{\int V^2dA}{\overline{V}^2A}$	h_L = head loss $\Sigma\dot{W}=\dot{W}_S+\dot{W}_{shear}+\dot{W}_I$ H_P = pump head $=\dot{W}_p/\dot{m}g$ H_T = turbine head $=\dot{W}_T/\dot{m}g$

[1]The control volume has one entrance (section 1) and one exit (section 2).

PROBLEMS

Basic Laws

2.1 **(a)** State the necessary conditions for momentum of a system to remain constant.
(b) State the necessary conditions for energy of a system to remain constant.
(c) Show the detailed steps and state the assumptions that allow Eq. 2.2.3 to be reduced to $\Sigma \mathbf{F} = m\mathbf{a}$.

2.2 Make a list of five extensive properties that are of interest in fluid mechanics. Also, list their associated intensive properties. In addition, list five other intensive properties.

2.3 A control volume is identified as the volume interior to a balloon. At an instant the system is also identified as the air inside the balloon. Air escapes during a short time increment Δt. Sketch the system and the control volume at t and at $t + \Delta t$.

2.4 At an instant in time the control volume and the system occupy the volume inside the pump shown in Figure P2.4 and a few diameters of the pipe at the inlet side. Sketch the system and the control volume at times t and $t + \Delta t$.

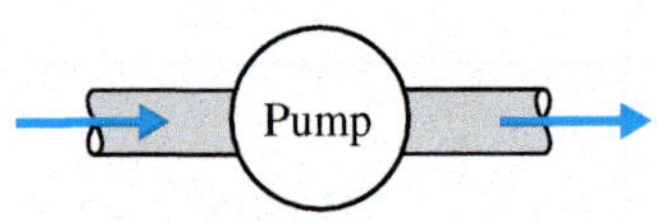

Figure P2.4

2.5 Indicate which fundamental equation would be most useful in determining the following quantity:
(a) The horsepower output of a pump
(b) The mass flux from closing baffles
(c) The drag force on an airfoil
(d) The head loss in a pipeline
(e) The rotational speed of a wind machine

2.6 Sketch the unit vector $\hat{\mathbf{n}}$ and the velocity vector $\mathbf{V}$ on each of the areas listed:
(a) The exit area of a fireman's nozzle
(b) The inlet area of a pump
(c) The wall area of a pipe
(d) The porous bottom area of a river into which a small amount of water flows
(e) The cylindrical exit area of a rotating impeller

2.7 Fluid flows through the enlargement shown in Figure P2.7 with a velocity distribution $v_1(r)$ at the inlet and $v_2(r)$ at the exit. Sketch a control volume showing $\mathbf{V}$ and $\hat{\mathbf{n}}$ at selected locations on the control volume. Include locations on the lateral sides as well as the ends.

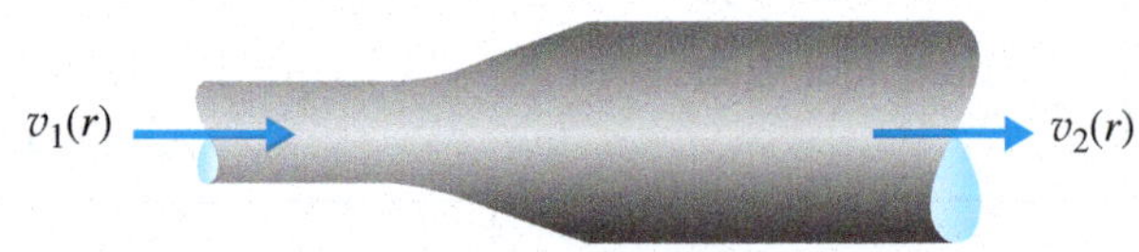

Figure P2.7

2.8 Sketch the unit vector $\hat{\mathbf{n}}$ and the velocity vector $\mathbf{V}$ at several positions on a rectangular box surrounding the airfoil shown in Figure P2.8.

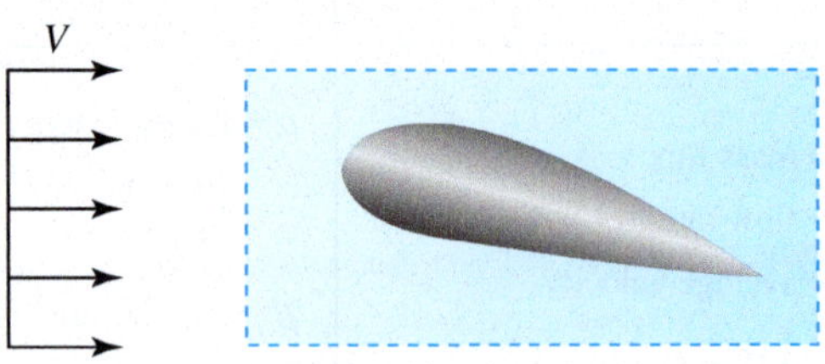

Figure P2.8

System-to-Control-Volume Transformation

2.9 Assume that $V_1 = V_2 = V_3 = 10$ m/s for the control volume shown in Figure P2.9. Write $\hat{\mathbf{n}}_1$, $\hat{\mathbf{n}}_2$, and $\hat{\mathbf{n}}_3$ in terms of $\hat{\mathbf{i}}$, $\hat{\mathbf{j}}$, and $\hat{\mathbf{k}}$, and calculate the normal component of the velocity vector on each of the three plane areas. The volume is of uniform depth in the z-direction.

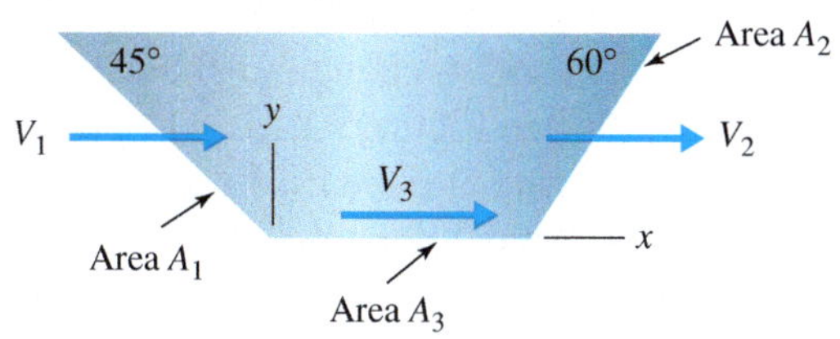

Figure P2.9

2.10 Write an expression for the flux of a property across each of the three areas of the control volume of Problem 2.9 if η and ρ are both constant over the entire control volume. Let the cross sectional area (normal to the xy-plane) be A. Use $V_1 = V_2 = V_3 = 10$ m/s.

2.11 Show that $(\mathbf{B} \cdot \hat{\mathbf{n}})A$ is the volume of the parallelepiped of depth 12 cm (Figure P2.11). Note that $\hat{\mathbf{n}}$ is normal to the area A.

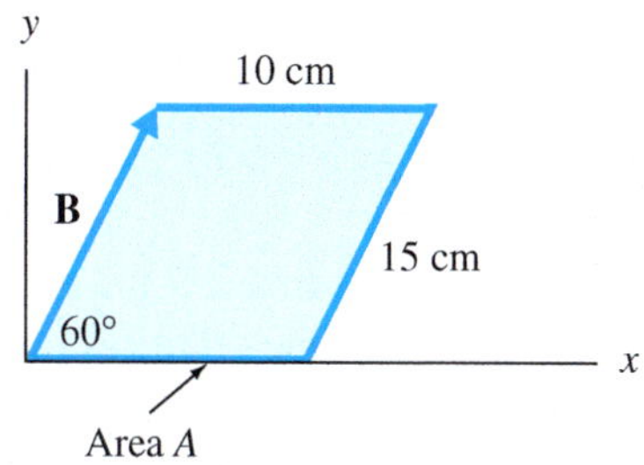

Figure P2.11

2.12 We recognize that

$$\frac{d}{dt}\int_{\text{c.v.}} \rho\eta \, d\mathcal{V} = \int_{\text{c.v.}} \frac{\partial}{\partial t} \rho\eta \, d\mathcal{V}$$

What condition allows this equivalence? Why is an ordinary derivative used on the left and a partial derivative on the right?

2.13 A spray can for mosquitos is activated at $t = 0$. Select the chemical inside the can as the system and sketch the system at $t = \Delta t$. Select a control volume and sketch the control volume at $t = \Delta t$.

2.14 The air inside the lungs at the end of an inhaled breath is identified as the system at $t = 0$. Select a control volume and sketch both the system and control volume at $t = \Delta t$ if air is exhaled through the nose only.

2.15 The control volume selected to analyze flow around an airfoil is the rectangular box as sketched in the figure of Problem 2.8. The system occupies the box at t. Sketch the system at $t + \Delta t$.

Conservation of Mass

2.16 Show that Eq. 2.4.5 results from Eq. 2.4.4 by assuming one inlet and one exit and uniform flow.

2.17 An incompressible fluid enters a volume filled with an absorbing material with a mass flux $\dot{m}$ and leaves the volume with a flow rate Q. Determine an expression for the rate of change of mass in the volume.

2.18 A liquid of density ρ flows into a volume filled with a sponge at a flow rate Q_1. It exits one area with a mass flux $\dot{m}_2$ and a second area A_3 with an average velocity V_3 as shown in Figure P2.18. Write an expression for dm_{sponge}/dt, the rate of change of mass in the sponge.

Figure P2.18

2.19 Water is flowing in a 2.5-in.-diameter pipe at 60 ft/s. If the pipe enlarges to a diameter of 5 in., calculate the reduced velocity. Also, calculate the mass flux and the flow rate. In addition, express the answers using SI units.

2.20 Water flows in the 5-cm-diameter pipe shown in Figure P2.20 with an average velocity of 10 m/s. It turns a 90° angle and flows radially between two parallel plates. What is the velocity at a radius of 60 cm? What are the mass flux and the discharge?

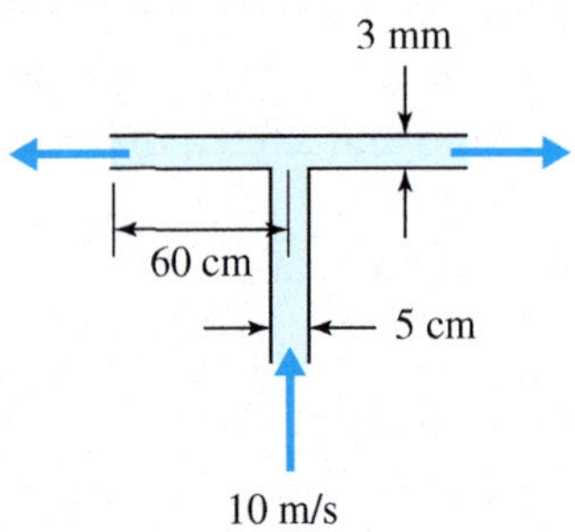

Figure P2.20

2.21 A pipe transports 200 kg/s of water. The pipe tees into a 5-cm-diameter pipe and a 7-cm- diameter pipe (Figure P2.21). If the average velocity in the smaller-diameter pipe is 25 m/s, calculate the flow rate in the larger pipe.

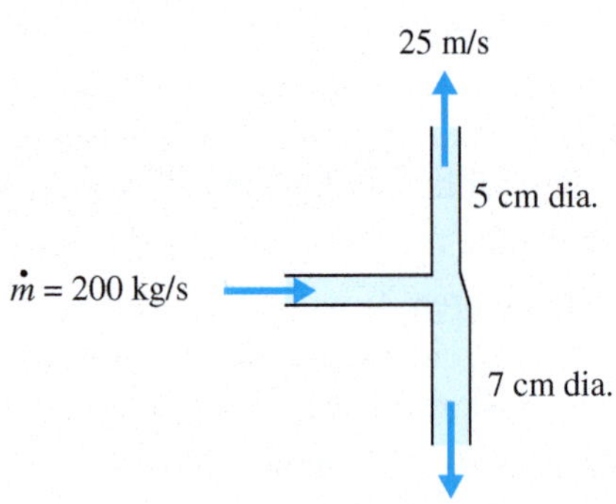

Figure P2.21

2.22 Air at 60°F and 40 psia flows in a 4-in.-diameter pipe with a mass flux of 0.2 slug/s. The pipe undergoes a conversion to a 2 in. by 3 in. rectangular duct in which $T = 150$°F and $p = 7$ psia. Calculate the velocity in each section.

2.23 Air at 120°C and 500 kPa absolute is flowing in a pipe at 600 m/s and suddenly undergoes an abrupt change to 249°C and 1246 kPa absolute at a location where the diameter is 10 cm. Calculate the velocity after the sudden change (a shock wave) illustrated in Figure P2.23. Also, calculate the mass flux and the flow rates before and after the abrupt change.

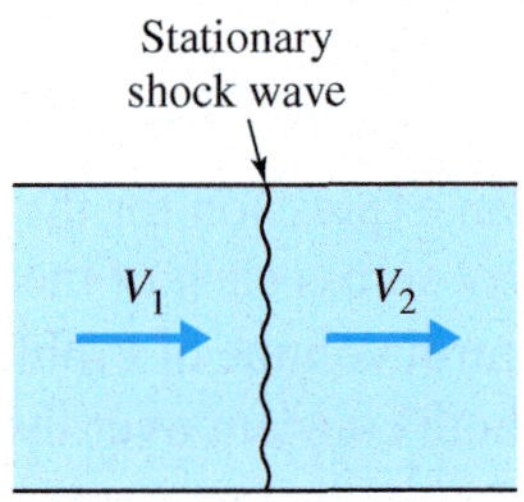

Figure P2.23

2.24 A laser velocimeter is used to measure velocities of 40 m/s and 120 m/s before and after an abrupt change in the diameter of a pipe from 10 cm to 6 cm, respectively. The pressure in the air before and after the change is measured to be 200 kPa and 120 kPa, respectively. If the temperature before the change is 20°C, what is the temperature after the change?

2.25 Water flows with a velocity of 3 m/s at a depth of 1.5 m in a trapezoidal channel with a 2-m base and sides sloping at 45°. It empties into a circular pipe and flows at 2 m/s. What is the diameter if:
(a) The pipe flows full?
(b) The pipe flows half full?
(c) The water in the pipe flows at a depth of one-half the radius?

2.26 Water flows in a 8-cm-diameter pipe with the profiles shown in Figure P2.26. Find the average velocity, the mass flux, and the flow rate for each.

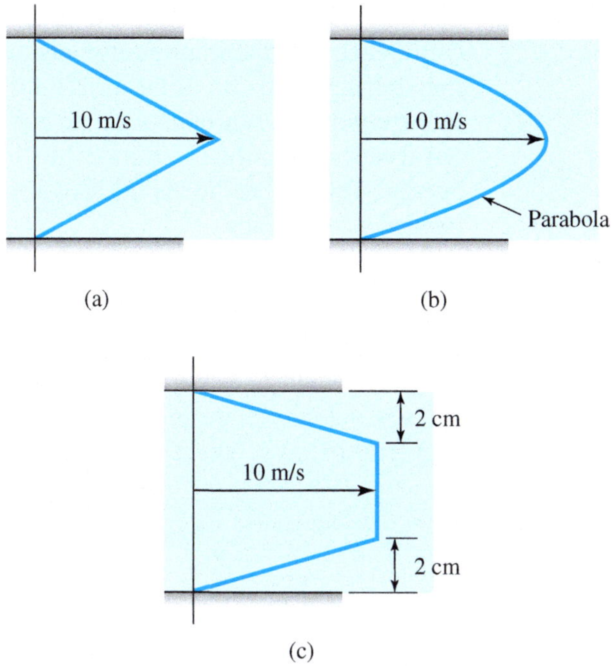

Figure P2.26

2.27 The profiles in Figure P2.26 are assumed to exist in a wide rectangular channel, 8 cm high and 80 cm wide. Find the average velocity, the mass flux, and the flow rate for each.

2.28 A constant-density fluid flows as shown in Figure P2.28. Find the equation of the parabola if the conduit is:
- **(a)** A pipe with $d = 1$ in. and $V = 6$ fps
- **(b)** A wide rectangular channel with $d = 1$ in. and $V = 6$ fps
- **(c)** A pipe with $d = 2$ cm and $V = 2$ m/s
- **(d)** A wide rectangular channel with $d = 2$ cm and $V = 2$ m/s

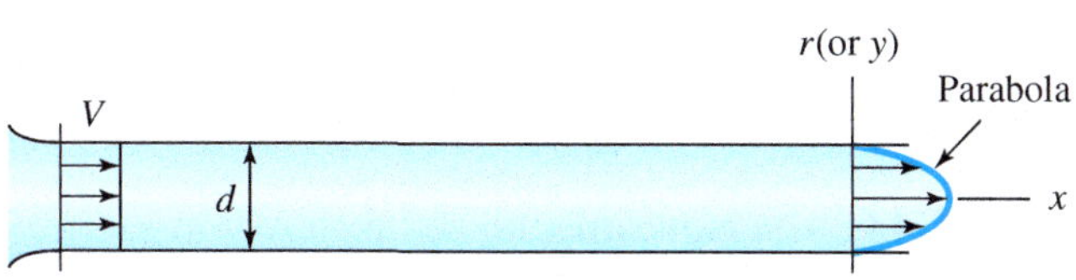

Figure P2.28

2.29 In Example 2.2 let $\dot{m}_2$ be unknown and V_1 and Q_3 be as shown in the figure. Calculate $\dot{m}_2$ so that dm/dt of the device is zero.

2.30 A parabolic profile exists in a 10-mm-diameter pipe. The pipe contracts to a 5-mm-diameter pipe in which the velocity profile is essentially uniform at 2 m/s. Write the equation for the parabola. Assume an incompressible flow.

2.31 As air flows as shown in Figure P2.31 over a flat plate, the velocity is reduced to zero at the wall. If $u(y) = 10(20\,y - 100\,y^2)$ m/s, find the mass flux $\dot{m}$ through a surface parallel to the plate and 0.2 m above the plate. The plate is 2 m wide and $\rho = 1.23$ kg/m^3.

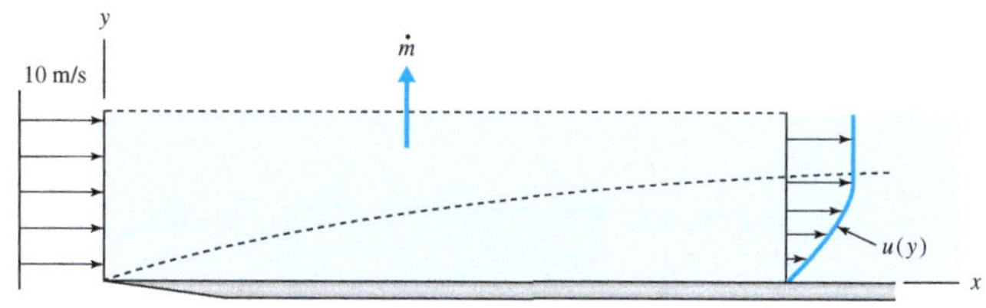

Figure P2.31

2.32 A streamline is 5 cm above the plate shown in Figure P2.31 at the leading edge. How far from the plate is that same streamline at the location of the profile $u(y) = 10(20\,y - 100\,y^2)$?

2.33 Stratified salt water flows at a depth of 4 in. in a channel with a velocity distribution $2(6y - 9y^2)$ ft/s, where y is measured in feet. If the density varies linearly from 2.2 slug/ft^3 at the bottom to clear water at the top, find $\dot{m}$. Also, show that $\dot{m} \neq \bar{\rho}\bar{V}A$. The channel is 5 ft wide.

2.34 Water flows as shown in Figure P2.34. Calculate V_2.

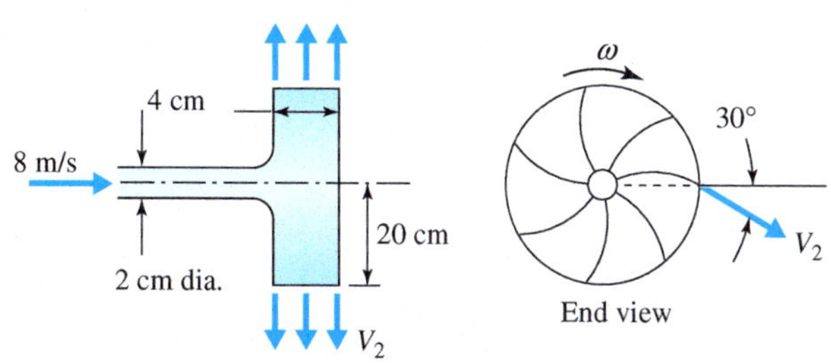

Figure P2.34

2.35 Rain is falling vertically downward with an average velocity of 5.0 m/s onto a 9000-m^2 parking lot. All of the water flows from the lot in an open rectangular ditch with an average velocity of 1.5 m/s. Estimate the depth of flow in the 1.5-m-wide ditch if two thousand 3-mm-diameter drops of water are contained in each cubic meter of rain.

2.36 Air at 37 psi gage and 60°F is being forced into a tire, which has a volume of 17 ft^3, at a velocity of 180 ft/s through a 1/4-in.-diameter opening. Determine the rate of change of density in the tire.

2.37 In Figure P2.37, if the mass of the control volume is not changing, find $\overline{V}_3$.

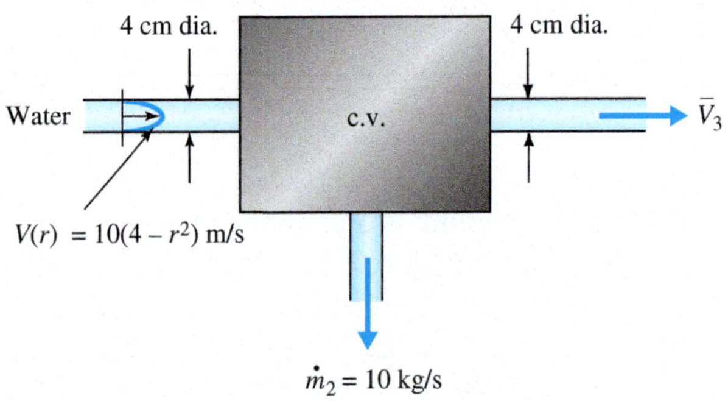

Figure P2.37

2.38 The average velocity $\overline{V}_3 = 10$ m/s in Problem 2.37. Find the rate at which the mass of the control volume is changing.

2.39 Find the velocity of the gas-fuel interface shown in Figure P2.39. Use $R_{gas} =$ 0.28 kJ/kg · K and $d_e = 30$ cm.

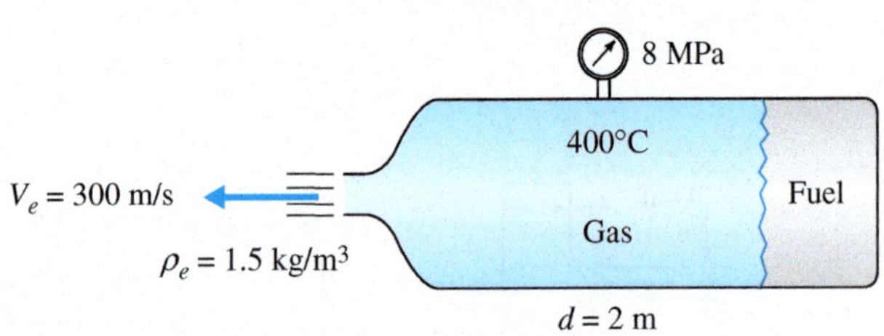

Figure P2.39

2.40 Conditioned air is delivered to a large conference room through four inlets, each transferring 1500 cfm. If the air is returned to the conditioner through a single 2 ft × 4 ft rectangular duct, estimate the average velocity in the duct. Make any needed assumptions.

2.41 A rectangular scoop, 80 cm deep, collects air as shown in Figure P2.41, and delivers it through a 30-cm-diameter pipe with a uniform velocity profile. Estimate the average velocity of air in the pipe if $u(y) = 20\, y^{1/5}$ m/s, where y is in meters.

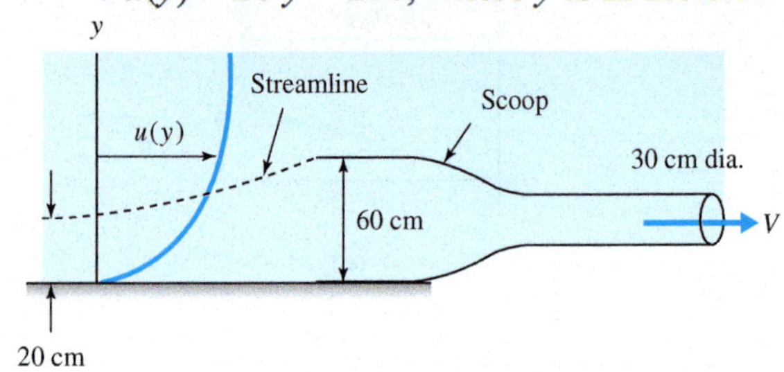

Figure P2.41

2.42 The jet pump operates by inducing a flow due to the high velocity in the 5-cm-diameter pipe as shown in Figure P2.42. The velocity in the small pipe is $200\,[1 - (r/R)^2]$. Estimate the average velocity at the exit.

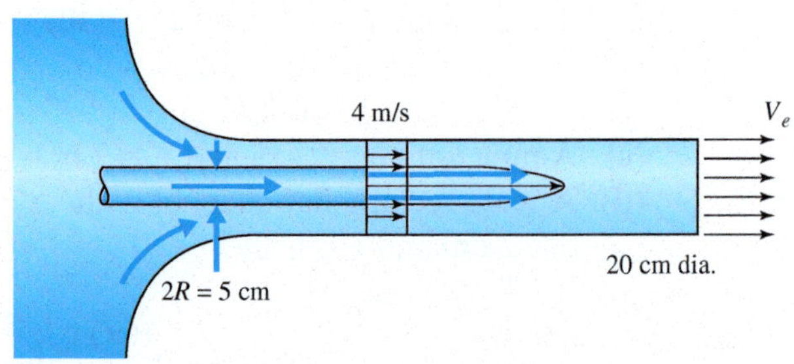

Figure P2.42

2.43 The experimental setup shown in Figure P2.43 is used to provide liquid for the tissue. Derive an expression for the storage rate of liquid in the tissue in terms of the relevant information.

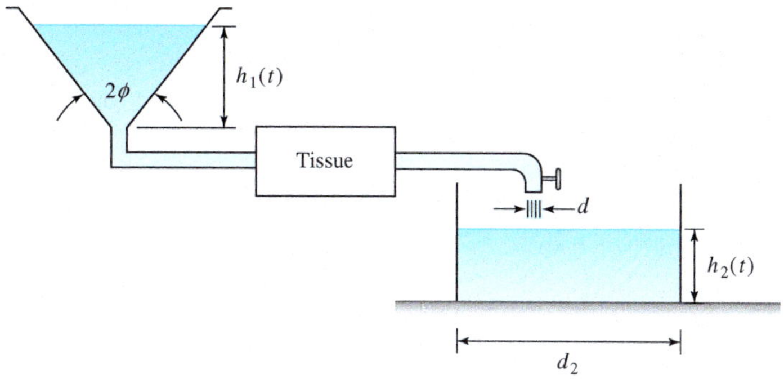

Figure P2.43

2.44 The water is 4 m deep behind a sluice gate in a rectangular channel that is suddenly opened (Figure P2.44). Find the initial dh/dt if $V_2 = 8$ m/s and $V_1 = 0.2$ m/s. The upstream channel length is 100 m.

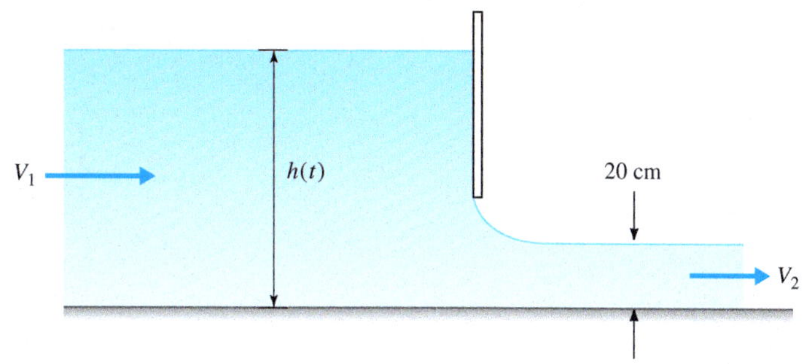

Figure P2.44

2.45 Water enters a kidney through a tube at 10 mL/min. It exits through a 6-mm-diameter tube at 20 mm/s. What is the rate of change of mass of water in the kidney?

2.46 The solid fuel in a rocket is burning at the rate of $400\,e^{-t/100}$ cm^3/s (Figure P2.46). If the density of the fuel is 900 kg/m^3, estimate the exit velocity V_e when $t = 10$ s assuming the density of the exiting gases to be 0.2 kg/m^3.

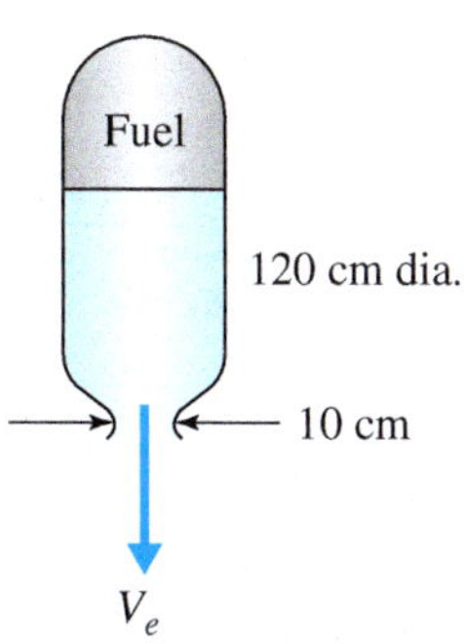

Figure P2.46

2.47 In Figure P2.47, find the rate of change of $h(t)$ if water is the fluid at all locations:

(a) $V_1 = 0.2$ m/s, $\dot{m}_2 = 10$ kg/s, $Q_3 = 600$ L/min

(b) $V_1 = 0$, $\dot{m}_2 = 20$ kg/s, $Q_3 = 10$ L/s

(c) $V_1 = 5$ m/s, $\dot{m}_2 = 10$ kg/s, $Q_3 = 1000$ L/min

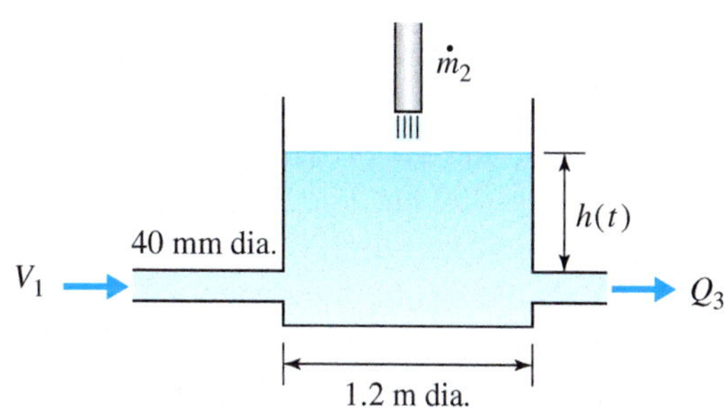

Figure P2.47

2.48 A1-m-diameter cylindrical tank initially contains liquid fuel and has a 2-cm rubber plug at the bottom, as shown in Figure P2.48. If the plug is removed, how long will it take to empty the tank. The initial height of liquid in the tank is 1.5 m. Bernoulli's equation is required.

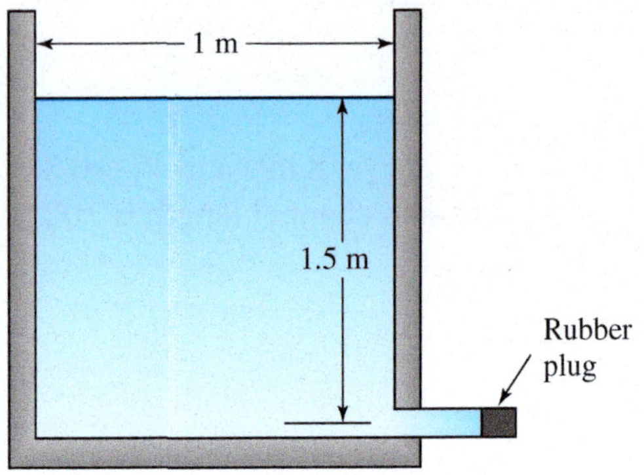

Figure P2.48

2.49 A 1-m^3-rigid tank initially contains compressed air at 15°C. Air is being removed from the tank through a 3-cm-diameter pipe, as shown in Figure P2.49. If the velocity and density of air in the pipe are 200 m/s and 1.8 kg/m^3, respectively, determine the initial rate at which the pressure inside the tank will vary assuming a constant temperature.

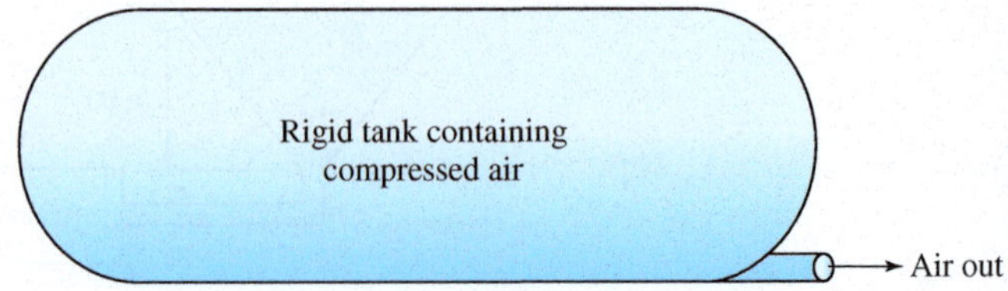

Figure P2.49

2.50 The velocity of the water entering the volume shown in Figure P2.50 is $V(t) = 10e^{-t/10}$ m/s. Assuming $h(0) = 0$, find $h(t)$ if the volume is:

(a) A cone. The inlet is 40 mm in diameter.

(b) A 10-m-long trough. The inlet is 4 cm high.

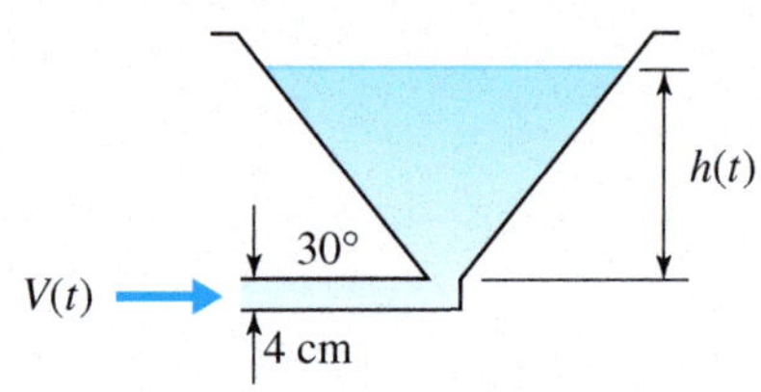

Figure P2.50

Energy Equation

2.51 In Figure P2.51, determine the work rate by the air at the instant shown if V_{piston} = 10 m/s, the torque $T = 20$ N·m, the velocity gradient at the belt surface is 100 s^{-1}, and the pressure acting on the piston is 400 Pa. The belt is 80 cm × 50 cm and the 40-cm-high rectangular piston is 50 cm deep (into the paper).

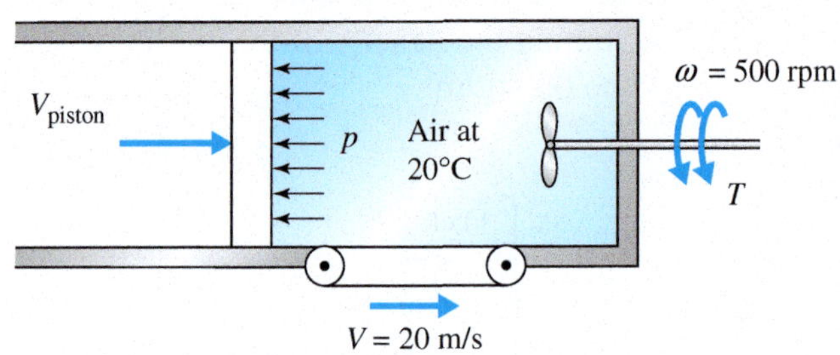

Figure P2.51

2.52 Assuming that the internal energy of natural gas depends on temperature only, what happens to the losses as natural gas is pumped from Texas to Michigan? The temperature remains essentially constant. Refer to Eq. 2.5.14.

2.53 An insulated water pump requires 500 W while pumping 0.02 m^3/s with an efficiency of 80%. What is the temperature rise of the water from inlet to exit of the pump assuming equal inlet and exit areas? The specific heat of water is 4.18 kJ/kg°C .

2.54 A water pump requires 5 hp to create a pump head of 20 m. If its efficiency is 87%, what is the flow rate of water?

2.55 An 89% efficient hydro turbine operates with a turbine head of 40 m. What is the turbine output if the mass flux is:

(a) 200 kg/s? **(b)** 90 000 kg/min?

(c) 8×10^6 kg/h?

2.56 The desired output from a set of 89% efficient turbines on a river is 10 MW. If the maximum turbine head attainable is 50 m, determine the average velocity at a location where the river is 60 m wide and 3 m deep.

2.57 Water flows in an open rectangular channel at a depth of 3 ft with a velocity of 12 ft/s. The bottom of the channel drops over a short length a distance of 3 ft. Calculate the two possible depths of flow after the drop. Neglect all losses.

2.58 If the head loss in Problem 2.57 across the channel drop is 0.6 ft, determine the two possible depths of flow.

2.59 Find the velocity V_1 of the water in the vertical pipe shown in Figure P2.59. Assume no losses.

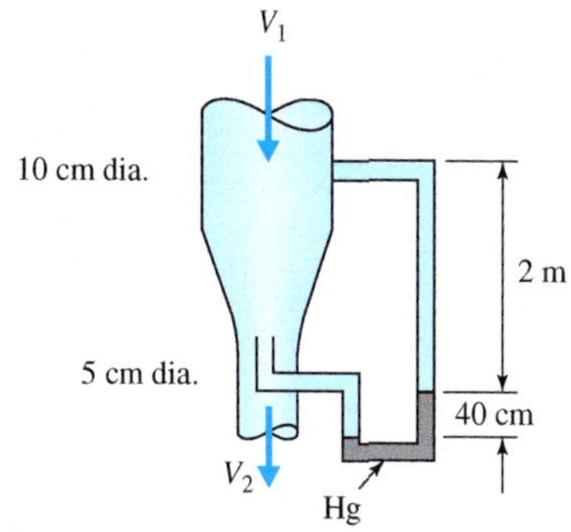

Figure P2.59

2.60 If the head-loss coefficient (based on V_2) between sections 1 and 2 of Problem 2.59 is 0.05, determine V_1 of the water.

2.61 The flow rate of water in a 2-in.-diameter horizontal pipeline at a pressure of 60 psi is 120 gal/min. If the pipeline increases to 3 in. diameter, calculate the increased pressure after the expansion if the loss coefficient (based on V_1) is 0.37.

2.62 Water flows at the rate of 600 L/min in a 4-cm-diameter horizontal pipeline with a pressure of 690 kPa. If the pressure after an enlargement to 6 cm diameter is measured to be 700 kPa, calculate the head loss across the enlargement.

2.63 Calculate the pressure p_1 shown in Figure P2.63 needed to maintain a flow rate of 0.08 m^3/s of water in a 6-cm-diameter horizontal pipe leading to a nozzle if a loss coefficient based on V_1 is 0.2 between the pressure gage and the exit.

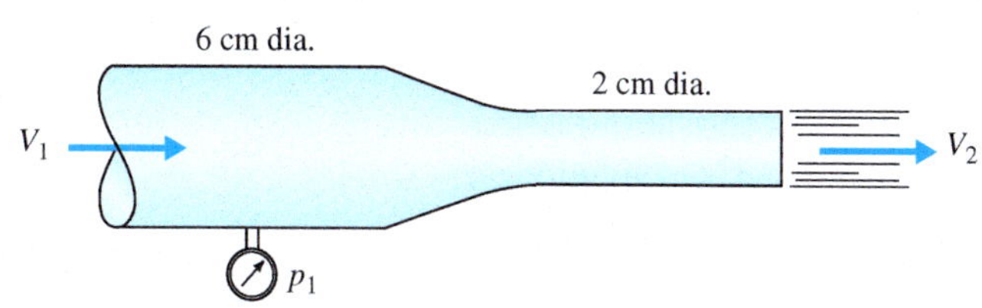

Figure P2.63

2.64 In Figure P2.64, neglect all losses and predict the value of H and p if:
(a) $h = 15$ cm **(b)** $h = 20$ cm

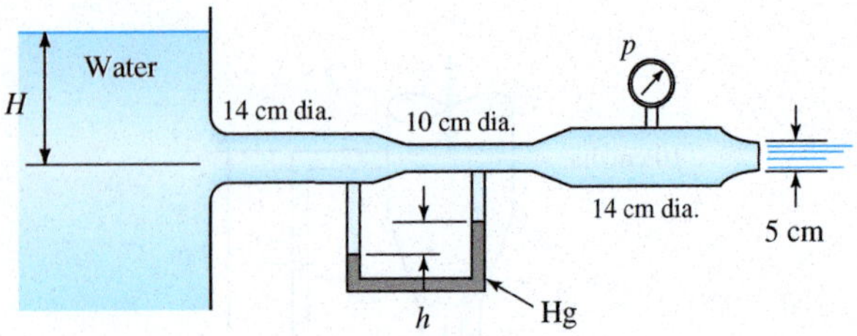

Figure P2.64

2.65 Water flows from the rectangular outlets shown in Figure P2.65. Estimate the flow rate per unit width for each if ($h = 80$ cm, $H = 2$ m). Neglect all losses;

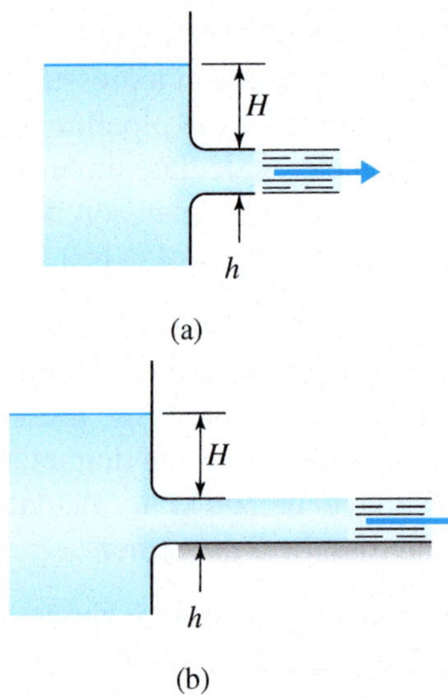

Figure P2.65

2.66 The overall loss coefficient for the pipe shown in Figure P2.66 is 5; up to A it is 0.8, from A to B it is 1.2, from B to C it is 0.8, from C to D it is 2.2. Estimate the flow rate and the pressures at A, B, C, and D. The elevations are shown.

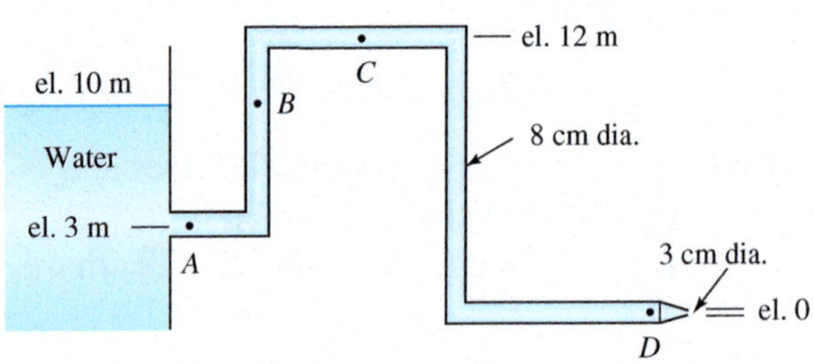

Figure P2.66

2.67 Water exits from a pressurized reservoir as shown in Figure P2.67. Calculate the flow rate if on section A we:
(a) Attach a nozzle with exit diameter 5 cm
(b) Attach a diffuser with exit diameter 18 cm
(c) Leave as an open pipe as shown
Neglect losses for all cases.

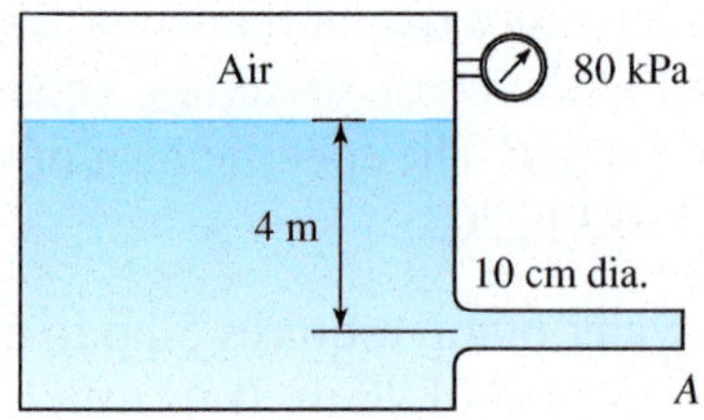

Figure P2.67

2.68 Rework Problem 2.67 assuming that $K_{pipe} = 1.5$, $K_{nozzle} = 0.04$ (based on V_1), and $K_{diffuser} = 0.8$ (based on V_1).

2.69 Relate the flow rate of the water through the venturi meter shown in Figure P2.69 to the diameter and the manometer reading. Assume no losses.

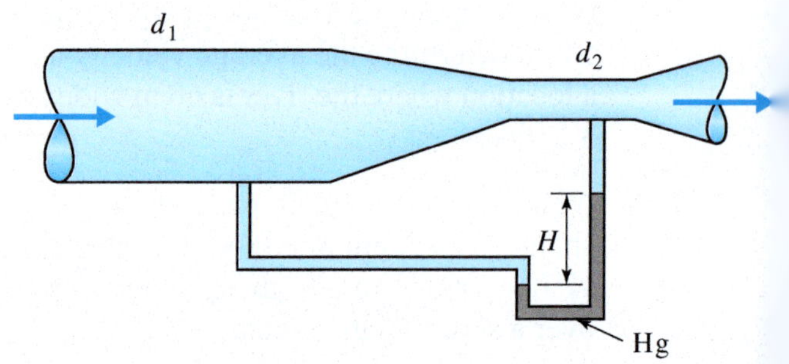

Figure P2.69

2.70 In the venturi meter of Figure P2.69, calculate the flow rate if:
(a) $H = 20$ cm, $d_1 = 2d_2 = 16$ cm
(b) $H = 40$ cm, $d_1 = 3d_2 = 24$ cm
(c) $H = 10$ in., $d_1 = 2d_2 = 6$ in
(d) $H = 15$ in., $d_1 = 3d_2 = 12$ in.

2.71 In Figure P2.71, determine the maximum possible height H if cavitation is to be avoided. Let:
(a) $d = 10$ cm and $T = 20°$C.
(b) $d = 4$ in. and $T = 70°$F.

Neglect all losses and assume $p_{atm} = 100$ kPa. (14.7 psi).

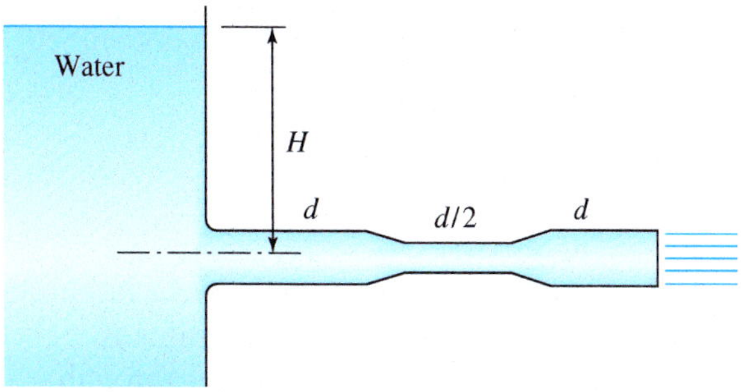

Figure P2.71

2.72 Cavitation is just observed in the small section of pipe of Figure P2.71 when $H = 65$ cm. Estimate the temperature of the water. Neglect all losses and assume $P_{atm} = 100$ kPa.
(a) $d = 10$ cm **(b)** $d = 12$ cm

2.73 In Figure P2.73, what is the minimum possible depth H if cavitation is to be avoided? Assume a vapor pressure of 6 kPa absolute and an overall loss coefficient of 8 based on V_2 and including the exit loss. Neglect losses up to the constriction.

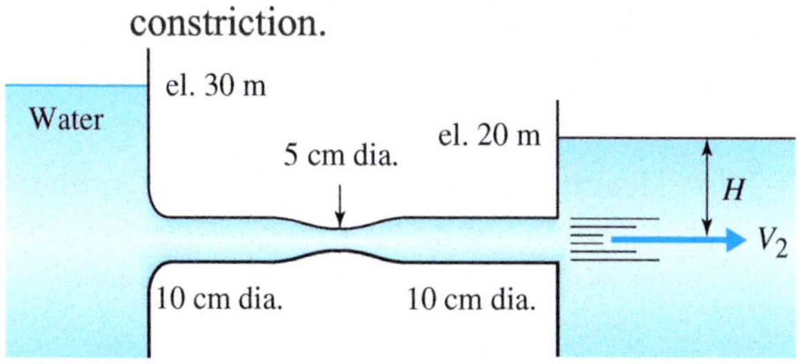

Figure P2.73

2.74 A contraction occurs in a 10-cm-diameter pipe to 6 cm followed by an enlargement back to 10 cm. The pressure upstream is measured to be 200 kPa when cavitation is first observed in the 20°C water. Calculate the flow rate. Neglect losses. Use $p_{atm} = 100$ kPa.

2.75 In Figure P2.75, calculate the maximum diameter D such that cavitation will be avoided if:
(a) $d = 20$ cm, $H = 5$ m, and $T_{water} = 20$°C.
(b) $d = 8$ in., $H = 15$ ft, and $T_{water} = 70$°F. Neglect all losses and use $p_{atm} = 100$ kPa (14.7 psi).

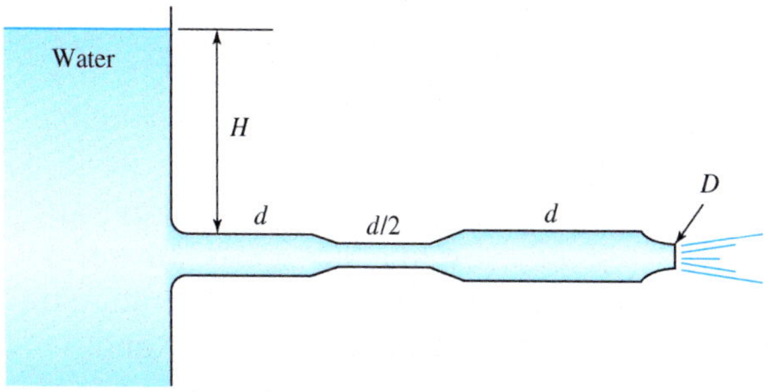

Figure P2.75

2.76 The overall loss coefficient in the siphon shown in Figure P2.76 is 4; up to section A it is 1.5. At what height H will the siphon cease to function?

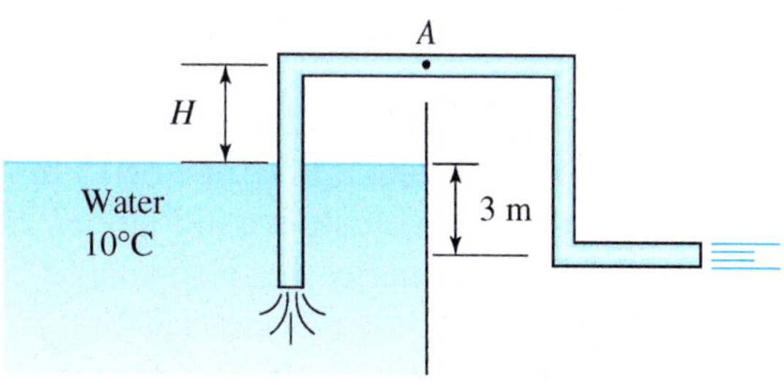

Figure P2.76

2.77 The pump shown in Figure P2.77 is 85% efficient. If the pressure rise is 120 psi, calculate the required energy input in horsepower.

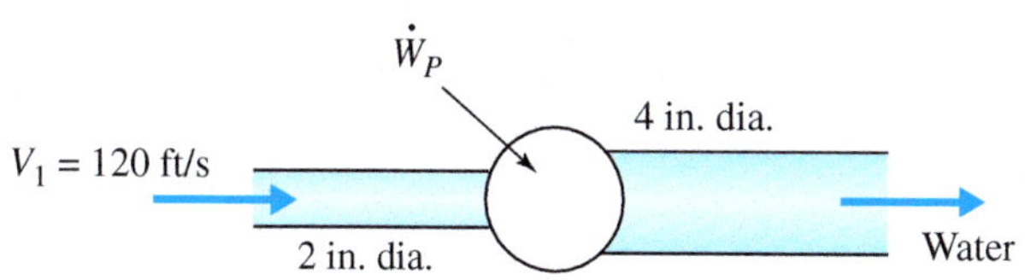

Figure P2.77

2.78 The pump shown in Figure P2.77 is powered by a 20-kW motor. If the pump is 82% efficient, determine the pressure rise.

2.79 A turbine, which is 87% efficient, accepts 2 m^3/s of water from a 50-cm-diameter pipe. The pressure drop is 600 kPa and the exit velocity is small. What is the turbine output?

2.80 A turbine receives 450 ft^3/s of water from 6-ft-diameter pipe at a pressure of 120 psi and delivers 10 000 kW. The pressure in the $7\frac{1}{2}$-ft-diameter exit pipe is 18 psi. Calculate the turbine efficiency.

2.81 Air enters a compressor with negligible velocity at 85 kPa absolute and 20°C. It leaves with a velocity of 200 m/s at 600 kPa absolute. For a mass flux of 5 kg/s, calculate the exit temperature if the power required is 1500 kW and:
(a) No heat transfer occurs
(b) The heat transfer loss is 60 kW

2.82 Air enters a compressor at standard conditions with negligible velocity. At the 1-in.-diameter exit the pressure, temperature, and velocity are 60 psia, 300°F, and 600 ft/s, respectively. If the heat transfer is 10 Btu/lb of air, find the power required by the compressor.

2.83 A small river with a flow rate of 15 m^3/s feeds the reservoir shown in Figure P2.83. Calculate the energy that is available continuously if the turbine is 80% efficient. The loss coefficient for the overall piping system is $K = 4.5$.

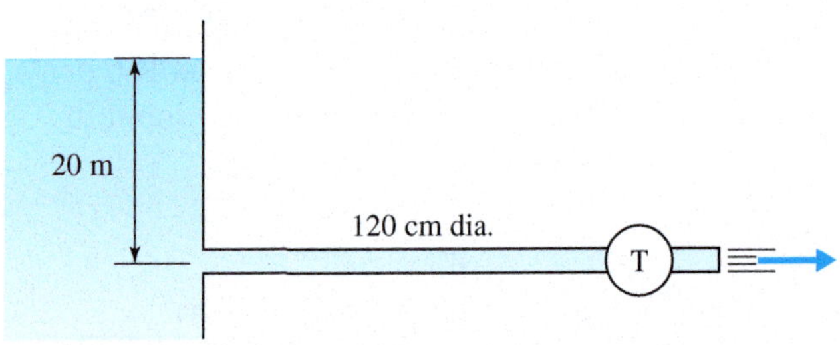

Figure P2.83

2.84 For the system shown in Figure P2.84 the average velocity in the pipe is 10 m/s. Up to point A, $K = 1.5$, from B to C, $K = 6.2$, and the pump is 80% efficient. If $p_C = 200$ kPa find p_A and p_B and the power required by the pump.

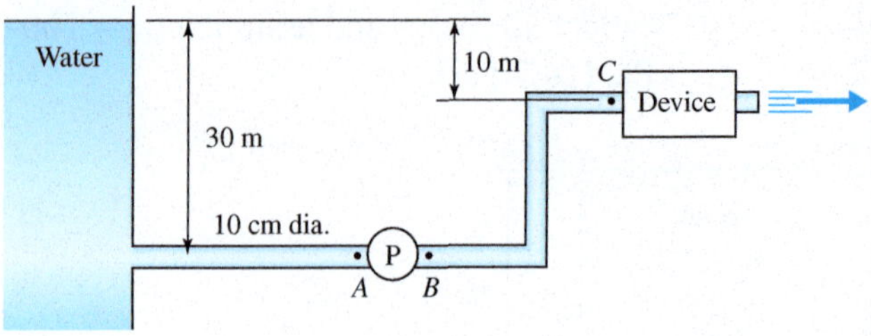

Figure P2.84

2.85 A gasoline-fueled engine uses a carburetor to mix fuel and air. The gasoline is pumped at a rate of 6.3 cm^3/s through 5-mm-diameter tubing from a tank maintained at 100 kPa, as shown in Figure P2.85. The gasoline discharges through a 0.8-mm nozzle in the carburetor to a pressure of 95 kPa. The total head loss in the system is $210V^2/2g$, where V is the average velocity in the pipe. If the pump has an efficiency of 75%, determine the power needed by the pump.

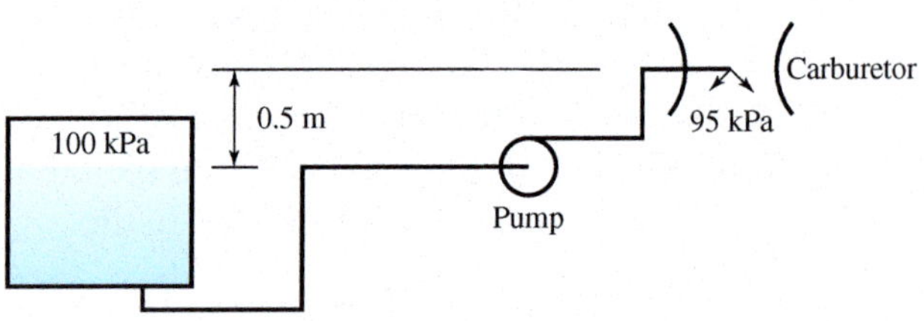

Figure P2.85

2.86 Determine the power output of the turbine shown in Figure P2.86 for a water flow rate of 18 ft^3/s. The turbine is 90% efficient.

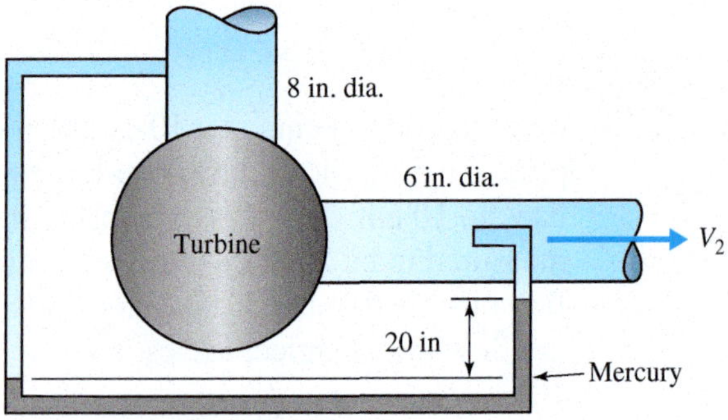

Figure P2.86

2.87 Find the power requirement of the 85%-efficient pump shown in Figure P2.87 if the loss coefficient up to A is 3.2, and from B to C, $K = 1.5$. Neglect the losses through the exit nozzle. Also, calculate p_A and p_B.

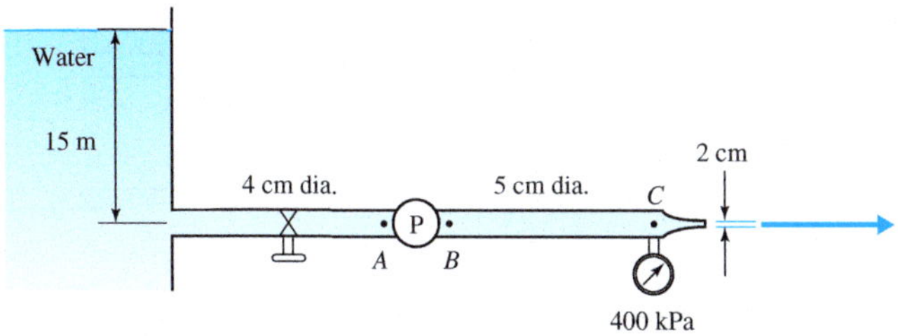

Figure P2.87

2.88 Water leaves a reservoir with a head of 10 m and flows through a horizontal 10-cm-diameter pipe, then exits to the atmosphere. At the end of the pipe a short length of smaller-diameter pipe is added. The loss coefficient, including the reduction, is 2.2, and after the reduction the losses are negligible. Calculate the minimum diameter of the smaller pipe if the flow rate is 0.02 m^3/s.

2.89 Neglect losses and find the depth of water on the raised section of the rectangular channel shown in Figure P2.89. Assume uniform velocity profiles.

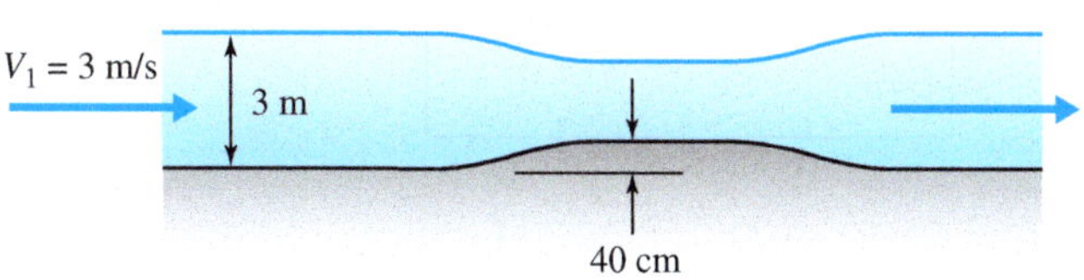

Figure P2.89

2.90 Determine the rate of kinetic energy loss, in watts, due to the cylinder shown in Figure P2.90. Assume a plane flow with $\rho = 1.23$ kg/m^3 and make the calculation per meter of cylinder length.

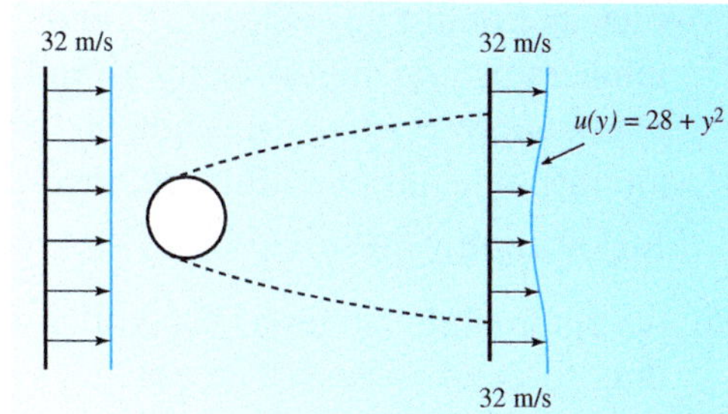

Figure P2.90

2.91 Calculate the head loss between the two sections shown in Figure P2.91.

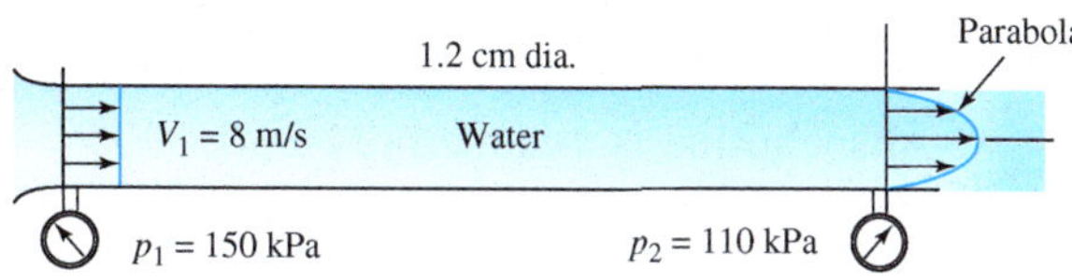

Figure P2.91

2.92 Calculate the kinetic-energy correction factor for the velocity profile at the downstream location in Figure P2.90.

2.93 Determine the kinetic-energy correction factor if:

(a) $u(r) = 10(1 - r^2/R^2)$ in a 2-cm-diameter pipe.

(b) $u(y) = 10(1 - y^2/h^2)$ in a 2-cm-high channel.

2.94 A turbulent velocity profile in a pipe is often written as $u(r) = u_{max}(1 - r/R)^{1/n}$, where n ranges between 5 and 9, 7 being the most common value. Calculate an expression for the kinetic energy passing a section of pipe and the kinetic-energy correction factor if:

(a) $n = 5$ **(b)** $n = 7$

(c) $n = 9$

2.95 A jet aircraft is flying with a velocity V_∞. Use the energy equation to relate the fuel consumption $\dot{m}_f$ to other flow variables such as exhaust gas velocity V_2 and temperature T_2, inlet velocity V_1 and temperature T_1, drag F_D acting on the aircraft, inlet air mass flux $\dot{m}$, and the fuel heating value q_f (kJ/kg).

2.96 An automobile moves at 100 km/hr with a drag force of 1340 N. The gas consumption is noted to be 5 km/L. If the engine efficiency is 15%, determine the energy released per kilogram of fuel. The fuel density is 680 kg/m^3.

2.97 A 180-m-long, 2-cm-diameter siphon provides water at 20°C from a reservoir to a field for irrigation. It exits 35 cm below the reservoir surface. The velocity distribution is assumed to be $u(r) = 2V(1 - r^2/r_0^2)$, where V is the average velocity. Determine the flow rate if the head loss is given by $32\nu LV/(D^2 g)$, where L is the siphon's length, D its diameter, and ν the kinematic viscosity of the water.

2.98 In the filtration system shown in Figure 2.98 water is circulated steadily through a filter using a pump. The pump characteristic curve is given by $H_P = 15 + 11Q - 150Q^2$, where H_P is in meters and Q is in m^3/s. A constant diameter pipe of 10 cm is used in the system. The total loss coefficient for the system based on the average velocity in the pipe is $K = 51$. Determine the flow rate and power input to the pump.

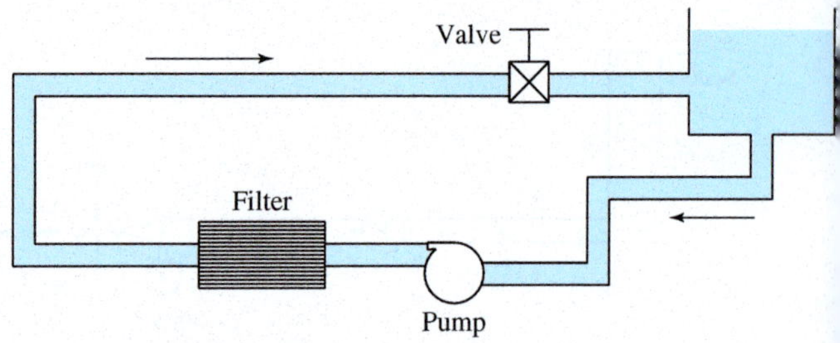

Figure P2.98

2.99 The manufacturer's pump curve for the pump in the flow system shown in Figure P2.99a is provided in Figure P2.99b. Estimate the flow rate. The overall loss coefficient is:
(a) $K = 5$ **(b)** $K = 20$

The solution involves a trial-and-error procedure, or the energy equation can be written as $H_P = H_P(Q)$ and plotted on the pump curve.

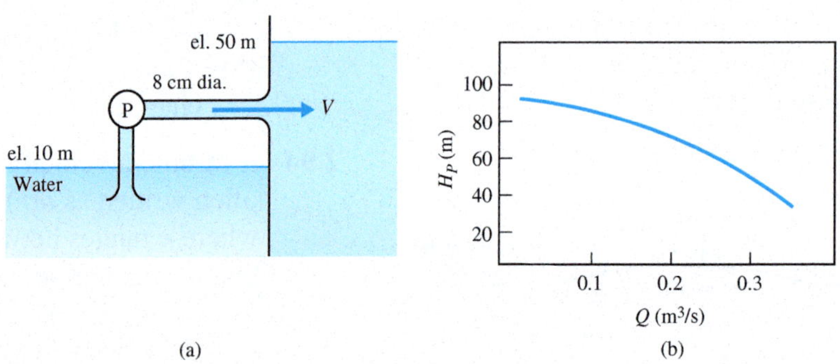

Figure P2.99

2.100 A water pump has one inlet and two outlets as shown in Figure P2.100, all at the same elevation. What pump power is required if the pump is 85% efficient? Neglect pipe losses.

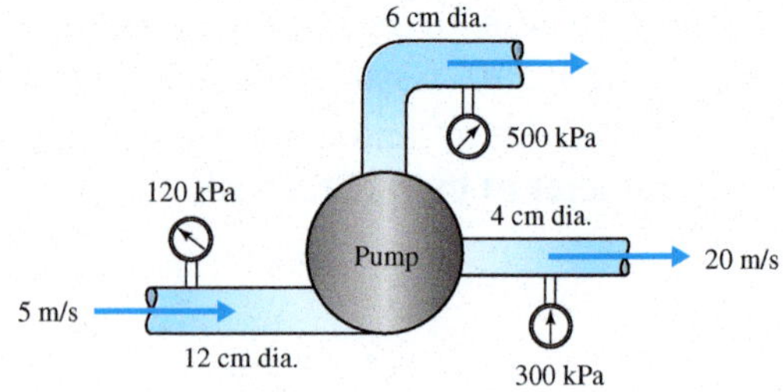

Figure P2.100

Momentum Equation

2.101 Water flows in a pipe of diameter d at a pressure p. It flows out a nozzle of diameter $d/2$ to the atmosphere. Calculate the force of the water on the nozzle if:
(a) $d = 6$ cm, $p = 200$ kPa
(b) $d = 6$ cm, $p = 400$ kPa
(c) $d = 12$ cm, $p = 200$ kPa
(d) $d = 3$ in., $p = 30$ psi
(e) $d = 3$ in., $p = 60$ psi
(f) $d = 6$ in., $p = 30$ psi

2.102 A nozzle and hose are attached to the ladder of a fire truck. What force is needed to hold a nozzle supplied by a 9-cm-diameter hose with a pressure of 2000 kPa? The nozzle outlet diameter is 3 cm.

2.103 Water flows in a 10-cm-diameter pipe at a pressure of 400 kPa out a straight nozzle. Calculate the force of the water on the nozzle if the outlet diameter is:
(a) 8 cm
(b) 6 cm
(c) 4 cm
(d) 2 cm

2.104 Find the horizontal force of the water on the horizontal bend shown in Figure P2.104.

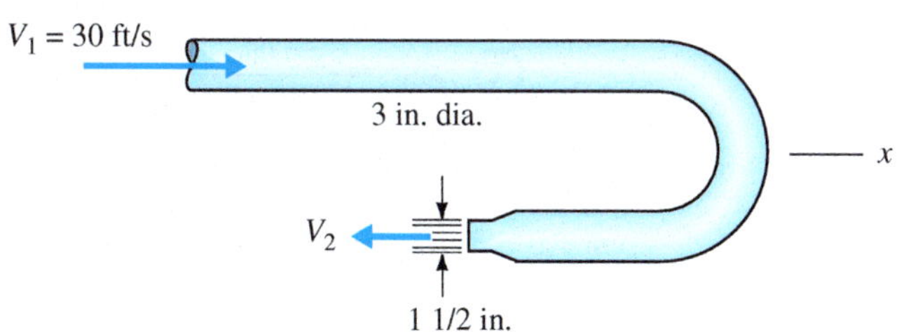

Figure P2.104

2.105 Find the horizontal force components of the water on the horizontal bend shown in Figure P2.105 if p_1 is:
(a) 200 kPa **(b)** 400 kPa
(c) 800 kPa

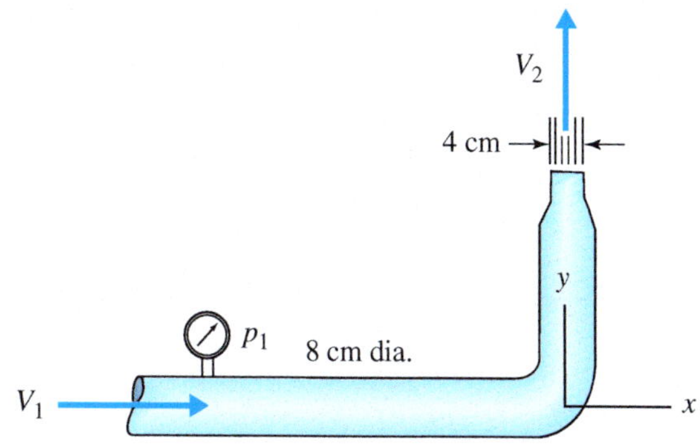

Figure P2.105

2.106 What is the net force needed to hold the orifice plate shown in Figure P2.106 onto the pipe?

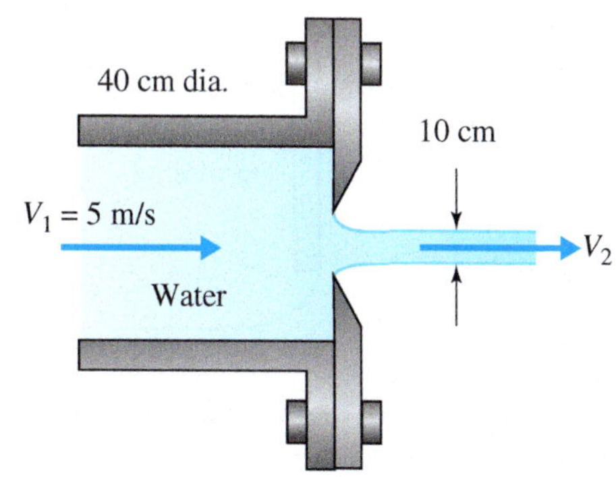

Figure P2.106

2.107 Assuming uniform velocity profiles, find F needed to hold the plug in the pipe shown in Figure P2.107. Neglect viscous effects.

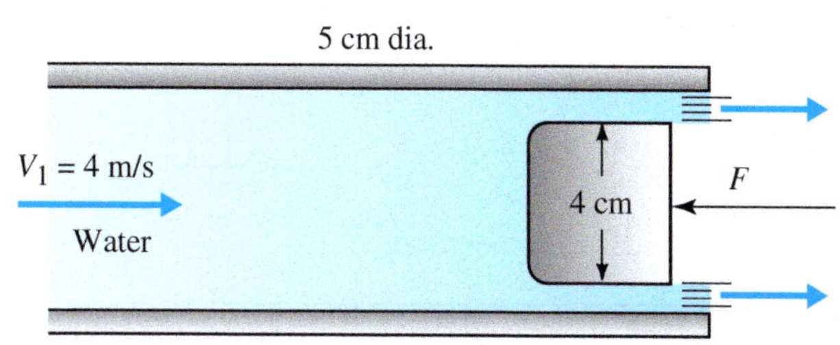

Figure P2.107

2.108 Neglect viscous effects, assume uniform velocity profiles, and find the horizontal force component acting on the obstruction shown in Figure P2.108.

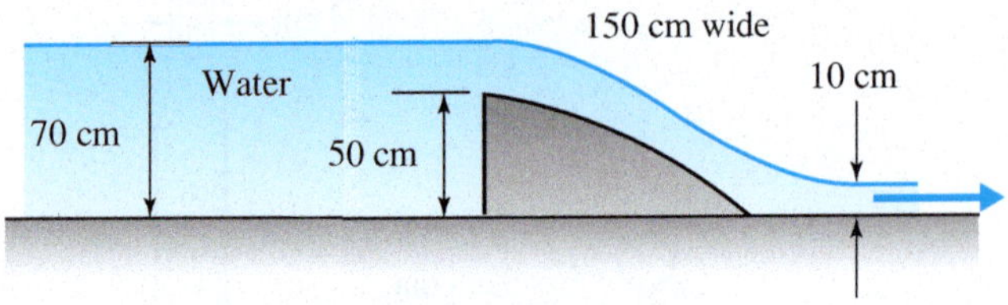

Figure P2.108

2.109 Assuming hydrostatic pressure distributions, uniform velocity profiles, and negligible viscous effects, find the horizontal force needed to hold the sluice gate in the position shown in Figure P2.109.

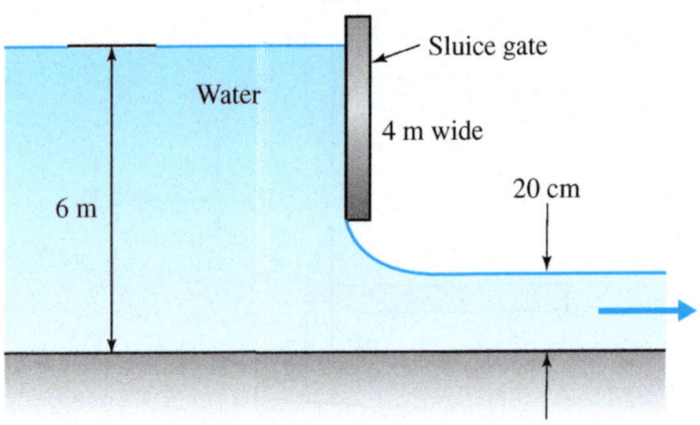

Figure P2.109

2.110 A sudden jump (a hydraulic jump) occurs in a rectangular channel as shown in Figure P2.110. Find y_2 and V_2 if:

(a) $V_1 = 8$ m/s, $y_1 = 60$ cm
(b) $V_1 = 12$ m/s, $y_1 = 40$ cm
(c) $V_1 = 20$ ft/s, $y_1 = 2$ ft
(d) $V_1 = 30$ ft/s, $y_1 = 3$ ft

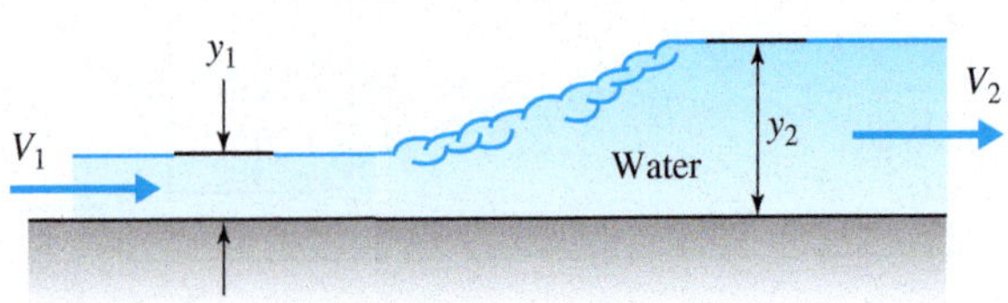

Figure P2.110

2.111 A hydraulic jump, as shown in Figure P2.110, occurs so that $V_2 = \frac{1}{4} V_1$. Find V_1 and y_2 if:

(a) $y_1 = 80$ cm **(b)** $y_1 = 2$ ft

2.112 For a flow rate of 9 m^3/s, find V_2 and y_2 for the hydraulic jump shown in Figure P2.112. The channel is 3 m wide. Neglect losses up to the jump.

Figure P2.112

2.113 The velocity is measured to be 10 ft/sec downstream of a hydraulic jump where the depth is 6 ft. Calculate the velocity and depth prior to the jump.

2.114 For the system shown in Figure P2.114, estimate the downstream pressure p_2 if $p_1 = 60$ kPa and $V_1 = 20$ m/s. Neglect losses.

(**Note:** The pressure immediately after the pipe expansion is p_1.)

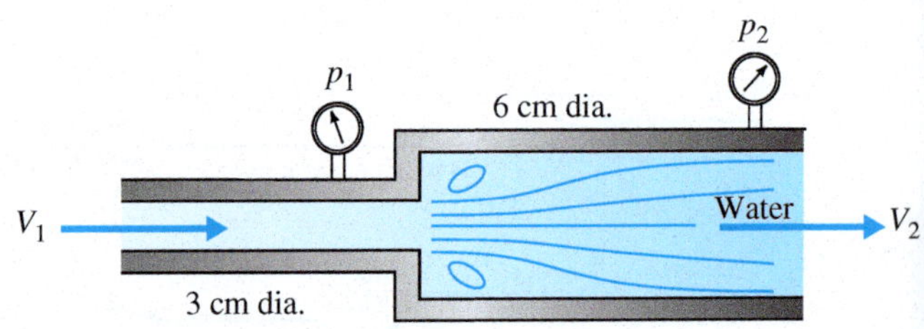

Figure P2.114

2.115 Water flows at 15 m/s in a 10-cm-diameter stem of a horizontal T-section that branches into 5-cm-diameter pipes. Find the force of the water on the T-section if the branches exit to the atmosphere. Neglect viscous effects.

2.116 Water flows steadily through the double elbow shown in Figure P2.116. Water flows into the elbow from the top at 5 m/s, and from the left at 15 m/s. Determine the vertical and horizontal components of the force needed to hold the elbow in place.

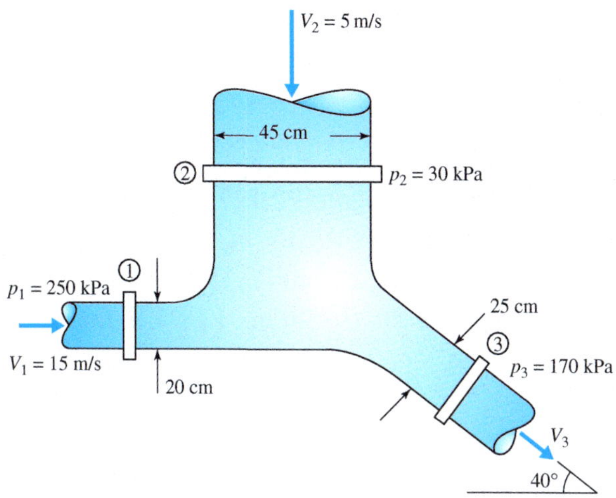

Figure P2.116

2.117 A horizontal 10-cm-diameter jet of water with $\dot{m} = 300$ kg/s strikes a vertical plate. Calculate:

(a) The force needed to hold the plate stationary

(b) The force needed to move the plate away from the jet at 10 m/s

(c) The force needed to move the plate into the jet at 10 m/s

2.118 A horizontal $2\frac{1}{2}$-in.-diameter jet of water strikes a vertical plate. Determine the velocity issuing from the jet if a force of 200 lb is needed to:

(a) Hold the plate stationary

(b) Move the plate away from the jet at 30 ft/s

(c) Move the plate into the jet at 30 ft/s

2.119 Determine the mass flux issuing from the jet shown in Figure P2.119 if a force of 700 N is needed to:

(a) Hold the cone stationary

(b) Move the cone away from the jet at 8 m/s

(c) Move the cone into the jet at 8 m/s

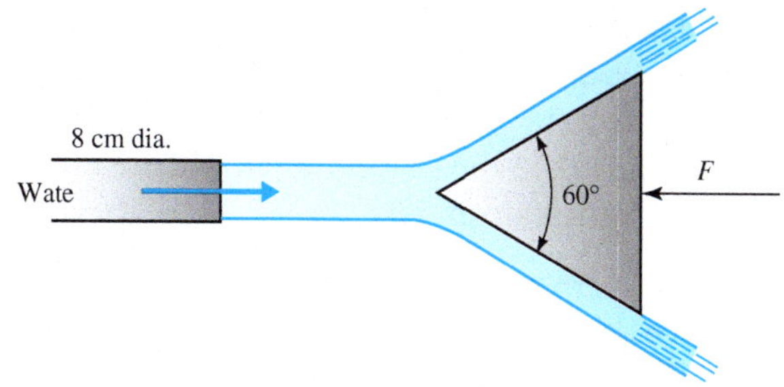

Figure P2.119

2.120 Calculate the force components of the water acting on the deflecting blade shown in Figure P2.120 if:

(a) The blade is stationary

(b) The blade moves to the right at 60 ft/s

(c) The blade moves to the left at 60 ft/s

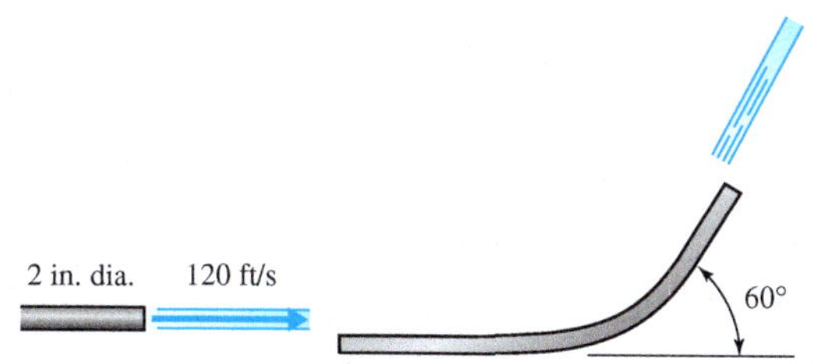

Figure P2.120

2.121 The blade of Problem 2.120 is one of a series of blades that are attached to a 50-cm-radius rotor that has a rotational speed of 30 rad/s. If there are 10 such water jets, find the power output.

2.122 Determine the force components of superheated steam acting on the blade shown in Figure P2.122 if:

(a) The blade is stationary

(b) The blade moves to the right at 100 m/s

(c) The blade moves to the left at 100 m/s

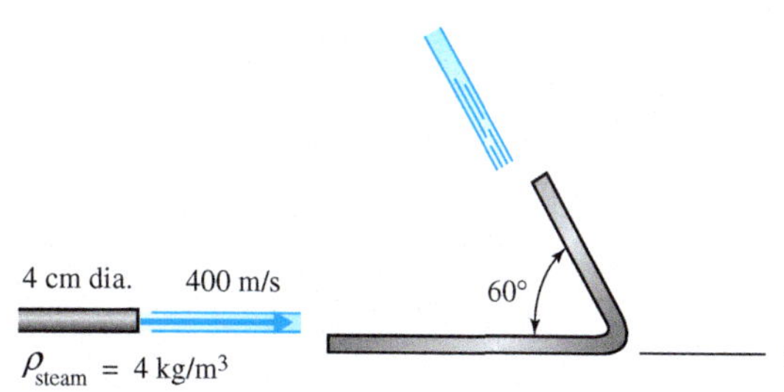

Figure P2.122

2.123 The blade of Problem 2.122 is one of a series of blades that are attached to a 1.2-m-radius rotor that rotates at 150 rad/s. Calculate the power output if there are 15 such steam jets.

2.124 Superheated steam jets strike the turbine blades shown in Figure P2.124. Find the power output of the turbine if there are 15 jets and α_1 is:
(a) 45° **(b)** 60° **(c)** 90°

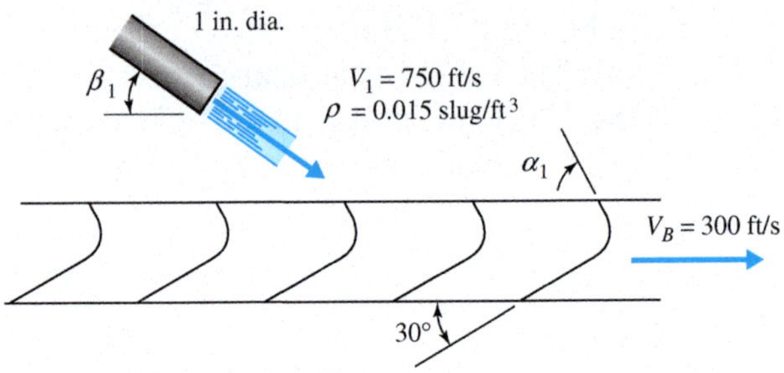

Figure P2.124

2.125 Twelve high-speed water jets strike the blades as shown in Figure P2.125. Find the power output and the blade angles if V_B is:
(a) 20 m/s **(b)** 40 m/s **(c)** 50 m/s

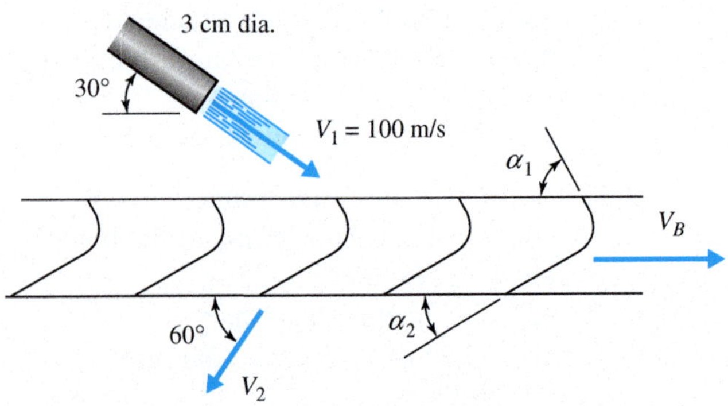

Figure P2.125

2.126 Fifteen water jets strike the blades of a turbine as shown in Figure P2.126. Calculate the power output and the blade angles if β_2 is:
(a) 60° **(b)** 70° **(c)** 80°

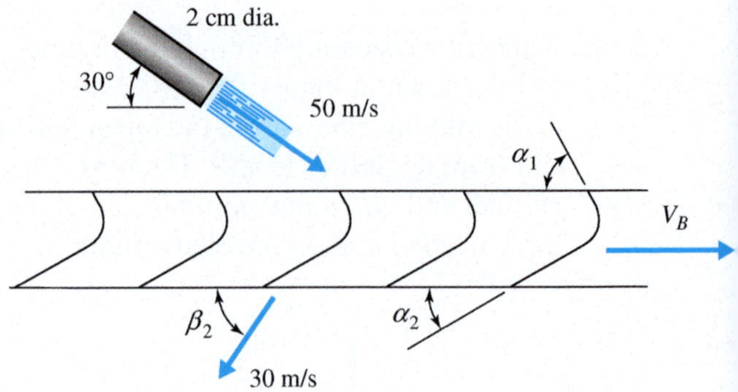

Figure P2.126

2.127 Water flows from the rectangular jet as shown in Figure P2.127. Find the force F and the mass fluxes $\dot{m}_2$ and $\dot{m}_3$ if
(a) $b = 20$, $h = 40$ cm, $V_1 = 40$ m/s
(b) $b = 20$, $h = 40$ in., $V_1 = 120$ fps

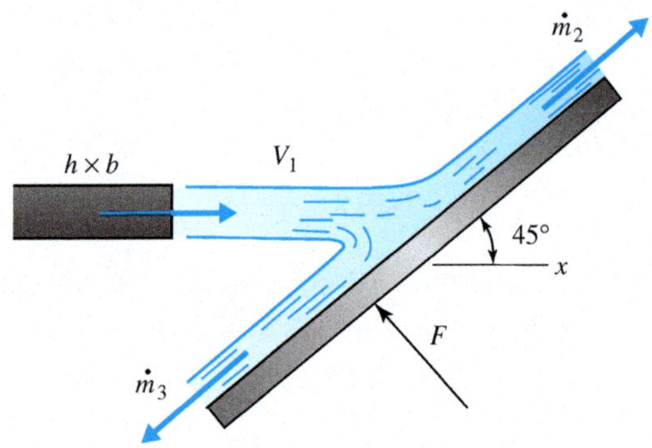

Figure P2.127

2.128 The plate of Problem 2.127a moves to the left at 20 m/s. Find the power required.

2.129 Calculate the velocity that the plate of Problem 2.127a must move with (in the x-direction) to produce the maximum power output.

2.130 A large vehicle with a mass of 100 000 kg is slowed by inserting a 180° deflector in a trough of water. If the 60-cm-wide deflector scoops off 10 cm of water, calculate the initial deceleration if the vehicle is traveling at 120 km/h. Also, find the time necessary to reach a speed of 60 km/h.

2.131 A 2.5-m-wide snowplow travels at 50 km/h plowing snow at a depth of 0.8 m. The snow leaves the blade normal to the direction of motion of the snowplow. What power does the plowing operation require if the density of the snow is 90 kg/m^3?

2.132 A vehicle with a mass of 5000 kg is traveling at 900 km/h. It is decelerated by lowering a 20-cm-wide scoop into water a depth of 6 cm (Figure P2.132). If the water is deflected through 180°, calculate the distance the vehicle must travel for the speed to be reduced to 100 km/h.

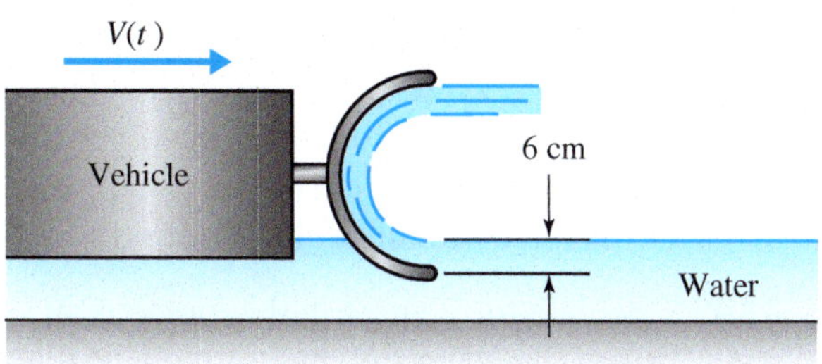

Figure P2.132

2.133 Modern jet engines have vanes that are deployed to produce a reverse thrust just after a plane touches down on a runway. The air flow rate through a particular engine is 100 kg/s, and the combustion gases exit the engine at a speed of 800 m/s relative to the engine. The fuel-air ratio is 1 to 40 in this case. Calculate the reverse thrust produced by the vanes for the configuration shown in Figure P2.133.

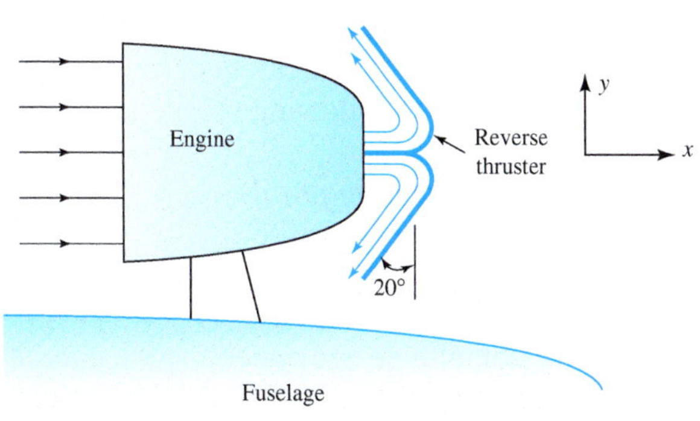

Figure P2.133

2.134 A 20-slug vehicle is required to have an initial acceleration of 6 ft/s^2. It is proposed that a $2\frac{1}{2}$-in.-diameter water jet strike a vane built in to the rear of the vehicle that will divert the water through an angle of 180°. What jet velocity is necessary? What speed will be achieved after 2 s?

2.135 The swamp boat shown in Figure P2.135 is propelled at 50 km/h by a 2-m-diameter propeller that requires a 20-kW engine. Calculate the thrust on the boat, the flow rate of air through the propeller, and the propeller efficiency.

Figure P2.135

2.136 An aircraft is propelled by a 2.2-m-diameter propeller at a speed of 200 km/h. The air velocity downstream of the propeller is 320 km/h relative to the aircraft. Determine the pressure difference across the propeller blades and the required power. Use $\rho = 1.2$ kg/m^3.

2.137 The 20-in.-diameter propeller on a boat moves at 20 mph by causing a velocity of 40 mph relative to the boat. Calculate the required power and the mass flux of water through the propeller.

2.138 A jet boat takes in 0.2 m^3/s of water and discharges it at a velocity of 20 m/s relative to the boat. If the boat travels at 10 m/s, calculate the thrust produced and the power required.

2.139 A 5-m-diameter wind turbine that operates in an 8 m/s wind at 100 kPa and 18°C is used to pump 10°C water from a well 18 m deep. If the overall efficiency of the turbine-pump system is 30%, calculate the amount of water pumped in cubic meters per day.

2.140 A 60-m-diameter wind turbine operates in a wind at 100 kPa and 18°C, and has a velocity pattern in one day that follows a full sinusoidal wave that has a mean velocity of 9 m/s and an amplitude of 4.5 m/s. Calculate, for the day:

(a) The total energy of the wind.

(b) The theoretical maximum energy developed by the turbine.

(c) The theoretical maximum energy developed by the turbine if the wind had a constant velocity equal to the mean value.

(d) The wind mean energy velocity. (see Example 2.19)

2.141 Repeat the previous problem but assume the wind turbine had cut-in and cut-out velocities of 3.6 m/s and 8.9 m/s, respectively.

2.142 The wind velocity flowing over a wind turbine is a function of elevation above ground due to the boundary layer effect. The ratio of wind velocity V at a height H to the reference velocity V_r at reference height H_r of 9.1 m is given by the empirical equation

$$\frac{V}{V_r} = \left(\frac{H}{H_r}\right)^a$$

where, $a = a_0\,(1 - \ln V_r/\ln V_0)$, $a_0 = (Z_0/H_r)^{0.2}$ and Z_0 is the surface roughness and a value of 0.4 m is used for evaluation purposes. V_0 has a fixed value of 67.1 m/s.

Wind at 100 kPa and 20°C has a mean velocity of 10 m/s measured at the reference height. Calculate:

(a) The wind power at the reference height.

(b) The power density for a 50 m diameter wind turbine with a hub 60 m above ground.

(c) The maximum theoretical power delivered by the turbine.

2.143 Consider a 10-m-diameter, horizontal-axis wind turbine whose rotor is stationary. A wind at 100 kPa and 20°C blows over the rotor at a speed of 15 m/s. At a location midway between the root and tip of the blade, the chord length is 0.5 m, the angle of attack is 17°, and the lift and drag coefficients are 0.651 and 0.274, respectively. Calculate the normal and tangential forces acting at that location.

2.144 A horizontal-axis wind turbine has a diameter of 10 m and is operating in a 100 kPa and 25°C wind at 10 m/s. The rotational speed is 60 rpm and the chord length is 0.5 m. At a location on the rotor blade where the radial distance is 50% of the blade length, the pitch angle is 6°, and the tangential and normal coefficients are 0.03 and 1.35, respectively. If the blade has a uniform chord length, airfoil profile, and pitch angle, calculate the torque and thrust forces acting at that location.

2.145 Calculate the change in the momentum flux of the water flowing through the plane contraction shown in Figure P2.145 if the flow rate is 0.2 m^3/s. The slope of the two profiles is the same. The upstream profile is created by a plate containing slots of various widths.

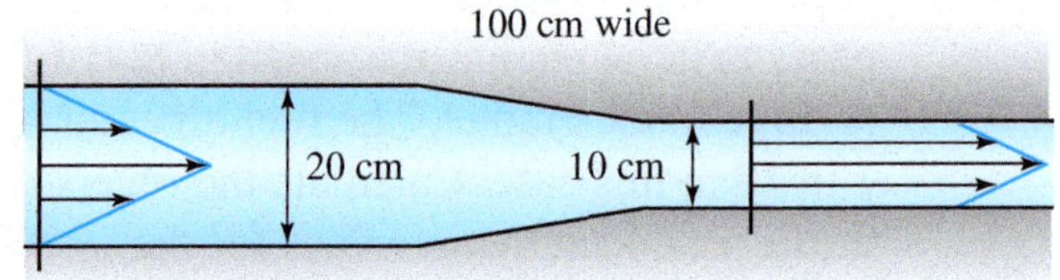

Figure P2.145

2.146 Determine the momentum correction factor for the following profile shown in Problem 2.145:

(a) The entering profile

(b) The exiting profile

2.147 Water at 60°F flows in a $1\frac{1}{2}$-in.-diameter horizontal pipe and experiences a pressure drop of 0.03 psi over a 30-ft length of pipe. Calculate the velocity gradient at the wall. Recall that $\tau = \mu|du/dr|$.

2.148 Find the drag force on the walls between the two sections of the horizontal pipe shown in Figure P2.145.

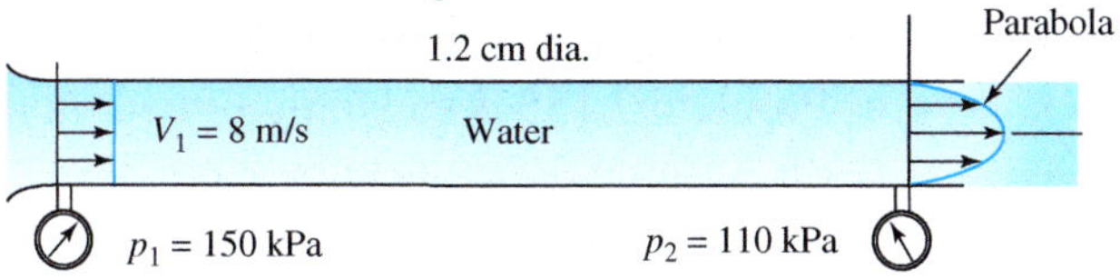

Figure P2.148

2.149 The velocity distribution downstream of a 10-m-long circular cylinder is as shown in Figure P2.149. Determine the force of the air on the cylinder. Use $\rho = 1.23$ kg/m^3.

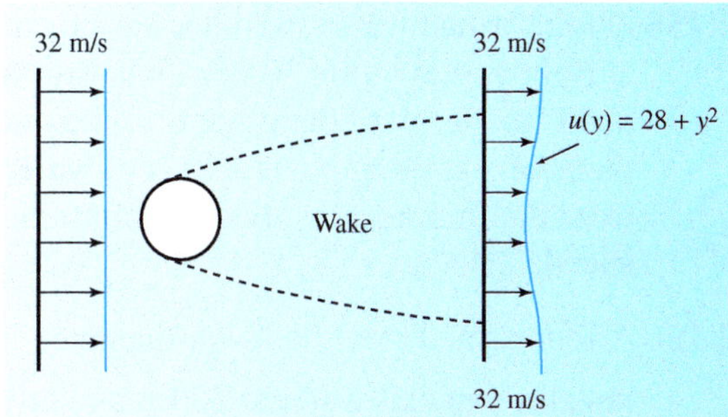

Figure P2.149

2.150 Calculate the drag force acting on the flat 2-m-wide plate shown in Figure P2.150. Outside the viscous region the velocity is uniform. Select:

(a) A rectangular control volume extending outside the viscous region (mass flux crosses the top).

(b) A control volume with the upper boundary a streamline (no mass flux crosses a streamline).

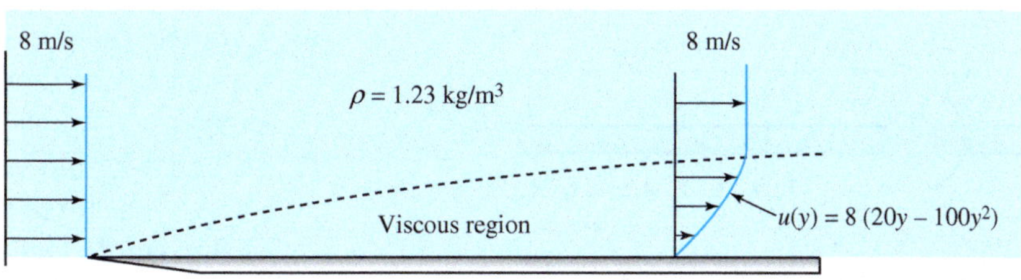

Figure P2.150

Momentum and Energy

2.151 Determine the power lost in the hydraulic jump of:

(a) Problem 2.110a **(b)** Problem 2.112
(c) Problem 2.113

2.152 Calculate the loss coefficient for the expansion of Problem 2.114. Base the coefficient on the velocity V_1.

2.153 Find a relationship between the acceleration of the cylindrical cart and the variables shown in Figure P2.153. Neglect friction. The initial mass of the cart and water is m_0.

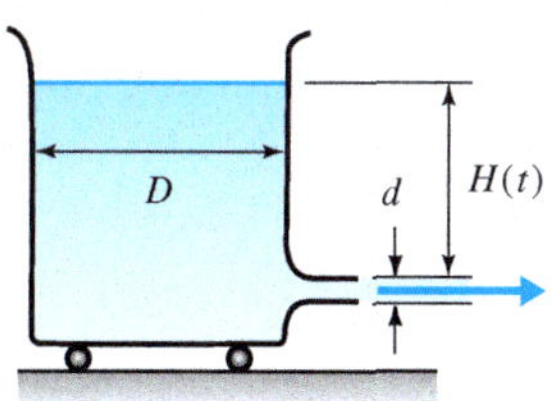

Figure P2.153

2.154 Set up the equations necessary to determine $H(t)$ for the toy air/water rocket shown in Figure P.2.154.

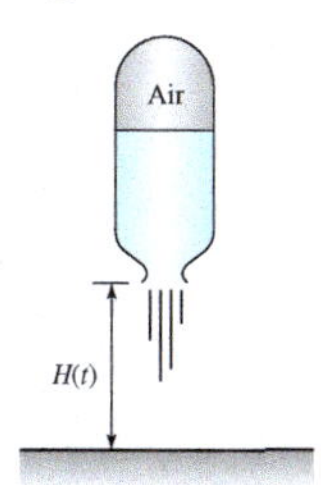

Figure P2.154

Moment of Momentum

2.155 A four-armed water sprinkler has nozzles at right angles to the 30-cm-long arms and at 45° angles with the ground. If the outlet diameters are 8 mm and 4 kg/s of water exits the four nozzles, find the rotational speed.

2.156 A four-armed rotor has $\frac{1}{2}$-in.-diameter nozzles that exit water at 200 ft/s relative to the arm. The nozzles are at right angles to the 10-in.-long arms and are parallel to the ground. If the rotational speed is 30 rad/s, find the power output. The arms are 1.5 in. in diameter.

2.157 Water flows out the 6-mm slots as shown in Figure P2.157. Calculate Ω if 20 kg/s is delivered by the two arms.

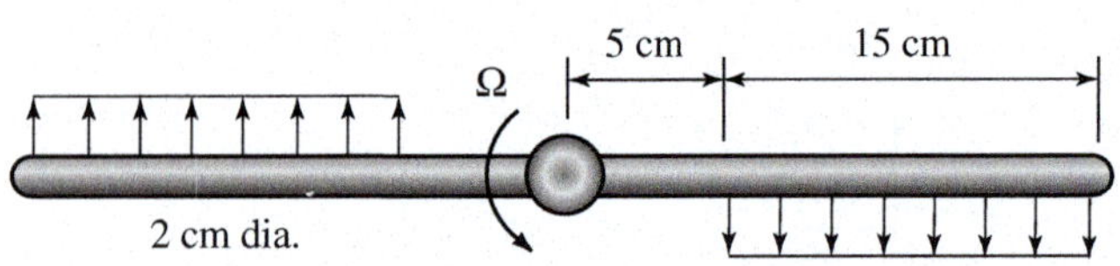

Figure P2.157

2.158 A 1-kW motor drives the rotor shown in Figure P2.158 at 500 rad/s. Determine the flow rate neglecting all losses. Use $\rho = 1.23\ \text{kg/m}^3$.

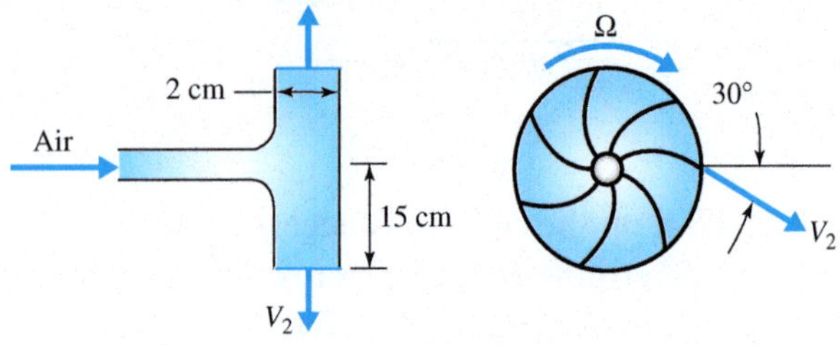

Figure P2.158

2.159 Find an expression for $\Omega(t)$ if the sprinkler of Problem 2.155 is suddenly turned on at $t = 0$. Assume the arms to be 2 cm in diameter.

2.160 Air enters the centrifugal-type air pump of a leaf blower through the blue area shown in Figure P2.160. The 10-cm-diameter 1.2-m-long tube has an attached nozzle with a 30-cm^2 exit area. The exit velocity is 240 km/h.

(a) Calculate the discharge.
(b) If the overall loss coefficient is 1.2, estimate the pump head.
(c) What power must the pump supply to the air?
(d) If the pump is 65% efficient, what is the required horsepower of the gasoline engine?
(e) Estimate the pressure at the tube entrance (just downstream of the pump).
(f) If the 10-kg blower hangs from a strap, what force must be applied at the handle located 30 cm above the nozzle? The center of gravity is 70 cm above and 120 cm to the left of the exit.

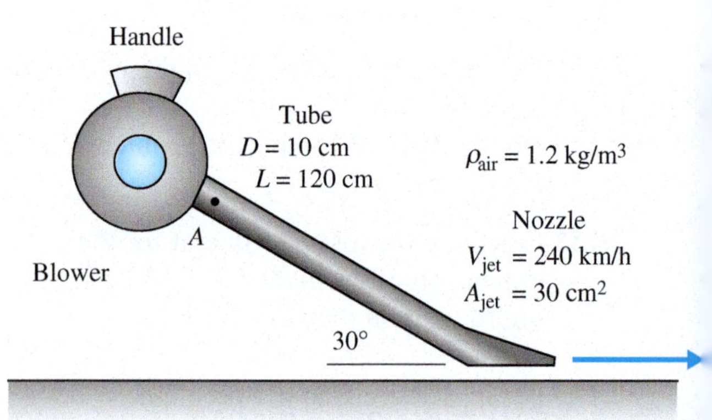

Figure P2.160

3 The Differential Forms of the Fundamental Laws

Outline

Chapter Objectives

The objectives of this chapter are to derive the differential equations and state the boundary and initial conditions needed to solve for the velocity and pressure fields in a fluid flow. The partial differential equations include:

- The continuity equation
- The momentum equations for inviscid flows (Euler's equations)
- The momentum equations for viscous flows (Navier–Stokes equations)
- The vorticity equations
- The energy equation

Numerous examples and problems will illustrate several applications of the differential equations along with how to simplify the equations depending on the flow situation.

3.1 INTRODUCTION

The subsequent chapters in this book have been designed to allow two possible routes: The general differential equations presented in this chapter may be used, or equations unique to a particular geometry may be derived without referring to these general equations.

In Chapter 2 the basic laws were expressed in terms of a fixed control volume, a finite volume in space. This is often described as the global approach to fluid mechanics. To find a solution using the control volume, it was necessary to assume an approximation to the integrands (primarily the velocity and pressure distributions), or expressions for the integrands were given. Suppose that we wish to find an integral quantity, such as the flow rate under a dam or the lift on an airfoil, and we cannot make a reasonable assumption for the velocity or pressure distribution. It is then necessary to determine the distributions that enter the integrands before the integral quantity can be found. This is done by solving the partial differential equations that express the basic laws.

M. C. Potter and D. C. Wiggert, *Applied Fluid Mechanics*,
https://doi.org/10.1007/978-3-031-65276-9_3

Not only do solutions of the differential forms of the basic laws allow us to determine integral quantities of interest, they often contain information in themselves. For example, we may wish to know the exact location of the minimum pressure on a body, or the region of separated flow from a surface may be of interest. So we often solve the differential equations to answer a specific question raised about a special flow.

There are two primary methods used in deriving the differential forms of the fundamental laws. One method involves the application of Gauss's theorem, which allows the area integrals of the basic equations of Chapter 2 to be transformed to volume integrals; the integrands are then collected under one integral, which can be set equal to zero. The integration is valid over any arbitrary control volume, and thus the integrand itself can be set equal to zero, providing us with the differential form of the basic law. The other approach, the one used in this book, is to identify an infinitesimal element in space and apply the basic laws directly to that element. Both methods result in the differential forms of the basic laws; the first method, however, demands the use of vector and tensor calculus. We introduce some vector calculus, but it will not be used at the operational level demanded by Gauss's theorem.

The conservation of mass, applied to an infinitesimal element, leads to the differential continuity equation; it relates the density and velocity fields.[1] Newton's second law (a vector relationship) results in three partial differential equations known as the Navier–Stokes equations; they relate the velocity, pressure, and density fields and include the viscosity and the gravity vector. The first law of thermodynamics provides us with the differential energy equation, which relates the temperature field to the velocity, density, and pressure fields and introduces the specific heat and thermal conductivity. Most problems considered are for isothermal, incompressible flows in which the temperature field does not play a role; for such flows the three Navier–Stokes equations and the continuity equation provide four partial differential equations that relate the three velocity components and the pressure. Thus the energy equation would not be needed. We will, however, derive the differential energy equation for use in a limited number of situations.

Partial differential equations require conditions that specify certain values for the dependent variables at particular values of the independent variables. If the independent variable is time, the conditions are called **initial conditions**; if the independent variable is a space coordinate, the conditions are **boundary conditions**. The total problem is referred to as an **initial-value problem** or a **boundary-value problem**.

In fluid mechanics the boundary conditions result from:

- The no-slip conditions for a viscous flow. The viscosity causes the fluid to stick to the wall, and thus the velocity of the fluid at the wall assumes the velocity of the wall. Usually the velocity of the wall is zero; this condition is used most often.
- The normal component of velocity in an inviscid flow, a flow in which viscous effects are negligible. Near a nonporous stationary wall the velocity vector must be tangent to the wall, demanding that the normal component be zero.

[1] When a dependent variable depends on more than one independent variable, it is referred to as a **field**; that is, $\mathbf{V}(x, y)$ is a velocity field and $p(x, y, z, t)$ is a pressure field. The partial differential equations describing the field quantities are often called **field equations**.

- The pressure in a flow involving a free surface. For problems with a free surface, such as a flow with a liquid-gas interface, the pressure is known on the interface. This would be the situation involving wave motion or in a separated flow involving cavitation.
- The temperature of the boundary or the temperature gradient at the boundary. If the temperature of the boundary is held constant, the temperature of the fluid next to the boundary will equal the boundary temperature. If the boundary is insulated, the temperature gradient will be zero at the boundary.

In an unsteady flow the initial condition demands that the three velocity components be specified at all points in the flow at a particular instant in time, usually taken as $t = 0$. This information would be very difficult to obtain in many situations, for example, the atmosphere; to obtain the three velocity components everywhere in the atmosphere at an instant is obviously an improbable task.

Differential equations take very different forms depending on the coordinate system chosen. We derive the equations using rectangular coordinates and then express the equations in vector form. Forms using both cylindrical and spherical coordinates are presented in Table 3.1 at the end of the chapter.

3.2 DIFFERENTIAL CONTINUITY EQUATION

Let us begin our quest for the partial differential equations that model the detailed motion of a fluid by applying the conservation of mass to a small volume in a fluid flow. Consider the mass flux through each face of the fixed infinitesimal control volume shown in Figure 3.1. We set the net flux of mass entering the element equal to the rate of change of the mass of the element; that is,

$$\dot{m}_{\text{in}} - \dot{m}_{\text{out}} = \frac{\partial}{\partial t} m_{\text{element}} \tag{3.2.1}$$

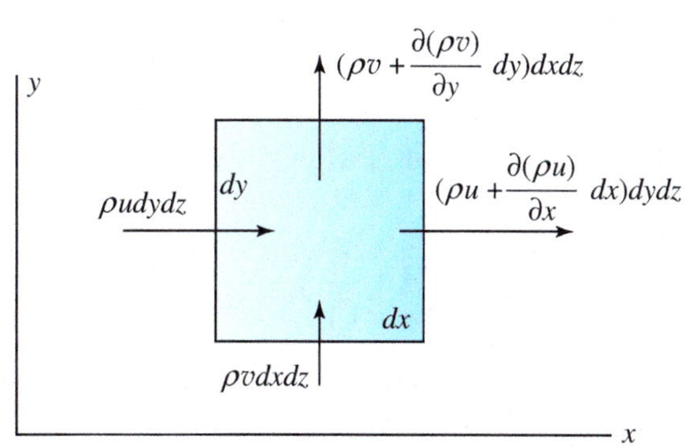

Figure 3.1 An infinitesimal control volume using rectangular coordinates.

Let us assume that the flow is only in the xy-plane so that no fluid flows in the z-direction. Then Eq. 3.2.1 is expressed as

$$\rho u dydz + \rho v dxdz - \left(\rho u + \frac{\partial(\rho u)}{\partial x}dx\right)dydz - \left(\rho v + \frac{\partial(\rho v)}{\partial y}dy\right)dxdz = \frac{\partial}{\partial t}(\rho dxdydz) \qquad (3.2.2)$$

where the density ρ is allowed to change[2] across the element. Simplifying the above, recognizing that the elemental control volume is fixed, results in

$$\frac{\partial(\rho u)}{\partial x} + \frac{\partial(\rho v)}{\partial y} = -\frac{\partial \rho}{\partial t} \qquad (3.2.3)$$

If an element included the z-component of velocity flowing through the side $dxdy$, the term $\partial(\rho w)/\partial z$ would be included on the left-hand side of Eq. 3.2.3 providing

$$\frac{\partial(\rho u)}{\partial x} + \frac{\partial(\rho v)}{\partial y} + \frac{\partial(\rho w)}{\partial z} = -\frac{\partial \rho}{\partial t} \qquad (3.2.4)$$

Differentiate the products and include the variation in the z-direction. Then the differential continuity equation can be put in the form

$$\frac{\partial \rho}{\partial t} + u\frac{\partial \rho}{\partial x} + v\frac{\partial \rho}{\partial y} + w\frac{\partial \rho}{\partial z} + \rho\left(\frac{\partial u}{\partial x} + \frac{\partial v}{\partial y} + \frac{\partial w}{\partial z}\right) = 0 \qquad (3.2.5)$$

The first four terms form the material derivative (see Eq. 1.2.11) so Eq. 3.2.5 becomes

$$\frac{D\rho}{Dt} + \rho\left(\frac{\partial u}{\partial x} + \frac{\partial v}{\partial y} + \frac{\partial w}{\partial z}\right) = 0 \qquad (3.2.6)$$

providing the most general form of the **differential continuity equation** expressed using rectangular coordinates.

We can introduce the **gradient operator**, called "del," which, in rectangular coordinates, is

$$\nabla = \frac{\partial}{\partial x}\hat{\mathbf{i}} + \frac{\partial}{\partial y}\hat{\mathbf{j}} + \frac{\partial}{\partial z}\hat{\mathbf{k}} \qquad (3.2.7)$$

[2] The product ρu could have been included as $\left(\rho + \frac{\partial \rho}{\partial x}dx\right)\left(u + \frac{\partial u}{\partial x}dx\right)$ on the right-hand-side of the element but the above is equivalent.

The continuity equation can then be written in the form

$$\frac{D\rho}{Dt} + \rho\nabla \cdot \mathbf{V} = 0 \tag{3.2.8}$$

where $\mathbf{V} = u\,\hat{\mathbf{i}} + v\,\hat{\mathbf{j}} + w\,\hat{\mathbf{k}}$; the dot product $\nabla \cdot \mathbf{V}$ is called the **divergence** of the velocity. This form of the continuity equation does not refer to any particular coordinate system, and thus is used to express the continuity equation using various coordinate systems.

For the case of **incompressible flow**, a flow in which the density of a fluid particle does not change as it travels along, we see that

$$\frac{D\rho}{Dt} = \frac{\partial \rho}{\partial t} + u\frac{\partial \rho}{\partial x} + v\frac{\partial \rho}{\partial y} + w\frac{\partial \rho}{\partial z} = 0 \tag{3.2.9}$$

Note that this is less restrictive than the assumption of constant density, which would require each term in Eq. 3.2.9 to be zero. Incompressible flows that have density gradients are sometimes referred to as **stratified flows** or **nonhomogeneous flows**; atmospheric and oceanic flows are examples of such flows. Using Eq. 3.2.6, the continuity equation for an incompressible flow takes the form

$$\frac{\partial u}{\partial x} + \frac{\partial v}{\partial y} + \frac{\partial w}{\partial z} = 0 \tag{3.2.10}$$

or, in vector form,

$$\nabla \cdot \mathbf{V} = 0 \tag{3.2.11}$$

The divergence of the velocity vector is zero for an incompressible flow even if the flow is unsteady.

In cylindrical and spherical coordinates the continuity equation for an incompressible flow is presented in Table 3.1. The expression for *D*/*Dt* in cylindrical and spherical coordinates can also be found in Table 3.1.

Example 3.1

The x-component velocity is given by $u(x, y) = Ay^2$ in an incompressible plane flow. Determine $v(x, y)$ if $v(x, 0) = 0$, as would be the case in flow between parallel plates.

Solution

The differential continuity equation for an incompressible, plane flow is

$$\frac{\partial u}{\partial x} + \frac{\partial v}{\partial y} = 0$$

since in a plane flow the two velocity components depend only on x and y. Using the given $u(x, y)$ we find that

$$\frac{\partial v}{\partial y}+\frac{\partial u}{\partial x}=\frac{\partial}{\partial y}(Ay^2)=0$$

Since this is a partial differential equation, its solution is

$$v(x,y)=f(x)$$

But $v(x, 0) = 0$ requiring that $f(x) = 0$. Consequently,

$$\underline{v(x,y)=0}$$

is the y-component velocity demanded by the conservation of mass. In order for $v(x, y)$ to be nonzero, $u(x, y)$ would have to vary with x or $v(x, 0)$ would have to be nonzero.

Example 3.2

Air flows in a pipe and the velocity at three neighboring points A, B, and C, 4-in. apart, is measured to be 274, 285, and 291 ft/s, respectively, as shown in Figure E3.2. The temperature and pressure are 50°F and 50 psia, respectively, at point B. Approximate $d\rho/dx$ at that point, assuming steady, uniform flow.

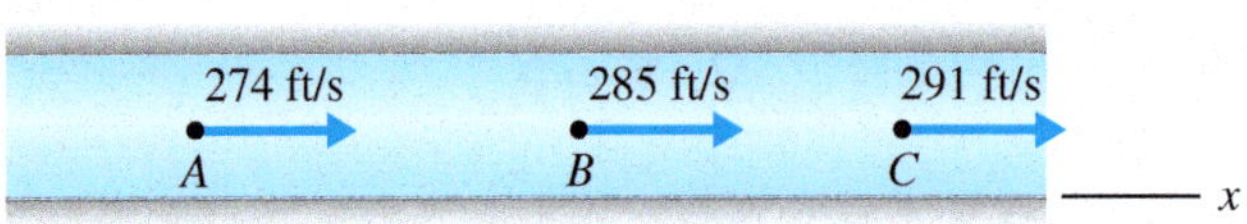

Figure E3.2

Solution

The continuity equation (3.2.5) for this steady $\left(\frac{\partial}{\partial t}=0\right)$, uniform $\left(\frac{\partial}{\partial y}=\frac{\partial}{\partial z}=0\right)$ flow reduces to

$$u\frac{d\rho}{dx}+\rho\frac{du}{dx}=0$$

We used ordinary derivatives since u and ρ depend only on x. The velocity derivative is approximated by

$$\frac{du}{dx}\simeq\frac{\Delta u}{\Delta x}=\frac{291-274}{8/12}=25.5\frac{\text{ft/s}}{\text{ft}}$$

where the more accurate central difference has been used.[3] The density is

$$\rho = \frac{p}{RT} = \frac{50\ \text{lb/in}^2 \times 144\ \text{in}^2/\text{ft}^2}{1716\ \text{ft-lb/slug-°R} \times (50 + 460)\text{°R}} = 0.00823\ \text{slug/ft}^3$$

where absolute pressure and temperature are used. The density derivative is then approximated to be

$$\begin{aligned} \frac{d\rho}{dx} &= -\frac{\rho}{u}\frac{du}{dx} \\ &= -\frac{0.00823\ \text{slug/ft}^3}{285\ \text{ft/s}} \times 25.5\ \text{s}^{-1} = \underline{-0.000736\ \text{slug/ft}^4} \end{aligned}$$

Example 3.3

The x-component of velocity at points A, B, C, and D, which are 10 mm apart, is measured to be 5.76, 6.72, 7.61, and 8.47 m/s, respectively, in the plane steady, symmetrical, incompressible flow shown in Figure E3.3 in which $w = 0$. Approximate the x-component acceleration at C and the y-component of velocity 6 mm above B.

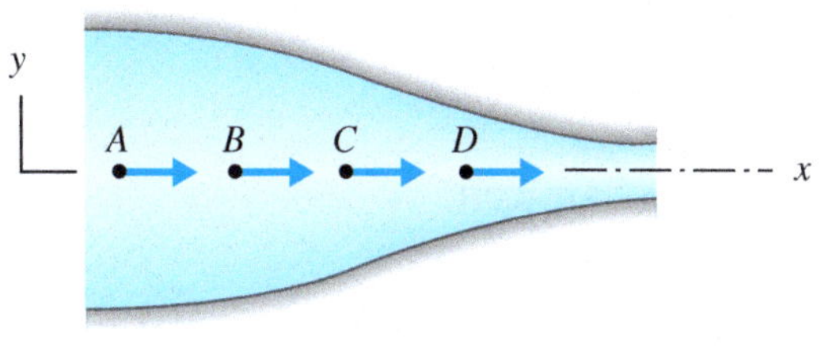

Figure E3.3

Solution

The desired acceleration component on the centerline is found from Eq. 1.2.9 to be

$$\begin{aligned} a_x &= \cancelto{0}{\frac{\partial u}{\partial t}} + u\frac{\partial u}{\partial x} + \cancelto{0}{v\frac{\partial u}{\partial y}} + \cancelto{0}{w\frac{\partial u}{\partial z}} \\ &\cong u\frac{\Delta u}{\Delta x} = 7.61\frac{8.47 - 6.72}{0.02} = 666\ \text{m/s}^2 \end{aligned}$$

[3] Using a forward difference $\Delta u/\Delta x = (291 - 285)/0.333 = 18$; a backward difference provides $\Delta u/\Delta x = (285 - 274)/0.333 = 33$. These two approximations are less accurate than the central difference used in the example. A sketch of a general curve $u(x)$ could graphically show this by displaying three points at $x - \Delta x$, x, and $x + \Delta x$ and sketching the slopes.

where we have assumed a symmetrical flow so that v along the centerline is zero. We have used central differences to approximate $\partial u/\partial x$ at point C, as done in Example 3.2 (see footnote 3).

The y-component of velocity 6 mm above B is found using the continuity equation (3.2.10) as follows:

$$\frac{\partial v}{\partial y} = -\frac{\partial u}{\partial x} \quad (\text{with } w = 0)$$

$$\frac{\Delta v}{\Delta y} \simeq -\frac{\Delta u}{\Delta x} = -\frac{7.61 - 5.76}{0.02} = -92.5$$

$$\therefore \Delta v = -92.5\ \Delta y = -92.5 \times 0.006 = -0.555 \text{ m/s}$$

We know that $v = 0$ at B; hence at the desired location, with $\Delta v = v - v_B$, there results

$$v = \underline{-0.555 \text{ m/s}}$$

Example 3.4

The continuity equation can be used to change the form of an expression. Write the expression $\rho D\tilde{u}/Dt + p\nabla \cdot \mathbf{V}$, which appears in the differential energy equation, in terms of enthalpy h rather than internal energy $\tilde{u}$. Recall that $h = \tilde{u} + p/\rho$.

Solution

Using the definition of enthalpy, we can write

$$\frac{D\tilde{u}}{Dt} = \frac{Dh}{Dt} - \frac{1}{\rho}\frac{Dp}{Dt} + \frac{p}{\rho^2}\frac{D\rho}{Dt}$$

where we used

$$\frac{D}{Dt}\left(\frac{p}{\rho}\right) = \frac{1}{\rho}\frac{Dp}{Dt} - \frac{p}{\rho^2}\frac{D\rho}{Dt}$$

The desired expression is then

$$\rho\frac{D\tilde{u}}{Dt} + p\nabla \cdot \mathbf{V} = \rho\frac{Dh}{Dt} - \frac{Dp}{Dt} + \frac{p}{\rho}\frac{D\rho}{Dt} + p\nabla \cdot \mathbf{V}$$

The continuity equation (3.2.8) is introduced resulting in

$$\rho\frac{D\tilde{u}}{Dt}+p\nabla\cdot\mathbf{V}=\rho\frac{Dh}{Dt}-\frac{Dp}{Dt}+\frac{p}{\rho}\frac{D\rho}{Dt}+p\left(-\frac{1}{\rho}\frac{D\rho}{Dt}\right)$$
$$=\rho\frac{Dh}{Dt}-\frac{Dp}{Dt}$$

and enthalpy has been introduced.

3.3 DIFFERENTIAL MOMENTUM EQUATION

3.3.1 General Formulation

Suppose that we do not know the velocity field or the pressure field in an incompressible[4] flow of interest and we wish to solve differential equations to provide us with that information. The differential continuity equation is one differential equation to help us toward this end; however, it has three unknowns, the three velocity components. The differential momentum equation is a vector equation and thus provides us with three scalar equations. These component equations will aid us in our attempt to determine the velocity and pressure fields. There is a difficulty in deriving these equations, however, since we must use the stress components to determine the forces required in the momentum equation. Let us identify these stress components.

There are nine stress components that act at a particular point in a fluid flow. They are the nine components of the stress tensor τ_{ij}. We will not study the properties of a stress tensor in detail in this study of fluid mechanics since we do not have to maximize or minimize the stress (as would be required in a solid mechanics course); we must, however, use the nine stress components in our derivations, then relate the stress components to the velocity and pressure fields with the appropriate equations. The stress components that act at a point are displayed on two- and three-dimensional rectangular elements in Figure 3.2. These elements are considered to be an exaggerated point, in this case a cubical point; the stress components act in the positive direction on a positive face (a normal vector points in the positive coordinate direction) and in the negative direction on a negative face (a normal vector points in the negative coordinate direction). The first subscript on a stress component denotes the face upon which the component acts, and the second subscript denotes the direction in which it acts; the component τ_{xy} acts in the positive y-direction on a positive x-face and in the negative y-direction on a negative x-face, as displayed in Figure 3.2a. A stress component that acts perpendicular to a face is referred to as a **normal stress**; the components σ_{xx}, σ_{yy}, and σ_{zz} are normal stresses. A stress component that acts tangential to a face is called a **shear stress**; the components τ_{xy}, τ_{yx}, τ_{xz}, τ_{zx}, τ_{yz}, and τ_{zy} are the shear stress components.

---

[4]An incompressible flow, when referred to in a general discussion such as in this section, will generally refer to a constant-density flow. This is true in most fluid mechanics literature, including textbooks on the subject. Two exceptions are atmospheric flows and flows in a river near the ocean where fresh water mixes with salt water.

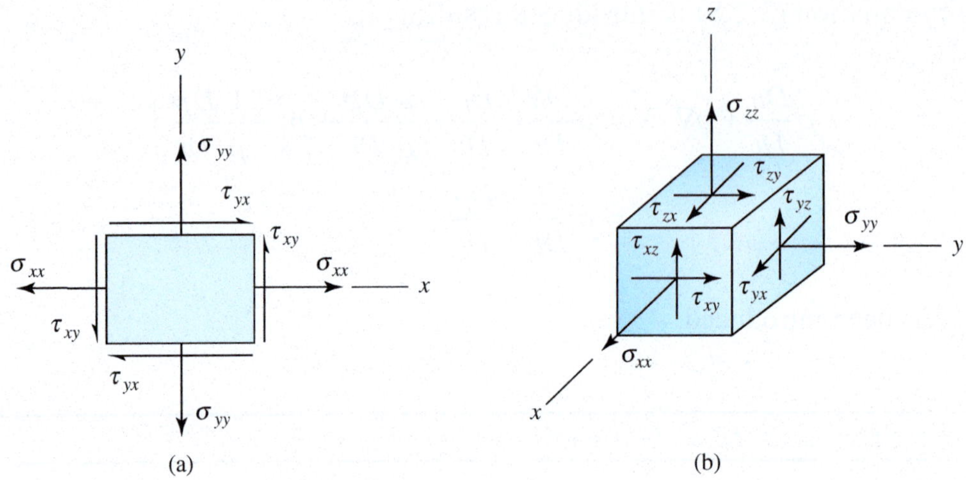

Figure 3.2 Stress components in rectangular coordinates: (a) two-dimensional stress components; (b) three-dimensional stress components.

There are nine stress components that act at each point in a fluid flow. If moments are taken about the axes passing through the centers of the faces of the elements in Figure 3.2, respectively, they would show that

$$\tau_{yx} = \tau_{xy} \qquad \tau_{zx} = \tau_{xz} \qquad \tau_{zy} = \tau_{yz} \tag{3.3.1}$$

showing that the stress tensor is symmetric. It is often displayed as

$$\tau_{ij} = \begin{pmatrix} \sigma_{xx} & \tau_{xy} & \tau_{xz} \\ \tau_{xy} & \sigma_{yy} & \tau_{yz} \\ \tau_{xz} & \tau_{yz} & \sigma_{zz} \end{pmatrix} \tag{3.3.2}$$

in which the subscripts i and j take on numerical values 1, 2, or 3. The term τ_{12} represents the element τ_{xy} in the first row, second column. Consequently, there are six independent stress components that result in the infinitesimal forces acting on the fluid contained in the element of Figure 3.2b.

Let us now apply Newton's second law to the element of Figure 3.3 assuming no shear stresses act in the z-direction (those will be added in later) and that gravity acts in the z-direction only:

$$\begin{aligned} &\left(\sigma_{xx} + \frac{\partial \sigma_{xx}}{\partial x} dx\right) dydz - \sigma_{xx} dydz + \left(\tau_{xy} + \frac{\partial \tau_{xy}}{\partial y} dy\right) dxdz - \tau_{xy} dxdz = \rho dxdydz \frac{Du}{Dt} \\ &\left(\sigma_{yy} + \frac{\partial \sigma_{yy}}{\partial y} dy\right) dxdz - \sigma_{yy} dxdz + \left(\tau_{yx} + \frac{\partial \tau_{yx}}{\partial x} dx\right) dydz - \tau_{yx} dydz = \rho dxdydz \frac{Dv}{Dt} \end{aligned} \tag{3.3.3}$$

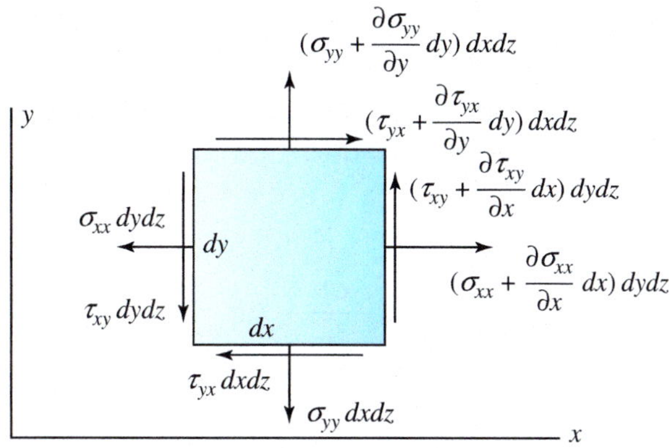

Figure 3.3 Rectangular stress components on a fluid element.

Equations (3.3.3) simplify to

$$\frac{\partial \sigma_{xx}}{\partial x} + \frac{\partial \tau_{xy}}{\partial y} = \rho \frac{Du}{Dt}$$
$$\frac{\partial \sigma_{yy}}{\partial y} + \frac{\partial \tau_{yx}}{\partial x} = \rho \frac{Dv}{Dt} \tag{3.3.4}$$

If the z-direction components are included, the differential equations become

$$\frac{\partial \sigma_{xx}}{\partial x} + \frac{\partial \tau_{xy}}{\partial y} + \frac{\partial \tau_{xz}}{\partial z} = \rho \frac{Du}{Dt}$$
$$\frac{\partial \sigma_{yy}}{\partial y} + \frac{\partial \tau_{yx}}{\partial x} + \frac{\partial \tau_{yz}}{\partial z} = \rho \frac{Dv}{Dt} \tag{3.3.5}$$
$$\frac{\partial \sigma_{zz}}{\partial z} + \frac{\partial \tau_{zx}}{\partial x} + \frac{\partial \tau_{zy}}{\partial y} - \rho g = \rho \frac{Dw}{Dt}$$

assuming the gravity term, $\rho g dxdydz$, acts in the negative z-direction.

Let us now apply these equations first to an inviscid flow in which the effects of viscosity are negligible and then to a flow in which viscous effects are included.

3.3.2 Euler's Equations

Good approximations to the components of the stress tensor for many flows, especially for flow away from a boundary (flow around an airfoil) or in regions of sudden change (flow through a contraction) are displayed by the stress tensor

$$\tau_{ij} = \begin{pmatrix} -p & 0 & 0 \\ 0 & -p & 0 \\ 0 & 0 & -p \end{pmatrix} \tag{3.3.6}$$

For such flows, we have assumed the shear stress components that result from viscous effects to be negligibly small and the normal stress components to be equal to the negative of the pressure; this is precisely what we did in Figure 1.17 when deriving Bernoulli's equation. If these stress components are introduced back into Eqs. 3.3.5 there results, for this frictionless flow,

$$\begin{aligned} \rho\frac{Du}{Dt} &= -\frac{\partial p}{\partial x} + \rho g_x \\ \rho\frac{Dv}{Dt} &= -\frac{\partial p}{\partial y} + \rho g_y \\ \rho\frac{Dw}{Dt} &= -\frac{\partial p}{\partial z} + \rho g_z \end{aligned} \tag{3.3.7}$$

The scalar equations above can then be written as the vector equation

$$\rho\frac{D}{Dt}(u\hat{\mathbf{i}} + v\hat{\mathbf{j}} + w\hat{\mathbf{k}}) = -\left(\frac{\partial p}{\partial x}\hat{\mathbf{i}} + \frac{\partial p}{\partial y}\hat{\mathbf{j}} + \frac{\partial p}{\partial z}\hat{\mathbf{k}}\right) - \rho\mathbf{g} \tag{3.3.8}$$

In vector form, we have the well-known **Euler's equation**

$$\rho\frac{D\mathbf{V}}{Dt} = -\nabla p - \rho\mathbf{g} \tag{3.3.9}$$

If we assume a constant-density, steady flow, Eq. 3.3.9 can be integrated along a streamline to yield Bernoulli's equation, a result that does not surprise us since the same assumptions were imposed when deriving Bernoulli's equation in Chapter 1; this will be illustrated in Example 3.6.

With the differential momentum equations in the form of Eqs. 3.3.7, we have added three additional equations to the continuity equation to give four equations and four unknowns, u, v, w, and p. With the appropriate boundary and initial conditions, a solution, yielding the velocity and pressure fields for this inviscid, incompressible flow, would be possible.

Example 3.5

A velocity field is proposed to be

$$u = \frac{10y}{x^2 + y^2} \qquad v = -\frac{10x}{x^2 + y^2} \qquad w = 0$$

(a) Is this a possible incompressible flow? (b) If so, find the pressure gradient ∇p assuming a frictionless air flow with the z-axis vertical. Use $\rho = 1.23\ \text{kg/m}^3$.

Solution

(a) The continuity equation (3.2.10) is used to determine if the velocity field is possible. For this incompressible flow we have

$$\frac{\partial u}{\partial x}+\frac{\partial v}{\partial y}+\cancelto{0}{\frac{\partial w}{\partial z}}=0$$

Substituting in the velocity components, we have

$$\frac{\partial}{\partial x}\left(\frac{10y}{x^2+y^2}\right)+\frac{\partial}{\partial y}\left(-\frac{10x}{x^2+y^2}\right)=\frac{-10y(2x)}{(x^2+y^2)^2}-\frac{-10x(2y)}{(x^2+y^2)^2}=\frac{1}{(x^2+y^2)^2}[-20xy+20xy]=0$$

The quantity in brackets is obviously zero; hence, the velocity field given is a possible incompressible flow.

(b) The pressure gradient is found using Euler's equation. In component form we have the following:

$$\rho\frac{Du}{Dt}=-\frac{\partial p}{\partial x}+\cancelto{0}{\rho g_x}$$

$$\therefore \frac{\partial p}{\partial x}=-\rho\left[u\frac{\partial u}{\partial x}+v\frac{\partial u}{\partial y}+w\cancelto{0}{\frac{\partial u}{\partial z}}+\cancelto{0}{\frac{\partial u}{\partial t}}\right]$$

$$=-1.23\left[\frac{10y}{x^2+y^2}\frac{-20xy}{(x^2+y^2)^2}+\frac{-10x}{x^2+y^2}\frac{(x^2+y^2)10-10y(2y)}{(x^2+y^2)^2}\right]=\frac{123x}{(x^2+y^2)^2}$$

$$\rho\frac{Dv}{Dt}=-\frac{\partial p}{\partial y}+\cancelto{0}{\rho g_y}$$

$$\therefore \frac{\partial p}{\partial y}=-\rho\left[u\frac{\partial v}{\partial x}+v\frac{\partial v}{\partial y}+w\cancelto{0}{\frac{\partial v}{\partial z}}+\cancelto{0}{\frac{\partial v}{\partial t}}\right]$$

$$=-1.23\left[\frac{10y}{x^2+y^2}\frac{(x^2+y^2)(-10)+10x(2x)}{(x^2+y^2)^2}+\frac{-10x}{x^2+y^2}\frac{20xy}{(x^2+y^2)^2}\right]=\frac{123y}{(x^2+y^2)^2}$$

$$\rho\cancelto{0}{\frac{Dw}{Dt}}=-\frac{\partial p}{\partial z}+\rho g_z$$

$$\therefore \frac{\partial p}{\partial z}=\rho g_z=1.23\ \text{kg/m}^3\times(-9.81)\ \text{m/s}^2=-12.07\ \text{N/m}^3$$

Thus, $$\nabla p=\frac{\partial p}{\partial x}\hat{\mathbf{i}}+\frac{\partial p}{\partial y}\hat{\mathbf{j}}+\frac{\partial p}{\partial z}\hat{\mathbf{k}}=\frac{123}{(x^2+y^2)^2}(x\hat{\mathbf{i}}+y\hat{\mathbf{j}})-12.07\hat{\mathbf{k}}\ \text{N/m}^3$$

Example 3.6

Assume a steady, constant-density flow and integrate Euler's equation along a streamline in a plane flow.

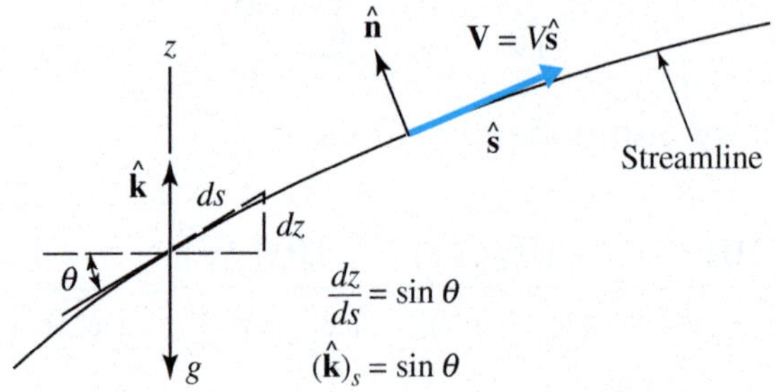

Figure E3.6

Solution

First, let us express the substantial derivative in streamline coordinates. Since the velocity vector is tangent to the streamline, we can write

$$\mathbf{V} = V\,\hat{\mathbf{s}}$$

where $\hat{\mathbf{s}}$ is the unit vector tangent to the streamline and V is the magnitude of the velocity, as shown in Figure E3.6. The substantial derivative is then, for this plane flow,

$$\frac{D\mathbf{V}}{Dt} = \frac{\partial \mathbf{V}}{\partial t} + V\frac{\partial(V\,\hat{\mathbf{s}})}{\partial s} + \overset{0}{\cancel{V_n}}\frac{\partial \mathbf{V}}{\partial n} = \frac{\partial \mathbf{V}}{\partial t} + V\frac{\partial V}{\partial s}\hat{\mathbf{s}} + V^2\frac{\partial \hat{\mathbf{s}}}{\partial s}$$

The quantity $\partial\hat{\mathbf{s}}/\partial s$ results from the change of the unit vector $\hat{\mathbf{s}}$; the unit vector cannot change magnitude (it must always have a magnitude of 1), it can only change direction. Hence the derivative $\partial\hat{\mathbf{s}}/\partial s$ is in a direction normal to the streamline and does not enter the streamwise component equation. For a steady flow $\partial\mathbf{V}/\partial t = 0$. Consequently, in the streamwise direction, Euler's equation (3.3.9) takes the form

$$\rho V\frac{\partial V}{\partial s} = -\frac{\partial p}{\partial s} - \rho g\frac{\partial z}{\partial s}$$

recognizing that the component of $\hat{\mathbf{k}}$ along the streamline can be expressed as $(\hat{\mathbf{k}})_s = \partial z/\partial s$ (see the sketch above). Note that we use partial derivatives in this equation since velocity and pressure also vary with the normal coordinate.

The equation above can be written, assuming constant density so that $\partial\rho/\partial s = 0$, as

$$\frac{\partial}{\partial s}\left(\rho\frac{V^2}{2} + p + \rho g z\right) = 0$$

Integrating along the streamline results in

$$\rho\frac{V^2}{2}+p+\rho gz=\text{const.}$$

or

$$\frac{V^2}{2}+\frac{p}{\rho}+gz=\text{const.}$$

This is, of course, Bernoulli's equation. We have integrated along a streamline assuming constant density, steady flow, negligible viscous effects, and an inertial reference frame, so it is to be expected that Bernoulli's equation will result.

3.3.3 Navier–Stokes Equations

Many fluids exhibit a linear relationship between the stress components and the velocity gradients. Such fluids are called **Newtonian fluids** and include common fluids such as water, oil, and air. If in addition to linearity, we require that the fluid be **isotropic**,[5] it is possible to relate the stress components and the velocity gradients using only two fluid properties, the *viscosity* μ and the **second coefficient of viscosity** λ. The stress–velocity gradient relations, often referred to as the **constitutive equations**,[6] are stated as follows:

$$\begin{aligned}\sigma_{xx}&=-p+2\mu\frac{\partial u}{\partial x}+\lambda\nabla\cdot\mathbf{V} & \tau_{xy}&=\mu\left(\frac{\partial u}{\partial y}+\frac{\partial v}{\partial x}\right)\\ \sigma_{yy}&=-p+2\mu\frac{\partial v}{\partial y}+\lambda\nabla\cdot\mathbf{V} & \tau_{xz}&=\mu\left(\frac{\partial u}{\partial z}+\frac{\partial w}{\partial x}\right)\\ \sigma_{zz}&=-p+2\mu\frac{\partial w}{\partial z}+\lambda\nabla\cdot\mathbf{V} & \tau_{yz}&=\mu\left(\frac{\partial v}{\partial z}+\frac{\partial w}{\partial y}\right)\end{aligned} \tag{3.3.10}$$

For most gases, and for monatomic gases exactly, the second coefficient of viscosity is related to the viscosity by

$$\lambda=-\frac{2}{3}\mu \tag{3.3.11}$$

[5] The condition of isotropy exists if the fluid properties at a point are independent of direction. Polymers are examples of anisotropic fluids.

[6] Details of the development of the constitutive equations can be found in any textbook on the subject of continuum mechanics.

a condition that is known as **Stokes's hypothesis**. With this relationship the negative average of the three normal stresses is equal to the pressure; that is,

$$-\frac{1}{3}(\sigma_{xx}+\sigma_{yy}+\sigma_{zz})=p \tag{3.3.12}$$

Using Eqs. 3.3.10, this can be shown to always be true for a liquid in which $\nabla \cdot \mathbf{V}=0$, and with Stokes's hypothesis it is also true for a gas.

If we substitute the constitutive equations into the differential momentum equations (3.3.5), there results, using Stokes's hypothesis,

$$\begin{aligned}
\rho\frac{Du}{Dt}&=-\frac{\partial p}{\partial x}+\rho g_x+\mu\left(\frac{\partial^2 u}{\partial x^2}+\frac{\partial^2 u}{\partial y^2}+\frac{\partial^2 u}{\partial z^2}\right)+\frac{\mu}{3}\frac{\partial}{\partial x}\left(\frac{\partial u}{\partial x}+\frac{\partial v}{\partial y}+\frac{\partial w}{\partial z}\right)\\
\rho\frac{Dv}{Dt}&=-\frac{\partial p}{\partial y}+\rho g_y+\mu\left(\frac{\partial^2 v}{\partial x^2}+\frac{\partial^2 v}{\partial y^2}+\frac{\partial^2 v}{\partial z^2}\right)+\frac{\mu}{3}\frac{\partial}{\partial y}\left(\frac{\partial u}{\partial x}+\frac{\partial v}{\partial y}+\frac{\partial w}{\partial z}\right)\\
\rho\frac{Dw}{Dt}&=-\frac{\partial p}{\partial z}+\rho g_z+\mu\left(\frac{\partial^2 w}{\partial x^2}+\frac{\partial^2 w}{\partial y^2}+\frac{\partial^2 w}{\partial z^2}\right)+\frac{\mu}{3}\frac{\partial}{\partial z}\left(\frac{\partial u}{\partial x}+\frac{\partial v}{\partial y}+\frac{\partial w}{\partial z}\right)
\end{aligned} \tag{3.3.13}$$

where we have assumed a **homogeneous fluid**, that is, fluid properties (e.g., the viscosity) are independent of position.

For an incompressible flow the continuity equation $(\nabla \cdot \mathbf{V}=0)$ allows the equations above to be reduced to

$$\begin{aligned}
\rho\frac{Du}{Dt}&=-\frac{\partial p}{\partial x}+\rho g_x+\mu\left(\frac{\partial^2 u}{\partial x^2}+\frac{\partial^2 u}{\partial y^2}+\frac{\partial^2 u}{\partial z^2}\right)\\
\rho\frac{Dv}{Dt}&=-\frac{\partial p}{\partial y}+\rho g_y+\mu\left(\frac{\partial^2 v}{\partial x^2}+\frac{\partial^2 v}{\partial y^2}+\frac{\partial^2 v}{\partial z^2}\right)\\
\rho\frac{Dw}{Dt}&=-\frac{\partial p}{\partial z}+\rho g_z+\mu\left(\frac{\partial^2 w}{\partial x^2}+\frac{\partial^2 w}{\partial y^2}+\frac{\partial^2 w}{\partial z^2}\right)
\end{aligned} \tag{3.3.14}$$

These are the **Navier–Stokes equations**, named after Louis M. H. Navier (1785–1836) and George Stokes (1819–1903); with these three differential equations and the differential continuity equation we have four equations and four unknowns, u, v, w, and p. The viscosity and density are fluid properties that are assumed to be known. With the appropriate boundary and initial conditions the equations can hopefully be solved. Several relatively simple geometries allow for analytical solutions; some of the solutions are presented in Chapter 5. Numerical solutions have also been determined for many flows of interest. Because the equations are nonlinear partial differential equations (the acceleration terms cause the equations to be nonlinear as observed in Eqs. 1.2.9), we cannot be assured that the solution we find will actually be realized in the laboratory; that is, the solutions are not unique. For example, a laminar flow and a turbulent flow may have the identical initial and boundary conditions, yet the two flows (the two solutions) are very different.

The Navier–Stokes equations have not been solved for any turbulent flow. All turbulent flows are unsteady and three-dimensional and hence the time-derivative terms must be retained. This requires an initial condition on all dependent variables; that is, u, v, w, and p must be known at all points in the flow field at $t = 0$. Such information would be extremely difficult, if not impossible, to obtain. To avoid this situation, time-averaged quantities are introduced for turbulent flows. This subject will be studied in a later chapter.

We can express the Navier–Stokes equations in vector form by multiplying Eqs. 3.3.14 by $\hat{\mathbf{i}}$, $\hat{\mathbf{j}}$, and $\hat{\mathbf{k}}$, respectively, and adding. We recognize that

$$\begin{aligned} \frac{Du}{Dt}\hat{\mathbf{i}} + \frac{Dv}{Dt}\hat{\mathbf{j}} + \frac{Dw}{Dt}\hat{\mathbf{k}} &= \frac{D\mathbf{V}}{Dt} \\ \frac{\partial p}{\partial x}\hat{\mathbf{i}} + \frac{\partial p}{\partial y}\hat{\mathbf{j}} + \frac{\partial p}{\partial z}\hat{\mathbf{k}} &= \nabla p \\ \nabla^2 u\hat{\mathbf{i}} + \nabla^2 v\hat{\mathbf{j}} + \nabla^2 w\hat{\mathbf{k}} &= \nabla^2 \mathbf{V} \end{aligned} \tag{3.3.15}$$

where we have introduced the Laplacian

$$\nabla^2 = \frac{\partial^2}{\partial x^2} + \frac{\partial^2}{\partial y^2} + \frac{\partial^2}{\partial z^2} \tag{3.3.16}$$

Combining the above, the Navier–Stokes equations (3.3.14) take the vector form

$$\rho\frac{D\mathbf{V}}{Dt} = -\nabla p + \rho\mathbf{g} + \mu\nabla^2\mathbf{V} \tag{3.3.17}$$

Using this vector form we can express the Navier–Stokes equations using other coordinate systems. The equations are listed for cylindrical and spherical coordinates in Table 3.1.

Example 3.7

Simplify the x-component Navier–Stokes equation for steady flow in a horizontal, rectangular channel assuming all streamlines parallel to the walls. Let the x-direction be in the direction of flow (Figure E3.7).

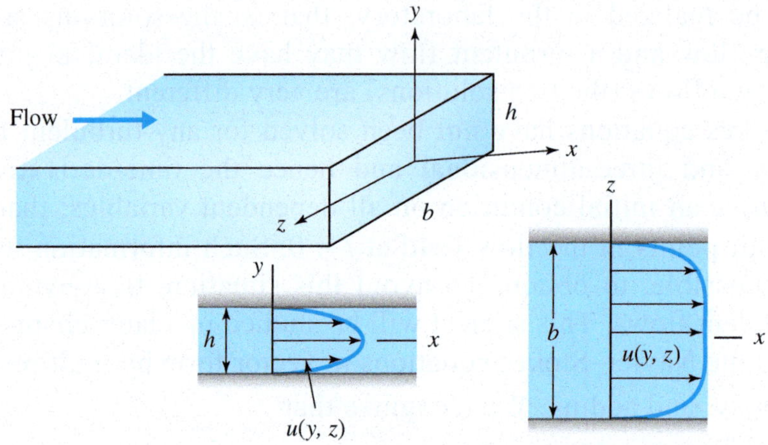

Figure E3.7

Solution

If the streamlines are parallel to the walls, only the x-component of velocity will be nonzero. Letting $v = w = 0$ the continuity equation (3.2.10) for an incompressible flow becomes

$$\frac{\partial u}{\partial x} = 0$$

showing that $u = u(y, z)$. The acceleration is then

$$\frac{Du}{Dt} = \overset{0}{\cancel{\frac{\partial u}{\partial t}}} + u\overset{0}{\cancel{\frac{\partial u}{\partial x}}} + \overset{0}{\cancel{v}}\frac{\partial u}{\partial y} + \overset{0}{\cancel{w}}\frac{\partial u}{\partial z} = 0$$

The x-component momentum equation (3.3.14) then simplifies to

$$0 = -\frac{\partial p}{\partial x} + \overset{0}{\cancel{\rho g_x}} + \mu\left(\overset{0}{\cancel{\frac{\partial^2 u}{\partial x^2}}} + \frac{\partial^2 u}{\partial y^2} + \frac{\partial^2 u}{\partial z^2}\right)$$

since $\partial^2 u/\partial x^2 = 0$ and $g_x = 0$ for a horizontal channel. We then have

$$\frac{\partial p}{\partial x} = \mu\left(\frac{\partial^2 u}{\partial y^2} + \frac{\partial^2 u}{\partial z^2}\right)$$

With the appropriate boundary conditions (the no-slip conditions), a solution to the foregoing equation could be sought. It would provide the velocity profiles sketched in Figure E3.7.

3.3.4 Vorticity Equations

There are certain fluid flow phenomena that cannot be explained or understood without reference to the vorticity equations, equations that are derived from the Navier–Stokes equations (Example 3.8 provides such a phenomenon). In addition to providing insight into such phenomena, the vorticity equations do not contain the pressure or gravity terms found in the Navier–Stokes equations but contain terms involving the velocity only. Since boundary conditions most often involve only the velocity, the vorticity equations are often the equations of choice for numerical solutions.

To derive the vorticity equations, we take the curl of Eq. 3.3.17, the vector form of the Navier–Stokes equations. This is a difficult task, so we will not display all the steps here but simply outline the process. First, let us define the vorticity of Eq. 1.2.18 in vector form using the del operator; using Eq. 3.2.6, we see that the three scalar equations of Eq. 1.2.18 can be written as the single vector equation

$$\boldsymbol{\omega} = \nabla \times \mathbf{V} \tag{3.3.18}$$

where $\nabla \times \mathbf{V}$ is the curl of the velocity. The **curl** is the cross-product of the del operator and a vector function. Second, let us write the acceleration in vector form as

$$\mathbf{a} = \frac{D\mathbf{V}}{Dt} = \frac{\partial \mathbf{V}}{\partial t} + (\mathbf{V} \cdot \nabla)\mathbf{V} \tag{3.3.19}$$

where we have used Eqs. 1.2.8, 1.2.10, and 3.2.7. Finally, let us take the curl of the Navier–Stokes vector equation (3.3.17):

$$\nabla \times \left[\rho \frac{\partial \mathbf{V}}{\partial t} + \rho (\mathbf{V} \cdot \nabla)\mathbf{V} \right] = -\cancelto{0}{\nabla \times \nabla p} + \cancelto{0}{\rho \nabla \times g} + \mu \nabla \times \nabla^2 \mathbf{V} \tag{3.3.20}$$

The curl of the gradient of a scalar function and the curl of a constant are zero. Also, since we can interchange differentiation, we can write

$$\begin{aligned} \nabla \times \frac{\partial \mathbf{V}}{\partial t} &= \frac{\partial}{\partial t} \nabla \times \mathbf{V} = \frac{\partial \boldsymbol{\omega}}{\partial t} \\ \nabla \times \nabla^2 \mathbf{V} &= \nabla^2 (\nabla \times \mathbf{V}) = \nabla^2 \boldsymbol{\omega} \end{aligned} \tag{3.3.21}$$

The difficult step, which we will leave as a homework problem, comes in showing that

$$\nabla \times [(\mathbf{V} \cdot \nabla)\mathbf{V}] = (\mathbf{V} \cdot \nabla)\boldsymbol{\omega} - (\boldsymbol{\omega} \cdot \nabla)\mathbf{V} \tag{3.3.22}$$

Equation 3.3.20 then becomes, assuming that ρ and μ are constants ($\nu = \rho/\mu$), the **vorticity equation**,

$$\frac{D\boldsymbol{\omega}}{Dt} = (\boldsymbol{\omega} \cdot \boldsymbol{\nabla})\mathbf{V} + \nu\nabla^2\boldsymbol{\omega} \tag{3.3.23}$$

The vorticity equation can be written as three scalar equations. Using rectangular coordinates, the three vorticity equations are

$$\begin{aligned} \frac{D\omega_x}{Dt} &= \omega_x \frac{\partial u}{\partial x} + \omega_y \frac{\partial u}{\partial y} + \omega_z \frac{\partial u}{\partial z} + \nu\nabla^2\omega_x \\ \frac{D\omega_y}{Dt} &= \omega_x \frac{\partial v}{\partial x} + \omega_y \frac{\partial v}{\partial y} + \omega_z \frac{\partial v}{\partial z} + \nu\nabla^2\omega_y \\ \frac{D\omega_z}{Dt} &= \omega_x \frac{\partial w}{\partial x} + \omega_y \frac{\partial w}{\partial y} + \omega_z \frac{\partial w}{\partial z} + \nu\nabla^2\omega_z \end{aligned} \tag{3.3.24}$$

Since the vorticity is the curl of the velocity, note that all of the terms in the vorticity equations involve only the velocity and its derivatives. Consequently, the vorticity equations often become the equations of choice when solving problems requiring the differential equations of motion. In fact, we refer to vortex lines and vortex tubes as being similar to streamlines and streamtubes. A **vortex line** is a line to which the vorticity vector is tangent. A **vortex tube**, or simply a **vortex**, is a tube whose walls contain vortex lines. A tornado is an example.

An interesting conclusion can be made by considering the vorticity equation (3.3.23). If an inviscid flow is everywhere irrotational (i.e., $\boldsymbol{\omega} = 0$ at all points in the flow), it must remain irrotational since $D\boldsymbol{\omega}/Dt = 0$. This is referred to as the **persistence of irrotationality**. Also, if a uniform flow approaches an object, vorticity (rotation of fluid particles) is introduced to the flow only by the action of viscosity. Without viscous effects, vorticity cannot be created in an oncoming irrotational flow.

Plane flows are often of particular interest. If $w = 0$, $u = u(x, y)$, and $v = v(x, y)$ the only nonzero vorticity component is ω_z. For such a flow, the vorticity equation takes the simplified form

$$\frac{D\omega_z}{Dt} = \nu\nabla^2\omega_z \tag{3.3.25}$$

We observe that viscous effects are needed to cause vorticity changes in a plane flow.

Example 3.8

In a snowstorm, the snow is actually scooped out in front of a tree, or post, as shown in Figure E3.8a. Explain this phenomenon by referring to the vorticity equations.

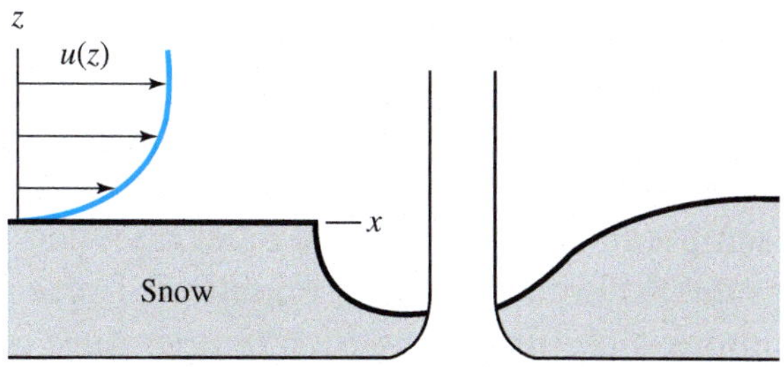

(a)

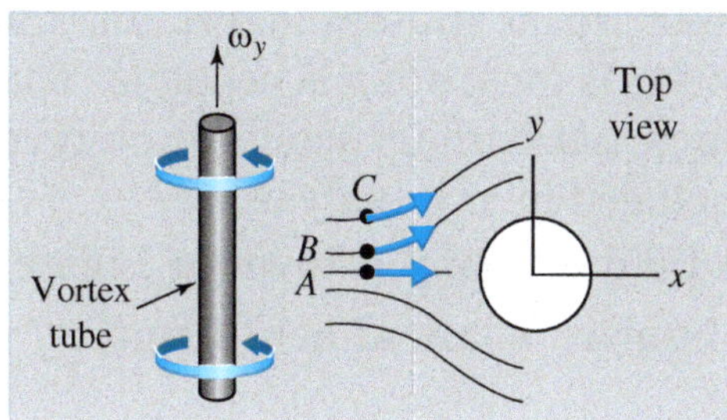

(b)

Figure E3.8

Solution

Let the velocity approaching the tree be in the x-direction with a velocity gradient $\partial u/\partial z$ near the ground. The vorticity components are then (refer to Eqs. 1.2.18)

$$\omega_x = 0 \qquad \omega_y = \frac{\partial u}{\partial z} \qquad \omega_z = 0$$

The y-component vorticity equation (3.3.24), ignoring viscous effects over the short flow length, reduces to

$$\frac{D\omega_y}{Dt} = \omega_y \frac{\partial v}{\partial y}$$

Observe from Figure E3.8b that in the vicinity of the tree $\partial v/\partial y$ is positive since $v_C > v_B > v_A$. ($\partial v/\partial y$ can be shown to be positive for negative y also.) Since ω_y and $\partial v/\partial y$ are both positive, $D\omega_y/Dt$ is positive and ω_y increases as the vortex tubes approach the tree. This increased vorticity creates a stronger vortex in front of the tree resulting in the snow being scooped out, as shown. This same phenomenon occurs in a sandstorm or in a water flow around a post in a riverbed. Without the use of the vorticity equation, it would be very difficult to explain.

3.4 DIFFERENTIAL ENERGY EQUATION

Most problems of interest in fluid mechanics do not involve temperature gradients; they involve flows in which the temperature is assumed to be everywhere constant. For such flows it is not necessary to introduce the differential energy equation. There are situations, however, for both compressible and incompressible flows in which temperature gradients are important, and for such flows the differential energy equation may be needed. We will derive the differential energy equation assuming negligible viscous effects, an assumption that significantly simplifies the derivation. Since the shear stresses that result from viscosity are quite small for many applications, this assumption is most often acceptable. These shear stresses do, however, account for the high temperatures that burn up satellites on reentry to the atmosphere; if they are significant, they must be included in any analysis.

Consider the infinitesimal fluid element[7], shown in Figure 3.4. The heat transfer rate $\dot{Q}$ through an area A is given by **Fourier's law of heat transfer**, named after Jean B. J. Fourier (1768–1830):

$$\dot{Q} = -KA\frac{\partial T}{\partial n} \tag{3.4.1}$$

where n is the direction normal to the area, T is the temperature, and K is the **thermal conductivity**, assumed constant. The work rate done by a force is the magnitude of the force multiplied by the velocity in the direction of the force; for a pressure force pA it is,

$$\dot{W} = pAV \tag{3.4.2}$$

where V is the velocity in the direction of the pressure force. The first law of thermodynamics applied to a fluid particle is

$$\dot{Q} - \dot{W} = \frac{DE}{Dt} \tag{3.4.3}$$

[7]We have used the lower corner at (x, y, z) from which to measure the various quantities. To be more exact mathematically, we would let Δx, Δy, and Δz approach zero so it matters not if we measure from a corner or from the center of the element.

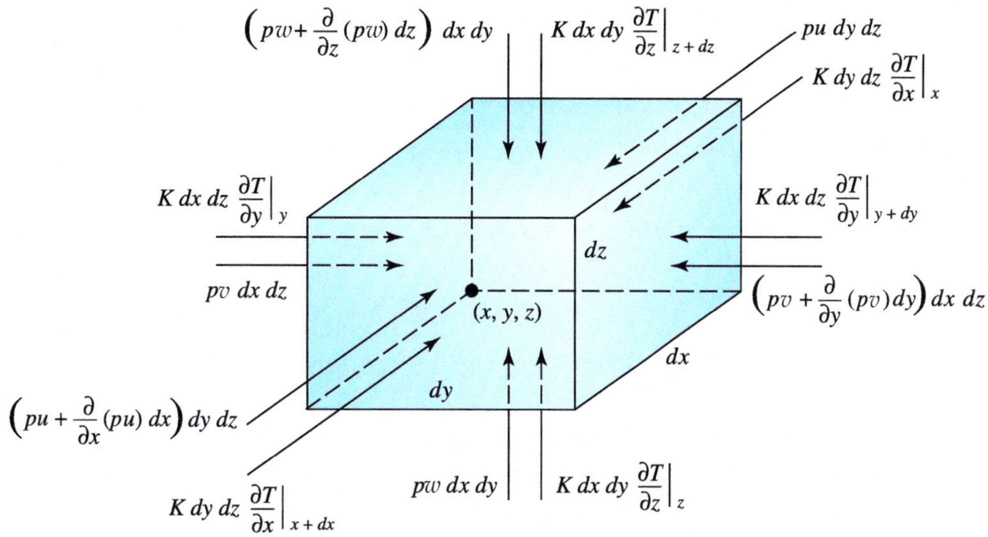

Figure 3.4 Rate of heat transfer and work rate on an infinitesimal fluid element. Shear stresses on the sides are assumed to be negligible.

where D/Dt is used since we are following a fluid particle at the instant shown. For the particle occupying the infinitesimal element of Figure 3.4, the relationships above allow us to write

$$
\begin{aligned}
&K\,dy\,dz\left(\frac{\partial T}{\partial x}\Big|_{x+dx} - \frac{\partial T}{\partial x}\Big|_{x}\right) - \frac{\partial}{\partial x}(pu)\,dx\,dy\,dz \\
&\quad + K\,dx\,dz\left(\frac{\partial T}{\partial x}\Big|_{y+dy} - \frac{\partial T}{\partial y}\Big|_{y}\right) - \frac{\partial}{\partial y}(pv)\,dx\,dy\,dz \\
&\quad + K\,dx\,dy\left(\frac{\partial T}{\partial z}\Big|_{z+dz} - \frac{\partial T}{\partial z}\Big|_{z}\right) - \frac{\partial}{\partial z}(pw)\,dx\,dy\,dz \\
&\quad = \rho\,dx\,dy\,dz\frac{D}{Dt}\left(\frac{u^2+v^2+w^2}{2} + gz + \tilde{u}\right)
\end{aligned}
\tag{3.4.4}
$$

where E has included kinetic, potential, and internal energy, and the z-axis is assumed vertical. Also, since the mass of a fluid particle is constant, $\rho dxdydz$ is outside the D/Dt-operator. Divide both sides by $dxdydz$. The result is[8]

$$
K\left(\frac{\partial^2 T}{\partial x^2} + \frac{\partial^2 T}{\partial y^2} + \frac{\partial^2 T}{\partial z^2}\right) - \frac{\partial}{\partial x}(pu) - \frac{\partial}{\partial y}(pv) - \frac{\partial}{\partial z}(pw) = \rho\frac{D}{Dt}\left(\frac{u^2+v^2+w^2}{2} + gz + \tilde{u}\right) \tag{3.4.5}
$$

[8]The second derivative is $(\partial T/\partial x|_{x+dx} - \partial T/\partial x|_{x})/dx = \partial^2 T/\partial x^2$.

This can be rearranged as follows:

$$K\left(\frac{\partial^2 T}{\partial x^2}+\frac{\partial^2 T}{\partial y^2}+\frac{\partial^2 T}{\partial z^2}\right)-p\left(\frac{\partial u}{\partial x}+\frac{\partial v}{\partial y}+\frac{\partial w}{\partial z}\right)-u\frac{\partial p}{\partial x}-v\frac{\partial p}{\partial y}-w\frac{\partial p}{\partial z}$$
$$=\rho u\frac{Du}{Dt}+\rho v\frac{Dv}{Dt}+\rho w\frac{Dw}{Dt}+\rho g\frac{Dz}{Dt}+\rho\frac{D\tilde{u}}{Dt} \tag{3.4.6}$$

Euler's scalar equations (3.3.7) are applicable for this inviscid flow; hence the last three terms on the left equal the first four terms on the right if we recognize that

$$\frac{Dz}{Dt}=\overset{0}{\cancel{\frac{\partial z}{\partial t}}}+u\overset{0}{\cancel{\frac{\partial z}{\partial x}}}+v\overset{0}{\cancel{\frac{\partial z}{\partial y}}}+w\frac{\partial z}{\partial z}=w \tag{3.4.7}$$

since x, y, z, t are all independent variables. The simplified energy equation then takes the form

$$\rho\frac{D\tilde{u}}{Dt}=K\left(\frac{\partial^2 T}{\partial x^2}+\frac{\partial^2 T}{\partial y^2}+\frac{\partial^2 T}{\partial z^2}\right)-p\left(\frac{\partial u}{\partial x}+\frac{\partial v}{\partial y}+\frac{\partial w}{\partial z}\right) \tag{3.4.8}$$

Introducing the $\boldsymbol{\nabla}$ operator, this is expressed as

$$\rho\frac{D\tilde{u}}{Dt}=K\nabla^2 T-p\boldsymbol{\nabla}\cdot\mathbf{V} \tag{3.4.9}$$

Before we simplify this equation for incompressible gas flow, let us write it in terms of enthalpy rather than internal energy. Using

$$\tilde{u}=h-\frac{p}{\rho} \tag{3.4.10}$$

the energy equation becomes, using Eq. 3.2.8,

$$\rho\frac{Dh}{Dt}=K\nabla^2 T+\frac{Dp}{Dt} \tag{3.4.11}$$

See Example 3.4 for the details of this conversion.

We have two special cases to be considered. First, for a liquid flow we can use $\nabla \cdot \mathbf{V} = 0$ and with $\tilde{u} = c_p T$, where c_p is the specific heat;[9] Eq. 3.4.9 simplifies to

$$\frac{DT}{Dt} = \alpha \nabla^2 T \tag{3.4.12}$$

where we have introduced the **thermal diffusivity** α defined by

$$\alpha = \frac{K}{\rho c_p} \tag{3.4.13}$$

For an incompressible gas flow, an interesting result occurs. In Example 3.10 we will show that

$$\left|\frac{Dp}{Dt}\right| \ll | p \nabla \cdot \mathbf{V} | \tag{3.4.14}$$

Thus, when comparing Eqs. 3.4.9 and 3.4.11, it is Eq. 3.4.11 that simplifies to

$$\rho c_p \frac{DT}{Dt} = K \nabla^2 T \tag{3.4.15}$$

for an incompressible gas flow if we make the ideal-gas assumption that

$$dh = c_p dT \tag{3.4.16}$$

If viscous effects are not negligible, the derivation would include the work input due to the shear stress components. This would add a term to the right-hand side of the differential energy equations (3.4.9) above; this term is called the **dissipation function** Φ, which, in rectangular coordinates, is

$$\Phi = 2\mu \left[\left(\frac{\partial u}{\partial x}\right)^2 + \left(\frac{\partial v}{\partial y}\right)^2 + \left(\frac{\partial w}{\partial z}\right)^2 + \frac{1}{2}\left(\frac{\partial u}{\partial y} + \frac{\partial v}{\partial x}\right)^2 + \frac{1}{2}\left(\frac{\partial v}{\partial z} + \frac{\partial w}{\partial y}\right)^2 + \frac{1}{2}\left(\frac{\partial u}{\partial z} + \frac{\partial w}{\partial x}\right)^2 \right] \tag{3.4.17}$$

[9]The specific heat for a liquid is often listed in tables as c_p. The specific heat at constant volume for a liquid is approximately equal to c_p. Hence we often simply drop the subscript and let $c_p = c$ for a liquid. It is assumed to be a constant, but it does depend slightly on temperature. Here we will use c_p.

With the addition of the energy equation, problems involving temperature variations in a flow can now be considered. Such problems are present in compressible flows in which pressure and density are related to temperature by an equation of state. In incompressible gas flows (Mach number < 0.3) and liquid flows, temperature variations are often negligible so that the differential energy equation is not of interest. If, however, a temperature field does exist in a liquid flow or an incompressible gas flow (heat exchangers, atmospheric flows, lake inversions, lubrication flows, free convection flows), the energy equation provides an additional equation relating the quantities of interest. For liquid flows involving temperature gradients it is often necessary to assume that $\mu = \mu T)$; in free convection flows we must assume that $\rho = \rho(T)$. In incompressible gas flows we can usually assume viscosity to be constant since the temperature variation is quite small.

Example 3.9

A constant-density liquid flows into a wide, rectangular horizontal channel, the walls of which are maintained at a higher temperature than the liquid, as shown in Figure E3.9. Assume a variable μ, include viscous dissipation, and write the three describing differential equations for a steady flow.

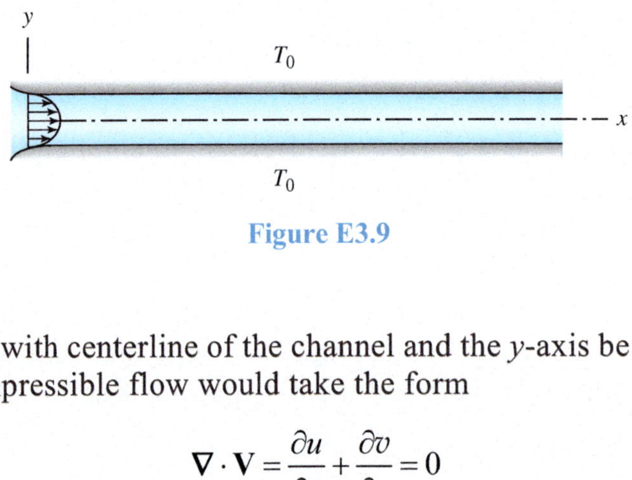

Figure E3.9

Solution

Let the x-axis coincide with centerline of the channel and the y-axis be vertical. The continuity equation for this incompressible flow would take the form

$$\nabla \cdot \mathbf{V} = \frac{\partial u}{\partial x} + \frac{\partial v}{\partial y} = 0$$

since $w = 0$ for the wide channel.

The flow will be primarily in the x-direction, but we must allow for variation of the y-component v. There will be no variation in the z-direction. The accelerations for this steady flow will be

$$\frac{Du}{Dt} = u\frac{\partial u}{\partial x} + v\frac{\partial u}{\partial y}$$

$$\frac{Dv}{Dt} = u\frac{\partial v}{\partial x} + v\frac{\partial v}{\partial y}$$

The stress terms contained in Eqs. 3.3.5 using Eqs. 3.3.10 with $\nabla \cdot \mathbf{V} = 0$, assuming a variable μ, become

$$\frac{\partial \sigma_{xx}}{\partial x} + \frac{\partial \tau_{xy}}{\partial y} = -\frac{\partial p}{\partial x} + \mu\left(\frac{\partial^2 u}{\partial x^2} + \frac{\partial^2 u}{\partial y^2}\right) + 2\frac{\partial \mu}{\partial x}\frac{\partial u}{\partial x} + \frac{\partial \mu}{\partial y}\left(\frac{\partial u}{\partial y} + \frac{\partial v}{\partial x}\right)$$
$$\frac{\partial \tau_{xy}}{\partial x} + \frac{\partial \sigma_{yy}}{\partial y} = -\frac{\partial p}{\partial y} + \mu\left(\frac{\partial^2 v}{\partial x^2} + \frac{\partial^2 v}{\partial y^2}\right) + 2\frac{\partial \mu}{\partial y}\frac{\partial v}{\partial y} + \frac{\partial \mu}{\partial x}\left(\frac{\partial u}{\partial y} + \frac{\partial v}{\partial x}\right)$$

The differential momentum equations (3.3.5) are then

$$\rho\left(u\frac{\partial u}{\partial x} + v\frac{\partial u}{\partial y}\right) = -\frac{\partial p}{\partial x} + \mu\left(\frac{\partial^2 u}{\partial x^2} + \frac{\partial^2 u}{\partial y^2}\right) + 2\frac{\partial \mu}{\partial x}\frac{\partial u}{\partial x} + \frac{\partial \mu}{\partial y}\left(\frac{\partial u}{\partial y} + \frac{\partial v}{\partial x}\right)$$
$$\rho\left(u\frac{\partial v}{\partial x} + v\frac{\partial v}{\partial y}\right) = -\frac{\partial p}{\partial y} + \mu\left(\frac{\partial^2 v}{\partial x^2} + \frac{\partial^2 v}{\partial y^2}\right) + 2\frac{\partial \mu}{\partial y}\frac{\partial u}{\partial y} + \frac{\partial \mu}{\partial x}\left(\frac{\partial u}{\partial y} + \frac{\partial v}{\partial x}\right)$$

The energy equation simplifies to

$$u\frac{\partial T}{\partial x} + v\frac{\partial T}{\partial y} = \alpha\left(\frac{\partial^2 T}{\partial x^2} + \frac{\partial^2 T}{\partial y^2}\right) + \frac{2\mu}{c_p}\left[\left(\frac{\partial u}{\partial x}\right)^2 + \left(\frac{\partial v}{\partial y}\right)^2 + \frac{1}{2}\left(\frac{\partial u}{\partial y} + \frac{\partial v}{\partial x}\right)^2\right]$$

where we have assumed K to be constant. The nonlinear, partial differential equations above, although quite formidable when attempting an analytic solution, could be solved numerically with the appropriate boundary conditions, and for a sufficiently low flow rate so that laminar flow exists (a turbulent flow is always unsteady and three-dimensional).

Example 3.10

Show that for an ideal gas, $Dp/Dt \,|\ll|\, p\nabla \cdot \mathbf{V}\,|$ in a low-speed flow, thereby concluding that Eq. 3.4.15 is the appropriate equation.

Solution

Let us consider a steady, uniform flow in a pipe so that $|\mathrm{V}| = u$ and $Dp/Dt = u\partial p/\partial x$. Then the problem can be stated as follows: Show that

$$\left|u\frac{\partial p}{\partial x}\right| \ll \left|p\frac{\partial u}{\partial x}\right|$$

Viscous effects are small and would not change the conclusion, so we can ignore any possible viscous effects. Then Euler's equation (3.3.7) allows us to use

$$\frac{\partial p}{\partial x} = -\rho u \frac{\partial u}{\partial x}$$

Using the speed of sound equation $c = \sqrt{kRT}$ and the equation of state, we see that

$$c = \sqrt{\frac{kp}{\rho}} \quad \text{or} \quad p = c^2 \frac{\rho}{k}$$

Thus,

$$p \frac{\partial u}{\partial x} = \frac{c^2}{k} \rho \frac{\partial u}{\partial x}$$

Our problem can now be stated: Show that

$$\left| \rho u^2 \frac{\partial u}{\partial x} \right| \ll \left| \frac{c^2}{k} \rho \frac{\partial u}{\partial x} \right|$$

Or, more simply, is it true that

$$u^2 \ll \frac{c^2}{k}?$$

This can be seen to be true since we have assumed for a low-speed gas flow that the speed of the gas is much less than the speed of sound (e.g., $u < 0.3c$ or $M < 0.3$). We know that k is of order unity ($k = 1.4$ for air), so it will not affect our conclusion that

$$\left| \frac{Dp}{Dt} \right| \ll \left| p \nabla \cdot \mathbf{V} \right|$$

3.5 SUMMARY

We have now completed our derivation of the partial differential equations that are used in describing flows of interest. Let's summarize the equations in vector form for an incompressible flow:

Continuity: $$\nabla \cdot \mathbf{V} = 0 \tag{3.5.1}$$

Momentum: $$\rho \frac{D\mathbf{V}}{Dt} = -\nabla p + \rho g + \mu \nabla^2 \mathbf{V} \tag{3.5.2}$$

Energy: $$\frac{DT}{Dt} = \alpha\nabla^2 T + \Phi/\rho c_p \qquad \text{Liquids} \tag{3.5.3}$$

Energy: $$\rho c_p \frac{DT}{Dt} = K\nabla^2 T + \Phi \qquad \text{Incompressible gases} \tag{3.5.4}$$

To express these equations in the forms above, we have assumed:

- A Newtonian fluid (a linear relationship between the stress components and the velocity gradients).
- An isotropic fluid (the fluid properties are independent of direction).
- A homogeneous fluid (the fluid properties μ, c_p, and K do not depend on position).
- An incompressible flow (the density of a particle is constant, that is, $D\rho/Dt = 0$; we do not demand that ρ = constant. For a gas flow we require that $\mathrm{M} < 0.3$).
- An inertial reference frame.

The vorticity equation is also of interest. It is

Vorticity: $$\frac{D\boldsymbol{\omega}}{Dt} = (\boldsymbol{\omega}\cdot\boldsymbol{\nabla})\mathbf{V} + \nu\nabla^2\boldsymbol{\omega} \tag{3.5.5}$$

Numerical methods often make use of this vorticity equation.

In the derivations of the differential equations in this chapter we have made no mention of laminar or turbulent flow. The equations are applicable to either class of flow. Some laminar flows in relatively simple geometries have been solved analytically, and many others have been solved numerically. Turbulent flows, however, have not been solved even for the simplest geometry. A turbulent flow is always an unsteady flow, and the presence of the time-derivative terms demands initial conditions; that is, at time $t = 0$ we must specify u, v, w and p at all points in the region of interest, information that is difficult, if not impossible, to obtain even in a simple pipe flow. Turbulent flow will be considered in later chapters.

Table 3.1 Fundamental Laws for Incompressible Flows

Continuity

Rectangular

$$\frac{\partial u}{\partial x}+\frac{\partial v}{\partial y}+\frac{\partial w}{\partial z}=0$$

Cylindrical

$$\frac{1}{r}\frac{\partial}{\partial r}(rv_r)+\frac{1}{r}\frac{\partial v_\theta}{\partial \theta}+\frac{\partial v_z}{\partial z}=0$$

Spherical

$$\frac{1}{r^2}\frac{\partial}{\partial r}(r^2 v_r)+\frac{1}{r\sin\theta}\frac{\partial}{\partial\theta}(v_\theta\sin\theta)+\frac{1}{r\sin\theta}\frac{\partial v_\phi}{\partial\phi}=0$$

Momentum

Rectangular

$$\frac{Du}{Dt}=-\frac{1}{\rho}\frac{\partial p}{\partial x}+g_x+\nu\nabla^2 u$$

$$\frac{Dv}{Dt}=-\frac{1}{\rho}\frac{\partial p}{\partial y}+g_y+\nu\nabla^2 v$$

$$\frac{Dw}{Dt}=-\frac{1}{\rho}\frac{\partial p}{\partial z}+g_z+\nu\nabla^2 w$$

$$\frac{D}{Dt}=\frac{\partial}{\partial t}+u\frac{\partial}{\partial x}+v\frac{\partial}{\partial y}+w\frac{\partial}{\partial z}$$

$$\nabla^2=\frac{\partial^2}{\partial x^2}+\frac{\partial^2}{\partial y^2}+\frac{\partial^2}{\partial z^2}$$

Spherical

$$\frac{Dv_r}{Dt}-\frac{v_\theta^2+v_\phi^2}{r}=-\frac{1}{\rho}\frac{\partial p}{\partial r}+g_r+\nu\left(\nabla^2 v_r-\frac{2v_r}{r^2}-\frac{2}{r^2}\frac{\partial v_\theta}{\partial\theta}-\frac{2v_\theta\cot\theta}{r^2}-\frac{2}{r^2\sin\theta}\frac{\partial v_\phi}{\partial\phi}\right)$$

$$\frac{Dv_\theta}{Dt}+\frac{v_r v_\theta-v_\phi^2\cot\theta}{r}=-\frac{1}{\rho r}\frac{\partial p}{\partial\theta}+g_\theta+\nu\left(\nabla^2 v_\theta+\frac{2}{r^2}\frac{\partial v_r}{\partial\theta}-\frac{v_\theta}{r^2\sin^2\theta}-\frac{2\cos\theta}{r^2\sin^2\theta}\frac{\partial v_\phi}{\partial\phi}\right)$$

$$\frac{Dv_\phi}{Dt}+\frac{v_\phi v_r}{r}+\frac{v_\theta v_\phi\cot\theta}{r}=-\frac{1}{\rho r\sin\theta}\frac{\partial p}{\partial\phi}+g_\phi+\nu\left(\nabla^2 v_\phi-\frac{v_\phi}{r^2\sin^2\theta}+\frac{2}{r^2\sin^2\theta}\frac{\partial v_r}{\partial\phi}+\frac{2\cos\theta}{r^2\sin^2\theta}\frac{\partial v_\theta}{\partial\phi}\right)$$

$$\frac{D}{Dt}=\frac{\partial}{\partial t}+v_r\frac{\partial}{\partial r}+\frac{v_\theta}{r}\frac{\partial}{\partial\theta}+\frac{v_\phi}{r\sin\theta}\frac{\partial}{\partial\phi}$$

$$\nabla^2=\frac{1}{r^2}\frac{\partial}{\partial r}\left(r^2\frac{\partial}{\partial r}\right)+\frac{1}{r^2\sin\theta}\frac{\partial}{\partial\theta}\left(\sin\theta\frac{\partial}{\partial\theta}\right)+\frac{1}{r^2\sin^2\theta}\frac{\partial^2}{\partial\phi^2}$$

Spherical

$$\rho\frac{Dh}{Dt}=K\nabla^2 T+2\mu\left[\left(\frac{\partial v_r}{\partial r}\right)^2+\left(\frac{1}{r}\frac{\partial v_\theta}{\partial\theta}+\frac{v_r}{r}\right)^2+\left(\frac{1}{r\sin\theta}\frac{\partial v_\phi}{\partial\phi}+\frac{v_r}{r}+\frac{v_\theta\cot\theta}{r}\right)^2\right]$$
$$+\mu\left[\left(\frac{1}{r\sin\theta}\frac{\partial v_\theta}{\partial\phi}+\frac{\sin\theta}{r}\frac{\partial}{\partial\theta}\left(\frac{v_\theta}{\sin\theta}\right)\right)^2+\left(\frac{1}{r\sin\theta}\frac{\partial v_r}{\partial\phi}+r\frac{\partial}{\partial r}\left(\frac{v_\phi}{r}\right)\right)^2+\left(r\frac{\partial}{\partial r}\left(\frac{v_\theta}{r}\right)+\frac{1}{r}\frac{\partial v_r}{\partial\theta}\right)^2\right]$$

Stresses

Rectangular

$$\sigma_{xx}=-p+2\mu\frac{\partial u}{\partial x}\qquad \tau_{xy}=\mu\left(\frac{\partial u}{\partial y}+\frac{\partial v}{\partial x}\right)$$

$$\sigma_{yy}=-p+2\mu\frac{\partial v}{\partial y}\qquad \tau_{yz}=\mu\left(\frac{\partial v}{\partial z}+\frac{\partial w}{\partial y}\right)$$

$$\sigma_{zz}=-p+2\mu\frac{\partial w}{\partial z}\qquad \tau_{xz}=\mu\left(\frac{\partial u}{\partial z}+\frac{\partial w}{\partial x}\right)$$

Cylindrical

$$\frac{Dv_r}{Dt}-\frac{v_\theta^2}{r}=-\frac{1}{\rho}\frac{\partial p}{\partial r}+g_r+\nu\left(\nabla^2 v_r-\frac{v_r}{r^2}-\frac{2}{r^2}\frac{\partial v_\theta}{\partial \theta}\right)$$

$$\frac{Dv_\theta}{Dt}+\frac{v_r v_\theta}{r}=-\frac{1}{\rho r}\frac{\partial p}{\partial \theta}+g_\theta+\nu\left(\nabla^2 v_\theta+\frac{2}{r^2}\frac{\partial v_r}{\partial \theta}-\frac{v_\theta}{r^2}\right)$$

$$\frac{Dv_z}{Dt}=-\frac{1}{\rho}\frac{\partial p}{\partial z}+g_z+\nu\nabla^2 v_z$$

$$\frac{D}{Dt}=\frac{\partial}{\partial t}+v_r\frac{\partial}{\partial r}+\frac{v_\theta}{r}\frac{\partial}{\partial \theta}+v_z\frac{\partial}{\partial z}$$

$$\nabla^2=\frac{\partial^2}{\partial r^2}+\frac{1}{r}\frac{\partial}{\partial r}+\frac{1}{r^2}\frac{\partial^2}{\partial \theta^2}+\frac{\partial^2}{\partial z^2}$$

Energy

Rectangular

$$\rho\frac{Dh}{Dt}=K\nabla^2 T+2\mu\left[\left(\frac{\partial u}{\partial x}\right)^2+\left(\frac{\partial v}{\partial y}\right)^2+\left(\frac{\partial w}{\partial z}\right)^2 +\frac{1}{2}\left(\frac{\partial u}{\partial y}+\frac{\partial v}{\partial x}\right)^2+\frac{1}{2}\left(\frac{\partial v}{\partial z}+\frac{\partial w}{\partial y}\right)^2 +\frac{1}{2}\left(\frac{\partial u}{\partial z}+\frac{\partial w}{\partial x}\right)^2\right]$$

Cylindrical

$$\rho\frac{Dh}{Dt}=K\nabla^2 T+2\mu\left[\left(\frac{\partial v_r}{\partial r}\right)^2+\left(\frac{1}{r}\frac{\partial v_\theta}{\partial \theta}+\frac{v_r}{r}\right)^2+\left(\frac{\partial v_z}{\partial z}\right)^2 +\frac{1}{2}\left(\frac{1}{r}\frac{\partial v_z}{\partial \theta}+\frac{\partial v_\theta}{\partial z}\right)^2+\frac{1}{2}\left(\frac{\partial v_r}{\partial z}+\frac{\partial v_z}{\partial r}\right)^2 +\frac{1}{2}\left(\frac{1}{r}\frac{\partial v_r}{\partial \theta}+\frac{\partial v_\theta}{\partial r}-\frac{v_\theta}{r}\right)^2\right]$$

Cylindrical

$$\sigma_{rr}=-p+2\mu\frac{\partial v_r}{\partial r}\qquad \tau_{r\theta}=\mu\left[r\frac{\partial}{\partial r}\left(\frac{v_\theta}{r}\right)+\frac{1}{r}\frac{\partial v_r}{\partial \theta}\right]$$

$$\sigma_{\theta\theta}=-p+2\mu\left(\frac{1}{r}\frac{\partial v_\theta}{\partial \theta}+\frac{v_r}{r}\right)\qquad \tau_{\theta z}=\mu\left[\frac{\partial v_\theta}{\partial z}+\frac{1}{r}\frac{\partial v_z}{\partial \theta}\right]$$

$$\sigma_{zz}=-p+2\mu\frac{\partial v_z}{\partial z}\qquad \tau_{rz}=\mu\left[\frac{\partial v_r}{\partial z}+\frac{\partial v_z}{\partial r}\right]$$

Spherical

$$\sigma_{rr}=-p+2\mu\frac{\partial v_r}{\partial r}$$

$$\sigma_{\theta\theta}=-p+2\mu\left(\frac{1}{r}\frac{\partial v_\theta}{\partial \theta}+\frac{v_r}{r}\right)$$

$$\sigma_{\phi\phi}=-p+2\mu\left(\frac{1}{r\sin\theta}\frac{\partial v_\phi}{\partial \phi}+\frac{v_r}{r}+\frac{v_\theta\cot\theta}{r}\right)$$

$$\tau_{r\theta}=\mu\left[r\frac{\partial}{\partial r}\left(\frac{v_\theta}{r}\right)+\frac{1}{r}\frac{\partial v_r}{\partial \theta}\right]$$

$$\tau_{\theta\phi}=\mu\left[\frac{\sin\theta}{r}\frac{\partial}{\partial \theta}\left(\frac{v_\phi}{\sin\theta}\right)+\frac{1}{r\sin\theta}\frac{\partial v_\theta}{\partial \phi}\right]$$

$$\tau_{r\phi}=\mu\left[\frac{1}{r\sin\theta}\frac{\partial v_r}{\partial \phi}+r\frac{\partial}{\partial r}\left(\frac{v_\phi}{r}\right)\right]$$

PROBLEMS

Differential Continuity Equation

3.1 The divergence theorem (also knows as Gauss's theorem) states that

$$\int_A \mathbf{V}\cdot\hat{\mathbf{n}}\, dA = \int_{\forall} \nabla\cdot\mathbf{V}\, d\forall$$

where **V** represents any vector and A completely surrounds the volume $\forall$. Apply theorem to the integral continuity equation (2.4.3) and derive the differential continuity equation (3.2.6).

3.2 Use the infinitesimal element shown in Figure P3.2 and derive the differential continuity equation in cylindrical coordinates. The velocity vector is $\mathbf{V} = (v_r, v_\theta, v_z)$.

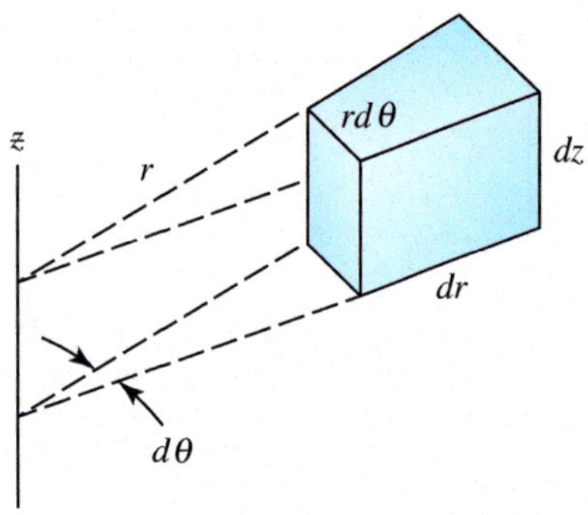

Figure P3.2

3.3 Use the infinitesimal elements shown in Figure P3.3 and derive the differential continuity equation in spherical coordinates. The velocity vector is $\mathbf{V} = (v_r, v_\theta, v_\phi)$.

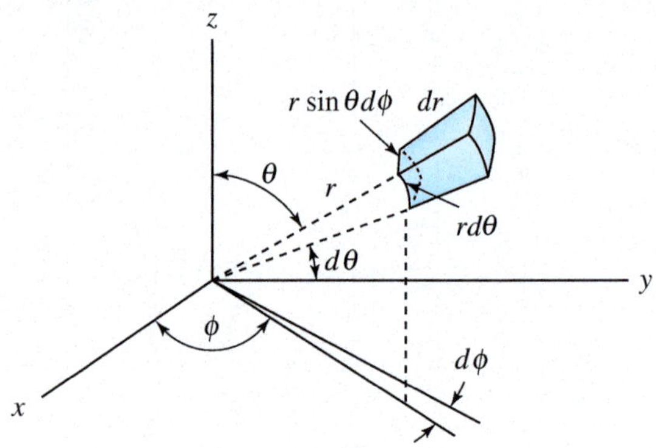

Figure P3.3

3.4 A uniform compressible flow occurs in a constant-diameter pipe. Write the simplified differential continuity equation for the steady flow.

3.5 An incompressible flow of air over the mountain range shown in Figure P3.5 can be approximated by a plane, steady flow. If the z-axis is vertical and density is allowed to vary, write the differential equations that result from conservation of mass considerations.

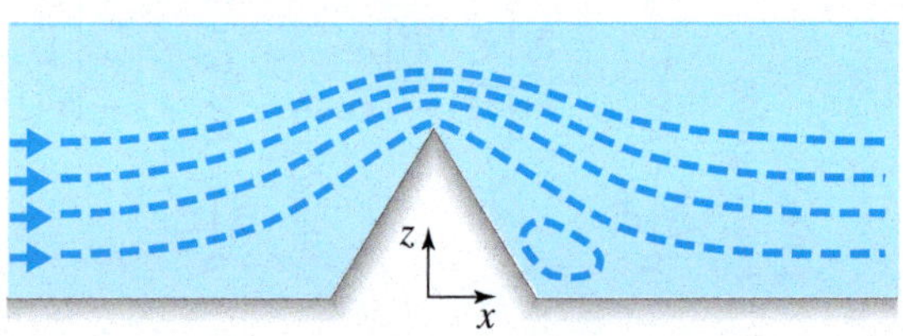

Figure P3.5

3.6 A stratified flow of salt water in which the density increases with depth occurs over an obstruction in the bottom of the channel of Figure P3.6. Assuming a plane steady flow with the z-axis vertical, write the equations that result from the differential continuity equation.

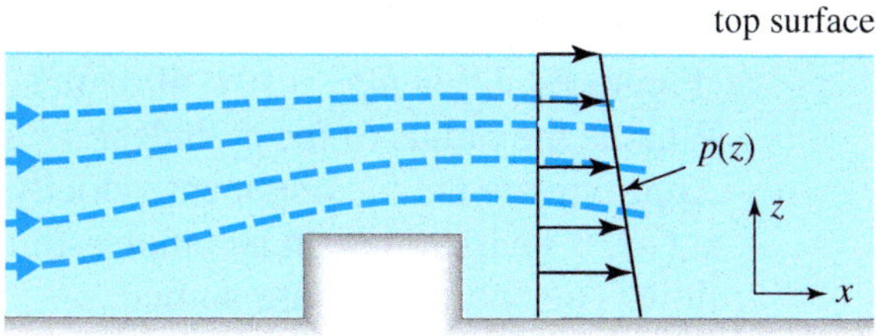

Figure P3.6

3.7 Show that for an isothermal compressible flow,

$$\frac{1}{p}\frac{Dp}{Dt} = -\nabla \cdot \mathbf{V}$$

3.8 An incompressible fluid is flowing radially into a sink (treated as a line or a point at the origin). Determine an expression for the radial velocity component if it is:
(a) A line sink **(b)** A point sink

3.9 Consider steady, inviscid, incompressible uniform $[u = u(x)]$ flow in a two-dimensional diverging channel as shown in Figure P3.9. The inclined walls of the channel are straight, and the fluid enters the diverging section with velocity $V_1 = 40$ m/s. Given $H = 1$ m, and assume unit width.
(a) Determine an expression for the velocity component u as a function of position x along the channel. (u does not depend on y.)
(b) Determine an expression for the acceleration of the fluid in the x-direction.

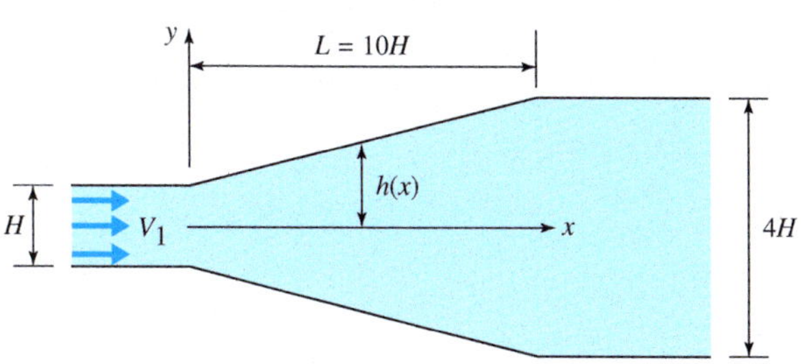

Figure P3.9

3.10 For the same conditions given in Problem 3.9, determine:
(a) An expression for the velocity component v.
(b) An expression for the acceleration in the y-direction.

3.11 A compressible flow occurs such that

$$u = 200\,xy \quad v = 200(x^2 + y^2) \quad w = 0 \text{ m/s}$$

Find the rate at which the density is changing at the point (2 m, 1 m) where $\rho = 2.3$ kg/m^3.

3.12 If, in an incompressible plane flow, the velocity component u = const, what can we say about the y-component of velocity? About the density?

3.13 In an incompressible flow we know that u and v are both nonzero but constant in magnitude. What can we infer about w from the differential continuity equation? About the density?

3.14 In an incompressible plane flow $u = Ax$. Find $v(x, y)$ if $v(x, 0) = 0$.

3.15 If the velocity component u is given by

$$u(x, y) = 10 + \frac{5x}{x^2 + y^2}$$

in an incompressible plane flow, determine $v(x, y)$. Let $v(x, 0) = 0$.

3.16 The θ-component of velocity is given by

$$v_\theta = -\left(10 + \frac{0.4}{r^2}\right)\cos\theta$$

Find the r-component of velocity for the incompressible plane flow if $v_r\,(0.2, \theta) = 0$.

3.17 In an incompressible plane flow

$$v_\theta = 20\left(1 + \frac{1}{r^2}\right)\sin\theta - \frac{40}{r}$$

Find $v_r\,(r, \theta)$ if $v_r\,(1, \theta) = 0$.

3.18 In an incompressible axisymmetric flow ($v_\phi = 0$) the velocity component v_θ is given by

$$v_\theta = -\left(10 + \frac{40}{r^3}\right)\sin\theta$$

Find $v_r(r, \theta)$ if $v_r(2, \theta) = 0$.

3.19 The velocity of air in a pipe is measured at points 2 in. apart to be 453, 486, and 526 ft/s, respectively. At the middle point the temperature is 40°F and the pressure is 18 psia. Find $d\rho/dx$ at the middle point of this steady, uniform flow.

3.20 The x-component velocity on the x-axis (Figure P3.20) is proposed to be $u(x) = -20\,(1 - e^{-x})$ m/s. Approximate the y-component velocity at the point (2, 0.2) in this plane incompressible flow. The coordinates are in meters.

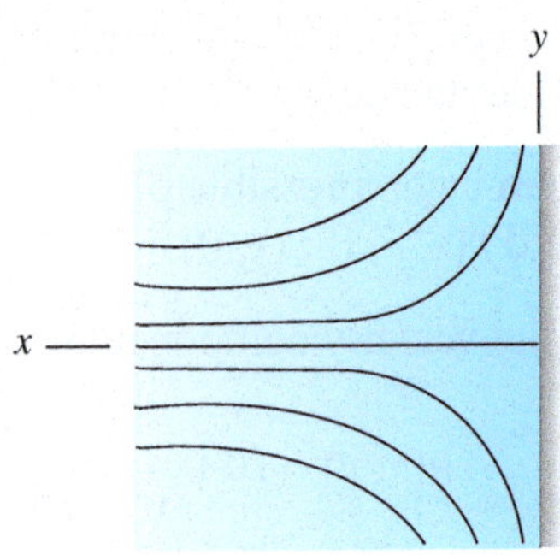

Figure P3.20

3.21 Assume the flow in Problem 3.20 is axisymmetric and replace y with r and x with z. Approximate the r-component velocity at (2, 0.2). The coordinates are in meters.

3.22 The velocity component along the x-axis (Figure P3.22) is $u(x) = 10 - 40/x^2$ m/s. What is the radius of the cylinder? Approximate the y-component velocity at (–3, 0.1) assuming an incompressible flow. The coordinates are in meters.

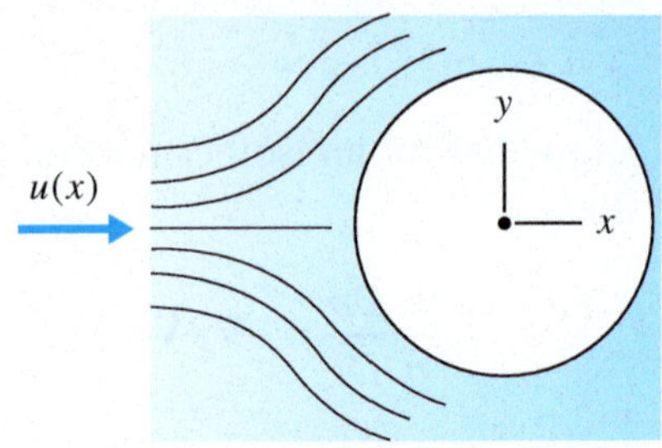

Figure P3.22

3.23 Assume the flow in Problem 3.22 represents flow around a sphere and let $v_r(r) = (40/r^2) - 10$ m/s along the negative x-axis. What is the radius of the sphere? Approximate the θ-component of velocity at (–3, 0.1) assuming an incompressible flow. The coordinates are in meters.

3.24 The x-component of the velocity vector is measured, at points A, B, and C, 5 mm apart, as 11.3, 12.6, and 13.5 m/s, respectively, in the incompressible, steady plane flow shown in Figure P3.24. Estimate:

(a) The y-component of velocity 4 mm above point B.

(b) The acceleration at point B.

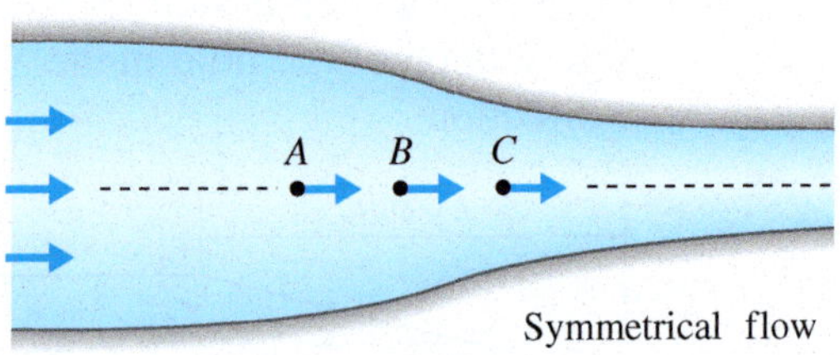

Figure P3.24

Differential Momentum Equation

3.25 Sum forces on the element of Figure 3.3 in the y-direction and show that Eq. 3.3.5b results.

3.26 Does the velocity field

$$u=\frac{10x}{x^2+y^2} \qquad v=\frac{10y}{x^2+y^2} \qquad w=0$$

represent a possible incompressible flow? If so, find the pressure gradient ∇p assuming a frictionless flow with negligible body forces.

3.27 Does the velocity field

$$v_r=10\left(1-\frac{1}{r^2}\right)\cos\theta$$

$$v_\theta=-10\left(1+\frac{1}{r^2}\right)\sin\theta$$

$$v_z=0$$

represent a possible incompressible flow? If so, find the pressure gradient Δp assuming a frictionless flow with negligible body forces.

3.28 Consider the velocity field

$$v_r=10\left(1-\frac{8}{r^3}\right)\cos\theta$$

$$v_\theta=-10\left(1+\frac{4}{r^3}\right)\sin\theta$$

$$v_\phi=0$$

Does it represent an incompressible flow? If so, find the pressure gradient ∇p assuming a frictionless flow, neglecting the body force.

3.29 For a steady plane flow shown in Figure P3.29, find an expression for $D\mathbf{V}/Dt$ in terms of coordinates (s, n) tangential and normal to a streamline. Let R be the radius of curvature.

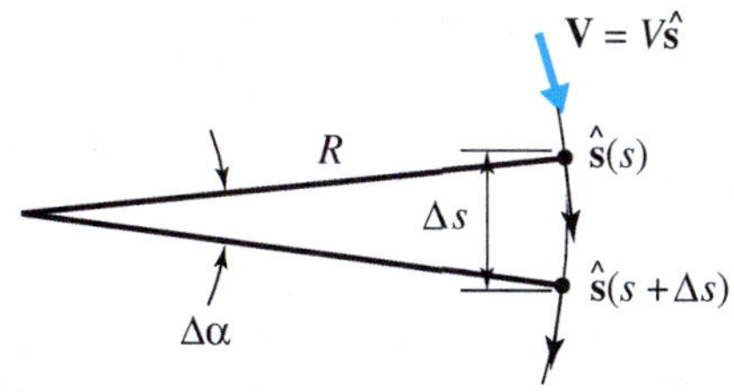

Figure P3.29

3.30 Write Euler's equation if the velocity is referred to a reference frame that is rotating with constant angular velocity.

3.31 A velocity field is given by $u = 30\,(y - 24y^2)$ ft/s, $v = 0$, and $w = 0$. Display the stress components at $y = 0.1$ in. using $\mu = 10^{-5}$ lb-s/ft^2 and $p = 30$ psi. Find the ratio τ_{xy}/σ_{xx}.

3.32 The velocity field near a surface is approximated by $u = 10\,(2y/\delta - y^2/\delta^2)$, where $\delta = Cx^{4/5}$. If $\delta = 8$ m at $x = 1000$ m, find $v(x, y)$ assuming that $w = 0$ and $v(x, 0) = 0$. Also, display the stress components at $(1000, 0)$ using $\mu = 2 \times 10^{-5}$ N · s/m^2 and $p = 100$ kPa. Assume an incompressible flow.

3.33 Show that for a steady flow Du/Dt can be written as $(\mathbf{V}\cdot\nabla)\,u$, and that $D\mathbf{V}/Dt = (\mathbf{V}\cdot\nabla)\mathbf{V}$. Verify using rectangular coordinates.

3.34 Write the compressible flow differential momentum equations (3.3.13) as one equation in vector form.

3.35 Simplify the Navier–Stokes equations for incompressible steady flow between horizontal parallel plates assuming that $u = u(y)$; $w = 0$. Write all three equations.

3.36 Consider two long, horizontal, parallel plates with a viscous incompressible fluid placed between them. The two plates move in two opposite directions with two different constant velocities, as shown in Figure P3.36. There is no pressure gradient and the only body force is due to the weight. Starting with the Navier–Stokes equations, determine an expression for the velocity profile for laminar flow between the two plates.

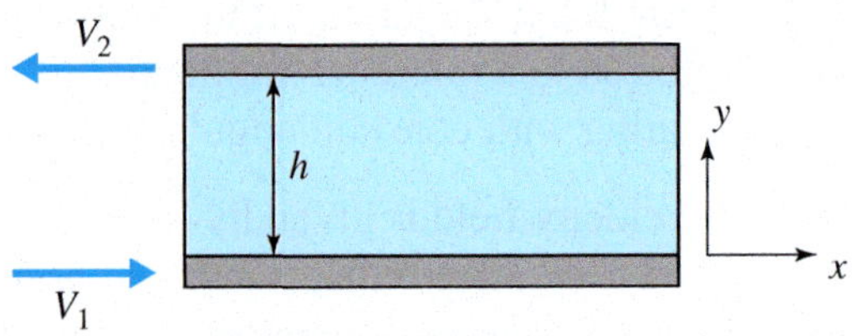

Figure P3.36

3.37 Consider steady, laminar, and incompressible fluid flow between two fixed vertical parallel plates. If the fluid is moving downward between the two plates, use the Navier–Stokes equations to determine an expression for the velocity profile.

3.38 Simplify the Navier–Stokes equations for incompressible steady flow in a horizontal pipe assuming that $v_z = v_z(r)$, $v_\theta = 0$. Write all three equations.

3.39 An incompressible fluid flows steadily between two infinitely long concentric cylinders with the outer cylinder fixed and the inner cylinder moving with a velocity V_c, as shown in Figure P3.39. For fully developed, laminar, axisymmetric flow between the two horizontal cylinders, determine an expression for the velocity V_c needed to produce zero drag on the inner cylinder.

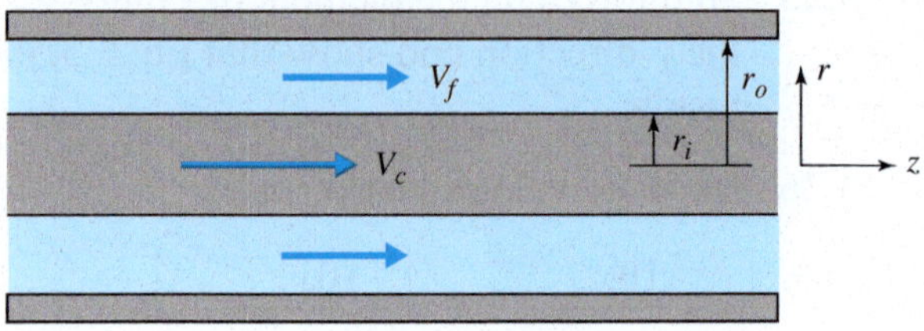

Figure P3.39

3.40 Fluid flows in the small gap between concentrically rotating spheres such that $v_\theta = v_\theta(r)$ and $v_\phi = 0$. Simplify the Navier–Stokes equations neglecting gravity for steady incompressible flow.

3.41 Consider viscous fluid between two infinitely long concentric vertical cylinders with the inner cylinder, of radius r_i, fixed. The outer cylinder has a radius r_0 and rotates with angular velocity ω. For steady, incompressible, axisymmetric laminar flow, determine an expression for the velocity profile using the Navier–Stokes equations.

3.42 Substitute the constitutive equations (3.3.10) for incompressible flow into the differential momentum equations (3.3.5) and derive the Navier–Stokes equations (3.3.14).

3.43 In Eqs. 3.3.14 the viscosity is assumed to be constant. If the temperature is not constant, as in a liquid flow with temperature gradients, we must let $\mu = \mu(T)$ so that $\mu = \mu(x, y, z)$ since $T = T(x, y, z)$. Modify Eqs. 3.3.14 to account for variable viscosity.

3.44 A large flat plate oscillates beneath a liquid, as shown in Figure P3.44. Write the differential equation that describes the motion if the plane laminar flow moves only parallel to the plate. Assume that $\mu = \text{const}$.

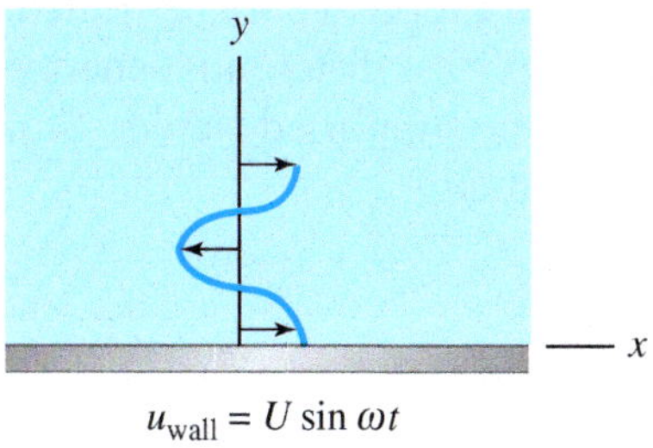

Figure P3.44

3.45 For a gas flow in which Stokes's hypothesis is not applicable, the negative average of the three normal stresses, denoted $\bar{p}$, may be different from the pressure p. Find an expression for $(\bar{p} - p)$.

Vorticity

3.46 Verify that the relationship (3.3.22) is indeed true by using rectangular coordinates.

3.47 A uniform flow moves over the flat plate of Figure P3.47 oriented parallel to the flow. The plate has a very sharp leading edge. Identify the term that is responsible for the creation of vorticity.

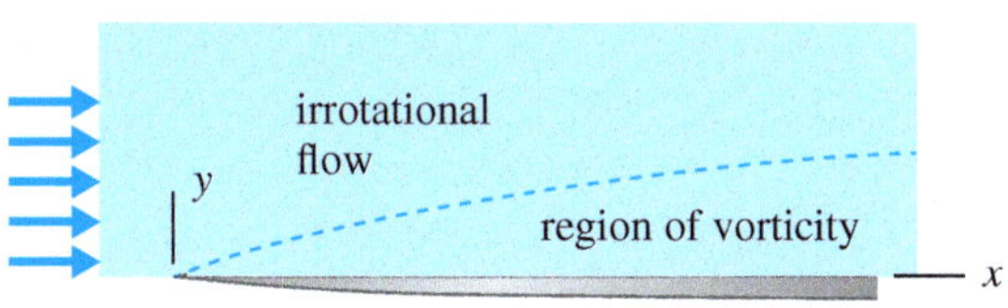

Figure P3.47

3.48 Neglect viscous effects and determine the velocity profile just downstream of the contraction at section 2 in the plane flow shown in Figure P3.48. *Hint:* In an inviscid flow, the fluid does not stick to the wall.

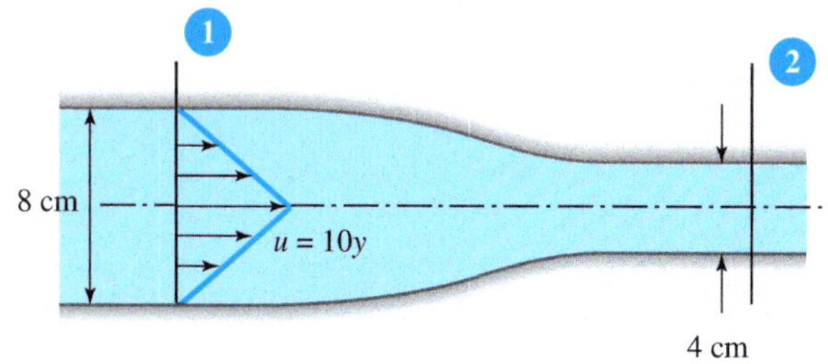

Figure P3.48

3.49 In Example 3.8, does ω_x remain zero as the vortex tube nears the tree? If not, explain why.

Differential Energy Equation

3.50 Derive the incompressible differential energy equation by applying Gauss's theorem (see Problem 3.1) to the integral energy equation (2.5.13) assuming that $\dot{W}_s = \dot{W}_{shear} = \dot{W}_I = 0$ and using

$\dot{Q}_s = \int_{c.s.} K\nabla T \cdot \hat{\mathbf{n}} dA$ assuming no fiscous effects.

3.51 Verify that Eq. 3.4.5 follows from Eq. 3.4.4. Recall the definition of a second derivative.

3.52 Verify that Eq. 3.4.11 follows from Eq. 3.4.9.

3.53 Simplify the differential energy equation for a liquid flow in which the temperature gradients are quite large and the velocity components are very small, such as in a lake heated from above.

3.54 Explain which term in the differential energy equation accounts for the extremely high temperatures that exist on satellites during reentry.

3.55 The velocity distribution in a 2.0-cm-diameter pipe is given by $u(r) = 10$ $(1 - 10\,000r^2)$ m/s. Find the magnitude of the dissipation function at the wall, at the centerline, and halfway between for air at 20°C.

3.56 Let the plate in Problem 3.44 be heated. Write the simplified differential energy equation and the simplified Navier–Stokes equation assuming:

(a) $\mu = \text{const.}$

(b) $\mu = \mu(T)$

4 Dimensional Analysis and Similitude

Outline

Chapter Objectives

The objectives of this chapter are to:

- Establish the parameters necessary to guide experimental studies.
- Present the technique used to apply the results of model studies to prototypes for a variety of flow situations.
- Extract the flow parameters from the differential equations and boundary conditions used to guide computational studies.
- Provide examples and problems that illustrate how dimensionless flow parameters are used, how model studies allow us to predict quantities of interest on a prototype, and how normalized differential equations are utilized.

4.1 INTRODUCTION

There are many problems of interest in the field of fluid mechanics in the real world of design that cannot be solved using the differential and integral equations only. It is often necessary to resort to experimental methods to establish relationships between the variables of interest. Since experimental studies are usually quite expensive, it is necessary to keep the required experimentation to a minimum. This is done using a technique called *dimensional analysis*, which is based on the notion of **dimensional homogeneity**—that all terms in an equation must have the same dimensions. For example, if we write Bernoulli's equation in the form

M. C. Potter and D. C. Wiggert, *Applied Fluid Mechanics*,
https://doi.org/10.1007/978-3-031-65276-9_4

$$\frac{V_1^2}{2g}+\frac{p_1}{\gamma}+z_1=\frac{V_2^2}{2g}+\frac{p_2}{\gamma}+z_2 \tag{4.1.1}$$

we note that the dimension of each term is length. Furthermore, if we factored out z_1 from the left-hand side and z_2 from the right-hand side, we would have

$$\frac{V_1^2}{2gz_1}+\frac{p_1}{\gamma z_1}+1=\left(\frac{V_2^2}{2gz_2}+\frac{p_2}{\gamma z_2}+1\right)\frac{z_2}{z_1} \tag{4.1.2}$$

In this form of Bernoulli's equation, the terms are all dimensionless and we have written the equation as a combination of dimensionless parameters, the basic idea in dimensional analysis that will be presented in the next section.

Often in experimental work we are required to perform experiments on objects that are quite large, too large to experiment with for a reasonable cost. This would include flows over weirs and dams; wave interactions with piers and breakwaters; flows around submarines and ships; subsonic and supersonic flows around aircraft; flows around stadiums and buildings, as shown in Figure 4.1; flows through large pumps and turbines; and flows around automobiles and trucks. Such flows are usually studied in laboratories using models that are smaller than the prototype, the actual device. This substantially reduces the costs when compared with full-scale studies and allows for the study of various configurations or flow conditions.

Figure 4.1 Scale model of the large buildings in a city. Air flow around the buildings is studied. The roughening elements on the floor generate the desired wall turbulence. (Source: IR Stone/shutterstock.com)

There are also flows of interest that involve rather small dimensions, such as flow around a turbine blade, flow into a capillary tube, flow around a microorganism, flow through a small control valve, and flow around and inside a falling droplet. These flows would require that the model be larger than the prototype so that observations could be made with an acceptable degree of accuracy.

Similitude is the study of predicting prototype conditions from model observations. This will be presented following dimensional analysis. Similitude involves the use of the dimensionless parameters obtained in dimensional analysis.

There are two approaches that can be used in the study of dimensional analysis. We will present both approaches. First, we use the **Buckingham π-theorem**, which organizes the steps to ensure dimensional homogeneity; it requires some knowledge of the phenomenon being studied in order that the appropriate quantities of interest are included. Second, we extract the dimensionless parameters that influence a particular flow situation from the differential equations and boundary conditions that are needed to describe the phenomenon being investigated.

4.2 DIMENSIONAL ANALYSIS

4.2.1 Motivation

In the study of phenomena involving fluid flows, either analytically or experimentally, there are invariably many flow and geometric parameters involved. In the interest of saving time and money, the fewest possible combinations of parameters should be utilized. For example, consider the pressure drop across the slider valve of Figure 4.2. We may suspect that the pressure drop depends on such parameters as pipe mean velocity V, the density ρ of the fluid, the fluid viscosity μ, the pipe diameter d, and the gap height h. This could be expressed as

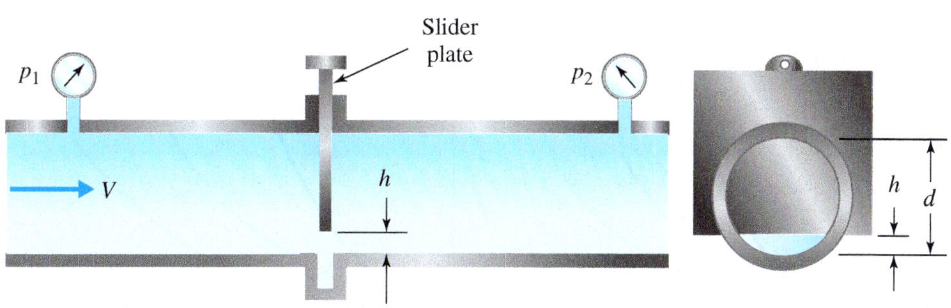

Figure 4.2 Flow around a slider valve.

$$\Delta p = f(V, \rho, \mu, d, h) \tag{4.2.1}$$

where $f(\ )$ simply means "a function of."

Now, if we attempt an experimental study of this problem, consider the strategy for finding the dependence of the pressure drop on the parameters involved. We could fix all parameters except the velocity and investigate the dependence of the pressure drop on the average velocity. Then the diameter could be changed and the experiment repeated. This would lead to the set of results shown in Figure 4.3a. Following that set of experiments the gap height h could be changed, leading to the curves of Figure 4.3b. Again, different fluids could be studied, leading to curves with ρ and μ changing values.

Consider next the notion that any equation that relates a certain set of variables, such as Eq. 4.2.1, can be written in terms of dimensionless parameters, as was done with the Bernoulli equation (4.1.2). We can organize the variables of Eq. 4.2.1 into dimensionless parameters (the steps needed to do this will be presented in a subsequent section) as follows:

$$\frac{\Delta p}{\rho V^2} = f\left(\frac{V\rho d}{\mu}, \frac{h}{d}\right) \tag{4.2.2}$$

Obviously, this is a much simpler relationship. We could perform an experiment with a fixed h/d (say, $h/d = 0.1$) by varying $V\rho d/\mu$ (this is done by simply varying V), resulting in the curve shown in Figure 4.4. The quantity h is changed so that $h/d = 0.5$ and the testing is repeated. Finally, the entire experiment is presented in one figure, as in Figure 4.4. This has greatly reduced the effort and the cost in determining the actual form of $f(V\rho d/\mu, h/d)$; we would use only one pipe and one valve, and we would use only one fluid.

It is not always clear, however, which parameters should be included in an equation such as 4.2.1. The selection of these parameters requires a detailed understanding of the physics involved. In the selection of the parameters that affect the pressure drop across the slider valve, it was assumed that density and viscosity are important parameters, while parameters such as slider plate thickness, pipe pressure, and gravity are not. It should be kept in mind that the selection of the proper parameters is a first crucial step in the application of dimensional analysis.

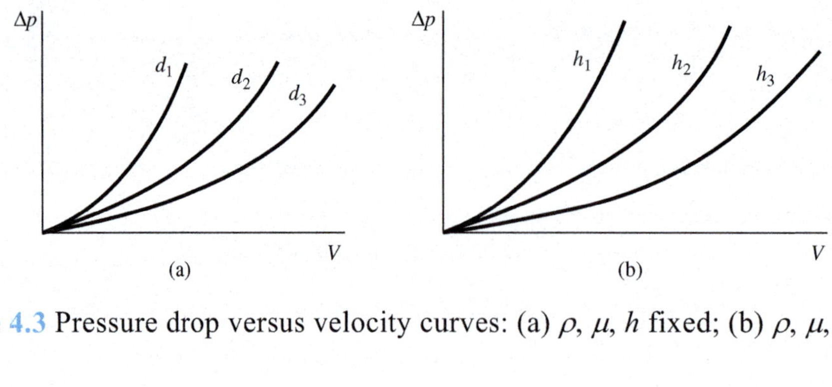

Figure 4.3 Pressure drop versus velocity curves: (a) ρ, μ, h fixed; (b) ρ, μ, d fixed.

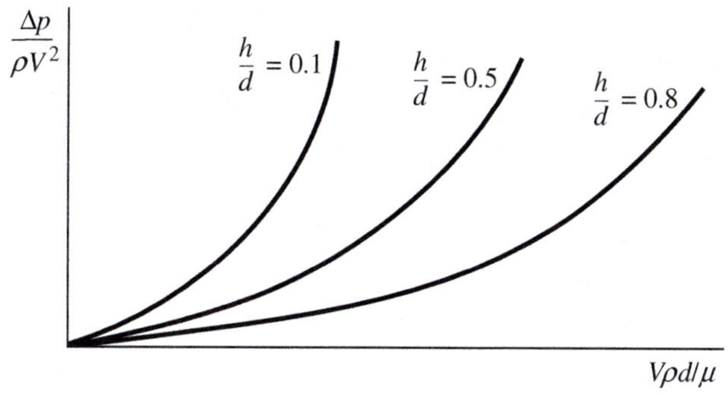

Figure 4.4 Dimensionless pressure drop versus dimensionless velocity.

4.2.2 Review of Dimensions

Before we present the dimensional analysis technique let's review the dimensions of the quantities of interest. All of the quantities have some combination of dimensions of length, time, mass, and force which are related by Newton's second law,

$$\Sigma \mathbf{F} = m\mathbf{a} \tag{4.2.3}$$

In terms of dimensions, it is written as

$$F = \frac{ML}{T^2} \tag{4.2.4}$$

where F, M, L, and T are the dimensions of force, mass, length, and time, respectively. Thus we see that it is sufficient to use only three basic dimensions. We will choose the M-L-T system because we can eliminate the force dimension with Eq. 4.2.4.

If we were considering more complicated flow situations such as those involving electromagnetic field interactions, or those involving temperature gradients, we would need to include the appropriate additional dimensions. However, in this book such phenomena will not be introduced, except for the compressible flow of an ideal gas; for that case an equation of state relates the thermal effects to the dimensions above. That is,

$$p = \rho RT \tag{4.2.5}$$

where T represents temperature. This allows us to write

$$[RT] = [p/\rho] = \frac{F}{L^2} \cdot \frac{L^3}{M} = \frac{ML/T^2}{L^2} \cdot \frac{L^3}{M} = \frac{L^2}{T^2} \tag{4.2.6}$$

where the brackets represent "the dimensions of." Notice that the equation of state does not introduce additional dimensions.

Table 4.1 Symbols and Dimensions of Quantities Used in Fluid Mechanics

Quantity	*Symbol*	*Dimensions*
Length	l	L
Time	t	T
Mass	m	M
Force	F	ML/T^2
Velocity	V	L/T
Acceleration	a	L/T^2
Frequency	ω	T^{-1}
Gravity	g	L/T^2
Area	A	L^2
Flow rate	Q	L^3/T
Mass flux	$\dot{m}$	M/T
Pressure	p	M/LT^2
Stress	τ	M/LT^2
Density	ρ	M/L^3
Specific weight	γ	M/L^2T^2
Viscosity	μ	M/LT
Kinematic viscosity	ν	L^2/T
Work	W	ML^2/T^2
Power, heat flux	$\dot{W}, \dot{Q}$	ML^2/T^3
Surface tension	σ	M/T^2
Bulk modulus	B	M/LT^2

The quantities of interest in fluid mechanics are listed with their respective dimensions in Table 4.1. Reference to this table will simplify writing the dimensions of the quantities introduced in the problems.

4.2.3 Buckingham π-Theorem

In a given physical problem the dependent variable x_1 can be expressed in terms of the independent variables as

$$x_1 = f(x_2,\ x_3,\ x_4,\ \ldots,\ x_n) \tag{4.2.7}$$

where n represents the total number of variables. Referring to Eq. 4.2.1, Δp is the dependent variable and V, ρ, μ, d, and h are the independent variables. The **Buckingham π-theorem**, named after Edgar Buckingham (1867–1940), states that $(n - m)$ dimensionless groups of variables, called π-terms, where m is the number[1] of basic dimensions included in the variables, can be related by

$$\pi_1 = f_1(\pi_2,\ \pi_3,\ \ldots,\ \pi_{n-m}) \tag{4.2.8}$$

[1]There are situations where m is less than the number of basic dimensions. Example 4.2 will illustrate.

where π_1 includes the dependent variable and the remaining π-terms include only independent variables, as in Eq. 4.2.2.

Further, it is noted that a requirement for a successful application of dimensional analysis is that a dimension must occur at least twice or not at all. For example, the equation $\Delta p = f(V, l, d)$ is ill-stated since pressure involves the dimensions of force and V, l, and d do not contain such a dimension.

The procedure used in applying the π-theorem is summarized as follows:

1. Write the functional form of the dependent variable depending on the $(n - 1)$ independent variables. This step requires knowledge of the phenomenon being studied. All variables that effect the dependent variable must be included. These include geometric variables, fluid properties, and external effects that influence the variable being studied. Quantities that have no influence on the dependent variable must not be included. Also, do not include variables that depend on each other; for example, both radius and diameter would not be included. The variables on the right-hand side of Eq. 4.2.7 should be independent.
2. Identify m **repeating variables**, variables that will be combined with each remaining variable to form the π-terms. The repeating variables selected from the independent variables must include all of the basic dimensions, but they must not form a π-term by themselves. An angle cannot be a repeating variable since it is dimensionless and forms a π-term itself.
3. Form the π-terms by combining the repeating variables with each of the remaining variables.
4. Write the functional form of the $(n - m)$ dimensionless π-terms.

Step 3 can be accomplished by an algebraic procedure; we will also illustrate a procedure in the examples that utilizes observation.

The algebraic procedure will now be illustrated with an example. Suppose that we desire to combine the variables surface tension σ, velocity V, density ρ, and length l into a π-term; this can be written as

$$\pi = \sigma^a V^b \rho^c l^d \tag{4.2.9}$$

The objective is to determine a, b, c, and d so that the grouping is dimensionless. In terms of dimensions (refer to Table 4.1), Eq. 4.2.9 is

$$M^0 L^0 T^0 = \left(\frac{M}{T^2}\right)^a \left(\frac{L}{T}\right)^b \left(\frac{M}{L^3}\right)^c L^d \tag{4.2.10}$$

Equate the exponents on each of the basic dimensions:

$$\begin{aligned} M:&\quad 0 = a + c \\ L:&\quad 0 = b - 3c + d \\ T:&\quad 0 = -2a - b \end{aligned} \tag{4.2.11}$$

The three algebraic equations are solved simultaneously to yield

$$a = -c \qquad b = 2c \qquad d = c \tag{4.2.12}$$

so that the π-term becomes

$$\pi = \left(\frac{\rho l V^2}{\sigma}\right)^c \tag{4.2.13}$$

A dimensionless parameter raised to any power remains dimensionless; consequently, we can select c to be any number other than zero. It is usually selected as $c = 1$, depending on the ratio desired. Selecting $c = 1$, the π-term is

$$\pi = \frac{\rho l V^2}{\sigma} \tag{4.2.14}$$

Actually, we could have selected $c = 1$ in Eq. 4.2.9 and proceeded with only three unknowns. Or if it were desired to have σ in the numerator to the first power, we could have set $a = 1$ and let b, c, and d be unknowns.

A final note: If only one π-term results, the functional form would state that the π-term must be a constant since the right-hand side of Eq. 4.2.8 would contain no additional π-terms. This would result in an expression that includes an arbitrary constant that could be determined through analysis or experimentation. Problem 4.12 illustrates this situation.

Example 4.1

The drag force F_D on a cylinder of diameter d and length l is to be studied. What functional form relates the dimensionless variables if a fluid with velocity V flows normal to the cylinder?

Solution

First, we must determine the variables that have some influence on the drag force. If we include variables that do not influence the drag force, we would have additional π-terms that experimentation would show to be unimportant; if we do not include a variable that does influence the drag force, experimentation would also reveal that problem. Experience is essential in choosing the correct variables; in this example we will include as influential variables the free stream velocity V, the viscosity μ, the density ρ of the fluid, in addition to the diameter d and the length l of the cylinder, resulting in $n = 6$ variables. This is written as

$$F_D = f(d,\ l,\ V,\ \mu,\ \rho)$$

The variables are observed to include $m = 3$ dimensions. Refer to Table 4.1 and write

$$[F_D] = \frac{ML}{T^2} \qquad [V] = \frac{L}{T} \qquad [\mu] = \frac{M}{LT} \qquad [d] = L \qquad [l] = L \qquad [\rho] = \frac{M}{L^3}$$

Consequently, we can expect $n - m = 6 - 3 = 3$ π-terms.

We choose repeating variables with the simplest combinations of dimensions such that they do not form a π-term by themselves (we could not include both d and l as repeating variables); the repeating variables are chosen to be d, V, and ρ. These three variables are combined with each of the remaining variables to form the π-terms. Rather than writing equations similar to Eq. 4.2.9 for the π-terms, let us form the π-terms by inspection. When the repeating variables are combined with F_D, we observe that only F_D and ρ have the mass dimension; hence F_D must be divided by ρ. Only F_D and V have the time dimension; hence, F_D must be divided by V^2. Thus F_D divided by ρ has L^4 in the numerator; when divided by V^2 this results in L^2 remaining in the numerator. Hence we must have d^2 in the denominator resulting in

$$\pi_1 = \frac{F_D}{\rho V^2 d^2}$$

When d, V, and ρ are combined with l there results

$$\pi_2 = \frac{l}{d}$$

The last π-term results from combining μ with d, V, and ρ. The mass dimension disappears if we divide μ by ρ. The time dimension disappears if we divide μ by V. This leaves one length dimension in the numerator; hence d is needed in the denominator resulting in

$$\pi_3 = \frac{\mu}{\rho V d}$$

The dimensionless, functional relationship relating the π-terms is

$$\pi_1 = f_1(\pi_2,\ \pi_3) \qquad \text{or} \qquad \frac{F_D}{\rho V^2 d^2} = f_1\left(\frac{l}{d},\ \frac{\mu}{\rho V d}\right)$$

where $f_1(\ )$ simply means it's different from $f(\ )$ above.

Rather than the original relationship of six variables we have reduced the relationship to one involving three π-terms, a much simpler expression. To determine the particular form of the functional relationship above, we would actually have to solve the problem; experimentation would be needed if analytical or numerical methods were not available. This is often the case in fluid mechanics.

Note that we could have included several additional variables in our original list, such as gravity g, the angle θ that the velocity makes with the cylinder, and the roughness e of the cylinder surface. To not include variables that are significant, or to include variables that are not significant is a matter of experience. The novice must learn how to identify significant variables; however, even the experienced researcher is often at a loss to correlate certain phenomena; much experimentation is often needed to discover the appropriate parameters.

Example 4.2

The rise of liquid in a capillary tube is to be studied. It is anticipated that the rise h will depend on surface tension σ, tube diameter d, liquid specific weight γ, and angle β of attachment between the liquid and tube. Write the functional form of the dimensionless variables.

Solution

The expression relating the variables is

$$h = f(\sigma, d, \gamma, \beta)$$

The dimensions of the variables are

$$[h] = L \qquad [\gamma] = \frac{M}{L^2T^2} \qquad [\beta] = 1 \text{ (dimensionless)} \qquad [\sigma] = \frac{M}{T^2} \qquad [d] = L$$

By observation we see that M/T^2 occurs as that combination in both σ and γ, hence M and T are not independent dimensions in this problem. There are only two independent groupings of basic dimensions, L and M/T^2. Thus $m = 2$, and we choose σ and d as the repeating variables. When combined with h, the first π-term is

$$\pi_1 = \frac{h}{d}$$

When σ and d are combined with γ, the second π-term is

$$\pi_2 = \frac{\gamma d^2}{\sigma}$$

Finally, since the angle β is dimensionless, it forms a π-term by itself; that is,

$$\pi_3 = \beta$$

The final functional form relating the π-terms is

$$\pi_1 = f_1(\pi_2, \pi_3) \qquad \text{or} \qquad \frac{h}{d} = f_1\left(\frac{\gamma d^2}{\sigma}, \beta\right)$$

Note: In this example we could not have chosen the angle β as a repeating variable since it already is a dimensionless π-term. Also, we could not have chosen three repeating variables since M and T were not independent.

Also, note that we may have thought that gravity should have been included in the problem. If it had been included above, it would not have appeared in any of the π-terms, indicating that it should not have been included. If density and gravity, rather than specific weight, had been included, the relationship above would have resulted since $\gamma = \rho g$; this, by the way, would have avoided the necessity of observing that M/T^2 was a dimensional grouping.

A final note regarding the functional form of the π-terms: The relationship above could equally have been written as

$$\frac{h}{d} = f_1\left(\frac{\sigma}{\gamma d^2}, \beta\right)$$

Also, occasionally a different set of repeating variables could be selected. This simply expresses the final functional equation in a different but equivalent form. Actually, a second form can be shown to be a combination of the π-terms from an initial form.

4.2.4 Common Dimensionless Parameters

Consider a relatively general relationship between the pressure drop Δp, a characteristic length l, a characteristic velocity V, the density ρ, the viscosity μ, the gravity g, the surface tension σ, the speed of sound c, and an angular frequency ω, written as

$$\Delta p = f(l, V, \rho, \mu, g, c, \omega, \sigma) \tag{4.2.15}$$

The π-theorem applied to this problem, with l, V, and p as repeating variables, results in

$$\frac{\Delta p}{\rho V^2} = f_1\left(\frac{V\rho l}{\mu}, \frac{V^2}{lg}, \frac{V}{c}, \frac{l\omega}{V}, \frac{V^2 \rho l}{\sigma}\right) \tag{4.2.16}$$

Each of the π-terms in this expression is a common dimensionless parameter that appears in numerous fluid flow situations. They are identified as follows:

$$\begin{aligned}
&\text{Euler number, Eu} = \frac{\Delta p}{\rho V^2}\\
&\text{Reynolds number, Re} = \frac{V\rho l}{\mu}\\
&\text{Froude number}^2\text{, Fr} = \frac{V}{\sqrt{lg}}\\
&\text{Mach number, M} = \frac{V}{c}\\
&\text{Strouhal number}^2\text{, St} = \frac{l\omega}{V}\\
&\text{Weber number}^2\text{, We} = \frac{V^2 l\rho}{\sigma}
\end{aligned} \tag{4.2.17}$$

The physical significance of each parameter can be determined by observing that each dimensionless number can be written as the ratio of two forces. The forces are observed to be

$$\begin{aligned}
F_p &= \text{pressure force} = \Delta pA \sim \Delta p l^2\\
F_I &= \text{inertial force} = mV\frac{dV}{ds} \sim \rho l^3 V\frac{V}{l} = \rho l^2 V^2\\
F_\mu &= \text{viscous force} = \tau A = \mu\frac{du}{dy}A \sim \mu\frac{V}{l}l^2 = \mu lV\\
F_g &= \text{gravity force} = mg \sim \rho l^3 g\\
F_B &= \text{compressibility force} = BA \sim \rho\frac{dp}{d\rho}l^2 = \rho c^2 l^2\\
F_\omega &= \text{centrifugal force} = mr\omega^2 \sim \rho l^3 l\omega^2 = \rho l^4\omega^2\\
F_\sigma &= \text{surface tension force} = \sigma l
\end{aligned} \tag{4.2.18}$$

[2]The Froude, Strouhal, and Weber numbers were named after William Froude (1810–1879), Vincenz Strouhal (1850–1922), and Moritz Weber (1871–1951), respectively.

Thus, we see that

$$\begin{aligned}
\text{Eu} &\propto \frac{\text{pressure force}}{\text{inertial force}} \\
\text{Re} &\propto \frac{\text{inertial force}}{\text{viscous force}} \\
\text{Fr} &\propto \frac{\text{inertial force}}{\text{gravity force}} \\
\text{M} &\propto \frac{\text{inertial force}}{\text{compressibility force}} \\
\text{St} &\propto \frac{\text{centrifugal force}}{\text{inertial force}} \\
\text{We} &\propto \frac{\text{inertial force}}{\text{surface tension force}}
\end{aligned} \tag{4.2.19}$$

Thinking of the dimensionless parameters in terms of the ratios of forces allows us to anticipate the significant parameters in a particular flow of interest. If viscous forces are important, such as in the pipe flow of Figure 1.8 or the boundary layer flow of Figure 1.10, we know that the Reynolds number is a significant dimensionless parameter. If surface tension forces are instrumental in affecting the flow, as in droplet formation or flow over a weir with a small head, we expect the Weber number to be important. Similar analysis can be applied to other fluid flow situations.

Obviously, all of the effects included in the general relationship (4.2.16) would not be of interest in any one situation. It would be very unlikely that both compressibility effects and surface tension effects would influence a flow simultaneously. In addition, there is often more than one length of importance, thereby introducing additional geometric, dimensionless ratios. We have, however, introduced the more common dimensionless flow parameters of interest in fluid mechanics. Table 4.2 summarizes this section.

Table 4.2 Common Dimensionless Parameters in Fluid Mechanics

Parameter	*Expression*	*Flow situations where parameter is important*
Euler number	$\frac{\Delta p}{\rho V^2}$	Flows in which pressure drop is significant: most flow situations
Reynolds number	$\frac{\rho l V}{\mu}$	Flows that are influenced by viscous effects: internal flows, boundary layer flows
Froude number	$\frac{V}{\sqrt{lg}}$	Flows that are influenced by gravity: primarily free surface flows
Mach number	$\frac{V}{c}$	Compressibility is important in these flows, usually if $V > 0.3\ c$
Strouhal number	$\frac{l\omega}{V}$	Flow with an unsteady component that repeats itself periodically
Weber number	$\frac{V^2 l \rho}{\sigma}$	Surface tension influences the flow; flow with an interface may be such a flow

4.3 SIMILITUDE

4.3.1 General Information

As stated in the introduction, **similitude** is the study of predicting prototype conditions from model observations. When an analytical or numerical solution is not practical, or when calculations are based on flow simplifications so that uncertainty is introduced, it is usually advisable to perform tests on a model if testing is not practical on a full-scale prototype, be it too large or too small.

If it is decided that a model study is to be performed, it is necessary to develop the means whereby a quantity measured on the model, signified with a subscript *m*, can be used to predict the associated quantity on the prototype, signified by a subscript *p*. We can develop such a means if we have **dynamic similarity** between model and prototype, that is, if the forces which act on corresponding masses in the model flow and the prototype flow are in the same ratio throughout the entire flow fields. Suppose that inertial forces, pressure forces, viscous forces, and gravity forces are present; then dynamic similarity requires that, at corresponding points in the flow fields,

$$\frac{(F_I)_m}{(F_I)_p} = \frac{(F_P)_m}{(F_P)_p} = \frac{(F_\mu)_m}{(F_\mu)_p} = \frac{(F_g)_m}{(F_g)_p} = \text{const.} \tag{4.3.1}$$

These can be rearranged to read

$$\left(\frac{F_I}{F_P}\right)_m = \left(\frac{F_I}{F_P}\right)_p \qquad \left(\frac{F_I}{F_\mu}\right)_m = \left(\frac{F_I}{F_\mu}\right)_p \qquad \left(\frac{F_I}{F_g}\right)_m = \left(\frac{F_I}{F_g}\right)_p \tag{4.3.2}$$

which, in the preceding section, have been shown to be

$$\text{Eu}_m = \text{Eu}_p \qquad \text{Re}_m = \text{Re}_p \qquad \text{Fr}_m = \text{Fr}_p \tag{4.3.3}$$

If the forces above were the only forces present, we could write

$$F_I = f(F_p,\ F_\mu,\ F_g) \tag{4.3.4}$$

Recognizing that there is only one basic dimension, namely force, dimensional analysis would allow us to write (see Eq. 4.2.8) the equation above in terms of force ratios or

$$\text{Eu} = f(\text{Re},\ \text{Fr}) \tag{4.3.5}$$

Hence we could conclude that if the Reynolds number and the Froude number are the same on the model and prototype, the Euler number must also be the same. Thus dynamic similarity between model and prototype is guaranteed by equating the Reynolds number and the Froude number of the model to those on the prototype, respectively. If compressibility forces were included here, the analysis above would result in the Mach number being included in Eq. 4.3.5.

We can write the inertial force ratio as

$$\frac{(F_I)_m}{(F_I)_p} = \frac{a_m m_m}{a_p m_p} = \text{const.} \tag{4.3.6}$$

showing that the acceleration ratio between corresponding points on the model and prototype is a constant, provided that the mass ratio of corresponding fluid elements is a constant. We can write the acceleration ratio as

$$\frac{a_m}{a_p} = \frac{V_m^2/l_m}{V_p^2/l_p} = \text{const.} \tag{4.3.7}$$

showing that the velocity ratio between corresponding points is a constant providing the length ratio is a constant. The velocity ratio being a constant between all corresponding points in the flow fields is the statement of **kinematic similarity**. This would result in the streamline pattern around the model being the same as that around the prototype except for a scale factor. The length ratio being constant between all corresponding points in the flow fields is the demand of **geometric similarity**, which results in the model having the same shape as the prototype. Hence, to ensure complete similarity between model and prototype, we demand that:

- Geometric similarity be satisfied
- The mass ratio of corresponding fluid elements be a constant
- The appropriate dimensionless parameters of Eq. 4.2.17 be equal

Assuming that complete similarity between model and prototype exists, we can now predict quantities of interest on a prototype from measurements on a model. If we measure a drag force F_D on a model and wish to predict the corresponding drag on the prototype, we could equate the ratio of the drag forces to the ratio of the inertial forces (see Eq. 4.2.18) as

$$\frac{(F_D)_m}{(F_D)_p} = \frac{(F_I)_m}{(F_I)_p} = \frac{\rho_m V_m^2 l_m^2}{\rho_p V_p^2 l_p^2} \tag{4.3.8}$$

If we measure the power input to a model and wish to predict the power requirement of the prototype, we would recognize that power is force times velocity and write

$$\frac{\dot{W}_m}{\dot{W}_p} = \frac{(F_I)_m V_m}{(F_I)_p V_p} = \frac{\rho_m V_m^2 l_m^2 V_m}{\rho_p V_p^2 l_p^2 V_p} \tag{4.3.9}$$

Hence, we can predict a prototype quantity if we select the model fluid (this provides ρ_m/ρ_p), the scale ratio (this provides l_m/l_p), and the appropriate dimensionless number from Table 4.2 (this provides V_m/V_p). Examples will illustrate.

4.3.2 Confined Flows

A confined flow is a flow that has no free surface (a liquid-gas surface) or interface (two different liquids forming an interface). It is confined to move within a specified region; such flows include external flows around objects, such as aircraft, buildings, and submarines, as well as internal flows in pipes and conduits.

Gravity does not influence the flow pattern in confined flows; that is, if gravity could be changed in magnitude, the flow pattern and associated flow quantities would not change. The dominant effect is that of viscosity in incompressible confined flows (all liquid flows and gas flows in which $M < 0.3$). Surface tension is obviously not a factor, as it would be in bubble formation, and for steady flows there would be no unsteady effects due to oscillations in the flow. The three relevant forces are pressure forces, inertial forces, and viscous forces. Therefore, in confined flows dynamic similarity is achieved if the ratios of viscous forces, inertial forces, and pressure forces between model and prototype are the same. This leads to the conclusion (see Eq. 4.3.5) that $\mathrm{Eu} = f(\mathrm{Re})$, so that it is only necessary to consider the Reynolds number as the dominant dimensionless parameter in a confined incompressible flow. If compressibility effects are significant, the Mach number would also be important.

Example 4.3

A test is to be performed on a proposed design for a large pump that is to deliver 1.5 m^3/s of water from a 40-cm-diameter impeller with a pressure rise of 400 kPa. A model with an 8-cm-diameter impeller is to be used. What flow rate should be used and what pressure rise is to be expected? The model fluid is water at the same temperature as the water in the prototype.

Solution

For similarity to exist in this confined incompressible flow problem, the Reynolds numbers must be equal; that is,

$$\mathrm{Re}_m = \mathrm{Re}_p$$

$$\frac{V_m d_m}{\nu_m} = \frac{V_p d_p}{\nu_p}$$

Recognizing that $\nu_m = \nu_p$ if the temperatures are equal, we see that

$$\frac{V_m}{V_p} = \frac{d_p}{d_m}$$

$$= \frac{0.4}{0.08} = 5$$

The ratio of flow rates is found recognizing that $Q = VA$:

$$\frac{Q_m}{Q_p} = \frac{V_m d_m^2}{V_p d_p^2}$$
$$= 5 \times \left(\frac{1}{5}\right)^2 = \frac{1}{5}$$

Thus we find that

$$Q_m = \frac{Q_p}{5} = \frac{1.5}{5} = \underline{0.3\ \text{m}^3/\text{s}}$$

The dimensionless pressure rise is found using the Euler number:

$$\left(\frac{\Delta p}{\rho V^2}\right)_m = \left(\frac{\Delta p}{\rho V^2}\right)_p$$

Hence the pressure rise for the model is

$$\Delta p_m = \Delta p_p \frac{\rho_m}{\rho_p} \frac{V_m^2}{V_p^2}$$
$$= 400 \times 1 \times 5^2 = \underline{10\ 000\ \text{kPa}}$$

Note that in this example we see that the velocity in the model is equal to the velocity in the prototype multiplied by the length ratio, and the pressure rise in the model is equal to the pressure rise in the prototype multiplied by the length ratio squared. If the length ratio were very large, it is obvious that to maintain Reynolds number equivalence would be quite difficult, indeed. This observation is discussed in more detail in Section 4.3.4.

4.3.3 Free-Surface Flows

A free-surface flow is one in which part of the boundary involves a pressure boundary condition. This includes flows over weirs and dams, as shown in Figure 4.5; flows in channels and spillways; flows involving two fluids separated by an interface; and flows around floating objects with waves and around submerged objects with cavitation present. In all of these flows the location of the free surface is unknown and the velocity at the free surface is unknown; it is the pressure that must be the same[3] on either side of the interface. In free-surface flows, gravity controls both the location and the motion of the free surface. This introduces the Froude

[3]The surface tension, if significant, results in a pressure difference across the interface.

Figure 4.5 The Bonneville lock and dam on the Columbia River. (Courtesy U.S. Army Corps of Engineers)

number because of the influence of the gravity forces. If we consider flows that do not exhibit periodic motions, which have negligible surface tension and compressibility effects, we can ignore the influence of St, M, and We. That leaves only the viscous effects to be considered. There are many free-surface flows in which viscous effects are significant. Consider, however, that in most model studies water is the only economical fluid to use; if the prototype fluid is also water, as it often is, we would find from the Froude numbers

$$\frac{V_m^2}{l_m g_m} = \frac{V_p^2}{l_p g_p} \qquad \therefore \frac{V_m}{V_p} = \left(\frac{l_m}{l_p}\right)^{1/2} \tag{4.3.10}$$

assuming that $g_m = g_p$. Match the Reynolds numbers (using $\nu_m = \nu_p$):

$$\frac{V_m l_m}{\nu_m} = \frac{V_p l_p}{\nu_p} \qquad \therefore \frac{V_m}{V_p} = \frac{l_p}{l_m} \tag{4.3.11}$$

Thus, we have a conflict. If we use the same fluid in the model study as in the prototype flow, we cannot satisfy both the Froude number criterion and the Reynolds number criterion. If we demand that both criteria be satisfied by using different fluids for model and prototype ($\nu_m \neq \nu_p$), we must select a model fluid with a viscosity $\nu_m = \nu_p(l_m/l_p)^{3/2}$ (this results from a matching of the Froude and Reynolds numbers). A fluid with this viscosity is probably either an impossibility or an impracticality. Hence, when modeling free-surface flows in which viscous effects are important, we equate the Froude numbers and include viscous effects by some other technique. For example, if we measured the total drag on the model of a ship, we would approximate the viscous drag (using some technique not included here) and subtract it from the total drag, leaving the drag due to wave resistance. The wave drag on the prototype would then be predicted using similitude, and the approximated viscous drag would be added to the wave drag, yielding the expected drag on the ship. For the better designs, the viscous drag on ship hulls can be of the same order of magnitude as the wave drag. The viscous drag on objects involving free surfaces will not be considered in this text.

Example 4.4

A 1:20 scale model of a surface vessel is used to test the influence of a proposed design on the wave drag. A wave drag of 25 N is measured at a model speed of 2.5 m/s. What speed does this correspond to on the prototype, and what wave drag is predicted for the prototype? Neglect viscous effects, and assume the same fluid for model and prototype.

Solution

The Froude number must be equated for both model and prototype. Thus

$$\mathrm{Fr}_m = \mathrm{Fr}_p \qquad \frac{V_m}{\sqrt{l_m g}} = \frac{V_p}{\sqrt{l_p g}}$$

This yields, recognizing that g does not vary significantly on the surface of the earth,

$$V_p = V_m \left(\frac{l_p}{l_m}\right)^{1/2} = 2.5\sqrt{20} = \underline{11.2\ \mathrm{m/s}}$$

To find the wave drag on the prototype, we equate the drag ratio to the inertia force ratio:

$$\frac{(F_D)_m}{(F_D)_p} = \frac{\rho_m V_m^2 l_m^2}{\rho_p V_p^2 l_p^2}$$

This allows us to calculate the wave drag on the prototype as, using $\rho_p = \rho_m$,

$$\begin{aligned}(F_D)_p &= (F_D)_m \frac{\rho_p V_p^2 l_p^2}{\rho_m V_m^2 l_m^2} \\ &= 25 \times \frac{11.2^2}{2.5^2} \times 20^2 = \underline{201\ \mathrm{kN}}\end{aligned}$$

Note: We could have used the gravity force ratio rather than the inertial force ratio, but we could not have used the viscous force ratio since viscous forces were assumed negligible.

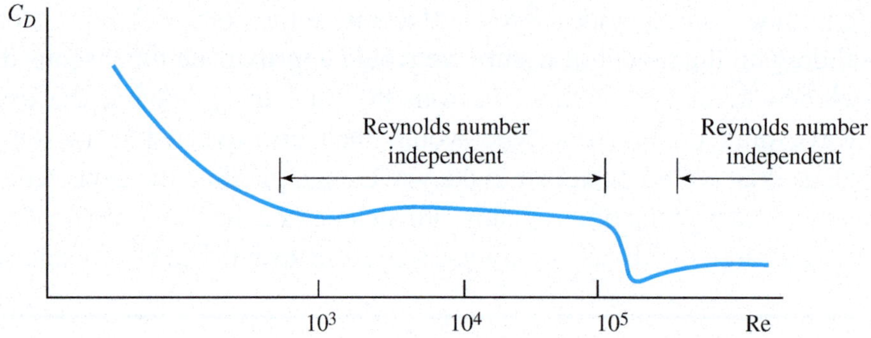

Figure 4.6 Drag coefficient versus Reynolds number for typical blunt object.

4.3.4 High-Reynolds-Number Flows

In a confined flow in which the Reynolds number is the dimensionless parameter that guarantees dynamic similarity, we note that if the same fluid is used in model and prototype, the velocity in the model study is $V_m = V_p l_p / l_m$; the velocity in the model is the velocity in the prototype multiplied by the scale factor. This often results in velocities that are prohibitively large in the model study. Also, the pressures encountered in the model study are large, as shown in Example 4.3, and the energy consumption is also very large. Because of these problems the Reynolds numbers may not be matched in studies involving large Reynolds numbers.

There is, however, some justification for not matching the Reynolds number in model studies. Consider a typical drag coefficient C_D versus Reynolds number curve, as shown in Figure 4.6 for a blunt object (the complete curve is presented in Figure 6.9). The drag coefficient is a dimensionless drag, defined as $C_D = \text{drag} / \frac{1}{2} \rho V^2 A$. At a sufficiently high Reynolds number, between about 10^3 and 10^5, the flow is insensitive to changes in the Reynolds number; note that the drag coefficient is essentially constant and independent of Re. That implies that the flow field is similar at $\text{Re} = 10^3$ to that at $\text{Re} = 10^5$. Thus if $\text{Re} = 10^5$, it is only necessary that $10^3 < \text{Re}_m < 10^5$ for the viscous effects to have the same effect on model and prototype. This often allows another parameter of interest to be matched, such as the Froude number or the Mach number. There are, however, high-Reynolds-number flows in which compressibility effects and free-surface effects are negligible, so that neither the Froude number nor the Mach number are applicable. Examples include flow around automobiles, large smoke stacks, and dirigibles. For such flows we must only ensure that the Reynolds number be within the range where the drag coefficient is constant. We should note that for very large Reynolds numbers, as is often the case ($\text{Re} > 5 \times 10^5$ in Figure 4.6), the flow may also become Reynolds number independent. If that is the case, it is only necessary that Re_m be sufficiently large.

Example 4.5

A 1:10 scale model of an automobile is used to measure the drag on a proposed design. It is to simulate a prototype speed of 90 km/h. What speed should be used in the wind tunnel if Reynolds numbers are equated? For this condition, what is the ratio of drag forces?

Solution

The same fluid exists on model and prototype; thus, equating the Reynolds numbers results in

$$\frac{V_m l_m}{\nu_m} = \frac{V_p l_p}{\nu_p} \qquad \therefore V_m = V_p \frac{l_p}{l_m}$$

$$= 90 \times 10 = \underline{900 \text{ km/h}}$$

This speed would, of course, introduce compressibility effects, effects that do not exist in the prototype. Hence the proposed model study would be inappropriate.

If we did use this velocity in the model, the drag force ratio would be

$$\frac{(F_D)_p}{(F_D)_m} = \frac{\rho_p V_p^2 l_p^2}{\rho_m V_m^2 l_m^2}. \qquad \therefore \frac{(F_D)_p}{(F_D)_m} = 1$$

Thus, we see that the drag force on the model is the same as the drag force on the prototype if the same fluids are used when we equate Reynolds numbers.

Example 4.6

In Example 4.5, if the Reynolds numbers were equated, the velocity in the model study was observed to be in the compressible flow regime (i.e., M > 0.3 or $V_m > 360$ km/h). To conduct an acceptable model study, could we use a velocity of 90 km/h on a model with a characteristic length of 10 cm? Assume that the drag coefficient ($C_D = F_D / \frac{1}{2} \rho V^2 A$, where A is the projected area) is independent of Re for Re > 10^5. If so, what drag force on the prototype would correspond to a drag force of 1.2 N measured on the model?

Solution

The proposed model study in a wind tunnel is to be conducted with V_m = 90 km/h and l_m = 0.1 m. Using $\nu = 1.6 \times 10^{-5}$ m^2/s, the Reynolds number is

$$\text{Re}_m = \frac{V_m l_m}{\nu_m} = \frac{(90 \times 1000/3600) \times 0.1}{1.6 \times 10^{-5}} = 1.56 \times 10^5$$

This Reynolds number is greater than 10^5, so we will assume that similarity exists between model and prototype. The velocity of 90 km/h is sufficiently high.

The drag force on the prototype traveling at 90 km/h corresponding to 1.2 N on the model is found from

$$\frac{(F_D)_p}{(F_D)_m} = \frac{\rho_p V_p^2 l_p^2}{\rho_m V_m^2 l_m^2} \qquad \therefore (F_D)_p = (F_D)_m \frac{\rho_p}{\rho_m} \frac{\overset{1}{\cancel{V_p^2}}}{\cancel{V_m^2}} \frac{\overset{1}{\cancel{l_p^2}}}{\cancel{l_m^2}} = 1.2 \times 10^2 = \underline{120 \text{ N}}$$

Note that in this example we have assumed that the drag coefficient is independent of Re for Re > 10^5. If the drag coefficient continued to vary above Re = 10^5 (this would be evident from experimental data), the foregoing analysis would have to be modified accordingly.

4.3.5 Compressible Flows

For most compressible flow situations the Reynolds number is so large (refer to Figure 4.6) that it is not a parameter of significance; the compressibility effects lead to the Mach number as the primary dimensionless parameter for model studies. Thus, for a particular model study we require

$$\mathrm{M}_m = \mathrm{M}_p \qquad \text{or} \qquad \frac{V_m}{c_m} = \frac{V_p}{c_p} \tag{4.3.12}$$

If the model study is carried out in a wind tunnel and the prototype fluid is air, we can assume that $c_m \simeq c_p$ if the temperature is about the same in the two flows. For such a case, the velocity in the model study is equal to the velocity associated with the prototype. Of course, if the speeds of sound are different, the velocity ratio will be different from unity, accordingly.

Example 4.7

The pressure rise from free stream to the nose of a fusilage section of an aircraft is measured in a wind tunnel at 20°C to be 34 kPa with a wind-tunnel airspeed of 900 km/h. If the test is to simulate flight at an elevation of 12 km, what is the prototype velocity and the expected nose pressure rise?

Solution

To find the prototype velocity corresponding to a wind-tunnel airspeed of 900 km/h, we equate the Mach numbers

$$\mathrm{M}_m = \mathrm{M}_p \qquad \text{or} \qquad \frac{V_m}{\sqrt{kRT_m}} = \frac{V_p}{\sqrt{kRT_p}}$$

Thus,

$$V_p = V_m \left(\frac{kRT_p}{kRT_m} \right)^{1/2} = 900 \left(\frac{216.7}{293} \right)^{1/2} = \underline{774 \text{ km/h}}$$

The pressure at the nose of the prototype fusilage is found using the Euler number as follows:

$$\frac{\Delta p_m}{\rho_m V_m^2} = \frac{\Delta p_p}{\rho_p V_p^2}$$

$$\therefore \Delta p_p = \Delta p_m \frac{\rho_p}{\rho_m} \frac{V_p^2}{V_m^2}$$

$$= 34 \times \frac{0.3119}{1.225} \times \frac{774^2}{900^2} = \underline{6.4 \text{ kPa}}$$

The densities and temperature T_p were found in Appendix B.

4.3.6 Periodic Flows

In many flow situations there are regions of the flows in which periodic motions occur. Such flows include the periodic fluid motion (in Section 4.3.2 this is called **vortex shedding**) that takes place when a fluid flows past a cylindrical object such as a bridge, a TV tower, a cable, or a tall building; the flow past a wind machine; and the flow through turbomachinery. In flows such as these, it is necessary to equate the Strouhal numbers, which can be written as

$$\frac{V_m}{\omega_m l_m} = \frac{V_p}{\omega_p l_p} \qquad (4.3.13)$$

in order that the periodic motion be properly modeled.

In addition to the Strouhal number, there may be additional dimensionless parameters that must be equated: in viscous flows, the Reynolds number; in free-surface flows, the Froude number; and in compressible flows, the Mach number.

Example 4.8

A large wind turbine, designed to operate at 50 km/h, is to be tested in a laboratory by constructing a 1:15 scale model. What airspeed should be used in the wind tunnel, what angular velocity should be used to simulate a prototype angular speed of 5 rpm, and what power output is expected from the model if the prototype output is designed to be 500 kW?

Solution

The speed in the wind tunnel can be any speed above that needed to provide a sufficiently large Reynolds number. Let us select the same speed with which the prototype is to operate, namely,

50 km/h, and calculate the minimum characteristic length that a Reynolds number of 10^5 would demand; this gives

$$\mathrm{Re} = \frac{Vl}{\nu} \qquad 10^5 = \frac{(50 \times 1000/3600) \times l}{1.6 \times 10^{-5}} \qquad \therefore l = 0.12 \text{ m}$$

Obviously, in a reasonably large wind tunnel we can maintain a characteristic length (e.g., the blade length) that large.

The angular velocity is found by equating the Strouhal numbers. There results

$$\frac{V_m}{\omega_m l_m} = \frac{V_p}{\omega_p l_p} \qquad \therefore \omega_m = \omega_p \frac{V_m}{V_p} \frac{l_p}{l_m} = 5 \times 1 \times 15 = \underline{75 \text{ rpm}}$$

assuming that the wind velocities are equal.

The power is found by observing that power is force times velocity:

$$\frac{\dot{W}_m}{\dot{W}_p} = \frac{\rho_m V_m^3 l_m^2}{\rho_p V_p^3 l_p^2}$$

or

$$\dot{W}_m = \dot{W}_p \frac{\rho_m}{\rho_p} \frac{V_m^3}{V_p^3} \frac{l_m^2}{l_p^2} = 500 \times \left(\frac{1}{15}\right)^2 = \underline{2.22 \text{ kW}}$$

4.4 NORMALIZED DIFFERENTIAL EQUATIONS

In Chapter 3 we derived the partial differential equations used to describe flows of interest that involve isotropic, homogeneous, Newtonian fluids. Such flows could be laminar or turbulent, steady or unsteady, compressible or incompressible, confined flows or free-surface flows, flows with surface tension effects or ones in which surface tension is negligible. Most often, when utilizing the differential equations, we express them in dimensionless, or normalized, form. This form of the equations provides us with information not contained in the dimensional form, information similar to that provided by dimensional analysis. Let us normalize the differential equations that would describe the motion of an incompressible, homogeneous flow. The energy equation will not be needed.

Before we normalize the equations, let's review them. In vector form the continuity equation and the Navier–Stokes equation are

$$\begin{aligned} \boldsymbol{\nabla} \cdot \mathbf{V} &= 0 \\ \rho \frac{D\mathbf{V}}{Dt} &= -\boldsymbol{\nabla} p - \rho g \boldsymbol{\nabla} h + \mu \nabla^2 \mathbf{V} \end{aligned} \tag{4.4.1}$$

where we have assumed h to be vertical. Because the first two terms on the right are of the same form (the gradient of a scalar function) we can combine them as follows:

$$\nabla p + \rho g \nabla h = \nabla p_k \tag{4.4.2}$$

where the **kinetic pressure** p_k is defined to be

$$p_k = p + \rho g h \tag{4.4.3}$$

In terms of the kinetic pressure the Navier–Stokes equation becomes

$$\rho \frac{D\mathbf{V}}{Dt} = -\nabla p_k + \mu \nabla^2 \mathbf{V} \tag{4.4.4}$$

So, as the fluid motion ceases, p_k goes to zero.

If the pressure p is never used in a boundary condition, we can retain the kinetic pressure in the equations, thereby "hiding" the gravity effects. If the pressure enters a boundary condition, we must return to Eq. 4.4.3, and gravity becomes important.

To normalize the differential equations and boundary conditions, we must select characteristic quantities that best describe the problem of interest. For example, consider a flow that has an average velocity V and a primary dimension l. The dimensionless velocities and coordinate variables are

$$u^* = \frac{u}{V} \qquad v^* = \frac{v}{V} \qquad w^* = \frac{w}{v} \qquad x^* = \frac{x}{l} \qquad y^* = \frac{y}{l} \qquad z^* = \frac{z}{l} \tag{4.4.5}$$

where asterisks are used to denote dimensionless quantities. The characteristic pressure will be twice the inviscid pressure rise between free stream and the stagnation point (i.e., ρV^2); the characteristic time, provided that no characteristic time such as a period of oscillation exists in the problem, will be the time it takes a fluid particle to travel the distance l at the velocity V (i.e., l/V). Hence,

$$p^* = \frac{p}{\rho V^2} \qquad t^* = \frac{t}{l/V} \tag{4.4.6}$$

Using the dimensionless quantities of Eq. 4.4.5, we see that

$$\mathbf{V}^* = u^*\hat{\mathbf{i}} + v^*\hat{\mathbf{j}} + w^*\hat{\mathbf{k}} = \frac{u}{V}\hat{\mathbf{i}} + \frac{v}{V}\hat{\mathbf{j}} + \frac{w}{V}\hat{\mathbf{k}} = \frac{\mathbf{V}}{V}$$
$$\nabla^* = \frac{\partial}{\partial x^*}\hat{\mathbf{i}} + \frac{\partial}{\partial y^*}\hat{\mathbf{j}} + \frac{\partial}{\partial z^*}\hat{\mathbf{k}} = l\frac{\partial}{\partial x}\hat{\mathbf{i}} + l\frac{\partial}{\partial y}\hat{\mathbf{j}} + l\frac{\partial}{\partial z}\hat{\mathbf{k}} = l\nabla \tag{4.4.7}$$

Introducing the above into the Navier–Stokes equation and the continuity equation, we have

$$\rho\frac{V}{l/V}\frac{D^*V^*}{Dt^*} = -\frac{\rho V^2}{l}\nabla^* p_k^* + \mu\frac{V}{l^2}\nabla^{*2}\,\mathbf{V}^* \quad \frac{V}{l}\nabla^*.\mathbf{V}^* = 0 \tag{4.4.8}$$

where D^*/Dt^* represents the dimensionless material derivative. Finally, in dimensionless form,

$$\frac{D^*V^*}{Dt^*} = -\nabla^* p_k^* + \frac{\mu}{\rho V l}\nabla^{*2}\,\mathbf{V}^*$$
$$\nabla^*.\mathbf{V}^* = 0 \tag{4.4.9}$$

Note that we have introduced the Reynolds number

$$\mathrm{Re} = \frac{\rho V l}{\mu} \tag{4.4.10}$$

as a parameter in the normalized Navier–Stokes equation.

Consider now the boundary conditions. The no-slip condition on a fixed boundary $\mathbf{V}^* = 0$ introduces geometric parameters that are necessary to specify the geometry of the fixed boundary. These could include, for example, a roughness parameter e/l, where e is the average height of the roughness elements; a thickness parameter t/l, where t is the maximum thickness of a turbine blade; and a nose radius parameter r_0/l, where r_0 is the nose radius of an airfoil.

Another boundary condition would be the entering velocity distribution, for example, $u(y)/V$. This may introduce parameters having to do with the profile and the turbulent structure of the entering flow. Such features could be extremely significant for a particular flow.

Also, a portion of the boundary may be oscillating, like that of a rotating component. On such a boundary we would require that the velocity of the fluid be the same as the velocity of the rotating part, that is, $v = r\,\omega$. This would introduce the Strouhal number St,

$$v^* = \mathrm{St}\, r^* \tag{4.4.11}$$

where

$$\mathrm{St} = \frac{\omega l}{V} \tag{4.4.12}$$

Another boundary condition that introduces an additional parameter is that of a free surface. It is required that, neglecting surface tension effects, the pressure be constant across a free surface. On an air-water surface this requires that $p = 0$ on the free surface. Returning to Eq. 4.4.3, and normalizing, there results

$$\rho V^2 p_k^* = \rho V^2 p^* + \rho g l h^* \qquad \text{or} \qquad p_k^* = p^* + \frac{gl}{V^2} h^* \tag{4.4.13}$$

where we have introduced the Froude number

$$\mathrm{Fr} = \frac{V}{\sqrt{lg}} \tag{4.4.14}$$

Finally, a boundary condition may involve surface tension, as in droplet formation problems. Such a boundary condition would involve the two radii of curvature leading to the normalized equation

$$\Delta p^* = \frac{\sigma}{l \rho V^2} \left(\frac{1}{r_1^*} + \frac{1}{r_2^*} \right) \tag{4.4.15}$$

where we have introduced the Weber number

$$\mathrm{We} = \frac{V^2 l \rho}{\sigma} \tag{4.4.16}$$

Note that we have introduced all of the dimensionless parameters discussed in Section 4.3 with the exception of the Mach number. It could be introduced in a normalized energy equation, a task that is left to the home problems.

Obviously, if we can write the differential equations that describe a particular flow, those equations and the boundary conditions contain all the parameters of interest; hence the Buckingham π-theorem is not actually necessary for flow situations in which the equations and boundary conditions are known.

In addition, we can obtain the similitude conditions from the normalized equations and boundary conditions. The idea is to make the normalized differential equations and boundary conditions the same for both model and prototype. If they are the same, they should possess the same solution.[4] The normalized boundary conditions being identical would demand that geometric similarity be satisfied. In order that the equations and boundary conditions be identical on both model and prototype, it is necessary that the dimensionless parameters be the same for both model and prototype. Thus, similitude conditions are also included in the normalized problem.

[4]Questions of uniqueness, which we will not consider, enter here. For example, it is possible to have identical conditions and equations, yet in one case a laminar flow could result and in the other it could be a turbulent flow or a second laminar flow different from the first; that is, different solutions would exist.

4.5 SUMMARY

Experimental studies are greatly simplified by reducing the number of variables that influence the phenomenon being studied. We demonstrated how the relationship

$$\Delta p = f(V, \rho, \mu, d, h) \tag{4.5.1}$$

could be written in the simplified form,

$$\frac{\Delta p}{\rho V^2} = f\left(\frac{V\rho d}{\mu}, \frac{h}{d}\right) \tag{4.5.2}$$

The most common flow parameters are, using l as the characteristic length,

$$\text{Re} = \frac{V\rho l}{\mu}, \quad \text{Fr} = \frac{V}{\sqrt{lg}}, \quad M = \frac{V}{c}, \quad \text{Eu} = \frac{\Delta p}{\rho V^2}, \quad \text{St} = \frac{l\omega}{V}, \quad \text{We} = \frac{V^2 l\rho}{\sigma} \tag{4.5.3}$$

The parameters, used to guide model studies that are of primary significance in a particular flow are identified as follows:

Confined flows: $$\text{Re} = \frac{V\rho l}{\mu} \tag{4.5.4}$$

Free-surface flows: $$\text{Fr} = \frac{V}{\sqrt{lg}} \tag{4.5.5}$$

High-Reynolds-number flows: $$\text{Re} > (\text{Re})_{\text{minimum}} \tag{4.5.6}$$

Compressible flows: $$\text{M} = \frac{V}{c} \tag{4.5.7}$$

Periodic flows: $$\text{St} = \frac{V}{l\omega} \tag{4.5.8}$$

The Navier–Stokes equation is written using dimensionless variables as

$$\frac{D\mathbf{V}}{Dt} = -\boldsymbol{\nabla} p_k + \frac{1}{\text{Re}}\nabla^2\mathbf{V} \tag{4.5.9}$$

where p_k is the kinetic pressure. We know that all variables are dimensionless since the Reynolds number appears in the equation. We often write equations in dimensionless form without the asterisks for simplicity.

PROBLEMS

4.1 Write Bernoulli's equation in dimensionless form by dividing Eq. 4.1.1 by V_1^2 and multiplying by g. Express the equation in a form similar to that of Eq. 4.1.2.

4.2 If the F-L-T system of units were used, what would the dimensions be on each of the following?
(a) Mass flux **(b)** Pressure
(c) Density **(d)** Viscosity
(e) Work **(f)** Power
(g) Surface tension

Dimensional Analysis

4.3 Verify that the dimensions of power are ML^2/T^3 as listed in Table 4.1.

4.4 If the velocity V in a fluid flow depends on a dimension l, the fluid density ρ, and the viscosity μ, show that this implies that the Reynolds number $Vl\rho/\mu$ is constant.

4.5 If the velocity V in a fluid depends on the surface tension σ, the density ρ, and a diameter d, show that this implies that the Weber number $V^2 d\rho/\sigma$ is constant.

4.6 Assume that the velocity V of fall of an object depends on the height H through which it falls, the gravity g, and the mass m of the object. Find an expression for V.

4.7 Include the density ρ and viscosity μ of the surrounding fluid and repeat Problem 4.6. This would account for the resistance (drag) of the fluid.

4.8 Select l, V, and ρ as the repeating variables in Example 4.1 and find an expression for F_D. Show that this is an equivalent form to that of Example 4.1.

4.9 Select d, μ, and V as the repeating variables in Example 4.1 and find an expression for F_D. Show that this is equivalent to the expression of Example 4.1.

4.10 Include gravity g in the list of variables of Example 4.2 and determine the final expression that results for h.

4.11 Find an expression for the centrifugal force F_c if it depends on the mass m, the angular velocity ω, and the radius R of an impeller.

4.12 The normal stress σ in a beam depends on the bending moment M, the distance y from the neutral axis, and the moment of inertia I. Relate σ to the other variables if we know that σ varies linearly with y.

4.13 Find an expression for the average velocity in a smooth pipe if it depends on the viscosity, the diameter, and the pressure gradient $\partial p/\partial x$.

4.14 It is suggested that the velocity of the water flowing from a hole in the side of an open tank depends on the height H of the hole from the surface, the gravity, and the density of the water. What expression relates the variables?

4.15 Derive an expression for the velocity of liquid issuing from a hole in the side of an open tank if the velocity depends on the height H of the hole from the free surface, the fluid viscosity and density, gravity, and the hole diameter.

4.16 The pressure drop Δp in the pipe of Figure P4.16 depends on the average velocity, the pipe diameter, the kinematic viscosity, the length L of pipe, the wall roughness height e, and the fluid density. Find an expression for Δp.

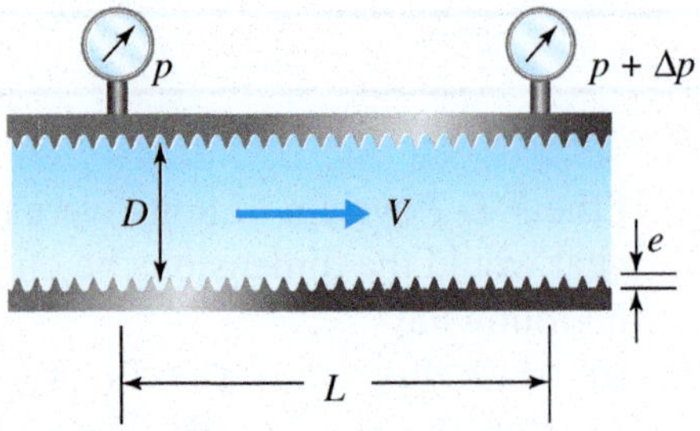

Figure P4.16

4.17 Select an appropriate set of variables that influence the drag force F_D on an airfoil (Figure P4.17) and write the final form in terms of dimensionless parameters.

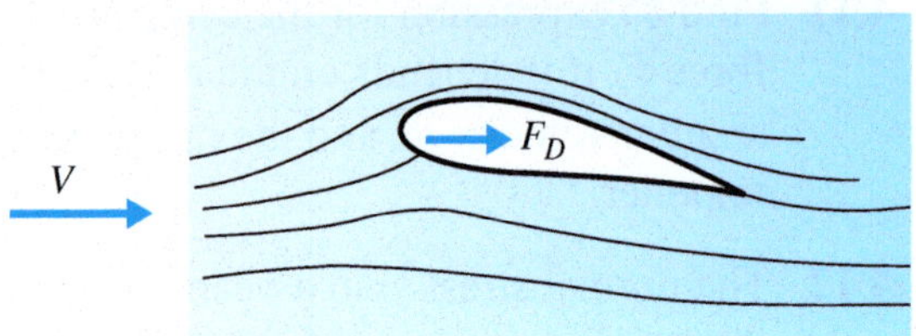

Figure P4.17

4.18 The flow rate Q in an open channel depends on the hydraulic radius R, the cross-sectional area A, the wall roughness height e, gravity g, and the slope S. Relate Q to the other variables using

(a) the M-L-T system

(b) the F-L-T system

4.19 The velocity of propagation of ripples in a shallow liquid depends on the liquid depth h, gravity g, surface tension σ, and the liquid density ρ. Find an expression for the propagation velocity V. See Figure P4.19.

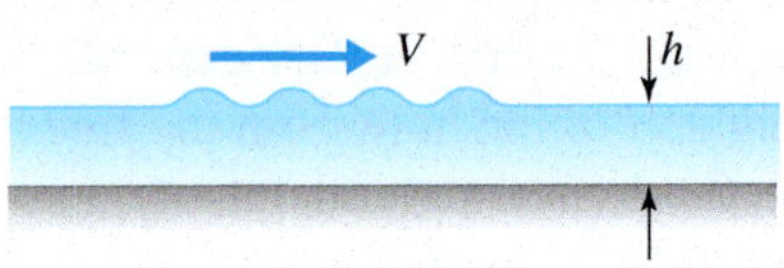

Figure P4.19

4.20 The drag force F_D on a sphere depends on the velocity V, the viscosity μ, the density ρ, the surface roughness height e, the freestream fluctuation intensity I (a dimensionless quantity), and the diameter D. Find an expression for F_D.

4.21 The drag force acting on a ship is considered to be a function of the fluid density ρ, viscosity μ, gravity g, ship velocity V, and a characteristic length l. Determine a set of suitable dimensionless numbers to describe the relationship $F_D = f(\rho, \mu, g, V, l)$.

4.22 The thrust force T, in newtons, from an airplane propeller is a function of propeller diameter D, air density ρ, air viscosity μ, and airplane speed V. Determine a suitable set of dimensionless numbers to describe the relationship $T = f(\rho, D, \mu, V)$.

4.23 The drag force F_D on the smooth sphere of Figure P4.23 falling in a liquid depends on the constant sphere speed V, the solid density ρ_s, the liquid density ρ and viscosity μ, the sphere diameter D, and gravity g. Find an expression for F_D using

(a) the M-L-T system

(b) the F-L-T system

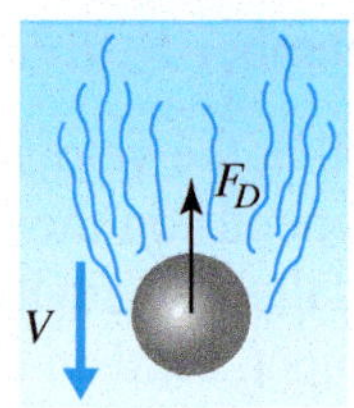

Figure P4.23

4.24 The drag force F_D on a golf ball depends on the velocity, the viscosity, the density, the diameter, the dimple depth, the dimple radius, and the concentration C of dimples measured by the number of dimples per unit area. What expression relates F_D to the other variables?

4.25 The frequency f with which vortices are shed from a cylinder depends on the viscosity, density, velocity, and diameter. Derive an expression for f.

4.26 The lift F_L on an airfoil depends on the aircraft speed V, the speed of sound c, the fluid density, the airfoil chord length l_c, the airfoil thickness t, and the angle of attack α. Find an expression for F_L.

4.27 Derive an expression for the torque T needed to rotate a disk of diameter d at the angular velocity ω in a fluid with density ρ and viscosity μ if the disk is a distance t from a wall. Also, find an expression for the power requirement.

4.28 The cables holding a suspension bridge (Figure P4.28) are observed to undergo large vibrations under certain wind conditions. Select a set of variables that influence the periodic force acting on a cable and write a simplified relationship of dimensionless parameters.

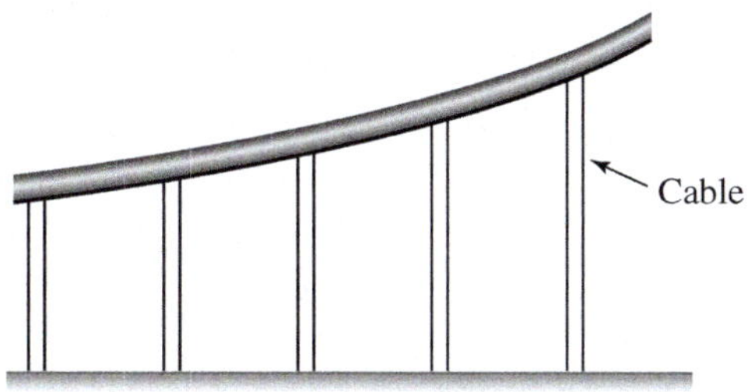

Figure P4.28

4.29 A vacuum cleaner creates a pressure drop Δp across its fan. Relate this pressure drop to the impeller diameter D and width h, its rotational speed ω, the air density ρ, and the inlet and outlet diameters d_i and d_o. Also, find an expression for the power requirement of the fan.

4.30 Derive an expression for the maximum torque needed to rotate an agitator if it depends on the frequency of oscillation, the angular velocity with which the agitator rotates during an oscillation, the diameter of the agitator, the nominal height of the paddles, the length of the paddles, the number of paddles, the depth of liquid, and the liquid density.

4.31 The flow rate of water over a weir depends on the head of water, the width of the weir, gravity, viscosity, density, and surface tension. Relate the flow rate to the other variables.

4.32 The size of droplets in the spray of a fruit sprayer depends on the air velocity; the spray jet velocity, the spray jet exit diameter; the liquid spray surface tension, density, and viscosity; and the air density. Relate the droplet diameter to the other variables.

4.33 Relate the torque T to the other variables shown in Figure P4.33.

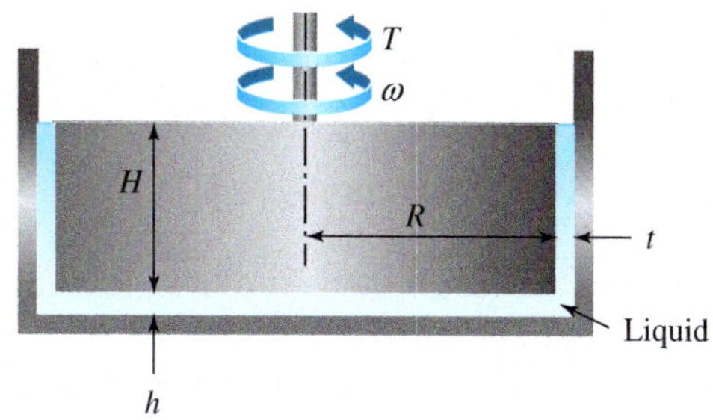

Figure P4.33

4.34 A viscometer is composed of an open tank in which the liquid level is held constant. A small-diameter tube empties into a calibrated volume. Find an expression for the viscosity using the relevant parameters.

4.35 Derive an expression for the torque T necessary to rotate the shaft of Figure P4.35.

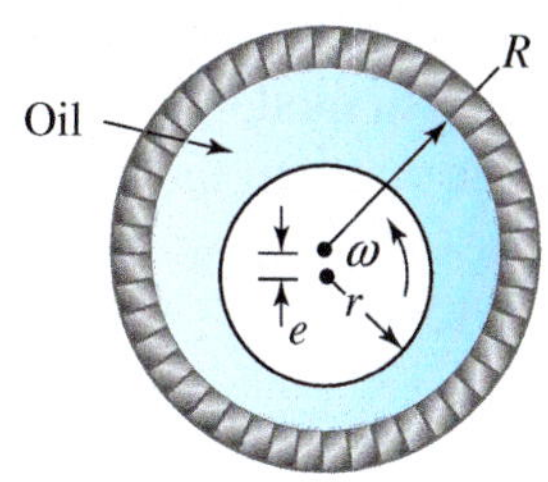

Figure P4.35

4.36 Derive an expression for the depth y_2 in the hydraulic jump of Figure P4.36.

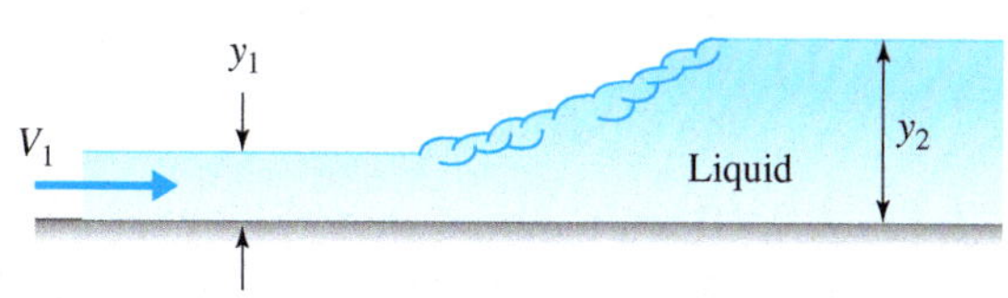

Figure P4.36

4.37 Derive an expression for the frequency with which a cylinder, suspended by a string in a liquid flow, oscillates.

Similitude

4.38 After a model study has been performed, it is necessary to predict quantities of interest that are to be expected on the prototype. Write expressions, in terms of density, velocity, and length, for the ratio of model to prototype of each of the following: flow rate Q, pressure drop Δp, pressure force F_p, shear stress τ_0, torque T, and heat transfer rate $\dot{Q}$.

4.39 A 1:7 scale model simulates the operation of a large turbine that is to generate 200 kW with a flow rate of 1.5 m^3/s. What flow rate should be used in the model, and what power output is expected?
- **(a)** Water at the same temperature is used in both model and prototype.
- **(b)** The model water is at 25°C and the prototype water is at 10°C.

4.40 A 1:5 scale model of a large pump is used to test a proposed change. The prototype pump produces a pressure rise of 600 kPa at a mass flux of 800 kg/s. Determine the mass flux to be used in the model and the expected pressure rise.
- **(a)** Water at the same temperature is used in both model and prototype.
- **(b)** The water in the model study is at 30°C and the water in the prototype is at 15°C.

4.41 A force on a component of a 1:10 scale model of a large pump is measured to be 10 lb. What force is expected on the prototype component if water is used for both model and prototype:
- **(a)** With the water at the same temperature?
- **(b)** With the prototype water at 50°F and the model water at 70°F?

4.42 A 1:10 scale model study of an automobile is proposed. A prototype speed of 100 km/h is desired. What wind-tunnel speed should be selected for the model study? Comment on the advisability of such a speed selection.

4.43 A 1:10 scale model study of a torpedo is proposed. Prototype speeds of 90 km/h are to be studied. Should a wind tunnel or a water facility be used?

4.44 A 1:10 scale model of an airplane propeller is to be tested in a laboratory using a wind tunnel. The prototype on which we are to estimate the thrust moves at V = 200 km/hr at 8000 m altitude. Assuming the prototype and model have geometric similarity, use the result from Problem 4.22 to achieve dynamic similarity.
- **(a)** What air speed should be used in the wind tunnel using standard air?
- **(b)** What is the corresponding thrust for the prototype, if a thrust of 10 N is measured on the model?

4.45 It is proposed that a model study be carried out on a proposed low-speed airfoil that is to fly at low altitudes at a speed of 50 m/s. If a 1:10 scale model is to be constructed, what velocity should be used in a wind tunnel? Comment as to the advisability of such a test. Would it be more advisable to perform the test in a 20°C water channel? If a water channel study were undertaken, calculate the drag ratio between the model and prototype.

4.46 A model study of oil (SAE-10W) at 30°F flowing in a 2.5-ft-diameter pipe is performed by using water at 70°F. Pipe of what diameter should be used if the average velocities are to be the same? What pressure drop ratio is expected? $S_{oil} = 0.9$.

4.47 A 0.025-mm-long microorganism moves through water at the rate of 0.1 body length per second. Could a model study be performed for such a prototype in a water channel or a wind tunnel?

4.48 A 1:30 scale model of a ship is to be tested with complete similarity. What should be the viscosity of the model fluid? Is such a liquid possible? What conclusion can be stated?

4.49 A 1:60 scale model of a ship is used in a water tank to simulate a ship speed of 10 m/s. What should be the model speed? If a towing force of 10 N is measured on the model, what force is expected on the prototype? Neglect viscous effects.

4.50 The flow rate over a weir is 2 m^3/s of water. A 1:10 scale model of the weir is tested in a water channel.
- **(a)** What flow rate should be used?
- **(b)** If a force of 12 N is measured on the model, what force would be expected on the prototype?

4.51 A 1:10 scale model of a hydrofoil is tested in a water channel with a force of 0.8 lb measured at a speed of 20 ft/s. Determine the speed and force predicted for the prototype. Neglect viscous effects.

4.52 The propeller of a ship is to be studied with a 1:10 scale model.
- **(a)** Assuming that the propeller operates close to the surface, select the propeller speed of the model if the prototype speed is 600 rpm.
- **(b)** What torque would be expected if 1.2 N · m is measured on the model?

4.53 The pontoon of a seaplane is to be studied in a water channel that has a channel speed capability of 6 m/s. If the seaplane is to lift off at 100 m/s, what model scale should be selected?

4.54 A 1:30 scale model study of a submarine is proposed in an attempt to study the influence of a suggested shape modification. The prototype is 2 m in diameter, and it is designed to travel at 15 m/s. The model is towed in a water tank at 2 m/s, and a drag force of 2.15 N is measured. Does similitude exist for this test? If so, predict the power needed for the prototype.

4.55 The smoke emanating from the stacks on a ship has a tendency to make its way to the deck down the outside of the stacks causing discomfort to the passengers. This problem is studied with a 1:20 scale model of a 4-m-diameter stack. The ship is traveling at 10 m/s. What range of wind-tunnel airspeeds could be used in this study?

4.56 A model study of a dirigible (a large balloon that travels through the air) is to be performed. The 10-m-diameter dirigible is to travel at 20 m/s. If a 40-cm-diameter model is proposed for use in a wind tunnel, or a 10-cm-diameter model for use in a 20°C water channel, which should be selected? Suppose that the wind-tunnel model is used with a speed of 15 m/s and a drag force of 3.2 N is measured. What force would be expected on the water-channel model with a speed of 2.4 m/s in the water channel? What horsepower would be predicted to overcome drag on the prototype? Assume that the flow is Reynolds number independent for $Re > 10^5$.

4.57 A model study is to be performed of a 1000-ft-tall, 45-ft-diameter smokestack of a power plant. It is known that the smokestack is submerged in a ground boundary layer that is 1200 ft thick. Could the study be performed in a wind tunnel that produces a 4-ft-thick boundary layer?

4.58 A 1:20 scale model of an aircraft is tested in a 23°C wind tunnel. A speed of 200 m/s is used in the model study. A drag force of 10 N is measured. What prototype speed and drag force does this simulate if the elevation is:
- **(a)** Sea level?
- **(b)** 5000 m?
- **(c)** 10 000 m?

4.59 A 1:10 scale model of an airfoil is tested in a wind tunnel that uses outside air at 0°C. The test is to simulate an aircraft speed of 250 m/s at an elevation of 10 000 m. What wind-tunnel speed should be used? If a velocity of 290 m/s and a pressure of 80 kPa absolute are measured on the model at a particular location at a 5° angle of attack, what speed and pressure are expected on the prototype at the corresponding location, and what is the angle of attack?

4.60 A 1:10 scale model of a propeller on a ship is to be tested in a water channel. What should be the rotational speed of the model if the rotational speed of the propeller is 2000 rpm and:
(a) The Froude number governs the study?
(b) The Reynolds number governs the study?

4.61 A torque of 12 N · m is measured on a 1:10 scale model of a large wind machine with a wind tunnel speed of 60 m/s. This is to simulate a wind velocity of 15 m/s since viscous effects are considered negligible. What torque is expected on the prototype? If the model rotates at 500 rpm, what prototype angular velocity is simulated?

4.62 An underwater study of a porpoise is to be performed by using a 1:10 scale model. A porpoise swimming at 10 m/s and making one swimming motion each second is to be simulated. What speed could be used in the water channel, and for that speed, how many swimming motions per second should be used?

4.63 A ship model test is used to predict the drag on a ship, which is very important in the design of ships. Model resistance tests are usually carried out in a towing tank. A ship model (1.2 to 2.2 m long for a small towing tank, 4 to 10 m for a large one) is towed at a constant velocity by a mechanically propelled towing carriage. The resistance of the model at constant velocity is recorded by the instruments on the carriage. Usually the test is carried out at a number of constant velocities, and a resistance curve is thus obtained. The drag coefficient $C_D = \rho V^2 A/2$, where A is the wetted area, is proportional to l^2. From dimensional analysis we can write $C_D = f(\text{Re}, \text{Fr})$. Since the drag force consists of two components, a friction drag and a wave drag, we write: $C_D = C_f + C_w$. Assuming that the two forces are independent, we can consider that the friction drag is only a function of Reynolds number, and the wave drag is only a function of Froude number; i.e., $C_f = f(\text{Re})$ and $C_w = f(\text{Fr})$. Since we cannot satisfy both Re and Fr similarity between model and prototype, we will use Fr to achieve dynamic similarity and use Re to determine the friction drag. Hence, from the model test data we can calculate

$$(C_D)_m = \left(\frac{F_D}{\frac{1}{2}\rho V^2 A} \right)_m = (C_f)_m + (C_w).$$

Assuming friction resistance on the ship is similar to that of a flat plate, we calculate $(C_f)_m$ from Re_m and an appropriate equation. The wave drag coefficient is determined using $(C_w)_m = (C_D)_m - (C_f)_m$. Using $(\text{Fr})_p = (\text{Fr})_m$ to determine $(C_w)_p = (C_w)_m$ we then calculate $(C_f)_p$ from Re_p. Finally, we calculate the drag coefficient for the prototype using $(C_D)_p = (C_f)_p + (C_w)_p$. Test data for the model are shown in Tables P4.63(1) and P4.63(2) below.

Table P4.63(1)

Model Features	*Model Data (1:20 scale)*
Wetted area (m^2)	12.205
Model length (m)	7.11
Water temperature (°C)	12.6

Table P4.63(2) Test Data for Model

V(m/s)	F_D(N)
0.805	17.189
0.995	25.49
1.199	35.80
1.399	49.450
1.599	69.66
1.759	92.18
1.899	132.98

If the prototype is used in seawater at 19.2°C, determine the following:

(a) Using this data, calculate the drag coefficient for the prototype.
(b) Plot the drag coefficient versus Re.
(c) Plot the wave drag coefficient versus Fr.

Normalized Differential Equations

4.64 Normalize the continuity equation

$$\frac{\partial \rho}{\partial t}+\frac{\partial}{\partial x}(\rho u)+\frac{\partial}{\partial y}(\rho v)=0$$

using a characteristic velocity V, length l, density ρ_0, and time f^{-1}, where f is the frequency. What dimensionless parameter is introduced?

4.65 Normalize Euler's equation

$$\frac{\partial \mathbf{V}}{\partial t}+u\frac{\partial \mathbf{V}}{\partial x}+v\frac{\partial \mathbf{V}}{\partial y}+w\frac{\partial \mathbf{V}}{\partial z}=-\frac{\nabla p}{\rho}$$

using a characteristic velocity U, length l, pressure ρU^2, and time f^{-1}, where f is the frequency. Find any dimensionless parameters that are introduced.

4.66 Normalize Euler's equation

$$\rho\frac{D\mathbf{V}}{Dt}=-\nabla p-\rho g\nabla h$$

using a characteristic velocity U, length l, pressure ρU^2, and time l/U. Determine any dimensionless parameters introduced in the normalized equation.

4.67 A fluid is at rest between the large horizontal plates shown in Figure P4.67. The top plate is suddenly given a velocity U. Show that the Navier–Stokes equation that describes the resulting motion simplifies to

$$\frac{\partial u}{\partial t}=\nu\frac{\partial^2 u}{\partial y^2}$$

Nondimensionalize this equation using characteristic velocity U, length h, and times **(a)** h/U and **(b)** h^2/ν. Identify any dimensionless parameters that result.

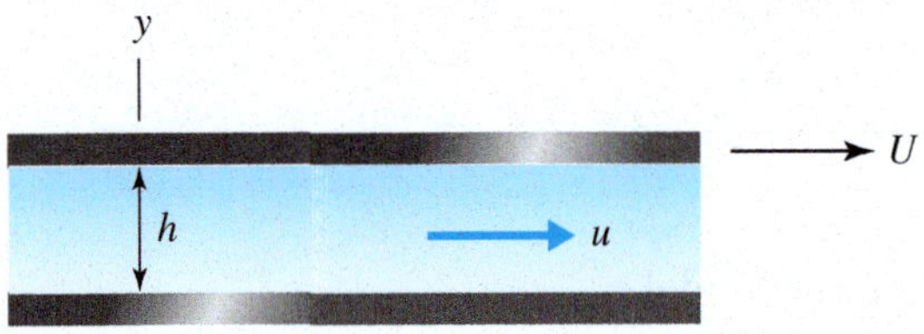

Figure P4.67

4.68 A fluid flows in a horizontal pipe of diameter d (Figure P4.68). The flow is suddenly increased to an average velocity V. Show that the appropriate Navier–Stokes equation, using the coordinates shown, simplifies to

$$\rho\frac{\partial u}{\partial t}=-\frac{\partial p}{\partial x}+\mu\left(\frac{\partial^2 u}{\partial r^2}+\frac{1}{r}\frac{\partial u}{\partial r}\right)$$

Normalize this equation using characteristic velocity V, length d, and time **(a)** d/V and **(b)** d^2/ν. Identify any dimensionless parameters that result.

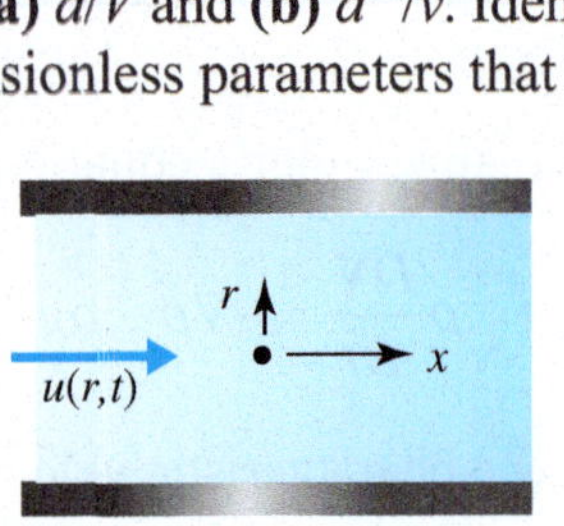

Figure P4.68

4.69 A highly viscous liquid such as honey is flowing down a vertical flat surface. Its thickness decreases as it progresses down the surface (Figure P4.69). Show that the steady flow is described by

$$\rho u\frac{\partial u}{\partial x}=\mu\left(\frac{\partial^2 u}{\partial x^2}+\frac{\partial^2 u}{\partial y^2}\right)+\rho g$$

where we have neglected the y-component of velocity. Normalize this equation using characteristic length h (measured at $x = 0$) and velocity V (the average velocity). Identify any dimensionless parameters that result.

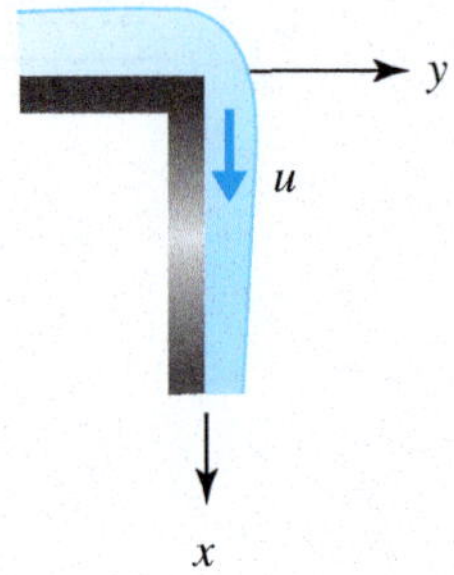

Figure P4.69

4.70 Nondimensionalize the energy equation

$$\rho c_p\left(u\frac{\partial T}{\partial x}+v\frac{\partial T}{\partial y}\right)=K\nabla^2 T$$

using a characteristic velocity U, length l, and temperature T_0. Express the dimensionless parameter that results in terms of the Prandtl number

$$\Pr=\mu\frac{c_p}{K}$$

4.71 Nondimensionalize the compressible flow differential momentum equation

$$\rho\frac{D\mathbf{V}}{Dt}=-\nabla p+\mu\nabla^2\mathbf{V}+\frac{\mu}{3}\nabla(\nabla\cdot\mathbf{V})$$

and the energy equation

$$\rho c_v\frac{DT}{Dt}=K\nabla^2 T-p\nabla\cdot\mathbf{V}$$

using characteristic quantities ρ_0, p_0, T_0, U, and l. The characteristic time is l/U. The speed of sound is $c=\sqrt{kRT_0}$. Find any dimensionless parameters that result if $\Pr = \mu c_p/K$.

5 Internal Flows

Outline

Chapter Objectives

The objectives of this chapter are to:

- Establish the length of the entrance region for both laminar and turbulent flows.
- Determine the laminar flow solution for pipes, parallel plates, and rotating cylinders.
- Present the quantities of interest for turbulent pipe flow, with losses being of particular interest.
- Calculate the flow rate with a centrifugal pump in a single piping system.
- Determine flow rates in open channels.
- Provide numerous examples and problems that demonstrate entrance and developed flow solutions for both laminar and turbulent flows, including losses due to wall friction and various devices. Open channel flows are also analyzed.

M. C. Potter and D. C. Wiggert, *Applied Fluid Mechanics*,
https://doi.org/10.1007/978-3-031-65276-9_5

5.1 INTRODUCTION

In this chapter the effects of viscosity on an incompressible, internal flow will be studied. Such flows are of particular importance to engineers. Flow in a circular pipe is undoubtedly the most common internal fluid flow; it is encountered in the veins and arteries in a body, in a city's water system, in a farmer's irrigation system, in the piping systems transporting fluids in a factory, in the hydraulic lines of an aircraft, and in the ink jet of a computer's printer. Flows in noncircular ducts and open channels must also be included in our study. In Chapter 4 we discovered that the viscous effects in a flow resulted in the introduction of the Reynolds number,

$$\mathrm{Re} = \frac{V\rho l}{\mu} \tag{5.1.1}$$

The Reynolds number was observed to be a ratio of the inertial force to the viscous force. Hence, when this ratio becomes large, it is expected that the inertial forces may dominate the viscous forces. This is usually true when short, sudden geometric changes occur; for long reaches of pipe or open channels, this is not the situation. When the surface areas, such as the wall area of a pipe, are relatively large, viscous effects become quite important and must be included in our study.

Internal flow between parallel plates, in a pipe, between rotating cylinders, and in an open channel will be considered in detail. For a sufficiently low Reynolds number ($\mathrm{Re} < 2000$ in a pipe and $\mathrm{Re} < 1500$ in a wide channel) a laminar flow results, and at a sufficiently high Reynolds number a turbulent flow occurs. Review Section 3.3.3 for a more detailed discussion. We consider laminar flow first and then turbulent flow.

5.2 ENTRANCE FLOW AND DEVELOPED FLOW

When considering internal flows we are interested primarily in developed flows in conduits. A **developed laminar flow** results when the velocity profile ceases to change in the flow direction. Let us first focus our attention on a laminar flow. In the **entrance region** of a laminar flow the velocity profile changes in the flow direction, as shown in Figure 5.1. The idealized flow from a reservoir begins at the inlet as a uniform flow (in reality, there is a thin viscous layer on the wall, as shown); the viscous wall layer then grows over the **inviscid core length** L_i until the viscous stresses dominate the entire cross section; the profile then continues to change in the **profile development region** due to viscous effects until a developed flow is achieved. The inviscid core length is one-fourth to one-third of the **entrance length** L_E, depending on the conduit geometry, shape of the inlet velocity profile, and the Reynolds number.

For a laminar flow in a circular pipe with a uniform profile at the inlet, the entrance length is given by

$$\frac{L_E}{D} = 0.065\,\mathrm{Re} \qquad \mathrm{Re} = \frac{VD}{\nu} \tag{5.2.1}$$

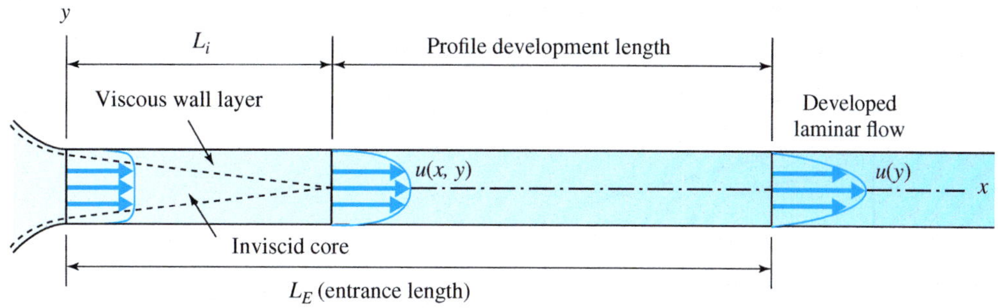

Figure 5.1 Entrance region of a laminar flow in a pipe or a wide rectangular channel.

where the Reynolds number is based on the average velocity and the pipe diameter. Laminar flow in a pipe has been observed at Reynolds numbers in excess of 40 000 for carefully controlled conditions in a smooth pipe. However, for engineering applications a value of about 2000 is the highest Reynolds number for which laminar flow is assured; this is due to vibrations of the pipe, fluctuations in the flow, and/or roughness elements on the pipe wall.

For a laminar flow in a high-aspect-ratio channel (the aspect ratio is the width divided by the distance between the top and bottom plates) with a uniform profile at the inlet the entrance length is

$$\frac{L_E}{h} = 0.04\,\text{Re} \qquad \text{Re} = \frac{Vh}{\nu} \tag{5.2.2}$$

where the Reynolds number is based on the average velocity and the distance h between the plates. The inviscid core length is approximately one-third of the entrance length. A laminar flow cannot exist over Re = 7700; for engineering situations a value of 1500 is often used as the upper limit for laminar flow.

For a turbulent flow the situation is slightly different, as shown in Figure 5.2 for flow in a pipe. A developed flow results when all characteristics of the flow cease to change in the flow direction; this includes details of the turbulence that will be introduced later in this chapter. The inviscid core exists followed by the velocity profile development region, which terminates at $x = L_d$. An additional length is needed, however, for the detailed structure of the turbulent flow to develop. The detailed structure is important in certain calculations such as accurate estimates

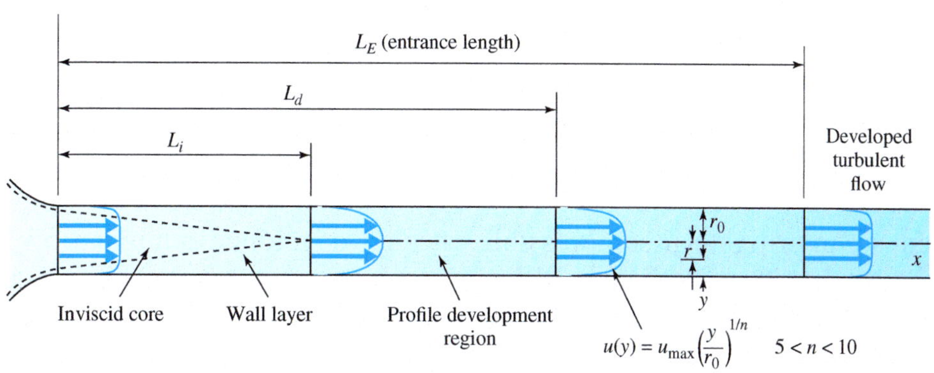

Figure 5.2 Velocity profile development in a turbulent pipe flow.

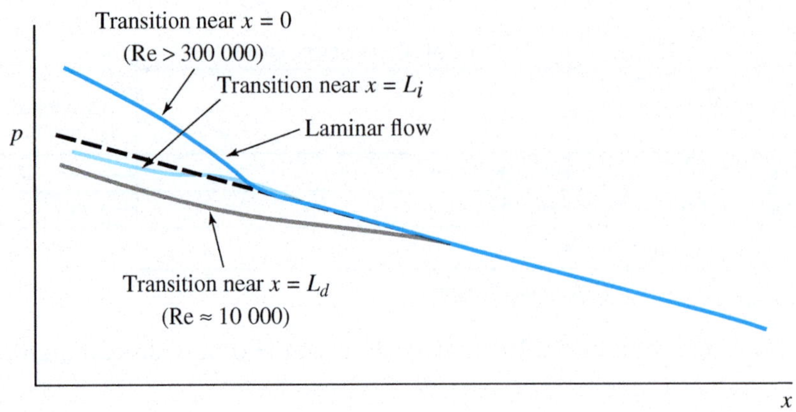

Figure 5.3 Pressure variation in a horizontal pipe flow for both laminar and turbulent flows. (From PhD Thesis of Dr. Jack Backus, Michigan State University)

of wall heat transfer. For large Reynolds number flow ($\text{Re} > 10^5$) in a pipe, where the turbulence fluctuations initiate near $x = 0$, tests have yielded

$$\frac{L_i}{D} \simeq 10 \qquad \frac{L_d}{D} \simeq 40 \qquad \frac{L_E}{D} \simeq 120 \tag{5.2.3}$$

For turbulent flow with Re = 4000, the foregoing developmental lengths would be substantially higher, perhaps five times the values listed. This is true since, for low-Re turbulent flows, transition to a turbulent flow occurs in the profile development region; hence much of the entry region is laminar with relatively low wall shear. Experimental data are not available for low-Reynolds-number turbulent flow in the entrance region.

In Figure 5.3 the pressure variation is sketched. In the flow beyond a sufficiently large x, it is noted that the pressure variation decreases linearly with x. If transition to turbulent flow occurs near the origin, the linear pressure variation begins near L_i and the pressure gradient [the slope of the $p(x)$ curve] in the inlet region is higher than in the developed flow region; if transition occurs near L_d, as it does for low Re, the linear variation begins at the end of the transition process and the pressure gradient in the inlet region is less than that of developed flow.

For a laminar flow, the pressure variation qualitatively resembles that associated with a large Reynolds number. The pressure gradient is higher than in the developed flow region due to the higher shear stress at the wall and the increasing momentum flux.

5.3 LAMINAR FLOW IN A PIPE

In this section we investigate incompressible, steady, developed laminar flow in a pipe, as sketched in Figure 5.4. Two methods will be used: an elemental approach in Sec. 5.3.1 and a direct solution of the appropriate Navier–Stokes equation in Sec. 5.3.2. Both develop the same equations so either can be used.

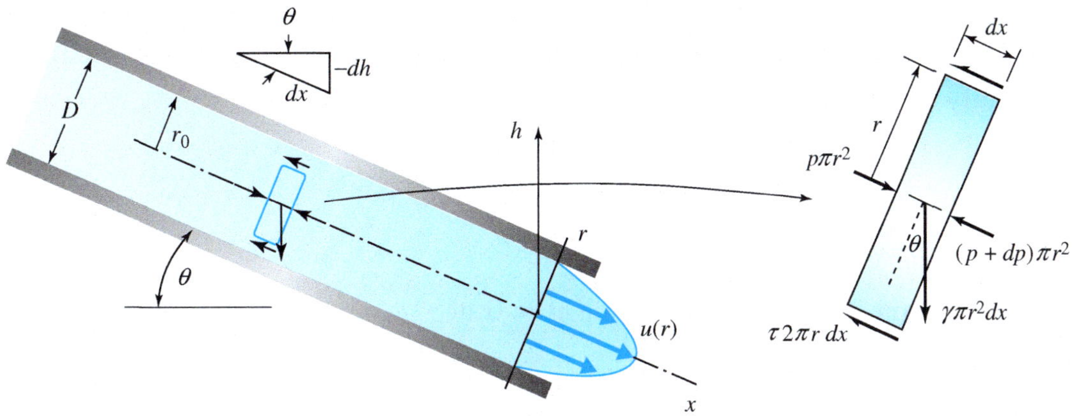

Figure 5.4 Developed flow in a circular pipe.

5.3.1 Elemental Approach

An elemental volume of the fluid is shown in Figure 5.4. This volume can be considered an infinitesimal control volume into which and from which fluid flows, or it can be considered an infinitesimal fluid mass upon which forces are acting. If it is considered a control volume, we would apply the momentum equation (2.6.5); if it is a fluid mass, we would apply Newton's second law. Since the velocity profile does not change in the x-direction, the momentum flux in equals the momentum flux out and the resultant force is zero; since there is no acceleration of the mass element, the resultant force must also be zero. Consequently, a force balance in the x-direction yields

$$p\pi r^2 - (p+dp)\pi r^2 - \tau 2\pi r\, dx + \gamma \pi r^2 dx \sin\theta = 0 \tag{5.3.1}$$

which can be simplified to

$$\tau = -\frac{r}{2}\frac{d(p+\gamma h)}{dx} \tag{5.3.2}$$

where we have used $\sin\theta = -dh/dx$, the vertical direction being denoted by h. Note that Eq. 5.3.2 can be applied to either a laminar or turbulent flow. The shear stress[1] in this laminar flow is related to the velocity gradient and the viscosity giving

$$-\mu\frac{du}{dr} = -\frac{r}{2}\frac{d(p+\gamma h)}{dx} \tag{5.3.3}$$

Since $u = u(r)$ for this developed flow, we can conclude from the above that $d/dx(p + \gamma h)$ must

[1]The shear stress is considered a positive quantity, as displayed in Figure 5.4. The minus sign in $\tau = -\mu\, du/dr$ is necessary since du/dr is negative.

be at most a constant; it cannot depend on x, and from fluid statics $(p + \gamma h)$ does not depend on x since there is no acceleration in the radial direction. Equation (5.3.3) can then be integrated to give the velocity distribution,

$$u(r) = \frac{r^2}{4\mu}\frac{d(p+\gamma h)}{dx} + A \tag{5.3.4}$$

where A is a constant of integration. Using $u = 0$ at $r = r_0$, we can evaluate A and find the velocity distribution to be

$$u(r) = \frac{1}{4\mu}\frac{d(p+\gamma h)}{dx}\left(r^2 - r_0^2\right) \tag{5.3.5}$$

a parabolic profile. It is often referred to as **Poiseuille flow**, named after Jean L. Poiseuille (1799–1869).

The foregoing result can also be obtained by integrating the appropriate Navier–Stokes equation, as illustrated in the following section. If such an exercise is not of interest, go to Section 5.3.3.

5.3.2 Solving the Navier–Stokes Equations

For developed flow in a circular pipe the streamlines are parallel to the wall with no swirl, so that $v_r = v_\theta = 0$ and $u = u(r)$ only. Referring to the momentum equations in cylindrical coordinates of Table 3.1 (note that $u = v_z$ and the z-coordinate has been replaced by x) the x-component Navier–Stokes equation is

$$\rho\left(\overset{\text{steady}}{\frac{\partial u}{\partial t}} + \overset{\text{streamlines} \parallel \text{wall}}{v_r\frac{\partial u}{\partial r}} + \overset{\text{no swirl}}{\frac{v_\theta}{r}\frac{\partial u}{\partial \theta}} + \overset{\text{developed flow}}{u\frac{\partial u}{\partial x}}\right)$$
$$= -\frac{\partial p}{\partial x} + \gamma\sin\theta + \mu\left(\frac{\partial^2 u}{\partial r^2} + \frac{1}{r}\frac{\partial u}{\partial r} + \underset{\text{symmetric flow}}{\frac{1}{r^2}\frac{\partial^2 u}{\partial \theta^2}} + \underset{\text{developed flow}}{\frac{\partial^2 u}{\partial x^2}}\right) \tag{5.3.6}$$

Note that there is no acceleration (the left-hand side is zero) of the fluid particles as they move in the pipe. Equation (5.3.6) simplifies to, using $\sin\theta = -dh/dx$ (see Figure 5.4),

$$\frac{1}{\mu}\frac{\partial(p+\gamma h)}{\partial x} = \frac{1}{r}\frac{\partial}{\partial r}\left(r\frac{\partial u}{\partial r}\right) \tag{5.3.7}$$

where the first two terms in parentheses on the right-hand side of Eq. 5.3.6 have been combined. We see that the left-hand side is at most a function of x and the right-hand side is at most a function of r. Since x and r can be varied independently, we must have

$$\frac{1}{r}\frac{d}{dr}\left(r\frac{du}{dr}\right)=\lambda \tag{5.3.8}$$

where λ is a constant[2] and we have used ordinary derivatives since u depends on only one variable r. Multiply both sides by r and integrate:

$$r\frac{du}{dr}=\frac{\lambda}{2}r^2+A \tag{5.3.9}$$

Divide both sides by r and integrate:

$$u(r)=\frac{\lambda}{4}r^2+A\ln r+B \tag{5.3.10}$$

The velocity must remain finite at $r = 0$; hence $A = 0$. Also, at $r = r_0$, $u = 0$; thus B can be evaluated and we have

$$\begin{aligned} u(r) &= \frac{\lambda}{4}\left(r^2-r_0^2\right) \\ &= \frac{1}{4\mu}\frac{d(p+\gamma h)}{dx}\left(r^2-r_0^2\right) \end{aligned} \tag{5.3.11}$$

where we have used λ as the left-hand side of Eq. 5.3.7. This is the parabolic velocity distribution for flow in a pipe, often referred to as **Poiseuille flow**.

5.3.3 Pipe Flow Quantities

For steady, laminar, developed flow in a circular pipe, the velocity distribution bas been shown to be

$$u(r)=\frac{1}{4\mu}\frac{d(p+\gamma h)}{dx}\left(r^2-r_0^2\right) \tag{5.3.12}$$

[2]If λ were a function of x, u would depend on x, which is not acceptable for this developed flow. If λ were a function of r, $p + \gamma h$ would depend on r, also unacceptable. Hence it is at most a constant.

The average velocity V is found to be

$$V = \frac{Q}{A} = \frac{1}{\pi r_0^2}\int_0^{r_0} u(r)\, 2\pi r\, dr$$
$$= \frac{2}{r_0^2}\int_0^{r_0} \frac{1}{4\mu}\frac{d(p+\gamma h)}{dx}\left(r^2 - r_0^2\right) r\, dr = -\frac{r_0^2}{8\mu}\frac{d(p+\gamma h)}{dx} \quad (5.3.13)$$

Or, expressing the pressure drop Δp in terms of the average velocity, we have, for a horizontal[3] pipe,

$$\Delta p = \frac{8\mu VL}{r_0^2} \quad (5.3.14)$$

where we have used $\Delta p/L = -dp/dx$ since dp/dx is a constant for developed flow. Note that the pressure drop is a positive quantity, whereas the pressure gradient is negative.

The maximum velocity at $r = 0$ from Eq. 5.3.12 is

$$u_{\max} = -\frac{r_0^2}{4\mu}\frac{d(p+\gamma h)}{dx} \quad (5.3.15)$$

so that (see Eq. 5.3.13)

$$V = \frac{1}{2}u_{\max} \quad (5.3.16)$$

The shearing stress is determined to be a linear distribution given by

$$\tau = -\mu\frac{du}{dr}$$
$$= -\frac{r}{2}\frac{d(p+\gamma h)}{dx} \quad (5.3.17)$$

Letting $\tau = \tau_0$ at $r = r_0$, we see that the pressure drop Δp over a length L of a horizontal section of pipe is

$$\Delta p = \frac{2\tau_0 L}{r_0} \quad (5.3.18)$$

where we have again used $dp/dx = -\Delta p/L$.

[3]For a pipe on an incline, simply replace p with $(p + \gamma h)$.

If we introduce the **friction factor** f, a quantity of substantial interest in pipe flow, a dimensionless wall shear, defined by

$$f = \frac{\tau_0}{\frac{1}{8}\rho V^2} \tag{5.3.19}$$

we see that

$$\frac{\Delta p}{\gamma} = h_L = f\frac{L}{D}\frac{V^2}{2g} \tag{5.3.20}$$

where h_L is the head loss (see Eq. 2.5.17) with dimension of length. This equation is often referred to as the **Darcy–Weisbach equation**,[4] named after Henri P. G. Darcy (1803–1858) and Julius Weisbach (1806–1871). Combining Eqs. 5.3.14 and 5.3.20, we find that

$$f = \frac{64}{\text{Re}} \tag{5.3.21}$$

for laminar flow in a pipe. Substituting this back into Eq. 5.3.20, we see that

$$h_L = \frac{32\mu LV}{\gamma D^2} \tag{5.3.22}$$

The head loss is directly proportional to the average velocity (and hence also the discharge) to the first power, a result that generally is applied to developed, laminar flows in conduits, including conduits of shapes other than circular.

Example 5.1

A small-diameter horizontal tube is connected to a supply reservoir as shown in Figure E5.1. If 6600 mm^3 is captured at the outlet in 10 s, estimate the viscosity of the water.

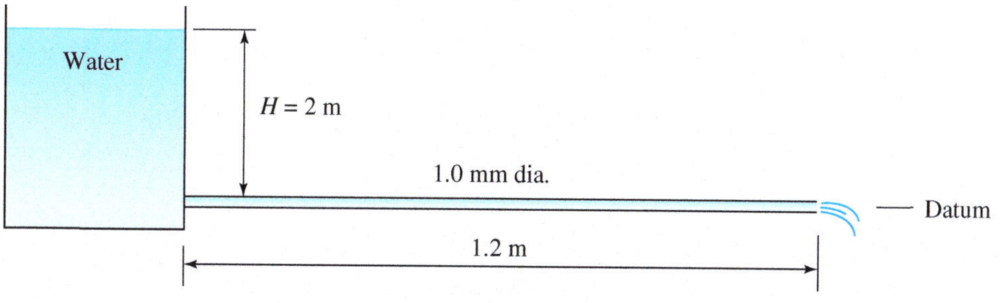

Figure E5.1

[4]This equation will be shown to be valid for both laminar and turbulent flows.

Solution

The tube is very small, so we expect viscous effects to limit the velocity to a small value. Using Bernoulli's equation from the surface to the entrance to the tube, and neglecting the velocity head, we have, letting 0 represent a point on the reservoir surface,

$$\cancelto{0}{\frac{p_0}{\gamma}} + H = \cancelto{0}{\frac{V^2}{2g}} + \frac{p}{\gamma}$$

where we have used gage pressure with $p_0 = 0$. This becomes from fluid statics, assuming $V^2/2g \cong 0$ at the tube's entrance,

$$p = \gamma H = 9800 \text{ N/m}^3 \times 2 \text{ m} = 19\ 600 \text{ Pa}$$

At the exit of the tube the pressure is zero; hence

$$\frac{\Delta p}{L} = \frac{19\,600}{1.2} = 16\,300 \text{ Pa/m (N/m}^3\text{)}$$

The average velocity is found to be

$$V = \frac{Q}{A} = \frac{6600 \times 10^{-9}/10}{\pi \times 0.001^2/4} = 0.840 \text{ m/s}$$

Check to make sure the velocity head is negligible: $V^2/2g = 0.036$ m compared with $p/\gamma = 2$ m, so the assumption of negligible velocity head is valid and our pressure calculation is acceptable. Using Eq. 5.3.14, we can find the viscosity of this assumed laminar flow to be

$$\mu = \frac{r_0^2}{8V}\frac{\Delta p}{L} = \frac{0.0005^2 \text{ m}^2}{8 \times 0.84 \text{ m/s}}(16\,300 \text{ N/m}^3) = \underline{6.06 \times 10^{-4} \text{ N} \cdot \text{s/m}^2}$$

We should check the Reynolds number to determine if our assumption of a laminar flow is acceptable. It is

$$\text{Re} = \frac{\rho V D}{\mu} = \frac{1000 \text{ kg/m}^3 \times 0.84 \text{ m/s} \times 0.001 \text{ m}}{6.06 \times 10^{-4} \text{N} \cdot \text{s/m}^2} = 1390$$

where we use kg/m^3 = N·s^2/m^4. This is obviously a laminar flow since Re < 2000, so the calculations are valid providing the entrance length is not too long. It is

$$L_E = 0.068 \text{Re} \times D = 0.065 \times 1390 \times 0.001 = 0.09 \text{ m}$$

This is approximately 8% of the total length, a sufficiently small quantity; hence, the calculation for viscosity is assumed acceptable.

Example 5.2

Derive an expression for the velocity distribution and the flow rate between horizontal, concentric pipes for a steady, incompressible developed flow (Figure E5.2).

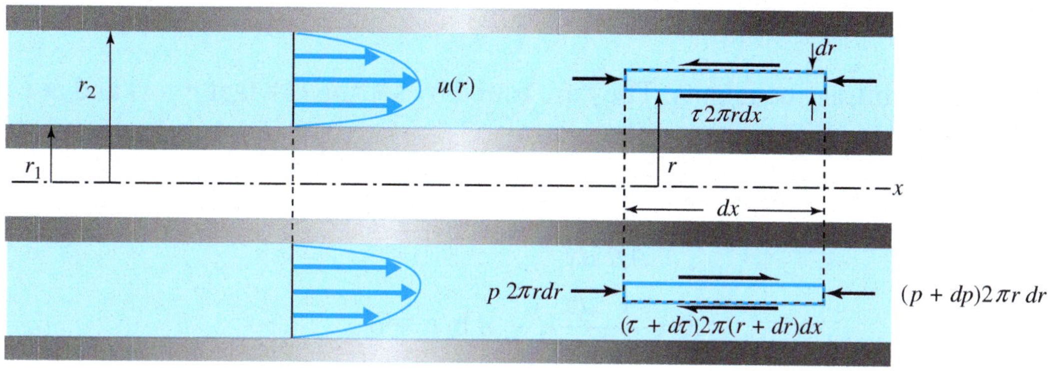

Figure E5.2

Solution

Let us use an elemental approach. The element is a hollow cylindrical shell as sketched in the figure. If we sum forces, we obtain

$$p2\pi r\,dr - (p+dp)2\pi r\,dr + \tau 2\pi r\,dx - (\tau + d\tau)2\pi(r+dr)dx = 0$$

Simplifying, there results, neglecting the term of differential magnitude,

$$\frac{dp}{dx} = -\frac{\tau}{r} - \frac{d\tau}{dr} - \frac{d\tau}{r}$$

neglect

Substituting $\tau = -\mu\, du/dr$ (du/dr is negative near the outer wall where the element is sketched) we have

$$\frac{dp}{dx} = \mu\left(\frac{1}{r}\frac{du}{dr} + \frac{d^2u}{dr^2}\right)$$
$$= \frac{\mu}{r}\frac{d}{dr}\left(r\frac{du}{dr}\right)$$

Multiply both sides by rdr and divide by μ, then integrate:

$$r\frac{du}{dr}=\frac{1}{2\mu}\frac{dp}{dx}r^2+A$$

Multiply both sides by dr/r and integrate again:

$$u(r)=\frac{1}{4\mu}\frac{dp}{dx}r^2+A\ln r+B$$

where A and B are arbitrary constants. They are found by setting $u = 0$ at $r = r_1$ and at $r = r_2$; that is,

$$0=\frac{1}{4\mu}\frac{dp}{dx}r_1^2+A\ln r_1+B$$

$$0=\frac{1}{4\mu}\frac{dp}{dx}r_2^2+A\ln r_2+B$$

Solve for A and B:

$$A=\frac{1}{4\mu}\frac{dp}{dx}\frac{r_1^2-r_2^2}{\ln(r_2/r_1)}$$

$$B=-A\ln r_2-\frac{r_2^2}{4\mu}\frac{dp}{dx}$$

Thus,

$$u(r)=\frac{1}{4\mu}\frac{dp}{dx}\left[r^2-r_2^2+\frac{r_2^2-r_1^2}{\ln(r_1/r_2)}\ln(r/r_2)\right]$$

This is integrated to give the flow rate:

$$Q=\int_{r_1}^{r_2}u(r)2\pi r\,dr$$

$$=-\frac{\pi}{8\mu}\frac{dp}{dx}\left[r_2^4-r_1^4-\frac{\left(r_2^2-r_1^2\right)^2}{\ln(r_2/r_1)}\right]$$

As $r_1 \to 0$ the velocity distribution approaches the parabolic distribution of pipe flow. As $r_1 \to r_2$ this distribution approaches that of parallel-plate flow. These two conclusions are not obvious and are presented as Problem 5.36 at the end of this chapter.

5.4 LAMINAR FLOW BETWEEN PARALLEL PLATES

Consider the incompressible, steady, developed flow of a fluid between parallel plates, with the upper plate moving with velocity U, as shown in Figure 5.5. We will derive the velocity distribution by two methods; either can be used.

5.4.1 Elemental Approach

Consider an elemental volume of unit depth in the z-direction, as sketched in Figure 5.5. If we sum forces in the x-direction, we can write

$$p\,dy - (p + dp)\,dy - \tau\,dx + (\tau + d\tau)dx + \gamma dx\,dy \sin\theta = 0 \tag{5.4.1}$$

The force sum is set equal to 0 since there is no acceleration. The pressure is assumed to depend only on x; variation with y is assumed to be negligible since the dimension a is quite small for most applications. After dividing by $dx\,dy$, the above is simplified to

$$\frac{d\tau}{dy} = \frac{dp}{dx} - \gamma \sin\theta \tag{5.4.2}$$

Since this is one-dimensional flow, the shear stress is

$$\tau = \mu \frac{du}{dy} \tag{5.4.3}$$

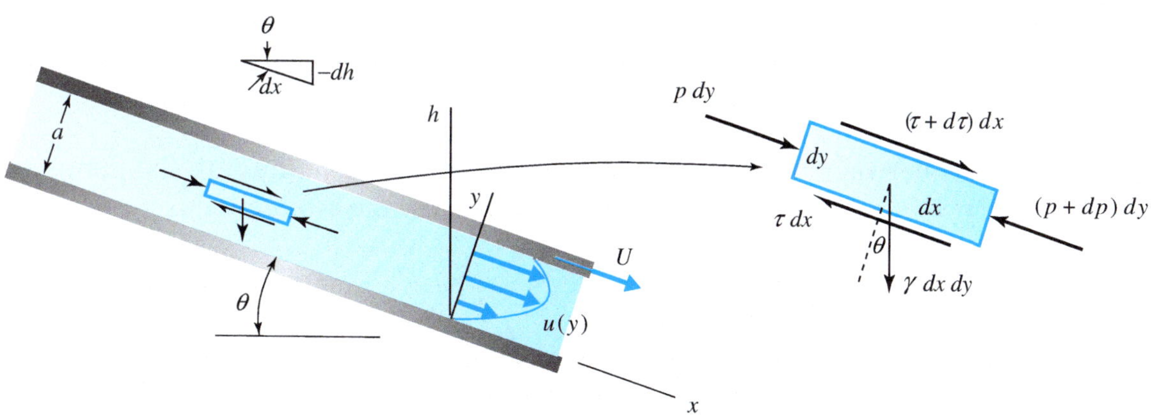

Figure 5.5 Developed flow between parallel plates.

The element is sketched near the lower plate where $du/dy > 0$. Using this and $\sin\theta = -dh/dx$, there results

$$\mu\frac{d^2u}{dy^2} = \frac{dp}{dx} + \gamma\frac{dh}{dx} = \frac{d(p+\gamma h)}{dx} \tag{5.4.4}$$

Since $u = u(y)$ for this developed flow, the left-hand side is only a function of y; because the right-hand side is a function of x we conclude that it must be a constant. Hence, it can be integrated twice to yield (first, divide by μ)

$$u(y) = \frac{y^2}{2\mu}\frac{d(p+\gamma h)}{dx} + Ay + B \tag{5.4.5}$$

where A and B are integration constants. If we require $u = 0$ at $y = 0$ and $u = U$ at $y = a$, we have

$$A = \frac{U}{a} - \frac{a}{2\mu}\frac{d(p+\gamma h)}{dx} \qquad \text{and} \qquad B = 0 \tag{5.4.6}$$

Thus, the velocity distribution is the parabola

$$u(y) = \frac{1}{2\mu}\frac{d(p+\gamma h)}{dx}(y^2 - ay) + \frac{U}{a}y \tag{5.4.7}$$

If the motion is due to the motion of the plate only (a linear profile), the flow is called a **Couette flow**; if the motion is due to the pressure gradient only, that is, $U = 0$, it is Poiseuille flow.

Rather than summing forces on an element we can integrate the appropriate Navier–Stokes equation, as follows. If such an exercise is not of interest, go to Section 5.4.3.

5.4.2 Solving the Navier–Stokes Equations

For developed flow between parallel plates the streamlines are parallel to the plates so that $u = u(y)$ only and $v = w = 0$. The Navier–Stokes equation for the x-direction is (see Eq. 3.3.14)

$$\rho\left(\underbrace{\frac{\partial u}{\partial t}}_{\text{steady}} + u\underbrace{\frac{\partial u}{\partial x}}_{\text{developed flow}} + \underbrace{v}_{0}\frac{\partial u}{\partial y} + \underbrace{w}_{0}\frac{\partial u}{\partial z}\right)$$

$$= -\frac{\partial \rho}{\partial x} + \mu\left(\underbrace{\frac{\partial^2 u}{\partial x^2}}_{\text{developed flow}} + \frac{\partial^2 u}{\partial y^2} + \underbrace{\frac{\partial^2 u}{\partial z^2}}_{\text{wide channel}}\right) + \gamma\sin\theta \tag{5.4.8}$$

If the parallel plates are the top and bottom of a channel, the channel must be wide; the analysis then applies to the midsection away from the side walls. Equation (5.4.8) reduces to, using $\sin\theta = -dh/dx$

$$\frac{\partial^2 u}{\partial y^2} = \frac{1}{\mu}\frac{d(p+\gamma h)}{dx} \tag{5.4.9}$$

The left-hand side is at most a function of y, and the right-hand side is at most a function of x. Hence, we must have

$$\frac{\partial^2 u}{\partial y^2} = \lambda \tag{5.4.10}$$

where λ is a constant, since x and y are independent variables. This can be integrated twice to yield

$$u(y) = \frac{\lambda}{2}y^2 + Ay + B \tag{5.4.11}$$

where A and B are arbitrary constants of integration. We demand that $u = 0$ at $y = 0$ and $u = U$ at $y = a$. This gives

$$A = \frac{U}{a} - \frac{\lambda a}{2} \qquad \text{and} \qquad B = 0 \tag{5.4.12}$$

Using these constants, we can write Eq. 5.4.11 as

$$\begin{aligned} u(y) &= \frac{\lambda}{2}(y^2 - ay) + \frac{U}{a}y \\ &= \frac{1}{2\mu}\frac{d(p+\gamma h)}{dx}(y^2 - ay) + \frac{U}{a}y \end{aligned} \tag{5.4.13}$$

where we have used λ as the right-hand side of Eq. 5.4.9. If the flow is due to the motion of the plate only (a linear profile), the flow is called a **Couette flow;** if the motion is due to the pressure gradient only, that is, $U = 0$, it is Poiseuille flow.

5.4.3 Simplified Flow Situation

Using Eq. 5.4.13, we can write the expression for the velocity distribution between fixed plates (let $U = 0$) as

$$u(y) = \frac{1}{2\mu}\frac{d(p+\gamma h)}{dx}(y^2 - ay) \tag{5.4.14}$$

Using this distribution, we can find the flow rate per unit width to be

$$\begin{aligned} Q &= \int u\, dA \\ &= \int_0^a \frac{1}{2\mu}\frac{d(p+\gamma h)}{dx}(y^2 - ay)\, dy = -\frac{a^3}{12\mu}\frac{d(p+\gamma h)}{dx} \end{aligned} \tag{5.4.15}$$

The average velocity V is found to be

$$\begin{aligned} V &= \frac{Q}{a \times 1} \\ &= -\frac{a^2}{12\mu}\frac{d(p+\gamma h)}{dx} \end{aligned} \tag{5.4.16}$$

This can be expressed as the pressure drop in terms of the average velocity; for a horizontal[5] channel we have

$$\Delta p = \frac{12\mu VL}{a^2} \tag{5.4.17}$$

where we have used $\Delta p/L = -dp/dx$ since dp/dx is constant for developed flow.

Observe that the maximum velocity occurs at $y = a/2$ and from Eq. 5.4.14 is

$$u_{\max} = -\frac{a^2}{8\mu}\frac{dp}{dx} \tag{5.4.18}$$

Thus the average velocity is related to the maximum velocity by

$$V = \frac{2}{3}u_{\max} \tag{5.4.19}$$

The shearing stress is found to be

$$\tau = \mu\frac{du}{dy} = \frac{1}{2}\frac{dp}{dx}(2y - a) \tag{5.4.20}$$

[5]For plates on an incline, simply replace p with $(p + \gamma h)$.

At the wall where $y = 0$ there results

$$\tau_0 = -\frac{a}{2}\frac{dp}{dx} \tag{5.4.21}$$

The pressure drop Δp over a length L of horizontal channel is found to be

$$\Delta p = \frac{2\tau_0}{a} L \tag{5.4.22}$$

recognizing that $dp/dx = -\Delta p/L$. This can be expressed in a more convenient form as

$$\frac{\Delta p}{\gamma} = f \frac{L}{2a}\frac{V^2}{2g} \tag{5.4.23}$$

if we introduce the **friction factor** f, defined by

$$f = \frac{\tau_0}{\frac{1}{8}\rho V^2} \tag{5.4.24}$$

In terms of the head loss (see Eq. 2.5.17), Eq. 5.4.23 becomes

$$h_L = f \frac{L}{2a}\frac{V^2}{2g} \tag{5.4.25}$$

If we combine Eqs. 5.4.17, 5.4.21, and 5.4.24, we find that

$$f = \frac{8}{\rho V^2}\left(-\frac{a}{2}\frac{dp}{dx}\right) = \frac{8}{\rho V^2}\left(-\frac{a}{2}\right)\left(-\frac{12\mu V}{a^2}\right) = \frac{48\mu}{\rho a V} = \frac{48}{\text{Re}} \tag{5.4.26}$$

where the Reynolds number $\text{Re} = \rho a V/\mu$, has been introduced. If this is substituted into Eq. 5.4.25, we see that

$$h_L = \frac{12\mu L V}{r a^2} \tag{5.4.27}$$

The head loss is directly proportional to the average velocity, a conclusion that is true for laminar flows, in general.

We have calculated most of the quantities of interest for laminar flow of an incompressible, steady flow between parallel plates. This would, of course, be an acceptable approximation for a wide channel, one for which the width exceeds the height by at least a factor of 8. For channels with smaller aspect ratios, the edge effects become significant and must be accounted for by adding an additional shear stress to the sides of the element in Figure 5.5, or by maintaining $\partial^2 u/\partial z^2$ in Eq. 5.4.8. The solution is beyond the scope of this book.

Example 5.3

Water at 60°F flows with a Reynolds number of 1500 between the 20-inch-wide, horizontal plates shown in Figure E5.3. Calculate

(a) the flow rate,
(b) the wall shear stress,
(c) the pressure drop over 10 ft, and
(d) the velocity at $y = 0.2$ in.

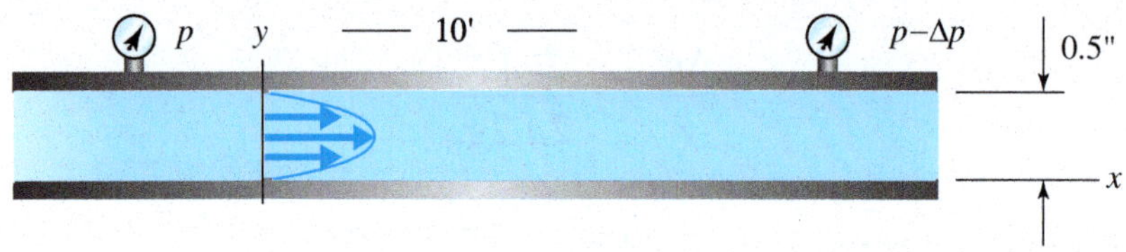

Figure E5.3

Solution

Since the Reynolds number is 1500, the laminar flow equations are assumed applicable.

(a) Using the definition of the Reynolds number (Eq. 7.2.1), the average velocity is found as follows:

$$1500 = \frac{Va}{\nu}$$

$$\therefore V = \frac{1500\nu}{a} = \frac{1500 \times 1.22 \times 10^{-5}}{0.5/12} = 0.439 \text{ ft/s}$$

Thus

$$Q = VA = 0.439 \times \frac{0.5 \times 20}{144} = \underline{0.0305 \text{ ft}^3/\text{s}}$$

(b) Using Eq. 5.4.17, the pressure gradient is

$$\frac{\Delta p}{L} = \frac{12\mu V}{a^2} = \frac{12 \times 2.36 \times 10^{-5} \text{ lb-s/ft}^2 \times 0.439 \text{ ft/s}}{(0.5/12)^2 \text{ ft}^2} = 0.0716 \text{ psf/ft}$$

The shearing stress at the wall is found, using Eq. 5.4.22, to be

$$\tau_0 = \frac{a}{2}\frac{\Delta p}{L} = \frac{0.5/12}{2} \times 0.0716 = \underline{0.00149 \text{ psf}}$$

(c) The pressure drop over 10 ft is found to be

$$\Delta p = 0.0716L = 0.0716 \times 10 = \underline{0.716 \text{ psf}}$$

(d) The velocity distribution of Eq. 5.4.14 is

$$u(y) = \frac{1}{2\mu}\frac{dp}{dx}(y^2 - ay)$$
$$= \frac{1}{2 \times 2.36 \times 10^{-5}}(-0.0716)\left(y^2 - \frac{0.5}{12}y\right) = -1517(y^2 - 0.0417y)$$

where we have used $dp/dx = -\Delta p/L$. At $y = 0{,}2$ in, The velocity is

$$u = -1517\left[\left(\frac{0.2}{12}\right)^2 - 0.0417 \times \frac{0.2}{12}\right] = \underline{0.633 \text{ ft/s}}$$

We have used three significant digits since the fluid properties are assumed known to three significant digits.

Example 5.4

Find an expression for the pressure gradient between two parallel plates with a moving top plate that results in a zero shear stress at the lower wall, where $y = 0$; also, sketch the velocity profiles for a top plate speed of U with various pressure gradients. Assume horizontal plates.

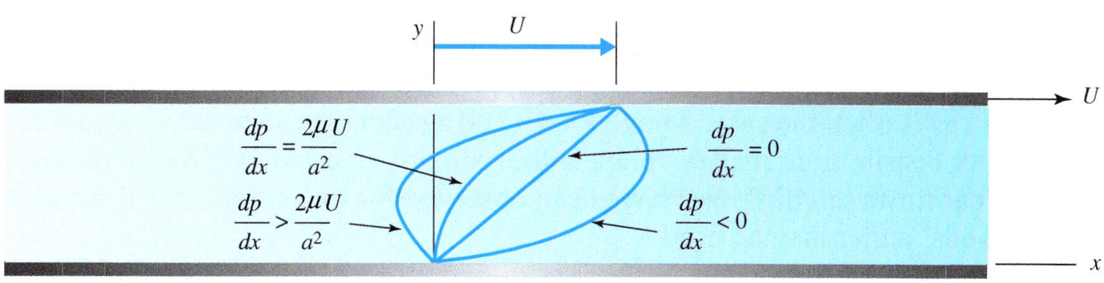

Figure E5.4

Solution

The velocity distribution for plates with the top plate moving with velocity U is given by Eq. 5.4.17. Letting $dh/dx = 0$, we have

$$u(y) = \frac{1}{2\mu}\frac{dp}{dx}(y^2 - ay) + \frac{U}{a}y$$

The shear stress is

$$\tau = \mu \frac{du}{dy}$$
$$= \frac{1}{2}\frac{dp}{dx}(2y-a) + \mu \frac{U}{a}$$

If $\tau = 0$ at $y = 0$, then $du/dy = 0$ at $y = 0$ and the pressure gradient is

$$\frac{dp}{dx} = \frac{2\mu U}{a^2}$$

If dp/dx is greater than this value, the slope du/dy at $y = 0$ is negative and thus the velocity u will be negative near $y = 0$. If $dp/dx = 0$, we observe that a linear velocity distribution results, namely,

$$u(y) = \frac{U}{a} y$$

If dp/dx is negative, $u(y)$ is greater at each y-location than the linear distribution since $(y^2 - ay)$ is a negative quantity for all y's of interest.

All of the results above can be qualitatively displayed on a sketch of $u(y)$ for several dp/dx as shown in Figure E5.4.

5.5 LAMINAR FLOW BETWEEN ROTATING CYLINDERS

Fully developed, steady flow between concentric, rotating cylinders, as shown in Figure 5.6, is another flow that yields a rather simple solution. It has particular application in the field of lubrication, where the fluid may be oil and the inner cylinder a rotating shaft. We will again use two approaches to find the velocity distribution. The laminar flow solution we will find will be valid up to a Reynolds number[6] of 1700, providing the angular velocity of the outer cylinder $\omega_2 = 0$, as is often the case. Above Re = 1700 a secondary laminar flow (a flow with a different velocity distribution) may develop, and eventually a turbulent flow develops. In fact, numerous laminar flows (all different) have been observed for Re > 1700. The flow is analyzed using two methods; either may be used.

5.5.1 Elemental Approach

Body forces will be neglected in this derivation, that is, the cylinders will be assumed to be vertical. Since pressure does not vary with θ, an element in the form of a thin cylindrical shell will be used, as shown in Figure 5.6b. The resultant torque acting on this element is zero because it has no angular acceleration; this is expressed as

$$\tau 2\pi r L \times r - (\tau + d\tau) 2\pi (r + dr) L \times (r + dr) = 0 \tag{5.5.1}$$

[6]The Reynolds number is defined as $\text{Re} = \omega_1 r_1 \delta / \nu$, where $\delta = r_2 - r_1$.

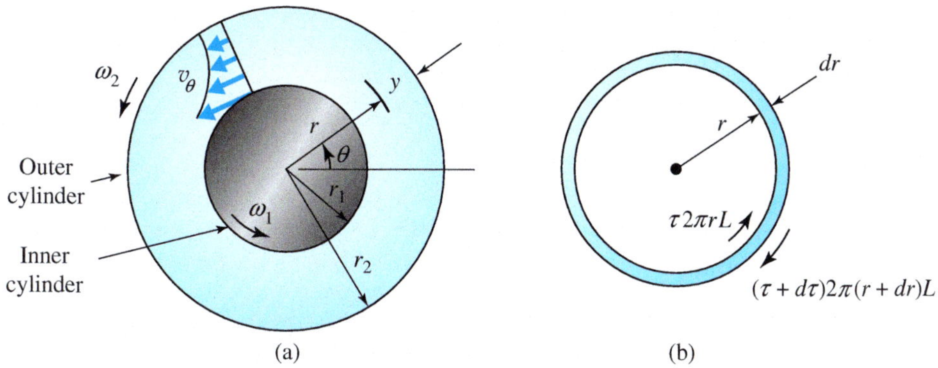

Figure 5.6 Flow between concentric vertical cylinders: (a) basic flow variables; (b) element from between the cylinders.

where L, the length of the cylinders, should be large with respect to the gap width $(r_2 - r_1)$ to avoid three-dimensional end effects. Equation (5.5.1) reduces to (neglecting the three higher-order terms which approach zero as $dr \to 0$)

$$2\tau + r\frac{d\tau}{dr} = 0 \tag{5.5.2}$$

The one-dimensional constitutive equation (see Table 3.1), recognizing that $\tau = -\tau_{r\theta}$, provides the shear stress:

$$\tau = -\mu r \frac{d}{dr}\left(\frac{v_\theta}{r}\right) \tag{5.5.3}$$

There results

$$-2\mu r \frac{d}{dr}\left(\frac{v_\theta}{r}\right) - r\mu \frac{d}{dr}\left[r\frac{d}{dr}\left(\frac{v_\theta}{r}\right)\right] = 0 \tag{5.5.4}$$

Divide by μr, multiply by dr, and integrate to find

$$2\frac{v_\theta}{r} + r\frac{d}{dr}\left(\frac{v_\theta}{r}\right) = A \tag{5.5.5}$$

This can be rearranged as $\left[\text{perform the differentiation}: \frac{d}{dr}\left(\frac{v_\theta}{r}\right) = \frac{1}{r}\frac{dv_\theta}{dr} - \frac{v_\theta}{r^2}\right]$

$$\frac{dv_\theta}{dr} + \frac{v_\theta}{r} = A \tag{5.5.6}$$

or

$$\frac{1}{r}\frac{d}{dr}(rv_\theta) = A \tag{5.5.7}$$

Multiply by $r\,dr$ and integrate again to find that

$$v_\theta(r) = \frac{A}{2}r + \frac{B}{r} \tag{5.5.8}$$

The boundary conditions are $v_\theta = r_1\omega_1$ at $r = r_1$, and $v_\theta = r_2\omega_2$ at $r = r_2$. These conditions allow the constants to be evaluated as

$$A = 2\frac{\omega_2 r_2^2 - \omega_1 r_1^2}{r_2^2 - r_1^2} \qquad B = \frac{r_1^2 r_2^2(\omega_1 - \omega_2)}{r_2^2 - r_1^2} \tag{5.5.9}$$

We can obtain the same result by integrating the appropriate Navier–Stokes equation, or if that is not of interest, go directly to Section 5.5.3.

5.5.2 Solving the Navier–Stokes Equations

For steady, laminar flow between concentric cylinders we assume the streamlines to be circular so that $v_r = v_z = 0$, $v_\theta = v_\theta(r)$ only, and $\partial p/\partial\theta = 0$. The θ-component Navier–Stokes equation from Table 3.1 is

$$\left(\overset{\text{steady}}{\cancel{\frac{\partial v_\theta}{\partial t}}} + \overset{0}{\cancel{v_r}}\frac{\partial v_\theta}{\partial r} + \frac{v_\theta}{r}\overset{\text{symmetric flow}}{\cancel{\frac{\partial v_\theta}{\partial \theta}}} + v_z\overset{0}{\cancel{\frac{\partial v_\theta}{\partial z}}} + \overset{0}{\cancel{\frac{v_r v_\theta}{r}}}\right)$$

$$= -\frac{1}{r}\underset{0}{\cancel{\frac{\partial p}{\partial \theta}}} + \mu\left[\frac{\partial^2 v_\theta}{\partial r^2} + \frac{1}{r}\frac{\partial v_\theta}{\partial r} + \frac{1}{r^2}\underset{\text{symmetric flow}}{\cancel{\frac{\partial^2 v_\theta}{\partial \theta^2}}} + \underset{\text{long cylinders}}{\cancel{\frac{\partial^2 v_\theta}{\partial z^2}}} + \frac{2}{r^2}\underset{\text{symmetric flow}}{\cancel{\frac{\partial v_\theta}{\partial \theta}}} - \frac{v_\theta}{r^2}\right] \tag{5.5.10}$$

This reduces to

$$0 = \frac{\partial^2 v_\theta}{\partial r^2} + \frac{1}{r}\frac{\partial v_\theta}{\partial r} - \frac{v_\theta}{r^2} \tag{5.5.11}$$

which can be written as, recognizing that v_θ depends only on r,

$$\frac{d^2 v_\theta}{dr^2} + \frac{d}{dr}\left(\frac{v_\theta}{r}\right) = 0 \tag{5.5.12}$$

Integrating once, there results

$$\frac{dv_\theta}{dr} + \frac{v_\theta}{r} = A \tag{5.5.13}$$

or

$$\frac{1}{r}\frac{d}{dr}(r\, v_\theta) = A \tag{5.5.14}$$

A second integration yields

$$v_\theta = \frac{A}{2} r + \frac{B}{r} \tag{5.5.15}$$

Applying the boundary conditions $v_\theta = r_1 \omega_1$ at $r = r_1$ and $v_\theta = r_2 \omega_2$ at $r = r_2$, we find that

$$A = 2\frac{r_2^2 \omega_2 - r_1^2 \omega_1}{r_2^2 - r_1^2} \qquad B = \frac{r_1^2 r_2^2 (\omega_1 - \omega_2)}{r_2^2 - r_1^2} \tag{5.5.16}$$

5.5.3 Flow with the Outer Cylinder Fixed

For a number of situations, such as a shaft rotating in a bearing, the outer cylinder is fixed. Setting $\omega_2 = 0$, the velocity distribution is

$$v_\theta = \frac{r_1^2 \omega_1}{r_2^2 - r_1^2} \left(\frac{r_2^2}{r} - r \right) \tag{5.5.17}$$

The shearing stress τ_1 on the inner cylinder (see Table 3.1 and let $\tau = -\tau_{r\theta}$) is found to be

$$\begin{aligned} \tau_1 &= -\left[\mu r \frac{d}{dr}\left(\frac{v_\theta}{r} \right) \right]_{r=r_1} \\ &= \mu \frac{2}{r_1^2} \frac{r_1^2 r_2^2 \omega_1}{r_2^2 - r_1^2} = \frac{2\mu r_2^2 \omega_1}{r_2^2 - r_1^2} \end{aligned} \tag{5.5.18}$$

The torque T necessary to rotate the inner cylinder of length L is

$$\begin{aligned} T &= \tau_1 A_1 r_1 \\ &= \frac{2\mu r_2^2 \omega_1}{r_2^2 - r_1^2} 2\pi r_1 L r_1 = \frac{4\pi\mu r_1^2 r_2^2 L \omega_1}{r_2^2 - r_1^2} \end{aligned} \tag{5.5.19}$$

The power $\dot{W}$ necessary to rotate the shaft is found by multiplying the torque by the rotational speed ω_1; it is

$$\begin{aligned}\dot{W} &= T\omega_1 \\ &= \frac{4\pi\mu r_1^2 r_2^2 L\omega_1^2}{r_2^2 - r_1^2}\end{aligned} \tag{5.5.20}$$

This power is necessary to overcome the resistance of viscosity and results in an increase in the internal energy and thus the temperature of the fluid. The removal of this energy from the fluid often requires special heat exchangers.

Example 5.5

Estimate the viscosity of an oil contained in the annulus between two 25-cm-long cylinders, as shown in Figure E5.6. The outer stationary cylinder is 8 cm in diameter. The 7.8-cm-diameter inner cylinder rotates at 3800 rpm when a torque of 1.2 N · m is applied. The specific gravity of the oil is 0.85. Neglect any torque due to the cylinder ends.

Solution

Assuming that the Reynolds number is less than 1700, Eq. 5.5.19 provides

$$\begin{aligned}\mu &= \frac{T(r_2^2 - r_1^2)}{4\pi r_1^2 r_2^2 L\omega_1} \\ &= \frac{1.2\ \text{N}\cdot\text{m}\,(0.04^2 - 0.039^2)\ \text{m}^2}{4\pi \times 0.04^2\ \text{m}^2 \times 0.039^2\ \text{m}^2 \times 0.25\ \text{m} \times (3800 \times 2\pi/60)\ \text{rad/s}} = \underline{0.0312\ \text{N}\cdot\text{s/m}^2}\end{aligned}$$

Check the Reynolds number using $\nu = \mu/\rho$:

$$\text{Re} = \frac{\omega_1 r_1 \delta}{\nu} = \frac{(3800 \times 2\pi/60)\ \text{rad/s} \times 0.039\ \text{m} \times 0.002/2\ \text{m}}{0.0312/(1000 \times 0.85)\ \text{m}^2/\text{s}} = 423$$

This is less than 1700 so the calculation is acceptable.

Example 5.6

Show that as the inner cylinder radius of Figure E5.6 approaches the outer cylinder radius the velocity distribution approaches the linear distribution between parallel plates with one plate moving and a zero pressure gradient. This is Couette flow.

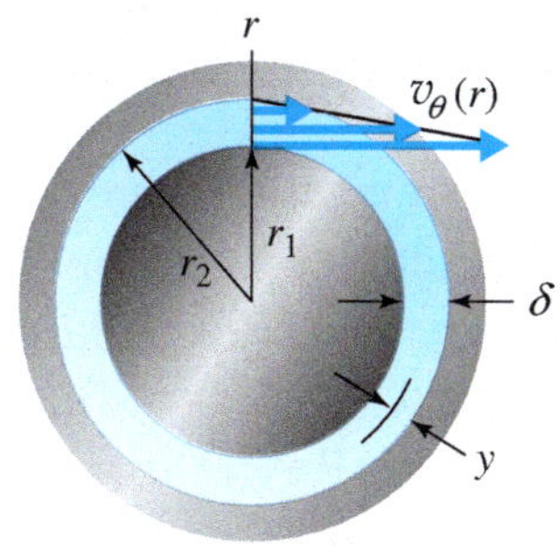

Figure E5.6

Solution

For this problem we will let $\omega_2 = 0$; the velocity distribution (5.5.17) is

$$v_\theta(r) = \frac{r_1^2\omega_1}{r_2^2 - r_1^2}\left(\frac{r_2^2}{r} - r\right)$$

$$= \frac{r_1^2\omega_1}{r_2^2 - r_1^2}\frac{r_2^2 - r^2}{r} = \frac{r_1^2\omega_1}{r_2^2 - r_1^2}(r_2 - r)\frac{r_2 + r}{r}$$

Introduce the independent variable y, measured from the outer cylinder defined by $r + y = r_2$ (see Figure E5.6); let $\delta = r_2 - r_1$. Then the above can be written as

$$v_\theta(r) = \frac{r_1^2\omega_1(r_2 - r)}{(r_2 - r_1)(r_2 + r_1)}\frac{r_2 + r}{r}$$

$$= \frac{r_1^2\omega_1 y}{\delta(r_2 + r_1)}\frac{2r_2 - y}{r_2 - y}$$

As the inner radius approaches the outer radius we can write $r_1 \simeq r_2$. Letting $r_2 \simeq r_1 \simeq R$ we have $r_2 + r_1 \simeq 2R$ and

$$\frac{2R - y}{R - y} \simeq 2$$

since $y << R$. The velocity distribution then simplifies to

$$v_\theta(y) = \frac{R^2\omega_1 y}{\delta 2R} \times 2 = \frac{R\omega_1}{\delta}y$$

This is a linear distribution and is a good approximation to the flow whenever $\delta << R$.

5.6 TURBULENT FLOW IN A PIPE

The study of developed turbulent flow in a circular pipe is of substantial interest in actual flows since most flows encountered in practical applications are turbulent flows in pipes. Even though for carefully controlled laboratory conditions, laminar flows have been observed up to Reynolds numbers of 40 000 in a pipe flow, turbulent flow is assumed to occur in a pipe under standard operating conditions whenever the Reynolds number

$$\mathrm{Re} = \frac{VD}{\nu}$$

exceeds 2000; between 2000 and 4000 it is possible for the flow to randomly oscillate between being laminar and turbulent. Consider, for example, water at 20°C flowing in a rather small 5-mm-diameter pipe; the average velocity need only be 0.8 m/s (2 ft/s in a $\frac{1}{4}$-in. pipe) for turbulent flow to occur. This is the situation whenever we drink water from a drinking fountain. For larger-diameter pipes the average velocity is sufficiently large so that a turbulent flow is produced in most situations.

In a turbulent flow all three velocity components are nonzero. If we measure the components as a function of time, graphs similar to those shown in Figure 5.7 would result for a flow in a pipe where u, v, w are in the x-, r-, and θ-directions, respectively. There is seldom any interest (to the engineer) in the details of the randomly fluctuating velocity components; hence we introduce the notion of a time-average quantity. The velocity components (u, v, w) are written as,

$$u = \overline{u} + u' \qquad v = \overline{v} + v' \qquad w = \overline{w} + w' \tag{5.6.1}$$

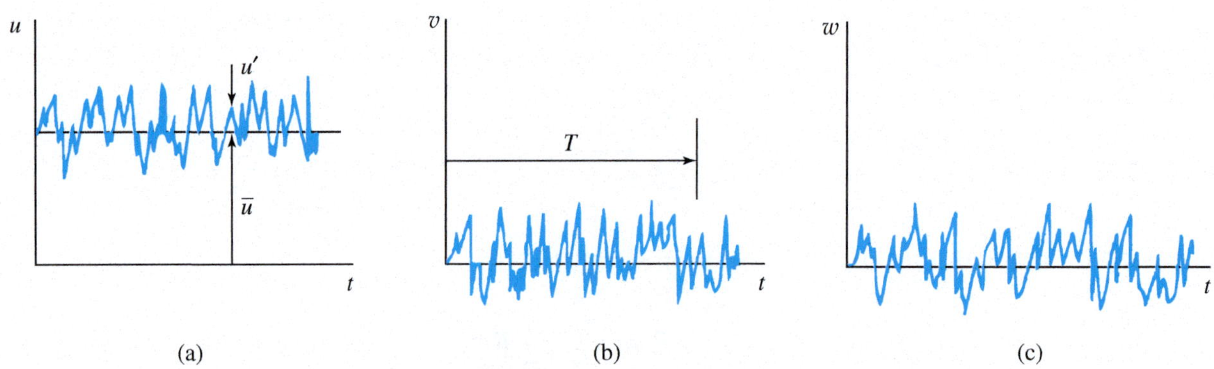

Figure 5.7 Velocity components in a steady turbulent pipe flow: (a) x-component velocity u; (b) r-component velocity v; (c) θ-component velocity w.

where a bar over a quantity denotes a time average and a prime denotes the fluctuating part. Using the component u as an example, the **time average** is defined as,

$$\overline{u} = \frac{1}{T}\int_0^T u(t)\,dt \tag{5.6.2}$$

where T is a time increment large enough[7] to eliminate all time dependence from $\overline{u}$. In a developed pipe flow $\overline{u}$ would be nonzero and $\overline{v} = \overline{w} = 0$, as is observed in Figure 5.7.

Example 5.7

Show that $\overline{u'} = 0$ and $\overline{\frac{\partial u}{\partial y}} = \frac{\partial \overline{u}}{\partial y}$ for a turbulent flow.

Solution

To show that $\overline{u'} = 0$ we simply substitute the expression (5.6.1) for $u(t)$ into Eq. 5.6.2 and obtain

$$\begin{aligned}\overline{u} &= \frac{1}{T}\int_0^T (\overline{u} + u')dt \\ &= \frac{1}{T}\int_0^T \overline{u}dt + \frac{1}{T}\int_0^T u'\,dt \\ &= \overline{u}\frac{1}{T}\int_0^T dt + \overline{u'} \\ &= \overline{u} + \overline{u'}\end{aligned}$$

Subtracting $\overline{u}$ from both sides results in

$$\overline{u'} = 0$$

Now, let us time average the derivative $\partial u/\partial y$. We have

$$\overline{\frac{\partial u}{\partial y}} = \frac{1}{T}\int_0^T \frac{\partial u}{\partial y}dt = \frac{\partial}{\partial y}\left(\frac{1}{T}\int_0^T u\,dt\right) = \frac{\partial}{\partial y}\overline{u}$$

since T is a constant. Thus

$$\overline{\frac{\partial u}{\partial y}} = \frac{\partial \overline{u}}{\partial y}$$

[7]At a given location in a pipe this could be checked experimentally as follows: Start with a relatively large value of T and then decrease T for subsequent runs. If $\overline{u}$ remains unchanged as T decreases, T is sufficiently large. If T is too small, $\overline{u}$ will be different for each run.

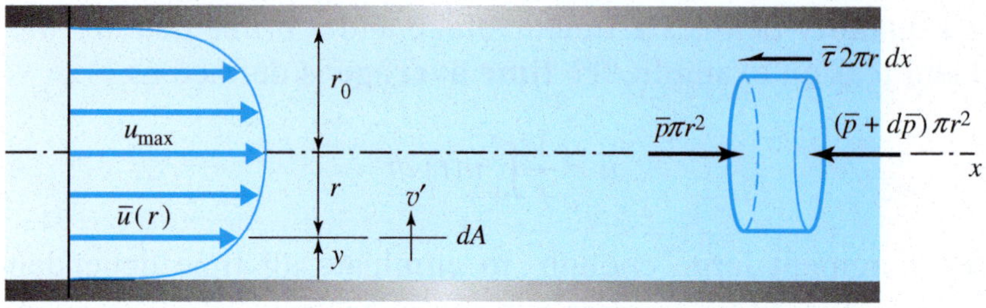

Figure 5.8 Turbulent flow in a horizontal pipe.

5.6.1 Differential Equation

The differential equation that must be solved to provide us with the time-average velocity distribution is derived by using a particle approach. The Navier–Stokes equations could be time averaged, arriving at the same differential equation; that approach is included in the problems.

Consider the situation for a turbulent flow in a horizontal pipe, as shown in Figure 5.8. We use velocity components u and v in the x- and y-directions, respectively. Fluid particles[8] move randomly throughout the flow. At an instant in time a fluid particle moves through the incremental area dA due to the velocity fluctuation v'; it enters a neighboring layer of fluid which is moving with a higher x-component velocity, thereby providing a retarding effect on the neighboring layer. A fluid particle that moves to a neighboring layer that is traveling with a lower x-component velocity would tend to accelerate the slower moving fluid. The x-component force that results due to the random motion of a fluid particle passing through the incremental area dA of Figure 5.8 would be (see Eq. 2.6.6)

$$dF = -\rho v'\,dA\,u' \tag{5.6.3a}$$

where u' is the negative change in x-component velocity due to the momentum exchange and $\rho v' dA$ is the mass flux through the area; the negative sign provides a positive dF. If we divide both sides by the area dA, we obtain a "stress" that we call the **turbulent shear stress**. It is

$$\tau_{\text{turb}} = \frac{dF}{dA} = -\rho u'v' \tag{5.6.3b}$$

where we know that $(u'v')$ is, on the average, a negative quantity since a positive v' produces a negative u'. This "shear stress" is actually a momentum exchange but since it has the same effect as a stress, we call it a shear stress.

The time-average turbulent shear stress, often called the **apparent shear stress**, which is of primary interest, is

$$\overline{\tau}_{\text{turb}} = -\rho\overline{u'v'} \tag{5.6.4}$$

[8]A particle is a relatively small mass of the fluid, consisting of a large number of molecules, that tends to move as a unit.

where Eq. 5.6.2 would be used for the time averages. Note that $\overline{u'w'}$ would be zero since a w'-component (in the θ-direction) would not move a fluid particle into a layer of higher or lower x-component velocity. Also, $\overline{v'w'} = 0$ using the same logic.

The total shear stress at a particular location would be due to both the viscosity and the momentum exchange described by Eq. 5.6.4; that is,

$$\begin{aligned}\bar{\tau} &= \bar{\tau}_{\text{lam}} + \bar{\tau}_{\text{turb}} \\ &= \mu\frac{\partial \bar{u}}{\partial y} - \rho\overline{u'v'}\end{aligned} \tag{5.6.5}$$

The shear stress can be related to the pressure gradient by summing forces on the horizontal cylindrical element shown in Figure 5.8. There results

$$\bar{\tau} = -\frac{r}{2}\frac{d\bar{p}}{dx} = \frac{r\Delta\bar{p}}{2L} \tag{5.6.6}$$

which shows that the shear stress distribution is linear for a turbulent flow as well as a laminar flow (see Eq. 5.3.17). The turbulent shear obviously goes to zero at the wall since the velocity perturbations are zero at the wall, and the total shear is zero at the centerline where $r = 0$ or $y = r_0$, as shown in Figure 5.9. The viscous shear is nonzero only in a very thin viscous wall layer, of thickness δ_v, near the wall, as shown exaggerated in part (b). Note that the turbulent shear reaches a maximum near the wall in the viscous wall layer.

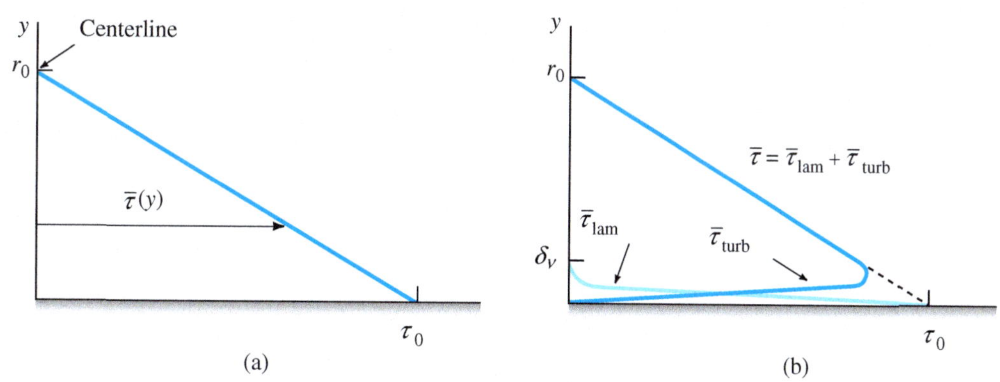

Figure 5.9 Shear stress distributions in a developed pipe flow.

The differential equation that must be solved if the time-average velocity distribution is to be determined is found by combining the two preceding equations; it is

$$\frac{r}{2}\frac{d\bar{p}}{dx} = \rho\overline{u'v'} + \mu\frac{d\bar{u}}{dr} \tag{5.6.7}$$

where we have used $r + y = r_0$ so that $dy = -dr$. For developed flow we know that $d\overline{p}/dx =$ const.; hence, if we know how $\overline{u'v'}$ varied with r, the differential equation can be solved. The quantity $\overline{u'v'}$ cannot be determined analytically, however, so the solution to Eq. 5.6.7 cannot be attempted until an empirical expression is found for $\overline{u'v'}$. Rather than finding an empirical expression for $\overline{u'v'}$ and then solving the differential equation for u, we will simply present the empirical results obtained for the velocity profile $u(y)$. However, before we do this in the next section we should introduce the eddy viscosity, the mixing length, and the correlation coefficient.

Instead of using the quantity $\overline{u'v'}$ as the unknown in Eq. 5.6.7, we often introduce the **eddy viscosity** η, defined by the relationship

$$\overline{u'v'} = \eta \frac{d\overline{u}}{dy} \tag{5.6.8}$$

Note that it has the same dimensions as the kinematic viscosity. In terms of the eddy viscosity the differential equation becomes

$$\frac{r}{2}\frac{d\overline{p}}{dx} = \rho(\nu + \eta)\frac{d\overline{u}}{dr} \tag{5.6.9}$$

If we view the turbulent process as the random and chaotic mixing of particles of fluid, we may choose to introduce the **mixing length** l_m, the distance a particle travels before interacting with another particle. Based on reasoning related to momentum interchange we relate the eddy viscosity to the mixing length with

$$\eta = l_m^2 \left| \frac{d\overline{u}}{dy} \right| \tag{5.6.10}$$

The **correlation coefficient** K_{uv}, a normalized turbulent shear stress, often used when describing turbulent motions, has limits of ± 1 and is defined by

$$K_{uv} = \frac{\overline{u'v'}}{\sqrt{\overline{u'^2}}\sqrt{\overline{v'^2}}} \tag{5.6.11}$$

where the time-average quantities would be defined by Eq. 5.6.2.

The quantities η, l_m, and K_{uv} are functions of r (or y) that simply replace the variable $\overline{u'v'}$. They do not simplify the differential equation (5.6.7); they allow the equation to take slightly different forms. Since we cannot derive an expression for η, l_m, or K_{uv} we cannot find $\overline{u}(r)$ using analytical methods. We must rely on experimental data. It follows in the next section.

Example 5.8

Note that in Figure 5.9b there is a region near the wall where the turbulent shear is near its maximum and is relatively constant, as shown in the expanded view of Figure E5.8, and the viscous shear is quite small. Assume that the mixing length is directly proportional to the distance from the wall. With this assumption show that $\bar{u}(y)$ is logarithmic in this region near the wall.

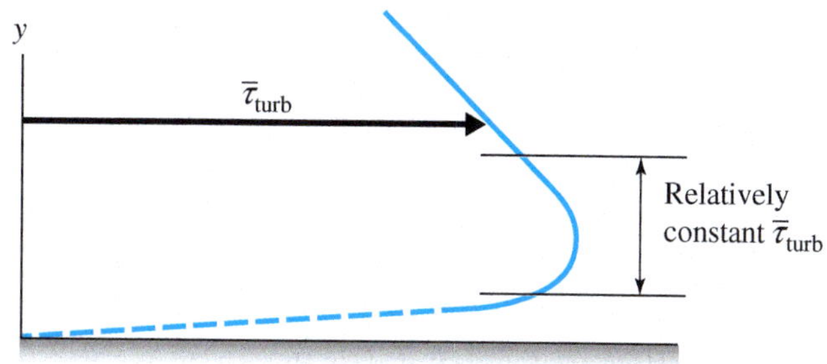

Figure E5.8

Solution

If the viscous shear is negligible (as it is away from the thin wall layer), we have, combining Eqs. 5.6.5 and 5.6.8, and 5.6.10,

$$\bar{\tau}_{turb} = \rho\eta\frac{d\bar{u}}{dy} = \rho l_m^2\left(\frac{d\bar{u}}{dy}\right)^2$$

Now, if $\bar{\tau}_{\text{turb}}$ = const. = c_1 and we assume, as given in the problem statement, that

$$l_m = c_2 y$$

there results

$$c_1 = \rho c_2^2 y^2\left(\frac{du}{dy}\right)^2$$

or

$$y\frac{d\bar{u}}{dy} = c_3$$

where $c_3 = \sqrt{c_1/\rho c_2^2}$. The differential equation is integrated to yield

$$\bar{u}(y) = c_3 \ln y + c_4$$

Hence, with the foregoing assumptions we see that a logarithmic profile is predicted for the region of constant turbulent shear near the wall. This is, in fact, observed from experimental data; so we conclude that the above assumptions are reasonable for a turbulent flow in a pipe.

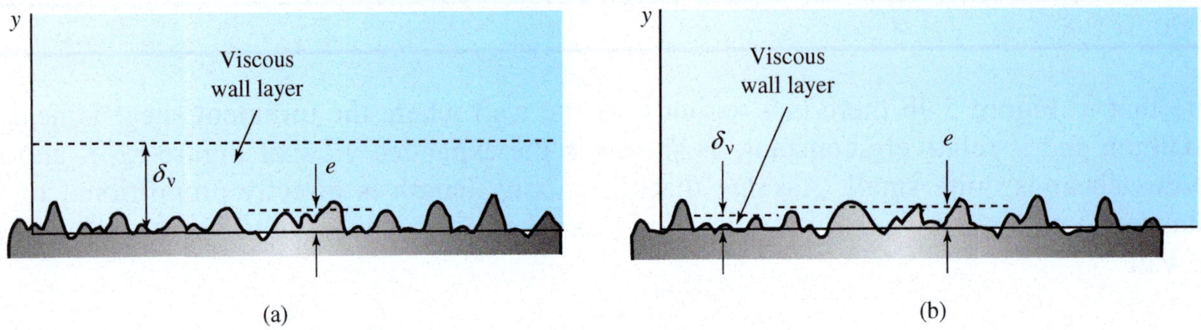

Figure 5.10 (a) A smooth wall and (b) a rough wall.

5.6.2 Velocity Profile

The time-average velocity profile in a pipe is quite sensitive to the magnitude of the average wall roughness height e, as sketched in Figure 5.10. Most materials are "rough" when viewed with sufficient magnification, although glass and plastic are assumed to be smooth with $e = 0$. Values for e are listed in Figure 5.13. As noted in the preceding section, the laminar shear is significant only near the wall in the viscous wall layer with thickness δ_v. If the thickness δ_v, is sufficiently large, it submerges the wall roughness elements, as in Figure 5.10a, so that they have negligible effect on the flow; it is as if the wall were smooth. Such a condition is often referred to as being **hydraulically smooth**. If the viscous wall layer is relatively thin, the roughness elements protrude out of this layer, as in Figure 5.10b, and the wall is rough. The **relative roughness** e/D and the Reynolds number can be used to determine if a pipe is smooth or rough. This will be observed from the friction factor data presented in the next section.

We do not solve for the velocity distribution in a developed turbulent flow since $\overline{u'v'}$ cannot be determined analytically. Thus we will present the empirical data for $\bar{u}(y)$ directly and not solve Eq. 5.6.7. The two most common expressions for turbulent flow are now presented.

The first method of empirically expressing the velocity distribution involves flows with smooth walls and flows with rough walls. If the flow has a smooth wall, as in Figure 5.10a, we identify two regions of the flow, the **wall region** and the **outer region**. In the wall region the characteristic velocity and length are the **shear velocity** u_τ defined by[9] $u_\tau = \sqrt{\tau_0/\rho}$ and the **viscous length** ν/u_τ. The dimensionless velocity distribution in the wall region, for a smooth pipe, is

$$\frac{\bar{u}}{u_\tau} = \frac{u_\tau y}{\nu} \text{ (viscous wall layer)} \qquad 0 \le \frac{u_\tau y}{\nu} \lesssim 5 \tag{5.6.12}$$

[9]The shear velocity u_τ is a fictitious velocity and has no relationship to an actual velocity in the flow. It is a quantity with dimensions of velocity that allows the experimental data to be presented in dimensionless form as **universal profiles** (profiles that are valid for all turbulent pipe flows). The viscous length ν/u_τ is also a fictitious length.

$$\frac{\bar{u}}{u_\tau} = 2.44 \ln \frac{u_\tau y}{\nu} + 4.9 \text{ (turbulent region)} \qquad 30 < \frac{u_\tau y}{\nu}, \ \frac{y}{r_0} < 0.15 \tag{5.6.13}$$

In the interval $5 < u_\tau y/\nu < 30$, the **buffer zone**, the experimental data do not fit either of the curves above but merge the two curves as shown in Figure 5.11a. The viscous wall layer has thickness δ_ν and it is in the viscous layer that turbulence is thought to be initiated; this layer possesses a time-average, linear velocity distribution, but instantaneously the layer is very time dependent. The outer edge of the wall region is quite dependent on the Reynolds number, as shown; for low Re it may be located near $u_\tau y/\nu = 3000$.

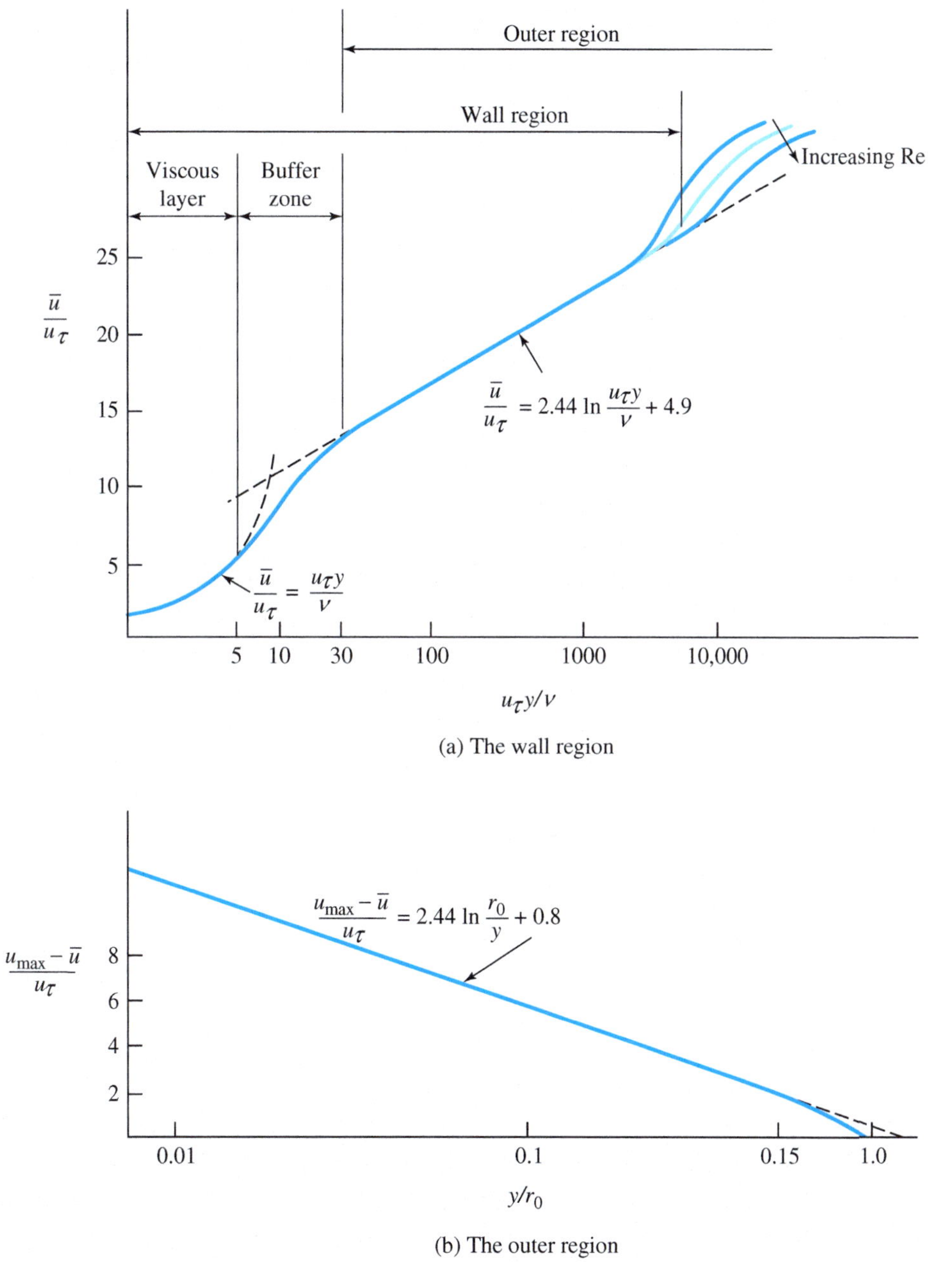

Figure 5.11 Empirical relations for turbulent flow in a smooth pipe: (a) wall region; (b) outer region.

For rough pipes the viscous wall layer does not play an important role since turbulence initiates from the protruding wall elements, so only a logarithmic profile is necessary in the wall region. The characteristic length is the average roughness height e; the dimensionless velocity profile for the rough pipe is

$$\frac{\overline{u}}{u_\tau} = 2.44 \ln \frac{y}{e} + 8.5 \qquad \frac{y}{r_0} < 0.15 \qquad \text{(wall region)} \tag{5.6.14}$$

where the constants 8.5 and 2.44 allow a good fit with experimental data.

In the outer region, sketched in Figure 5.11b, the characteristic length is r_0; the velocity defect ($u_{max} - \overline{u}$) is normalized with u_τ, and the empirical relation, for both smooth and rough pipes, is

$$\frac{u_{max} - \overline{u}}{u_\tau} = 2.44 \ln \frac{r_0}{y} + 0.8 \frac{y}{r_0} \leq 0.15 \qquad \text{(outer region)} \tag{5.6.15}$$

An additional empirical equation is needed to complete the profile for $0.15 < y/r_0 \leq 1$.

The wall region and the outer region overlap as shown in Figure 5.11a. In this overlap region we can combine the equations above to obtain an expression for the maximum velocity; for a smooth pipe it is

$$\frac{u_{max}}{u_\tau} = 2.44 \ \ln \ \frac{u_\tau r_0}{\nu} + 5.7 \qquad \text{(smooth pipes)} \tag{5.6.16}$$

and for a rough pipe we find that

$$\frac{u_{max}}{u_\tau} = 2.44 \ \ln \ \frac{r_0}{e} + 9.3 \qquad \text{(rough pipes)} \tag{5.6.17}$$

Although we do not often desire the actual time-average velocity at a specified radial position in a pipe, the distributions above are of occasional use and are presented for completeness. Note, however, that before u_{max} can be found we must know u_τ; before u_τ can be found we must know τ_0. To find τ_0 we use the pressure gradient,

$$\tau_0 = -\frac{r_0}{2} \frac{dp}{dx} \tag{5.6.18}$$

or the friction factor with Eq. 5.3.19. If neither dp/dx nor f is known, we may use the power-law form of the profile, described in the following paragraph, to approximate f.

An alternative, and simpler form that adequately describes the turbulent flow velocity distribution in a pipe is the **power-law profile**, namely,

$$\frac{\bar{u}}{u_{\max}} = \left(\frac{y}{r_0}\right)^{1/n} \tag{5.6.19}$$

where y is measured from the pipe wall and n is an integer between 5 and 10. Using this distribution the average velocity is found to be

$$V = \frac{1}{\pi r_0^2}\int_0^{r_0} \bar{u}(r) 2\pi r\, dr = \frac{2n^2}{(n+1)(2n+1)} u_{\max} \tag{5.6.20}$$

This distribution is compared with a laminar profile in Figure 5.12.

The value of n in the exponent is related to the friction factor f by the empirical expression

$$n = \frac{1}{\sqrt{f}} \tag{5.6.21}$$

The constant n varies from 5 to 10 depending on the Reynolds number and the pipe wall roughness e/D. For smooth pipes, the exponent n is related to the Reynolds number as shown in Table 5.1.

The power-law profile cannot be used to obtain the slope at the wall since it will always yield $(d\bar{u}/dy)_{\text{wall}} = \infty$ for all n. Thus it cannot be used to predict the wall shear stress. The wall shear is found by combining Eqs. 5.6.21 and 5.3.19. In addition, it gives a positive slope $(d\bar{u}/dy)$ at the centerline of the pipe, where the slope should be zero, so it is not valid near the centerline.

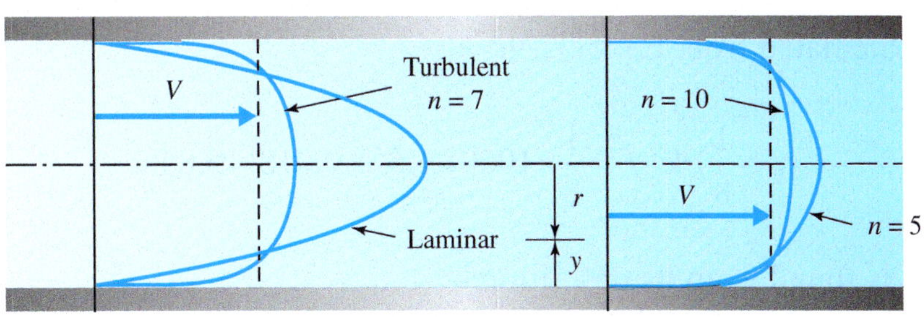

Figure 5.12 Turbulent velocity profile.

Table 5.1 Exponent n for Smooth Pipes

$\text{Re} = VD/\nu$	4×10^3	10^5	10^6	$> 2 \times 10^6$
n	6	7	9	10

It should be noted that the kinetic-energy-correction factor α (see Eq. 4.5.27 in pipes is 1.11, 1.06, and 1.03 for $n = 5, 7$, and 10, respectively. Because it is close to unity for $n > 7$, it is often set equal to unity in the energy equation when working problems involving turbulent flow.

Example 5.9

Water at 20°C flows in a 10-cm-diameter pipe at an average velocity of 1.6 m/s. If the roughness elements are 0.046 mm high, would the wall be rough or smooth? Refer to Figure 5.10.

Solution

To determine if the wall is rough or smooth, we must compare the viscous wall layer thickness with the height of the roughness elements. So, let's find the viscous wall layer thickness. From Figure 5.11 the viscous layer thickness is determined by letting $u_\tau y/\nu = 5$, where $y = \delta_\nu$. First, we must find u_τ. The Reynolds number is

$$\text{Re} = \frac{VD}{\nu} = \frac{1.6 \times 0.1}{10^{-6}} = 1.6 \times 10^5$$

From Table 5.1 $n \simeq 7.5$, so that, from Eq. 5.6.21,

$$f = \frac{1}{n^2} = \frac{1}{7.5^2} = 0.018$$

The wall shear is calculated from Eq. 5.3.19:

$$\tau_0 = \frac{1}{8}\rho V^2 f = \frac{1}{8} \times 1000 \times 1.6^2 \times 0.018 = 5.8 \text{ Pa}$$

The shear velocity is found from its definition:

$$u_\tau = \sqrt{\frac{\tau_0}{\rho}} = \sqrt{\frac{5.8}{1000}} = 0.076 \text{ m/s}$$

This allows us to calculate the viscous wall layer thickness using $y = \delta_v$:

$$\delta_v = \frac{5\nu}{u_\tau} = \frac{5\times10^{-6}\ \text{m}^2/\text{s}}{0.076\ \text{m/s}} = 6.6\times10^{-5}\ \text{m} \quad \text{or} \quad 0.066\ \text{mm}$$

Since the roughness elements are only 0.046 mm high, they are submerged in the viscous wall layer. Consequently, the wall is smooth (see Figure 5.10a). If the pipe were made of cast iron with $e = 0.26$ mm, the wall would be rough.

Note that the viscous wall layer, even at this relatively low velocity, is about 0.1% of the radius. The viscous wall layer is usually extremely thin.

Example 5.10

The 4-cm-diameter smooth, horizontal pipe of Figure E5.10 transports 0.004 m^3/s of water at 20°C. Using the power-law profile, approximate (a) the friction factor, (b) the maximum velocity, (c) the radial position where $u = V$, (d) the wall shear, (e) the pressure drop over a 10-m length, and (f) the maximum velocity using Eq. 5.6.16.

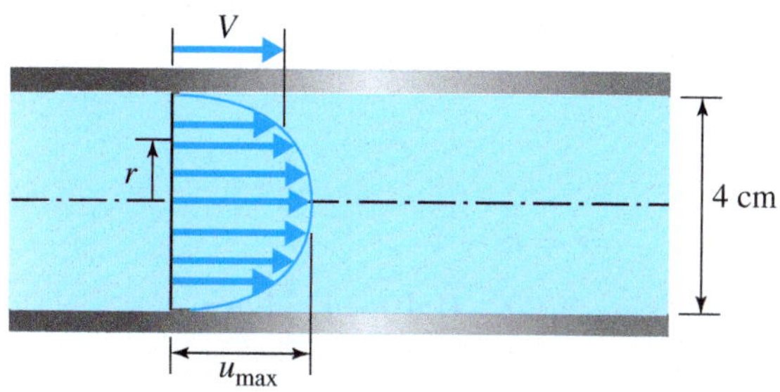

Figure E5.10

Solution

(a) The average velocity is calculated to be

$$V = \frac{Q}{A} = \frac{0.004}{10^{-6}} = 3.18\ \text{m/s}$$

The Reynolds number is

$$\text{Re} = \frac{VD}{\nu} = \frac{3.18\times0.04}{\pi\times0.02^2} = 1.27\times10^5$$

From Table 5.1 we see that $n \cong 7.5$ and from Eq. 5.6.21,

$$f = \frac{1}{n^2} = \frac{1}{7.5^2} = \underline{0.018}$$

(b) The maximum velocity is found using Eq. 5.6.20 to be

$$\begin{aligned} u_{\max} &= \frac{(n+1)(2n+1)}{2n^2} V \\ &= \frac{8.5 \times 16}{2 \times 7.5^2} \times 3.18 = \underline{3.84 \text{ m/s}} \end{aligned}$$

(c) The distance from the wall where $u = V = 3.18$ m/s is found using Eq. 5.6.19 as follows:

$$\frac{\bar{u}}{u_{\max}} = \left(\frac{y}{r_0}\right)^{1/7.5}$$

$$\therefore y = r_0 \left(\frac{u}{u_{\max}}\right)^{7.5} = 2\left(\frac{3.18}{3.84}\right)^{7.5} = 0.49 \text{ cm}$$

The radial position is thus

$$\begin{aligned} r &= r_0 - y \\ &= 2 - 0.49 = \underline{1.51 \text{ cm}} \end{aligned}$$

(d) The wall shear is found using Eq. 5.3.19 and is

$$\begin{aligned} \tau_0 &= \frac{1}{8} \rho V^2 f \\ &= \frac{1}{8} \times 1000 \times 3.18^2 \times 0.018 = \underline{23 \text{ Pa}} \end{aligned}$$

(e) The pressure drop is calculated using Eq. 5.6.18 with $\Delta p/L = -dp/dx$ to be

$$\begin{aligned} \Delta p &= \frac{2\tau_0 L}{r_0} \\ &= \frac{2 \times 23 \times 10}{0.02} = 23\ 000 \text{ Pa} \quad \text{or} \quad \underline{23 \text{ kPa}} \end{aligned}$$

(f) To use Eq. 5.6.16 we must know the shear velocity. It is

$$u_\tau = \sqrt{\frac{\tau_0}{\rho}}$$
$$= \sqrt{\frac{23}{1000}} = 0.152 \text{ m/s}$$

We then find u_{max} to be, using $\nu = 10^{-6}$ m^2/s,

$$u_{max} = 0.152\left(2.44 \ln \frac{0.152 \times 0.02}{10^{-6}} + 5.7\right) = \underline{3.84 \text{ m/s}}$$

This is the same as that given by the power-law formula in part (b). This answer is considered to be more accurate if it differs from that of Eq. 5.6.20. Note that the experimental data do not allow for accuracy in excess of three significant digits, and often to only two significant digits.

5.6.3 Losses in Developed Pipe Flow

Perhaps the most calculated quantity in pipe flow is the head loss. If the head loss is known in a developed flow, the pressure change can be calculated which allows pumps to be selected; for developed flow in a pipe the energy equation (2.5.17) provides us with

$$h_L = \frac{\Delta(p + \gamma h)}{\gamma} \tag{5.6.22}$$

The head loss that results from the wall shear in a developed flow is related to the friction factor (see Eq. 5.3.20) by the Darcy–Weisbach equation, namely,

$$h_L = f \frac{L}{D} \frac{V^2}{2g} \tag{5.6.23}$$

Consequently, if the friction factor is known, we can find the head loss and then the pressure drop.

The friction factor f depends on the various quantities that affect the flow, written as

$$f = f(\rho, \mu, V, D, e) \tag{5.6.24}$$

where the average wall roughness height e accounts for the influence of the wall roughness elements. A dimensional analysis, following the steps of Section 4.2, provides us with

$$f = f\left(\frac{\rho V D}{\mu}, \frac{e}{D}\right) \tag{5.6.25}$$

where e/D is the relative roughness.

Experimental data that relate the friction factor to the Reynolds number have been obtained for fully developed pipe flow over a wide range of wall roughnesses. The results of these data are presented in Figure 5.13, which is commonly referred to as the **Moody diagram**, named after Lewis F. Moody (1880–1953). There are several features of the Moody diagram that should be noted.

- For a given wall roughness, measured by the relative roughness e/D, there is a sufficiently large value of Re above which the friction factor is constant, thereby defining the **completely turbulent regime**. The average roughness element size e is substantially greater than the viscous wall layer thickness δ_v so that viscous effects are not significant; the resistance to the flow is produced primarily by the drag of the roughness elements that protrude into the flow.
- For the smaller relative roughness e/D values it is observed that, as Re decreases, the friction factor increases in the **transition zone** and eventually becomes the same as that of a smooth pipe. The roughness elements become submerged in the viscous wall layer so that they produce little effect on the main flow.
- For Reynolds numbers less than 2000, the friction factor of laminar flow is shown. The **critical zone** couples the turbulent flow to the laminar flow and may represent an oscillatory flow that alternately exists between turbulent and laminar flow. (Problem 5.130 illustrates the phenomenon.)
- The e values in this diagram are for new pipes. With age a pipe will corrode and become fouled, changing both the roughness and the pipe diameter, with a resulting change in the friction factor. Such factors should be included in design considerations; they will not be reviewed here.

The following empirical equations represent the Moody diagram for Re > 4000: Smooth pipe flow:

Smooth pipe flow:
$$\frac{1}{\sqrt{f}} = 0.86 \ln \mathrm{Re}\sqrt{f} - 0.8 \tag{5.6.26}$$

Completely turbulent zone:
$$\frac{1}{\sqrt{f}} = -0.86 \ln \frac{e}{3.7D} \tag{5.6.27}$$

Transition Zone:
$$\frac{1}{\sqrt{f}} = -0.86 \ln\left(\frac{e}{3.7D} + \frac{2.51}{\mathrm{Re}\sqrt{f}}\right) \tag{5.6.28}$$

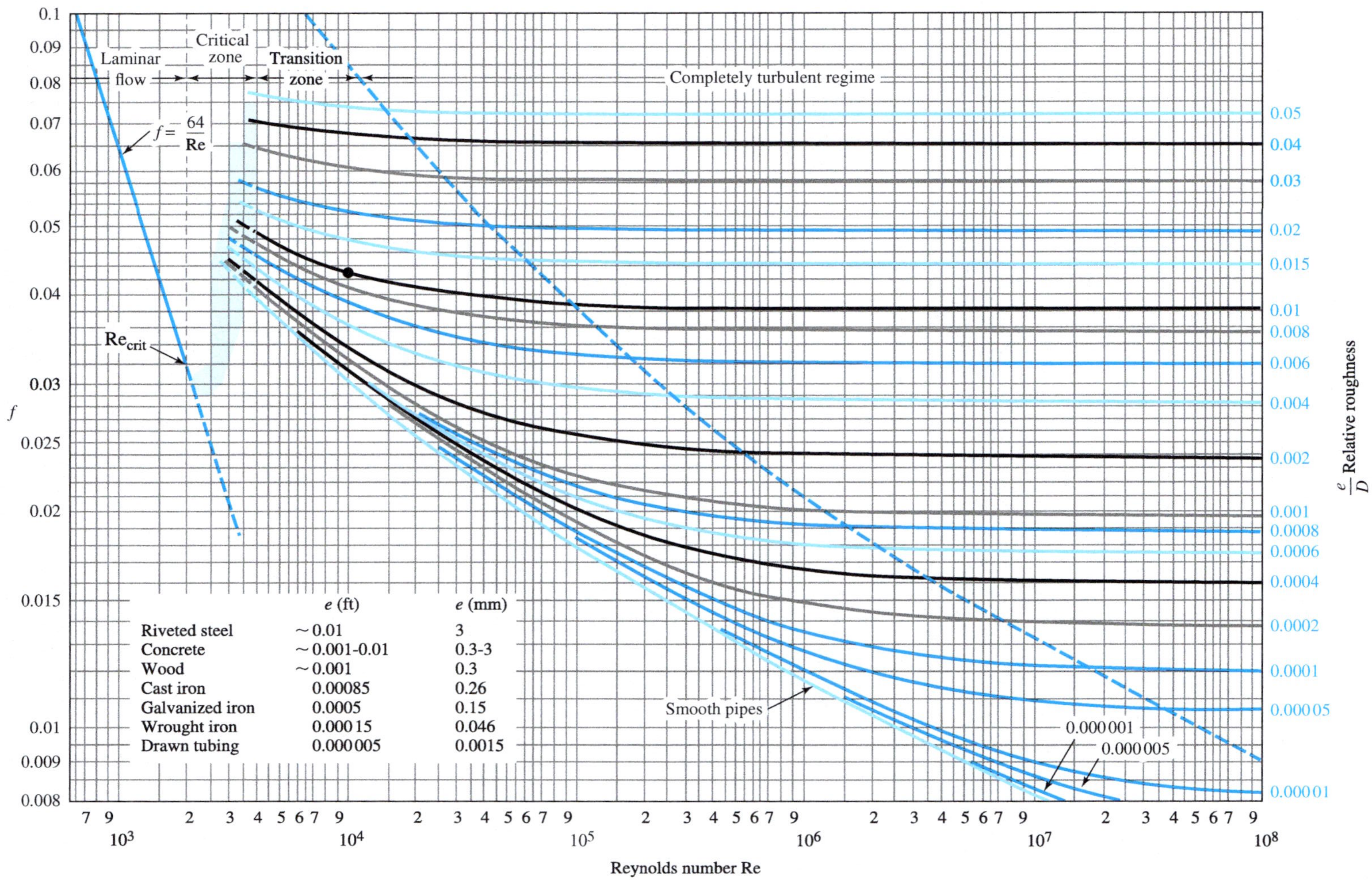

Figure 5.13 Moody diagram. (From L. F. Moody, Trans. ASME, Vol. 66, 1944. Reproduced with permission of ASME.) (Note: If $e/D = 0.01$ and $Re = 10^4$, the dot locates $f = 0.043$.)

The transition zone equation (5.6.28) that couples the smooth pipe equation to the completely turbulent regime equation is known as the **Colebrook equation**. Note that Eq. 5.6.26 is the Colebrook equation with $e = 0$, and Eq. 5.6.27 is the Colebrook equation with $\mathrm{Re} = \infty$.

Three categories of problems can be identified for developed turbulent flow in a pipe of length L:

Category	*Known*	*Unknown*
1	Q, D, e, ν	h_L
2	D, e, ν, h_L	Q
3	Q, e, ν, h_L	D

A Category 1 problem is straightforward and requires no iteration procedure when using the Moody diagram. Category 2 and 3 problems are more like problems encountered in engineering design situations and require an iterative trial-and-error process when using the Moody diagram. Each of these types will be illustrated with an example.

An alternative to using the Moody diagram that avoids any trial-and-error process is made possible by empirically derived formulas. Perhaps the best of such formulas were presented by Swamee and Jain (1976) for pipe flow; an explicit expression that provides an approximate value for the unknown in each category above is as follows:

$$h_L = 1.07\frac{Q^2L}{gD^5}\left\{\ln\left[\frac{e}{3.7D} + 4.62\left(\frac{\nu D}{Q}\right)^{0.9}\right]\right\}^{-2} \qquad \begin{matrix} 10^{-6} < e/D < 10^{-2} \\ 3000 < \mathrm{Re} < 3\times 10^8 \end{matrix} \qquad (5.6.29)$$

$$Q = -0.965\left(\frac{gD^5h_L}{L}\right)^{0.5}\ln\left[\frac{e}{3.7D} + \left(\frac{3.17\nu^2L}{gD^3h_L}\right)^{0.5}\right] \qquad \mathrm{Re} > 2000 \qquad (5.6.30)$$

$$D = 0.66\left[e^{1.25}\left(\frac{LQ^2}{gh_L}\right)^{4.75} + \nu Q^{9.4}\left(\frac{L}{gh_L}\right)^{5.2}\right]^{0.04} \qquad \begin{matrix} 10^{-6} < e/D < 10^{-2} \\ 5000 < \mathrm{Re} < 3\times 10^8 \end{matrix} \qquad (5.6.31)$$

One may use either English or SI units in the equations above.

Equation 5.6.30 is as accurate as the Moody diagram, and Eqs. 5.6.29 and 5.6.31 are accurate to within approximately 2% of the Moody diagram. These tolerances are acceptable for engineering calculations. It is important to realize that the Moody diagram is based on experimental data that likely is accurate to within no more than 5%. Hence the foregoing three formulas of Swamee and Jain, are often used by design engineers. The following examples also illustrate the use of these approximate formulas.

Example 5.11

Water at 74°F is transported for 1500 ft in a 1½-in.-diameter wrought iron horizontal pipe with a flow rate of 0.1 ft^3/s. Calculate the head loss and the pressure drop over the 1500-ft length of pipe, using (a) the Moody diagram and (b) the alternate method.

Solution

(a) The average velocity is

$$V = \frac{Q}{A} = \frac{0.1}{\pi \times 0.75^2/144} = 8.15 \text{ ft/s}$$

The Reynolds number is

$$\text{Re} = \frac{VD}{\nu} = \frac{8.15 \times 1.5/12}{10^{-5}} = 1.02 \times 10^5$$

Obtaining e from Figure 5.13, we have, using D = 1.5/12 ft,

$$\frac{e}{D} = \frac{0.00015}{0.125} = 0.0012$$

The friction factor is read from the Moody diagram to be

$$f = 0.023$$

The head loss is calculated as

$$\begin{aligned} h_L &= f\frac{L}{D}\frac{V^2}{2g} \\ &= 0.023 \frac{1500}{1.5/12} \frac{8.15^2 \text{ ft}^2/\text{s}^2}{2 \times 32.2 \text{ ft/s}^2} = \underline{280 \text{ ft}} \end{aligned}$$

This answer is given to two significant numbers since the friction factor is known to at most two significant numbers. The pressure drop is found by Eq. 5.6.22 to be

$$\begin{aligned} \Delta p &= \gamma h_L \\ &= 62.4 \text{ lb/ft}^3 \times 280 \text{ ft} = 17{,}500 \text{ psf} \quad \text{or} \quad \underline{120 \text{ psi}} \end{aligned}$$

(b) The alternate method for this Category 1 problem uses Eq. 5.6.29, with $D = 1.5/12 = 0.125$ ft:

$$h_L = 1.07\frac{0.1^2 \times 1500}{32.2 \times 0.125^5}\left\{\ln\left[\frac{0.0012}{3.7} + 4.62\left(\frac{10^{-5} \times 0.125}{0.1}\right)^{0.9}\right]\right\}^{-2}$$
$$= 1.07 \times 15{,}265 \times 0.01734 = \underline{280 \text{ ft}}$$

This much simpler method provides the same value as that found using the Moody diagram.

Example 5.12

A pressure drop of 700 kPa is measured over a 300-m length of horizontal, 10-cm-diameter wrought iron pipe that transports oil ($S = 0.9$, $\nu = 10^{-5}$ m^2/s). Calculate the flow rate using (a) the Moody diagram, and (b) the alternate method.

Solution

(a) The relative roughness is

$$\frac{e}{D} = \frac{0.046}{100} = 0.00046$$

Assuming that the flow is completely turbulent (Re is not needed), the Moody diagram gives

$$f = 0.0165$$

The head loss is found to be

$$h_L = \frac{\Delta p}{\gamma_{\text{oil}}} = \frac{700\,000 \text{ N/m}^2}{9800 \text{ N/m}^3 \times 0.9} = 79.4 \text{ m}$$

The velocity is calculated from Eq. 5.6.23 to be

$$V = \left(\frac{2gDh_L}{fL}\right)^{1/2} = \left(\frac{2 \times 9.8 \text{ m/s}^2 \times 0.1 \text{ m} \times 79.4 \text{ m}}{0.0165 \times 300 \text{ m}}\right)^{1/2} = 5.61 \text{ m/s}$$

This provides us with a Reynolds number of

$$\text{Re} = \frac{VD}{\nu} = \frac{5.61 \text{ m/s} \times 0.1 \text{ m}}{10^{-5} \text{ m}^2/\text{s}} = 5.61 \times 10^4$$

Using this Reynolds number and $e/D = 0.00046$, the Moody diagram gives the friction factor as

$$f = 0.023$$

This corrects the original value for f. The velocity is recalculated to be

$$V = \left(\frac{2 \times 9.8 \times 0.1 \times 79.4}{0.023 \times 300} \right)^{1/2} = 4.75 \text{ m/s}$$

The Reynolds number is then

$$\text{Re} = \frac{4.75 \times 0.1}{10^{-5}} = 4.75 \times 10^4$$

From the Moody diagram $f = 0.023$ appears to be satisfactory. Thus the flow rate is

$$Q = VA = 4.75 \times \pi \times 0.05^2 = \underline{0.037 \text{ m}^3\text{/s}}$$

Only two significant numbers are given since f is known to at most two significant numbers.

(b) The alternative method for this Category 2 problem uses the explicit relationship (5.6.30). We can directly calculate Q to be

$$Q = -0.965 \left(\frac{9.8 \times 0.1^5 \times 79.4}{300} \right)^{0.5} \ln \left[\frac{0.00046}{3.7} + \left(\frac{3.17 \times 10^{-10} \times 300}{9.8 \times 0.1^3 \times 79.4} \right)^{0.5} \right]$$
$$= -0.965 \times 5.096 \times 10^{-3} \times (-7.655) = \underline{0.038 \text{ m}^3\text{/s}}$$

This much simpler method produces a value essentially the same as that obtained using the Moody diagram.

Example 5.13

Drawn tubing of what diameter should be selected to transport 0.002 m^3/s of 20°C water over a 400-m length so that the head loss does not exceed 30 m? (a) Use the Moody diagram and (b) the alternative method.

Solution

(a) In this problem we do not know D. Thus, a trial-and-error solution is anticipated. The average velocity is related to D by

$$V = \frac{Q}{A} = \frac{0.002}{\pi D^2/4} = \frac{0.00255}{D^2}$$

The friction factor and D are related as follows:

$$h_L = f\frac{L}{D}\frac{V^2}{2g}$$

$$30 = f\frac{400}{D}\frac{(0.00255/D^2)^2}{2\times 9.8}$$

$$\therefore D^5 = 4.42\times 10^{-6} f$$

The Reynolds number is

$$\text{Re} = \frac{VD}{\nu} = \frac{0.00255D}{D^2\times 10^{-6}} = \frac{2550}{D}$$

Now, let us simply guess a value for f and check with the relations above and the Moody diagram. The first guess is f = 0.03, and the correction is listed in the following table.

Note: The second guess is the value for f found from the calculations of the first guess.

f	D(m)	Re	e/D	f (Figure 5.13)
0.03	0.0421	6.06×10^4	0.000036	0.02
0.02	0.0388	6.57×10^4	0.000039	0.02

The value of f = 0.02 is acceptable, yielding a diameter of 3.88 mm. Since this diameter would undoubtedly not be standard, a diameter of

$$D = \underline{4\text{ cm}}$$

would be the tube size selected. This tube would have a head loss less than the limit of h_L = 30 m imposed in the problem statement. Any larger-diameter tube would also satisfy this criterion but would be more costly, so it should not be selected.

(b) The alternative method for this Category 3 problem uses the explicit relationship (5.6.31). We can directly calculate D to be

$$D = 0.66\left[(1.5\times 10^{-6})^{1.25}\left(\frac{400\times 0.002^2}{9.81\times 30}\right)^{4.75} + 10^{-6}\times 0.002^{9.4}\left(\frac{400}{9.81\times 30}\right)^{5.2}\right]^{0.04}$$

$$= 0.66[5.163\times 10^{-33} + 2.102\times 10^{-31}]^{0.04} = \underline{0.039\text{ m}}$$

Hence, D = 4 cm would be selected, the same size as that selected using the Moody diagram.

5.6.4 Losses in Noncircular Conduits

Good approximations can be made for the head loss in conduits with noncircular cross sections by using the **hydraulic radius** R, defined by

$$R = \frac{A}{P} \tag{5.6.32}$$

where A is the cross-sectional area and P is the **wetted perimeter**, that perimeter where the fluid is in contact with the solid boundary. For a circular pipe flowing full the hydraulic radius is $R = r_0/2$. Hence we simply replace the radius r_0 with $2R$ and use the Moody diagram with

$$\text{Re} = \frac{4VR}{\nu} \qquad \text{relative roughness} = \frac{e}{4R} \tag{5.6.33}$$

The head loss becomes

$$h_L = f\frac{L}{4R}\frac{V^2}{2g} \tag{5.6.34}$$

To use this hydraulic radius technique the cross section should be fairly "open," such as a rectangle with aspect ratio less than 4:1, an equilateral triangle, or an oval. For some other shapes, such as an annulus, the error would be significant.

Example 5.14

Air at standard conditions is to be transported through 500 m of a smooth, horizontal, 30 cm × 20 cm rectangular duct at a flow rate of 0.24 m^3/s. Calculate the pressure drop.

Solution

The hydraulic radius is

$$R = \frac{A}{P} = \frac{0.3 \times 0.2}{(0.3 + 0.2) \times 2} = 0.06 \text{ m}$$

The average velocity is

$$V = \frac{Q}{A} = \frac{0.24}{0.3 \times 0.2} = 40 \text{ m/s}$$

This gives a Reynolds number of

$$\mathrm{Re} = \frac{4VR}{\nu} = \frac{4 \times 4 \times 0.06}{1.5 \times 10^{-5}} = 6.4 \times 10^4$$

Using the smooth pipe curve of the Moody diagram, there results

$$f = 0.0196$$

Hence,

$$h_L = f \frac{L}{4R} \frac{V^2}{2g} = 0.0196 \frac{500 \text{ m}}{4 \times 0.06 \text{ m}} \frac{4^2 \text{m}^2/\text{s}^2}{2 \times 9.8 \text{ m/s}^2} = 33.3 \text{ m}$$

The pressure drop is

$$\Delta p = \rho g h_L = 1.23 \times 9.8 \times 33.3 = \underline{400 \text{ Pa}}$$

5.6.5 Minor Losses in Pipe Flow

We now know how to calculate the losses due to a developed flow in a pipe. Pipe systems do, however, include valves, elbows, enlargements, contractions, inlets, outlets, bends, and other fittings that cause additional losses, referred to as **minor losses**, even though such losses can exceed the frictional losses of Eq. 5.6.23. Each of these elements causes a change in the magnitude and/or the direction of the velocity vectors and hence results in a loss. In general, if the flow is gradually accelerated by an element, the losses are very small; relatively large losses are associated with sudden enlargements or contractions because of the separated regions that result (a separated flow occurs when the primary flow separates from the wall).

A minor loss is expressed in terms of a loss coefficient K, defined by

$$h_L = K \frac{V^2}{2g} \tag{5.6.35}$$

Values of K have been determined experimentally for the various fittings and geometry changes of interest in piping systems. One exception is the sudden expansion from area A_1 to area A_2, for which the loss can be calculated; this was done in Example 2.14, where we found that

$$h_L = \left(1 - \frac{A_1}{A_2}\right)^2 \frac{V_1^2}{2g} \tag{5.6.36}$$

Thus, for the sudden expansion

$$K = \left(1 - \frac{A_1}{A_2}\right)^2 \tag{5.6.37}$$

If A_2 is extremely large (e.g., a pipe exiting into a reservoir), $K = 1.0$, since the entire kinetic energy is lost.

A pipe bend or elbow fitting has a relatively large loss coefficient with no change in cross-sectional area. This results primarily from the secondary flow caused by the fluid flowing from the high-pressure region to the low-pressure region (see Eq. 1.4.15), as shown in Figure 5.14; this secondary flow is eventually dissipated after the fluid leaves the long sweep bend or elbow. In addition, a separated region occurs at the sharp corner in a standard elbow. Energy is needed to maintain a secondary flow and the flow in the separated region. This wasted energy is measured in terms of the loss coefficient.

The loss coefficients for various geometries are presented in Table 5.2 and Figure 5.15. A globe valve may be used to control the flow rate by introducing large losses caused by partially closing the valve. Most other types of valves should not be used to control the flow; damage could result.

Loss coefficients for sudden contractions and orifice plates can be approximated by neglecting the losses in the converging flow up to the vena contracta and calculating the losses in the diverging flow using the loss coefficient for a sudden expansion. Figure 5.16 provides the information necessary to establish the area A_c of the **vena contracta**, the minimum area; this minimum area occurs where the converging streamlines begin to expand to fill the downstream area.

It is often the practice to express a loss coefficient as an **equivalent length** L_e of pipe. This is done by equating Eq. 5.6.35 to Eq. 5.6.23:

$$K\frac{V^2}{2g} = f\frac{L_e}{D}\frac{V^2}{2g} \tag{5.6.38}$$

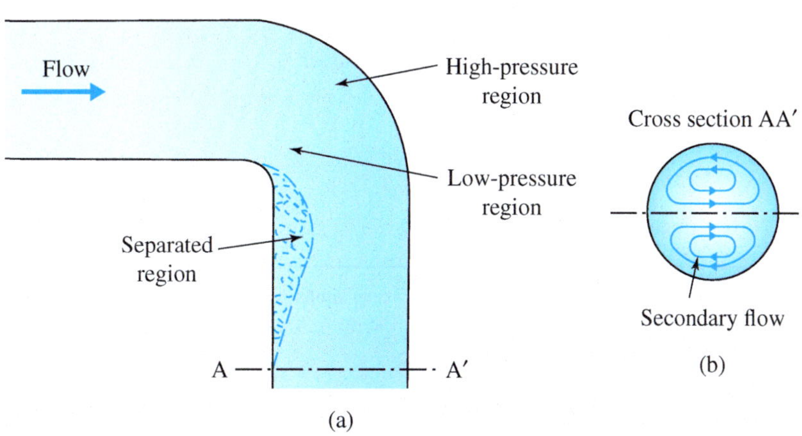

Figure 5.14 Flow in an elbow.

Table 5.2 Nominal Loss Coefficients K (Turbulent Flow)[a]

Type of fitting		*Screwed*			*Flanged*		
Diameter		2.5 cm	5 cm	10 cm	5 cm	10 cm	20 cm
Globe valve	(fully open)	8.2	6.9	5.7	8.5	6.0	5.8
	(half open)	20	17	14	21	15	14
	(one-quarter open)	57	48	40	60	42	41
Angle valve (fully open)		4.7	2.0	1.0	2.4	2.0	2.0
Swing check valve (fully open)		2.9	2.1	2.0	2.0	2.0	2.0
Gate valve (fully open)		0.24	0.16	0.11	0.35	0.16	0.07
Return bend		1.5	0.95	0.64	0.35	0.30	0.25
Tee (branch)		1.8	1.4	1.1	0.80	0.64	0.58
Tee (line)		0.9	0.9	0.9	0.19	0.14	0.10
Standard elbow		1.5	0.95	0.64	0.39	0.30	0.26
Long sweep elbow		0.72	0.41	0.23	0.30	0.19	0.15
45° elbow		0.32	0.30	0.29			
Square-edged entrance				0.5			
Reentrant entrance				0.8			
Well-rounded entrance				0.03			
Pipe exit				1.0			
Sudden contraction[b]	Area ratio						
	2:1			0.25			
	5:1			0.41			
	10:1			0.46			
Orifice plate	Area ratio A/A_0						
	1.5:1			0.85			
	2:1			3.4			
	4:1			29			
	$\geq 6:1$			$2.78\left(\frac{A}{A_0}-0.6\right)^2$			
Sudden enlargement[c]	A_1			$\left(1-\frac{A_1}{A_2}\right)^2$			
90° miter bend (without vanes)				1.1			
(with vanes)				0.2			
General contraction	(30° included angle)			0.02			
	(70° included angle)			0.07			

[a]Values for other geometries can be found in *Technical Paper 410*, The Crane Company, 1957.
[b]Based on exit velocity V_2.
[c]Based on entrance velocity V_1.

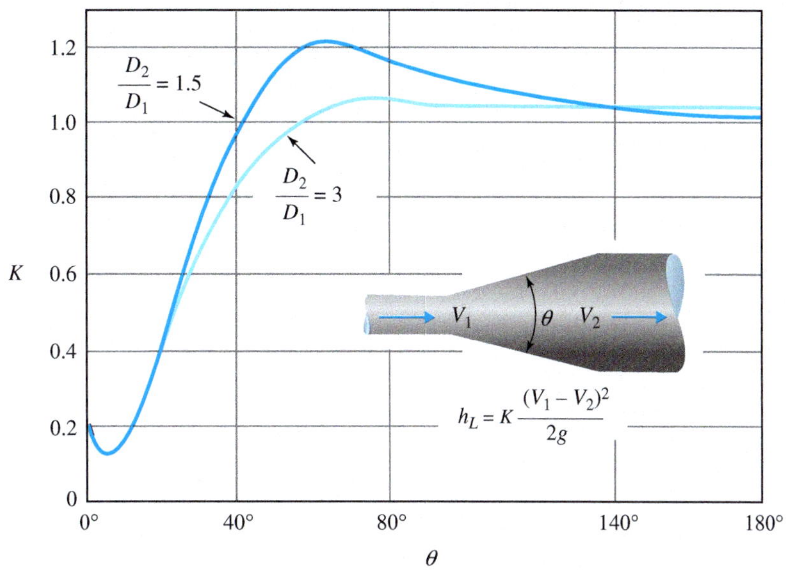

Figure 5.15 Loss coefficients in a conical expansion.

giving the relationship

$$L_e = K\frac{D}{f} \tag{5.6.39}$$

Hence, the square-edged entrance of a 20-cm-diameter pipe with a friction factor of $f = 0.02$ could be replaced by an equivalent pipe length of $L_e = 5$ m.

Finally, a comment should be made concerning the magnitude of the minor losses. In piping systems involving intermediate lengths (i.e., about 100 diameters of length) of pipe between minor losses, the minor losses may be of the same order of magnitude as the frictional losses; for relatively short lengths the minor losses may be substantially greater than the frictional losses; and for long lengths (i.e., 1000 diameters) of pipe, the minor losses are usually neglected.

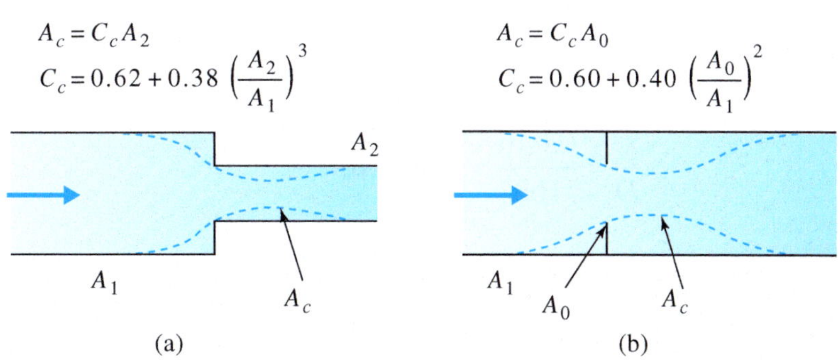

Figure 5.16 Vena contractas in contractions and orifices: (a) sudden contraction; (b) concentric orifice.

Example 5.15

If the flow rate through a 10-cm-diameter wrought iron pipe (Figure E5.15) is 0.04 m^3/s, find the difference in elevation H of the two reservoirs.

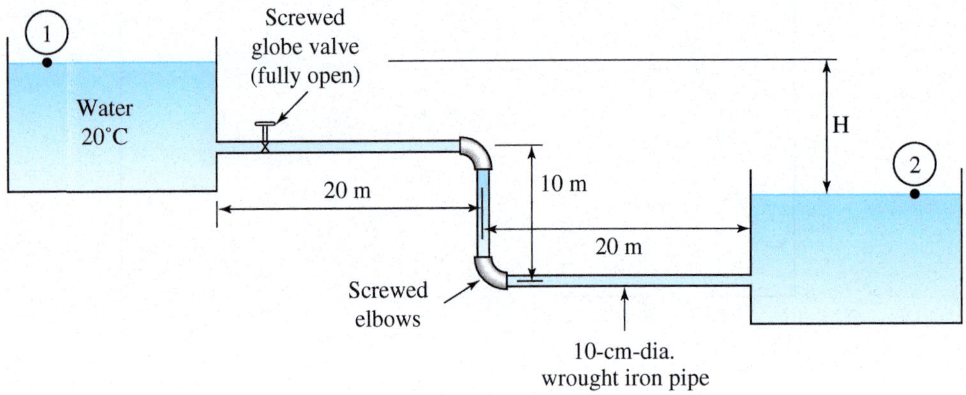

Figure E5.15

Solution

The energy equation written for a control volume that contains the two reservoir surfaces (see Eq. 2.5.17), where $V_1 = V_2 = 0$ and $p_1 = p_2 = 0$, is

$$0 = z_2 - z_1 + h_L$$

Thus, letting $z_1 - z_2 = H$, we have, using Eqs. 5.6.35 and 5.6.23,

$$H = (K_{\text{entrance}} + K_{\text{valve}} + 2K_{\text{elbow}} + K_{\text{exit}})\frac{V^2}{2g} + f\frac{L}{D}\frac{V^2}{2g}$$

The average velocity, Reynolds number, and relative roughness are

$$V = \frac{Q}{A} = \frac{0.04}{\pi \times 0.05^2} = 5.09 \text{ m/s}$$

$$\text{Re} = \frac{VD}{\nu} = \frac{5.09 \times 0.1}{10^{-6}} = 5.09 \times 10^5$$

$$\frac{e}{D} = \frac{0.046}{100} = 0.00046$$

From the Moody diagram we find that

$$f = 0.0173$$

Using the loss coefficients from Table 5.2 for an entrance, a globe valve, screwed 10-cm-diameter standard elbows, and an exit there results

$$H = (0.5 + 5.7 + 2 \times 0.64 + 1.0)\frac{5.09^2}{2 \times 9.8} + 0.0173\frac{50}{0.1}\frac{5.09^2}{2 \times 9.8}$$
$$= 11.2 + 11.4 = \underline{22.6 \text{ m}}$$

Note: The minor losses are about equal to the frictional losses as expected, since there are five minor loss elements in 500 diameters of pipe length.

Example 5.16

Approximate the loss coefficient for the sudden contraction $A_1/A_2 = 2$ by neglecting the losses in the contracting portion up to the vena contracta and assuming that all the losses occur in the expansion from the vena contracta to A_2 (see Figure 5.16). Compare with that given in Table 5.2.

Solution

The head loss from the vena contracta to area A_2 is (see Table 5.2, sudden enlargement)

$$h_L = \left(1 - \frac{A_c}{A_2}\right)^2 \frac{V_c^2}{2g}$$

Continuity allows us to write (V_c is the velocity at the area A_c)

$$V_c = \frac{A_2}{A_c} V_2$$

Thus, the head loss based on V_2 is

$$h_L = \left(1 - \frac{A_c}{A_2}\right)^2 \left(\frac{A_2}{A_c}\right)^2 \frac{V_2^2}{2g}$$

so the loss coefficient of Eq. 5.6.35 is

$$K = \left(1 - \frac{A_c}{A_2}\right)^2 \left(\frac{A_2}{A_c}\right)^2$$

Using the expression of C_c given in Figure 5.16, we have

$$\frac{A_c}{A_2} = C_c = 0.62 + 0.38\left(\frac{1}{2}\right)^3 = 0.67$$

Finally,

$$K = (1 - 0.67)^2 \frac{1}{0.67} = \underline{0.24}$$

This compares favorably with the value of 0.25 given in Table 5.2.

5.6.6 Hydraulic and Energy Grade Lines

When the energy equation is written with $\gamma_1 = \gamma_2$ in the form of Eq. 2.5.17, that is,

$$-\frac{\dot{W}_S}{\dot{m}g} - \frac{V_2^2 - V_1^2}{2g} + \frac{p_2 - p_1}{\gamma} + z_2 - z_1 + h_L \qquad (5.6.40)$$

the terms have dimensions of length. This has led to the conventional use of the hydraulic grade line and the energy grade line. The **hydraulic grade line** (HGL), the dashed line in Figure 5.17, in a piping system is formed by the locus of points located a distance p/γ above the center of the pipe, or $p/\gamma + z$ above a preselected datum; the liquid in a piezometer tube would rise to the HGL. The **energy grade line** (EGL), the solid line in Figure 5.17, is formed by the locus of points a distance $V^2/2g$ above the HGL, or the distance $V^2/2g + p/\gamma + z$ above the datum; the liquid in a pitot tube would rise to the EGL. The following points are noted relating to the HGL and the EGL:

- As the velocity goes to zero, the HGL and the EGL approach each other. Thus, in a reservoir, they are identical and lie on the surface (see Figure 5.17).
- The EGL and, consequently, the HGL slope downward in the direction of the flow due to the head loss in the pipe. The greater the loss per unit length, the greater the slope. As the average velocity in the pipe increases, the loss per unit length increases.
- A sudden change occurs in the HGL and the EGL whenever a loss occurs due to a sudden geometry change, as represented by the valve or the sudden enlargement of Figure 5.17.
- A jump occurs in the HGL and the EGL whenever useful energy is added to the fluid, as occurs with a pump, and a drop occurs if useful energy is extracted from the flow, as occurs with a turbine.
- At points where the HGL passes through the centerline of the pipe, the pressure is zero. If the pipe lies above the HGL, there is a vacuum in the pipe, a condition that is often avoided, if possible, in the design of piping systems; an exception would be in the design of a siphon.

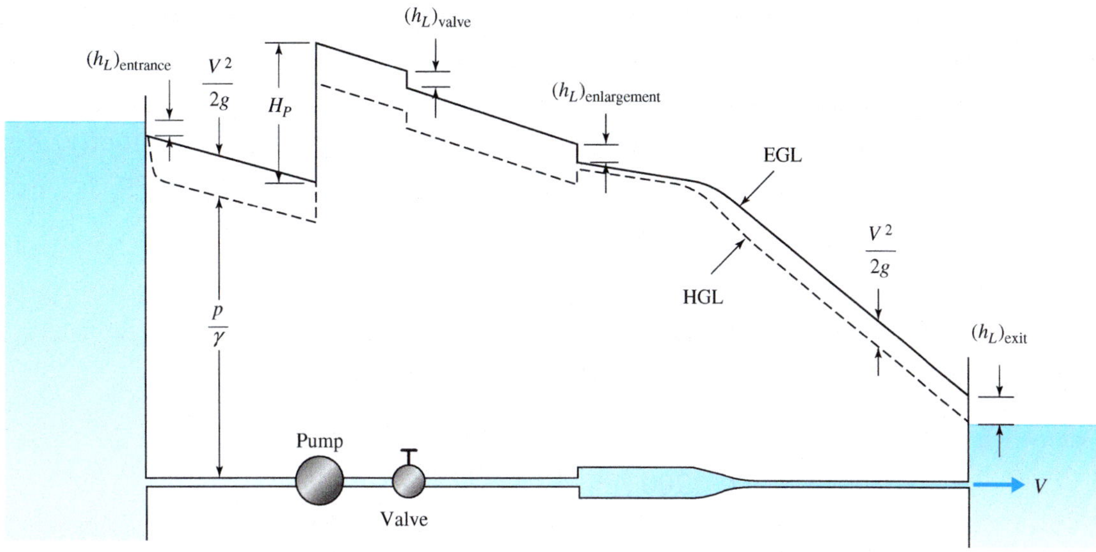

Figure 5.17 Hydraulic grade line (HGL) and energy grade line (EGL) for a piping system.

The concepts of energy grade line and hydraulic grade line may also be applied to open-channel flows. The HGL coincides with the free surface and the EGL is a distance $V^2/2g$ above the free surface. Uniform flows in open channels will be discussed in the following section, and Chapter 8 is devoted primarily to nonuniform open-channel flow.

Example 5.17

Water at 20°C flows between two reservoirs at the rate of 0.06 m^3/s as shown in Figure E5.17. Sketch the HGL and the EGL. What is the minimum diameter D_B allowed to avoid the occurrence of cavitation?

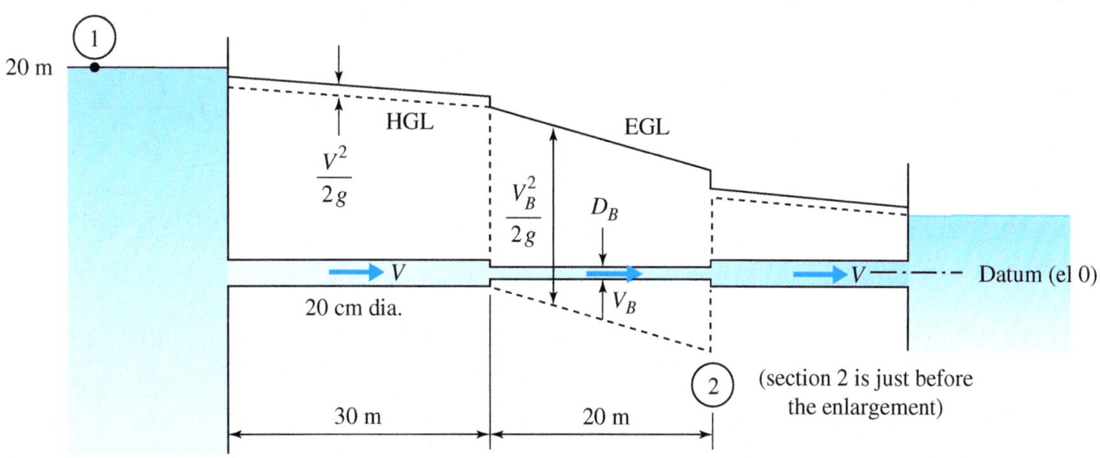

Figure E5.17

Solution

The EGL and the HGL are sketched on the figure, including sudden changes at the entrance, contraction, enlargement, and the exit. Note the large velocity head (the difference between the EGL and the HGL) in the smaller pipe because of the high velocity. The velocity, Reynolds number, and relative roughness in the 20-cm-diameter pipe are calculated to be

$$V = \frac{Q}{A} = \frac{0.06}{\pi \times 0.20^2/4} = 1.91 \text{ m/s}$$

$$\text{Re} = \frac{VD}{\nu} = \frac{1.91 \times 0.2}{10^{-6}} = 3.8 \times 10^5$$

$$\frac{e}{D} = \frac{0.26}{200} = 0.0013$$

Thus $f = 0.022$ from Figure 5.13. The velocity, Reynolds number, and relative roughness in the smaller pipe are

$$V_B = \frac{0.06}{\pi D_B^2/4} = \frac{0.0764}{D_B^2}$$

$$\text{Re}_B = \frac{0.0764 \times D_B}{D_B^2 \times 10^{-6}} = \frac{76400}{D_B}$$

$$\frac{e}{D_B} = \frac{0.00026}{D_B}$$

The minimum possible diameter is established by recognizing that the water vapor pressure (2450 Pa absolute) at 20°C is the minimum allowable pressure. Since the distance between the pipe and the HGL is an indication of the pressure in the pipe, we can conclude that the minimum pressure will occur at section 2. Hence the energy equation applied between section 1, the reservoir surface, and section 2 gives

$$\overset{0}{\cancel{\frac{V_1^2}{2g}}} + \frac{p_1}{\gamma} + z_1 = \frac{V_B^2}{2g} + \frac{p_2}{\gamma} + \overset{0}{\cancel{z_2}} + K_{\text{ent}}\frac{V_A^2}{2g} + K_{\text{cont}}\frac{V_B^2}{2g} + f_A \frac{L_A}{D_A}\frac{V_A^2}{2g} + f_B \frac{l_B}{D_B}\frac{V_B^2}{2g}$$

where the subscript A refers to the 20-cm-diameter pipe. This simplifies, using absolute pressure, to

$$\frac{101\,000}{9810} + 20 = \frac{(0.0764/D_B^2)^2}{2 \times 9.81}\left(1 + 0.25 + f_B \frac{20}{D_B}\right) + \frac{2450}{9810} + \left(0.5 + 0.022\frac{30}{0.2}\right)\frac{1.91^2}{2 \times 9.81}$$

$$98\,600 = \frac{1.25}{D_B^4} + f_B \frac{20}{D_B^5}$$

where we have used $K_{\text{ent}} = 0.5$ and assumed that $K_{\text{cont}} = 0.25$. This requires a trial-and-error solution. The following illustrates the procedure.

Let $D_B = 0.1$ m. Then $e/D_B = 0.0026$ and $\text{Re}_B = 7.6 \times 10^5$. Therefore, $f = 0.026$:

$$98\,600 \stackrel{?}{=} 12\,500 + 52\,000$$

Let $D_B = 0.09$ m. Then $e/D_B = 0.0029$ and $\text{Re}_B = 8.4 \times 10^5$. Therefore, $f = 0.027$:

$$98\,600 \stackrel{?}{=} 19\,000 + 91\,000$$

We see that 0.1 m is too large and 0.09 m is too small. In fact, the value of 0.09 m is only slightly too small. Consequently, to be safe we must select the next larger pipe size of 0.1 m diameter. If there were a pipe size of 9.5 cm diameter, that could be selected. Assuming that that size is not available, we select

$$D_B = \underline{10 \text{ cm}}$$

Note that the assumption of a 2:1 area ratio for the contraction is too small. It is actually 4:1. This would give $K_{\text{cont}} \simeq 0.4$. After a quick check we conclude that this value does not significantly influence the result.

5.6.7 Simple Pipe System with a Pump

The problems we have considered thus far in this section have not involved a pump. If a centrifugal pump is included in the piping system and the flow rate is specified, the solution is straightforward using the techniques we have already developed. If, on the other hand, the flow rate is not specified, as is often the case, a trial-and-error solution involving the centrifugal pump results since the head produced by a centrifugal pump and its efficiency η_P (see Eq. 2.5.26) depend on the discharge, as shown by the pump characteristic curves, the solid curves in Figure 5.18. Companies furnish such characteristic curves for each centrifugal pump manufactured. Such a curve provides one equation relating the flow rate Q and pump head H_P. The other equation is provided by the energy equation, which can typically be written as (see the energy equation in the following example)

$$H_P = c_1 + c_2 Q^2 \tag{5.6.41}$$

This is the **system demand curve**. This along with the characteristic curve must be solved simultaneously to yield the desired flow rate. To determine the power requirement of the pump, the efficiency η_P must be used.

Note that for the piping system, the required pump energy head H_P, demanded by the energy equation, increases with Q and from the pump characteristic curve we see that H_P decreases with Q; hence the two curves will intersect at a point, called the **operating point** of the system. An example will illustrate the solution technique.

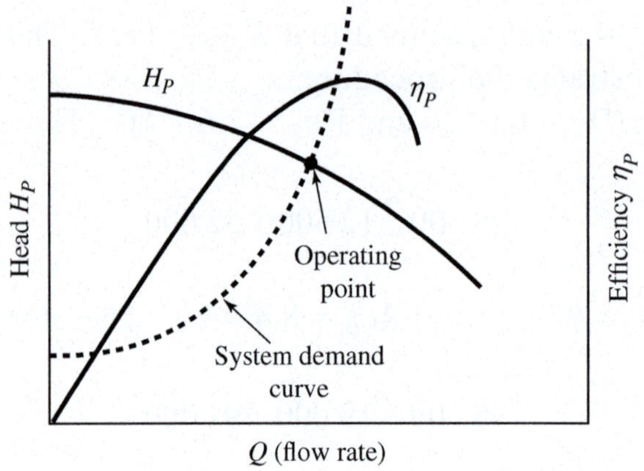

Figure 5.18 Pump characteristic curves and the system demand curve.

Example 5.18

Estimate the flow rate in the simple piping system of Figure E5.18a if the pump characteristic curves are as shown in Figure E5.18b. Also, find the pump power requirement.

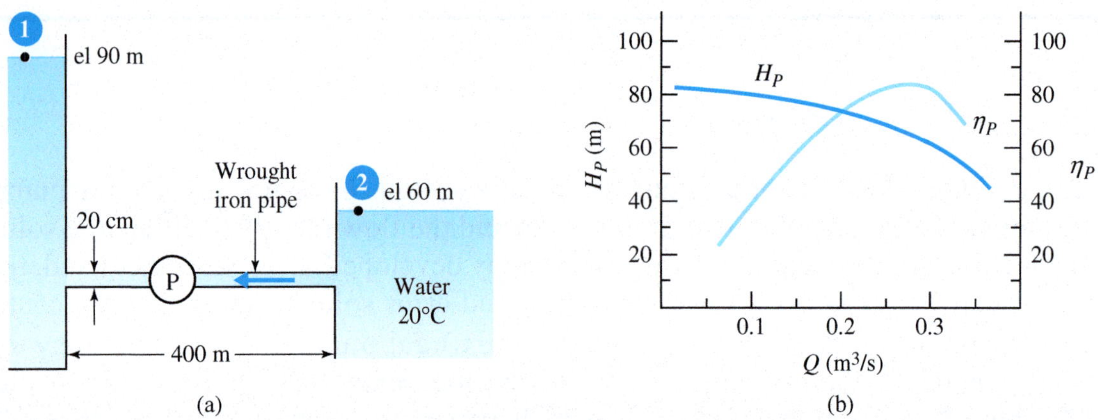

Figure E5.18

Solution

We will assume that the Reynolds number is sufficiently large that the flow is completely turbulent. So, using $e/D = 0.046/200 = 0.00023$, the friction factor from the Moody diagram is

$$f = 0.014$$

The energy equation (see Eq. 5.6.40), with $H_P = -\dot{W}_s/\dot{m}g$, applied between the two surfaces, yields

$$H_P = \cancelto{0}{\frac{V_2^2 - V_1^2}{2g}} + z_2 - z_1 + \cancelto{0}{\frac{P_2 - P_1}{\gamma}} + h_L$$

or

$$
\begin{aligned}
H_P &= 90 - 60 + \left(K_{\text{entrance}} + K_{\text{exit}} + f\frac{L}{D} \right)\frac{V^2}{2g} \\
&= 30 + \left(0.5 + 1.0 + 0.014\frac{400}{0.2} \right)\frac{Q^2}{2 \times 9.8 \times \left[\pi \times 0.1^2\right]^2} \\
&= 30 + 1520Q^2
\end{aligned}
$$

This equation, the system demand curve, and the characteristic curve $H_P\,(Q)$ of the pump are now solved simultaneously by trial and error. Actually, the curve could be plotted on the same graph as the characteristic curve, and the point of intersection, the operating point, would provide Q. Try $Q = 0.2\ \text{m}^3/\text{s}$: $(H_P)_{\text{energy}} = 91$ m, $(H_P)_{\text{char}} \cong 75$ m. Try $Q = 0.15\ \text{m}^3/\text{s}$: $(H_P)_{\text{energy}} = 64$ m, $(H_P)_{\text{char}} \cong 75$ m. Try $Q = 0.17\ \text{m}^3/\text{s}$: $(H_P)_{\text{energy}} = 74$ m, $(H_P)_{\text{char}} \cong 76$ m. This is our solution. We have

$$
Q = \underline{0.17\ \text{m}^3/\text{s}}
$$

Check the Reynolds number: $\text{Re} = DQ/A\nu = 0.2 \times 0.17/(\pi \times 0.1^2 \times 10^{-6}) = 1.08 \times 10^6$. This is sufficiently large, but marginally so.

The power requirement of the pump is given by Eq. 2.5.26:

$$
\begin{aligned}
\dot{W}_P &= \frac{Q\gamma H_P}{\eta_P} \\
&= \frac{0.17\ \text{m}^3/\text{s} \times 9800\ \text{N/m}^3 \times 75\ \text{m}}{0.65} = 198\ 000\ \text{W} \quad \text{or} \quad \underline{200\ \text{kW}}
\end{aligned}
$$

where the efficiency $\eta_P = 0.65$ is found from the characteristic curve at $Q = 0.17\ \text{m}^3/\text{s}$.

Note: Since $L/D > 1000$, minor losses due to the entrance and exit could have been neglected.

5.7 UNIFORM TURBULENT FLOW IN OPEN CHANNELS

The last internal flow situation we consider in this chapter is that of steady uniform (constant depth) flow in an open channel, shown in Figure 5.19. We could already treat this flow by using the Darcy-Weisbach relation presented in Section 5.6.3. In fact, that technique predicts better results than does the more common method we present here. Both methods will be compared in two examples. However, unless otherwise stated, uniform flow in open, rough channels is commonly analyzed using the following less-complicated method.

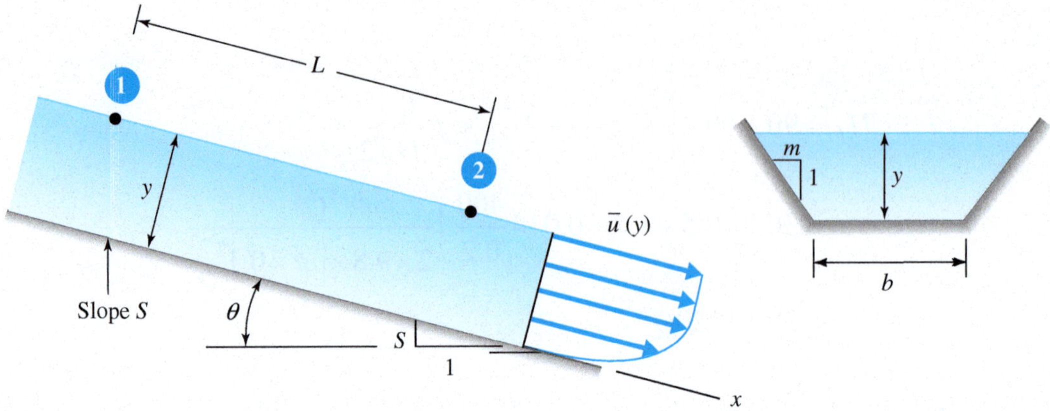

Figure 5.19 Uniform flow in an open channel.

If the energy equation is applied between two sections of the channel sketched in Figure 5.19, we obtain

$$0 = \frac{V_2^2 - V_1^2}{2g} + \frac{P_2 - P_1}{\gamma} + z_2 - z_1 + h_L \tag{5.7.1}$$

which shows that the head loss is

$$\begin{aligned} h_L &= z_1 - z_2 \\ &= L \sin\theta = LS \end{aligned} \tag{5.7.2}$$

where L is the length of the channel between the two sections and S is the slope of the channel, which is assumed small, so that $\sin\theta \simeq S$. (Do not confuse S with specific gravity.)

The Darcy–Weisbach equation (5.6.34), with $h_L = LS$ from Eq. 5.7.2, takes the form

$$LS = f\frac{L}{4R}\frac{V^2}{2g} \qquad \therefore \mathrm{RS} = \frac{f}{8g}V^2 \tag{5.7.3}$$

where R is the hydraulic radius. Since open channels are usually quite large with large Reynolds numbers, the friction factor is invariably in the completely turbulent regime. Thus the equation above is written as

$$V = C\sqrt{RS} \tag{5.7.4}$$

where the coefficient C is a dimensional constant; the equation above is referred to as the **Chezy equation**, named after Antoine Chezy (1718–1798). The Chezy coefficient is related to the channel roughness and the hydraulic radius (much like f is in a pipe) by

$$C = \frac{c_1}{n}R^{1/6} \tag{5.7.5}$$

where the dimensional constant c_1 has a value of 1.0 if SI units are used and 1.49 if English units are used. The dimensionless constant n is directly related to the wall roughness; it is called the **Manning** n, named after Robert Manning (1816–1897). Values for various wall materials are listed in Table 5.3.

The flow rate, which is of primary interest in open-channel flow problems, is found to be

$$Q = \frac{c_1}{n} A R^{2/3} S^{1/2} \qquad c_1 = \begin{cases} 1.0 \text{ for SI units} \\ 1.49 \text{ for English units} \end{cases} \tag{5.7.6}$$

This is the **Chezy–Manning equation**.

For smooth-surfaced channels, the use of the Chezy–Manning equation is discouraged since it implicitly assumes a rough wall. Calculations for smooth-surfaced channels such as glass or plastic should be based on the Darcy–Weisbach relation using f from Figure 5.13; see Section 5.6.3.

Table 5.3 Average Values[a] of the Manning n

Wall material	*Manning n*
Planed wood	0.012
Unplaned wood	0.013
Finished concrete	0.012
Unfinished concrete	0.014
Sewer pipe	0.013
Brick	0.016
Cast iron, wrought iron	0.015
Concrete pipe	0.015
Riveted steel	0.017
Earth, straight	0.022
Corrugated metal flumes	0.025
Rubble	0.03
Earth with stones and weeds	0.035
Mountain streams	0.05

[a]The values in this table result in flow rates too large for hydraulic radii greater than about 3 m (10 ft). The Manning n should be increased by 10 to 15% for such large conduits.

Example 5.19

The depth of water at 60°F flowing in a 12-ft-wide rectangular, finished concrete channel is measured to be 4 ft. The slope is measured to be 0.0016. Estimate the flow rate using (a) the Chezy–Manning equation and (b) the Darcy–Weisbach equation.

Solution

The hydraulic radius is calculated to be

$$R = \frac{A}{P} = \frac{yb}{2y+b} = \frac{4\times 12}{2\times 4\times 12} = 2.4 \text{ ft}$$

(a) Using the Chezy–Manning equation, with $n = 0.012$ from Table 5.3 and $c = 1.49$, we have

$$\begin{aligned} Q &= \frac{1.49}{n} A R^{2/3} S^{1/2} \\ &= \frac{1.49 \text{ ft}^{1/3}/\text{s}}{0.012} \times (4\times 12)\text{ ft}^2 \times (2.4 \text{ ft})^{2/3} \times 0.0016^{1/2} = \underline{427 \text{ ft}^3/\text{s}} \end{aligned}$$

(b) The relative roughness is, using a low value $e = 0.0015$ ft (it is finished concrete) shown on the Moody diagram:

$$\frac{e}{4R} = \frac{0.0015}{4\times 2.4} = 0.00016$$

Assuming a completely turbulent flow, the Moody diagram gives the friction factor as

$$f = 0.013$$

The Darcy–Weisbach equation (5.7.3) then yields the velocity as follows:

$$\begin{aligned} \therefore V &= \left(\frac{8RgS}{f}\right)^{1/2} \\ &= \left(\frac{8\times 2.4 \text{ ft}\times 32.2 \text{ ft/s}^2 \times 0.0016}{0.013}\right)^{1/2} = 8.72 \text{ ft/s} \end{aligned}$$

The flow rate is calculated as

$$Q = VA = 8.72\times 4\times 12 = \underline{419 \text{ ft}^3/\text{s}}$$

These two values are within 2%, an acceptable engineering tolerance for this type of problem. That found using the Moody diagram is considered to be more accurate, however.

Example 5.20

A 1.0-m-diameter concrete pipe transports 20°C water at a depth of 0.4 m. If the slope is 0.001, find the flow rate using (a) the Chezy–Manning equation and (b) the Darcy–Weisbach equation.

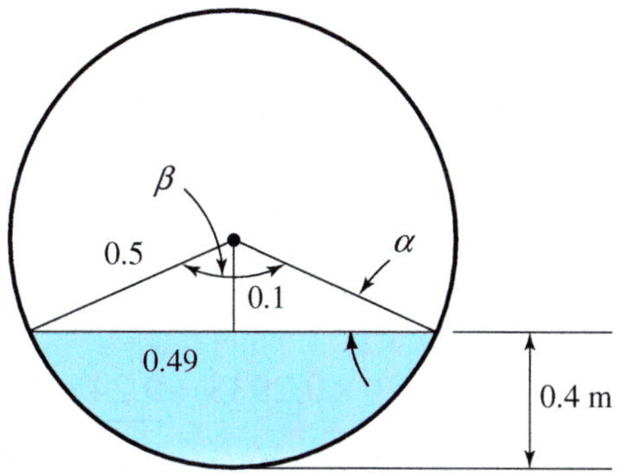

Figure E5.20

Solution

From the sketch of the pipe in Figure E5.20 the following are calculated:

$$\alpha = \sin^{-1}\frac{0.1}{0.5} = 11.54°$$

$$\therefore \beta = 180 - 2\times 11.54 = 156.9°$$

$$\therefore A = \pi \times 0.5^2 \times \frac{156.9}{360} - 0.49\times 0.1 = 0.2933\ \text{m}^2$$

$$P = 2\pi \times 0.5 \times \frac{156.9}{360} = 1.369\ \text{m}$$

The hydraulic radius is found, using the above calculations, to be

$$R = \frac{A}{P} = \frac{0.2933}{1.369} = 0.2142\ \text{m}$$

(a) The Chezy–Manning equation yields, with n from Table 5.3 and $c_1 = 1.0\ \text{m}^{1/3}/\text{s}$,

$$Q = \frac{1.0}{n} AR^{2/3}S^{1/2} = \frac{1.0}{0.015}\times 0.2933\times 0.2142^{2/3}\times 0.001^{1/2} = \underline{0.22\ \text{m}^3/\text{s}}$$

(b) The relative roughness is, using a relatively rough value for concrete pipe from Figure 5.13, as suggested by Table 5.3 of e = 20 mm,

$$\frac{e}{4R} = \frac{2}{4\times 214.2} = 0.0023$$

Assuming completely turbulent flow, the Moody diagram yields

$$f = 0.025$$

The Darcy–Weisbach equation (5.7.3) then gives the following:

$$\therefore V = \left(\frac{8RgS}{f}\right)^{1/2} = \left(\frac{8 \times 0.2142 \times 9.81 \times 0.001}{0.025}\right)^{1/2} = 0.820 \text{ m/s}$$

The flow rate is

$$Q = VA = 0.820 \times 0.2933 = \underline{0.24 \text{ m}^3/\text{s}}$$

This is within 8% of the result above, an acceptable tolerance for this type of problem. The second method, which is more difficult to apply, is considered to be more accurate, however.

5.8 SUMMARY

The laminar entrance lengths for a pipe and wide channel are, respectively,

$$\frac{L_E}{D} = 0.065 \text{ Re} \qquad \frac{L_E}{h} = 0.04 \text{ Re} \tag{5.8.1}$$

For a high-Reynolds-number turbulent pipe flow the entrance length is

$$\frac{L_E}{D} = 120 \tag{5.8.2}$$

For laminar flow in a pipe and a wide channel the pressure drop and friction factor are, respectively,

$$\Delta p = \frac{8\mu VL}{r_0^2} \qquad f = \frac{64}{\text{Re}} \qquad \text{Pipe} \tag{5.8.3}$$

$$\Delta p = \frac{12\mu VL}{a^2} \qquad f = \frac{48}{\text{Re}} \qquad \text{channel} \tag{5.8.4}$$

where a is the channel height.

The torque required to rotate an inner cylinder with the outer cylinder fixed is

$$T = \frac{4\pi\mu r_1^2 r_2^2 L\omega_1}{r_2^2 - r_1^2} \tag{5.8.5}$$

In a turbulent flow the head loss is calculated using

$$h_L = f\frac{L}{D}\frac{V^2}{2g} \tag{5.8.6}$$

where f is found using the Moody diagram of Figure 5.13. Minor losses are included using

$$h_L = K\frac{V^2}{2g} \tag{5.8.7}$$

where many loss coefficients K are listed in Table 5.2.

To include a pump in a piping system when the flow rate is not known, it is necessary to have pump characteristic curves, such as those of Figure E5.18. Example 5.18 illustrated the procedure.

The flow rate in an open channel is most often estimated using the equation

$$Q = \frac{c_1}{n}AR^{2/3}S^{1/2} \qquad c_1 = \begin{cases} 1.0\ \text{m}^{1/3}/\text{s for SI units} \\ 1.49\ \text{ft}^{1/3}/\text{s for English units} \end{cases} \tag{5.8.8}$$

where n is found in Table 5.3.

PROBLEMS

Laminar or Turbulent Flow

5.1 Calculate the maximum average velocity V with which 20°C water can flow in a pipe in the laminar state if the critical Reynolds number ($Re = VD/\nu$) at which transition occurs is 2000; the pipe diameter is:
(a) 2 m **(b)** 2 cm
(c) 2 mm

5.2 Water at 20°C flows in a wide river. Using a critical Reynolds number ($Re = Vh/\nu$) at which transition occurs of 1500, calculate the average velocity V that will result in a laminar flow if the depth h of the river is:
(a) 4 m **(b)** 1 m
(c) 0.3 m

5.3 Water at 50°F is flowing in a thin sheet in a parking lot at a depth of 0.2 in. with an average velocity of 1.5 ft/s. Is the flow laminar or turbulent?

5.4 Water is flowing, apparently quite placidly, in a river that is 20 m wide and 1.4 m deep. A leaf floating in the river is observed to move 1 m in 2 s. Is the flow laminar or turbulent? See Problem 5.2 for the definition of the Reynolds number.

5.5 Flow occurs in a 20-cm-diameter pipe. What is the maximum velocity that can occur for water at 20°C for a laminar flow if:
(a) Re = 2000? **(b)** Re = 40 000?

Entrance and Developed Flow

5.6 Calculate the laminar entrance length in a 4-cm-diameter pipe if 2×10^{-4} m^3/s of water is flowing at:
(a) 10°C **(b)** 20°C
(c) 40°C **(d)** 80°C

5.7 A laminar flow is to occur in an experimental facility with 20°C air flowing through a 4-cm-diameter pipe. Calculate the average velocity, the inviscid core length, and the entrance length if the Reynolds number is:
(a) 1000 **(b)** 80 000

5.8 A 6-cm-diameter pipe originates in a tank and delivers 0.025 m^3/s of water at 20°C to a receiver 50 m away. Is the assumption of developed flow acceptable?

5.9 A laboratory experiment is designed to create a laminar flow in a 2-mm-diameter tube shown in Figure P5.9. Water flows from a reservoir through the tube. If 18 L is collected in 2 hr, can the entrance length be neglected?

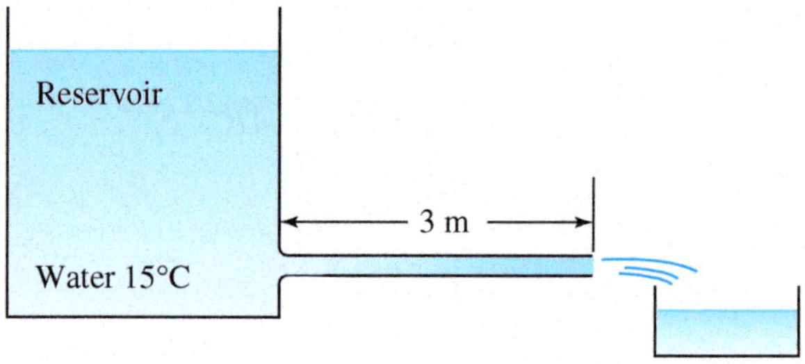

Figure P5.9

5.10 Air at 23°C is used as the working fluid in a parallel-plate research project. If the plates are 1.2 cm apart, how long is the longest possible entrance length for laminar flow? What is the shortest entrance length?

5.11 Air at 25°C can exist in either the laminar state or the turbulent state (a trip wire near the entrance is used to make it turbulent) for flow in a 6-cm-diameter pipe in a research lab. If the average velocity is 5 m/s, compare the length of the entrance region for the laminar flow with that of the turbulent flow.

5.12 Water at 20°C flows with an average velocity of 0.2 m/s from a reservoir through a 4-cm-diameter pipe. Estimate the inviscid core length and the entrance length if the flow is:
(a) Laminar
(b) Turbulent

5.13 Draw an incremental control volume with length Δx and radius r_0 and show that for a laminar flow $(\Delta p/\Delta x)_{\text{entrance}} > (\Delta p/\Delta x)_{\text{developed}}$.

5.14 Explain the pressure variations observed for a turbulent flow in Figure 5.3 for:
(a) A high-Re flow (Re > 300 000)
(b) A low-Re flow (Re ≃ 10 000)
(c) An intermediate-Re flow

Laminar Flow in a Pipe

5.15 Define $p_k = p + \gamma h$ as the kinetic pressure and write Eq. 5.3.5 or 5.3.11 in terms of p_k. Can we let $dp_x/dx = \Delta p_k/L$, where L is the length over which Δp_k is measured? If so, express $u(r)$ in terms of $\Delta p_k/L$.

5.16 Verify that Eq. 5.3.13 is, in fact, correct.

5.17 A pressure drop of 0.07 psi occurs over a section of 0.8-in.-diameter pipe transporting water at 70°F. Determine the length of the horizontal section if the Reynolds number is 1600. Also, find the shear stress at the wall and the friction factor.

5.18 Find the angle θ of the 10-mm-diameter pipe of Figure P5.18 in which water at 40°C is flowing with Re = 1500 such that no pressure drop occurs. Also, find the flow rate.

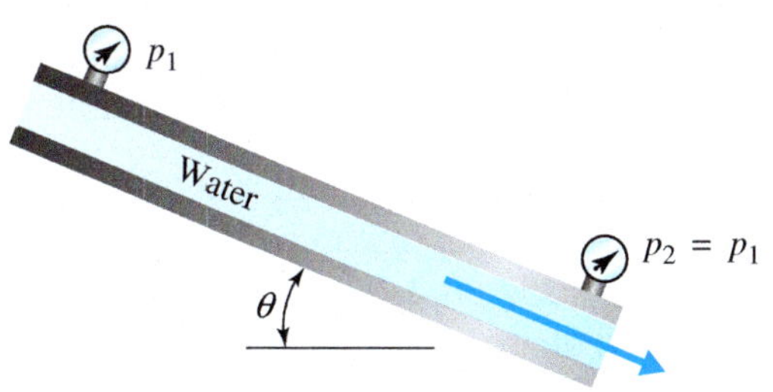

Figure P5.18

5.19 A liquid is pumped through a 2-cm-diameter pipe at a flow rate of 12 L/min. Calculate the pressure drop assuming laminar flow, in a 10-m horizontal section if the liquid is:
(a) SAE-10W oil at 20°C
(b) Water at 20°C
(c) Glycerine at 40°C

Is the assumption of laminar flow acceptable?

5.20 A liquid flows with no pressure drop in a vertical 2-cm-diameter pipe. Find the flow rate if, assuming a laminar flow, the liquid is:
(a) Water at 5°C
(b) SAE-30W oil at 25°C
(c) Glycerine at 20°C

Is the assumption of laminar flow acceptable?

5.21 A laminar flow is to exist in a pipe transporting 012 ft^3/sec of SAE-10W oil at 70°F. What is the maximum allowable diameter? What is the pressure drop over 30 ft of horizontal pipe for this diameter?

5.22 Estimate the flow rate through the smooth pipe shown in Figure P5.22. How long is the entrance region? Assume a laminar flow.

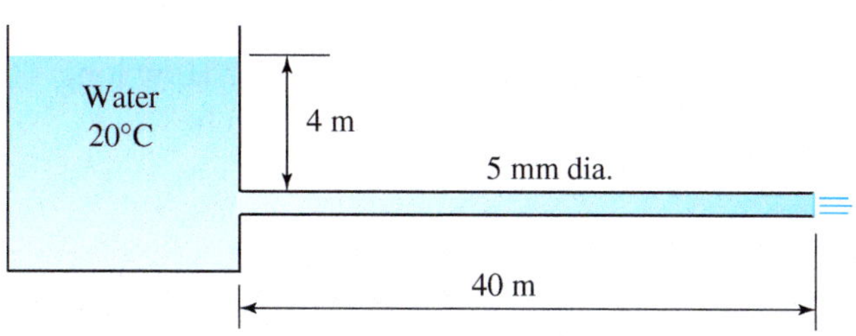

Figure P5.22

5.23 A manufacturer of small-diameter tubes wishes to know if the diameters are, in fact, accurate. An experimental setup, like that in Figure P5.22, is used with a 4-m-long horizontal tube transporting 20°C water with a head of 4 m. If 3.4 L of water is collected in 60 min, what is the inside diameter of the tube, neglecting the effect of the entrance region? Is the effect of entrance region actually negligible?

5.24 Air at 70°F flows in a horizontal 0.08-in.-diameter pipe. Calculate the maximum pressure drop in a 30-ft section for a laminar flow. Assume that $\rho = 0.0024$ slug/ft^3.

5.25 Water at 20°C flows in the 4-mm-diameter pipe of Figure P5.25. The pressure rise over the 10-m section is 6 kPa. Find the Reynolds number of the flow and the wall shear stress. Assume laminar flow.

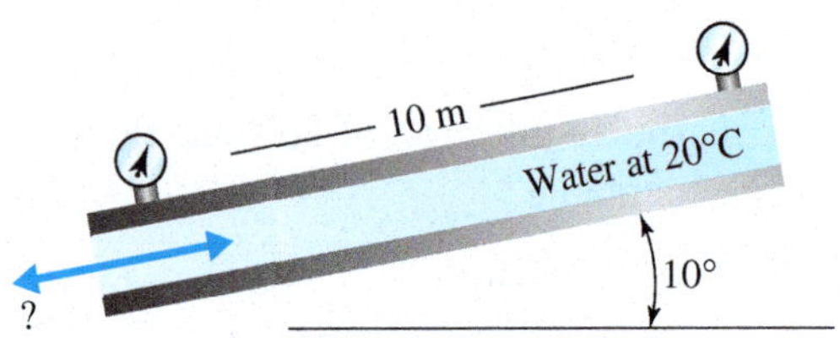

Figure P5.25

5.26 A research experiment requires a laminar flow of air at 20°C through a 10-cm-diameter pipe at a Reynolds number of 40 000. What maximum velocity is to be expected? What would be the pressure drop over a 10-m horizontal length of developed flow? How long would the entrance region be? Use $\rho = 1.2$ kg/m^3.

5.27 Calculate the radius where a pitot probe must be placed in the laminar liquid flow of Figure P5.27 so that the flow rate is given by $\pi R^2 \sqrt{2gH}$.

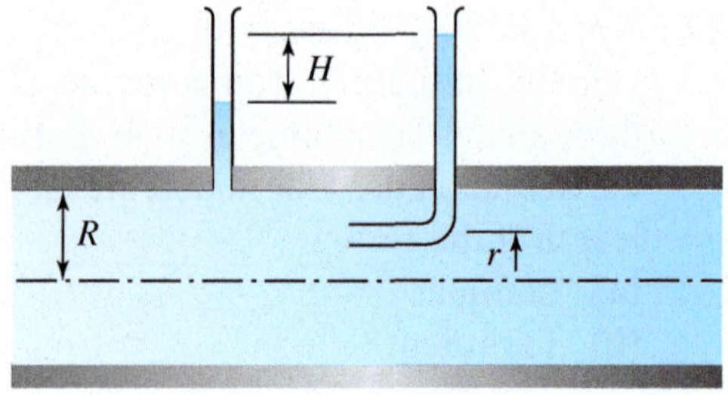

Figure P5.27

5.28 A laminar flow of water at 20°C occurs in a vertical 2-mm-diameter pipe. Calculate the flow rate if the pressure is constant. Is it reasonable to assume a laminar flow?

5.29 Water flows downward at a rate of 4.0 liters/min in a 40-mm-diameter vertical pipe.

(a) Determine the pressure drop over a distance of 10 meters.

(b) Calculate the friction head loss per unit length.

(c) What is the shear stress at the pipe wall? Use $\mu = 1.14 \times 10^{-6}$ N· s/m^2.

5.30 Find the radius in a developed lammar flow in a pipe where:

(a) The velocity is equal to the average velocity.

(b) The shear stress is equal to one-half the wall shear stress.

5.31 Find the ratio of the total flow rate through a pipe of radius r_0 to the flow rate through an annulus with inner and outer radii of $r_0/2$ and r_0. Assume developed laminar flow with the same pressure gradient.

5.32 A laminar flow of water at 60°F is obtained in a research lab at Re = 20,000 in a horizontal 2-in.-diameter pipe. Calculate the head loss in a 30-ft section of developed flow, the wall shear stress, and the length of the entrance region.

5.33 Water at 20°C flows between the two concentric horizontal pipes of Figure P5.33 with diameters of 2 cm and 3 cm. A pressure drop of 100 Pa is measured over a 10-m section of developed laminar flow. Find the flow rate and the shear stress on the inner pipe.

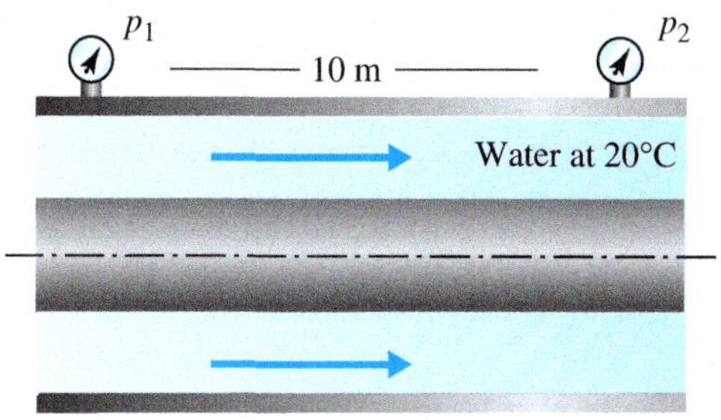

Figure P5.33

5.34 Air at 20°C is to flow in the annulus between two concentric, horizontal pipes, with respective diameters of 2 cm and 3 cm, such that a pressure drop of 10 Pa occurs over a 10-m length. Find the average velocity and the shear stress at the inner pipe. Assume a developed laminar flow.

5.35 Fluid flows in the annulus between two concentric horizontal pipes. The inner pipe is maintained at a higher temperature than the outer pipe so that the viscosity in the annulus cannot be assumed to be constant but $\mu = \mu(T)$. What differential equation would be solved to yield $u(r)$ assuming a developed laminar flow?

5.36 Show that the velocity distribution of Example 5.2 approaches that of pipe flow as $r_1 \to 0$ and approaches that of parallel-plate flow as $r_1 \to r_2$.

Laminar Flow Between Parallel Plates

5.37 A flow occurs in a horizontal channel ½-in. × 20 in. with Re = 2000. Calculate the flow rate if the fluid is:
(a) Water at 60°F
(b) Atmospheric air at 60°F

5.38 A board 1 m × 1 m that weighs 40 N moves down the incline shown in Figure P5.38 with a velocity of $V = 0.2$ m/s. Estimate the viscosity of the fluid if θ is:

(a) 20° **(b)** 30°

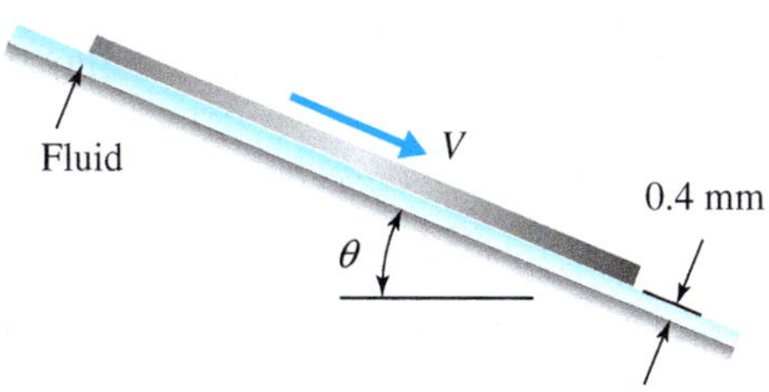

Figure P5.38

5.39 Water at 20°C exists between the plate and the surface of Problem 5.38. Calculate the velocity of the plate for an angle θ of:
(a) 20°
(b) 30°

5.40 Water at 20°C flows down a 20° incline at a depth of 6 mm and a width of 50 m. Calculate the flow rate and the Reynolds number assuming laminar flow. Also, find the maximum velocity and the wall shear.

5.41 Water at 20°C flows down a 100-m-wide parking lot on a slope of 0.00015 at a depth of 10 mm. Determine the flow rate and the Reynolds number assuming laminar flow. Also, calculate the friction factor and the wall shear.

5.42 A pressure drop of 50 Pa is measured over a 60-m length of a 90 cm × 2 cm rectangular horizontal channel transporting 20°C air. Calculate the maximum flow rate and the associated Reynolds number. Use $\rho = 1.2$ kg/m^3.

5.43 In Figure P5.43, a pressure difference $p_A - p_B$ is measured to be 96 kPa. Find the friction factor for the wide channel assuming laminar flow. The flow direction is unknown.

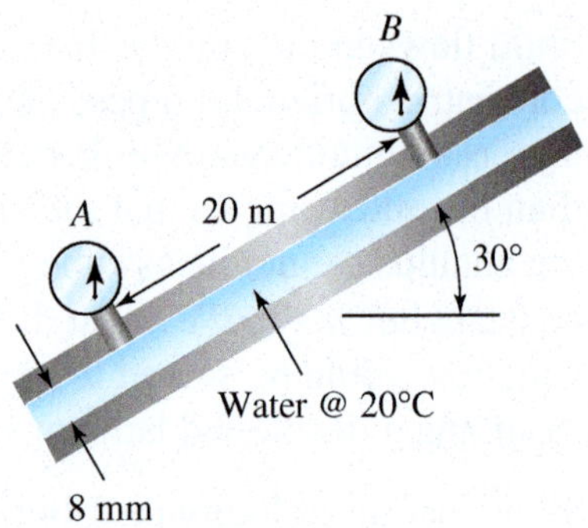

Figure P5.43

5.44 A slit with dimensions 0.02 in. × 4 in. exists in the 2-in.-thick side of a pressure vessel containing SAE-10W oil at 80°F and 600 psi. What is the maximum flow rate that can exist from the slit? Assume developed laminar flow.

5.45 Air flows between the parallel plates as shown in Figure P5.45. Find the pressure gradient such that:

(a) The shear stress at the upper surface is zero.

(b) The shear stress at the lower surface is zero.

(c) The flow rate is zero.

(d) The velocity at $y = 2$ mm is 4 m/s.

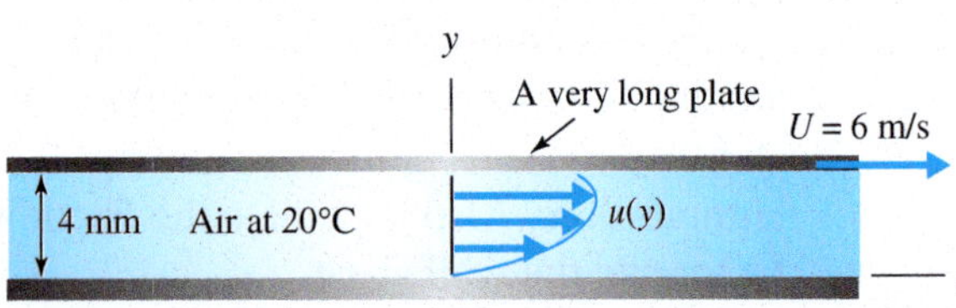

Figure P5.45

5.46 A pressure gradient of –20 Pa/m exists in 50°C air flowing between horizontal parallel plates spaced 6 mm apart. Find the velocity of the upper plate so that:

(a) The shear stress at the upper plate is zero.

(b) The shear stress at the lower plate is zero.

(c) The flow rate is zero.

(d) The velocity at $y = 2$ mm is 2 m/s.

5.47 Using the Navier–Stokes equations (a) determine an expression for the velocity profile of a pressure driven flow between two parallel horizontal plates placed 10-mm apart. The lower plate is fixed, and the upper plate is moving with a constant velocity of 2 m/s. Assume steady, laminar incompressible flow of oil with viscosity of 0.4 N · s/m^2. (b) Based on your answer in part (a), determine the pressure gradient needed to cause a zero velocity midway between the plates.

5.48 Consider steady, laminar, fully developed, incompressible flow between two inclined parallel plates separated by a distance h. The upper plate of Figure P5.48 is moving upward with a constant velocity U, and the lower plate is fixed. Starting with the Navier–Stokes equations, obtain an expression for the velocity in the fluid between the two plates. Consider that dp/dx = constant with the fluid flowing downward.

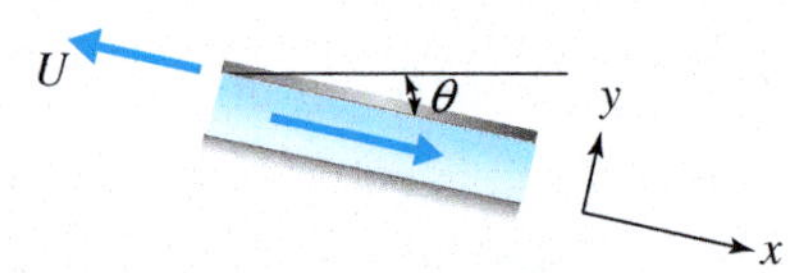

Figure P5.48

5.49 Oil with $\mu = 10^{-4}$ lb-s/ft^2 fills the concentric space between the rod and the surface shown in Figure P5.49. Find the force F if $V = 45$ ft/s. Assume that $dp/dx = 0$.

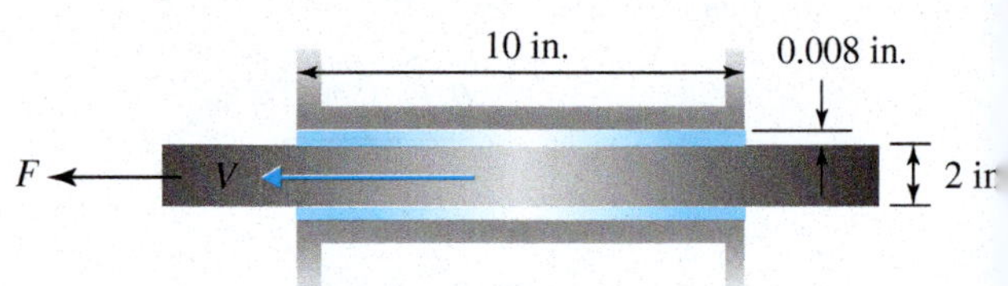

Figure P5.49

5.50 Calculate the torque T necessary to rotate the rod shown in Figure P5.50 at 30 rad/s if the fluid filling the gap is SAE-10W oil at 20°C. Assume a linear velocity profile.

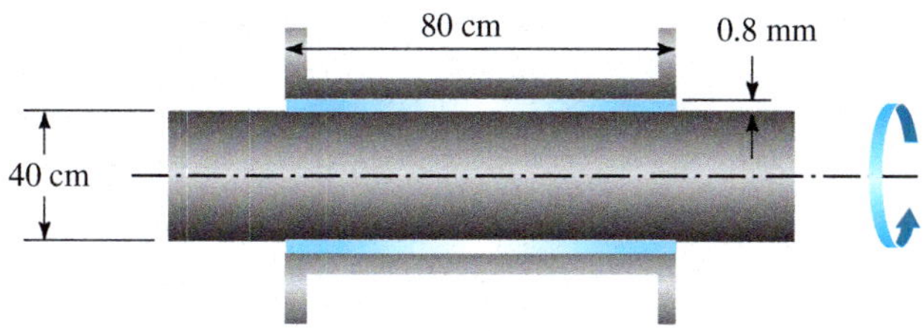

Figure P5.50

5.51 Oil with $\mu = 0.01 \text{ N} \cdot \text{s/m}^2$ fills the gap shown in Figure P5.51. Estimate the torque necessary to rotate the disk shown, assuming a linear velocity profile. Is the assumption of laminar flow valid? Use $S = 0.86$.

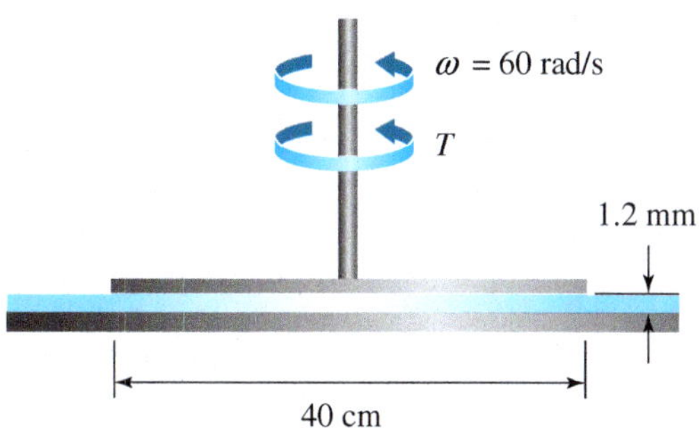

Figure P5.51

5.52 Approximate the torque necessary to rotate the inner 20-cm-diameter cylinder shown in Figure P5.52. SAE-30W oil at 20°C fills the gap. Assume a linear velocity profile.

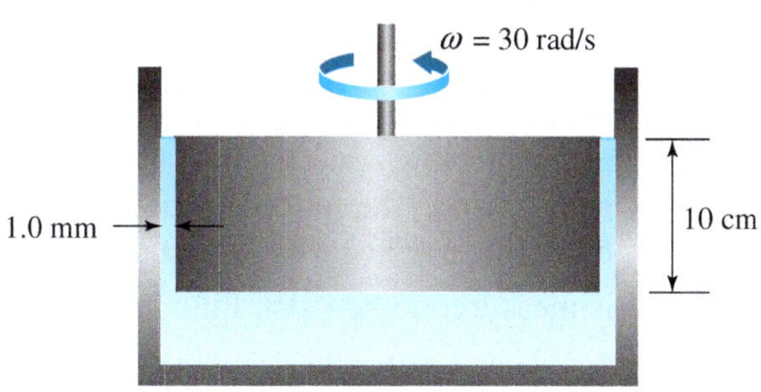

Figure P5.52

5.53 Find the torque needed to rotate the cone shown in Figure P5.53 if oil with $\mu = 0.01$ $\text{N} \cdot \text{s/m}^2$ fills the gap as shown. Assume a linear velocity profile.

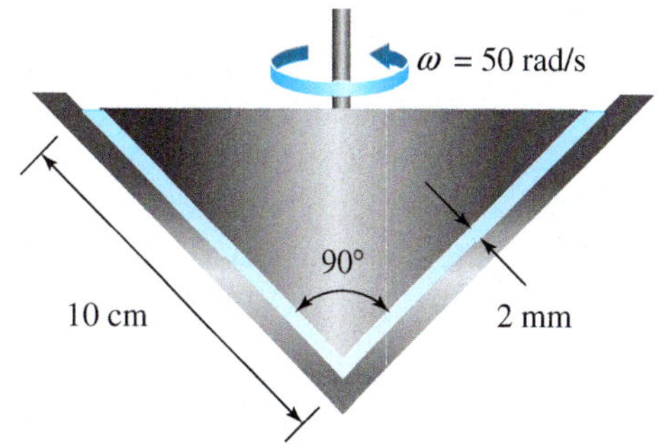

Figure P5.53

5.54 To create a high-Reynolds-number flow, the channel setup shown in Figure P5.54 was proposed by Prof. John Foss of Michigan State University. It is a pressurized channel, thereby avoiding fatal leaks that always are present in a suction channel. (An upstream fan would produce blade vortices that make a high Re impossible to attain.) Estimate the power requirement of the 70%-efficient fan if the channel is 1.2 m wide and Re = 7000.

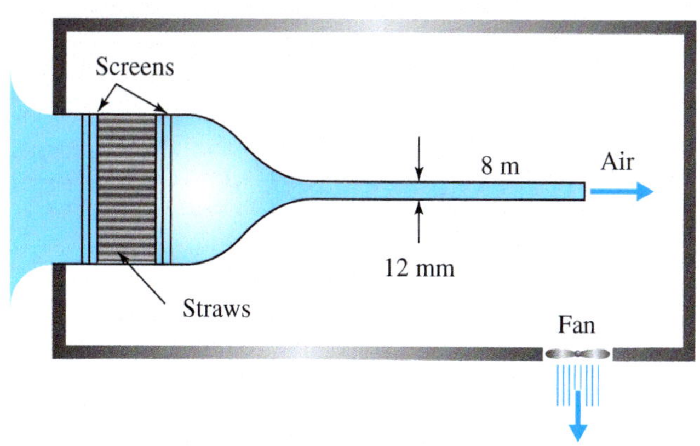

Figure P5.54

Laminar Flow Between Rotating Cylinders

5.55 A long cylinder of radius R rotates in a large container of liquid. What is the velocity distribution in the laminar flow? Calculate the torque needed to rotate a 40-in.-long, 2-in.-diameter cylinder at 1000 rpm if the liquid is water at 60°F. Assume a laminar flow.

5.56 SAE-10W oil at 40°C fills the gap between two concentric, 40-cm-long cylinders with respective radii of 2 cm and 3 cm.

What torque is necessary to rotate the inner cylinder at 3000 rpm if the outer cylinder is fixed? What power is required? Verify that Eq. 5.5.17 is, in fact, the velocity profile.

5.57 A torque of 0.015 N · m is required to rotate a 4-cm-radius cylinder inside a fixed 5-cm-radius cylinder at 40 rad/s. The concentric cylinders are 50 cm long. Calculate the viscosity of the fluid. Use $S = 0.9$. Verify that Eq. 5.5.17 is, in fact, the velocity profile.

5.58 Find an expression for the torque necessary to rotate the outer cylinder if the inner cylinder of Figure 5.6 is fixed.

5.59 Rework Problem 5.50 using the velocity distribution of Eq. 5.5.17 and calculate the percent error assuming the linear velocity profile.

Turbulent Flow

5.60 Time average the differential continuity equation for an incompressible flow and show that two continuity equations result: the instantaneous continuity equation

$$\frac{\partial u'}{\partial x}+\frac{\partial v'}{\partial y}+\frac{\partial w'}{\partial z}=0$$

and the time-average continuity equation

$$\frac{\partial \bar{u}}{\partial x}+\frac{\partial \bar{v}}{\partial y}+\frac{\partial \bar{w}}{\partial z}=0$$

5.61 Find an expression for the difference between the time-average acceleration $\overline{Du/Dt}$ and the quantity $\bar{D}\bar{u}/Dt$ using the fact that

$$\overline{u'\frac{\partial u'}{\partial x}}+\overline{v'\frac{\partial u'}{\partial y}}+\overline{w'\frac{\partial u'}{\partial z}}=\frac{\partial}{\partial x}\overline{u'^2}+\frac{\partial}{\partial y}\overline{u'v'}+\frac{\partial}{\partial z}\overline{u'w'}$$

5.62 Prove that the equation written in Problem 5.61 is indeed a fact. (*Hint:* Use the instantaneous continuity equation.)

5.63 Show that the time-averaged x-component Navier–Stokes equation results in

$$\rho\frac{\partial}{\partial y}\overline{u'v'}=-\frac{\partial\bar{p}}{\partial x}+\mu\frac{\partial^2\bar{u}}{\partial y^2}$$

for developed flow in a wide horizontal channel. Using $\bar{\tau}_{\text{lam}}=\mu(\partial\bar{u}/\partial y)$ and $\bar{\tau}_{\text{turb}}=-\rho\overline{u'v'}$, write the time-averaged Navier–Stokes equation in terms of stresses.

5.64 The velocity components at a point in a turbulent flow are given in Table 1. Find $\bar{u},\bar{v},\overline{u'^2},\overline{v'^2}$, and $\overline{u'v'}$ at that point.

Table 1

t(s)	u(m/s)	v(m/s)	t(s)	u(m/s)	v(m/s)
0.00	16.1	1.6	0.06	17.1	–1.4
0.01	25.7	–5.4	0.07	28.6	6.7
0.02	10.6	–8.6	0.08	6.7	–5.2
0.03	17.3	3.5	0.09	19.2	–8.2
0.04	5.2	4.1	0.10	21.6	1.5
0.05	10.2	–6.0			

5.65 Over a small radial distance in a developed turbulent flow, the time-average velocity is as given in the following table. The pressure drop in a 30-ft horizontal section is measured to be 8 psf. Find $\overline{u'v'}$ at $r = 0.69$ ft. Air with $\rho = 0.0035$ slug/ft^3 and $\nu = 1.6 \times 10^{-4}$ ft^2/s is flowing.

r (ft)	0.60	0.63	0.69	0.72
$\bar{u}$ (ft/s)	81.4	76.9	70.2	60.7

5.66 If at $r = 0.69$ ft for the data of Problem 5.65, we measure $\overline{u'^2} = 316\ \text{ft}^2/\text{s}^2$ and $\overline{v'^2} = 156\ \text{ft}^2/\text{s}^2$, what are the magnitudes of the eddy viscosity, the correlation coefficient, and the mixing length?

5.67 The velocity components are measured at a point in a laminar flow to be as shown in Figure P5.67. Find $\overline{u'v'}$, η, l_m, and K_{uv}, if $d\bar{u}/dy = -10\ \text{s}^{-1}$ at the point.

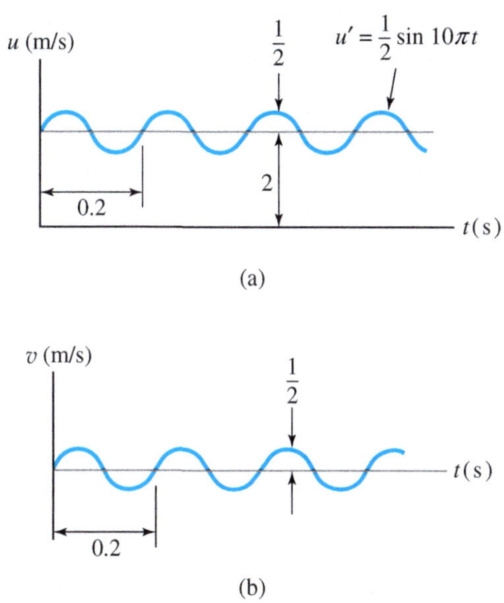

Figure P5.67

5.68 A 20-cm-diameter pipe with $e = 0.26$ mm transports water at 20°C. Determine if the pipe is smooth or rough if the average velocity is:
(a) 0.02 m/s
(b) 0.2 m/s
(c) 2 m/s

5.69 SAE-30 oil at 40°C is transported in a 10-cm-diameter pipe at an average velocity of 6 m/s. What is the largest size allowed for the roughness element if the pipe is hydraulically smooth?

5.70 Calculate the maximum velocity in the pipe of:
(a) Problem 5.68a
(b) Problem 5.68c

5.71 The velocity profile for water at 20°C in a turbulent flow in a 10-cm-diameter smooth pipe is approximated by $\bar{u} = 9.2\,y^{1/7}$ m/s. Find:
(a) The wall shear
(b) The velocity gradient $d\bar{u}/dy$ at the wall
(c) The pressure gradient
(d) The value of η at $r = 2.5$ cm

5.72 Water at 70°F flows in a 5-in.-diameter horizontal pipe at the rate of 2.5 ft^3/s. Find the constant n in the exponent of Eq. 5.6.19. What is the maximum velocity?

5.73 Show that the kinetic-energy correction factor is 1.10 for $n = 5$ and 1.03 for $n = 10$ using $u = u_{max}(y/r_0)^{1/n}$ in a circular pipe.

5.74 Water at 20°F flows in a 10-cm-diameter pipe with an average velocity of 10 m/s. Using $u = u_{max}(y/r_0)^{1/n}$ with $n = 7$, plot the viscous shear and the turbulent shear as a function of r. Also, find $d\bar{p}/dx$.

5.75 SAE-10W oil at 10°C is transported in a smooth 80-cm-diameter pipe at the rate of 1.2 m^3/s.
(a) Find the Reynolds number.
(b) Find the friction factor.
(c) Find the maximum velocity using Eq. 5.6.20.
(d) Find the viscous wall layer thickness.
(e) Compare part (*c*) with the solution using the logarithmic velocity profile.

5.76 Let the pipe in Problem 5.75 be a cast iron pipe (Figure 5.13 provides a value for *e*). Estimate the maximum velocity using the logarithmic velocity profile.

5.77 A pressure drop of 1.5 psi is measured with gages placed 15 ft apart on a smooth horizontal 4-in.-diameter pipe transporting water at 100°F. Estimate:
(a) The wall shear
(b) The maximum velocity
(c) The average velocity
(d) The Reynolds number
(e) The flow rate

5.78 A 12-cm-diameter horizontal pipe transports SAE-10 oil at 10°C. Calculate the wall shear, the average velocity, and the flow rate if the pressure drop over a 10-m section of pipe is measured to be:
(a) 5 kPa
(b) 20 kPa
(c) 200 kPa

5.79 Make a linear plot (not a semilog plot) of the velocity profile of the flow in Problem 5.77 using:
(a) The log profile
(b) The power-law profile

Turbulent Flow in Pipes and Conduits

5.80 Water at 20°C flows in an 8-cm-diameter plastic pipe with a flow rate of 20 L/s. Determine the friction factor using (a) the Moody diagram, and (b) Eq. 5.6.26.

5.81 A flow rate of 0.03 m^3/s of 15°C water occurs in a 10-cm-diameter cast iron pipe. Determine the friction factor using (a) the Moody diagram, and (b) one of the equations (5.6.26) through (5.6.28).

5.82 Water at 20°C flows in a 4-cm-diameter cast iron pipe. Determine the friction factor, using the Moody diagram, if the average velocity is:
(a) 0.025 m/s **(b)** 0.25 m/s
(c) 2.5 m/s **(d)** 25 m/s

5.83 A flow rate of 0.02 m^3/s occurs in a 10-cm-diameter wrought iron pipe. Using the Moody diagram, calculate the pressure drop over a 100-m horizontal section if the pipe transports:
(a) Water at 20°C
(b) Glycerine at 60°C
(c) SAE-30W at 30°C
(d) Kerosene at 10°C

Compare each answer with that using Eq. 5.6.29.

5.84 Water at 60°F flows in a 1.5-in.-diameter pipe with a flow rate of 0.06 ft^3/s. Using the Moody diagram, determine the head loss in a 600-ft section if the pipe is:
(a) Cast iron **(b)** Galvanized iron
(c) Wrought iron **(d)** Plastic

5.85 A mass flux of 1.2 kg/s occurs in a 10-cm-diameter plastic pipe at 20°C and 500 kPa absolute. Assume an incompressible flow and, using the Moody diagram, find the pressure drop in a 100-m section of pipe if the fluid flowing is:
(a) Air
(b) Carbon dioxide
(c) Hydrogen

Compare each answer with that using Eq. 5.6.29.

5.86 SAE-30W oil flows at the rate of 0.08 m^3/s in a 15-cm-diameter horizontal galvanized iron pipe. Find the pressure drop in 100 m if the temperature of the oil is:
(a) 0°C **(b)** 30°C
(c) 60°C **(d)** 90°C

Compare each answer with that using Eq. 5.6.29.

5.87 Select the material, listed in Figure 5.13, from which each of the following pipes is probably constructed. Each 5-cm-diameter pipe is tested with water at 20°C using a flow rate of 400 L/min. The following pressure drops are measured over a 10-m length of horizontal pipe:
(a) Pipe 1:36 kPa **(b)** Pipe 2:24 kPa
(c) Pipe 3:19 kPa

5.88 Water at 50°F flows up a 50° incline in a 2½-in.-diameter plastic pipe with a flow rate of 0.3 ft^3/s. Find the pressure change over a 300-ft length of pipe.

5.89 Water at 40°C flows in a horizontal section of 5-cm-diameter wrought iron pipe with a flow rate of 0.02 m^3/s. Does the pipe behave as a smooth pipe, or is the roughness significant?

5.90 An 80-cm-diameter concrete pipe is to transport storm water at 20°C at a rate of 5 m^3/s. What pressure drop is to be expected over a 100-m section of horizontal pipe?

5.91 A pressure drop of 500 kPa is not to be exceeded over a 200-m length of horizontal 10-cm-diameter cast iron pipe. Calculate the maximum flow rate if the fluid is:
(a) Water at 20°C
(b) Glycerine at 20°C
(c) SAE-10W oil at 20°C
(d) Kerosene at 20°C

5.92 A pressure drop of 200 kPa is not to be exceeded over a 100-m length of horizontal 4-cm-diameter pipe. Estimate the maximum flow rate if water at 20°C is transported and the pipe is:
(a) Cast iron **(b)** Wrought iron
(c) Plastic

5.93 Neglecting all losses except that due to wall friction, estimate the flow rate through the pipe shown in Figure P5.93 if the diameter is:
(a) 4 cm **(b)** 8 cm
(c) 12 cm **(d)** 16 cm

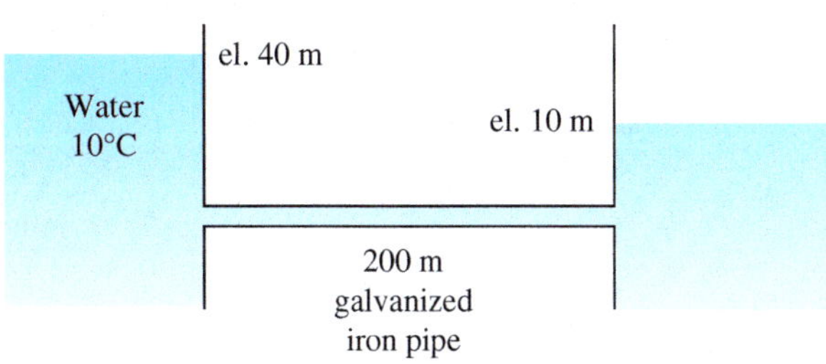

Figure P5.93

5.94 A pressure drop of 400 Pa is allowable in gas flow in a 400-m horizontal section of 12-cm-diameter wrought iron pipe. If the temperature and pressure are 40°C and 200 kPa absolute, find the maximum mass flux if the gas is:
(a) Air **(b)** Carbon dioxide
(c) Hydrogen

5.95 A pressure drop of 30 psi is not to be exceeded over a 600-ft length of horizontal 4-ft-diameter concrete pipe transporting water at 60°F. What flow rate can be accommodated? Use:
(a) The Moody diagram
(b) Equation 5.6.30

5.96 Estimate the size of plastic tubing that should be selected if 0.002 m^3/s of fluid is to be transported such that the pressure drop does not exceed 200 kPa in a 100-m horizontal section. The fluid is:
(a) Water at 20°C
(b) Glycerine at 60°C
(c) Kerosene at 20°C
(d) SAE-10W oil at 40°C

5.97 Select a concrete pipe size that will transport 5 m^3/s of 20°C water so that the head loss does not exceed 20 m in a 300-m horizontal pipe section. Use:
(a) The Moody diagram
(b) Equation 5.6.31

5.98 A farmer wishes to siphon 10°C water from a lake 1200 m from a field positioned a distance of 3 m below the surface of the lake. What size of drawn tubing should be selected if 400 L of water is desired each minute? Use:
(a) The Moody diagram
(b) Equation 5.6.31

Neglect all losses except that due to wall friction. Is the exiting kinetic energy also negligible?

5.99 Atmospheric air at 30°C is to be transported through a square sheet metal (smooth) conduit at the rate of 4 m^3/s. What should be the conduit dimensions so that the head loss does not exceed 10 m over a horizontal length of 200 m?

5.100 Water at 20°C is transported through a 2 cm × 4 cm smooth conduit and experiences a pressure drop of 80 Pa over a 2-m horizontal length. What is the flow rate?

5.101 A plastic conduit 4 cm × 10 cm transports water at 20°C. If a pressure drop of 100 Pa is measured by gauges spaced 5 m apart on a horizontal section, find the flow rate.

5.102 An open rectangular 1.2-m-wide concrete (use $e = 1.5$ mm) channel transports 20°C water from a reservoir to a location 10 000 m away. Using the Moody diagram, estimate the flow rate if the channel is on a slope of 0.0015 and the water depth is:
(a) 0.3 m **(b)** 0.6 m
(c) 0.9 m

Minor Losses

5.103 If the loss coefficient for a sudden expansion is based on the exit velocity V_2, determined the loss coefficient in terms of A_1 and A_2.

5.104 Explain, with reference to Eq. 1.4.15, why the high and low-pressure regions of Figure 5.14 exist. Also, sketch a velocity profile from the inside corner of the bend to the outside of the bend along a line from B to C as sketched in Figure P5.104. Explain why a secondary flow results after the bend.

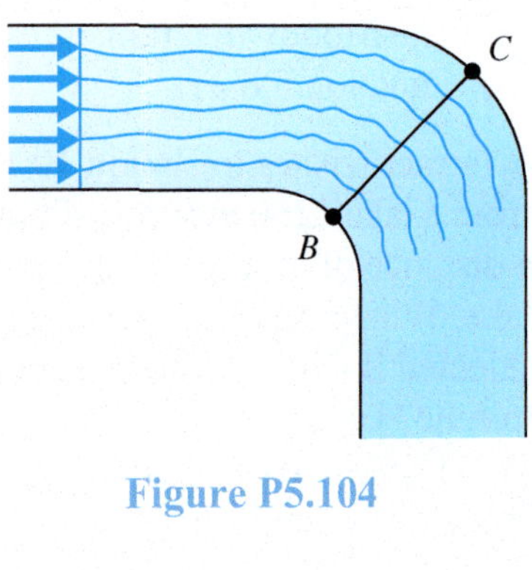

Figure P5.104

5.105 For each setup shown in Figure P5.105, find p_2 if $Q = 0.02$ m^3/s of air at 20°C and $p_1 = 50$ kPa.

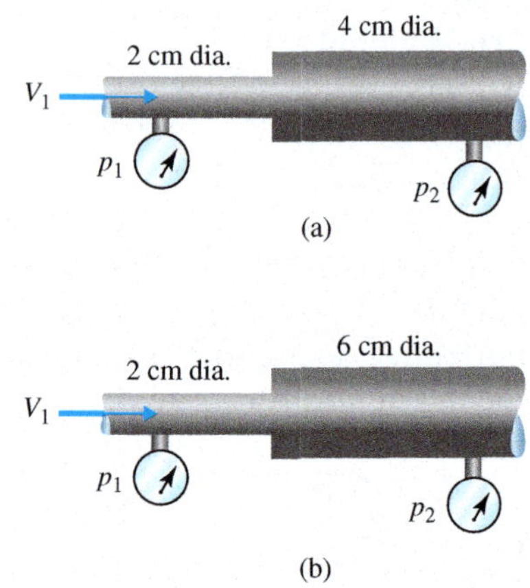

Figure P5.105

5.106 Replace the sudden enlargement of Problem 5.105 with a 20° expansion angle and rework the problem.

5.107 For each setup shown in Figure P5.107, estimate the loss coefficient based on V_2 using the data of Figure 5.16.

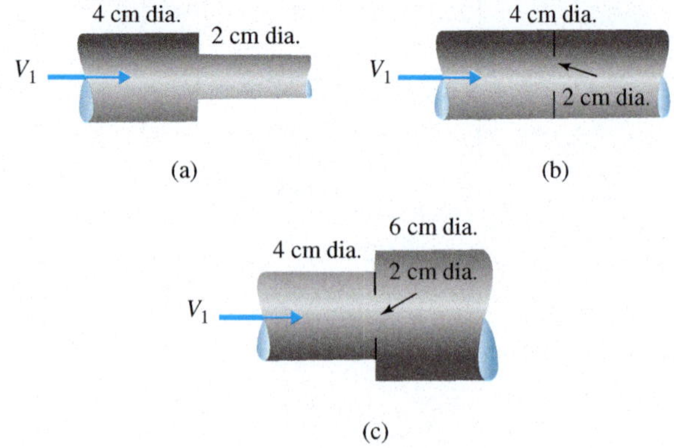

Figure P5.107

5.108 The flow rate is measured to be 0.12 ft^3/sec from the pipe shown in Figure P5.108. Find the loss coefficient of the valve. Neglect wall friction.

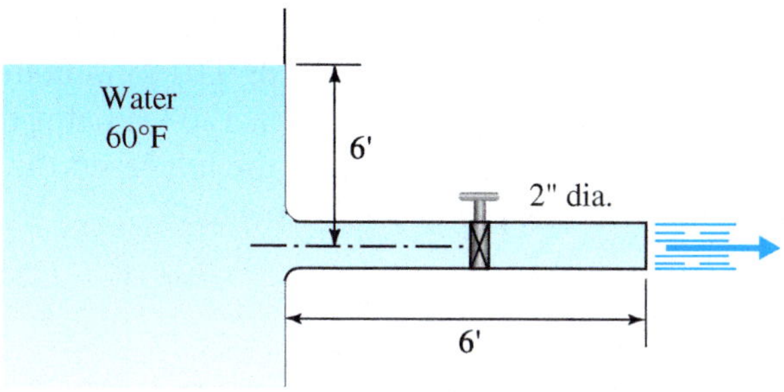

Figure P5.108

5.109 The flow rate is measured to be 6 L/s in the pipe shown in Figure P5.109. Find the loss coefficient of the valve if H is:
(a) 4 cm **(b)** 8 cm

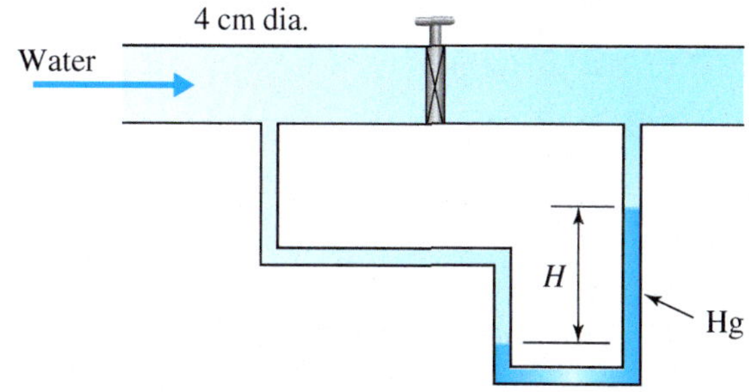

Figure P5.109

Simple Piping Systems

5.110 Find the flow rate from the pipe shown in Figure P5.110. Sketch the EGL and the HGL.

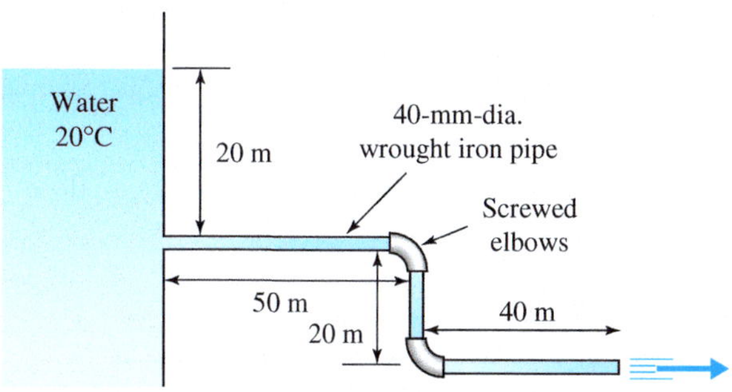

Figure P5.110

5.111 Water at 70°F flows from a 4-in.-diameter 1200-ft length of cast iron horizontal pipe that is attached to a reservoir with a square-edged entrance. A screwed globe valve that controls the flow is half open. Find the flow rate if the reservoir elevation above the pipe exit is:
(a) 15 ft
(b) 30 ft
(c) 60 ft

5.112 Estimate the flow rate to be expected through the plastic siphon shown in Figure P5.112 if the diameter is:
(a) 4 cm
(b) 8 cm
(c) 12 cm

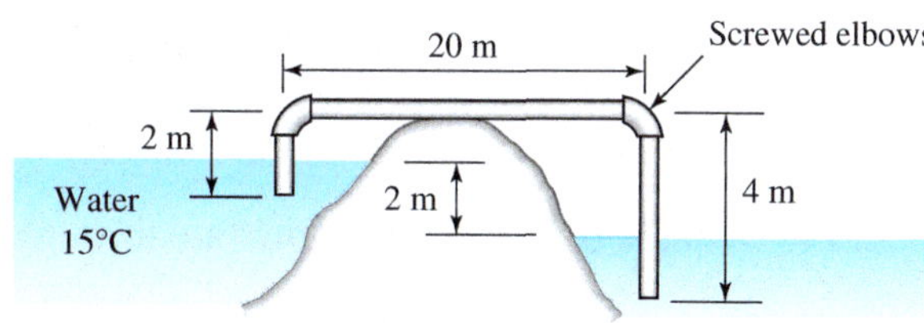

Figure P5.112

5.113 For the cast-iron piping shown in Figure P5.113, calculate the flow rate and minimum pressure and sketch the HGL and the EGL if:
(a) $H = 10$ m
(b) $H = 20$ m
(c) $H = 30$ m

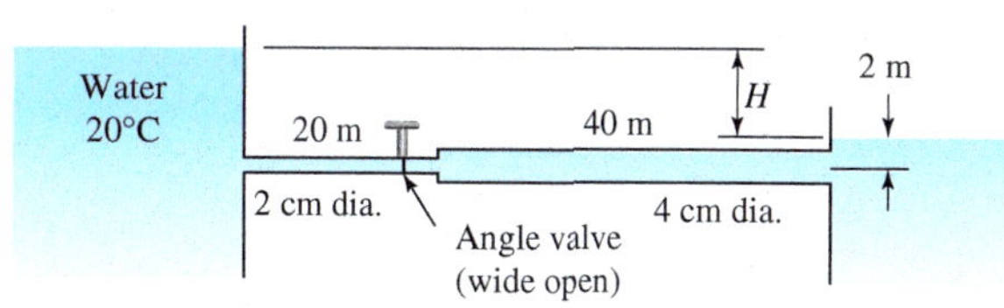

Figure P5.113

5.114 Water at 35°C flows from a bathtub through a 2.5-cm-diameter plastic pipe to a large air-filled drain pipe. There are two screwed elbows in the 10 m of pipe. If the drain pipe is 0.8 m below the water level in the tub, estimate how long it will take to drain 10 L of water.

5.115 A 3-cm-diameter plastic tube with screwed elbows is used to siphon the water as shown in Figure P5.115. Estimate the maximum height H for which the siphon will operate.

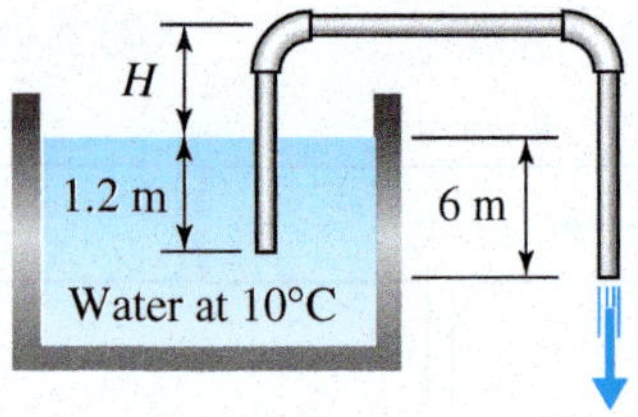

Figure P5.115

5.116 A 12-L lawn sprayer is filled with 8 L of 20°C water. It is 1.2 m high and has an 8-mm-diameter copper tube ($e \simeq 0$) that reaches the bottom (it is a little short). A 1.2-m-long 5-mm-diameter smooth hose is attached to the copper tube. The hose ends with a 2-mm-diameter nozzle. If the sprayer is pressurized to 100 kPa, estimate the initial velocity exiting the nozzle.

5.117 What is the maximum flow rate through the pipe shown in Figure P5.117 if the elevation difference of the reservoir surfaces is:
(a) 80 m
(b) 150 m
(c) 200 m

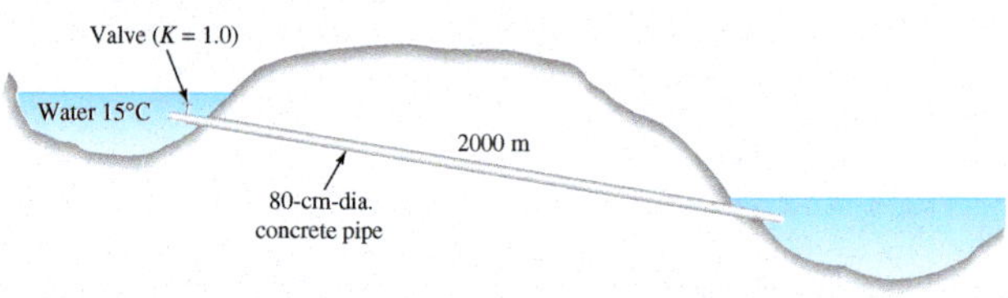

Figure P5.117

5.118 A 9.4-mm-diameter 60-m-long plastic pipe transports 20°C water from a spring to a pond 3 m below, similar to the situation shown in Problem 5.129. It is observed that the water alternates between a relatively fast moving stream to a relatively slow moving stream. Explain this phenomenon with supporting calculations.

5.119 Water at 60°F is to be pumped through 900 ft of cast iron pipe from a reservoir to a device that is 30 ft above the reservoir surface. It is to enter the device at 30 psi. Screwed components include two elbows, a square-edged entrance, and an angle valve. If the flow rate is to be 0.6 ft^3/s, what pump power is needed (assume 80% efficiency) if the pipe diameter is:
(a) 1.5 in. **(b)** 3 in **(c)** 4.5 in.

5.120 What pump power (85% efficient) is needed for a flow rate of 0.01 m^3/s in the pipe shown in Figure P5.120? What is the greatest distance from the left reservoir that the pump can be located?

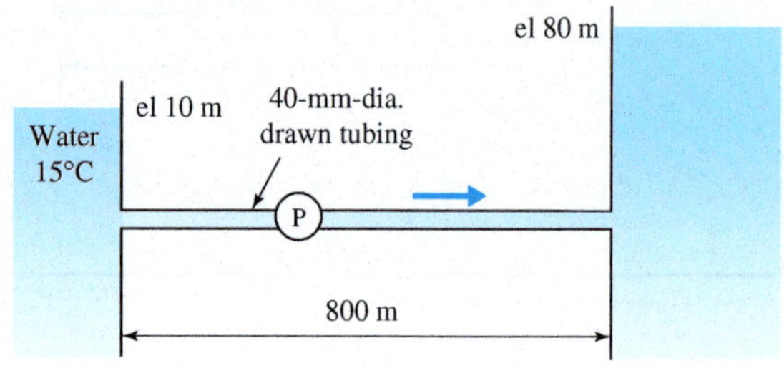

Figure P5.120

5.121 A flow rate of 2 m^3/s exists in the pipe shown in Figure P5.121. What is the expected power output of the turbine (85% efficient) if the elevation difference of the reservoir surfaces is:
(a) 20 m **(b)** 60 m **(c)** 100 m

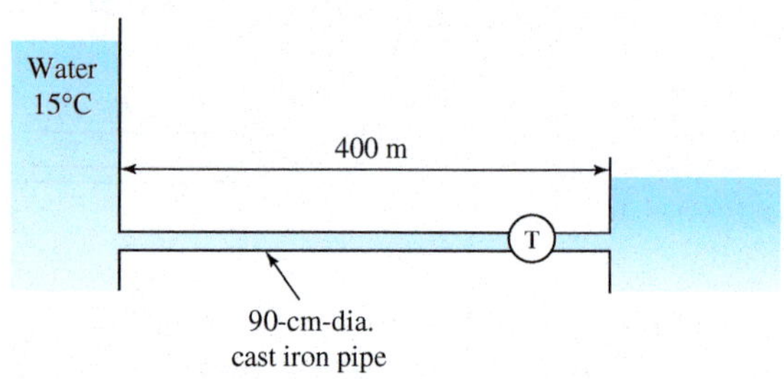

Figure P5.121

5.122 What pump power (75% efficient) is needed in the piping system shown in Figure P5.122? What is the greatest distance from the reservoir that the pump can be located?

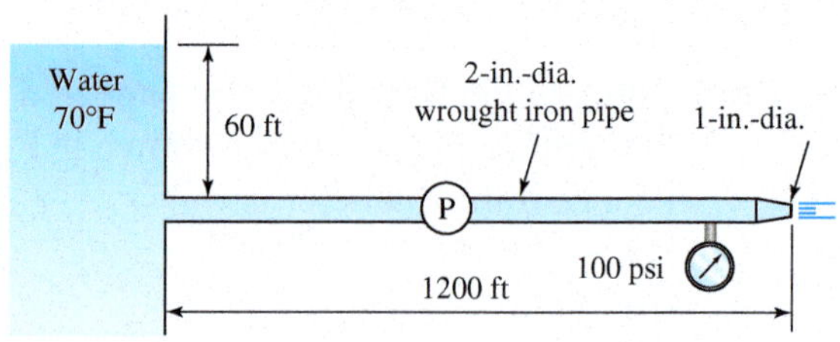

Figure P5.122

5.123 The pump shown in Figure P5.123 has the characteristic curves shown in Example 5.18.

(a) Estimate the flow rate and the power required by the pump.

(b) Sketch the EGL and the HGL.

(c) If cavitation is possible, determine the maximum distance from the reservoir to locate the pump.

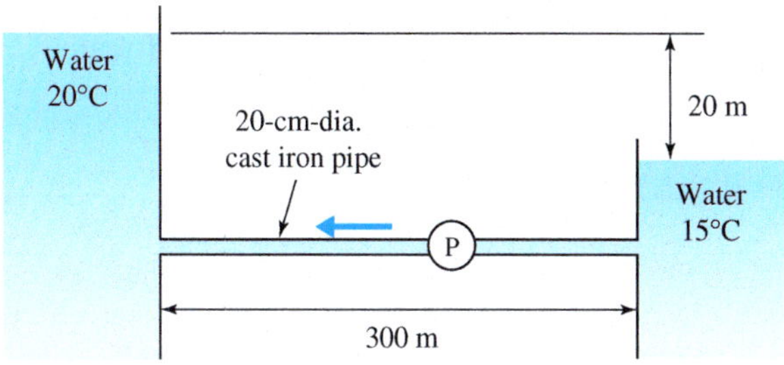

Figure P5.123

5.124 Reverse the flow direction in Problem 5.123 and redo the problem.

5.125 A turbine replaces the pump of Problem 5.123. Estimate the power output if $\eta_T = 0.88$. The turbine's characteristic curve is $H_T = 0.8Q$, where H_T is measured in meters and Q in L/s.

5.126 The pump shown in Figure P5.126 has characteristic curves shown in Example 5.18. Estimate the flow rate and

(a) Calculate the pump power requirement.

(b) Calculate the pressure at the pump inlet.

(c) Calculate the pressure at the pump outlet.

(d) Sketch the EGL and the HGL.

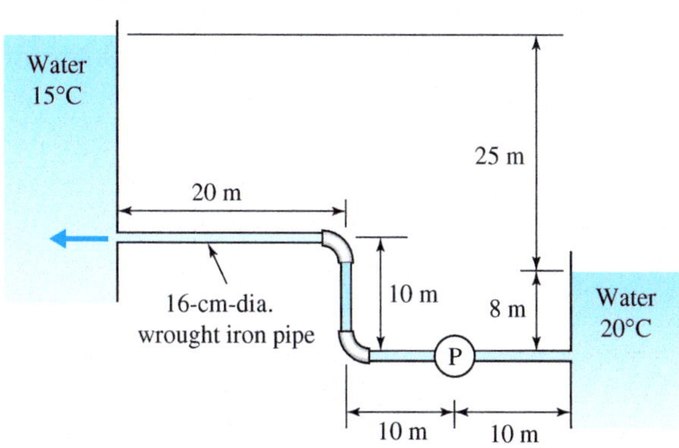

Figure P5.126

5.127 Reverse the flow direction in Problem 5.126 and redo the problem.

5.128 In the filtration system shown in Figure P5.128, water is circulated steadily through a filter using a pump. The pump characteristic curve is given by $H_p = 10 + 12Q - 150Q^2$, where H_p is in meters and Q is in m^3/s. A 10-cm-diameter pipe is used in the system. The total length of pipe used is 60 m, and the friction factor of the pipe is $f = 0.04$. Determine the flow rate and power input to the pump. All bends are 90-miter bends without vanes.

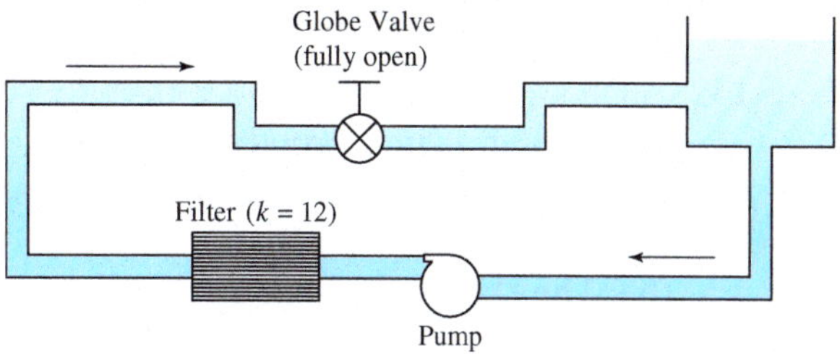

Figure P5.128

5.129 A turbine with a characteristic curve shown in Figure P5.141 is inserted in the pipeline. Calculate the turbine power output. Assume that $\eta_T = 0.90$.

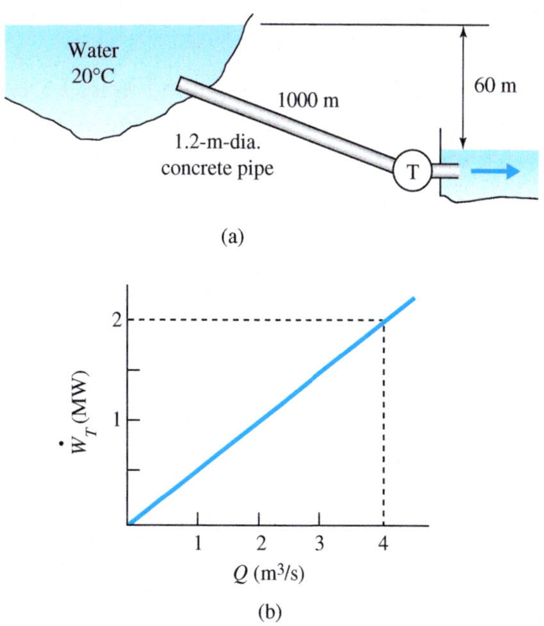

Figure P5.129

Open-Channel Flow

5.130 Using a control volume surrounding a finite length of the liquid in a channel flowing at constant depth, find the average shear stress on the walls if water flows in a 10-ft-wide rectangular channel at a depth of 6 ft on a 0.001 slope.

5.131 Using the approach mentioned in Problem 5.130, determine the average shear stress on the portion of the wall of a 40-cm-diameter circular conduit in contact with the water, which is flowing at a constant depth of 10 cm. The slope is 0.0016.

5.132 Calculate the flow rate in a 2-m-wide rectangular, planed wood channel on a slope of 0.001 if the depth is 60 cm. Use:
(a) The Chezy–Manning equation
(b) The Darcy–Weisbach equation

5.133 Water flows in a 6-ft-diameter finished concrete conduit on a 0.0012 slope. Predict the flow rate if the depth of flow is:
(a) 6 ft minus a little
(b) 5.7 ft
(c) 3 ft
(d) 1.5 ft
(e) 0.5 ft

5.134 For the channel shown in Figure P5.134, find the flow rate and average velocity if $S = 0.001$ using:
(a) The Chezy–Manning equation
(b) The Darcy–Weisbach equation

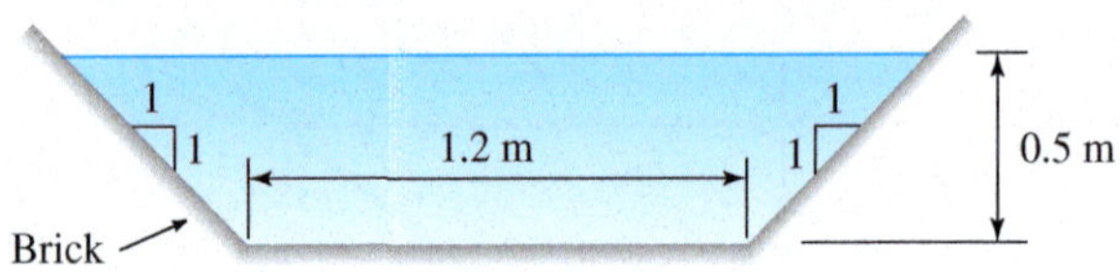

Figure P5.134

5.135 At what depth will 5 m^3/s of water flow in a 2-m-wide rectangular brick channel with $S = 0.001$? Use:
(a) The Chezy–Manning equation
(b) The Darcy–Weisbach equation

5.136 The cross section of a straight river is approximated as shown in Figure P5.136. At what depth will 100 m^3/s of water flow? The slope is 0.001.

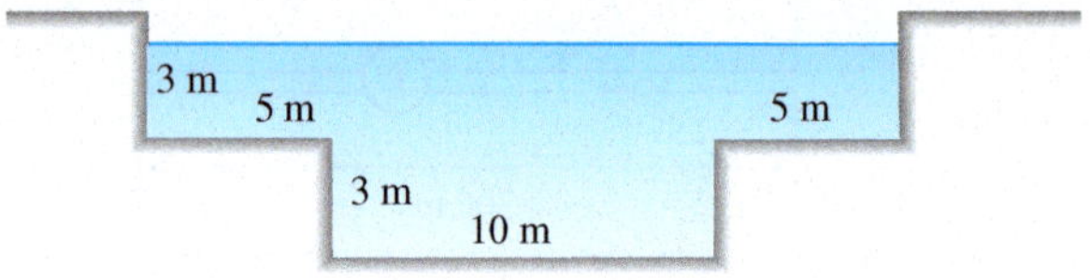

Figure P5.136

5.137 For the channel shown in Figure P5.137, if $S = 0.0016$, find the flow depth if $Q = 10$ m^3/s and the channel is constructed with:
(a) Planed wood
(b) Brick

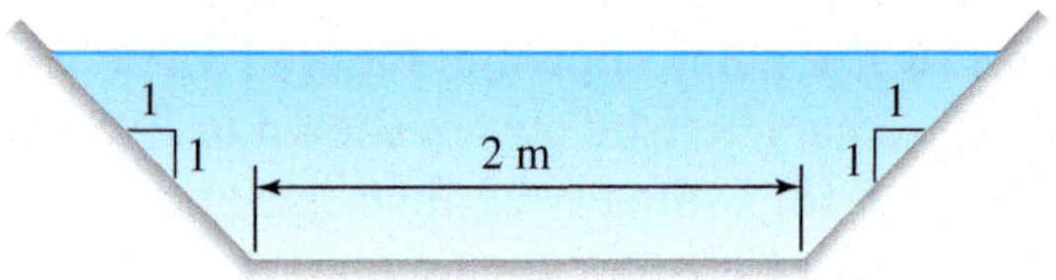

Figure P5.137

5.138 Water flows in a 4-ft-diameter sewer pipe at a flow rate of 24 ft^3/s. Estimate the depth if the slope is 0.001.

5.139 Water flows in a 80-cm-diameter sewer pipe at a flow rate of 0.2 m^3/s. Determine the depth if the slope is 0.001.

6 External Flows

Outline

Chapter Objectives

The objectives of this chapter are to:

- Discuss separated and attached flows.
- Introduce the lift and drag coefficients.
- Determine the drag on a variety of bodies.
- Study the influence of vortex shedding and streamlining.
- Determine when cavitation is present.
- Calculate the lift and drag on airfoils.
- Superimpose simple potential flows to construct flows of interest.
- Analyze both laminar and turbulent boundary layers on a flat plate.
- Provide numerous examples and problems demonstrating how the various quantities of interest are determined for the many external flows discussed in this chapter.

M. C. Potter and D. C. Wiggert, *Applied Fluid Mechanics*,
https://doi.org/10.1007/978-3-031-65276-9_6

6.1 INTRODUCTION

The study of external flows is of particular importance to the aeronautical engineer in the analysis of airflow around the various components of an aircraft. In fact, much of the present knowledge of external flows has been obtained from studies motivated by such aerodynamic problems. There is, however, substantial interest by other engineers in external flows; the flow of fluid around turbine blades, automobiles, buildings, athletic stadiums, smokestacks, spray droplets, bridge abutments, submarine pipelines, river sediment, and red blood cells includes a variety of phenomena that can be understood only from the perspective of external flows.

It is a difficult task to determine the flow field external to a body and the pressure distribution on the surface of the body, even for the simplest geometry. To discuss this subject, consider low-Reynolds-number flows ($Re < 5$, or so) and high-Reynolds-number flows ($Re > 1000$). Low-Reynolds-number flows, called **creeping flows** or **Stokes flows**, seldom occur in engineering applications (flow around spray droplets and red blood cells, lubrication in small gaps, and flow in porous media would be exceptions) and are not presented in this book; they are left to the specialist. We will direct our attention to high-Reynolds-number flows only. We do, however, show a Stokes flow in Figure 6.1.

High-Reynolds-number flows can be subdivided into three major categories: (1) incompressible immersed flows involving such objects as automobiles, helicopters, submarines, low-speed aircraft, take-off and landing of commercial aircraft, buildings, and turbine blades; (2) flows of liquids that involve a free surface as experienced by a ship or a bridge abutment; and (3) compressible flows involving high-speed objects ($V > 100$ m/s) such as aircraft, missiles, and bullets. We will focus our attention on the first category of flows in this chapter and consider cases in which the object is far from a solid boundary or other objects. The flow becomes significantly influenced by the presence of a boundary or another object, as shown in Figure 6.2; in part (d) the slender object must be at least five body lengths below the free surface before free surface effects can be neglected. Flows such as those shown in Figure 6.2 are not included in this presentation.

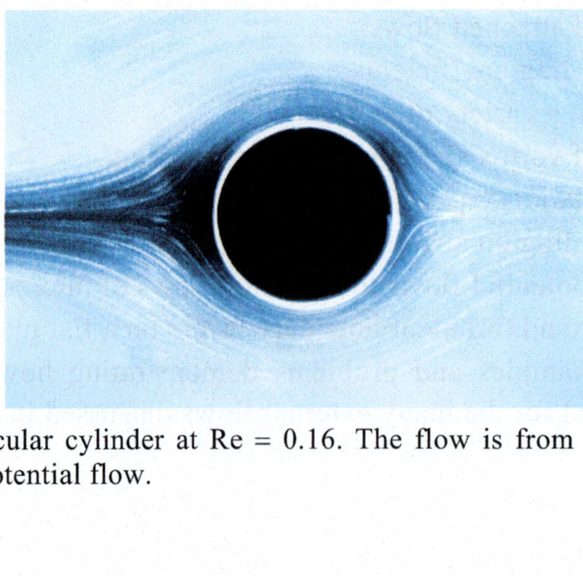

Figure 6.1 Flow past a circular cylinder at $Re = 0.16$. The flow is from left to right. It resembles superficially the pattern of potential flow.

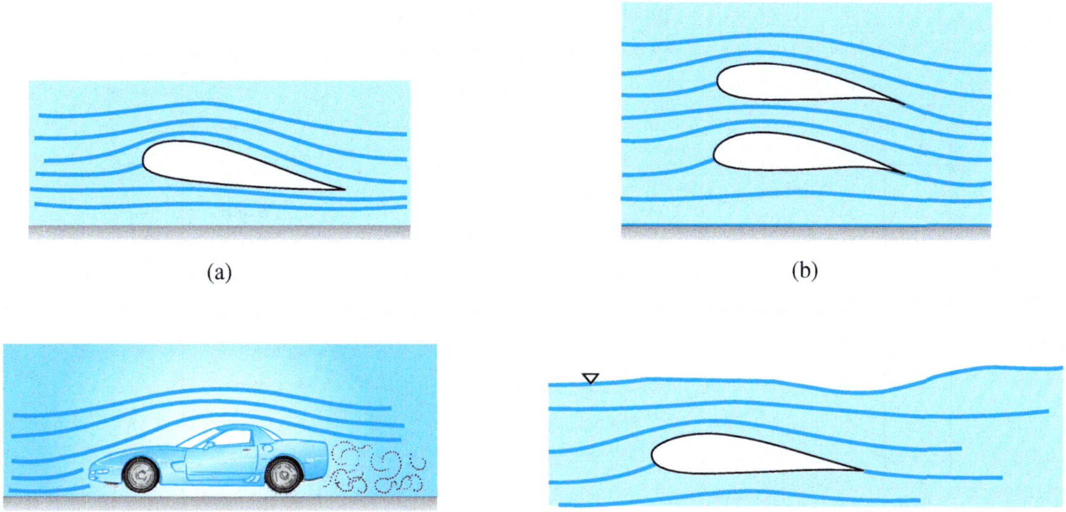

(a) (b) (c) (d)

Figure 6.2 Examples of complicated immersed flows: (a) flow near a solid boundary; (b) flow between two turbine blades; (c) flow around an automobile; (d) flow near a free surface.

High-Reynolds-number incompressible immersed flows are divided into two categories: flows around blunt bodies and flows around streamlined bodies, as displayed in Figure 6.3. The boundary layer (see Section 1.3.2) near the stagnation point is a laminar boundary layer and, for a sufficiently large Reynolds number, undergoes transition downstream to a turbulent boundary layer, as shown; the flow may separate from the body and form a **separated region**, a region of recirculating flow, as shown for the blunt body, or it simply leaves the streamlined body at the trailing edge (there may be a small separated region there). The **wake**, which is characterized by a **velocity defect**, is a growing (diffusion) region and trails each body, as shown. The boundaries of the wake, the separated region, and the turbulent boundary layer are quite time dependent; in the sketch the time-average location of the wake is shown by the dashed lines. Shear stresses due to viscosity are concentrated in the thin boundary layer, the separated region, and the wake; outside these regions the flow is approximated by an inviscid flow.

From Figure 6.3 it could be assumed that the separated region does not exchange mass with the free stream since mass does not cross a streamline. When viewed instantaneously, however, the separation streamline is highly time dependent, and due to this unsteady character, the separated region is able to exchange mass slowly with the free stream.

Several comments should be made regarding the separated region and the wake. The separated region eventually closes; the wake keeps diffusing into the main flow and eventually disappears as its area becomes exceedingly large (the fluid regains the free-stream velocity). Time-average streamlines do not enter a separated region; they do enter a wake. The separated region is submerged within the wake.

Flow around a blunt object is usually treated empirically, as was done for a turbulent flow in a conduit. We follow this procedure here. We are interested primarily in the **drag**, the force the flow exerts on the body in the direction of the flow. The **lift**, which acts normal to the direction of flow, will be of interest for airfoil shapes, as presented in Section 6.4. The actual details of the flow field are seldom of interest and are not presented in this introductory text.

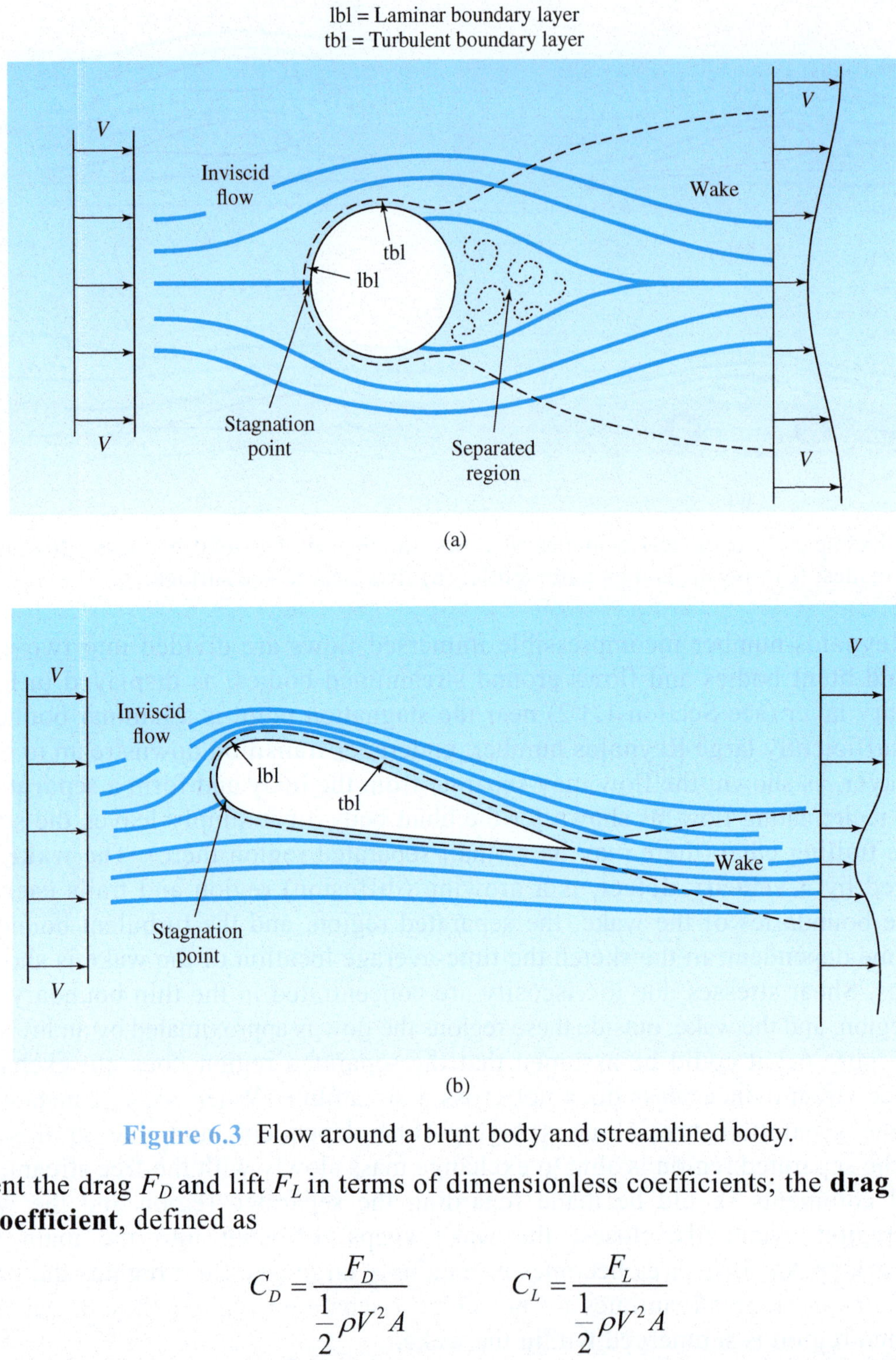

Figure 6.3 Flow around a blunt body and streamlined body.

We present the drag F_D and lift F_L in terms of dimensionless coefficients; the **drag coefficient** and **lift coefficient**, defined as

$$C_D = \frac{F_D}{\frac{1}{2}\rho V^2 A} \qquad C_L = \frac{F_L}{\frac{1}{2}\rho V^2 A} \tag{6.1.1}$$

where A is most often the projected area (projected on a plane normal to the direction of the flow); for airfoil shapes, the area is based upon the chord (see Figure 6.4). Drag coefficients for several common shapes are presented in Section 6.3.1. Since the drag on a blunt object is dominated by the flow in the separated region, there is little interest in studying the boundary layer growth on the front part of a blunt body and the associated viscous shear at the wall. Hence, interest is focused on the empirical data that provide the drag coefficient.

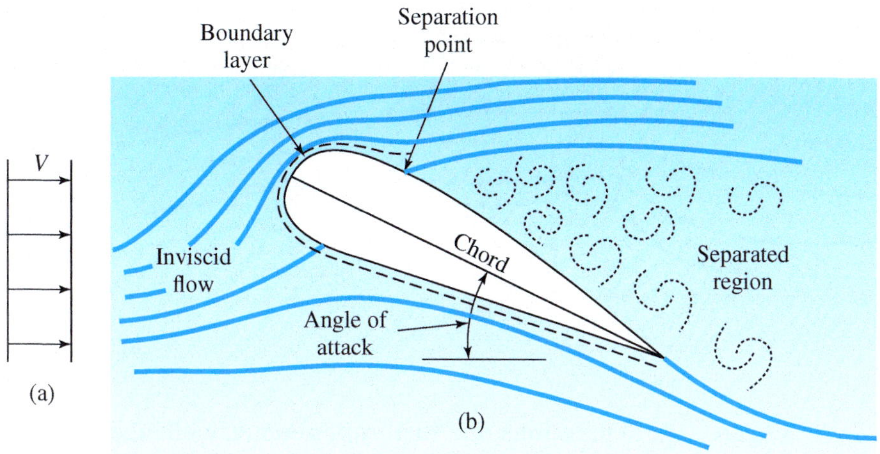

Figure 6.4 Streamlined body that is stalled.

Flow around a streamlined body—the separated region is insignificantly small or nonexistent—provides the motivation for detailed study of the laminar and turbulent boundary layers. A boundary layer that develops on a plane streamlined surface, such as an airfoil, is usually sufficiently thin that the curvature of the surface can be ignored and the problem can be treated as a boundary layer developing on a flat plate with a nonzero pressure gradient. We will provide a detailed study of flow on a flat plate with zero pressure gradient; once that problem is understood, the influence of a pressure gradient can be discussed. If the flow in the boundary layer on a streamlined body can be determined, the drag can be calculated, since the drag is a result of the shear stress and pressure force acting on the body surface.

Outside the boundary layer there exists an in Combine: inviscid, **free-stream flow**, as shown in Figures 6.3 and 6.4. Initially, we will assume that the free-stream flow is known. Before the velocity profile in the boundary layer can be determined, it is necessary that the inviscid flow solution be known. It is found by completely ignoring the boundary layer, since it is so thin, and solving the appropriate inviscid equations. The inviscid flow solution is then used to provide the lift on the body, and the two quantities used in the boundary layer flow solution: the pressure gradient and the velocity at the boundary. With the inviscid flow known and the boundary layer flow determined, quantities of interest can be obtained in the flow about a streamlined body. It should be emphasized that the inviscid flow will be assumed to be irrotational, that is, possessing no vorticity.

6.2 SEPARATION

Before we present the empirical data associated with the flow around blunt bodies, the general nature of separation will be discussed. Separation occurs when the main stream flow leaves the body, resulting in a separated region of flow, as sketched in Figure 6.4. When separation occurs on a streamlined body near the forward portion of an airfoil, as it will with a sufficiently large **angle of attack** (the angle the oncoming flow makes with the **chord**, a line connecting the trailing edge with the nose), the flow situation is referred to as **stall**, as shown in Figure 6.4. Stall is highly undesirable on aircraft at cruise conditions and leads to inefficiencies when it occurs on turbine blades. It is used, however, to provide the high drag

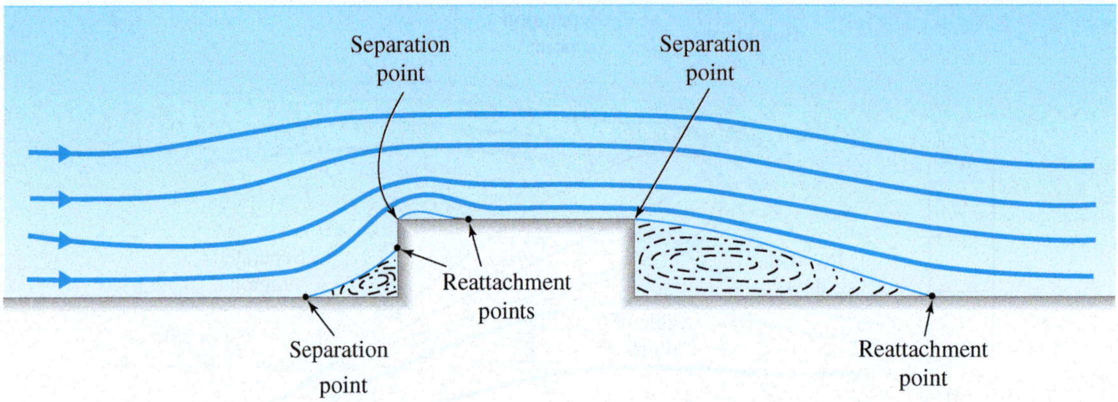

Figure 6.5 Separations due to abrupt geometry changes.

needed when landing an aircraft, or in certain maneuvers by stunt planes. For blunt bodies, however, separation is unavoidable at high Reynolds numbers and its effect must be understood.

The location of the separation point is dependent primarily on the geometry of the body; if the body has an abrupt change in geometry, such as that shown in Figure 6.5, separation will occur at, or near, the abrupt change; however, it will also occur upstream on the flat surface as shown. In addition, reattachment will occur at some locations, as shown. Let us establish the criterion used in predicting the location of the separation point on a surface with no abrupt geometry change.

Consider the flow on the flat surface just before the step of Figure 6.5. The region near the forward separation point is enlarged and shown in Figure 6.6; the y-coordinate is normal to the wall and the x-coordinate is measured along the wall. Downstream of the separation point the x-component velocity near the wall is in the negative x-direction and thus at the wall $\partial u/\partial y$ must be negative. Upstream of the separation point the x-component velocity near the wall is in the positive x-direction, demanding that $\partial u/\partial y$ at the wall be positive. Hence, we conclude that the separation point is defined as that point where $(\partial u/\partial y)_{\text{wall}} = 0$.

Observe that separation on the flat surface occurs as the flow is approaching a stagnation region where the velocity is low and the pressure is high. As the flow approaches the stagnation region the pressure increases, that is, $\partial p/\partial x > 0$; the pressure gradient is positive. Since separation

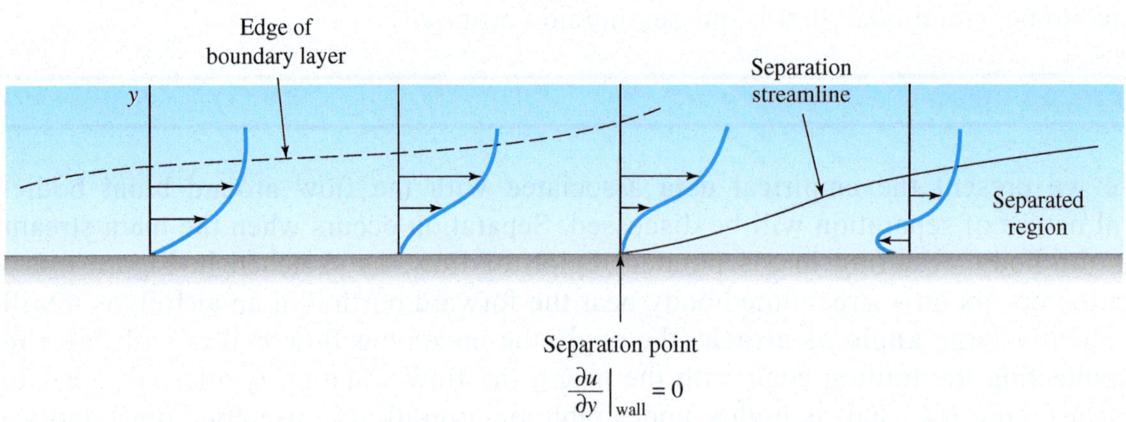

Figure 6.6 Flow separationd on a flat surface due to an adverse pressure gradient.

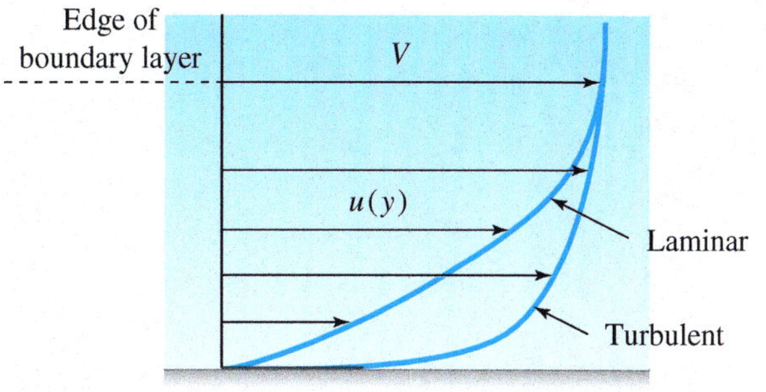

Figure 6.7 Comparison of laminar and turbulent velocity profiles.

is often undesirable, a positive pressure gradient is called an **adverse pressure gradient**; a negative gradient is a **favorable pressure gradient**. In general, the effect of an adverse pressure gradient results in decreasing velocities in the streamwise direction; if an adverse pressure gradient acts on a surface over a sufficient distance, separation may result. This is true even if the surface is a flat plate, such as the wall of a diffuser. See Section 6.6.7 for further discussion.

In addition to geometry and pressure gradient, several other parameters influence separation. These include the Reynolds number as a very important parameter, with the wall roughness, the **free-stream fluctuation intensity**[1] (the intensity of the disturbances that exist away from the boundary), and the wall temperature having less but occasionally significant influence.[2] Visualize, for example, flow around a sphere; at sufficiently low Reynolds numbers, no separation will occur. As the Reynolds number increases to a particular value, separation will occur over a small area at the rear; this area will become larger and larger as the Reynolds number increases until at a sufficiently large Reynolds number no additional increase in separation area will be observed. The boundary layer before separation will still be laminar. An interesting phenomenon takes place as the boundary layer before separation becomes turbulent; there is a sudden movement of the separation point to the rear of the sphere, resulting in a substantial reduction in the separation area and thus the drag. This phenomenon is explained by comparing the velocity profile of a laminar boundary layer with that of a turbulent boundary layer, as displayed in Figure, 6.7. As was true in a pipe flow, the turbulent profile has a much larger gradient near the wall (much larger wall shear stress) and thus the momentum of the fluid near the wall is substantially larger in the turbulent boundary layer. For a given geometry a greater distance is required to reduce the velocity near the wall to zero, resulting in the movement of the separation point to the rear, as can be observed in Figure 6.8, where both spheres are moving with the same velocity (the sphere in (b) has sandpaper in the nose region). In Figure 6.8a it is observed that separation occurs on the front half of the sphere, in a region of favorable pressure gradient. This separation is due to the centrifugal effects as the fluid moves around the sphere. This phenomenon of drag reduction is observed in the drop in the drag coefficient curves for a sphere and a cylinder, presented in the following section.

---

[1]Free-stream fluctuation intensity is defined as $\sqrt{\overline{u'^2}}/V$, where u' is the fluctuation. A value of 0.001 is quite low, and 0.1 quite high.

[2]If a fluid is flowing past a body that is rigidly supported, the vibration level of the support system will also influence the separation phenomenon. External sound waves can also be significant.

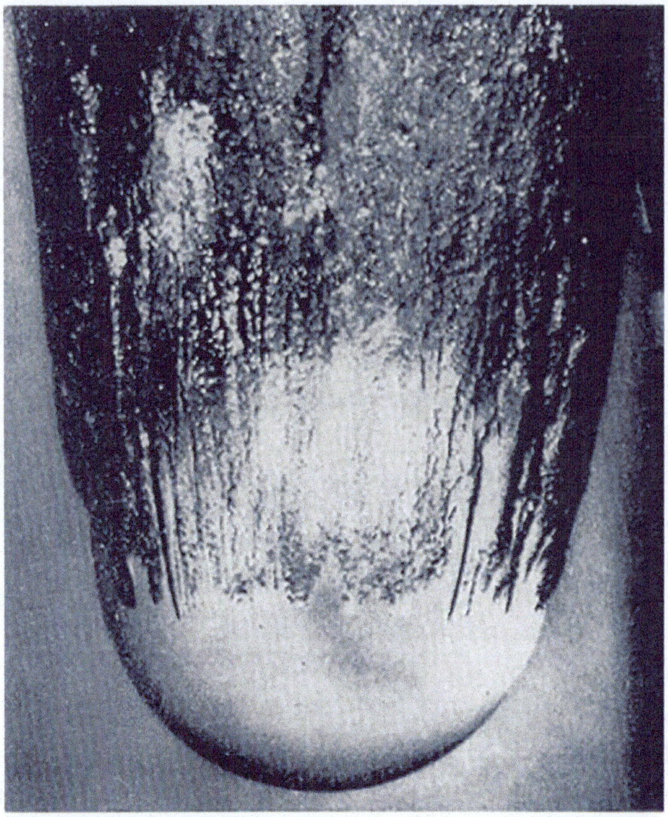

(a)

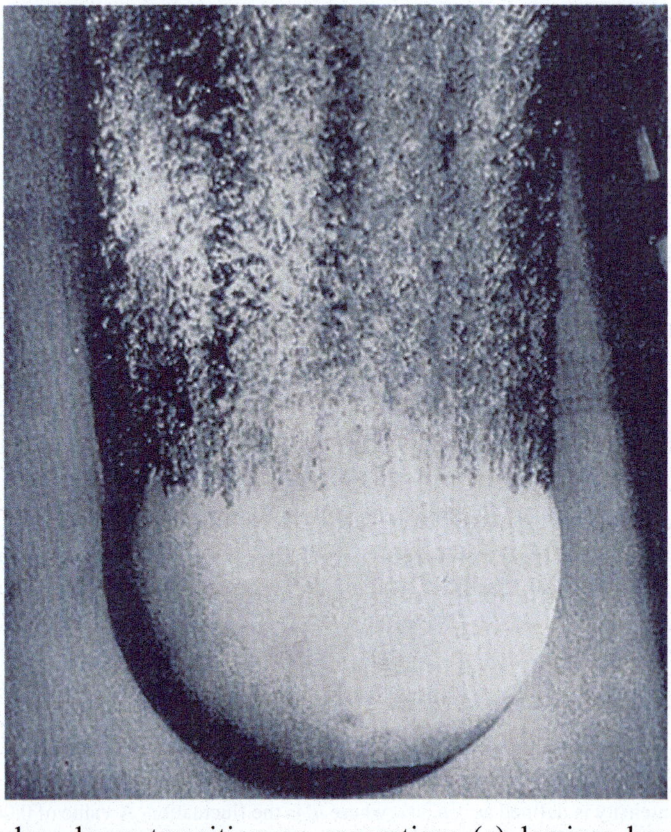

(b)

Figure 6.8 Effect of boundary layer transition on separation: (a) laminar boundary layer before separation; (b) turbulent boundary layer before separation. (Courtesy of the U.S. Navy)

6.3 FLOW AROUND IMMERSED BODIES

6.3.1 Drag Coefficients

From our study of dimensional analysis, recall that for a steady, incompressible flow in which gravity, thermal, and surface tension effects are neglected, the primary flow parameter that influences the flow is the Reynolds number; other occasionally important parameters include the relative wall roughness and the free-stream velocity fluctuation intensity.

The drag coefficient curves for two bodies that do not exhibit sudden geometric changes will be presented; the drag coefficients for the smooth sphere and the long smooth cylinder are shown in Figure 6.9 over a large range of Reynolds numbers. At Re < 1 creeping flow with no separation results. For the sphere, this creeping flow problem has been solved, with the result that

$$C_D = \frac{24}{\text{Re}} \qquad \text{Re} < 1 \tag{6.3.1}$$

Separation is observed at Re $\simeq$ 10 over a very small area on the rear of the body. The separated area increases as the Reynolds number increases until Re $\simeq$ 1000, where the separated region ceases to enlarge; during this growth of the separated region the drag coefficient decreases. At Re = 1000, 95% of the drag is due to form drag (the drag force due to the pressure acting on the body), and 5% is due to frictional drag (the drag force due to the shear stresses acting on the body).

The drag coefficient curve is relatively flat for smooth bodies over the range $10^3 < \text{Re} < 2 \times 10^5$.

The boundary layer before the point of separation is laminar and the separated region is as shown in Figure 6.8a. At Re $\simeq 2 \times 10^5$, for a smooth surface and with low free-stream fluctuation intensity, the boundary layer before separation undergoes transition to a turbulent state and the increased momentum in the boundary layer "pushes" the separation back, as shown in Figure 6.8b, with a substantial decrease (a 60 to 80% drop) in the drag. If the surface is rough (dimples on a golf ball) or the free stream has high free-stream fluctuation intensity, the drop in the C_D curve may occur at Re $\simeq 8 \times 10^4$. Since a lower drag is usually desirable, surface roughness is often added; the dimples on the golf ball may increase the flight distance by 50 to 100%.

After the sudden drop in drag, the C_D curve is observed to again increase with increased Reynolds number. Experimental data are not readily available for Re > 10^6 for a sphere and Re > 6×10^7 for a cylinder; however, a value of $C_D = 0.2$ for a sphere at large Reynolds number appears acceptable. Some engineers use $C_D = 0.4$ for cylinders at large Reynolds numbers; however, the data presented here suggests that is too low. Additional experimental data is needed.

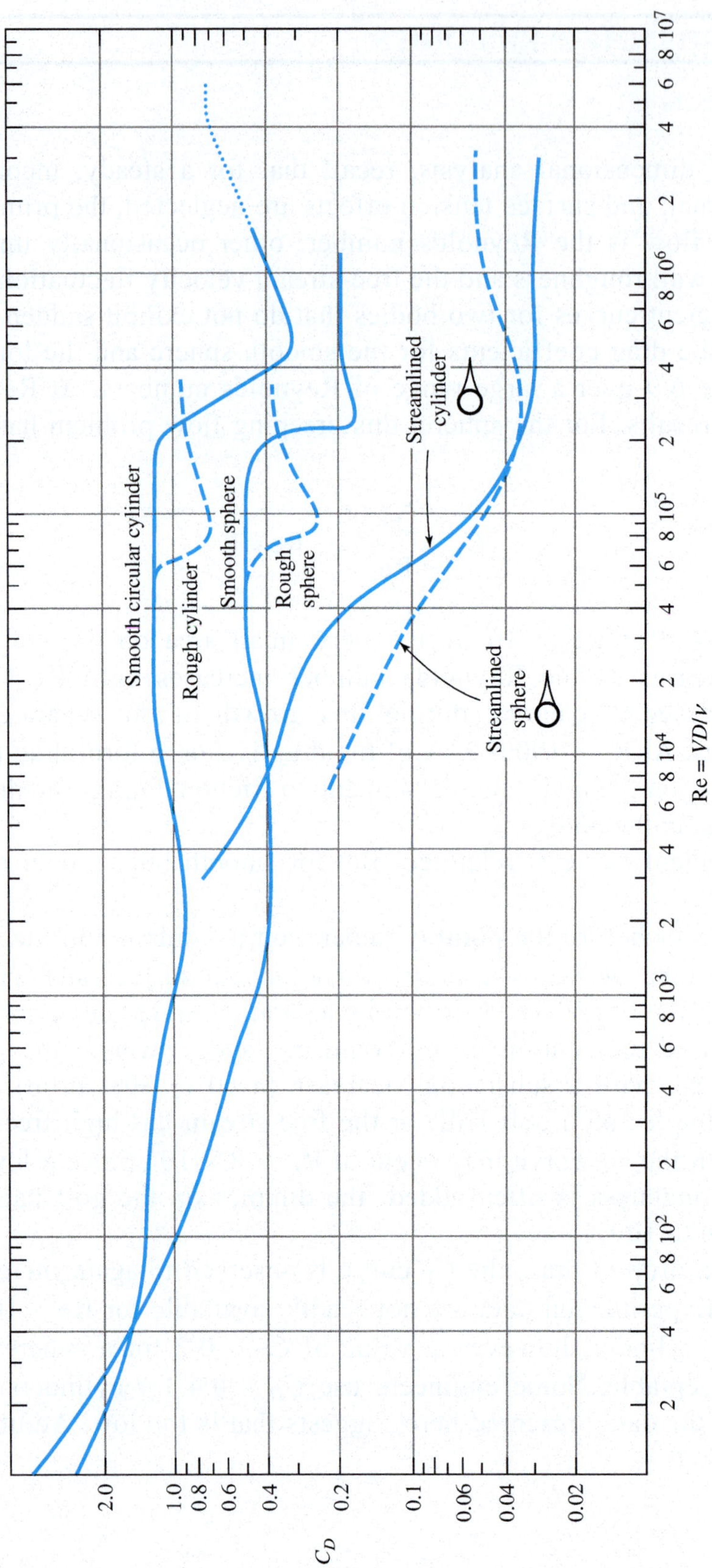

Figure 6.9 Drag coefficients for flow around a long cylinder and a sphere. (See E. Achenbach, *J Fluid Mech.*, Vol. 46, 1971, and Vol. 54, 1972.)

Table 6.1 Drag Coefficients of Finite-Length Circular Cylinders[a] with Free Ends[b] and of Infinite-Length Elliptic Cylinders

Circular cylinder		*Elliptic cylinder*[c]		
$\frac{Length}{Diameter}$	$\frac{C_D}{C_{D_\infty}}$	$\frac{Major\ axis}{Minor\ axis}$	Re	C_D
∞	1	2	4×10^4	0.6
40	0.82	4	10^5	0.46
20	0.76	4	2.5×10^4 to 10^5	0.32
10	0.68	8	2.5×10^4	0.29
5	0.62	8	2×10^5	0.20
3	0.62			
2	0.57			
1	0.53			

[a]C_{D_∞} is the drag coefficient for the infinite-length circular cylinder obtained in Figure 6.9.
[b]If one end is fixed to a solid surface, double the length of the cylinder.
[c]Flow is in the direction of the major axis.

For cylinders of finite length and for elliptic cylinders, the drag coefficients are presented in Table 6.1. The finite-length cylinders are assumed to have two free ends. If one end is fixed to a solid surface, its length must be doubled when using Table 6.1. Blunt objects with sudden geometry changes have separated regions that are relatively insensitive to the Reynolds number; the drag coefficients for some common shapes are given in Table 6.2.

Table 6.2 Drag Coefficient for Various Blunt Objects

Object		*Re*	C_D
Square cylinder →□ (L, w)	$L/w = \infty$	$>10^4$	2.0
	$L/w = 1$ (cube)	$>10^4$	1.1
rounded corners ($r = 0.2w$)	$L/w = \infty$	10^5	1.2
Rectangular plates →| w	$L/w = \infty$	$>10^3$	2.0
	$L/w = 20$	$>10^3$	1.5
	$L/w = 5$	$>10^3$	1.2
	$L/w = 1$	$>10^3$	1.1
Circular cylinder →□ (L, D)	$L/D = 0.01$ (disk)	$>10^3$	1.1
	$L/D = 4$	$>10^3$	0.9
	$L/D = 7$	$>10^3$	1.0
Semicircular cylinder → D		$>10^4$	2.2
→ ᗡ		$>10^4$	1.2
Semicircular shell →)		2×10^4	2.3
→ (		2×10^4	1.1
Equilateral cylinders → ◁	2.0	$>10^4$	2.0
→ ▷	1	$>10^4$	1.4
Cone → ◁ (α)	$\alpha = 30°$	$>10^4$	0.6
	$\alpha = 60°$	$>10^4$	0.8
	$\alpha = 90°$	$>10^4$	1.2
Solid hemisphere → ◗		$>10^4$	1.2
→ ◖		$>10^4$	0.4
Hollow hemisphere → D		$>10^4$	1.4
→ ᗡ		$>10^4$	0.4
Parachute		$>10^7$	1.4
Automobile 1920	—	$>10^5$	0.80
Modern, with square corners	—	$>10^5$	0.30
Modern, with rounded corners	—	$>10^5$	0.29
Van		$>10^5$	0.42
Bicycle, upright rider			1.1
racing, bent over			0.9
racing, drafting			0.5
Semitruck, standard			0.96
with streamlined deflector			0.76
with deflector and gap seal			0.70

Example 6.1

A square sign, 10 ft × 10 ft, is attached to the top of a 60-ft-high pole which is 12 in. in diameter (Figure E6.1). Approximate the maximum moment that must be resisted by the base for a wind speed of 100 ft/s.

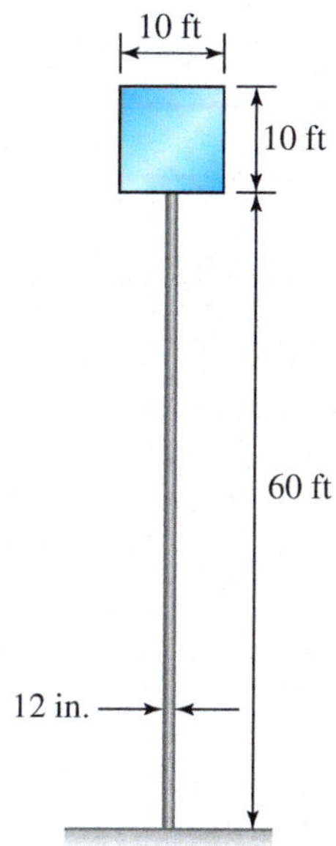

Figure E6.1

Solution

The maximum force F_1 acting on the sign occurs when the wind is normal to the sign; it is

$$F_1 = C_D \times \frac{1}{2}\rho V^2 A$$

$$= 1.1 \times \frac{1}{2} \times 0.0024 \times 100^2 \times 10^2 = 1320 \text{ lb}$$

where C_D is found in Table 6.2 and we use the standard value $\rho = 0.0024$ slug/ft^3 since it was not given (recall that slug = lb-s^2/ft). The force F_2 acting on the cylindrical pole is (using the projected area as $A = 60 \times 1$ ft^2)

$$F_2 = C_D \times \frac{1}{2}\rho V^2 A$$

$$= 0.8 \times \frac{1}{2} \times 0.0024 \times 100^2 \times 60 = 576 \text{ lb}$$

where C_D is found from Figure 6.9 with Re = 100 ×1/1.6 × 10^{-4} = 6.2 × 10^5, assuming a high-intensity fluctuation level (i.e., a rough cylinder); since neither end is free, we do not use the multiplication factor of Table 6.1.

The resisting moment that must be supplied by the supporting base is

$$\begin{aligned} M &= d_1F_1 + d_2F_2 \\ &= 65 \times 1320 + 30 \times 576 = \underline{103{,}000 \text{ ft-lb}} \end{aligned}$$

assuming that the forces act at the centers of their respective areas.

Example 6.2

Determine the terminal velocity of a 30-cm-diameter smooth sphere ($S = 1.02$) if it is released from rest in (a) air at 20°C and (b) water at 20°C.

Solution

(a) When terminal velocity is reached by a falling object, the weight of the object is balanced by the drag force acting on the object. Using $\Sigma F = 0$ and Eq. 6.1.1, we have

$$W = F_D$$

$$\therefore \gamma_{\text{sphere}} \times \frac{4}{3}\pi R^3 = C_D \times \frac{1}{2}\rho V^2 A$$

Using $\gamma_{\text{sphere}} = S\gamma_{\text{water}}$ and projected area $A = \pi R^2$, this become

$$S\gamma_{\text{water}} \times \frac{4}{3}\pi R^3 = C_D \times \frac{1}{2}\rho V^2 \pi R^2$$

The velocity can now be expressed as

$$V = \left(\frac{8RS\gamma_{\text{water}}}{3\rho C_D}\right)^{1/2} = \left(\frac{8 \times 0.15 \text{ m} \times 1.02 \times 9800 \text{ N/m}^3}{3 \times 1.20 \text{ kg/m}^3 \times C_D}\right)^{1/2} = \frac{57.7}{\sqrt{C_D}}$$

The Reynolds number should be quite large so $C_D = 0.2$ from Figure 6.9. Then

$$V = \frac{57.7}{\sqrt{0.2}} = \underline{129 \text{ m/s}}$$

We must check the Reynolds number to verify the C_D value assumed. It is

$$\text{Re} = \frac{VD}{\nu} = \frac{129 \times 0.3}{1.6 \times 10^{-5}} = 2.42 \times 10^6$$

This is beyond the end of the curve where data are unavailable; we will assume that the drag coefficient is unchanged at 0.2, so the terminal velocity is 129 m/s.

(b) For the sphere falling in water, we must include the buoyancy force B acting in the same direction as the drag force F_D. Hence the summation of forces yields

$$W = F_D + B$$

$$\therefore \gamma_{\text{sphere}} \times \frac{4}{3}\pi R^3 = C_D \times \frac{1}{2}\rho V^2 A + \gamma_{\text{water}} \times \frac{4}{3}\pi R^3$$

This gives

$$(S-1)\gamma_{\text{water}} \times \frac{4}{3}\pi R^3 = C_D \times \frac{1}{2}\rho V^2 \pi R^2$$

Using $\rho = 1000$ kg/m^3, there results

$$V = \left(\frac{8R(S-1)\gamma_{\text{water}}}{3\rho C_D}\right)^{1/2} = \left(\frac{8 \times 0.15 \times 0.02 \times 9800}{3 \times 1000 \times C_D}\right)^{1/2} = \frac{0.28}{\sqrt{C_D}}$$

We anticipate the Reynolds number being lower than in part (a), so let's assume that it is in the range $2 \times 10^4 < \text{Re} < 2 \times 10^5$. Then $C_D = 0.5$ and there results

$$V = \underline{0.40 \text{ m/s}}$$

This gives a Reynolds number of

$$\text{Re} = \frac{VD}{\nu} = \frac{0.40 \times 0.3}{10^{-6}} = 1.2 \times 10^5$$

This is in the required range, so the terminal velocity is expected to be 0.40 m/s. Of course, if the sphere were roughened (sand glued to the surface), the C_D value would be less and the velocity would be greater.

6.3.2 Vortex Shedding

Long blunt objects, such as circular cylinders, exhibit a particularly interesting phenomenon when placed normal to a fluid flow; vortices or eddies (regions of circulating fluid) are shed from the object, regularly and alternately from opposite sides, as shown in Figure 6.10. The resulting flow downstream is often referred to as a **Kármán vortex street**, named after Theodor von Kármán (1881–1963). The vortices are shed in the Reynolds number range $40 < \text{Re} < 10{,}000$, and are accompanied by turbulence above Re = 300. Images of high- and low-Reynolds-number vortex shedding are presented in Figure 6.11.

Dimensional analysis may be applied to find an expression for the shedding frequency. For high-Reynolds-number flows, that is, flows with insignificant viscous forces, the shedding frequency f, in hertz, depends only on the velocity and diameter. Thus $f = f(V, D)$. Using dimensional analysis we can show that fD/V = const. The shedding frequency, expressed as a dimensionless quantity, is expressed as the Strouhal number,

$$\text{St} = \frac{fD}{V} \tag{6.3.2}$$

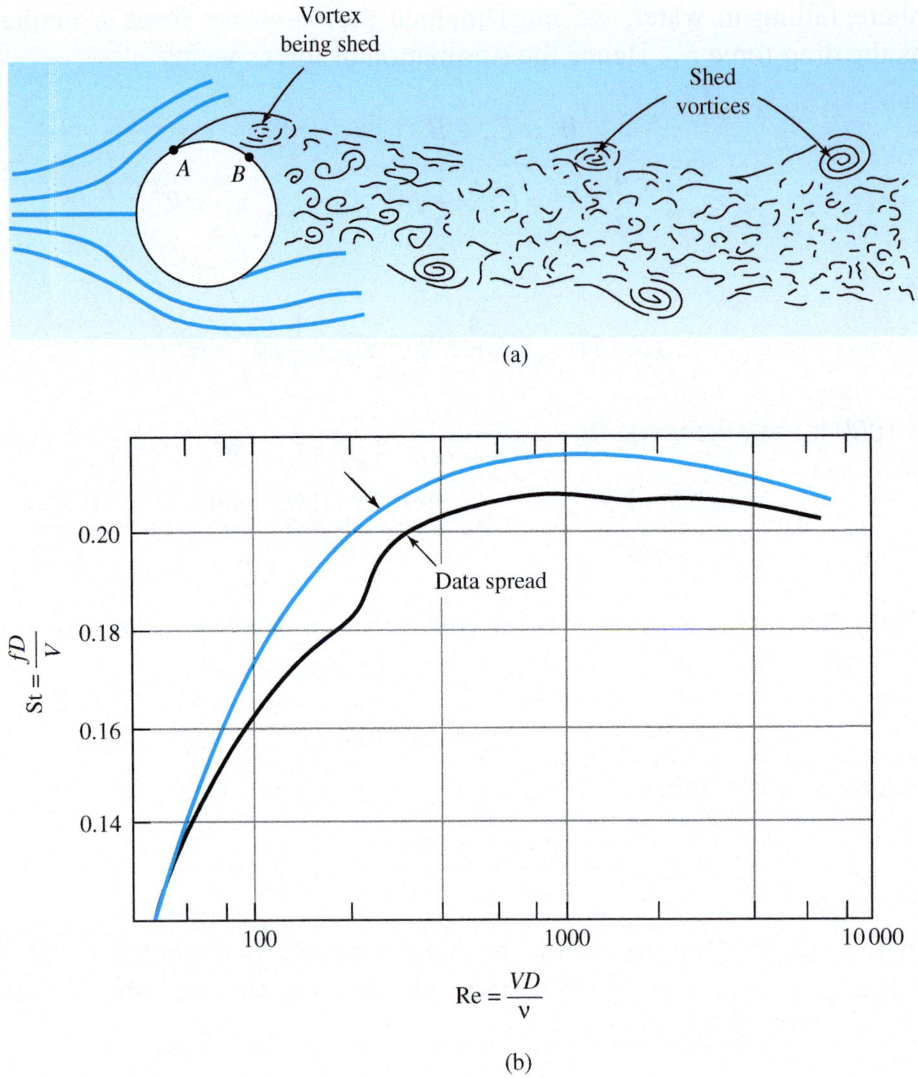

Figure 6.10 Vortex shedding from a cylinder: (a) vortex shedding; (b) Strouhal number versus Reynolds number. (Roshko, A. (1952-01-01) "On the development of turbulent wakes from vortex streets.")

From the experimental results of Figure 6.10, we observe that the Strouhal number is essentially constant (0.21) over the range $300 < Re < 10,000$; hence, the frequency is directly proportional to the velocity over this relatively large Reynolds number range.

The engineer or architect must be very careful when designing structures, such as towers and bridges that shed vortices. When a vortex is shed, a small force is applied to the structure; if the frequency of shedding is close to the natural frequency[3] (or one of the harmonics) of the structure, the phenomenon of resonance may occur in which the response to the applied force is multiplied by a large factor. For example, when resonance occurs on a television tower, the deflection of the tower due to the applied force may become so large that the supporting

[3]The natural frequency is that frequency with which a structure vibrates when given a "knock."

(a)

(b)

Figure 6.11 Vortex shedding at high and low Reynolds numbers: (a) Re = 10,000 (Photograph by Thomas Corke and Hassan Nagib.); (b) Re = 140.

cables fail, leading to collapse of the structure. This has occurred many times, leading to severe damage and numerous deaths and injuries. The collapse of the Tacoma Narrows suspension bridge is undoubtedly the most spectacular failure due to vortex shedding. "Galloping" power lines, in which a power line alternates between the usual catenary and an inverted catenary, is another example that may lead to significant damage; this may occur when a power line has iced up, providing a much larger cross-sectional area for the wind.

Example 6.3

The velocity of a slow-moving, 30°C air stream is to be measured using a cylinder and a pressure tap located between points A and B on the cylinder in Figure 6.10a. The velocity range is expected to be $0.1 < V < 1$ m/s. What size cylinder should be selected and what frequency would be observed by the pressure-measuring device for $V = 1$ m/s?

Solution

The Reynolds number should be in the vortex shedding range, say 4000. For the maximum velocity the diameter would be found as follows:

$$4000 = \frac{VD}{\nu}$$

$$= \frac{1.0\ \text{m/s} \times D}{1.6 \times 10^{-5}\ \text{m}^2/\text{s}}$$

$$\therefore\ D = 0.064\ \text{m} \quad \text{Select } D = \underline{6\ \text{cm}}$$

At $V = 0.1$ m/s the Reynolds number is $0.1 \times 0.06/1.6 \times 10^{-5} = 375$. Vortex shedding would occur, so this is acceptable. The expected vortex shedding frequency at $V = 1.0$ m/s is found using a Strouhal number from Figure 6.19b of 0.21. Hence

$$0.21 = \frac{fD}{V}$$

$$= \frac{f \times 0.06}{1.0}$$

$$\therefore f = \underline{3.5\ \text{hertz}}$$

6.3.3 Streamlining

If the flow is to remain attached to the surface of a blunt object, such as a cylinder or a sphere, it must move into regions of higher and higher pressure as it progresses to the rear stagnation point. At sufficiently high Reynolds numbers $(\text{Re} > 10)$ the slow-moving boundary layer flow near the surface is unable to make its way into the high-pressure region near the rear stagnation point, so it separates from the object. **Streamlining** reduces the high pressure at the rear of the object so that the slow-moving flow near the surface is able to negotiate its way into a slightly higher-pressure region. The fluid may not be able to make it all the way to the trailing edge of the streamlined object, but the separation region will be reduced to only a small percentage of the initial separated region on the blunt object. The included angle at the trailing edge must not be greater than about 20° or the separation region will be too large and the effect of streamlining will be negated. Drag coefficients for streamlined cylinders and spheres are displayed in Figure 6.9.

When a body is streamlined, the surface area is increased substantially. This eliminates the majority of the pressure drag but increases the shear drag on the surface. To minimize drag, the idea is to minimize the sum of the pressure drag and the shear drag. Consequently, the streamlined body cannot be so long that the shear drag is larger than the pressure drag plus the shear drag for a shorter body. An optimization procedure is required. Such a procedure would lead to a thickness-to-chord length ratio of about 0.25 for a strut.

Obviously, for a low-Reynolds-number flow (Re < 10) the drag is due primarily to shear drag and thus streamlining is unnecessary; it would undoubtedly lead to increased drag since the surface area would be increased.

Finally, it should be pointed out that another advantage of streamlining is that the periodic shedding of vortices is usually eliminated. The vibrations produced by vortex shedding are often undesirable, so streamlining not only decreases drag but can eliminate the vibrations.

Example 6.4

A strut on a stunt plane traveling at 60 m/s is 4 cm in diameter and 24 cm long. Calculate the drag force acting on the strut as a circular cylinder, and as a streamlined strut, as shown in Figure E6.4. Neglect any viscous drag. Would you expect vortex shedding from the circular cylinder?

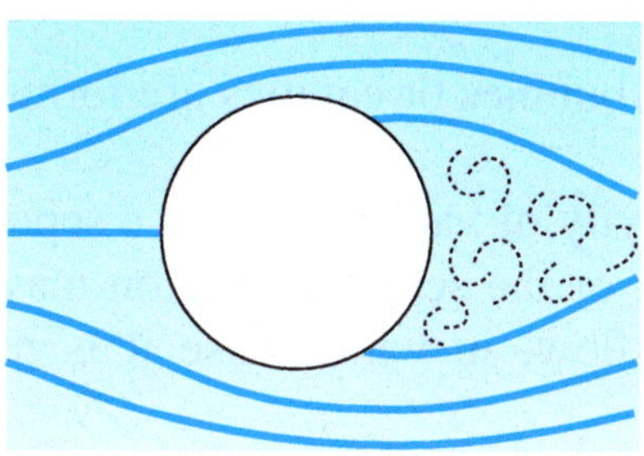

Figure E6.4(a)

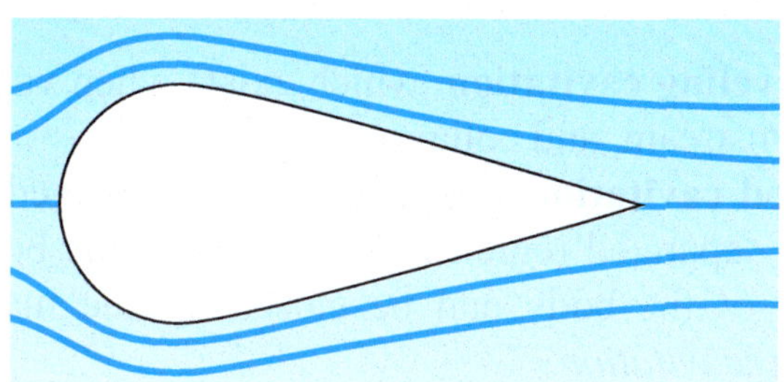

Figure E6.4(b)

Solution

The Reynolds number associated with the cylinder and the streamlined strut is, assuming air at $T = 20°C$,

$$\begin{aligned} \text{Re} &= \frac{VD}{\nu} \\ &= \frac{60 \times 0.04}{1.5 \times 10^{-5}} = 1.6 \times 10^5 \end{aligned}$$

Assuming a smooth surface as in Figure E6.4a, the drag coefficient is $C_D = 1.2$ from Figure 6.9. The drag force is then

$$\begin{aligned} F_D &= C_D \times \frac{1}{2} \rho V^2 A \\ &= 1.2 \times \frac{1}{2} \times 1.20 \text{ kg/m}^3 \times 60^2 \text{ m}^2/\text{s}^2 \times (0.24 \times 0.04) \text{ m}^2 = \underline{24.9 \text{ N}} \end{aligned}$$

For the streamlined strut of Figure E6.4b, Figure 6.9 yields $C_D = 0.04$. The drag force is

$$F_D = C_D \times \frac{1}{2}\rho V^2 A$$
$$= 0.04 \times \frac{1}{2} \times 1.20 \times 60^2 \times (0.24 \times 0.04) = \underline{0.82\,\text{N}}$$

This is a reduction of 97% in the drag, a rather substantial reduction.

Vortex shedding is not to be expected on the circular cylinder; the Reynolds number is too high. (See Figure 6.10.)

6.3.4 Cavitation

Cavitation is a very rapid change of phase from liquid to vapor which occurs in a liquid whenever the local pressure is equal to or less than the vapor pressure. The first appearance of cavitation is at the position of lowest pressure in a field of flow. Four types of cavitation have been identified:

1. **Traveling cavitation**, which exists when vapor bubbles or cavities are formed, are swept downstream, and collapse.
2. **Fixed cavitation**, which exists when a fixed cavity of vapor exists as a separated region. The separated region may reattach to the body, or the separated region may enclose the rear of the body and be closed by the main flow, in which case it is referred to as *supercavitation*.
3. **Vortex cavitation**, which is found in the high-velocity, and thus low-pressure, core of a vortex, often observed in the tip vortex leaving a propeller.
4. **Vibratory cavitation**, which may exist when a pressure wave moves in a liquid. A pressure wave consists of a pressure pulse, which has a high pressure followed by a low pressure. The low-pressure part of the wave (or vibration) can result in cavitation.

The first type of cavitation, in which vapor bubbles are formed and collapse, is associated with potential damage. The instantaneous pressures resulting from the collapse can be extremely high (perhaps 1400 MPa) and may cause damage to stainless steel components, as happens on the propellers of ships.

Cavitation occurs whenever the **cavitation number** $\boldsymbol{\sigma}$, defined by

$$\sigma = \frac{p_\infty - p_v}{\frac{1}{2}\rho V^2} \tag{6.3.3}$$

is less than the critical cavitation number σ_{crit}, which depends on the geometry of the body and the Reynolds number. Here p_∞ is the absolute pressure in the undisturbed free stream and

Table 6.3 Drag Coefficients for Zero Cavitation Number for Blunt Objects for Re $\cong 10^5$

Two-dimensional body				*Axisymmetric body*			
Geometry		θ	$C_D(0)$	*Geometry*		θ	$C_D(0)$
Flat plate	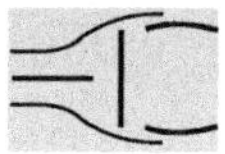	–	0.88	Disk		–	0.8
Circular cylinder		–	0.50	Sphere		–	0.30
		120	0.74	Cone		120	0.64
Wedge	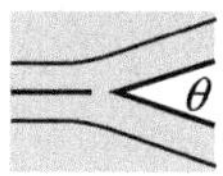	90	0.64		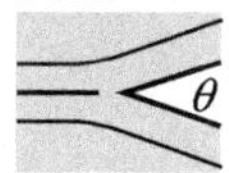	90	0.52
		60	0.49			60	0.38
		30	0.28			30	0.20

Table 6.4 Drag and Lift Coefficients and Critical Cavitation Number for a Typical Hydrofoil

Angle (°)	*Lift coefficient* C_L	*Drag coefficient* C_D	*Critical cavitation number* σ_{crit}
–2	0.2	0.014	0.5
0	0.4	0.014	0.6
2	0.6	0.015	0.7
4	0.8	0.018	0.8
6	0.95	0.022	1.2
8	1.10	0.03	1.8
10	1.22	0.04	2.5

p_v is the vapor pressure. As σ decreases below σ_{crit}, the cavitation increases in intensity, moving from traveling cavitation to fixed cavitation to supercavitation.

The drag coefficient of a body is dependent on the cavitation number and for small cavitation numbers is given by

$$C_D(\sigma) = C_D(0)(1 + \sigma) \tag{6.3.4}$$

where some values of $C_D(0)$ for common shapes are listed in Table 6.3 for Re $\simeq 10^5$.

The hydrofoil, an airfoil-type body that is used to lift a vessel out of the water, is a shape that is invariably associated with cavitation. Drag and lift coefficients and critical cavitation numbers are given in Table 6.4 for a typical hydrofoil with $10^5 < \text{Re} < 10^6$, where the Reynolds number is based on the chord length, and the area used with C_D and C_L is the chord times the length.

Example 6.5

A hydrofoil is to operate 20 in. below the surface of 60°F water at an angle of attack of 8° and travel at 45 ft/s. If its chord length is 24 in. and it is 6 ft long, calculate its lift and drag. Is cavitation present?

Solution

The absolute pressure p_∞ is

$$p_\infty = \gamma h + p_{\text{atm}}$$
$$= 62.4 \text{ lb/ft}^3 \times \frac{20}{12} \text{ ft} + 2117 = 2221 \text{ psf absolute}$$

The vapor pressure is $p_v = 0.256$ psia, so

$$\sigma = \frac{p_\infty - p_v}{\frac{1}{2}\rho V^2}$$
$$= \frac{2221 - 0.256 \times 144}{\frac{1}{2} \times 1.94 \times 45^2} = 1.11$$

Answering the last question first, we see that this is less than 1.8; hence cavitation exists.

The lift force is, finding C_L in Table 6.4,

$$F_L = C_L \times \frac{1}{2}\rho V^2 A$$
$$= 1.1 \times \frac{1}{2} \times 1.94 \text{ slug/ft}^3 \times 45^2 \text{ ft}^2/\text{s}^2 \times \left(\frac{24}{12} \times 6\right) \text{ ft}^2 = \underline{25{,}900 \text{ lb}}$$

The drag force is, finding C_D in Table 6.4,

$$F_D = C_D \times \frac{1}{2}\rho V^2 A$$
$$= 0.03 \times \frac{1}{2} \times 1.94 \times 45^2 (2 \times 6) = \underline{707 \text{ lb}}$$

6.3.5 Added Mass

The previous sections in this chapter have dealt with bodies moving at constant velocity. In this section we consider bodies accelerating from rest in a fluid. When a body accelerates, we say that an unbalanced force acts on the body; not only does the body accelerate but so does some of the fluid surrounding the body. The acceleration of the surrounding fluid requires an added force over and above the force required to accelerate only the body. A relatively simple way of accounting for the fluid mass being accelerated is to add a mass, called the **added mass** m_a, to the mass of the body. Summing forces in the direction of motion for a symmetrical body moving in the direction of its axis of symmetry we have, for horizontal motion,

$$F - F_D = (m + m_a)\frac{dV_B}{dt} \tag{6.3.5}$$

where V_B is the velocity of the body and F_D is the drag force. For an initial acceleration from rest F_D would be zero.

The added mass is related to the mass of the fluid m_f displaced by the body by the relationship

$$m_a = km_f \tag{6.3.6}$$

where k is an added mass coefficient. For a sphere $k = 0.5$; for an ellipsoid with major axis twice the minor axis and moving in the direction of the major axis, $k = 0.2$; for a long cylinder moving normal to its axis, $k = 1.0$. These values were calculated for inviscid flows and thus are applicable for motions starting from rest so that viscous forces are negligible.

For dense bodies accelerating in the atmosphere, the added mass is negligibly small and is typically ignored. Masses accelerating from rest in a liquid are more influenced by the added mass and it must usually be accounted for. Offshore structures that are subjected to oscillatory wave motions experience periodic forces the determination of which must include the added mass effect.

Example 6.6

A sphere with specific gravity 2.5 is released from rest in water. Calculate its initial acceleration. What is the percentage error if the added mass is ignored?

Solution

The summation of forces in the vertical direction, with zero drag, is

$$W - B = (m + m_a)\frac{dV_B}{dt}$$

Where B is the buoyant force. Substituting in the appropriate quantities gives, letting $\forall$ = sphere volume,

$$S\gamma_{\text{water}}\forall - \gamma_{\text{water}}\forall = (\rho_{\text{water}} S\forall + 0.5\rho_{\text{water}}\forall)\frac{dV_B}{dt}$$

This gives

$$g(S-1) = (S+0.5)\frac{dV_B}{dt}$$

Hence,

$$\frac{dV_B}{dt} = \frac{g(S-1)}{S+0.5} = \frac{9.8(2.5-1)}{2.5+0.5} = \underline{4.90 \text{ m/s}^2}$$

If we ignored the added mass, the acceleration would be

$$\frac{dV_B}{dt} = \frac{g(S-1)}{S} = \frac{9.8(2.5-1)}{2.5} = 5.88 \text{ m/s}^2$$

This is an error of 20%.

6.4 LIFT AND DRAG ON AIRFOILS

Separation occurs on a blunt body, such as a cylinder, due to the strong adverse pressure gradient in the boundary layer on the rear of the body. An airfoil is a streamlined body designed to reduce the adverse pressure gradient so that separation will not occur, usually with a small angle of attack, as shown in Figure 6.12. Without separation the drag is due primarily to the wall shear stress, which results from viscous effects in the boundary layer.

The boundary layer on an airfoil is very thin, and thus it can be ignored when solving for the flow field (the streamline pattern and the pressure distribution) surrounding the airfoil. Since the boundary layer is so thin, the pressure on the wall is not significantly influenced by the boundary layer's existence. Hence the lift on an airfoil can be approximated by integrating the pressure distribution as given by the inviscid flow solution on the wall. In the next section we will demonstrate how this is done; in this section we simply give empirical results.

The drag on an airfoil can be predicted by solving the boundary layer equations (simplified Navies–Stokes equations) for the shear stress on the wall and performing the appropriate integration. The inviscid flow field must be known before the boundary layer equations can be solved since the pressure gradient on the wall and the inviscid flow velocity at the wall[4] are needed as inputs in solving for the boundary layer flow. Boundary layer calculations will be presented in Section 6.6; in this section we present the empirical results for the drag.

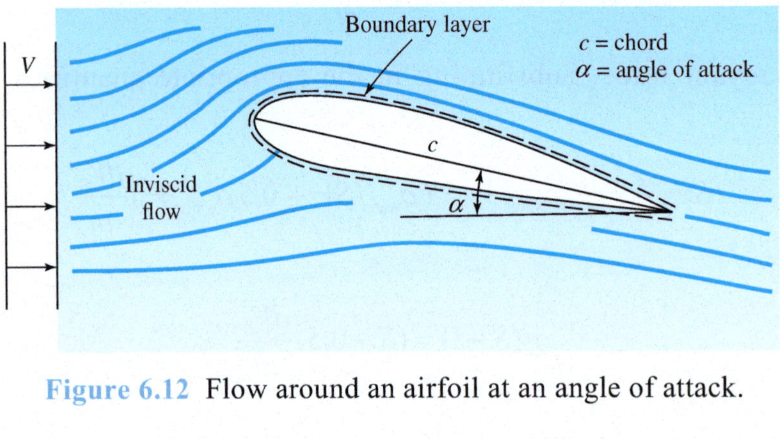

Figure 6.12 Flow around an airfoil at an angle of attack.

[4]Because the boundary layer is very thin, the velocity at its outer edge is taken to be the velocity at the wall from the inviscid flow solution.

The drag coefficient presented may seem quite low compared with the coefficients of the preceding section. For airfoils a much larger projected area is used, namely, the plan area, which is the chord c (see Figure 6.12) times the length L of the airfoil. Thus the drag and lift coefficients are defined as

$$C_D = \frac{F_D}{\frac{1}{2}\rho V^2 cL} \qquad C_L = \frac{F_L}{\frac{1}{2}\rho V^2 cL} \tag{6.4.1}$$

For a typical airfoil the lift and drag coefficients are given in Figure 6.13. For a specially designed airfoil the drag coefficient may be as low as 0.0035, but the maximum lift coefficient is about 1.5. The design lift coefficient (cruise condition) is about 0.3, which is near the minimum drag coefficient condition. This corresponds to an angle of attack of about 2°, far from the stall condition of about 16°.

Conventional airfoils are not symmetric; hence there is a positive lift coefficient at zero angle of attack. The lift is directly proportional to the angle of attack but deviates from the straight-line function just before stall. The drag coefficient also increases linearly up to an angle of attack of about five degrees for a conventional airfoil; then it increases in a nonlinear relation with angle of attack.

To take off and land at relatively low speeds, it is necessary to attain significantly higher lift coefficients than the maximum of 1.7 of Figure 6.13. Or if a relatively low lift coefficient is to be accepted, the area $c \times L$ must be enlarged. Both are actually accomplished. Flaps are moved out from a section of each airfoil, resulting in an increased chord, and the angle of

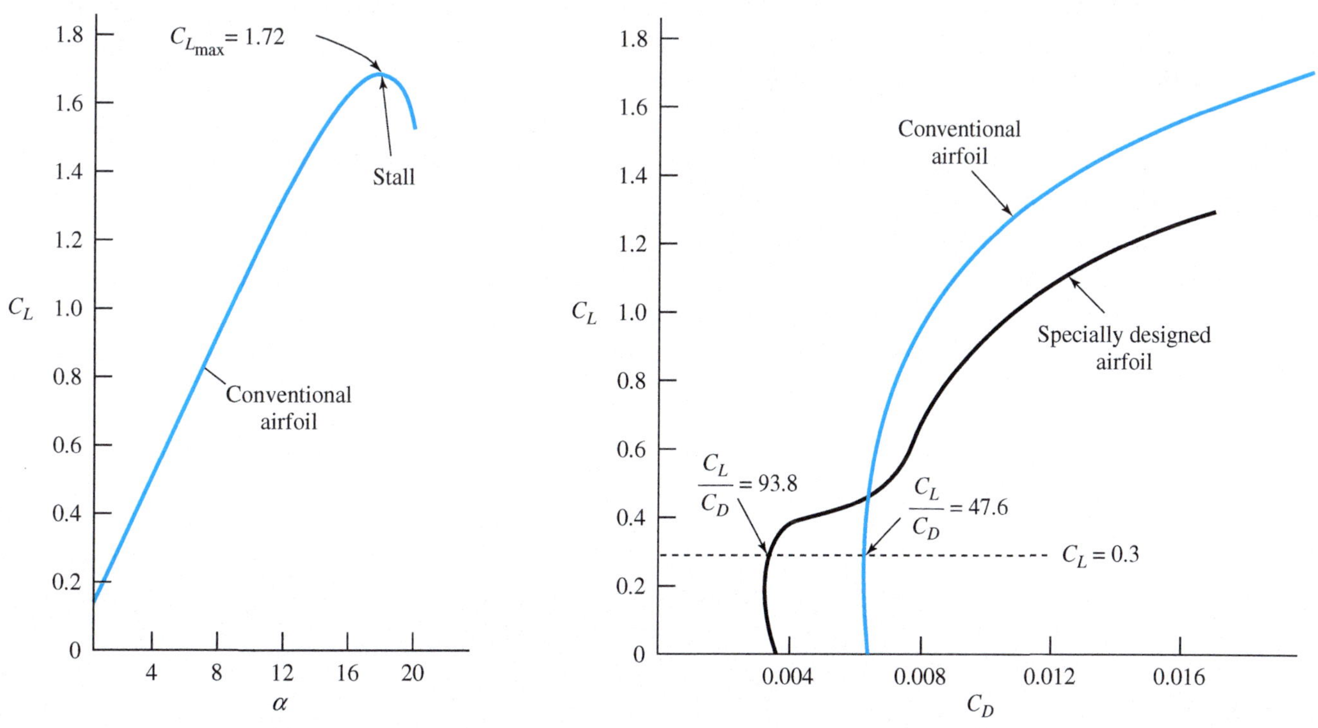

Figure 6.13 Lift and drag coefficients for airfoils with Re = $Vc/\nu \simeq 9 \times 10^6$ (c is the chord length).

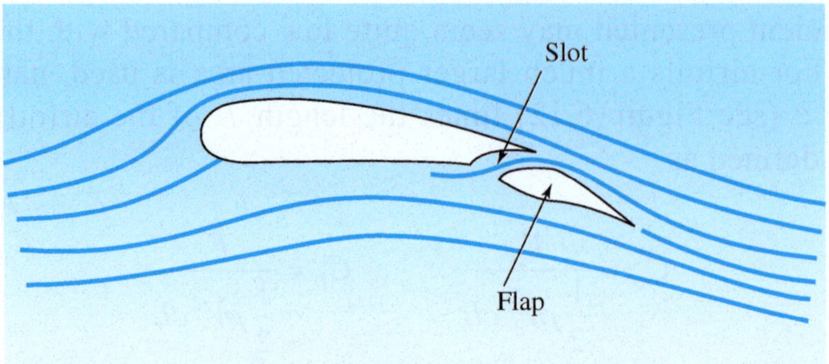

Figure 6.14 Flapped airfoil with slot for separation control.

attack of the flap is also increased. Slots are used to move high-pressure air from the underside into the relatively low momentum boundary layer flow on the top side, as shown in Figure 6.14; this prevents separation from the flap, thereby maintaining high lift. The lift coefficient can reach 2.5 with a single-slotted flap and 3.2 with a double-slotted flap. On some modern aircraft there may be three flaps in series with three slots along with a nose flap, to ensure that the boundary layer does not separate from the upper surface of the airfoil.

The total lift on an aircraft is supplied primarily by the airfoil. The effective length of the airfoil when calculating the lift is taken to be the tip-to-tip distance, the **wingspan**, since the fuselage acts to produce lift on the midsection of the aircraft. The drag calculation must include the shear acting on the airfoil, the fuselage, and the tail section.

The drag coefficient is essentially constant on airfoils up to a Mach number of about 0.75. Then a sudden rise occurs until the Mach number reaches unity; see Figure 6.15. The drag coefficient then slowly falls. Obviously, the condition of $M = 1$ is to be avoided. Thus aircraft fly at either $M < 0.75$ or $M > 1.5$ or so, to avoid the high drag coefficients near $M = 1$. Near $M = 1$ there are also regions of flow that oscillate from subsonic to supersonic. Such oscillations create forces that are best avoided.

It is useful to use swept-back airfoils since it is the component of velocity normal to the leading edge of the airfoil that must be used in calculating the Mach number in Figure 6.15. Cruise speeds at $M = 0.8$ with swept-back wings are not uncommon. It should be pointed out, though, that fuel consumption depends on power required, and power is drag force times velocity; hence fuel consumption depends on the velocity cubed since the drag force depends on the velocity squared, assuming all other parameters are constant. A lower velocity results in a fuel savings even though the engines must operate longer when traveling a fixed distance.

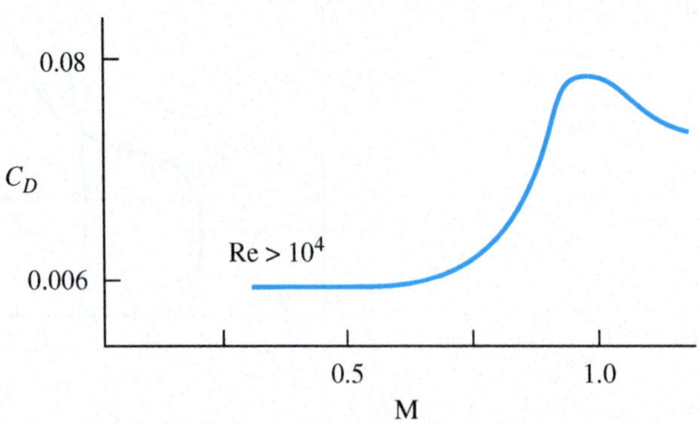

Figure 6.15 Drag coefficient as a function of Mach number (speed) for a typical un wept airfoil.

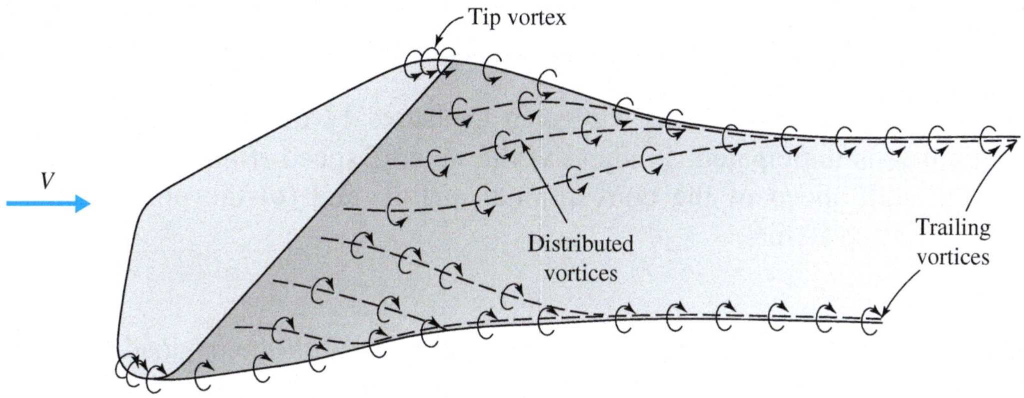

Figure 6.16 Trailing vortex.

A final comment on airfoils regards the influence of a finite airfoil. To understand flow around a finite airfoil we make reference to a vortex. Fluid particles rotate about the center of a vortex as they travel along in the flow field. There is a high pressure on the bottom and a low pressure on the top side of the airfoil sketched in Figure 6.16 with a model airfoil shown in Figure 6.17. This results in a movement of air from the bottom side to the top side around the ends of the airfoil, as shown, resulting in a strong tip vortex. Distributed vortices are also shed all along the airfoil, and they all collect into two large trailing vortices. On a clear day, the two trailing vortices may show up as visible white streaks of water vapor behind a high-flying aircraft. The trailing vortices persist for a considerable distance (perhaps 15 km) behind a large aircraft, and their 90-m/s velocities can cause a small trailing aircraft to flip over. Also, the trailing vortices induce a downwash, a downward velocity component, that must be accounted for in the design of the aircraft. The tail section is located up high to minimize the effect of this downwash.

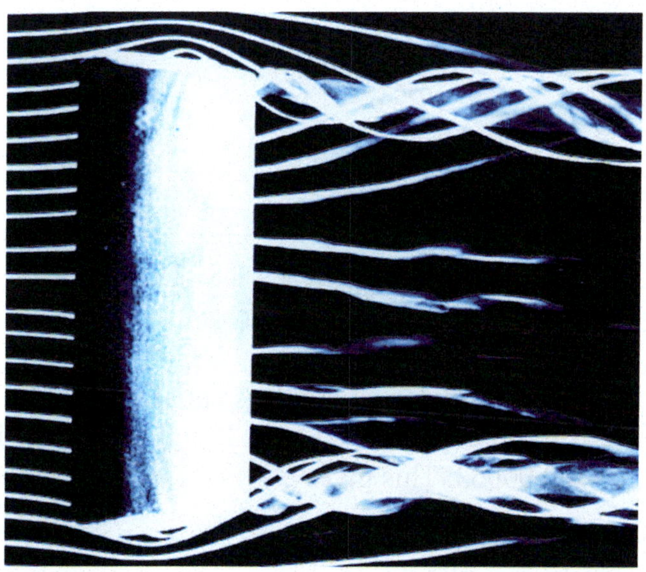

Figure 6.17 Trailing vortices from a rectangular wing. The flow remains attached over the entire wing surface. The centers of the vortex cores leave the trailing edge at the tips. The model is tested in a smoke tunnel at Reynolds number 100 000.

Example 6.7

A light airplane weighs 10 000 N its wingspan measures 12 m chord measures 1.8 m, and a payload of 2000 N is anticipated. Predict (a) the takeoff speed if an angle of attack of 8° is desired, (b) the stall speed of the conventional airfoil, and (c) the power required by the airfoil during cruise at 50 m/s.

Solution

(a) At takeoff the lift on an airplane is equal to its weight. With the payload the total weight is 12 000 N; hence, the lift-coefficient equation (6.4.1) gives the following:

$$V = \left(\frac{F_L}{\frac{1}{2}\rho C_L cL} \right)^{1/2}$$

$$= \left(\frac{12\ 000\,\text{N}}{\frac{1}{2} \times 1.20\ \text{kg/m}^3 \times 1.0 \times 1.8\ \text{m} \times 12\ \text{m}} \right)^{1/2} = \underline{30.4\ \text{m/s}}$$

where we have used $C_L = 1.0$ at $\alpha = 8°$ from Figure 6.13, and $\rho = 1.2\ \text{kg/m}^3$ since light aircraft take off at ground level, and the elevation is not given.

(b) The stall speed is found using a maximum lift coefficient of 1.72 from Figure 6.13:

$$V_{\text{stall}} = \left(\frac{F_L}{\frac{1}{2}\rho C_L cL} \right)^{1/2}$$

$$= \left(\frac{12\ 000}{\frac{1}{2} \times 1.20 \times 1.72 \times 1.8 \times 12} \right)^{1/2} = \underline{23.2\ \text{m/s}}$$

(c) The power demanded by the airfoil during cruise is equal to the drag force times the velocity. The design lift coefficient is assumed equal to 0.3, and thus from Figure 6.13, assuming a conventional airfoil, $C_D = 0.00063$. This gives

$$F_D = \frac{1}{2}\rho V^2 cLC_D$$

$$= \frac{1}{2} \times 1.20\ \text{kg/m}^3 \times 50^2\ \text{m}^2/\text{s}^2 \times 1.8\ \text{m} \times 12\ \text{m} \times 0.0063 = \underline{204\ \text{N}}$$

The power is then

$$\text{power} = F_D \times V$$
$$= 204 \text{ N} \times 50 \text{ m/s} = 10\,200 \text{ W} \quad \text{or} \quad \underline{13.7 \text{ hp}}$$

The total power would be significantly larger since the drag on the fuselage and tail section must be included.

6.5 POTENTIAL-FLOW THEORY

6.5.1 Basic Flow Equations

Inviscid flow exists outside the boundary layer and the wake in high-Reynolds-number flow around bodies. For an airfoil the boundary layer is quite thin, and the inviscid flow provides a good approximation to the actual flow; it is used to predict the pressure distribution on the surface, thereby giving a good estimate of the lift. It will also provide us with the velocity to be used as a boundary condition in the boundary layer solution of Section 6.6; from such a solution we can estimate the drag and predict possible points of separation. Consequently, the inviscid flow solution is very important in our study of external flows. Obviously, if we use the empirical results of previous sections, the details of the inviscid flow solution are unnecessary. If, on the other hand, we wish to predict quantities such as the lift and drag and locate possible points of separation by using the required differential equations, the inviscid flow solution is essential.

Consider a velocity field that is given by the gradient of a scalar function ϕ, that is,

$$\mathbf{V} = \nabla\phi \tag{6.5.1}$$

in which ϕ is called a **velocity potential function**. Such a velocity field is called a **potential flow** (or an **irrotational flow**) and possesses the property that the vorticity $\boldsymbol{\omega}$, which is the curl of the velocity vector, is zero; this is expressed by

$$\boldsymbol{\omega} = \nabla \times \mathbf{V} = 0 \tag{6.5.2}$$

The fact that the vorticity is zero for a potential flow can be shown by letting $\mathbf{V} = \nabla\phi$ and expanding Eq. 6.5.2 in rectangular coordinates. A fluid particle that does not possess vorticity (i.e., is not rotating) cannot obtain vorticity without the action of viscosity; normal pressure forces and body forces that act through the mass center cannot impart rotation to a fluid particle. This result is also observed from the **vorticity equation**, which is obtained by taking the curl of the Navier–Stokes equation (3.3.17) (see Section 3.3.4 if the details are of interest); the vorticity equation is

$$\frac{D\boldsymbol{\omega}}{Dt} = (\boldsymbol{\omega} \cdot \nabla)\mathbf{V} + \nu\nabla^2\boldsymbol{\omega} \tag{6.5.3}$$

Note that if $\boldsymbol{\omega} = 0$, the only way that $D\boldsymbol{\omega}/Dt$ can be nonzero is for viscous effects to act through the last term. If viscous effects are absent, as in an inviscid flow, then $D\boldsymbol{\omega}/Dt = 0$ and vorticity must remain zero.

With the velocity given by the gradient of a scalar function, the differential continuity equation (3.2.10), for an incompressible flow, gives

$$\nabla \cdot \nabla \phi = \nabla^2 \phi = 0 \tag{6.5.4}$$

which is known as **Laplace's equation**, named after Pierre S. Laplace (1749–1827). In rectangular coordinates this is

$$\frac{\partial^2 \phi}{\partial x^2} + \frac{\partial^2 \phi}{\partial y^2} + \frac{\partial^2 \phi}{\partial z^2} = 0 \tag{6.5.5}$$

With the appropriate boundary conditions this equation could be solved. However, three-dimensional problems are quite difficult, so we focus our attention on plane flows in which the velocity components u and v depend on x and y. This is acceptable for two-dimensional airfoils and other plane flows, such as flow around cylinders.

Before we attempt a solution to Eq. 6.5.5, let us define another scalar function that will aid us in our study of plane fluid flows. The continuity equation (3.2.9)

$$\frac{\partial u}{\partial x} + \frac{\partial v}{\partial y} = 0 \tag{6.5.6}$$

motivates the definition. If we let

$$u = \frac{\partial \psi}{\partial y} \quad \text{and} \quad v = -\frac{\partial \psi}{\partial x} \tag{6.5.7}$$

We observe that the continuity equation is automatically satisfied; the scalar function $\psi(x, y)$ is called the **stream function**. By using the mathematical description of a streamline, $\mathbf{V} \times d\mathbf{r} = 0$ we see that, for a plane flow, $u dy - v dx = 0$. Substituting from Eqs. 6.5.7, this becomes

$$\frac{\partial \psi}{\partial y} dy + \frac{\partial \psi}{dx} dx = 0 \tag{6.5.8}$$

This is, by definition, $d\psi = 0$. Thus ψ is constant along a streamline. Example 6.9 will show that the difference $(\psi_2 - \psi_1)$ between any two streamlines is equal to the flow rate per unit depth between the two streamlines.

The vorticity vector for a plane flow has only a z-component since $w = 0$ and there is no variation with z. The vorticity is

$$\boldsymbol{\omega}_z = (\nabla \times \mathbf{V})_z = \frac{\partial v}{\partial x} - \frac{\partial u}{\partial y} \tag{6.5.9}$$

For our potential flow we demand that the vorticity be zero, so Eq. 6.5.9 gives, using Eqs. 6.5.7,

$$\frac{\partial^2 \psi}{\partial x^2} + \frac{\partial^2 \psi}{\partial y^2} = 0 \tag{6.5.10}$$

So we see that both the stream function ψ and the potential function ϕ satisfy Laplace's equation for this plane flow.

Rather than attempt a solution of Laplace's equation for a particular flow of interest, we will use a different technique; we will identify some relatively simple functions that satisfy Laplace's equation and then superimpose these simple functions to create flows of interest. It is possible to generate any plane flow desired using this technique. Hence we will not actually solve Laplace's equation.

Before we present some simple functions, some additional observations concerning ϕ and ψ will be made. Using Eqs. 6.5.1 and 6.5.7. We see that

$$u = \frac{\partial\phi}{\partial x} = \frac{\partial\psi}{\partial y} \qquad v = \frac{\partial\phi}{\partial y} = -\frac{\partial\psi}{\partial x} \tag{6.5.11}$$

These relationships between the derivatives of ϕ and ψ are the famous **Cauchy–Riemann equations**, named after Augustin L. Cauchy (1789–1857) and Georg F. Riemann (1826–1866), from the theory of complex variables. The functions ϕ and ψ are harmonic functions since they satisfy Laplace's equation and form an **analytic function** $(\phi + \hat{\mathbf{i}}\psi)$ called the **complex velocity potential**. The theory of complex variables with all its powerful theorems is thus applicable to this restricted class of problcms: namely, plane, incompressible, potential flows. An example will show that the streamlines and constant-potential lines intersect each other at right angles.

Example 6.8

A scalar potential function is given by $\phi = A\tan^{-1}(y/x)$. Find the stream function $\psi(x, y)$.

Solution

The relationship between ϕ an ψ is given by Eq. 6.5.11. We have

$$\frac{\partial\psi}{\partial y} = \frac{\partial\phi}{\partial x} = \frac{\partial}{\partial x}\left(A\tan^{-1}\frac{y}{x}\right) = -\frac{Ay}{x^2 + y^2}$$

This can be integrated as follows:

$$\int\frac{\partial\psi}{\partial y}dy = -A\int\frac{y}{x^2 + y^2}dy$$

$$\therefore \psi = -\frac{A}{2}\ln\left(x^2 + y^2\right) + f(x)$$

A function of x rather than a constant must be added since partial derivatives are being used. Now, let us differentiate this expression with respect to x. There results

$$\frac{\partial\psi}{\partial x} = -A\frac{x}{x^2 + y^2} + \frac{df}{dx}$$

This must equal $(-\partial\phi/\partial y)$ as demanded by Eq. 6.5.11; that is,

$$-\frac{Ax}{x^2+y^2}+\frac{df}{dx}=-\frac{Ax}{x^2+y^2}$$

Thus

$$\frac{df}{dx}=0 \quad \text{or} \quad f=C$$

Since ϕ and ψ are used to find the velocity components by differentiation, the constant C is of no concern; it is usually set equal to zero. Hence

$$\psi=-\frac{A}{2}\ln\left(x^2+y^2\right)$$

Example 6.9

Show that the difference in the stream function between any two streamlines is equal to the flow rate per unit depth between the two streamlines. The flow rate per unit depth is denoted by q.

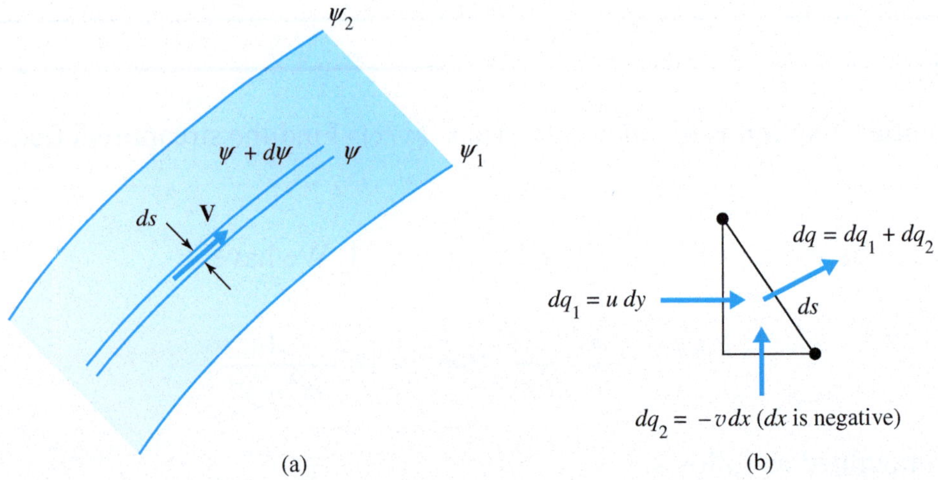

Figure E6.9

Solution

Consider the flow between two streamlines infinitesimally close, as shown in Figure E6.9a. The flow rate per unit depth through the elemental area is, referring to Figure E6.9b,

$$\begin{aligned} dq &= dq_1+dq_2 \\ &= udy-vdx \\ &= \frac{\partial\psi}{\partial y}dy+\frac{\partial\psi}{\partial x}dx=d\psi \end{aligned}$$

If this is integrated between two streamlines with $\psi = \psi_1$ and $\psi = \psi_2$, there results

$$q = \psi_2 - \psi_1$$

thereby proving the statement of the example.

Example 6.10

Show that the streamlines and equipotential lines of a plane, incompressible, potential flow intersect one another at right angles.

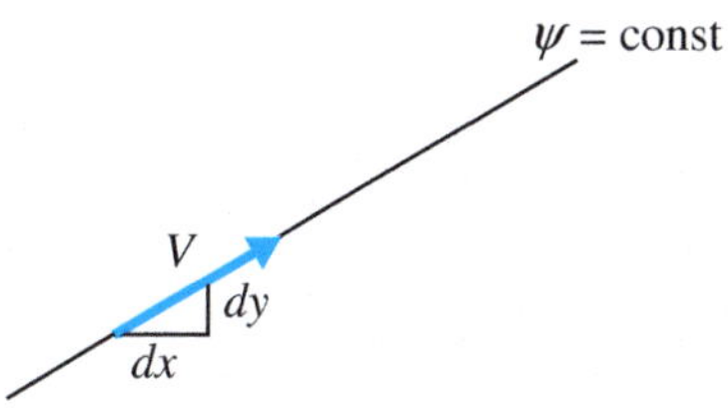

Figure E6.10a

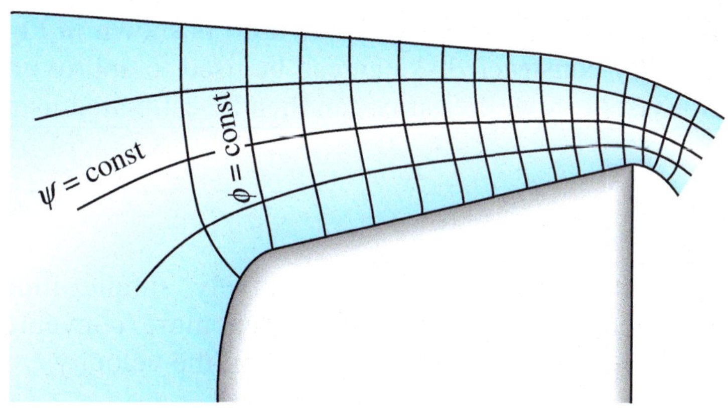

Figure E6.10b

Solution

If, at a point, the slope of a streamline is the negative reciprocal of the slope of an equipotential line, the two lines are perpendicular to each other. The slope of a streamline (see Figure E6.10a) is given by

$$\left.\frac{dy}{dx}\right|_{\psi = \text{const}} = \frac{v}{u}$$

The slope of an equipotential line is found from

$$d\phi = \frac{\partial \phi}{\partial \phi} dx + \frac{\partial \phi}{\partial y} dy = 0$$

since ϕ = constant along an equipotential line. This gives

$$\left.\frac{dy}{dx}\right|_{\phi\,=\,\text{const}} = -\frac{\partial\phi/\partial x}{\partial\phi/\partial y} = -\frac{u}{v}$$

Hence we see that the slope of the streamline is the negative reciprocal of the slope of the equipotential line; that is,

$$\left.\frac{dy}{dx}\right|_{\phi\,=\,\text{const}} = -\left(\left.\frac{dy}{dx}\right|_{\psi\,=\,\text{const}}\right)^{-1}$$

Thus, whenever the streamlines intersect the equipotential lines, they must do so at right angles. A sketch of the streamlines and equipotential lines (equally spaced at large distances from the body), known as a **flow net**, is shown in Figure E6.10b for flow over a weir. Such a carefully constructed sketch can be used to approximate the velocities at points of interest in an inviscid flow. Pressures can then be estimated using Bernoulli's equation.

6.5.2 Simple Solutions

Now, let us identify some relatively simple functions that satisfy Laplace's equation. However, before we do, it is often more convenient to use polar coordinates. Laplace's equation, the continuity equation, and the velocity components take the following forms

$$\nabla^2\psi = \frac{1}{r}\frac{\partial}{\partial r}\left(r\frac{\partial\psi}{\partial r}\right) + \frac{1}{r^2}\frac{\partial^2\psi}{\partial\theta^2} = 0 \tag{6.5.12}$$

$$\nabla\cdot\mathbf{V} = \frac{1}{r}\frac{\partial}{\partial r}(rv_r) + \frac{1}{r}\frac{\partial v_\theta}{\partial\theta} = 0 \tag{6.5.13}$$

$$v_r = \frac{1}{r}\frac{\partial\psi}{\partial\theta} = \frac{\partial\phi}{\partial r} \qquad v_\theta = -\frac{\partial\psi}{\partial r} = \frac{1}{r}\frac{\partial\phi}{\partial\theta} \tag{6.5.14}$$

We will introduce the names of four simple flows, sketched in Figure 6.18 and their corresponding functions, each of which satisfies Laplace's equation. The names and functions are:

Uniform flow: $\psi = U_\infty y \qquad \phi = U_\infty x$ (6.5.15)

Line source: $\psi = \dfrac{q}{2\pi}\theta \qquad \phi = \dfrac{q}{2\pi}\ln r$ (6.5.16)

Irrotational vortex: $\psi = \dfrac{\Gamma}{2\pi}\ln r \qquad \phi = \dfrac{\Gamma}{2\pi}\theta$ (6.5.17)

Doublet: $\psi = -\dfrac{\mu\sin\theta}{r} \qquad \phi = \dfrac{\mu\cos\theta}{r}$ (6.5.18)

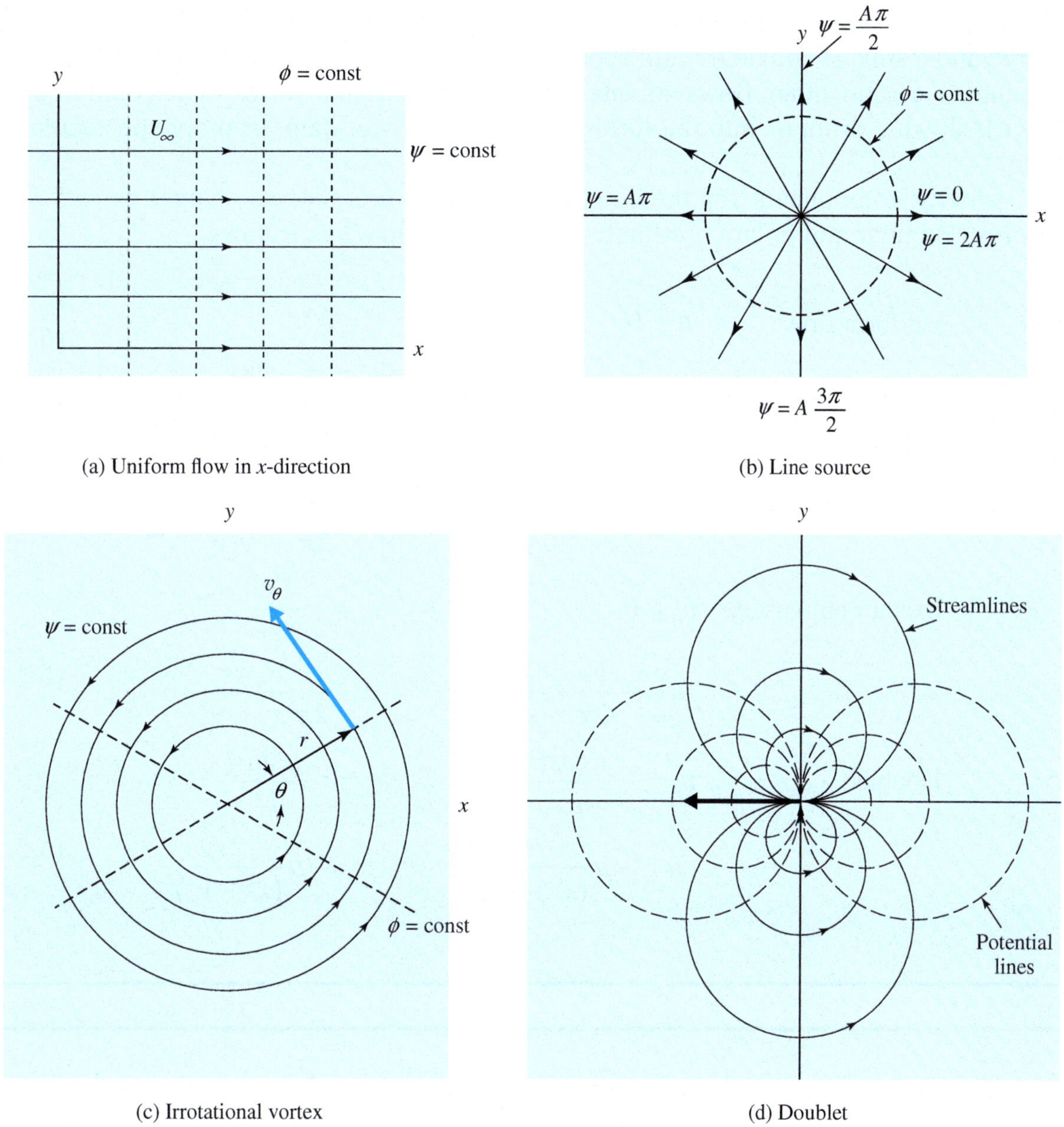

Figure 6.18 Four simple potential flows.

The uniform flow velocity U_∞ is assumed to be in the x-direction; if a y-component is desired, an appropriate term is added. The **source strength q** is the volume rate of flow per unit depth issuing from the source; a negative value would create a **sink**. The **vortex strength Γ** is the **circulation** about the origin, defined by

$$\Gamma = \oint_L \mathbf{V} \cdot d\mathbf{s} \tag{6.5.19}$$

where L must be a closed curve (a circle is usually used) around the origin and clockwise is positive. The **doublet strength μ** is for a doublet oriented in the negative x-direction; note the large arrow (in Figure 6.18d) showing the direction of the doublet. Doublets oriented in other directions are seldom of interest and are not considered here.

Of the four flows presented above, the doublet is rather mysterious; it can be visualized as a source and a sink of equal strength separated by a very small distance. Its usefulness is in the creation of certain other flows of interest. The irrotational vortex is encountered when water swirls down a drain or into the turbine of a hydropower dam, or more spectacularly, in a tornado.

The velocity components for the four simple flows are shown, using Eqs. 6.5.11 and 6.5.14 for rectangular and polar coordinates respectively, to be as follows:

$$\text{Uniform flow:} \quad u = U_\infty \qquad v = 0 \qquad v_r = U_\infty \cos\theta \qquad v_\theta = -U_\infty \sin\theta \tag{6.5.20}$$

$$\text{Line source:} \quad v_r = \frac{q}{2\pi r} \qquad v_\theta = 0 \qquad u = \frac{q}{2\pi}\frac{x}{x^2 + y^2} \qquad v = \frac{q}{2\pi}\frac{y}{x^2 + y^2} \tag{6.5.21}$$

$$\text{Irrotational vortex:} \quad v_r = 0 \qquad v_\theta = -\frac{\Gamma}{2\pi r} \qquad u = -\frac{\Gamma}{2\pi}\frac{y}{x^2 + y^2} \qquad v = \frac{\Gamma}{2\pi}\frac{x}{x^2 + y^2} \tag{6.5.22}$$

$$\text{Doublet:} \quad v_r = -\frac{\mu \cos\theta}{r^2} \qquad v_\theta = -\frac{\mu \sin\theta}{r^2} \qquad u = -\mu\frac{x^2 - y^2}{(x^2 + y^2)^2} \qquad v = -\mu\frac{2xy}{(x^2 + y^2)^2} \tag{6.5.23}$$

Example 6.11

The pressure far from an irrotational vortex (a simplified tornado) in the atmosphere is zero gage. If the velocity at $r = 20$ m is 20 m/s, estimate the velocity and the pressure at $r = 2$ m. (The irrotational vortex ceases to be a good model for a tornado when r is small. In the "eye" of the tornado the motion is approximated by rigid-body motion.)

Solution

For an irrotational vortex, we know that

$$v_\theta = -\frac{\Gamma}{2\pi r}$$

Hence

$$\begin{aligned}\Gamma &= -2\pi r v_\theta \\ &= -2\pi \times 20 \times 20 = -800\pi \text{ m}^2/\text{s}\end{aligned}$$

The velocity at $r = 2$ m is then

$$v_\theta = -\frac{-800\pi}{2\pi \times 2} = 200 \text{ m/s}$$

Bernoulli's equation for this incompressible, inviscid, steady flow gives the pressure as follows assuming a stagnant atmosphere away from the tornado:

$$\overset{0}{p_\infty} + \frac{\overset{0}{U_\infty^2}}{2}\rho = p + \frac{v_\theta^2}{2}\rho$$

$$\therefore p = -\frac{1}{2}\rho v_\theta^2$$

$$= -\frac{1}{2} \times 1.20 \times 200^2 = -24\,000 \text{ Pa or } \underline{-24 \text{ kPa}}$$

The negative sign denotes a vacuum. It is this vacuum that causes roofs of buildings to blow off during a tornado.

6.5.3 Superposition

The simple flows presented in Section 6.5.2 are of particular interest because they can be superimposed with each other to form more complicated flows of engineering importance. In fact, the most complicated, incompressible plane flow can be constructed using these simple flows. For example, suppose that the flow around an airfoil with a slotted flap is desired. We could divide the surface of the airfoil into a relatively large number (say, 200) of panels, locate a source or a sink (or alternatively, a doublet) at the center of each panel, add a uniform flow and an irrotational vortex, and then by adjusting[5] the panel source strengths, the desired inviscid flow could be created. The development of the model and the computer routine necessary to perform the calculations are considered beyond the scope of this book.

In this section we demonstrate superposition by creating flow around a circular cylinder with and without circulation. First, superimpose a uniform flow and a doublet; there results

$$\psi = U_\infty y - \frac{\mu \sin\theta}{r} \tag{6.5.24}$$

The velocity component v_r is (let $y = r\sin\theta$)

$$v_r = \frac{1}{r}\frac{\partial\psi}{\partial\theta}$$

$$= U_\infty \cos\theta - \frac{\mu}{r^2}\cos\theta \tag{6.5.25}$$

[5]The source strengths are adjusted so that the normal component of velocity at the center of each panel is zero. The vortex strength is adjusted so that the rear stagnation point occurs at the trailing edge.

Let us ask the question: Is there a radius r_c for which $v_r = 0$? If we set $v_r = 0$ we find that

$$r_c = \sqrt{\frac{\mu}{U_\infty}} \tag{6.5.26}$$

At this radius, v_r is identically zero for all θ and thus the circle $r = r_c$ must be a streamline.

The stagnation points are found by setting $v_\theta = 0$ on the circle $r = r_c$. There results

$$\begin{aligned} v_\theta &= -\frac{\partial \psi}{\partial r} \\ &= -U_\infty \sin\theta - \frac{\mu \sin\theta}{r_c^2} = -2U_\infty \sin\theta = 0 \end{aligned} \tag{6.5.27}$$

Thus we see that $v_\theta = 0$ at $\theta = 0°$ and 180°. The flow is as sketched in Figure 6.19a. We only have an interest in the flow external to the circular streamline $r = r_c$.

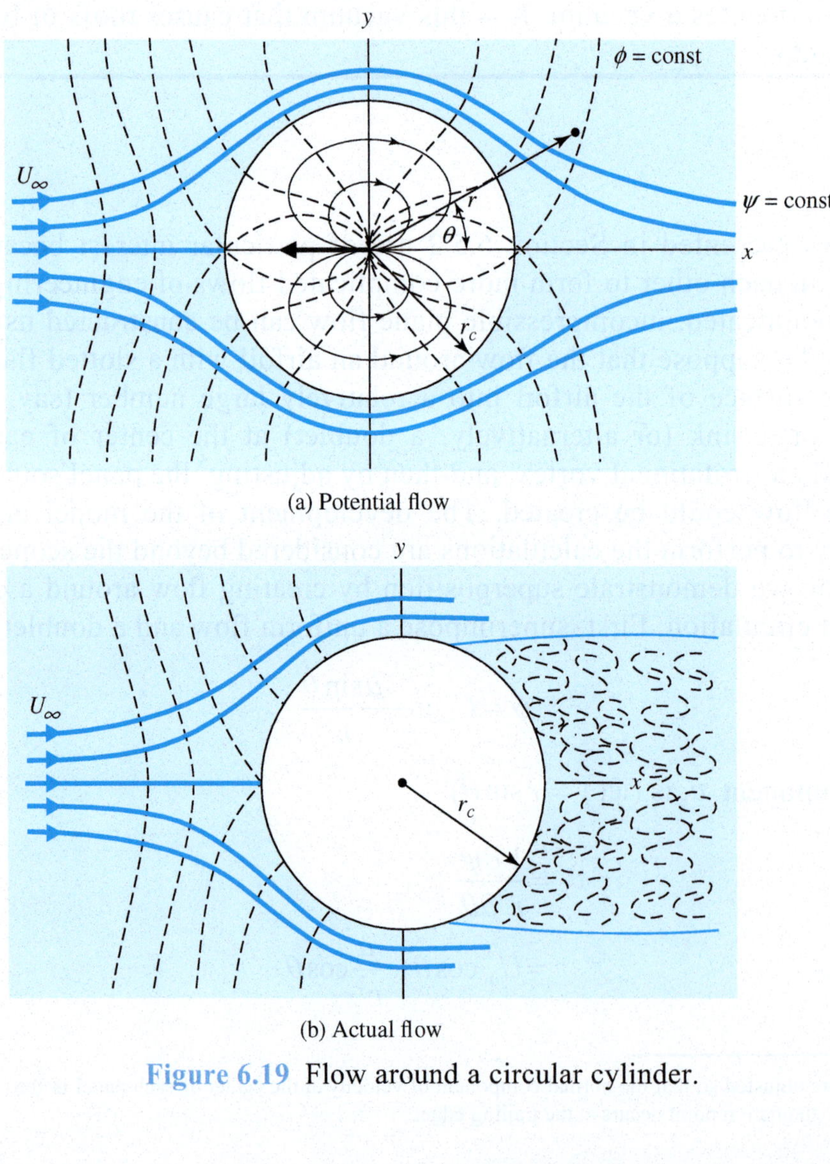

Figure 6.19 Flow around a circular cylinder.

If the pressure distribution were desired on the cylinder, Bernoulli's equation could be used between the stagnation point where $V = 0$ and $p = p_0$ and some arbitrary point on the cylinder to yield

$$\begin{aligned} p_c &= p_0 - \rho\frac{v_\theta^2}{2} \\ &= p_0 - 2\rho U_\infty^2 \sin^2\theta \end{aligned} \tag{6.5.28}$$

This provides a symmetric pressure distribution that yields zero drag and zero lift. The zero-lift prediction is acceptable for a real flow, but the zero-drag result is unacceptable. This may suggest that we ignore the inviscid flow solution; however, compared with the actual flow situation of Figure 6.19b, the measured pressure distribution on the cylinder up to the point of separation is very nearly the same as that predicted by the potential flow solution. Hence the potential flow solution is quite useful to us, even for blunt bodies that experience a separated flow. At low Reynolds numbers, viscous effects are not confined to a thin boundary layer, so potential-flow theory is not useful.

Let us now consider flow around a rotating cylinder. This is accomplished by adding an irrotational vortex to the doublet and uniform flow so that

$$\psi = U_\infty y - \frac{\mu \sin\theta}{r} + \frac{\Gamma}{2\pi}\ln r \tag{6.5.29}$$

Since the vortex flow, consisting of circular streamlines, does not influence the velocity component v_r the cylinder $r = r_c$ remains unchanged. The stagnation points, however, are changed and are located by setting $v_\theta = 0$ on $r = r_c$; that is, letting $\mu = U_\infty r_c^2$,

$$\begin{aligned} v_\theta &= -\frac{\partial\psi}{\partial r} \\ &= -2U_\infty \sin\theta - \frac{\Gamma}{2\pi r_c} = 0 \end{aligned} \tag{6.5.30}$$

This gives the location of the stagnation points as shown in Figure 6.20. In (a) the stagnation points are on the cylinder where $r = r_c$, but in (b) the circulation is sufficiently large that a single stagnation point is off the cylinder with $\theta = 270°$.

Bernoulli's equation gives the pressure distribution as

$$p_c = p_0 - \rho\frac{U_\infty^2}{2}\left(2\sin\theta + \frac{\Gamma}{2\pi r_c U_\infty}\right)^2 \tag{6.5.31}$$

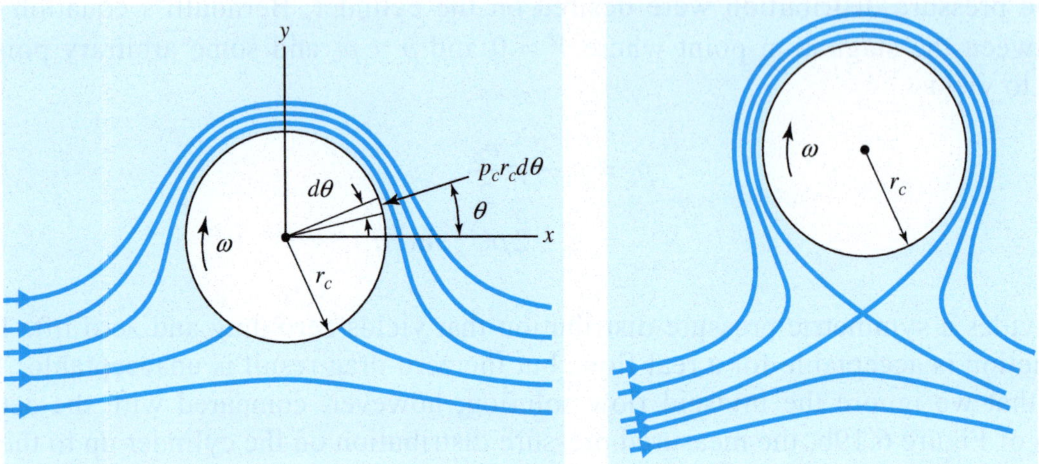

(a) $\Gamma < 4\pi U_\infty r_c$ (b) $\Gamma > 4\pi U_\infty r_c$

Figure 6.20 Flow around a circular cylinder with circulation.

This can be integrated to give drag = 0 and the lift per unit length as

$$\begin{aligned} F_L &= -\int_0^{2\pi} p_c \sin\theta \, r_c d\theta \\ &= \rho U_\infty \Gamma \end{aligned} \tag{6.5.32}$$

This expression for the lift provides an excellent approximation to the lift for all cylinders, including the airfoil. Along with the zero-drag conclusion, it forms the **Kutta–Joukowsky theorem**.

Other superpositions of the simple flows are included in the problems.

Example 6.12

An 8-in.-diameter cylinder rotates clockwise at 1000 rpm in a 60°F-atmospheric airstream flowing at 15 ft/s. Locate any stagnation points and find the minimum pressure on the cylinder.

Solution

The circulation is calculated (see Eq. 6.5.19) to be

$$\begin{aligned} \Gamma &= \oint_L \mathbf{V} \cdot d\mathbf{s} \\ &= 2\pi r_c^2 \omega = 2\pi \times \left(\frac{4}{12}\right)^2 \times \frac{1000 \times 2\pi}{60} = 73.1 \text{ ft}^2/\text{s} \end{aligned}$$

This is greater than $4\pi U_\infty r_c = 4\pi \times 15 \times 4/12 = 62.8$ ft^2/s; hence the stagnation point is off the cylinder (see Figure 6.20b) at $\underline{\theta = 270°}$ at a radius of

$$r_0 = -\frac{\Gamma}{4\pi U_\infty \sin 270°} = -\frac{73.1}{4\pi \times 15 \times (-1)} = \underline{0.388 \text{ ft}}$$

Only one stagnation point exists.

The minimum pressure is located on the top of the cylinder where $\theta = 90°$. Using Bernoulli's equation from the free stream to that point, we have, letting $p_\infty = 0$,

$$\overset{0}{\cancel{p_\infty}} + \rho\frac{U_\infty^2}{2} = p_{\min} + \frac{\rho}{2}(v_\theta)_{\max}^2$$

$$\therefore p_{\min} = \frac{\rho}{2}\left[U_\infty^2 - (v_\theta)_{\max}^2\right] = \frac{\rho}{2}\left[U_\infty^2 - \left(-2U_\infty \sin 90° - \frac{\Gamma}{2\pi r_c}\right)^2\right]$$

$$= \frac{0.0024 \text{ slug/ft}^3}{2}\left[15^2 - \left(2\times 15 + \frac{73.1}{2\pi \times 4/12}\right)^2\right]\text{ft}^2/\text{s}^2 = \underline{-4.78 \text{ psf}}$$

6.6 BOUNDARY-LAYER THEORY

6.6.1 General Background

In our study of high-Reynolds-number external flows we have observed that the viscous effects are confined to a thin layer of fluid, a boundary layer, next to the body and to the wake downstream of the body. For a streamlined body such as an airfoil, a good approximation to the drag can be obtained by integrating the viscous shear stress on the wall. To predict the wall shear, the velocity gradient at the wall must be known. This requires a complete solution of the flow field (i.e., a solution of the Navier–Stokes equations) inside the boundary layer. Such a solution also allows us to predict locations of possible separation. In this section we derive the integral and differential equations and provide solution techniques for a boundary–layer flow on a flat plate with a zero pressure gradient; this simplified flow has many applications. Nonzero-pressure-gradient flows on flat plates and flows on curved surfaces will not be considered in this introductory presentation.

Let us now discuss some of the characteristics of a boundary layer. The edge of the boundary layer, with thickness designated by $\delta(x)$, cannot be observed in an actual flow; we arbitrarily define it to be that locus of points where the velocity is equal to 99% of the free-stream velocity [the free-stream velocity is the inviscid flow wall-velocity $U(x)$], as shown in Figure 6.21. Since the boundary layer is thin, the pressure in the boundary layer is assumed to be the pressure $p(x)$ at the wall, as predicted by the inviscid flow solution.

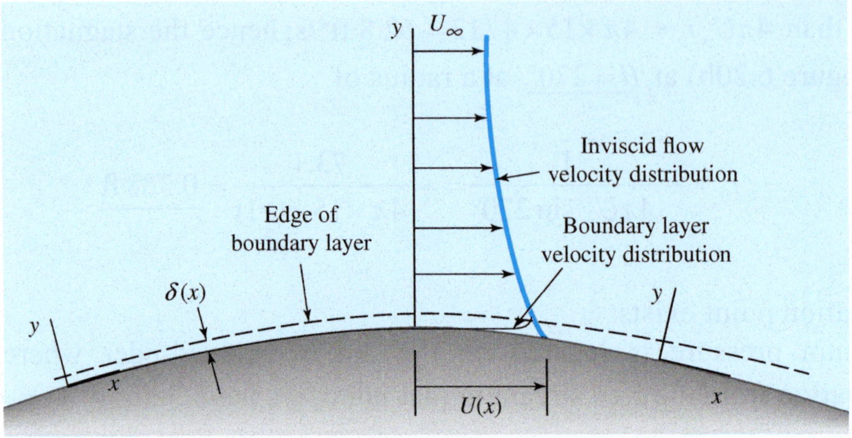

Figure 6.21 Boundry layer on a curved surface.

The boundary layer begins as a laminar flow with zero thickness at the leading edge of a flat plate, as shown in Figure 6.22, or with some finite thickness at the stagnation point of a blunt object or an airfoil (see Figure 6.3). After a distance x_T, which depends on the freestream velocity, the viscosity, the pressure gradient, the wall roughness, the free-stream fluctuation level, and wall rigidity, the laminar flow undergoes a transition process that results, after a short distance, in a turbulent flow, as sketched. For flow over a flat plate with zero pressure gradient, this transition process occurs when $U_\infty x_T/\nu = 3\times10^5$ for flow on rough plates or with high free-stream fluctuation intensity $\left(\sqrt{\overline{u'^2}}/U_\infty \simeq 0.1\right)$, or $U_\infty x_T/\nu = 5\times10^5$ for flow on smooth rigid plates with low free-stream fluctuation intensity. For extremely low fluctuation levels in research labs, laminar flows have been observed on smooth rigid plates with carefully designed leading edges up to $U_\infty x_T/\nu = 10^6$.

The quantity $U_\infty x/\nu$ is the **local Reynolds number** and $U_\infty x_T/\nu$ is the **critical Reynolds number**. For a smooth rigid flat plate and a very low free-stream fluctuation level, a zero pressure-gradient flow becomes unstable (i.e., small disturbances will grow) at a local

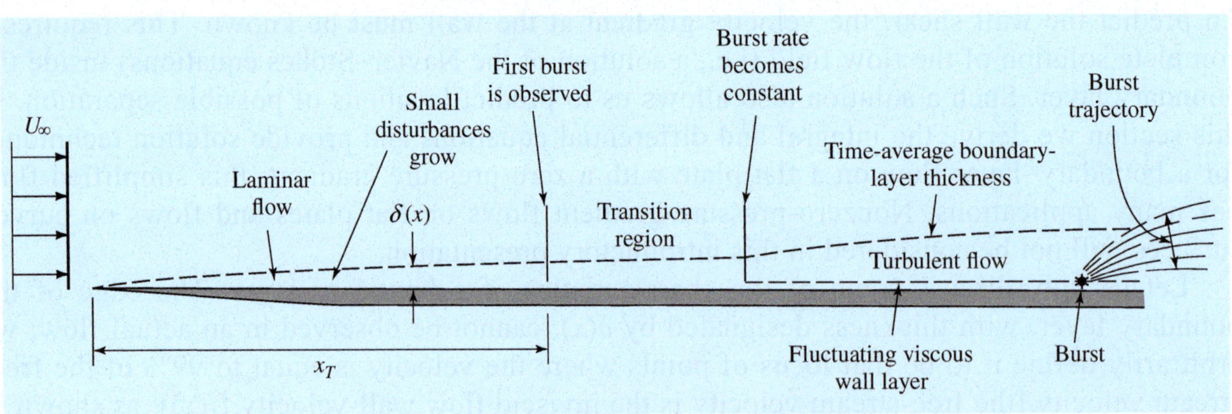

Figure 6.22 Boundary layer with transition.

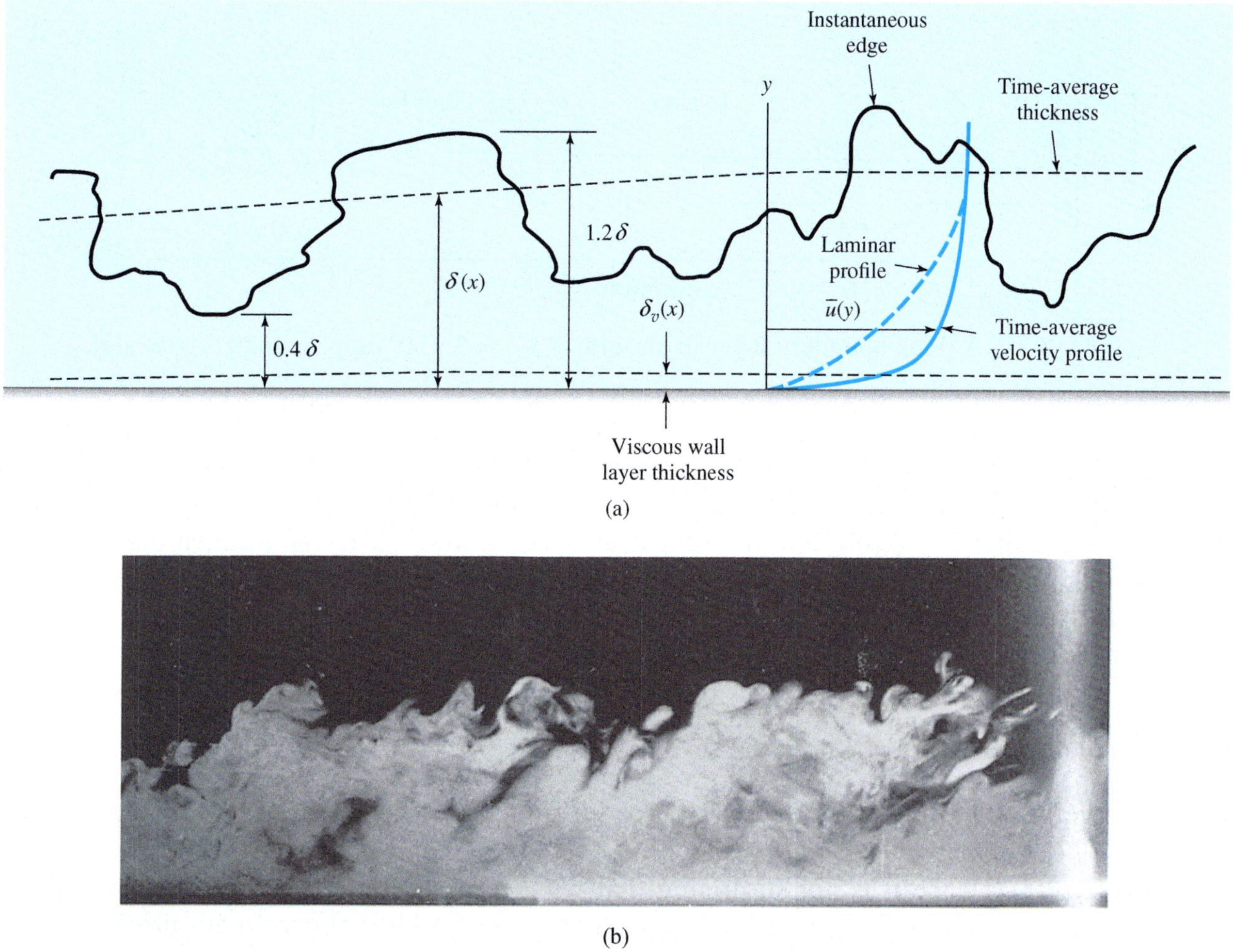

Figure 6.23 Turbulent boundary layer: (a) nomenclature sketch: (b) stream wise slice of the boundary layer. (Photograph by R. E. Falco)

Reynolds number of about 6×10^4. The small disturbances grow initially as a two-dimensional wave, then a three-dimensional wave, and finally burst into a turbulent spot; the initial burst forms the beginning of the transition region. The transition region is relatively short and is usually ignored in calculations. The flow up to x_T is assumed to be laminar, and the flow after x_T is considered to be turbulent.

The turbulent boundary layer thickens much more rapidly than the laminar layer. It also has a substantially greater wall shear. A sketch of a turbulent boundary layer with its submerged viscous wall layer is shown in Figure, 6.23a, and an actual photograph in Figure 6.23b. The time-average thickness is $\delta(x)$, and the average viscous wall-layer thickness is $\delta_\nu(x)$. Both layers are actually quite time dependent. The instantaneous boundary-layer thickness varies between 0.4δ and 1.2δ, as shown. The turbulent profile has a greater slope at the wall than does a laminar profile with the same boundary-layer thickness, as is sketched in Figure 6.23a.

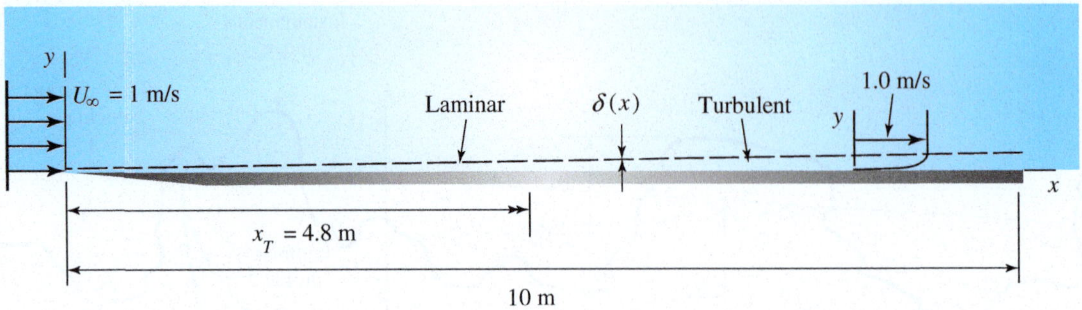

Figure 6.24 A thick boundary layer in air with $\text{Re}_{crit} = 3 \times 10^5$ (approximately to scale).

Finally, we should emphasize that the boundary layer is quite thin. A thick boundary layer ($U_\infty = 1$ m/s) is drawn to scale in Figure 6.24. We have assumed a laminar flow up to x_T and a turbulent flow thereafter. For higher velocities the boundary-layer thickness decreases. If we assumed that $U_\infty = 100$ m/s, the boundary layer would hardly be noticed drawn to the same scale, yet all the viscous effects are confined in that thin layer; the velocity is brought to rest with very large gradients. Viscous dissipative effects in this thin layer are large enough to cause sufficiently high temperatures that satellites burn up on reentry.

6.6.2 Von Kármán Integral Equation

From the velocity profile in the boundary layer of Figure 6.24, it is observed that the velocity goes from $u = 0.99U_\infty$ at $y = \delta$ to $u = 0$ at $y = 0$ over a very short distance (the thickness of the boundary layer). Hence it is not surprising that we can approximate the velocity profile, for both laminar and turbulent flow with a good deal of accuracy. If the velocity profile can be assumed known, the continuity and momentum integral equations will enable us to predict the boundary-layer thickness and the wall shear and thus the drag. So let us develop the integral equations for the boundary layer.

Consider an infinitesimal control volume, shown in Figure 6.25a. The integral continuity equation allows us to find $\dot{m}_{\text{top}}$ (see Figure 6.25b). It is, assuming unit depth,

$$\begin{aligned} \dot{m}_{\text{top}} &= \dot{m}_{\text{out}} - \dot{m}_{\text{in}} \\ &= \frac{\partial}{\partial x}\int_0^\delta \rho u \, dy \, dx \end{aligned} \tag{6.6.1}$$

The integral momentum equation takes the form

$$\Sigma F_x = m\dot{o}m_{\text{out}} - m\dot{o}m_{\text{in}} - m\dot{o}m_{\text{top}} \tag{6.6.2}$$

Where $m\dot{o}m$ represents the momentum flux in the x-direction. Referring to Figures 6.25c and 6.25d this becomes, neglecting high-order terms,

$$-\delta dp - \tau_0 dx = \frac{\partial}{\partial x}\int_0^\delta \rho u^2 dy \, dx - \left(\frac{\partial}{\partial x}\int_0^\delta \rho u \, dy \, dx\right) U(x) \tag{6.6.3}$$

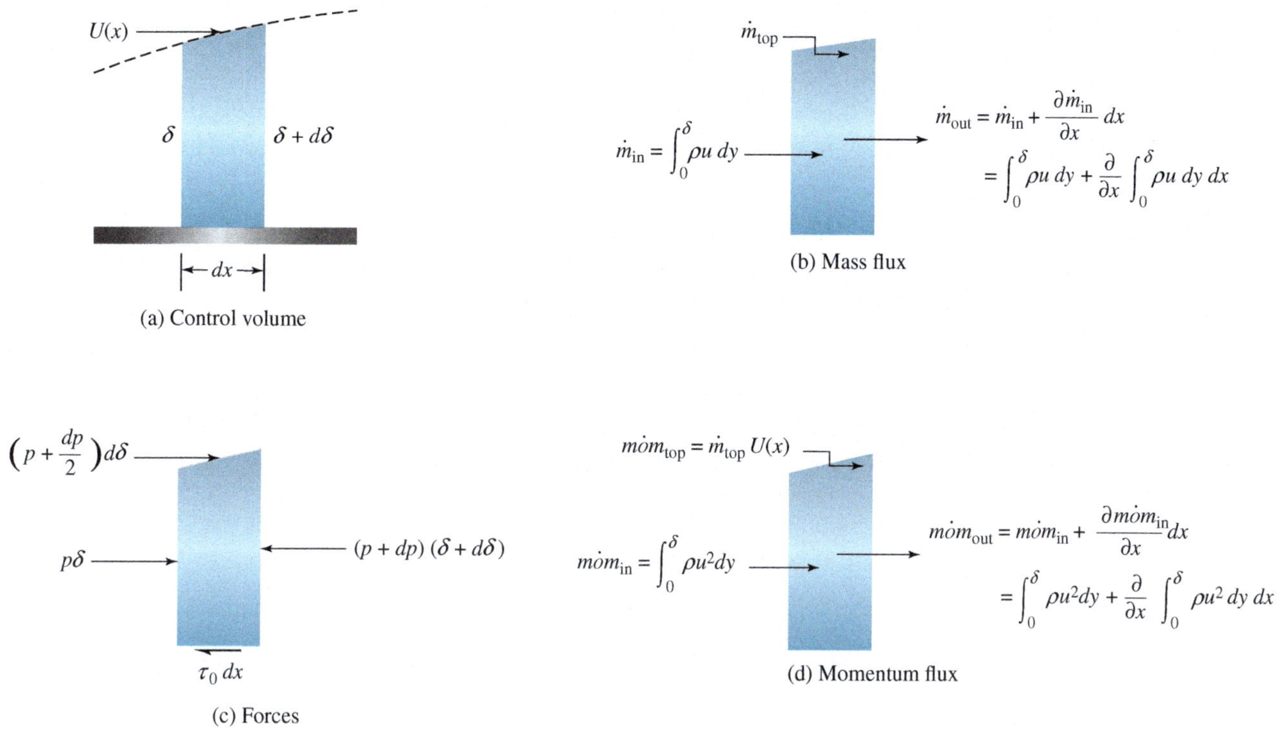

Figure 6.25 Control volume for a boundary layer with variable $U(x)$.

where $\dot{m}_{top}$ is given in Eq. 6.6.1. We have not assumed that $U(x)$ is constant. Divide through by $(-dx)$ and obtain

$$\tau_0 + \delta\frac{dp}{dx} = U(x)\frac{d}{dx}\int_0^\delta \rho u\,dy - \frac{d}{dx}\int_0^\delta \rho u^2\,dy \tag{6.6.4}$$

where we have used ordinary derivatives since the integrals are only functions of x. This equation is often referred to as the **von Kármán integral equation**.

For flow over a flat plate with zero pressure gradients, so that $dp/dx = 0$ and $U(x) = U_\infty$, von Kármán's integral equation takes the simplified form

$$\begin{aligned}\tau_0 &= \frac{d}{dx}\int_0^\delta \rho U_\infty\,dy - \frac{d}{dx}\int_0^\delta \rho u^2\,dy \\ &= \frac{d}{dx}\int_0^\delta \rho u(U_\infty - u)\,dy\end{aligned} \tag{6.6.5}$$

If the velocity profile can be assumed, this equation along with $\tau_0(x) = \mu\partial u/\partial y\big|_{y=0}$ allows us to solve for both $\delta(x)$ and $\tau_0(x)$. This will be demonstrated in the following sections for both a laminar and a turbulent boundary layer.

Before we do this, however, there are two additional lengths often used in boundary-layer theory. They are the **displacement thickness δ_d** and the **momentum thickness θ**, defined by

$$\delta_d = \frac{1}{U}\int_0^{\delta} (U - u)\, dy \tag{6.6.6}$$

$$\theta = \frac{1}{U^2}\int_0^{\delta} u(U - u)\, dy \tag{6.6.7}$$

The displacement thickness is the displacement of the streamlines in the free stream as a result of velocity deficits in the boundary layer, as can be shown by continuity considerations. The momentum-layer thickness is the equivalent thickness of a fluid layer with velocity U with momentum equal to the momentum lost due to friction; the momentum thickness is often used as a characteristic length in turbulent boundary-layer studies. The end-of-chapter problems will demonstrate the use of δ_d and θ. It should be noted, though, that von Kármán's integral equation (6.6.5) takes the form, assuming that $\rho = \text{const.}$,

$$\tau_0 = \rho U_\infty^2 \frac{d\theta}{dx} \tag{6.6.8}$$

6.6.3 Approximate Solution to the Laminar Boundary Layer

It is possible to use the von Kármán integral equation and obtain a fairly accurate approximation to the laminar boundary layer on a flat plate with zero pressure gradients. We have four conditions that a proposed velocity profile should satisfy:

$$\begin{aligned} u &= 0 \quad \text{at} \quad y = 0 \\ u &= U_\infty \quad \text{at} \quad y = \delta \\ \frac{\partial u}{\partial y} &= 0 \quad \text{at} \quad y = \delta \\ \frac{\partial^2 u}{\partial y^2} &= 0 \quad \text{at} \quad y = 0 \end{aligned} \tag{6.6.9}$$

The first three of these conditions are obvious from a sketch of a velocity profile; the fourth condition comes from the x component Navier–Stokes equation (3.3.14) since $u = v = 0$ at the wall, and $dp/dx = 0$ for the steady flow on the flat plate resulting in $\partial^2 u/\partial x^2 = 0$ on the wall.

A cubic polynomial can satisfy the four conditions above; let us assume that

$$\frac{u}{U_\infty} = A + By + Cy^2 + Dy^3 \tag{6.6.10}$$

where A, B, C, and D can be functions of x. Using the four conditions, we see that

$$A = 0 \qquad B = \frac{3}{2\delta} \qquad C = 0 \qquad D = -\frac{1}{2\delta^3} \tag{6.6.11}$$

Hence, a good approximation for the velocity profile in a laminar flow is

$$\frac{u}{U_\infty} = \frac{3}{2}\frac{y}{\delta} + \frac{1}{2}\left(\frac{y}{\delta}\right)^3 \tag{6.6.12}$$

Let us now use this velocity profile to find $\delta(x)$ and $\tau_0(x)$. Von Kármán's integral equation (6.6.5) gives

$$\tau_0 = \frac{d}{dx}\int_0^\delta \rho\left(\frac{3y}{2\delta} - \frac{y^3}{2\delta^3}\right)\left(1 - \frac{3y}{2\delta} - \frac{y^3}{2\delta^3}\right)U_\infty^2 dy$$

$$= 0.139\rho U_\infty^2 \frac{d\delta}{dx} \tag{6.6.13}$$

At the wall we know that $\tau_0 = \mu\,\partial u/\partial y\big|_{y=0}$, or using the cubic profile (6.6.12),

$$\tau_0 = \mu\left(U_\infty \frac{3}{2\delta}\right) \tag{6.6.14}$$

Equating the foregoing expressions for $\tau_0(x)$, we find that

$$\delta d\delta = \frac{\frac{3}{2}\mu U_\infty}{0.139\rho U_\infty^2}dx = 10.8\frac{\nu}{U_\infty}dx \tag{6.6.15}$$

Using $\delta = 0$ at $x = 0$ (the leading edge), Eq. 6.6.15 is integrated to give

$$\delta = 4.65\sqrt{\frac{\nu x}{U_\infty}} = 4.65\frac{x}{\sqrt{\mathrm{Re}_x}} \tag{6.6.16}$$

where Re_x is the local Reynolds number. This is substituted back into Eq. 6.6.14, giving the wall shear as

$$\tau_0 = 0.323\rho U_\infty^2\sqrt{\frac{\nu}{xU_\infty}}$$

$$= \frac{0.323\rho U_\infty^2}{\sqrt{\mathrm{Re}_x}} \tag{6.6.17}$$

The shearing stress is made dimensionless by dividing by $\frac{1}{2}\rho U_\infty^2$. The **local skin friction coefficient** c_f results; it is

$$c_f = \frac{\tau_0}{\frac{1}{2}\rho U_\infty^2}$$

$$= \frac{0.646}{\sqrt{U_\infty x/\nu}} = \frac{0.646}{\sqrt{\mathrm{Re}_x}} \tag{6.6.18}$$

If the wall shear is integrated over the length L, there results, per unit width,

$$F_D = \int_0^L \tau_0 dx = 0.646 \rho U_\infty \sqrt{U_\infty L \nu}$$

$$= \frac{0.646 \rho U_\infty^2 L}{\sqrt{\text{Re}_L}} \tag{6.6.19}$$

or in terms of the **skin friction coefficient** C_f,

$$C_f = \frac{F_D}{\frac{1}{2}\rho U_\infty^2 L}$$

$$= \frac{1.29}{\sqrt{U_\infty L/\nu}} = \frac{1.29}{\sqrt{\text{Re}_L}} \tag{6.6.20}$$

where Re_L is the Reynolds number at the end of the flat plate.

Note that the shearing stress τ_0 becomes unbounded as $x \to 0$. Hence we would not expect $\tau_0(x)$ be a very good approximation to the wall shear near the leading edge. The expression for the drag is acceptable, however.

The results above are quite good when compared with results from a solution of the differential equations (refer to Section 6.6.6). The boundary layer thickness is 7% too low; the constant in Eq. 6.6.16 should be 5 if an exact solution were obtained. The wall shear is less than 3% too low; the constant in Eq. 6.6.17 should be 0.332.

Example 6.13

Assume that the velocity profile in a boundary-layer flow can be approximated by a parabolic velocity profile. Calculate the boundary-layer thickness and the wall shear. Compare with those calculated above for the cubic profile.

Solution

The parabolic velocity profile is assumed to be

$$\frac{u}{U_\infty} = A + By + Cy^2$$

The fourth condition, which would be impossible to satisfy, of Eqs. (6.6.9) is omitted; this leaves

$$0 = A$$

$$1 = A + B\delta + C\delta^2$$

$$0 = B + 2C\delta$$

A simultaneous solution provides

$$A = 0 \qquad B = \frac{2}{\delta} \qquad C = -\frac{1}{\delta^2}$$

The velocity profile is then

$$\frac{u}{U_\infty} = 2\frac{y}{\delta} - \frac{y^2}{\delta^2}$$

This is substituted into von Kármán's integral equation (6.6.5) to obtain

$$\tau_0 = \frac{d}{dx}\int_0^\delta \rho U_\infty^2 \left(2\frac{y}{\delta} - \frac{y^2}{\delta^2}\right)\left(1 - \frac{2y}{\delta} + \frac{y^2}{\delta^2}\right)dy$$

$$= \frac{2}{15}\rho U_\infty^2 \frac{d\delta}{dx}$$

We also use $\tau_0 = \mu \partial u / \partial y\,|_{y=0}$; that is,

$$\tau_0 = \mu U_\infty \frac{2}{\delta}$$

Equating the two expressions above, we obtain

$$\delta\, d\delta = 15\frac{\nu}{U_\infty}dx$$

Using $\delta = 0$ at $x = 0$ this is integrated to

$$\delta = 5.48\sqrt{\frac{\nu x}{U_\infty}}$$

This is 18% higher than the value using the cubic but only 10% higher than the more accurate result of $5\sqrt{\nu x / U_\infty}$.

The wall shear is found to be

$$\tau_0 = \frac{2\mu U_\infty}{\delta}$$

$$= 0.365 \rho U_\infty^2 \sqrt{\frac{\nu}{x U_\infty}}$$

This is 13% higher than the value using the cubic and 10% higher than the more accurate value of 0.332 $\rho U_\infty^2 \sqrt{\nu / x U_\infty}$. Because the boundary layer is so thin, there is little difference between a cubic and a parabola or the actual profile; refer to the profile in Figure 6.24.

6.6.4 Turbulent Boundary Layer: Power-Law Form

For turbulent boundary-layer flow we have two methods for obtaining the information desired. Both methods utilize experimental data, but the one that we present in this section is the simpler of the two. The second method, to be presented in the next section, provides us with more information than we usually desire for most applications; however, it is more accurate.

In the method to be presented first we fit the data for the velocity profile with a power-law equation. The **power-law form** is

$$\frac{\bar{u}}{U_\infty} = \left(\frac{y}{\delta}\right)^{1/n} \qquad n = \begin{cases} 7 & \mathrm{Re}_x < 10^7 \\ 8 & 10^7 < \mathrm{Re}_x < 10^8 \\ 9 & 10^8 < \mathrm{Re}_x < 10^9 \end{cases} \tag{6.6.21}$$

where

$$\mathrm{Re}_x = \frac{U_\infty x}{\nu} \tag{6.6.22}$$

Von Kármán's integral equation can now be applied following the steps used for laminar flow, except when the shear stress at the wall is evaluated. The power-law form (6.6.21) yields $(\partial\bar{u}/\partial y)_{y=0} = \infty$; hence the profile gives poor results near the wall, especially for the shear stress. So rather than using $\tau_0 = (\mu\partial\bar{u}/\partial y)_{y=0}$, we use an empirical relation; the **Blasius formula**, named after Paul R.H. Blasius (1883–1970), which relates the local skin friction coefficient to the boundary-layer thickness, is

$$c_f = 0.046\left(\frac{\nu}{U_\infty \delta}\right)^{1/4} \tag{6.6.23}$$

Or, relating τ_0 to c_f using the definition of c_f in Eq. 6.6.18, the shear stress relation is

$$\tau_0 = 0.023\rho U_\infty^2\left(\frac{\nu}{U_\infty \delta}\right)^{1/4} \tag{6.6.23}$$

Von Kármán's integral equation provides us with a second expression for τ_0. Substitute the velocity profile of (6.6.21) with $\mathrm{Re}_x < 10^7$ into Eq. 6.6.5 and obtain

$$\tau_0 = \frac{d}{dx}\int_0^\delta \rho U_\infty^2 \left(\frac{y}{\delta}\right)^{1/7}\left[1-\left(\frac{y}{\delta}\right)^{1/7}\right]dy$$

$$= \frac{7}{72}\rho U_\infty^2 \frac{d\delta}{dx} \tag{6.6.25}$$

Combining the two expressions above for τ_0, we find that

$$\delta^{1/4} d\delta = 0.237\left(\frac{\nu}{U_\infty}\right)^{1/4} dx \tag{6.6.26}$$

Assuming a turbulent flow from the leading edge (the laminar portion is often quite short, i.e., $L \gg x_T$), there results

$$\begin{aligned}\delta &= 0.38x\left(\frac{V}{U_x x}\right)^{1/5} \\ &= 0.38x\,\mathrm{Re}_x^{-1/5} \qquad \mathrm{Re}_x < 10^7\end{aligned} \tag{6.6.27}$$

Substituting this expression for δ back into Eq.6.6.23, we find that

$$c_f = 0.059\,\mathrm{Re}_x^{1/5} \qquad \mathrm{Re}_x < 10^7 \tag{6.6.28}$$

and, performing the required integration for F_D, there results, with $n = 7$,

$$C_f = 0.073\,\mathrm{Re}_L^{-1/5} \qquad \mathrm{Re}_L < 10^7 \tag{6.6.29}$$

where $\mathrm{Re}_L = U_\infty L/\nu$. The relations above can be stretched to $\mathrm{Re}_x \simeq 10^8$ without substantial error.

If L is not much larger than x_T, say $L = 3x_T$, then there is a significant laminar portion on the leading part of a flat plate, and the skin friction coefficient can be modified as

$$C_f = 0.073\,\mathrm{Re}_L^{-1/5} - 1700\,\mathrm{Re}_L^{-1} \qquad \mathrm{Re}_L < 10^7 \tag{6.6.30}$$

This relationship is based on transition occurring at $\mathrm{Re}_{\mathrm{crit}} = 5\times10^5$. If $\mathrm{Re}_{\mathrm{crit}} = 3\times10^5$, the constant of 1700 is replaced by 1060; if $\mathrm{Re}_{\mathrm{crit}} = 6\times10^5$, it is replaced by 2080.

Finally, the displacement thickness and momentum thickness can be evaluated, using $n = 7$, to be

$$\begin{aligned}\delta_d &= 0.048x\,\mathrm{Re}_x^{-1/5} \\ \theta &= 0.037x\,\mathrm{Re}_x^{-1/5}\end{aligned} \tag{6.6.31}$$

Example 6.14

Estimate the boundary-layer thickness at the end of a 4-m-long flat surface if the free-stream velocity is $U_\infty = 5$ m/s. Use atmospheric air at 30°C. Also, predict the drag force if the surface is 5 m wide. (a) Neglect the laminar portion of the flow and (b) account for the laminar portion using $\mathrm{Re}_{\mathrm{crit}} = 5 \times 10^5$.

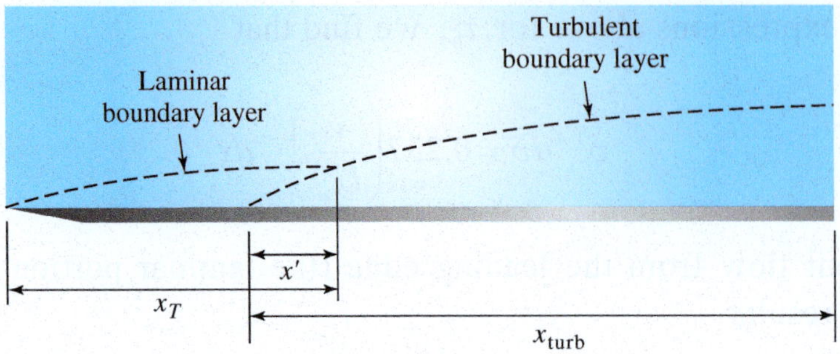

Figure E6.14

Solution

(a) Let us first assume turbulent flow from the leading edge. The boundary-layer thickness is given by Eq. 6.6.27. It is

$$\delta = 0.38x\,\mathrm{Re}_x^{-1/5}$$

$$= 0.38 \times 4 \times \left(\frac{5 \times 4}{1.6 \times 10^{-5}}\right)^{-1/5} = \underline{0.0917\ \mathrm{m}}$$

The drag force is, using Eq. 6.6.29,

$$F_D = C_f \times \frac{1}{2}\rho U_\infty^2\, Lw$$

$$= 0.073\left(\frac{5 \times 4}{1.6 \times 10^{-5}}\right)^{-1/5} \times \frac{1}{2} \times 1.16 \times 5^2 \times 4 \times 5 = \underline{1.28\ \mathrm{N}}$$

The predictions above assume that $\mathrm{Re}_L < 10^7$. The Reynolds number is

$$\mathrm{Re}_L = \frac{5 \times 4}{1.6 \times 10^{-5}} = 1.25 \times 10^6$$

Hence, the calculations are acceptable.

(b) Now let us account for the laminar portion of the boundary layer. Referring to Figure E6.14, the distance x_T is found as follows:

$$\mathrm{Re}_{\mathrm{crit}} = 5 \times 10^5 = \frac{U_\infty x_T}{\nu}$$

$$\therefore x_T = 5 \times 10^5 \times \frac{1.6 \times 10^{-5}}{5} = 1.6\ \mathrm{m}$$

The boundary-layer thickness at x_T is, replacing the constant of 4.65 in Eq. 6.6.16 with the more accurate value of 5,

$$\delta = 5\sqrt{\frac{x\nu}{U_\infty}}$$

$$= 5\sqrt{\frac{1.6\times 1.6\times 10^{-5}}{5}} = 0.0113 \text{ m}$$

The location of the fictitious origin of the turbulent flow (see Figure E6.14) is found using Eq. 6.6.27 to be

$$x'^{4/5} = \frac{\delta}{0.38}\left(\frac{U_\infty}{\nu}\right)^{1/5}$$

$$\therefore x' = \left(\frac{0.0113}{0.38}\right)^{3/4}\left(\frac{5}{1.6\times 10^{-5}}\right)^{1/4} = 0.292 \text{ m}$$

The distance x_{turb} is then $x_{\text{turb}} = 4 - 1.6 + 0.292 = 2.69$ m. Using Eq. 6.6.27, the thickness at the end of the surface is

$$\delta = 0.38x\left(\frac{\nu}{U_\infty x}\right)^{1/5}$$

$$= 0.38\times 2.69\times\left(\frac{1.6\times 10^{-5}}{5\times 2.69}\right)^{1/5} = \underline{0.067 \text{ m}}$$

The value of part (a) is 37% too high when compared with this more accurate value.

The more accurate drag force is found using Eq. 6.6.30 to be

$$F_D = C_f \times \frac{1}{2}\rho U_\infty^2 Lw$$

$$= \left[0.073\text{Re}_L^{-1/5} - 1700\text{Re}_L^{-1}\right]\times\frac{1}{2}\rho U_\infty^2 Lw$$

$$= \left[0.073\left(\frac{5\times 4}{1.6\times 10^{-5}}\right)^{-1/5} - 1700\left(\frac{5\times 4}{1.6\times 10^{-5}}\right)^{-1}\right]\times\frac{1}{2}\times 1.16\times 5^2\times 4\times 5$$

$$= \underline{0.88 \text{ N}}$$

The prediction of part (a) is 45% too high. For relatively short surfaces it is obvious that significant errors result if the Laminar portion with its smaller shear stress is neglected.

6.6.5 Turbulent Boundary Layer: Empirical Form

The second method of predicting turbulent flow quantities on a flat plate with zero pressure gradient is based entirely on data. It is more accurate than the power-law form but also more complicated. The time-average turbulent velocity profile can be divided into two regions, the **inner region** and the **outer region**, as shown in Figure 6.26. The inner region is characterized by the **self-similar** (the dimensionless dependent variable depends on only one dimensionless independent variable) relation,

$$\frac{\bar{u}}{u_\tau} = f\left(\frac{u_\tau y}{\nu}\right) \tag{6.6.32}$$

in which u_τ, is the shear velocity, given by[6]

$$u_\tau = \sqrt{\frac{\tau_0}{\rho}} \tag{6.6.33}$$

The velocity profile in the outer region is given by the self-similar relation

$$\frac{U_\infty - \bar{u}}{u_\tau} = f\left(\frac{y}{\delta}\right) \tag{6.6.34}$$

where $(U_\infty - \bar{u})$ is called the **velocity defect.**

The inner region has three distinct zones: the viscous wall layer, the buffer zone, and the turbulent zone, as shown in Figure 6.26b. The highly fluctuating **viscous wall layer** has a linear time-average profile given by

$$\frac{\bar{u}}{u_\tau} = \frac{u_\tau y}{\nu} \tag{6.6.35}$$

The quantity ν/u_τ is the characteristic length in the turbulent inner region; hence the dimensionless distance from the wall is denoted by

$$y^* = \frac{u_\tau y}{\nu} \tag{6.6.36}$$

The viscous wall layer is very thin, extending up to $y^* \simeq 5$. A logarithmic profile exists from $y^* \simeq 50$ to $y/\delta \simeq 0.15$. In this self-similar **turbulent zone,**

$$\frac{\bar{u}}{u_\tau} = 2.44 \ln \frac{u_\tau y}{\nu} + 4.9 \qquad 50 < \frac{u_\tau y}{\nu} \qquad \frac{y}{\delta} < 0.15 \tag{6.6.37}$$

---

[6] The shear velocity is a fictitious velocity and is defined because the quantity $\sqrt{\tau_0/\rho}$ occurs often in empirical relations in turbulent boundary-layer flows.

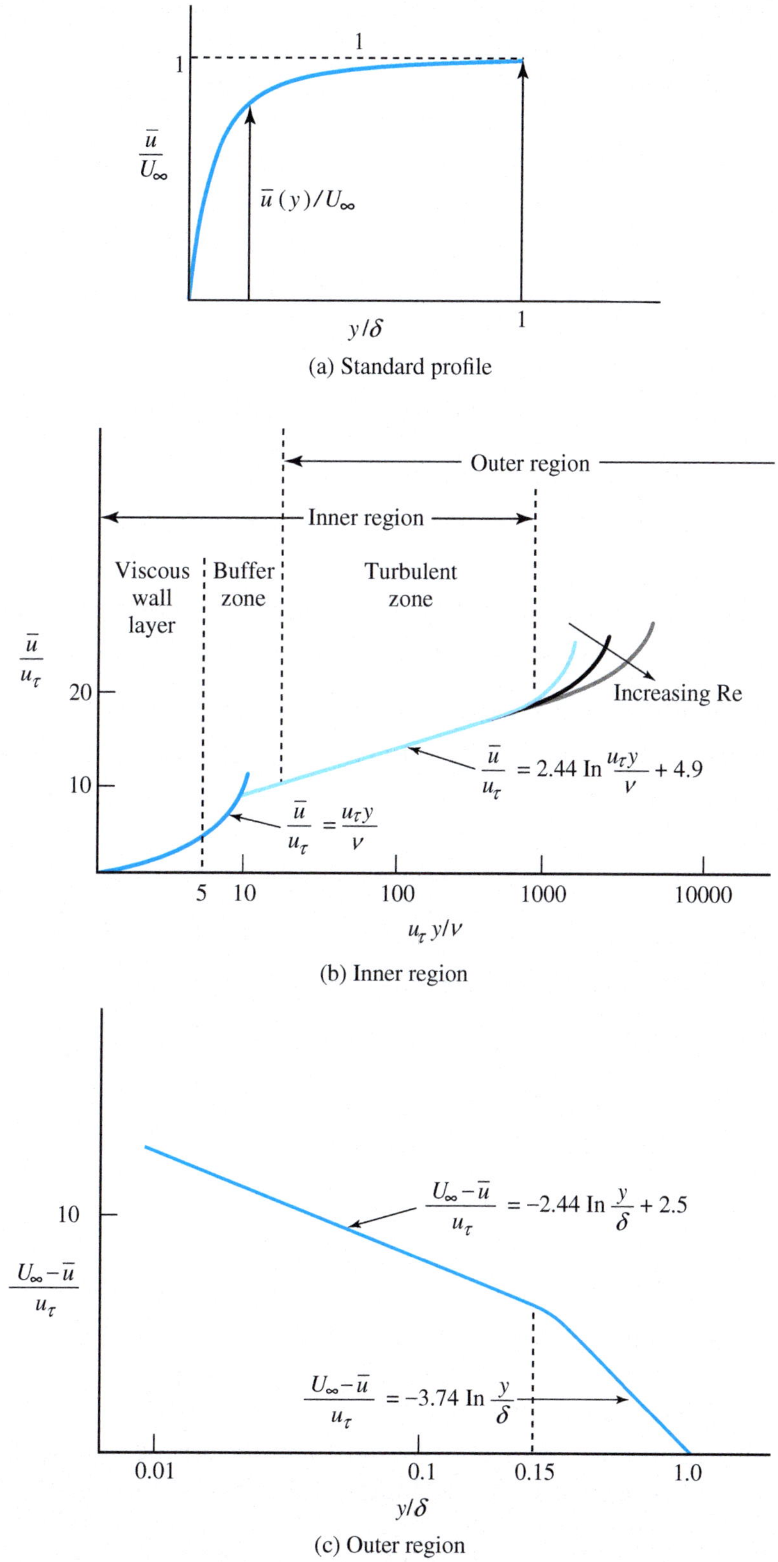

Figure 6.26 Velocity profile in a turbulent boundary layer.

The location of the outer edge of the turbulent zone is strongly dependent on the Reynolds number. The value of $u_\tau y/\nu$ that locates the outer edge increases as the Reynolds number increases, as shown. A **buffer zone**, with no specified velocity profile, connects the two self-similar zones.

The outer region relates the velocity defect to y/δ. In the turbulent zone the velocity-defect profile is sketched in Figure 6.26 and is

$$\frac{U_\infty - \bar{u}}{u_\tau} = -2.44 \ln \frac{y}{\delta} + 2.5 \qquad 50 < \frac{u_\tau y}{\nu} \qquad \frac{y}{\delta} < 0.15 \tag{6.6.38}$$

Between $y/\delta = 0.15$ and $y/\delta = 1$ researchers fit the data with several relations; the one selected here is

$$\frac{U_\infty - \bar{u}}{u_\tau} = -3.74 \ \ln \frac{y}{\delta} \qquad \frac{y}{\delta} > 0.15 \tag{6.6.39}$$

The equations above involve the shear velocity u_τ, which depends on the wall shear $\tau_0\,(x)$. The equation that is valid at the wall is Eq. 6.6.35, which, using $\tau_0 = \mu \partial u/\partial y\big|_{y=0}$, simply provides us with an identity. It does not allow us to calculate τ_0. Hence a relationship is necessary to provide us with τ_0 (or equally c_f). There are several such relationships used; one that gives excellent results is

$$c_f = \frac{0.455}{(\ln 0.06\,\mathrm{Re}_x)^2} \tag{6.6.40}$$

This local skin friction coefficient allows us to determine τ_0 and thus u_τ at any location of interest. The velocity profiles can be used to calculate quantities of interest, but u_τ must be known.

Assuming turbulent flow from the leading edge, the shear stress can be integrated to yield the drag. Then the skin friction coefficient becomes

$$C_f = \frac{0.523}{(\ln 0.06\,\mathrm{Re}_L)^2} \tag{6.6.41}$$

This relation is very good and can be used up to $\mathrm{Re}_L = 10^9$ with an error of 2% or less. Even at $\mathrm{Re}_L = 10^{10}$ the error is about 4%. To account for a laminar portion, the same term included in Eq. 6.6.30 can be subtracted from Eq. 6.6.41.

In concluding this section, a very useful relationship can be obtained by combining the two logarithmic profiles for the common turbulent zone. Substitute Eq. 6.6.37 into Eq. 6.6.38 and obtain

$$\frac{U_\infty}{u_\tau} = 2.44 \ln \frac{u_\tau \delta}{\nu} + 7.4 \tag{6.6.42}$$

This equation allows an easy calculation of δ knowing u_τ.

Example 6.15

Estimate the thickness δ_τ of the viscous wall layer and the boundary-layer thickness at the end of a 15-ft-long flat plate if $U_\infty = 100$ ft/s in 60°F-atmospheric air. Also, calculate the drag force on one side if the plate is 10 ft wide. Use the empirical data.

Solution

To find the viscous wall layer thickness we must know the shear velocity and hence the wall shear. The wall shear, using Eq. 6.6.40, and the shear velocity at $x = 15$ ft are

$$\tau_0 = \frac{1}{2}\rho U_\infty^2 c_f$$

$$= \frac{1}{2}\rho U_\infty^2 \frac{0.455}{(\ln 0.06\,\mathrm{Re}_x)^2}$$

$$= \frac{1}{2}\times 0.0024\ \frac{\text{slug}}{\text{ft}^3}\times 100^2\ \frac{\text{ft}^2}{\text{s}^2}\frac{0.455}{\left(\ln 0.06\dfrac{100\times 15}{1.6\times 10^{-4}}\right)^2} = 0.0311 \text{ psf}$$

$$u_\tau = \sqrt{\frac{\tau_0}{\rho}} = \sqrt{\frac{0.0311}{0.0024}} = 3.6 \text{ ft/s}$$

The viscous wall-layer thickness is determined using Eq. 6.6.36 with $y^* = 5$ as follows:

$$\frac{u_\tau \delta_\nu}{\nu} = 5$$

$$\delta_\nu = \frac{5\nu}{u_\tau} = \frac{5\times 1.6\times 10^{-4}}{3.6} = \underline{2.22\times 10^{-4}\text{ ft}}$$

The boundary-layer thickness is found using Eq. 6.6.42:

$$\frac{U_\infty}{u_\tau} = 2.44\ln\frac{u_\tau \delta}{\nu} + 7.4$$

$$\frac{100}{3.6} = 2.44\ln\frac{3.6\times\delta}{1.6\times 10^{-4}} + 7.4 \qquad \therefore \delta = \underline{0.188\text{ ft}}$$

The drag force is calculated using Eq. 6.6.41 to be

$$F_D = C_f \times \frac{1}{2}\rho U_\infty Lw$$

$$= \frac{0.523}{(\ln 0.06\,\mathrm{Re}_L)^2}\times\frac{1}{2}\rho U_\infty Lw$$

$$= \frac{0.523}{\left(\ln 0.06 \frac{100 \times 15}{1.6 \times 10^{-4}}\right)^2} \times \frac{1}{2} \times 0.0024 \times 100^2 (15 \times 10) = 5.4 \text{ lb}$$

The laminar portion of the boundary layer has been neglected.

Example 6.16

Estimate the maximum boundary-layer thickness and the drag due to friction on the side of a ship that measures 40 m long with a submerged depth of 8 m assuming the side of the ship is approximated as a flat plate. The ship travels at 10 m/s. (a) Use the empirical methods and (b) compare with the results using the power-law model.

Solution

(a) The boundary-layer thickness is found from Eq. 6.6.42. First we must find τ_0 from Eq. 6.6.40 and then u_τ as follows:

$$\tau_0 = \frac{1}{2} \rho U_\infty^2 \frac{0.455}{(\ln 0.06 \mathrm{Re}_L)^2}$$

$$= \frac{1}{2} \times 1000 \times 10^2 \frac{0.455}{\left(\ln 0.06 \frac{10 \times 40}{10^{-6}}\right)^2} = 78.8 \text{ Pa}$$

$$\therefore u_\tau = \sqrt{\frac{\tau_0}{\rho}}$$

$$= \sqrt{\frac{78.8}{1000}} = 0.28 \text{ m/s}$$

The maximum boundary-layer thickness is found using Eq. 6.6.42:

$$\frac{U_\infty}{u_\tau} = 2.44 \ln \frac{u_\tau \delta}{\nu} + 7.4$$

$$\frac{10}{0.28} = 2.44 \ln \frac{0.28 \delta}{10^{-6}} + 7.4 \qquad \therefore \delta = \underline{0.39 \text{ m}}$$

The drag is

$$F_D = C_f \times \frac{1}{2} \rho U_\infty^2 L w$$

$$= \frac{0.523}{\left(\ln 0.06 \frac{10 \times 40}{10^{-6}}\right)^2} \times \frac{1}{2} \times 1000 \times 10^2 \times 40 \times 8 = \underline{29\,000 \text{ N}}$$

(b) First, calculate the Reynolds number: $\mathrm{Re} = 10 \times 40/10^{-6} = 4 \times 10^8$. We select $n = 9$. Equation (6.6.25) becomes

$$\tau_0 = \frac{d}{dx}\int_0^\delta \rho U_\infty^2 \left(\frac{y}{\delta}\right)^{1/9}\left[1-\left(\frac{y}{\delta}\right)^{1/9}\right]dy$$

$$= \frac{9}{110}\rho U_\infty^2 \frac{d\delta}{dx}$$

Equating this to the τ_0 of Eq. 6.6.24, we find that

$$\delta^{1/4} d\delta = 0.281(\nu/U_\infty)^{1/4}\, dx$$

Assume $\delta = 0$ at $x = 0$ and integrate. This provides

$$\delta = 0.433x\ \mathrm{Re}_x^{-1/5}$$

$$= 0.433(40)\left(\frac{10\times 40}{10^{-6}}\right)^{-1/5} = \underline{0.33\ \mathrm{m}}$$

This value is 15% too low.

The drag force is found to be

$$F_D = 0.071\mathrm{Re}_L^{-1/5} \times \frac{1}{2}\rho U_\infty^2 Lw$$

$$= 0.071\left(\frac{10\times 40}{10^{-6}}\right)^{-1/5} \times \frac{1}{2} \times 1000 \times 10^2 \times 40 \times 8 = \underline{21\,600\ \mathrm{N}}$$

This value is 25% too low. Obviously, the power-law equations are insignificant error.

6.6.6 Laminar Boundary-Layer Equations

The solution presented in Section 6.6.3 for the laminar boundary layer was an approximate solution using a cubic polynomial to approximate the velocity profile. In this section we simplify the Navier–Stokes equations and present a more accurate solution for the laminar boundary layer on a flat plate with a zero pressure gradient.

The x-component Navier–Stokes equation for a steady, incompressible, plane flow is (see Eq. 3.3.14 and ignore the gravity term)

$$u\frac{\partial u}{\partial x} + v\frac{\partial u}{\partial y} = -\frac{1}{\rho}\frac{\partial p}{\partial x} + \nu\left(\frac{\partial^2 u}{\partial x^2} + \frac{\partial^2 u}{\partial y^2}\right) \qquad (6.6.43)$$

In boundary-layer theory the boundary layer is assumed to be very thin (see Figure 6.24), so there is no pressure variation in the y-direction in the boundary layer; that is, $p = p(x)$. In addition (this is a very important point), the pressure $p(x)$ is given by the inviscid flow solution as the wall pressure; hence the pressure is not an unknown. This leaves only two unknowns, u and v. Equation 6.6.43 provides us with one equation and the continuity equation

$$\frac{\partial u}{\partial x}+\frac{\partial v}{\partial y}=0 \tag{6.6.44}$$

provides us with the other. The y-component Navier–Stokes equation is not of use in boundary layer theory since all the terms are negligibly small ($v \ll u$ as inferred from Figure 6.24).

In addition to the simplification that a known pressure gradient provides, $\partial^2 u/\partial x^2$ is much, much less than the large gradients that exist in the y-direction (refer to the sketch of Figure 6.24); consequently, neglecting $\partial^2 u/\partial x^2$, the boundary-layer equation that must be solved is

$$u\frac{\partial u}{\partial x}+v\frac{\partial u}{\partial y}=-\frac{1}{\rho}\frac{dp}{dx}+\nu\frac{\partial^2 u}{\partial y^2} \tag{6.6.45}$$

where the pressure gradient dp/dx is assumed to be known from the inviscid flow solution. This is often referred to as the **Prandtl boundary-layer equation**, named after Ludwig Prandtl (1875–1953). Neither term on the left can be neglected; the y-component v may be small, but the velocity gradient $\partial u/\partial y$ is quite large; hence the product must be retained.

Let us now focus on flow over a flat plate with zero pressure gradient. In addition, let us introduce the stream function:

$$u=\frac{\partial \psi}{\partial y} \qquad v=-\frac{\partial \psi}{\partial x} \tag{6.6.46}$$

The boundary-layer equation in terms of the stream function, becomes

$$\frac{\partial \psi}{\partial y}\frac{\partial^2 \psi}{\partial x \partial y}-\frac{\partial \psi}{\partial x}\frac{\partial^2 \psi}{\partial y^2}=\nu\frac{\partial^3 \psi}{\partial y^3} \tag{6.6.47}$$

In this form the x and y dependence cannot be separated. If we transform this equation (such transformations are selected by trial and error and experience) by letting

$$\xi = x \qquad \eta = y\sqrt{\frac{U_\infty}{\nu x}} \tag{6.6.48}$$

there results

$$-\frac{1}{2\xi}\left(\frac{\partial \psi}{\partial \eta}\right)^2+\frac{\partial \psi}{\partial \eta}\frac{\partial^2 \psi}{\partial \xi \partial \eta}-\frac{\partial \psi}{\partial \xi}\frac{\partial^2 \psi}{\partial \eta^2}=\nu\frac{\partial^3 \psi}{\partial \eta^3}\sqrt{\frac{U_\infty}{\nu \xi}} \tag{6.6.49}$$

This equation may appear to be a more difficult equation to solve than Eq. 6.6.47, but by observing the position of ξ in this equation, we separate variables by letting

$$\psi(\xi,\eta) = \sqrt{U_\infty \nu \xi} F(\eta) \tag{6.6.50}$$

The velocity components can then be shown to be, using Eqs. 6.6.48 and 6.6.50,

$$u = \frac{\partial \psi}{\partial y} = U_\infty F'(\eta)$$

$$v = -\frac{\partial \psi}{\partial x} = \frac{1}{2}\sqrt{\frac{\nu U_\infty}{x}}(\eta F' - F) \tag{6.6.51}$$

Substitute Eq. 6.6.50 into Eq. 6.6.49 and an ordinary nonlinear, differential equation results; it is

$$F\frac{d^2F}{d\eta^2} + 2\frac{d^3F}{d\eta^3} = 0 \tag{6.6.52}$$

This equation replaces the partial differential equation (6.6.47). Now let us state the boundary conditions. The boundary conditions [$u(x, 0) = 0$, $v(x, 0) = 0$ and $u(x, y > \delta) = U_\infty$] take the form

$$F = F' = 0 \text{ at } \eta = 0 \qquad \text{and} \qquad F' = 1 \text{ at large } \eta \tag{6.6.53}$$

The boundary-value problem, consisting of the ordinary differential equation (6.6.52) and boundary conditions (6.6.53), can now be solved numerically. The results are tabulated in Table 6.5. The last two columns are used to provide v and τ_0, respectively.

Defining the boundary-layer thickness to be the point where $u = 0.99U_\infty$, we see from Table 6.5 that this occurs where $\eta = 5$. Hence, letting $\eta = 5$ and $y = \delta$ in Eq. 6.6.48, we have

$$\delta = 5\sqrt{\frac{\nu x}{U_\infty}} \tag{6.6.54}$$

Using

$$\frac{\partial u}{\partial y} = \frac{\partial u}{\partial \eta}\frac{\partial \eta}{\partial y} = U_\infty F''\sqrt{\frac{U_\infty}{\nu x}} \tag{6.6.55}$$

Table 6.5 Solution for the Laminar Boundary Layer with $dp/dx = 0$

$\eta = y\sqrt{\frac{U_\infty}{\nu x}}$	F	$F' = u/U_\infty$	$\frac{1}{2}(\eta F' - F)$	F''
0	0	0	0	0.3321
1	0.1656	0.3298	0.0821	0.3230
2	0.6500	0.6298	0.3005	0.2668
3	1.397	0.8461	0.5708	0.1614
4	2.306	0.9555	0.7581	0.0642
5	3.283	0.9916	0.8379	0.0159
6	4.280	0.9990	0.8572	0.0024
7	5.279	0.9999	0.8604	0.0002
8	6.279	1.0000	0.8605	0.0000

the wall shear in a laminar boundary layer with $dp/dx = 0$ is

$$\tau_0 = \mu \left.\frac{\partial u}{\partial y}\right|_{y=0} = 0.332 \rho U_\infty^2 \sqrt{\frac{\nu}{x U_\infty}} \tag{6.6.56}$$

The local skin friction coefficient is

$$c_f = \frac{0.664}{\sqrt{\mathrm{Re}_x}} \tag{6.6.57}$$

and the skin friction coefficient is

$$C_f = \frac{1.33}{\sqrt{\mathrm{Re}_L}} \tag{6.6.58}$$

Numerically integrating Eqs. 6.6.6 and 6.6.7, the displacement and momentum thicknesses are found to be

$$\delta_d = 1.72 \sqrt{\frac{\nu x}{U_\infty}} \qquad \theta = 0.644 \sqrt{\frac{\nu x}{U_\infty}} \tag{6.6.59}$$

Example 6.17

Atmospheric air at 30°C flows over a 8-m-long, 2-m-wide flat plate at 2 m/s. Assume that laminar flow exists in the boundary layer over the entire length. At $x = 8$ m, calculate (a) the maximum value of v, (b) the wall shear, and (c) the flow rate through the layer. (d) Also, calculate the drag force on the plate.

Solution

(a) The y-component of velocity has been assumed to be small in boundary-layer theory. Its maximum value at $x = 8$ m is found, using 6.6.51, to be

$$v = \sqrt{\frac{\nu U_\infty}{x}} \times \frac{1}{2}(\eta F' - F)$$

$$= \sqrt{\frac{1.6\times10^{-5}\times 2}{8}} \times 0.86 = \underline{0.00172 \text{ m/s}}$$

where 0.86 comes from Table 6.5. Compare v with $U_\infty = 2$ m/s.

(b) The wall shear at $x = 8$ m is found using Eq. 6.6.56 to be

$$\tau_0 = 0.332\rho U_\infty^2\sqrt{\frac{\nu}{U_\infty x}}$$

$$= 0.332\times1.16\times2^2\sqrt{\frac{1.6\times10^{-5}}{2\times8}}$$

$$= \underline{0.00154 \text{ Pa}}$$

(c) The flow rate through the boundary layer at $x = 8$ m is given by

$$Q = \int_0^\delta u \times w dy = w\sqrt{\frac{\nu x}{U_\infty}}\int_0^\delta U_\infty F' d\eta$$

where we have substituted for u and y from Eqs. 6.6.51 and 6.6.48. Recognizing that $\int F' d\eta = F$, the flow rate is

$$Q = wU_\infty\sqrt{\frac{\nu x}{U_\infty}}[F(5) - \overset{0}{F(0)}]$$

$$= 2\times2\sqrt{\frac{1.6\times10^{-5}\times8}{2}}\times3.28 = \underline{0.105 \text{ m}^3\text{/s}}$$

(d) The drag force is determined to be

$$F_D = \frac{1}{2}\rho U_\infty^2 LwC_f$$

$$= \frac{1}{2} \times 1.16 \times 2^2 \times 8 \times 2 \times \frac{1.33}{\sqrt{2 \times 8/1.6 \times 10^{-5}}}$$

$$= \underline{0.049 \text{ N}}$$

6.6.7 Pressure-Gradient Effects

In the foregoing sections we have concentrated our study of boundary layers on the flat plate with a zero pressure gradient. This is the simplest boundary-layer flow and allows us to model many flows of engineering interest. The inclusion of a pressure gradient, even though relatively low, markedly alters the boundary-layer flow. In fact, a strong negative pressure gradient (such as flow in a contraction) may relaminarize a turbulent boundary layer; that is, the turbulence production in the viscous wall layer which sustains the turbulence ceases and a laminar boundary layer is reestablished. A positive pressure gradient quickly causes the boundary layer to thicken and eventually to separate. These two effects are shown in the photographs of Figure 6.27.

The flow about any plane body with curvature, such as an airfoil, can be modeled as flow over a flat plate with a nonzero pressure gradient. The boundary-layer thickness is so much smaller than the radius of curvature that the additional curvature terms drop out of the differential equations. The inviscid flow solution at the wall provides the pressure gradient dp/dx and the velocity $U(x)$ at the edge of the boundary-layer. For axisymmetric flows, such as flow over the nose of an aircraft, boundary layer equations in cylindrical coordinates must be utilized.

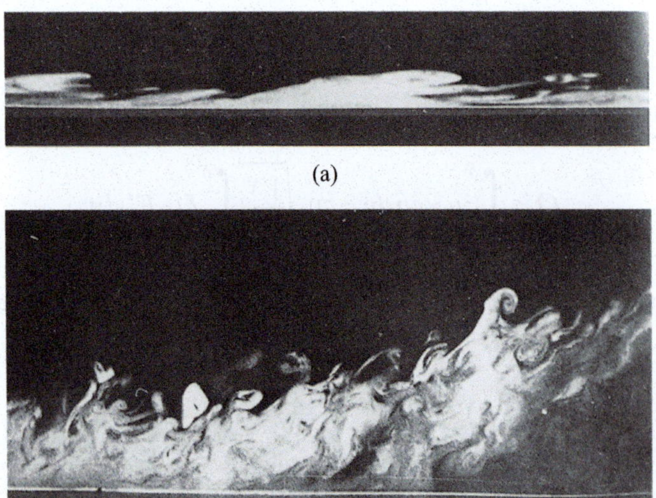

Figure 6.27 Influence of a strong pressure gradient on a turbulent flow: (a) a strong negative pressure gradient may relaminarize a flow; (b) a strong positive pressure gradient causes a boundary layer to thicken. (Photographs by R. E. Falco)

The pressure gradient determines the value of the second derivative $\partial^2 u/\partial y^2$ at the wall. From the boundary-layer equation (6.6.45) at the wall, $u = v = 0$, so that

$$\frac{dp}{dx} = \mu \left.\frac{\partial^2 u}{\partial y^2}\right|_{y=0} \tag{6.6.60}$$

for either a laminar or a turbulent boundary-layer flow. For a zero pressure gradient, the second derivative is zero at the wall; then since the first derivative is greatest at the wall and decreases as y increases, the second derivative must be negative for positive y. The profiles are sketched in Figure 6.28a.

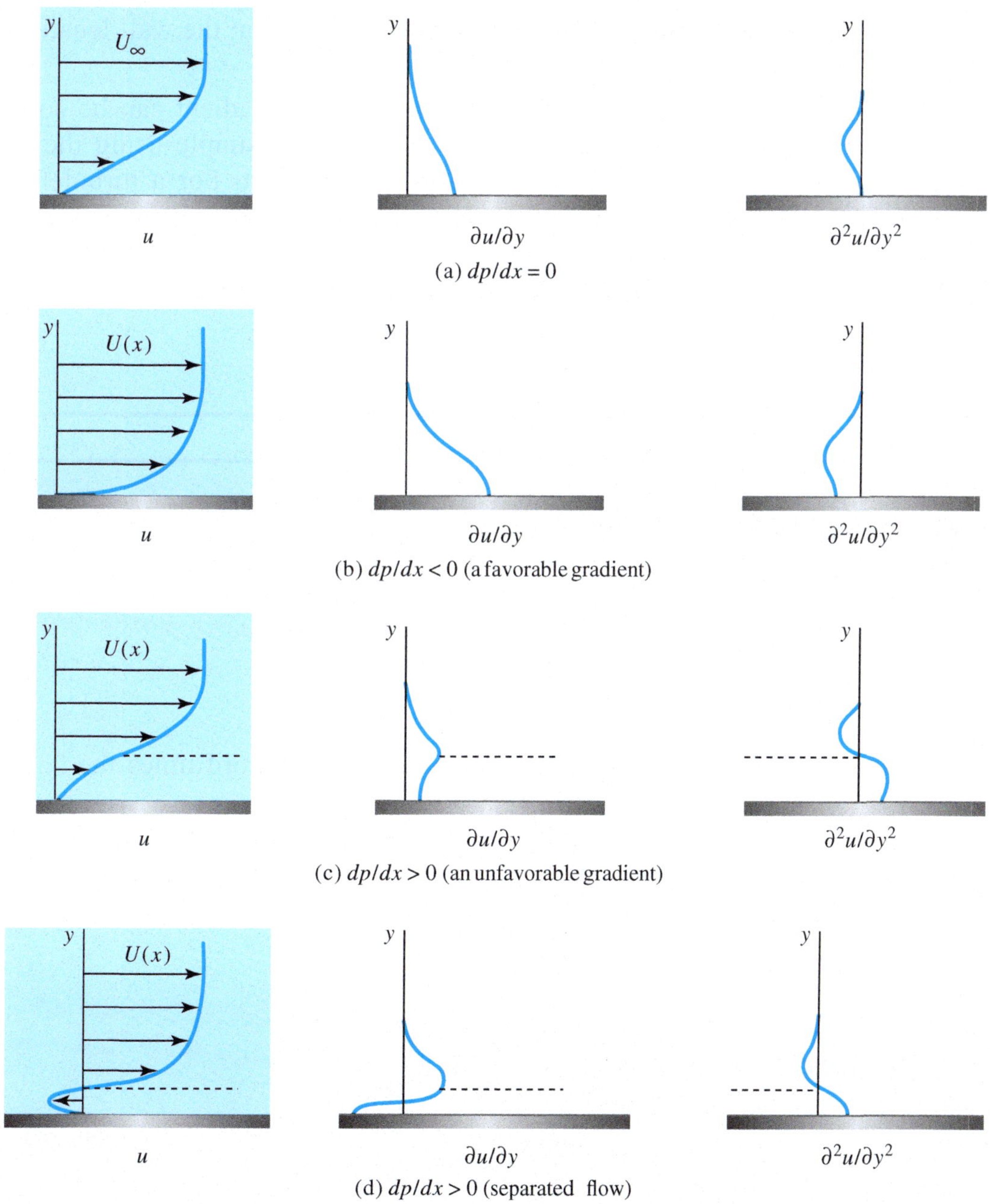

Figure 6.28 Influence of the pressure gradient.

For a negative (favorable) pressure gradient, the slope of the velocity profile near the wall is relatively large with a negative second derivative at the wall and throughout the layer. The momentum near the wall is larger than that of the zero pressure gradient flow, as shown in Figure 6.28b, and thus there is a reduced tendency for the flow to separate. Turbulence production is discouraged, and relaminarization may occur for a sufficiently large negative pressure gradient over a sufficient distance.

If a positive (unfavorable) pressure gradient is imposed on the flow, the second derivative at the wall will be positive and the flow will be as sketched in part (c) or (d). If the unfavorable pressure gradient acts over a sufficient distance, part (d) will probably represent the flow situation with the flow separated from the surface. Near the wall the higher pressure downstream will drive the low momentum flow near the wall in the upstream direction, resulting in flow reversal, as shown. The point at which $\partial u / \partial y = 0$ at the wall locates the point of separation.

The problem of a laminar boundary layer with a pressure gradient can be solved using conventional numerical techniques. The procedure is relatively simple using the simplified boundary-layer equation (6.6.45) with a known pressure gradient. For a turbulent flow the Reynolds stress term must be included; research continues to develop models of turbulent quantities that will result in acceptable numerical solutions. Experimental results are often necessary for turbulent-flow problems, as was the situation for internal flows.

6.7 SUMMARY

The drag and lift coefficients are defined as

$$C_D = \frac{\text{Drag}}{\frac{1}{2}\rho V^2 A} \qquad C_L = \frac{\text{Lift}}{\frac{1}{2}\rho V^2 A} \tag{6.7.1}$$

where the area is the projected area for blunt objects, and the chord times the length for an airfoil.

Vortex shedding occurs from a cylinder whenever the Reynolds number is in the range $300 < \text{Re} < 10\,000$. The frequency of shedding is found from the Strouhal number

$$\text{St} = \frac{fD}{V} \tag{6.7.2}$$

where f is the frequency, in hertz (cycles/s).

Plane potential flows are constructed by superimposing the following simple flows:

$$\begin{aligned}
&\text{Uniform flow:} && \psi = U_\infty y && \phi = U_\infty x \\
&\text{Line source:} && \psi = \frac{q}{2\pi\theta} && \phi = \frac{q}{2\pi}\ln r \\
&\text{Irrotational vortex:} && \psi = \frac{\Gamma}{2\pi}\ln r && \phi = \frac{\Gamma}{2\pi}\theta \\
&\text{Double:} && \psi = -\frac{\mu}{r}\sin\theta && \phi = -\frac{\mu}{r}\cos\theta
\end{aligned} \tag{6.7.3}$$

The stream function for the rotating cylinder is given by

$$\psi_{\text{cylinder}} = U_\infty y - \frac{\mu}{r}\sin\theta + \frac{\Gamma}{2\pi}\ln r \tag{6.7.4}$$

where the cylinder radius is

$$r_c = \sqrt{\frac{\mu}{U_\infty}} \tag{6.7.5}$$

The velocity components are

$$u = \frac{\partial\psi}{\partial y} \qquad v = -\frac{\partial\psi}{\partial x}$$
$$v_r = \frac{1}{r}\frac{\partial\psi}{\partial\theta} \qquad v_\theta = -\frac{\partial\psi}{\partial r} \tag{6.7.6}$$

For a laminar boundary layer on a flat plate with zero pressure gradient the exact solution provides.

$$\delta = 5\sqrt{\frac{\nu x}{U_\infty}} \qquad c_f = 0.664\sqrt{\frac{\nu}{xU_\infty}} \qquad C_f = 1.33\sqrt{\frac{\nu}{LU_\infty}} \tag{6.7.7}$$

For a turbulent flow from the leading edge, the power-law profile with $\eta = 7$ gives

$$\delta = 0.38x\left(\frac{\nu}{xU_\infty}\right)^{1/5} \qquad c_f = 0.059\left(\frac{xU_\infty}{\nu}\right)^{1/5} \qquad C_f = 0.073\left(\frac{\nu}{LU_\infty}\right)^{1/5} \tag{6.7.8}$$

where the wall shear and drag force per unit width are, respectively,

$$\tau_0 = \frac{1}{2}c_f\rho U_\infty^2 \qquad F_D = \frac{1}{2}C_f\rho U_\infty^2 L \tag{6.7.9}$$

PROBLEMS

Separated Flows

6.1 Sketch the flow over an airfoil at a large angle of attack for both attached flow and separated flow. Also, sketch the expected pressure distributions on the top and bottom surfaces for both flows. Identify regions of favorable and unfavorable pressure gradients.

6.2 A spherical particle is moving in 20°C atmospheric air at a velocity of 20 m/s. What must its diameter be for Re = 5 and Re = 10^5? Sketch the expected flow field at these Reynolds numbers. Identify all regions of the flow.

6.3 Sketch the flow that is expected over a semitruck (a tractor and trailer) where the trailer is substantially higher than the tractor with and without an air deflector attached to the roof of the tractor. Sketch a side view indicating any separated regions, boundary layers, and the wake.

6.4 Air is blowing by a long, rectangular building with the wind blowing parallel to the long sides. Sketch the top view showing the separated flow regions, the inviscid flow region, the boundary layers, and the wake.

6.5 A 0.8-in.-diameter sphere is to move with Re = 5. At what velocity does it travel if it is submerged in:
(a) Water at 60°F?
(b) Water at 180°F?
(c) Standard air at 60°F?

6.6 Air at 20°C flows around a cylindrical body at a velocity of 20 m/s. Calculate the Reynolds number if the body is:
(a) A 6-m-diameter smokestack
(b) A 6-cm-diameter flagpole
(c) A 6-mm-diameter wire
Use Re = VD/ν. Would a separated flow be expected?

6.7 The pressure distribution on the front of a 2-m-diameter disk (Figure P6.7) is approximated by $p(r) = p_0(1 - r^2)$. If $V = 20$ m/s in this 20°C-atmospheric airflow, estimate the drag force and the drag coefficient for this disk. Assume that the pressure on the back side is zero.

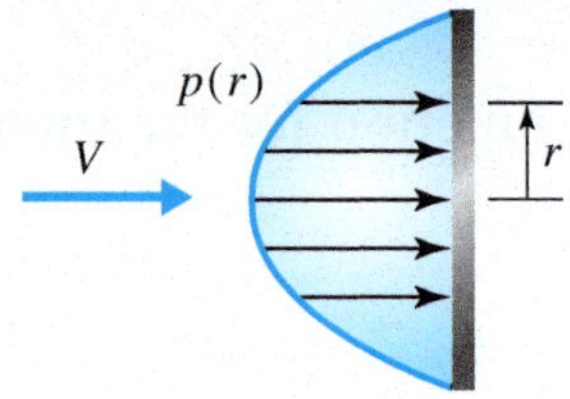

Figure P6.7

6.8 A flat plate 30 cm × 30 cm acts as a hydrofoil. If it is oriented at an angle of attack of 10°, estimate the lift and the drag if the pressure on the bottom side is 20 kPa and on the top side there is a vacuum of 10 kPa; neglect the effect of shear stress. Also, estimate the lift and drag coefficients if the hydrofoil velocity is 5 m/s. Use the surface area of the plate in the definition of the lift and drag coefficients.

6.9 The symmetric airfoil shown in Figure P6.9 flies at an altitude of 12 000 m with an angle of attack of 5°. If $p_l = 26$ kPa and $p_u = 8$ kPa, estimate the lift and drag coefficients neglecting shear stresses.

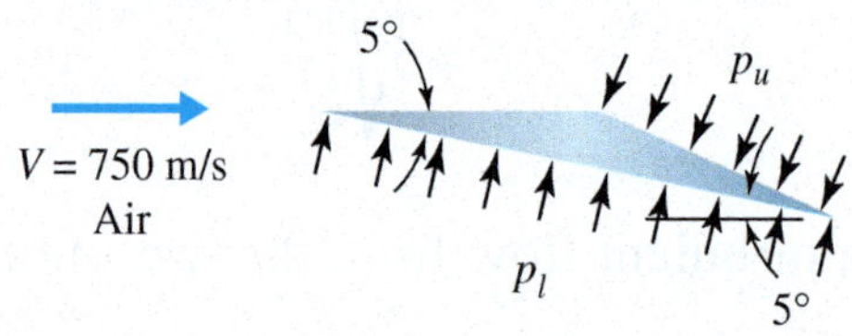

Figure P6.9

6.10 If the drag coefficient for a 10-cm-diameter sphere is given by $C_D = 1.0$, calculate the drag if the sphere is falling in the atmosphere:
(a) At sea level
(b) At 30 000 m
(c) In 10°C water

6.11 Calculate the drag on a smooth 50-cm-diameter sphere when subjected to a 20°C-atmospheric airflow of:
(a) 6 m/s
(b) 15 m/s

6.12 A 4.45-cm-diameter golf ball is roughened to reduce the drag during its flight. If the Reynolds number at which the sudden drop occurs is reduced from 3×10^5 to 6×10^4 by the roughening (dimples), would you expect this to lengthen the flight of a golf ball significantly? Justify your reasoning with appropriate calculations.

6.13 A 4-in.-diameter smooth sphere experiences a drag of 0.5 lb when placed in standard 60°F air.
(a) What is the velocity of the airstream?
(b) At what increased speed will the sphere experience the same drag?

6.14 A smooth 20-cm-diameter sphere experiences a drag of 4.2 N when placed in a 20°C water channel. Calculate the drag coefficient and the Reynolds number.

6.15 A 2-m-diameter smokestack stands 60 m high. It is designed to resist a 40-m/s wind. At this speed, what total force would be expected, and what moment would the base be required to resist? Assume atmospheric-20°C air.

6.16 A flagpole is composed of three sections: a 5-cm-diameter top section that is 10 m long, a 7.5-cm-diameter middle section 15 m long, and a 10-cm-diameter bottom section 20 m long.
Calculate the total force acting on the flagpole and the resisting moment provided by the base when subjected to a 25 m/s wind speed. Make the calculations for:
(a) A winter day at –30°C
(b) A summer day at 35°C

6.17 A drag force of 10 lb is desired at $Re = 10^5$ on a 6-ft-long cylinder in a 60°F-atmospheric airflow. What velocity should be selected, and what should be the cylinder diameter?

6.18 A 20-m-high structure is 2 m in diameter at the top and 8 m in diameter at the bottom as shown in Figure P6.18. If the diameter varies linearly with height, estimate the total drag force due to a 30-m/s wind. Use atmospheric air at 20°C.

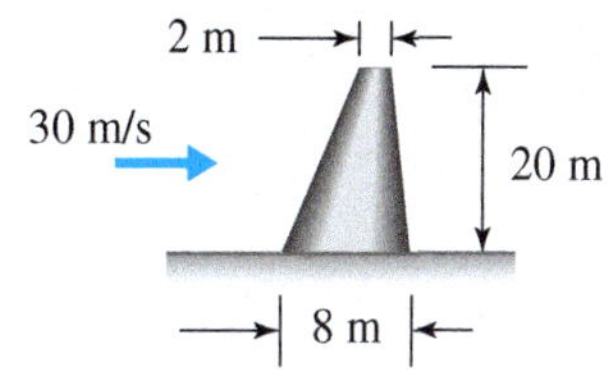

Figure P6.18

6.19 A steel sphere (S = 7.82) is dropped in water at 20°C. Calculate the terminal velocity if the diameter of the sphere is:
(a) 10 cm **(c)** 1 cm
(b) 5 cm **(d)** 2 mm

6.20 Estimate the terminal velocity of a 20-in.-diameter sphere as it falls in a 60°F atmosphere near the Earth if it has a specific gravity of:
(a) 0.005 **(b)** 0.02 **(c)** 1.0

6.21 Estimate the terminal velocity of a skydiver by making reasonable approximations of the arms, legs, head, and body. Assume air at 20°C.

6.22 Assuming that the drag on a modem automobile at high speeds is due primarily to form drag, estimate the power (horsepower) needed by an automobile with a 3.2-m^2 cross-sectional area to travel at:
(a) 80 km/h **(b)** 90 km/h
(c) 100 km/h

6.23 The 2 m × 3 m sign shown in Figure P6.23 weighs 400 N. What wind speed is required to blow over the sign?

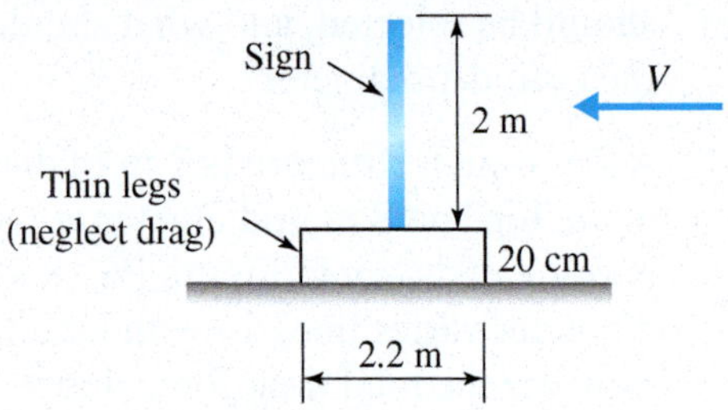

Figure P6.23

6.24 Calculate the drag force on a 60-cm-diameter, 6-m-long cylinder if 20°C-air is blowing normal to its axis at 40 km/h and the cylinder:

(a) Is a section of a very long cylinder
(b) Has both ends free
(c) Is fixed to the ground with the top free

6.25 An 80-kg paratrooper leaps from an elevation of 3000 m. Estimate the paratrooper's landing speed if she:

(a) Curls up as tightly as possible
(b) Uses an 3-m-diameter lightweight parachute
(c) Uses a 2-m-diameter safety parachute

6.26 A semitruck travels 200 000 km/h each year at an average speed of 90 km/ h. Estimate the savings if a streamlined deflector is added to reduce the drag coefficient. Fuel costs $0.40 per liter and the truck without the deflector averages 1.2 km per liter of fuel.

6.27 A rectangular car carrier has a cross section 6 ft × 2 ft. Estimate the minimum added power (horsepower) required to travel at 60 mph because of the car carrier.

6.28 Assume that the velocity at the corners of an automobile where the rear-view mirrors are located is 1.6 times the automobile's velocity. How much horse-power is required by the two 10-cm-diameter rear-view mirrors for an automobile speed of 100 km/h?

6.29 Atmospheric air at 25°C blows normal to a 4-m-long section of a cone, 30 cm in diameter at one end and 2 m in diameter at the other end, with a velocity of 20 m/s. Predict the drag on the object. Assume that $C_D = 0.4$ for each cylindrical element of the object.

6.30 An 80-cm-diameter balloon (Figure P6.30) that weighs 0.5 N is filled with 20°C helium at 20 kPa. Neglecting the weight of the string, estimate V if α equals:

(a) 80° **(c)** 60°
(b) 70° **(d)** 50°

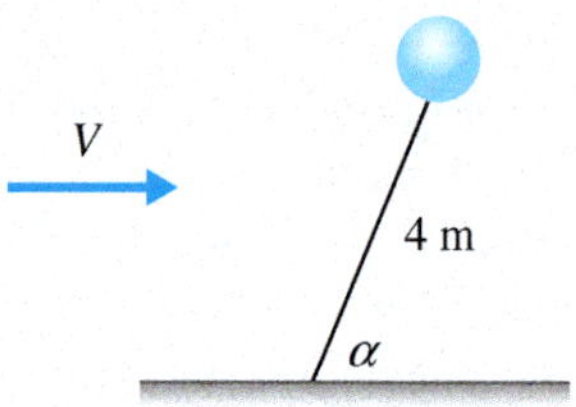

Figure P 6.30

6.31 For the newly planted tree shown in Figure P6.31, the soil-ball interface is capable of resisting a moment of 5000 N. m. Predict the minimum wind speed that might possibly tip the tree over. Assume that $C_D = 0.4$ for a cylinder in this air flow.

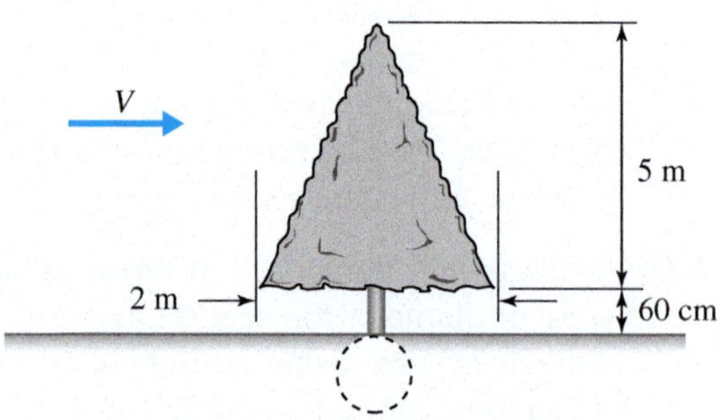

Figure P6.31

6.32 A sign 1.2 m × 0.6 m is attached to the top of a pizza delivery car. The car travels 10 hours a day, 6 days a week. Estimate the added cost in a year that the sign adds to the fuel used by the car. The average speed of the car is 40 km/h, fuel costs $0.60 per liter, the engine/drive train is 30% efficient, and the fuel contains 12 000 kJ/kg.

6.33 A bike rider is able to travel at an average speed of 25 mph while riding upright. It is determined that the rider's projected area is 0.56 m^2. If the rider assumes a racing position so that his projected area is 0.40 m^2, estimate his increased speed if his drag coefficient decreases 20%, assuming the same energy expenditure.

6.34 An automobile with a cross-sectional area of 3 m^2 is powered by a 40-hp engine. Estimate the maximum possible speed if the drive train is 90% efficient. (The engine is rated by the power produced before the transmission).

Vortex Shedding

6.35 Over what range of velocities would you expect vortex shedding from a telephone wire 3 mm in diameter? Could you hear any of the vortices being shed? (Human beings can hear frequencies between 20 and 20 000 Hz).

6.36 A wire is being towed through 60°F water normal to its axis at a speed of 6 ft/s. What diameter (both large and small) could the wire have so that vortex shedding would not occur?

6.37 It is quite difficult to measure low velocities. To determine the velocity in a low-speed airflow, the vorticies being shed from a 10-cm-diameter cylinder are observed to occur at 0.2 Hz. Estimate the airspeed if the temperature, is 20°C.

6.38 Motion pictures show that vortices are shed from a 2-m-diameter cylinder at 0.002 Hz while it is moving in 20°C water. What is the cylinder velocity?

6.39 The cables supporting a suspended bridge (Figure P6.39) have a natural frequency of $T/(\pi\rho L^2 d^2)$ hertz, where T is the tension, ρ the cable density, d its diameter, and L its length. The vortices shed from the cables may lead to resonance and possible failure. A certain 1.6-cm-diameter steel cable is subjected to a force of 30 000 N. What length cable would result in resonance in a 10-m/s wind? (**Note:** The third and fifth harmonics may also lead to resonance. Calculate all three lengths.)

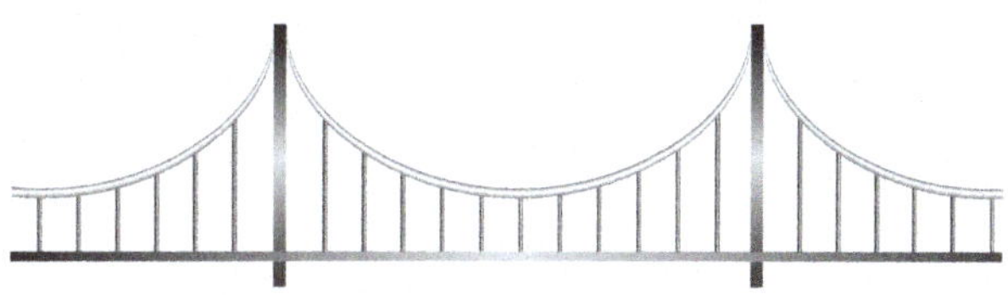

Figure P6.39

Streamlining

6.40 A 6-in.-diameter smokestack on a semitruck extends 6 ft straight up into the free stream. Estimate the horsepower needed because of the stack for a speed of 60 mp. If the stack were streamlined, estimate the reduced horsepower.

6.41 A wind speed of 3 m/s blows normal to an 8-cm-diameter smooth cylinder that is 2 m long. Calculate the drag force. The cylinder is now streamlined. What is the percentage reduction in the drag? Assume that $T = 20°C$.

6.42 Water flows by an 80-cm-diameter cylinder that protrudes 2 m up from the bottom of a river. For an average water speed of 2 m/s, estimate the drag on the cylinder. If the cylinder were streamlined, what would be the percentage reduction in drag?

6.43 Circular 2-cm-diameter tubes are used as supports on an ultra light airplane, designed to fly at 50 km/h. If there are 20 linear meters of the tubes, estimate the horsepower needed by the tubes. If the tubes were streamlined, estimate the reduced horsepower required of the tubes.

6.44 A bike racer is able to ride 50 km/h at top speed. Estimate the drag force just due to his head. If he wore a streamlined, tight-fitting helmet, estimate the reduced drag force.

Cavitation

6.45 The critical cavitation number for a streamlined strut is 0.7. Find the maximum velocity of the body to which the strut is attached if cavitation is to be avoided. The body is traveling 5 m beneath a water surface.

6.46 A lift force of 200 kN is desired at a speed of 12 m/s on the hydrofoil of Figure P6.46, designed to operate at a depth of 40 cm. It has a 40-cm chord and is 10 m long. Calculate the angle of attack and the drag force. Is cavitation present under these conditions?

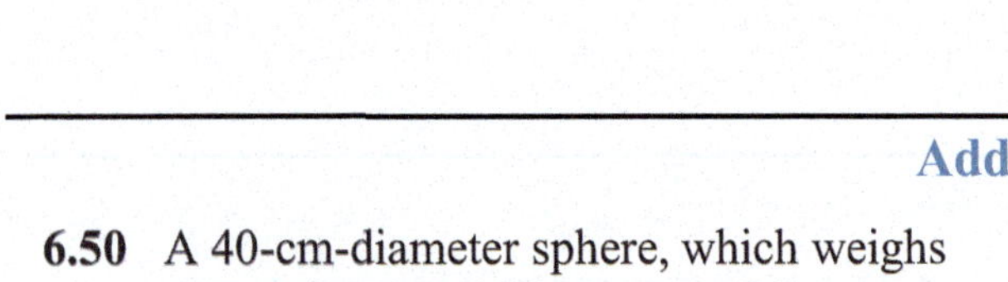

Figure P6.46

6.47 A hydrofoil, designed to operate at a depth of 16 in., has a 16-in. chord and is 30 ft long. A lift force of 50,000 lb is desired at a speed of 35 ft/s. Calculate the angle of attack and the drag force. Is cavitation present under these conditions?

6.48 A body resembling a sphere has a diameter of approximately 0.8 m. It is towed at a speed of 20 m/s, 5 m below the water surface. Estimate the drag acting on the body.

6.49 A 2200-kg mine sweeper is designed to ride above the water with hydrofoils at the four corners providing the lift. If the hydrofoils have a 40-cm chord length, what total hydrofoil length is required if they are to operate 60 cm below the surface at a 6° angle of attack? The vehicle is to travel at 50 m/s.

Added Mass

6.50 A 40-cm-diameter sphere, which weighs 400 N, is released from rest while submerged in water. Calculate its initial acceleration:

(a) Neglecting the added mass

(b) Including the added mass

6.51 A submersible, whose length is twice its maximum diameter, resembles an ellipsoid. If its added mass is ignored, what is the percentage error in a calculation of its initial acceleration if its specific gravity is 1.2?

Lift and Drag on Airfoils

6.52 A rectangular wing on a small airplane has a 1.3 m chord and a 10 m span. When flying in air at 220 km/h, the wing experiences a total aerodynamic force of 18 kN. If the lift to drag ratio is 3, determine the lift coefficient of the wing.

6.53 In a lift-and-drag experiment performed in a wind tunnel, an NACA0012 airfoil was used. The airfoil has a 15 cm chord and a span length of 45 cm. Using a force measuring device, a lift force of 60 N was measured on the airfoil for a Reynolds number of 4.586×10^5. The lift coefficient for this airfoil is given by $C_L = 2\pi \sin \alpha$, where α is the angle of attack. Under the given conditions, what is the angle of attack in degrees?

6.54 An aircraft with a mass, including payload, of 1000 kg is designed to cruise at a speed of 80 m/s at 10 km. The effective wing area is approximately 15 m^2. Determine the lift coefficient and the angle of attack. What power is required by the airfoil during cruise? Assume a conventional airfoil.

6.55 A 1500 kg aircraft is designed to carry a pay-load of 3000 N when cruising at 80 m/s at a height 10 km. The effective wing area is 20 m^2. Assuming a conventional airfoil, calculate:

(a) The takeoff speed if an angle of attack of 10° is desired
(b) The stall speed when landing
(c) The power required at cruise if 45% of the power is needed to move the airfoil

6.56 In Problem.6.55 it was necessary to assume that takeoff was at sea level. What would be the takeoff speed in Wyoming, where the elevation is 2000 m?

6.57 The aircraft of Problem 6.55 is flown at 2 km rather than 10 km. Estimate the percentage increase or decrease in power required at cruise.

6.58 An additional load of 6000 N is added to the aircraft of Problem 6.55. Estimate the takeoff speed if the angle of attack remains at 10°.

6.59 Estimate the minimum landing speed for a 250 000-kg aircraft in an emergency situation if the angle of attack is selected to be near stall and:

(a) No slotted flaps are used
(b) One slotted flap is used
(c) Two slotted flaps are used

Its wingspan is 60 m and the airfoil chord averages 8 m.

6.60 In Problem 6.59 it is assumed that the aircraft lands at standard sea-level conditions since neither elevation nor temperature is given. Calculate the percentage increase or decrease in the emergency landing speed if the aircraft is to land:

(a) In Denver where the elevation is 1600 m
(b) At sea level when the temperature is very cold at −40°C
(c) At sea level when the temperature is hot at 50°C

6.61 A proposed aircraft is to resemble a huge airfoil, a flying wing (Figure P6.61). Its wingspan will be 200 m and its chord will average 30 m. Estimate, assuming a conventional airfoil, the total mass of the aircraft, including payload, for a design speed of 800 km/h at an elevation of 8 km. Also, calculate the power requirement.

Figure P6.61

Vortices, Velocity Potential, and Stream Function

6.62 Take the curl of the Navier–Stokes equation and show that the vorticity equation (6.5.3) results. (See Eq. 3.3.20)

6.63 Write the three component vorticity equations contained in Eq. 6.5.3 using rectangular coordinates. Use $\boldsymbol{\omega} = \omega_x \hat{\mathbf{i}} + \omega_y \hat{\mathbf{j}} + \omega_z \hat{\mathbf{k}}$.

6.64 Simplify the vorticity equation 6.5.3 for a plane flow ($w = 0$ and $\partial/\partial z = 0$). Use $\omega = (\omega_x, \omega_y, \omega_z)$. What conclusion can be made about the magnitude of ω_z in an inviscid, plane flow (such as flow through a short contraction) that contains vorticity?

6.65 Consider the flow in the contracting channel shown in Figure P6.65. It is assumed that a vortex tube (y-vorticity) exists in the boundary layer on the upper and lower plates. Using the vorticity transport equation, explain the existence of x-vorticity downstream of the obstruction.

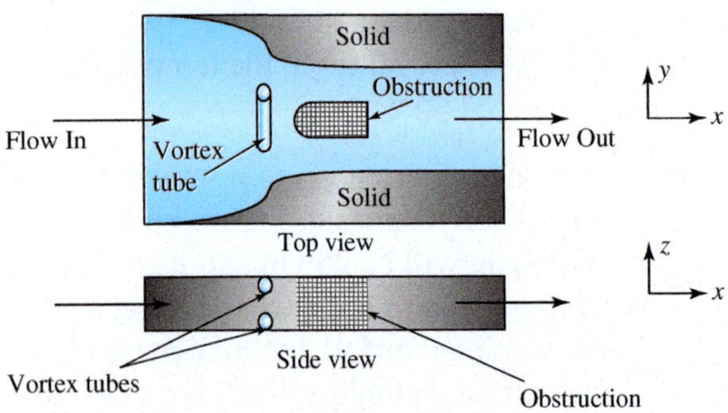

Figure P6.65

6.66 Determine which of the following flows are irrotational and incompressible and find the velocity potential function, should one exist for each incompressible flow.

(a) $V = 10x\hat{\mathbf{i}} + 20y\hat{\mathbf{j}}$

(b) $V = 8y\hat{\mathbf{i}} + 8x\hat{\mathbf{j}} - 6z\hat{\mathbf{k}}$

(c) $V = (x\hat{\mathbf{i}} + y\hat{\mathbf{j}})/\sqrt{x^2 + y^2}$

(d) $V = (x\hat{\mathbf{i}} + y\hat{\mathbf{j}})/(x^2 + y^2)$

6.67 An attempt is to be made to solve Laplace's equation for flow around a circular cylinder of radius r_c oriented in the center of a channel of height $2h$. The velocity profile far from the cylinder is uniform. State the necessary boundary conditions. Assume that $\psi = 0$ at $y = -h$. The origin of the coordinate system is located at the center of the cylinder.

6.68 State the stream function and velocity potential corresponding to a uniform velocity of $100\hat{\mathbf{i}} + 50\hat{\mathbf{j}}$ using rectangular coordinates.

6.69 A flow is represented by the stream function $\psi = 40 \tan^{-1}(y/x)$.

(a) Express the stream function in polar form

(b) Is this an incompressible flow? Show why

(c) Determine the velocity potential

(d) Find the radius where the acceleration is -10 m/s^2

6.70 The stream function for a flow is

$$\psi = 20 \ln(x^2 + y^2) \text{ m}^2/\text{s}$$

Determine the complex velocity potential for this incompressible flow. If the pressure at a large distance from the origin is 20 kPa, what is the pressure at the point (0, 20 cm) if water is flowing?

6.71 A stream function is given by

$$\psi = 10y - \frac{10y}{x^2 + y^2}$$

(a) Show that this satisfies $\nabla^2\psi = 0$

(b) Find the velocity potential $\phi = (x, y)$

(c) Assuming water to be flowing, find the pressure along the x-axis if $p = 50$ kPa at $x = -\infty$

(d) Locate any stagnation points

6.72 The velocity potential for a flow is

$$\phi = 10x + 5 \ln(x^2 + y^2)$$

(a) Show that this function satisfies Laplace's equation
(b) Find the stream function $\psi(x, y)$
(c) Assume that water is flowing and find the pressure along the x-axis if $p = 100$ kPa at $x = -\infty$
(d) Locate any stagnation points
(e) Find the acceleration at $x = -2$ m, $y = 0$

6.73 The velocity profile in a wide 0.2-m-high channel is given by $u(y) = y - y^2/0.2$. Determine the stream function for this flow. Calculate the flow rate by integrating the velocity profile and by using $\Delta\psi$. Explain why a velocity potential does not exist by referring to Eq. 6.5.2.

Superimposition of Simple Flows

6.74 The body formed by superimposing a source at the origin of strength 5π ft^2/s and a uniform flow of 30 ft/s is shown in figure P6.74.
(a) Locate any stagnation points
(b) Find the y-intercept y_B of the body
(c) Find the thickness of the body at $x = \infty$
(d) Find u at $x = -12$ in., $y = 0$

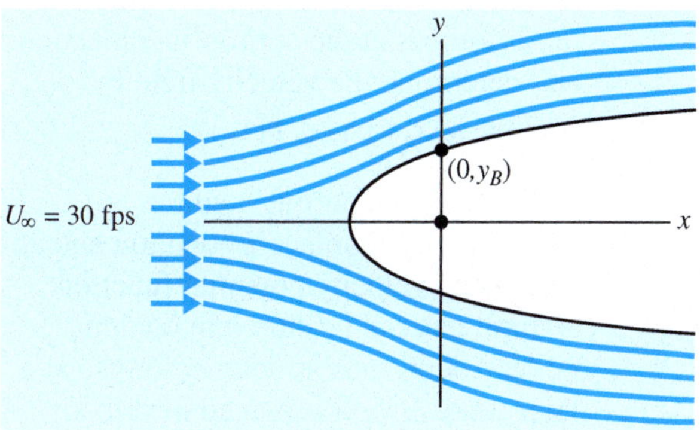

Figure P6.74

6.75 A source with strength π m^2/s and a sink of equal strength are located at (–1 m, 0) and (1 m, 0), respectively. They are combined with a uniform flow $U_\infty = 10$ m/s to form a **Rankine oval**. Calculate the length and maximum thickness of the oval. If $p = 10$ kPa at $x = -\infty$, find the minimum pressure if water is flowing.

6.76 An oval is formed from a source and sink of strengths 2π m^2/s located at (–1, 0) and (1, 0), respectively, combined with a uniform flow of 2 m/s. Locate any stagnation points, and find the velocity at (–4, 0) and (0, 4). Distances are in meters.

6.77 Two sources of strength 2π m^2/s are located at (0, 1) and (0, –1), respectively. Sketch the resulting flow and locate any stagnation points. Find the velocity at (1, 1). Distances are in meters.

6.78 The two sources of Problem 6.77 are superimposed with a uniform flow. Sketch the flow, locate any stagnation points, and find the y-intercept of the body formed if:
(a) $U_\infty = 10$ m/s
(b) $U_\infty = 1$ m/s
(c) $U_\infty = 0.2$ m/s

6.79 A doublet of strength 60 m^3/s superimposed with a uniform flow of 8 m/s of water. Calculate:
(a) The radius of the resulting cylinder
(b) The pressure increase from $x = -\infty$ to the stagnation point
(c) The velocity $v_\theta(\theta)$ on the cylinder
(d) The pressure decrease from the stagnation point to the point of minimum pressure on the cylinder

6.80 A sink of strength 4π m^2/s is superimposed with a vortex of strength 20π m^2/s.
(a) Sketch a pathline of a particle initially occupying the point ($x = 0$, $y = 1$ m). Use tangents at every 45° with straightline extensions
(b) Calculate the acceleration at (0, 1)
(c) If $p(10,10) = 20$ kPa, what is $p(0,0.1$ m$)$ if atmospheric air is flowing?

6.81 The cylinder shown in Figure 6.81 is formed by combining a doublet of strength 40 m^3/s with a uniform flow of 10 m/s

(a) Sketch the velocity along the y-axis from the cylinder to $y = \infty$

(b) Calculate the velocity at ($x = -4$ m, $y = 3$ m)

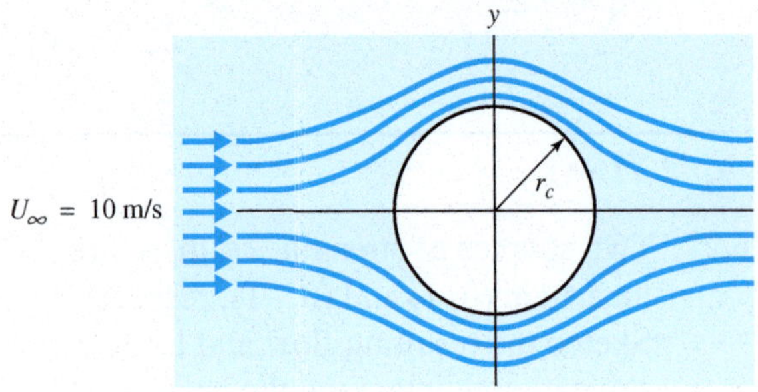

Figure P6.81

(c) Calculate the drag coefficient for the cylinder assuming potential flow over the front half and constant pressure over the back half

6.82 A 2-m diameter cylinder is placed in a uniform water flow of 4 m/s .

(a) Sketch the velocity along the x-axis from the cylinder to $x = -\infty$

(b) Find v_θ on the front half of the cylinder

(c) Find $p(\theta)$ on the front half of the cylinder if $p = 50$ kPa at $x = -\infty$

(d) Estimate the drag force on a 1-m length of the cylinder if the pressure over the rear half is constant and equal to the value at $\theta = 90°$

6.83 Superimpose a free stream $U_\infty = 30$ ft/s, a doublet $\mu = 400$ ft^2/s, and a vortex $\Gamma = 1000$ ft^2/s. Locate any stagnation points and predict the minimum and maximum pressure on the surface of the cylinder if $p = 0$ at $x = -\infty$ and atmospheric air is flowing.

6.84 A 0.8 m-diameter cylinder is placed in a 20-m/s atmospheric airflow. At what rotational speed should the cylinder rotate so that only one stagnation point exists on its surface? Calculate the minimum pressure acting on the cylinder if $p = 0$ at $x = -\infty$.

6.85 A 1.2-m-diameter cylinder rotates at 120 rpm in 3-m/s atmospheric airstream. Locate any stagnation points and calculate the minimum and maximum pressures on the cylinder if $p = 0$ at $x = -\infty$.

6.86 The circulation around a 60-ft-long airfoil (measured tip to tip) is calculated to have a value of 15000 ft^2/s. Estimate the lift generated by the airfoil if the aircraft is flying at 30 000 with a speed of 350 ft/s. Assume the flow to be incompressible.

6.87 The velocity field due to a source of strength 2π m^2/s per meter, located at (2 m, 2m) in a 90° corner, is desired. Use the **method of images**, that is, add one or more sources at the appropriate locations, and determine the velocity field by finding $u(x, y)$, and $v(x, y)$.

6.88 Flow through a porous medium is modeled with Laplace's equation and an associated velocity potential function. Natural gas can be stored in certain underground rock structures for use at a later time. A well is placed next to an impervious rock formation, as shown in Figure P6.88. If the well is to extract 0.2 m^3/s per meter, predict the velocity to be expected at the point (4 m, 3m). See Problem 6.87 for the method of images.

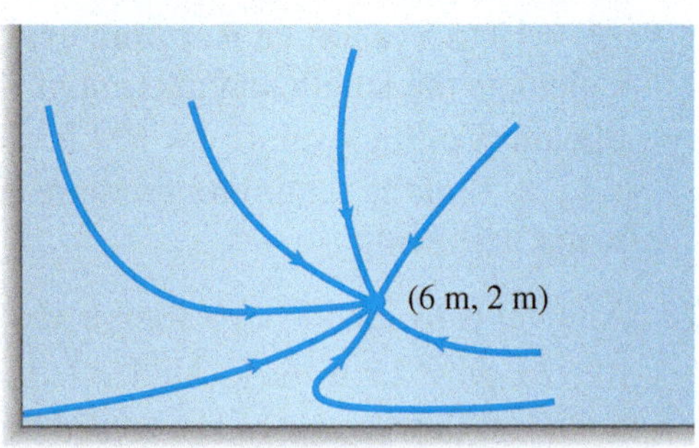

Figure P6.88

Boundary Layers

6.89 How far from the leading edge can turbulence be expected on an airfoil traveling at 300 ft/s if the elevation is:
(a) 0 ft? **(c)** 30, 000 ft?
(b) 12, 000 ft?

Use $Re_{crit} = 6\times10^5$ and assume a flat plate with zero pressure gradient.

6.90 An airfoil on a commercial jetliner can be crudely approximated by a flat plate. Determine the distance from the leading edge at which transition to turbulence can be expected. The jet is traveling at 800 km/h at an altitude of 9000 m, where the pressure and temperature are 31 kPa and –43°C, respectively.

6.91 The boundary layer on a flat plate with zero pressure gradient is to be studied in a tunnel transporting 20°C air. How far from the leading edge would you expect to find turbulent flow if $U_\infty = 10$ m/s and:
(a) The plate is held rigid with high free-stream, disturbance level?
(b) The plate is held rigid with low free-stream disturbance level?
(c) The plate is vibrated with low free-stream disturbance level?
(d) The plate is vibrated with high free-stream disturbance level?
(e) At what distance would you expect a small disturbance to grow for the flow of part (b)?

6.92 Repeat Problem 6.91 but place the flat plate in a channel transporting 20°C water.

6.93 A laminar region is desired to be at least 2 m long on a smooth rigid flat plate. A wind tunnel and a water channel are available. What maximum speed can be selected for each? Assume low free-stream fluctuation intensity and 20°C for each.

6.94 Determine the pressure $p(x)$ in the boundary layer and the velocity $U(x)$ at the edge of the boundary layer that would be present on the front of the cylinder of Problem 6.81. Let the pressure at the stagnation point be 20 kPa with water flowing. Let x be measured from the stagnation point; see Figure 6.19.

6.95 A boundary layer would develop as shown in Figure P6.95 from the front stagnation point of Problem 6.82. Determine $U(x)$ and $p(x)$ that would be needed to solve for the boundary-layer growth on the cylinder front. Measure x from the stagnation point.

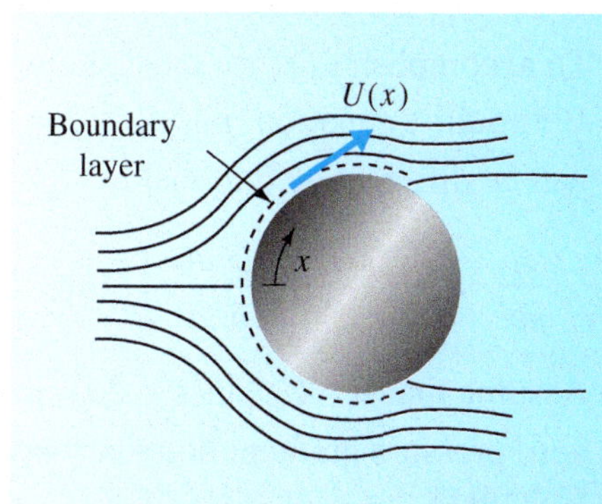

Figure P6.95

6.96 Assuming inviscid uniform flow of air through the contraction shown in Figure P6.96; estimate $U(x)$ and dp/dx, which are necessary to solve for the boundary-layer growth on the flat plate. Assume a one-dimensional flow with $\rho = 1.0$ kg/m^3.

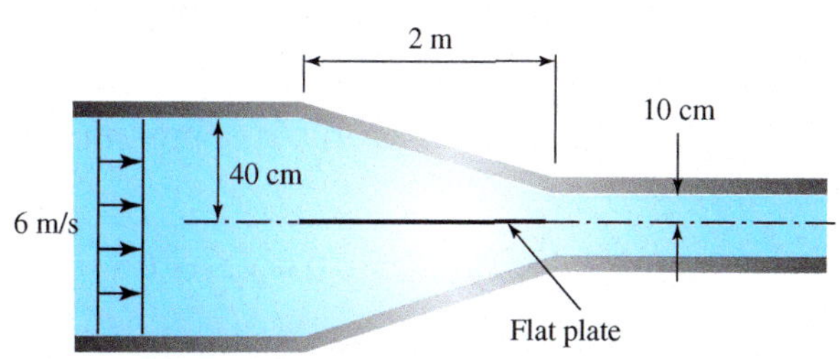

Figure P6.96

Von Kármán Integral Equation

6.97 Provide the detailed steps to the final form of Eq. 6.6.1 and Eq. 6.6.3. Refer to Figure 6.25.

6.98 Show that the von Kármán integral equation 6.6.4 can be put in the form

$$\tau_0 = -\delta\frac{dp}{dx} + \rho\frac{d}{dx}\int_0^\delta u(U-u)dy - \rho\frac{dU}{dx}\int_0^\delta udy$$

Note that the quantity $\int_0^\delta udy$ is only a function of x.

6.99 Show that the von Kármán integral equation of Problem 6.98 can be written as

$$\tau_0 = \rho\frac{d}{dx}(\theta U^2) + \rho\delta_d U\frac{dU}{dx}$$

To accomplish this, we must show that Bernoulli's equation $p + \rho U^2/2 =$ const. can be differentiated to yield

$$\frac{dp}{dx} = -\rho U\frac{dU}{dx} = -\frac{\rho}{\delta}\frac{dU}{dx}\int_0^\delta Udy$$

6.100 Assume that $u = U_\infty \sin(\pi y/2\delta)$ in a zero pressure gradient boundary layer. Calculate:

(a) $\delta(x)$

(b) $\tau_0(x)$

(c) v at $y=\delta$ and $x = 3$ m

6.101 Assume a linear velocity profile and find $\delta(x)$ and $\tau_0(x)$. Compute the percentage error when compared with the exact expressions for a laminar flow. Use $dp/dx = 0$.

6.102 A boundary-layer profile is approximated with:

$$u = 3U_\infty\frac{y}{\delta} \qquad 0 < y \le \delta/6$$

$$u = U_\infty\left(\frac{y}{\delta}+\frac{1}{3}\right) \qquad \delta/6 < y \le \delta/2$$

$$u = U_\infty\left(\frac{y}{3\delta}+\frac{2}{3}\right) \qquad \delta/2 < y \le \delta$$

Determine $\delta(x)$ and $\tau_0(x)$; compute the percentage error when compared with the exact expressions for a laminar flow.

6.103 If the walls in a wind-tunnel test section are parallel, the velocity in the center portion of the tunnel will accelerate as shown in Figure P6.103. To maintain a constant tunnel velocity so that $dp/dx = 0$, show that the walls should be displaced outward a distance $\delta_d(x)$. If a wind tunnel were square, how far should one wall be displaced outward for $dp/dx = 0$?

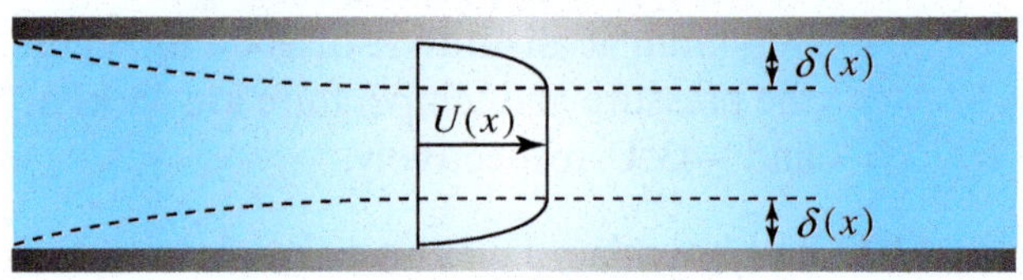

Figure P6.103

6.104 The velocity profile at a given x-location in the boundary layer (Figure P6.104) is assumed to be

$$u(y) - 10\left(2\frac{y}{\delta} = \frac{y^2}{\delta^2}\right)$$

A streamline is 2 cm from the flat plate at the leading edge. How far is it from the plate when $x = 3$ m (i.e., what is h)? Also, calculate the displacement thickness at $x = 3$ m. Compare the displacement thickness to $(h-2)$ cm.

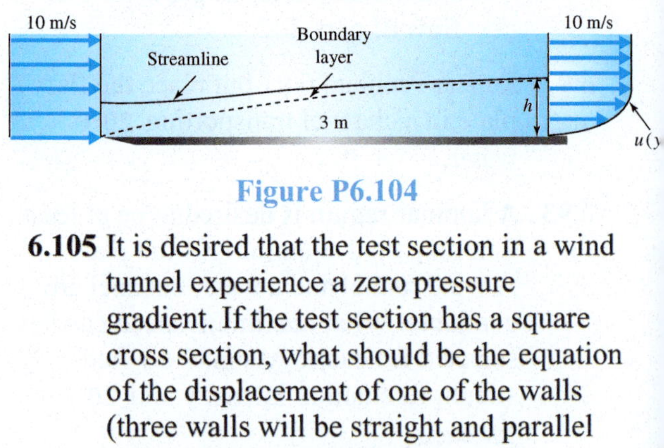

Figure P6.104

6.105 It is desired that the test section in a wind tunnel experience a zero pressure gradient. If the test section has a square cross section, what should be the equation of the displacement of one of the walls (three walls will be straight and parallel or perpendicular) if the 30°C-air is pressurized to 160 kPa absolute? Assume a boundary layer profile of $u/U_\infty = 2y/\delta - y^2/\delta^2$. Assume that $y = 0$ at $x = 0$ of the displacement equation $y(x)$.

6.106 Find δ_d and θ for a laminar boundary layer assuming:
(a) A cubic profile
(b) A parabolic profile
(c) That $u = U_\infty \sin(\pi y/2\delta)$
Compute percentage errors when compared with the exact values of $\delta_d = 1.72\sqrt{\nu x/U_\infty}$ and $\theta = 0.644\sqrt{\nu x/U_\infty}$.

6.107 A laminar flow is maintained in a boundary layer on a 20-ft-long, 15-ft-wide flat plate with 60°F-atmospheric air flowing at 12 ft/s. Assuming a parabolic profile, calculate:
(a) δ at $x = 20$ ft
(b) τ_0 at $x = 20$ ft
(c) The drag force on one side
(d) v at $y = \delta$ and $x = 10$ ft

6.108 Work Problem 6.107, but assume a cubic profile.

Laminar and Turbulent Boundary Layers

6.109 Atmospheric air at 20°C at 10 m/s 2-m-long 4-m-wide flat plate. Calculate the maximum boundary-layer thickness and the drag force on one side assuming:
(a) Laminar flow over the entire length
(b) Turbulent flow over the entire length

6.110 Fluid flows over a flat plate at 20 m/s Determine δ and τ_0 at $x = 6$ if the fluid is:
(a) Atmospheric air at 20°C
(b) Water at 20°C
Neglect the laminar portion

6.111 Assume a turbulent velocity profile $\bar{u} = U_\infty (y/\delta)^{1/7}$. Does this profile satisfy the conditions at $y = \delta$? Can it give the shear stress at the wall? Plot both a cubic laminar profile and the one-seventh power-law profile on the same graph assuming the same boundary-layer thickness.

6.112 Estimate the drag on one side of a 12-ft-long, 51-ft-wide flat plate if atmospheric air at 60° is flowing with a velocity of 20 ft/s. Assume that:
(a) $\text{Re}_{\text{crit}} = 3 \times 10^5$
(b) $\text{Re}_{\text{crit}} = 5 \times 10^5$
(c) $\text{Re}_{\text{crit}} = 6 \times 10^5$

6.113 A 1-m-long flat plate with a sharp leading edge that is 2 m wide is towed parallel to itself in 20°C water at 1.2 m/s. Estimate the total drag if:
(a) $\text{Re}_{\text{crit}} = 3 \times 10^5$
(b) $\text{Re}_{\text{crit}} = 6 \times 10^5$
(c) $\text{Re}_{\text{crit}} = 9 \times 10^5$

6.114 Air moving at 60 km/h is considered to have a zero boundary-layer thickness at a distance of 100 km from shore. At the beach, estimate the boundary-layer thickness and the wall shear using:
(a) A one-seventh power Jaw
(b) Empirical data
Use $T = 20$°C

6.115 For the conditions of Problem 6.114, calculate:
(a) The thickness of the viscous wall layer
(b) The displacement thickness at the beach

6.116 Atmospheric air at 60°F flows over a flat plate at 300 ft/s. At $x = 20$ ft, estimate:
(a) The local skin friction coefficient
(b) The wall shear
(c) The viscous wall-layer thickness
(d) The boundary-layer thickness

6.117 Water at 20°C flows over a flat plate at 10 m/s. At $x = 3$ m, estimate:

(a) The viscous wall layer thickness
(b) The velocity at the edge of the viscous wall layer
(c) The value of y at the outer edge of the turbulent zone
(d) The boundary-layer thickness

6.118 Estimate the total shear drag on a ship traveling at 10 m/s if the sides are flat plates 10 m × 100 m with zero pressure gradients. What is the maximum boundary-layer thickness?

6.119 The large 600-m-long, 100-m-diameter, cigar shaped dirigible of Figure P6.119 is planned for cruises for retired engineers. As a first estimate the drag is calculated assuming it to be a flat plate with zero pressure gradient, with the drag on the nose and rear areas neglected.

(a) Estimate the power required of each of four engines if it is to cruise at 15 m/s.
(b) Estimate the payload if one half of its volume is filled with helium, and its engines, equipment, and structure have a mass of 1.2×10^6 kg.

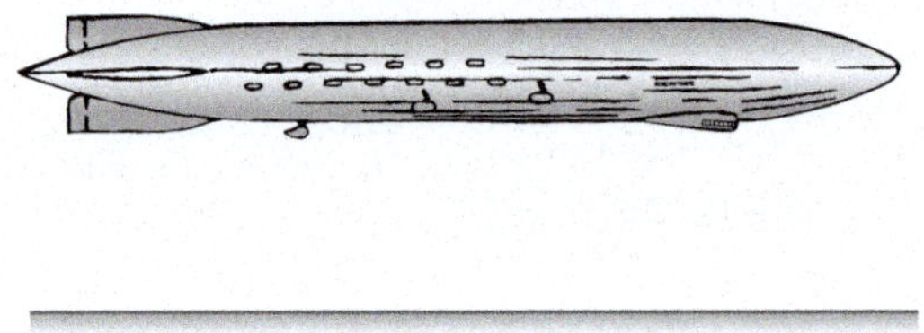

Figure P6.119

Laminar Boundary–Layer Equation

6.120 Assuming that $dp/dx = 0$, show that Eq. 6.6.47 follows from Eq. 6.6.45.

6.121 Recalling from calculus that

$$\frac{\partial \psi}{\partial y} = \frac{\partial \psi}{\partial \xi}\frac{\partial \xi}{\partial y} + \frac{\partial \psi}{\partial \eta}\frac{\partial \eta}{\partial y}$$

so that

$$\frac{\partial^2 \psi}{\partial y^2} = \frac{\partial}{\partial y}\left(\frac{\partial \psi}{\partial y}\right) = \frac{\partial(\partial \psi/\partial y)}{\partial \xi}\frac{\partial \xi}{\partial y} + \frac{\partial(\partial \psi/\partial y)}{\partial \eta}\frac{\partial \eta}{\partial y}$$

show that Eq. 6.6.49 follows from Eq. 6.6.47.

6.122 Show that Eq. 6.6.51 follow from the preceding equations.

6.123 Solve Eq. 6.6.52 with the appropriate boundary conditions, using a third-order Runge-Kutta scheme (or any other appropriate numerical algorithm) and verify the results of Table 6.5. (This was a Ph.D. thesis project for Blasius before the advent of the computer!)

6.124 A laminar boundary layer exists on a flat plate with atmospheric air at 20°C moving at 5 m/s. At $x = 2$ m, find:

(a) The wall shear
(b) The boundary-layer thickness
(c) The maximum value of v
(d) The flow rate through the boundary layer

6.125 A laminar boundary layer exist on a flat plate with atmospheric at 60°F moving at 15 ft/s. At $x = 6$ ft, find:

(a) The wall shear
(b) The boundary-layer thickness
(c) The maximum value of v
(d) The flow rate through the boundary layer

6.126 Water at 20°C flows over a flat plate with zero pressure gradient at 5 m/s. At $x = 2$ m, find:

(a) The wall shear
(b) The boundary-layer thickness
(c) The flow rate through the boundary-layer

6.127 If, when we defined the boundary-layer thikness, we defined δ to be that y-location where $u = 0.999U_\infty$, estimate the thickness of the boundary layer of:

(a) Problem 6.124
(b) Problem 6.125

6.128 Find the y-location where $u = 0.5U_\infty$ for the boundary layer of Problem 6.132. What is the value of v at that y-location? what is the shear stress there?

6.129 Assume that v remains unchanged between $y = \delta$ and $y = 10\delta$. Sketch $u(y)$ for $0 \leq y < 10\delta$ for a flat plate with zero pressure gradient. Now, assume that $v = 0$ at $y = 10\delta$. Again sketch $u(y)$. Explain by referring to appropriate equations.

6.130 Sketch the boundary-layer velocity profile near the end of the flat plate of Problem 6.115 and show a Blasius profile of the same sketch.

Pressure Gradient Effects

6.131 Sketch the velocity profiles near and normal to the cylinder's surface at each of the points indicated in Figure P6.131. The flow separates at C.

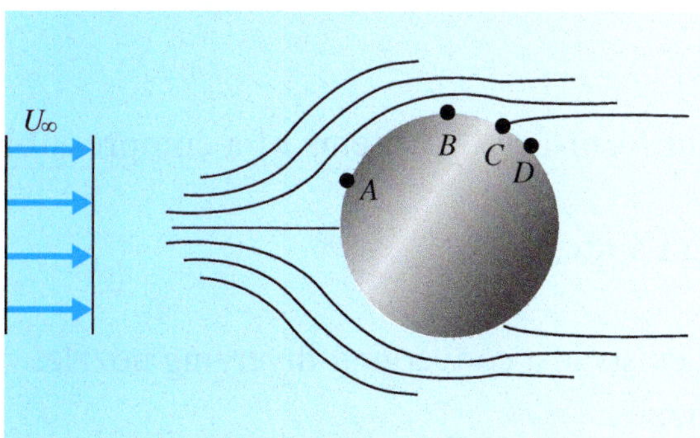

Figure P6.131

6.132 Sketch the expected boundary-layer profiles at each of the points indicated in Figure P6.132, showing relative thicknesses. The flow undergoes transition to turbulence just after point A. It separates at D. Show all profiles on the same plot. Indicate the sign of the pressure gradient at each point.

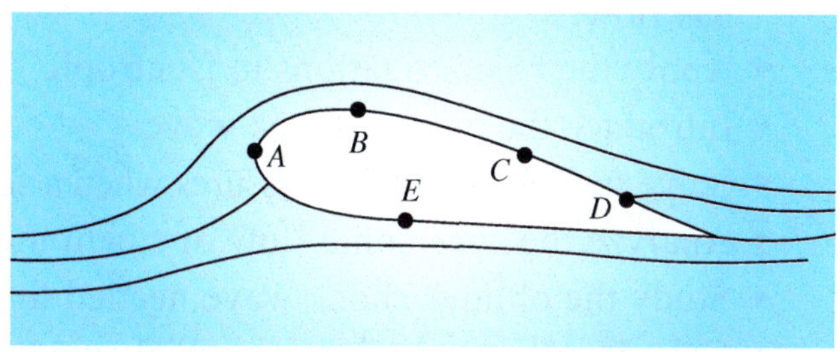

Figure P6.132

7 Compressible Flow

Outline

Chapter Objectives

The objectives of this chapter are to:

- Present the various equations needed to solve the uniform-flow problems of a compressible gas flow.
- Apply the basic equations to isentropic flow through a nozzle.
- Introduce the normal shock wave.
- Solve for velocities and pressures when a shock wave exists in a converging-diverging nozzle.
- Analyze the supersonic flow of steam through a nozzle.
- Study the oblique shock wave needed to turn a supersonic flow through an angle.
- Calculate the angle through which expansion waves can turn a flow around a convex corner.
- Provide numerous examples and problems that illustrate isentropic flow, flow through normal and oblique shock waves, and the turning of a supersonic flow around a convex corner.

7.1 INTRODUCTION

In this chapter we consider flows of gases in which the density changes significantly between points on a streamline; such flows are called **compressible flows**. We will consider those problems that can be solved using the integral equations: the continuity equation, the energy equation, and for some problems, the momentum equation.

Not all gas flows are compressible flows, neither are all compressible flows gas flows. At low speeds, less than a Mach-number $(\mathrm{M} = V/\sqrt{kRT})$ of about 0.3, gas flows may be treated as incompressible flows. This is justified because the density variations caused by the flow are insignificant (less than about 3%). Incompressible gas flows occur in a large number of situations of engineering interest; many of these have been considered in earlier chapters. There are many flows, however, in which the density variations must be accounted for.

M. C. Potter and D. C. Wiggert, *Applied Fluid Mechanics*,
https://doi.org/10.1007/978-3-031-65276-9_7

Included among these are airflows around commercial and military aircraft, airflow through jet engines, and the flow of a gas in compressors and turbines.

There are examples of compressible effects important in liquid flows; water hammer and compression waves due to underwater blasts are examples of compressible liquid flows. Compressibility of rock accounts for the propagation through the earth's surface of longitudinal waves due to an earthquake. In this chapter we are only concerned with compressibility effects in gas flows.

Let us introduce the effects of compressibility into the simplest of flow situations. The velocity at a given streamwise location in a conduit is assumed to be uniform and hence does not vary normal to the flow direction. For this simple uniform flow we recall that the continuity equation takes the form (see Eq. 2.4.5)

$$\dot{m} = \rho_1 A_1 V_1 = \rho_2 A_2 V_2 \tag{7.1.1}$$

The momentum equation for the uniform, compressible flow takes the form (see Eq. 2.6.5)

$$\sum \mathbf{F} = \dot{m}(\mathbf{V}_2 - \mathbf{V}_1) \tag{7.1.2}$$

The energy equation, neglecting potential energy changes, is written (see Eqs. 2.5.17 and 2.5.18)

$$\frac{\dot{Q} - \dot{W}_s}{\dot{m}} = \frac{V_2^2 - V_1^2}{2} + h_2 - h_1 \tag{7.1.3}$$

where we have used the enthalpy $h = \tilde{u} + p/\rho$. Assuming an ideal gas with constant specific heats, the energy equation takes the form

$$\frac{\dot{Q} - \dot{W}_s}{\dot{m}} = \frac{V_2^2 - V_1^2}{2} + c_p(T_2 - T_1) \tag{7.1.4}$$

or

$$\frac{\dot{Q} - \dot{W}_s}{\dot{m}} = \frac{V_2^2 - V_1^2}{2} + \frac{k}{k-1}\left(\frac{p_2}{\rho_2} - \frac{p_1}{\rho_1}\right) \tag{7.1.5}$$

where we have used the thermodynamic relations

$$h_2 - h_1 = c_p(T_2 - T_1) \tag{7.1.6a}$$

$$c_p = R + c_v \tag{7.1.6b}$$

$$k = \frac{c_p}{c_v} \tag{7.1.6c}$$

and the ideal gas law

$$p = \rho R T \tag{7.1.7}$$

If we have an interest in calculating the entropy change between two sections, we will use the definition of entropy as

$$\Delta S = \int \frac{\delta Q}{T}\Big|_{\text{reversible}} \tag{7.1.8}$$

where δQ represents the differential heat transfer. Using the first law, this becomes, for an ideal gas with constant specific heat,

$$\Delta s = c_p \ln \frac{T_2}{T_1} - R \ln \frac{p_2}{p_1} \tag{7.1.9}$$

If a process is adiabatic ($Q = 0$) and reversible (no losses), Eq. 7.1.8 shows that the entropy change is zero (i.e., the flow is **isentropic**). If the flow is isentropic, the relationship above can be used, along with the ideal gas law and Eqs. 7.1.6b and c, to show that

$$\frac{T_2}{T_1} = \left(\frac{p_2}{p_1}\right)^{(k-1)/k} \qquad \frac{p_2}{p_1} = \left(\frac{\rho_2}{\rho_1}\right)^{k} \tag{7.1.10}$$

Note: Temperatures and pressures must be measured on absolute scales.

Let us now introduce a parameter, the Mach number, that will be of special interest throughout our study of compressible flow.

7.2 SPEED OF SOUND AND THE MACH NUMBER

The speed of sound is the speed at which a pressure disturbance of small amplitude travels through a fluid. It is analogous to the small ripple, a gravity wave, that travels radially outward when a pebble is dropped into a pond. To determine the speed of sound consider a small pressure disturbance, called a **sound wave**, to be passing through a pipe, as shown in Figure 7.1. It travels with a velocity c relative to a stationary observer as shown in part (a); the pressure, density, and temperature will change by the small amounts Δp, $\Delta\rho$ and ΔT, respectively. There will also be an induced velocity ΔV in the fluid immediately behind the sound wave. To simplify the problem we will create a steady flow by having the observer travel at the speed of the wave so that the sound wave appears to be stationary. The flow will then approach the wave from the right, with the speed of sound c, as shown in Figure 7.1b. The flow properties will all change across the wave and the velocity in the downstream flow will be expressed as $c + \Delta V$, where ΔV is the small change in velocity. If the flow speed is reduced, the wave would move to the right, and if the flow speed is increased, it would move to the left. For the flow speed equal to the speed of sound, $V = c$, the sound wave would be stationary, as shown.

Let us apply the continuity equation and the energy equation to a small control volume enclosing the sound wave, shown in Figure 7.1c. The continuity equation (7.1.1) takes the form

$$\rho A c = (\rho + \Delta\rho) A (c + \Delta V) \tag{7.2.1}$$

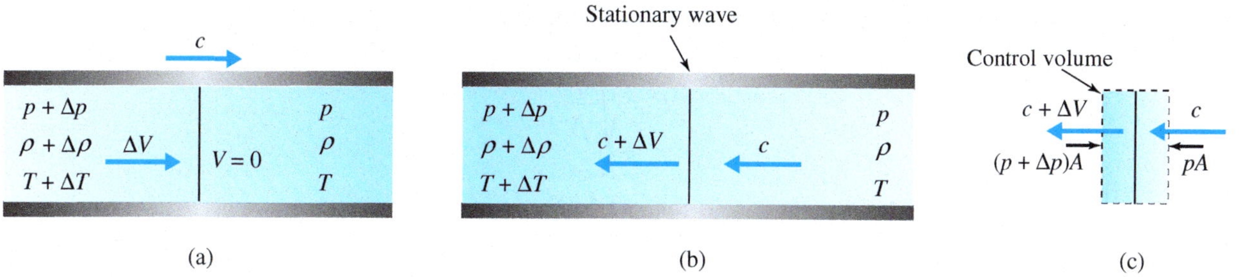

Figure 7.1 Sound wave: (a) stationary observer; (b) observer moving with the wave; (c) control volume enclosing the wave.

This can be rewritten as

$$\rho\Delta V = -c\Delta\rho \tag{7.2.2}$$

where we have neglected the higher-order term $\Delta\rho\Delta V$; that is, $\Delta\rho$ represents a small percentage change in ρ so that $\Delta\rho \ll \rho$; also, $\Delta V \ll c$.

The streamwise-component momentum equation (9.1.2) yields

$$pA - (p + \Delta p)A = \rho Ac(c + \Delta V - c) \tag{7.2.3}$$

which simplifies to,

$$-\Delta p = \rho c\Delta V \tag{7.2.4}$$

Combining this with Eq. 7.2.2 results in

$$c = \sqrt{\frac{\Delta p}{\Delta\rho}} \tag{7.2.5}$$

Since the changes Δp and $\Delta\rho$ are quite small, we can write

$$\frac{\Delta p}{\Delta\rho} \simeq \frac{dp}{d\rho} \tag{7.2.6}$$

Small-amplitude, moderate-frequency waves (up to about 18 000 Hz) travel with no change in entropy (isentropically), so that

$$\frac{p}{\rho^k} = \text{const.} \tag{7.2.7}$$

This can be differentiated to give

$$\frac{dp}{d\rho} = k\frac{p}{\rho} \tag{7.2.8}$$

Using this in Eq. 7.2.5, the speed of sound c is given by

$$c = \sqrt{\frac{kp}{\rho}} \tag{7.2.9}$$

or, using the ideal gas law,

$$c = \sqrt{kRT} \tag{7.2.10}$$

At high frequency, sound waves generate friction and the process ceases to be isentropic. It is better approximated by an isothermal process. For an ideal gas an isothermal approximation would lead to

$$c = \sqrt{RT} \tag{7.2.11}$$

For small waves traveling through liquids or solids the **bulk modulus** is used; it has dimensions of pressure and is equal to $\rho dp/d\rho$. For water it has a nominal value of 2110 MPa. It does vary slightly with temperature and pressure. Using Eq. 7.2.5, this leads to a speed of propagation of 1450 m/s for a small-amplitude pressure wave in water.

An important quantity used in the study of compressible flow is the dimensionless velocity called the **Mach number**, introduced in Eq.1.3.3 as

$$\mathrm{M} = \frac{V}{c} \tag{7.2.12}$$

If M < 1, the flow is a **subsonic** flow, and if M > 1, it is a **supersonic** flow. If M > 5 it may be referred to as **hypersonic** flow; such flows will not be considered in this book.

If a source of sound waves is at a fixed location, the waves travel radially away from the source with the speed of sound c. Figure 7.2a shows the position of the sound waves after a time increment Δt and after multiples of Δt. Part (b) shows a source moving with a velocity V

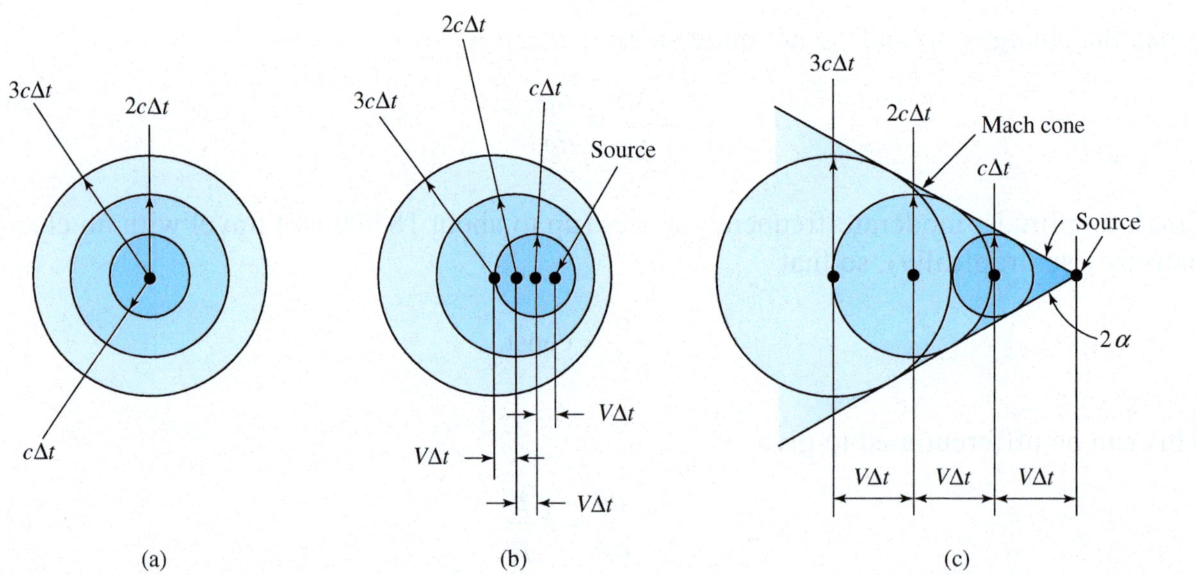

Figure 7.2 Sound waves propagating from a noise source: (a) stationary source; (b) moving source, $V < c$; (c) moving source, $V > c$.

which is less than the speed of sound. Note that the sound waves always propagate ahead of the source so that an airplane travelling at a speed less than the speed of sound will always "announce" its approach. This is not true, however, for an object travelling at a speed greater than the speed of sound, as shown in Figure 7.2c. The region outside the cone is a **zone of silence**, so that an approaching object moving at a supersonic speed could not be heard until it passed overhead and the **Mach cone**, the cone shown, intercepted the observer. From the figure the angle α of the Mach cone is given by

$$\alpha = \sin^{-1}\frac{c}{V} = \sin^{-1}\frac{1}{\mathrm{M}} \tag{7.2.13}$$

The discussion above is limited to small-amplitude sound waves, often called **Mach waves**. They are formed on the needle nose of an aircraft, or on the leading edge of an airfoil if the leading edge is sufficiently sharp. If the nose is blunt or if the leading edge is not sufficiently sharp, a supersonic aircraft will produce a large-amplitude wave called a **shock wave**. The shock wave will also form a zone of silence, but the initial angle at the source created by the shock wave will be larger than that of a Mach wave. Shock waves will be considered in subsequent sections.

Example 7.1

A needle-nose projectile traveling at a speed with M = 3 passes 200 m above the observer of Figure E7.1. Calculate the projectile's velocity and determine how far beyond the observer the projectile will first be heard.

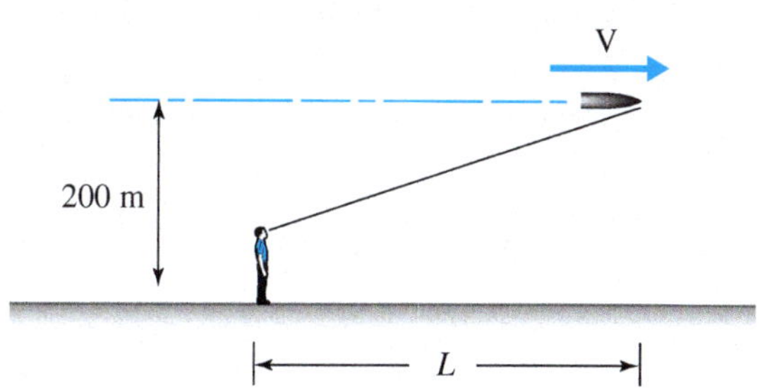

Figure E7.1

Solution

At a Mach number of 3 the velocity is

$$V = \mathrm{M}c = \mathrm{M}\sqrt{kRT}$$
$$= 3\sqrt{1.4 \times 287\,\mathrm{J/kg\cdot K} \times 288\,\mathrm{K}} = 1021\ \mathrm{m/s}$$

where a standard temperature of 15°C has been assumed since no temperature is given. (Remember, kg = N·s^2/m and J = N·m.) Using h as the height and L as the distance beyond the observer (refer to Figure 7.2c), we have

$$\sin\alpha = \frac{h}{\sqrt{L^2 + h^2}} = \frac{1}{\mathrm{M}}$$

With the information given,

$$\frac{200}{\sqrt{L^2+200^2}}=\frac{1}{3}$$

giving

$$L = \underline{566\ \text{m}}$$

Note: The units on kRT are $\frac{\text{N}\cdot\text{m}}{\text{kg}\cdot\text{K}}\times\text{K}=\frac{\text{N}\cdot\text{m}}{\text{N}\cdot\text{s}^2/\text{m}}=\frac{\text{m}^2}{\text{s}^2}$. The quantity k is dimensionless.

7.3 ISENTROPIC NOZZLE FLOW

There are many applications where gas flows through a section of tube or conduit that has a changing area in which a steady, uniform, isentropic flow is a good approximation to the actual flow situation. The diffuser near the front of a jet engine, exhaust gases passing through the blades of a turbine, the nozzles on a rocket engine, a broken natural gas line, and gas flow measuring devices are all examples of situations that can be modeled with a steady, uniform, isentropic flow. Consider the flow through the infinitesimal control volume shown in Figure 7.3. With a changing area the continuity equation

$$\rho AV = \text{const} \tag{7.3.1}$$

applied between two sections a distance dx apart takes the form

$$\rho AV = (\rho + d\rho)(A + dA)(V + dV) \tag{7.3.2}$$

Keeping only first-order terms in the differential quantities, Eq. 7.3.2 can be put in the form

$$\frac{dV}{V}+\frac{dA}{A}+\frac{d\rho}{\rho}=0 \tag{7.3.3}$$

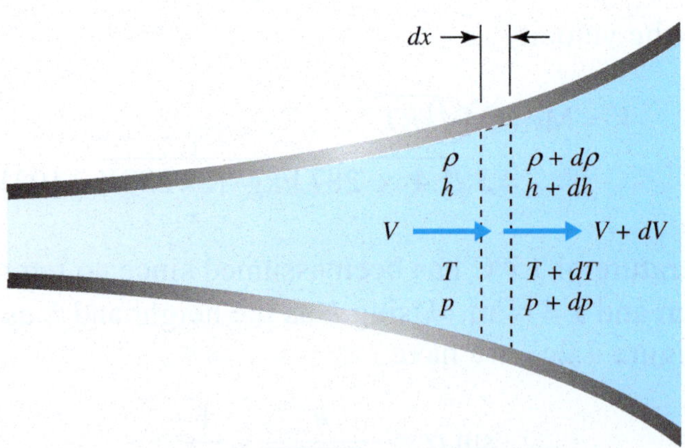

Figure 7.3 Uniform, isentropic flow.

The energy equation can be written (see Eq. 7.1.5)

$$\frac{V_1^2}{2}+\frac{k}{k-1}\frac{p_1}{\rho_1}=\frac{V_2^2}{2}+\frac{k}{k-1}\frac{p_2}{\rho_2} \tag{7.3.4}$$

For the present application, with state 2 an infinitesimal distance from state 1, we have

$$\frac{V^2}{2}+\frac{k}{k-1}\frac{p}{\rho}=\frac{(V+dV)^2}{2}+\frac{k}{k-1}\frac{p+dp}{\rho+d\rho} \tag{7.3.5}$$

or again, retaining only first-order terms,

$$V\,dV+\frac{k}{k-1}\frac{\rho dp-pd\rho}{\rho^2}=0 \tag{7.3.6}$$

For an isentropic process we use Eq. 7.2.8 and there results

$$V\,dV+k\frac{p}{\rho^2}d\rho=0 \tag{7.3.7}$$

Substituting for $d\rho/\rho$ from Eq. 7.3.3, the above becomes

$$\frac{dV}{V}\left(\frac{\rho V^2}{kp}-1\right)=\frac{dA}{A} \tag{7.3.8}$$

In terms of the speed of sound this is written as

$$\frac{dV}{V}\left(\frac{V^2}{c^2}-1\right)=\frac{dA}{A} \tag{7.3.9}$$

Introduce the Mach number and we have the very important relationship

$$\frac{dV}{V}(\mathrm{M}^2-1)=\frac{dA}{A} \tag{7.3.10}$$

for an isentropic uniform flow in a changing area.

From Eq. 7.3.10 we can make the following observations:

1. If the area is increasing, $dV > 0$, and M < 1, we see that dV must be negative, that is, $dV < 0$. The flow is decelerating for a subsonic flow.
2. If the area is increasing and M < 1, we see that $dV > 0$; hence the flow is accelerating in the diverging section for a supersonic flow.
3. If the area is decreasing and M < 1, then $dV > 0$, resulting is an accelerating flow.
4. If the area is decreasing and M < 1, then $dV > 0$, indicating a decelerating flow.
5. At a throat where $dA = 0$, either $dV > 0$ or M $= 1$, or possibly both.

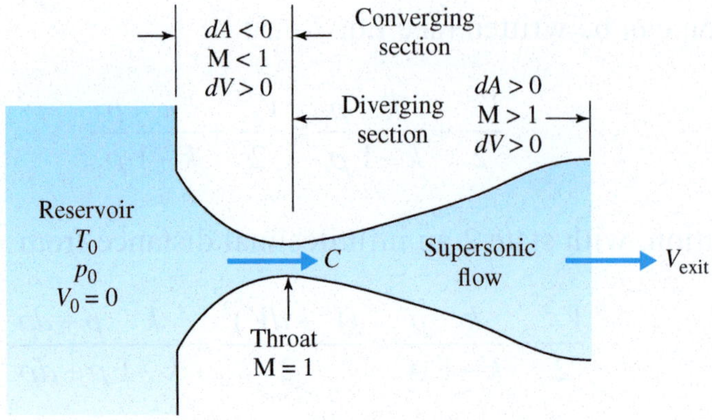

Figure 7.4 Supersonic nozzle.

If we define a **nozzle** as a device that accelerates the flow, we see that observations 2 and 3 describe a nozzle and observations 1 and 4 describe a **diffuser**, a device that decelerates the flow. The supersonic flow leads to rather surprising results: an accelerating flow in an enlarging area and a decelerating flow in a decreasing area. This is, in fact, the situation encountered in a traffic flow on a freeway; hence a supersonic flow might be used to model a traffic flow.

Note that the observations above prohibit a supersonic flow in a converging section attached to a reservoir. If a supersonic flow is to be generated by releasing a gas from a reservoir, there must be a converging section in which a subsonic flow accelerates to the throat where M = 1 followed by a diverging section in which the flow continues to accelerate with $\mathrm{M} > 1$, as shown in Figure 7.4. This is the type of nozzle observed on rockets used to place satellites into orbit.

The isentropic flow in the nozzle will now be considered in more detail. The energy equation between the reservoir where $V_0 = 0$ and any section can be written in the form

$$c_p T_0 = \frac{V^2}{2} + c_p T \tag{7.3.11}$$

Quantities with a zero subscript are often called **stagnation quantities** since they occur at a location where $V = 0$. Recognizing that $V = \mathrm{M}c$, $c_p/c_v = k$, $c_p = c_v + R$, and $c = \sqrt{kRT}$, we can write the energy equation as

$$\frac{T_0}{T} = 1 + \frac{k-1}{2}\mathrm{M}^2 \tag{7.3.12}$$

For our isentropic flow, the pressure and density ratios are expressed as

$$\begin{aligned} \frac{p_0}{p} &= \left(1 + \frac{k-1}{2}\mathrm{M}^2\right)^{k/(k-1)} \\ \frac{\rho_0}{\rho} &= \left(1 + \frac{k-1}{2}\mathrm{M}^2\right)^{1/(k-1)} \end{aligned} \tag{7.3.13}$$

If a supersonic flow occurs downstream of the throat, then M = 1 at the throat, and denoting this **critical area** with an asterisk (*) superscript, we have the following critical ratios:

$$\frac{T^*}{T_0}=\frac{2}{k+1}$$
$$\frac{p^*}{p_0}=\left(\frac{2}{k+1}\right)^{k/(k-1)} \tag{7.3.14}$$
$$\frac{\rho^*}{\rho_0}=\left(\frac{2}{k+1}\right)^{1/(k-1)}$$

We often make reference to the critical area even though an actual throat does not occur; we can imagine a throat occurring and call it the critical area. In fact, the isentropic flow table, Table D.1 in Appendix D, includes just such an area. For air with $k = 1.4$ the critical values are

$$p^* = 0.5283p_0 \qquad T^* = 0.8333T_0 \qquad \rho^* = 0.6340\rho_0 \tag{7.3.15}$$

We can determine an expression for the mass flux through the nozzle from the equation

$$\begin{aligned}\dot{m} &= \rho AV \\ &= \frac{p}{RT}A\mathrm{M}\sqrt{kRT}=\frac{p}{\sqrt{T}}\sqrt{\frac{k}{R}}A\mathrm{M}\end{aligned} \tag{7.3.16}$$

where we have used $V = \mathrm{M}c$. Using Eqs. 7.3.12 and 7.3.13, this can be expressed as

$$\begin{aligned}\dot{m} &= \frac{p_0\left(1+\frac{k-1}{2}\mathrm{M}^2\right)^{k/(1-k)}}{\sqrt{T_0}\left(1+\frac{k-1}{2}\mathrm{M}^2\right)^{-1/2}}\sqrt{\frac{k}{R}}A\mathrm{M} \\ &= p_0\sqrt{\frac{k}{RT_0}}\mathrm{M}A\left(1+\frac{k-1}{2}\mathrm{M}^2\right)^{(k+1)/2(1-k)}\end{aligned} \tag{7.3.17}$$

If we choose the critical area where M* = 1, we see that

$$\dot{m} = p_0A^*\sqrt{\frac{k}{RT_0}}\left(\frac{k+1}{2}\right)^{(k+1)/2(1-k)} \tag{7.3.18}$$

This shows that the mass flux in the nozzle is only dependent on the reservoir conditions and the critical area A^*. By combining Eqs. 7.3.17 and 7.3.18, the area ratio A/A^* can be written in terms of the Mach number as

$$\frac{A}{A^*}=\frac{1}{\mathrm{M}}\left[\frac{2+(k-1)\mathrm{M}^2}{k+1}\right]^{(k+1)/2(k-1)} \tag{7.3.19}$$

This ratio is included in Table D.1 for airflow.

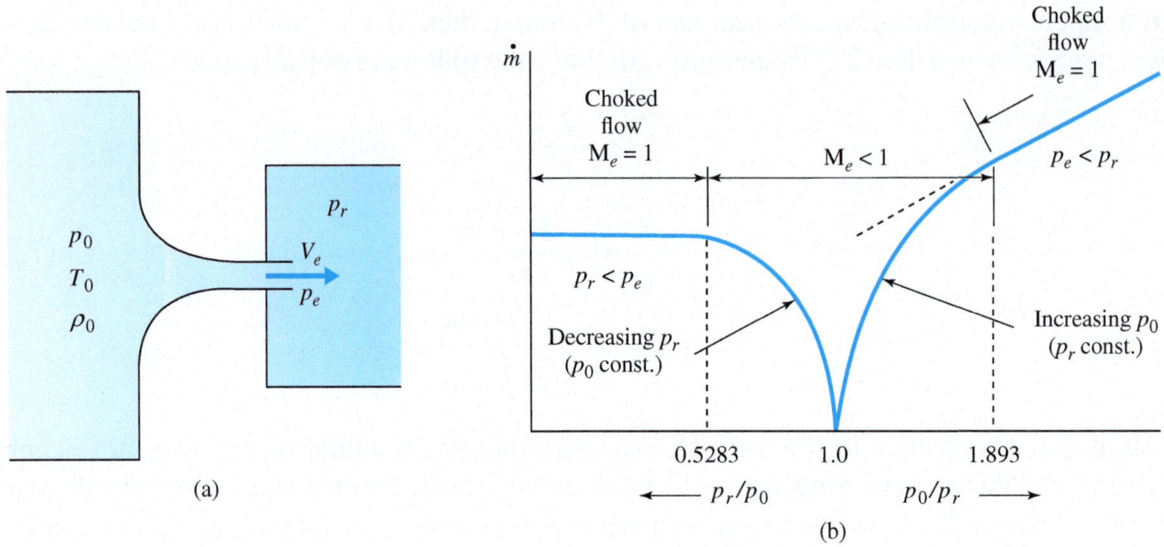

Figure 7.5 Converging nozzle.

As a final consideration in our study of isentropic nozzle flow, we will present the influence of reservoir pressure and receiver pressure on the mass flux. First, the converging nozzle will be presented; then the converging-diverging nozzle will follow. The converging nozzle is assumed to be attached to a reservoir, as shown in Figure 7.5a, with fixed conditions; the pressure in the receiver can be lowered to provide an increasing mass flux through the nozzle, as shown by the left curve in Figure 7.5b. When the receiver pressure p_r reaches the critical pressure (for air $p_r = 0.5283\ p_0$), the Mach number M_e at the throat (the exit) is unity. As p_r is reduced below this critical value the mass flux will not increase, and the condition of **choked flow** occurs, as shown by the left curve in Figure 7.5b. If the throat is a critical area, that is, $M = 1$ at the throat, then $\dot{m}$ is only dependent on the throat area A^* and the reservoir conditions, as indicated by Eq. 7.3.18. Hence, reducing p_r below the critical pressure has no effect on the upstream flow. This is reasonable since disturbances travel at the speed of sound; if p_r is reduced, the disturbances that would travel upstream, thereby changing conditions, cannot do so since the velocity of the stream at the exit is equal to the speed of sound thereby preventing any disturbances from propagating upstream.

If, in the converging nozzle, we keep p_r constant and increase the reservoir pressure (keep T_0 constant also), a choked flow again occurs when $M_e = 1$; however, when p_0 is increased still further, we see from Eq. 7.3.18 that the mass flux will increase, as shown by the right curve of Figure 7.5b. The pressure p_e is equal to p_r until the Mach number M_e is just equal to unity. This begins the condition of choked flow. The exit pressure p_e for the choked flow condition will be greater than the receiver pressure p_r, a condition that occurs when a gas line ruptures.

A note may be in order explaining how it is possible for the flow exit pressure p_e to exceed the receiver pressure p_r. If $p_e > p_r$, the flow exiting the nozzle is able to turn rather sharply (air is very light), causing a flow pattern sketched in Figure 7.6. This possible flow situation will be studied in Section 7.8.

Now, consider the converging–diverging nozzle, shown in Figure 7.7, with reservoir and receiver as indicated. We will only present the condition of constant reservoir pressure and

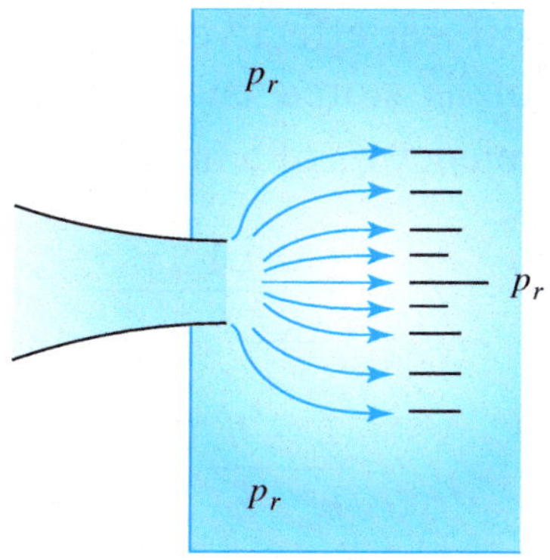

Figure 7.6 Nozzle exit flow for $p_e > p_r$.

reduced receiver pressure. For this nozzle we sketch the pressure ratio p/p_0 as a function of location in the nozzle for various receiver pressure ratios p_r/p_0. If $p_r/p_0 = 1$, no flow occurs, corresponding to curve A. If p_r is reduced a small amount, curve B results and a subsonic flow exists throughout the nozzle with a minimum pressure occurring at the throat. As the pressure is reduced still further, a pressure is reached that will result in the Mach number at the throat just being unity, as sketched by curve C; the flow remains subsonic throughout, however. Another particular receiver pressure exists, considerably below that of curve C, that will also produce an isentropic flow throughout; it results in curve D. Any receiver pressure in between these two particular pressures will result in a nonisentropic flow in the nozzle; a shock wave, to be studied later, occurs which renders our assumption of isentropic flow invalid. If the receiver pressure is below that associated with curve D, we again find that the nozzle exit pressure p_e is greater than the receiver pressure p_r. The mass flux in the nozzle increases from curve A to curve C; but as the receiver pressure is reduced below that of curve C, no increase in mass flux can occur since the throat condition will remain unchanged. The mass flux is given by Eq. 7.3.17.

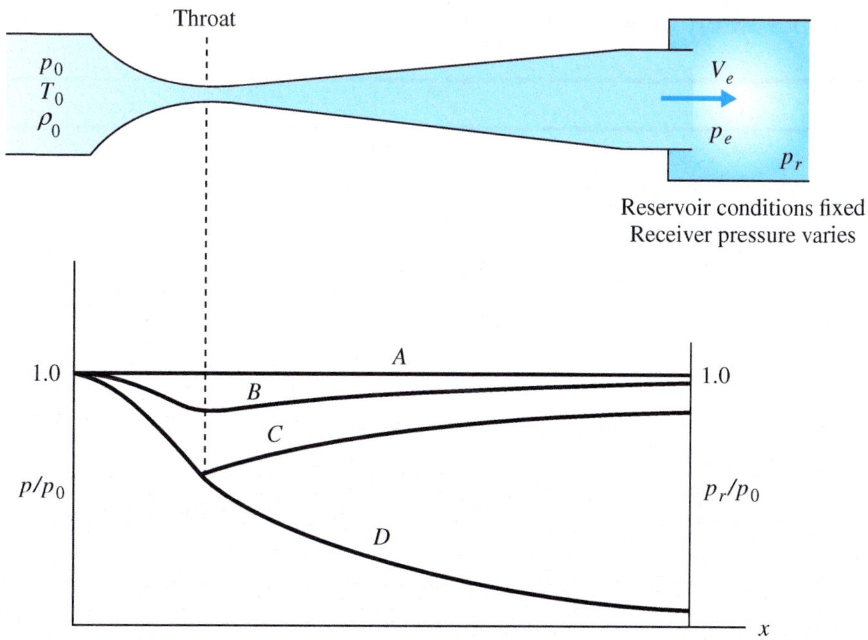

Figure 7.7 Converging-diverging nozzle.

Final notes regard nozzle and diffuser effectiveness. The purpose of a nozzle is to convert enthalpy (which can be thought of as stored energy) into kinetic energy as described by Eq. 7.1.3 with $\dot{Q} = \dot{W}s = 0$. The efficiency η_N of a nozzle is defined as

$$\eta_N = \frac{(\Delta \text{KE})_{\text{actual}}}{(\Delta \text{KE})_{\text{isentropic}}}$$

$$= \frac{h_0 - h_e}{h_0 - h_{es}} \tag{7.3.20}$$

where h_e is the actual exit enthalpy and h_{es} is the isentropic exit enthalpy. Efficiencies are between 90 and 99% with larger nozzles having the higher percentages because the viscous wall effects that account for most of the losses are relatively small with the larger nozzles.

The purpose of a diffuser is to slow down the fluid and recover the pressure. For a diffuser we define the **pressure recovery factor** C_p to be

$$C_p = \frac{\Delta p_{\text{actual}}}{\Delta p_{\text{isentropic}}} \tag{7.3.21}$$

Such factors vary between 40% when the flow actually separates from the wall (the included angle should be less than about 10° for a subsonic section to avoid this separation) to 85%. Vanes can be installed in wide-angled subsonic diffusers such that each passage between vanes expands at an angle of 10° or less. Viscous effects are significantly greater in a diffuser than in a nozzle because of the thicker viscous wall layers.

The reader may think that the flow in the supersonic diverging nozzle may also tend to separate from the wall; this is not the case. Expansion fans, a phenomenon provided by nature, to be studied in Section 7.8, allow the supersonic flow to turn rather sharp angles so that supersonic nozzles are constructed with large included angles, such as those attached to the rockets that propel satellites into orbit.

Example 7.2

Air exits from a reservoir maintained at 20°C and 500 kPa absolute into a receiver maintained at (a) 300 kPa absolute and (b) 200 kPa absolute. Estimate the mass flux if the exit area is 10 cm^2. Use the equations first and then the isentropic flow table, Table D.1. Refer to Figure 7.5.

Solution

To estimate the mass flux we will assume isentropic flow. For air the receiver pressure that would result in $\text{M}_e = 1$ is

$$p_r = 0.5283 p_0 = 0.5283 \times 500 = 264.2 \text{ kPa}$$

For part (a) $\text{M}_e < 1$ since $p_r < 264.2$ kPa, and for part (b) choked flow occurs and $\text{M}_e = 1$ since $p_r < 264.2$ kPa.

(a) To find the exit Mach number, Eq. 7.3.13 gives

$$1+\frac{k-1}{2}\mathrm{M}^2=\left(\frac{p_0}{p}\right)^{(k-1)/k}$$

or

$$\mathrm{M}_e^2=\frac{2}{k-1}\left(\frac{p_0}{p}\right)^{(k-1)/k}-\frac{2}{k-1}$$

$$=\frac{2}{0.4}\left(\frac{500}{300}\right)^{0.2857}-\frac{2}{0.4}=0.7857 \qquad \therefore \mathrm{M}_e=0.8864$$

The mass flux is given by Eq. 7.3.17 and is found to be

$$\dot{m}=p_0\sqrt{\frac{5}{RT_0}}\mathrm{M}A\left(1+\frac{k-1}{2}\mathrm{M}^2\right)^{(k+1)/2(1-k)}$$

$$=500\,000\sqrt{\frac{1.4}{287\times 93}}\times 0.8864\times 0.001\left(1+\frac{0.4}{2}\times 0.8864^2\right)^{-2.4/0.8}$$

$$=\underline{1.167 \text{ kg/s}}$$

(b) Choked flow occurs and thus $\mathrm{M}_e = 1$; Eq. 7.3.18 yields

$$\dot{m}=p_0A^*\sqrt{\frac{k}{RT_0}}\left(\frac{k+1}{2}\right)^{(k+1)/2(1-k)}$$

$$=500\,000\times 0.001\sqrt{\frac{1.4}{287\times 293}}\left(\frac{2.4}{2}\right)^{-2.4/0.8}=\underline{1.181 \text{ kg/s}}$$

Now, let us use the isentropic flow table (Table D.1) and solve parts (a) and (b).

(a) For a pressure ratio of $p/p_0 = 300/500 = 0.6$, we interpolate to find

$$\mathrm{M}_e=\frac{0.6041-0.6}{0.6041-0.5913}\times 0.02+0.88=0.886$$

$$\frac{T_e}{T_0}=\frac{0.6041-0.6}{0.6041-0.5913}(0.8606-0.8659)+0.8659=0.864$$

$$\therefore T_e=0.864\times 293=253\,\mathrm{K}$$

The velocity and density are, respectively,

$$V=\mathrm{M}c=0.886\sqrt{1.4\times 287\times 253}=282 \text{ m/s}$$

$$\rho=\frac{p}{RT}=\frac{300}{0.287\times 253}=4.13\,\mathrm{kg/m^3}$$

The mass flux is then

$$\dot{m} = \rho AV = 4.13 \times 0.001 \times 282 = \underline{1.165\,\text{kg/s}}$$

(b) For choked flow we know that $M_e = 1$. The table gives

$$\frac{T_e}{T_0} = 0.8333 \quad \text{and} \quad \frac{p_e}{p_0} = 0.5283$$

Thus the temperature, velocity, and density are, respectively,

$$T = 0.8333 \times 293 = 244.2\,\text{K}$$

$$V = \text{M}c = 1 \times \sqrt{1.4 \times 287 \times 244.2} = 313.2\,\text{m/s}$$

$$\rho = \frac{p}{RT} = \frac{0.5283 \times 500}{0.287 \times 244.2} = 3.769\,\text{kg/m}^3$$

The mass flux is calculated to be

$$\dot{m} = \rho AV = 3.769 \times 0.001 \times 313.2 = \underline{1.180\,\text{kg/s}}$$

The results using the equations are essentially the same as those using the tables.

Example 7.3

A converging–diverging nozzle, with an exit area of 40 cm^2 and a throat area of 10 cm^2, is attached to a reservoir with T = 20°C and p = 500 kPa absolute. Determine the two exit pressures that result in M = 1 at the throat for an isentropic flow. Also, determine the associated exit temperatures and velocities. See Figure 7.7.

Solution

The exit pressures we seek are associated with curves C and D of Figure 7.7. The area ratio is

$$\frac{A}{A^*} = \frac{40}{10} = 4$$

We could solve Eq. 7.3.19 for M using a trial-and-error technique; however, let us use the isentropic flow table, Table D.1. There are two entries for A/A^* = 4. Interpolation gives

$$\left(\frac{p}{p_0}\right)_C = \frac{4.182 - 4.0}{4.182 - 3.673}(0.9823 - 0.9864) + 0.9864 = 0.9849$$

$$\left(\frac{p}{p_0}\right)_D = \frac{4.0 - 3.999}{4.076 - 3.999}(0.02891 - 0.02980) + 0.02980 = 0.02979$$

Hence, the two exit pressures that will result in isentropic flow are

$$p_C = \underline{492.4 \text{ kPa}} \qquad \text{and} \qquad p_D = \underline{14.9 \text{ kPa}}$$

Note the very small pressure difference (7.6 kPa) between receiver and reservoir necessary to create the flow condition of curve *C* of Figure 7.7.

The exit temperature ratios and Mach numbers are interpolated to be

$$\left(\frac{T}{T_0}\right)_C = 0.3576(0.9949 - 0.9961) + 0.9961 = 0.9957$$

$$\left(\frac{T}{T_0}\right)_D = 0.01299(0.3633 - 0.3665) + 0.3665 = 0.3665$$

$$\mathrm{M}_C = 0.3576 \times 0.02 + 0.14 = 0.147$$

$$\mathrm{M}_D = 0.01299 \times 0.02 + 2.94 = 2.94$$

The exit temperatures associated with curves *C* and *D* are thus

$$T_C = 0.9957 \times 293 = \underline{291.7\,\mathrm{K}}$$

$$T_D = 0.3665 \times 293 = \underline{107.4\,\mathrm{K}}$$

The exit velocities are found from $V = \mathrm{M}c$ to be

$$V_C = 0.147\sqrt{1.4 \times 287 \times 291.7} = \underline{50.3\,\mathrm{m/s}}$$

$$V_D = 2.94\sqrt{1.4 \times 287 \times 107.4} = \underline{611\,\mathrm{m/s}}$$

Example 7.4

Gas flows are assumed to be incompressible flows at Mach numbers less than about 0.3. Determine the error involved in calculating the stagnation pressure for an airflow with $\mathrm{M} = 0.3$.

Solution

For an incompressible airflow the energy equation (2.5.20) with no losses would give

$$p_0 = p_1 + \rho\frac{V_1^2}{2}$$

where $V_0 = 0$ at the stagnation point. The isentropic flow equation (7.3.13), with $k = 1.4$, gives

$$p_0 = p_1\left(1 + 0.2\mathrm{M}_1^2\right)^{3.5}$$

Use the binomial theorem $(1+x)^n = 1+nx+n(n-1)x^2/2! + \ldots$ and express this as, letting $x = 0.2\text{M}_1^2$,

$$p_0 = p_1\left(1+0.7\text{M}_1^2+0.175\text{M}_1^4+\ldots\right)$$

This can be written as (see Eqs. 7.2.9 and 7.2.12)

$$\begin{aligned} p_0 - p_1 &= p_1\text{M}_1^2\left(0.7+0.175\text{M}_1^2+\ldots\right) \\ &= \frac{\rho_1 V_1^2}{1.4}\left(0.7+0.175\text{M}_1^2+\ldots\right) \\ &= \rho_1\frac{V_1^2}{2}\left(1+0.25\text{M}_1^2+\ldots\right) \end{aligned}$$

Substituting $\text{M}_1 = 0.3$, we see that

$$p_0 - p_1 = \rho_1\frac{V_1^2}{2}(1+0.0225+\ldots)$$

Comparing this with the incompressible flow equation, we see that the error is only slightly greater than 2%. Hence it is reasonable to approximate a gas flow below M = 0.3 ($V \cong 100$ m/s for air at standard conditions) with an incompressible flow for many engineering applications.

7.4 NORMAL SHOCK WAVE

Small-amplitude disturbances travel at the speed of sound, as was established in Section 7.2. In this section a large-amplitude disturbance will be studied. Its speed of propagation and its effect on other flow properties, such as pressure and temperature, will be considered, Large-amplitude disturbances occur in a number of situations. Examples include the flow in a gun barrel ahead of the projectile, the exit flow from a rocket or jet engine nozzle, the airflow around a supersonic aircraft, and the expanding front due to an explosion. Such large disturbances propagating through a gas are called **shock waves**. They can be oriented normal to the flow or at oblique angles. In this section we consider only the normal shock wave that occurs in a tube or directly in front of a blunt object; the image in Figure 7.8 shows the shock wave in front of a sphere.

The property changes that occur across a shock wave take place over an extremely short distance. For usual conditions the distance is only several mean free paths of the molecules, on the order of 10^{-4} mm. Phenomena such as viscous dissipation and heat conduction that occur within the shock wave will not be studied here. We will treat the shock wave as a zero-thickness discontinuity in the flow and allow our integral control volume equations to relate the global quantities of interest.

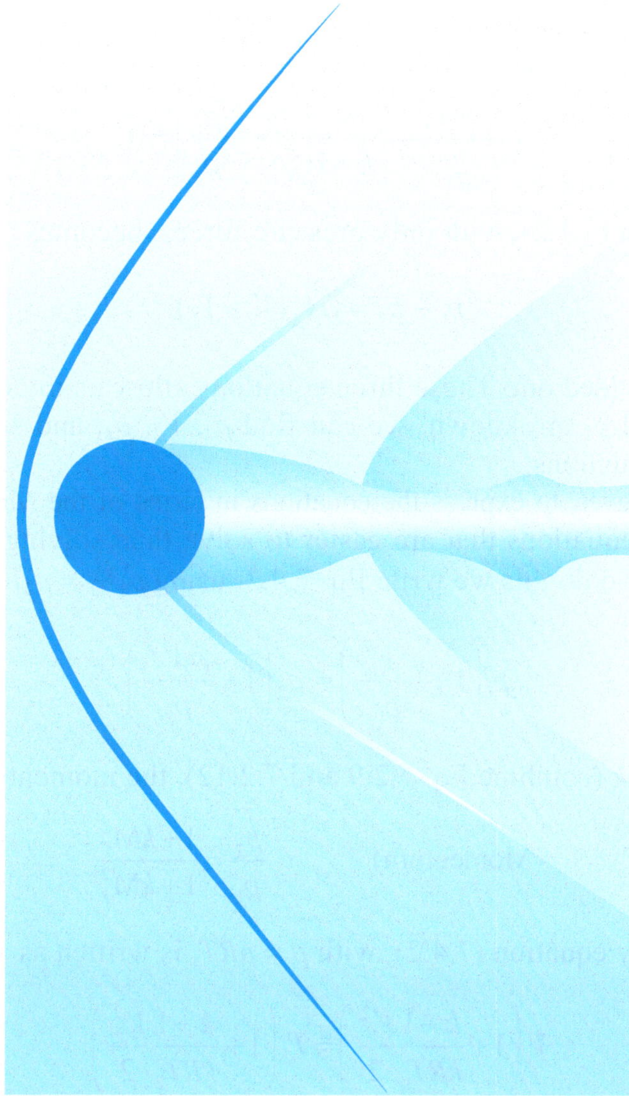

Figure 7.8 A shock wave is observed in front of a sphere at M = 1.53.

Consider a normal shock wave moving with velocity V_1. We can make it stationary by moving the flow in a tube at velocity V_1, as shown in Figure 7.9. The continuity equation, recognizing that $A_1 = A_2$, is

$$\rho_1 V_1 = \rho_2 V_2 \tag{7.4.1}$$

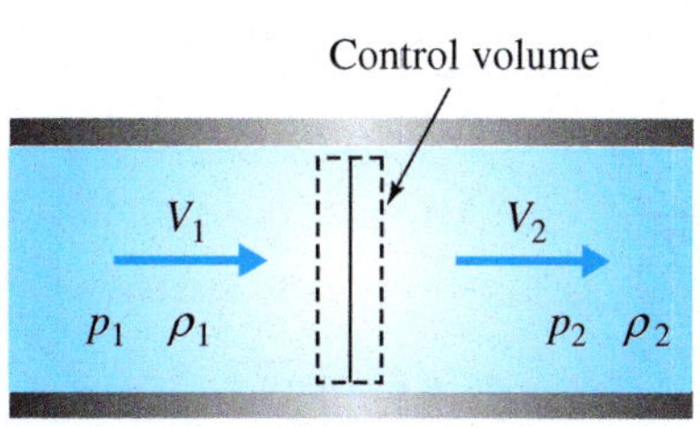

Figure 7.9 Stationary shock wave in a tube.

The energy equation (7.1.5), with $\dot{Q} = \dot{W}_S = 0$, is

$$\frac{V_2^2 - V_1^2}{2} + \frac{k}{k-1}\left(\frac{p_2}{\rho_2} - \frac{p_1}{\rho_1}\right) = 0 \tag{7.4.2}$$

The momentum equation (7.1.2), with only pressure forces, becomes

$$p_1 - p_2 = \rho_1 V_1 (V_2 - V_1) \tag{7.4.3}$$

where the areas have divided out. These three equations allow us to determine three unknown quantities; if ρ_1, V_1, and p_1 are known, we can find ρ_2, V_2, p_2, and subsequently T_2 and M_2, using the appropriate equations.

It is convenient, however, to express the equations in terms of the Mach numbers M_1 and M_2. This results in a set of equations that are easier to solve than solving the three simultaneous equations listed above. To do this we write Eq. 7.4.3, using $\rho_1 V_1 = \rho_2 V_2$, in the form

$$p_1\left(1 + \frac{\rho_1 V_1^2}{p_1}\right) = p_2\left(1 + \frac{\rho_2 V_2^2}{p_2}\right) \tag{7.4.4}$$

Introducing $\mathrm{M}^2 = V^2 \rho/pk$ (combine Eq. 7.2.9 and 7.2.12), the momentum equation becomes

$$\text{(Momentum)} \qquad \frac{p_2}{p_1} = \frac{1 + k\mathrm{M}_1^2}{1 + k\mathrm{M}_2^2} \tag{7.4.5}$$

Similarly, the energy equation (7.4.2), with $p = \rho RT$, is written as

$$T_1\left(1 + \frac{k-1}{kRT_1}\frac{V_1^2}{2}\right) = T_2\left(1 + \frac{k-1}{kRT_2}\frac{V_2^2}{2}\right) \tag{7.4.6}$$

or, substituting $\mathrm{M}^2 = V^2/kRT$, the energy equation takes the form

$$\text{(Energy)} \qquad \frac{T_2}{T_1} = \frac{1 + \frac{k-1}{2}\mathrm{M}_1^2}{1 + \frac{k-1}{2}\mathrm{M}_2^2} \tag{7.4.7}$$

If we substitute $\rho = p/RT$ into the continuity equation (7.4.1), we have

$$\frac{p_1 V_1}{RT_1} = \frac{p_2 V_2}{RT_2} \tag{7.4.8}$$

which becomes, using $V = \mathrm{M}\sqrt{kRT}$,

$$\frac{p_2}{p_1}\frac{\mathrm{M}_2}{\mathrm{M}_1}\sqrt{\frac{T_1}{T_2}} = 1 \tag{7.4.9}$$

Substituting for the pressure and temperature ratios from Eqs. 7.4.5 and 7.4.7, the continuity equation takes the form

$$\text{(Continuity)} \qquad \frac{M_1\left(1+\frac{k-1}{2}M_1^2\right)^{1/2}}{1+kM_1^2} = \frac{M_2\left(1+\frac{k-1}{2}M_2^2\right)^{1/2}}{1+kM_2^2} \tag{7.4.10}$$

Hence the downstream Mach number is related to the upstream Mach number (an algebraic challenge) by

$$M_2^2 = \frac{M_1^2 + \frac{2}{k-1}}{\frac{2k}{k-1}M_1^2 - 1} \tag{7.4.11}$$

This allows us to express the pressure and temperature ratios in terms of M_1 only. The momentum equation (7.4.5) takes the form

$$\frac{p_2}{p_1} = \frac{2k}{k+1}M_1^2 - \frac{k-1}{k+1} \tag{7.4.12}$$

and the energy equation becomes

$$\frac{T_2}{T_1} = \frac{\left(1+\frac{k-1}{2}M_1^2\right)\left(\frac{2k}{k-1}M_1^2 - 1\right)}{\frac{(k+1)^2}{2(k-1)}M_1^2} \tag{7.4.13}$$

For air, with $k = 1.4$, the preceding three equations reduce to

$$\begin{aligned} M_2^2 &= \frac{M_1^2+5}{7M_1^2-1} \\ \frac{p_2}{p_1} &= \frac{7M_1^2-1}{6} \\ \frac{T_2}{T_1} &= \frac{\left(M_1^2+5\right)\left(7M_1^2-1\right)}{36M_1^2} \end{aligned} \tag{7.4.14}$$

From the first of these three equations, we observe:

- If $M_1 = 1$, then $M_2 = 1$ and no shock wave exists.
- If $M_1 > 1$, then $M_2 < 1$ and the normal shock wave converts a supersonic flow into a subsonic flow.
- If $M_1 < 1$, then $M_2 > 1$ and a subsonic flow appears to be converted into a supersonic flow by the presence of a normal shock wave. This possibility is eliminated by the second law since it would demand a decrease in entropy by a process in an isolated system, an impossibility.

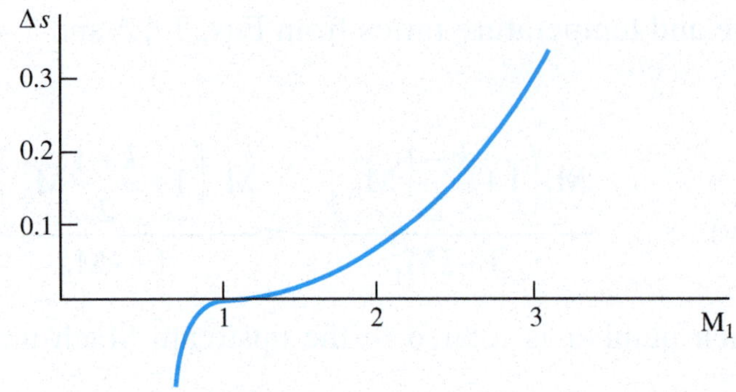

Figure 7.10 Entropy change for a normal shock in air.

The impossibility stated above is observed by considering the entropy increase, given by

$$
\begin{aligned}
s_2 - s_1 &= c_p \ln \frac{T_2}{T_1} - R \ln \frac{p_2}{p_1} \\
&= c_p \ln \frac{2+(k-1)M_1^2}{2+(k-1)M_2^2} - R \ln \frac{1+kM_1^2}{1+kM_2^2}
\end{aligned}
\qquad (7.4.15)
$$

For air, with $k = 1.4$, this is plotted in Figure 7.10, relating M_2 to M_1, with Eq. 7.4.11. Note the impossible negative entropy change whenever $M_1 < 1$.

The relationship among the thermodynamic properties, listed in the equations above, can be demonstrated graphically with reference to the *T*-*s* diagram in Figure 7.11. The conditions upstream of the normal shock are designated by state 1, and downstream by state 2. Note the dashed line from state 1 to state 2 for the irreversible process that occurs inside the shock wave.

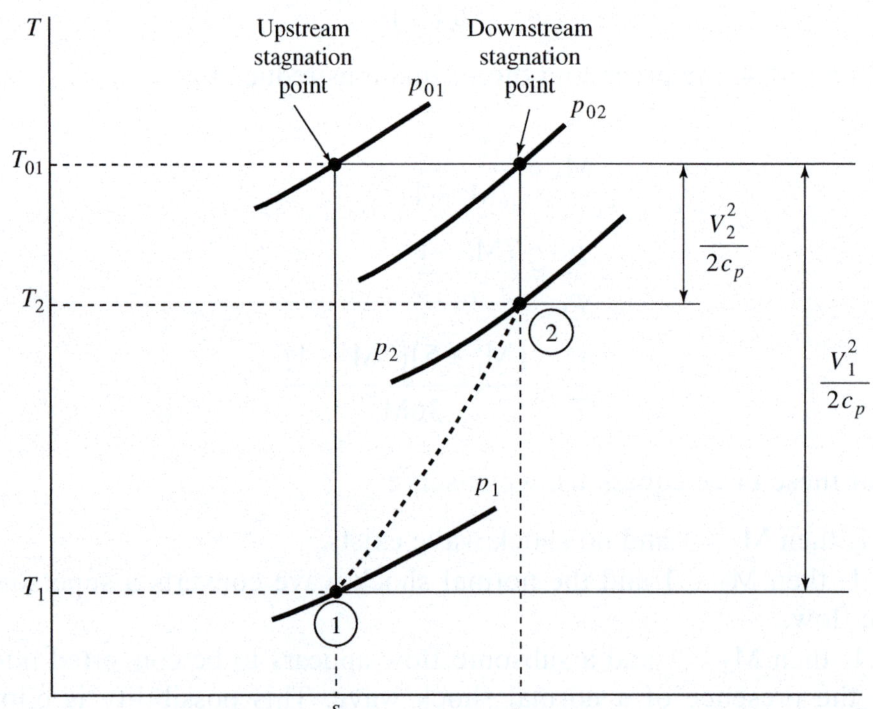

Figure 7.11 *T*-*s* diagram for a normal shock wave.

The energy equation, with $\dot{Q} = \dot{W}_S = 0$, can be written as

$$\frac{V_1^2}{2c_p} + T_1 = \frac{V_2^2}{2c_p} + T_2 \tag{7.4.16}$$

The stagnation temperature is defined as the temperature that would exist if the flow is brought to rest isentropically. Thus, the energy equation gives

$$T_{01} = T_{02} \tag{7.4.17}$$

as shown in the figure. The substantial decrease in stagnation pressure, $p_{02} < p_{01}$, is also observed in Figure 7.11. If the entropy increases from state 1 to state 2, as it must, the stagnation pressure p_{02} must decrease, as shown, if we are to maintain $T_{01} = T_{02}$.

Gas tables are available that give the pressure ratio, the temperature ratio, the downstream Mach number, and the stagnation pressure ratio as a function of the upstream Mach number. Table D.2 is such a table for $k = 1.4$ and includes the ratios given by Eqs. 7.4.14. Note that M_2 is always less than unity, p_2 is always greater than p_1, T_2 is always greater than T_1, and p_{02} is always less than p_{01}.

Example 7.5

A normal shock wave passes through stagnant air at 60°F and atmospheric pressure of 12 psia with a speed of 1500 ft/s. Calculate the pressure and temperature downstream of the shock wave. Use (a) the equations and (b) the gas tables.

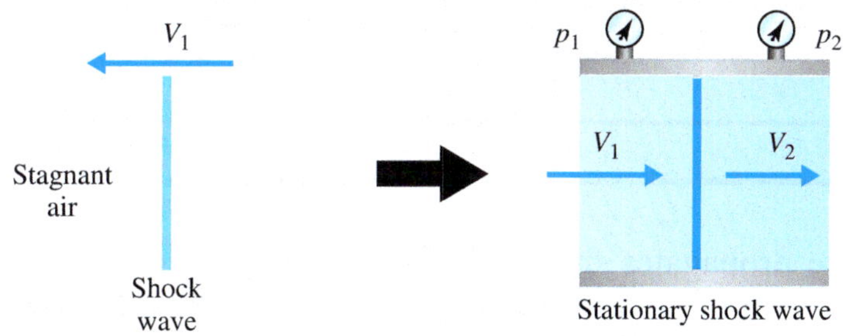

Figure E7.5

Solution

We consider the shock wave to be stationary with $V_1 = 1500$ ft/s and $p_1 = 12$ psia.

(a) To use the simplified equations (7.4.14) we must know the upstream Mach number. It is

$$\mathrm{M}_1 = \frac{V_1}{c_1} = \frac{V_1}{\sqrt{kRT_1}}$$

$$= \frac{1500\,\text{ft/s}}{\sqrt{1.4 \times 1716(\text{ft-lb/slug-°R}) \times 520\text{°R}}} = 1.342$$

The pressure and temperature are then found to be

$$p_2 = \frac{p_1\left(7\mathrm{M}_1^2 - 1\right)}{6}$$
$$= \frac{12(7 \times 1.342^2 - 1)}{6} = \underline{23.21\,\text{psia}}$$

$$T_2 = \frac{T_1\left(\mathrm{M}_1^2 + 5\right)\left(7\mathrm{M}_1^2 - 1\right)}{36\mathrm{M}_1^2}$$
$$= \frac{520(1.342^2 + 5)(7 \times 1.342^2 - 1)}{36 \times 1.342^2} = \underline{633.1°\text{R}}$$

(b) From part (a), we use $\mathrm{M}_1 = 1.342$. Interpolation in Table D.2 yields

$$\frac{p_2}{p_1} = \frac{1.342 - 1.34}{1.36 - 1.34}(1.991 - 1.928) + 1.928 = 1.934$$

$$\frac{T_2}{T_1} = \frac{1.342 - 1.34}{1.36 - 1.34}(1.229 - 1.216) + 1.216 = 1.217$$

Using the information given, we have

$$p_2 = 12 \times 1.934 = \underline{23.21\text{ psia}}$$

$$T_2 = 520 \times 1.217 = \underline{632.8°\text{R}}$$

Example 7.6

A normal shock wave propagates through otherwise stagnant air at standard conditions at a speed of 700 m/s. Determine the speed induced in the air immediately behind the shock wave as shown in Figure E7.6. Use the equations.

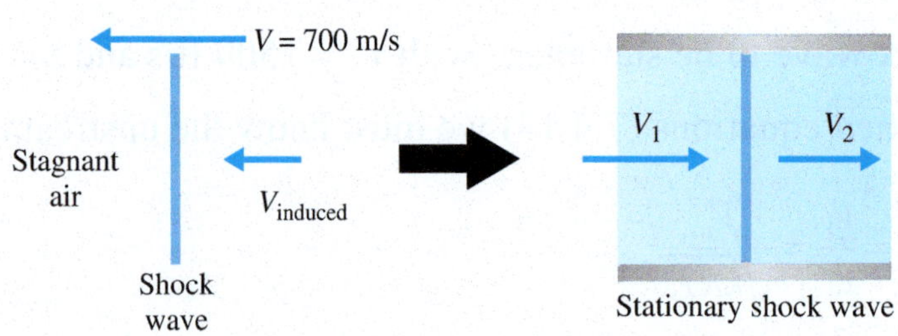

Figure E7.6

Solution

For standard conditions the temperature is 15°C. The upstream Mach number is thus

$$M_1 = \frac{V_1}{c_1} = \frac{V_1}{\sqrt{kRT_1}} = \frac{700}{\sqrt{1.4 \times 287 \times 288}} = 2.06$$

Using Eq. 7.4.14, we find that

$$M_2 = \left(\frac{M_1^2 + 5}{7M_1^2 - 1}\right)^{1/2} = \left(\frac{2.06^2 + 5}{7 \times 2.06^2 - 1}\right)^{1/2} = 0.567$$

$$T_2 = \frac{T_1\left(M_1^2 + 5\right)\left(7M_1^2 - 1\right)}{36M_1^2}$$

$$= \frac{288(2.06^2 + 5)(7 \times 2.06^2 - 1)}{36 \times 2.06^2} = 500.2 \text{ K}$$

This allows us to calculate

$$V_2 = M_2 c_2 = 0.567\sqrt{1.4 \times 287 \times 500.2} = 254.4 \text{ m/s}$$

This velocity assumes a flow with the shock wave stationary and the air approaching the shock wave at 700 m/s. If we superimpose a velocity of 700 m/s moving opposite to V_1, we find the induced velocity to be

$$V_{\text{induced}} = V_2 - V_1$$

$$= 254.4 - 700 = \underline{-446 \text{ m/s}}$$

where the negative sign means that the induced velocity would be moving to the left if V_1 is to the right. The induced velocity would be in the same direction as the propagation of the shock wave. Such large induced velocities are responsible for much of the damage away from the bomb center caused by explosions of high-power bombs.

7.5 SHOCK WAVES IN CONVERGING-DIVERGING NOZZLES

The converging-diverging nozzle has already been presented for an isentropic flow; for receiver-to-reservoir pressure ratios between those of curves *C* and *D* of Figures 7.7 and 7.12, shock waves exist in the flow either inside or outside the nozzle. If $p_r/p_0 = a$ (locate *a* on the vertical axis on the right of Figure 7.12), a normal shock wave would exist at an internal location in the diverging portion of the nozzle. Usually, the location of the shock wave is prescribed in student problems since a trial-and-error solution is necessary to locate the shock. When $p_r/p_0 = b$ the normal shock wave is located in the exit plane of the nozzle. For a pressure ratio less than *b* but greater than *e*, two types of oblique shock wave patterns are

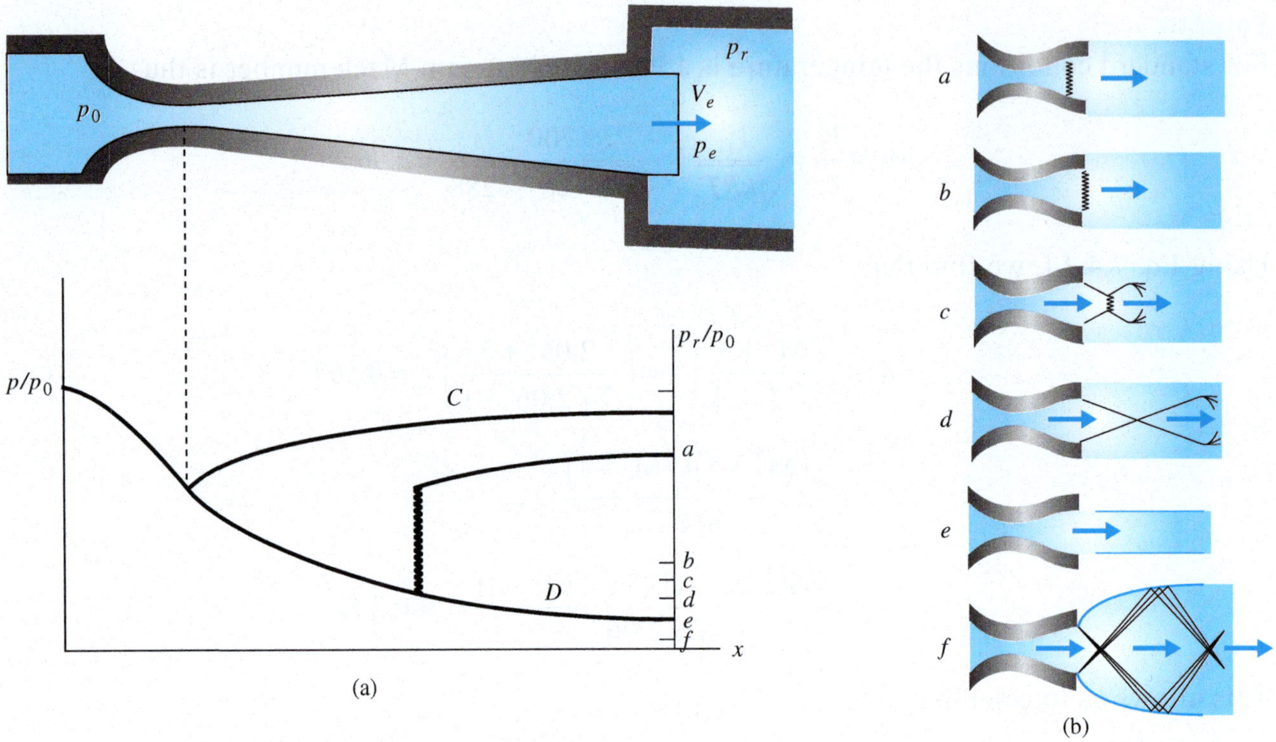

Figure 7.12 Converging-diverging nozzle.

observed, one with a central normal shock wave, as sketched for $p_r/p_0 = c$, and one with only oblique waves, as sketched for $p_r/p_0 = d$. The pressure ratios that result in oblique shock waves will not be considered here. As we move from *d* to *e*, the oblique shock waves become weaker and weaker until the isentropic flow is again realized at $p_r/p_0 = e$ with all shock waves absent. For pressure ratios below *e* a very complicated flow exists. The flow turns the corner at the nozzle exit rather abruptly due to expansion waves (isentropic waves to be considered in Section 7.8), then turns back due to the same expansion waves, resulting in a billowing out of the exhaust flow, as is visible from high-altitude satellite rocket engines. Let us work some examples for the converging-diverging nozzle; no new equations are necessary.

Example 7.7

A converging-diverging nozzle has a throat diameter of 5 cm and an exit diameter of 10 cm. The reservoir is the laboratory, maintained at atmospheric conditions of 20°C and 90 kPa absolute. Air is constantly pumped from a receiver so that a normal shock wave stands across the exit plane of the nozzle. Determine the receiver pressure and the mass flux.

Solution

Isentropic flow occurs from the reservoir, to the throat, to the exit plane in front of the normal shock wave at state 1. Supersonic flow occurs downstream of the throat making the throat the critical area. Hence

$$\frac{A_1}{A^*} = \frac{10^2}{5^2} = 4$$

Interpolation in the isentropic flow table (Table D.1) gives

$$M_1 = 2.94 \qquad \frac{p_1}{p_0} = 0.0298$$

Hence, the pressure in front of the normal shock is

$$p_1 = p_0 \times 0.0298$$
$$= 90 \times 0.0298 = 2.68 \text{ kPa}$$

From the normal shock table (Table D.2), using $M_1 = 2.94$, we find that

$$\frac{p_2}{p_1} = 9.918$$
$$\therefore p_2 = 9.918 \times 2.68 = \underline{26.6 \text{ kPa}}$$

This is the receiver pressure needed to orient the shock across the exit plane as shown for $P_r/p_0 = b$ in Figure 7.12.

To find the mass flux through the nozzle, we need only consider the throat. Recognizing that $M_t = 1$, so that $V_t = c_t$, we can write

$$\dot{m} = \rho_t A_t V_t = \frac{p_t}{RT_t} A_t \sqrt{kRT_t} = p_t A_t \sqrt{\frac{k}{RT_t}}$$

The isentropic flow table yields

$$\frac{p_t}{p_0} = 0.5283 \qquad \frac{T_t}{T_0} = 0.8333$$

Thus the mass flux becomes

$$\dot{m} = (0.5283 \times 90\,000) \times \frac{\pi \times 0.05^2}{4} \sqrt{\frac{1.4}{287 \times (0.8333 \times 293)}}$$
$$= \underline{0.417 \text{ kg/s}}$$

Remember, the pressure must be measured in pascals in the equation above. Check the units to make sure that the units on $\dot{m}$ are kg/s.

Example 7.8

Air flows from a reservoir at 20°C and 200 kPa absolute through a 5-cm-diameter throat and exits from a 10-cm-diameter nozzle, as shown in Figure E7.8. Calculate the exit pressure p_e needed to locate a normal shock wave at a position where the diameter is 7.5 cm.

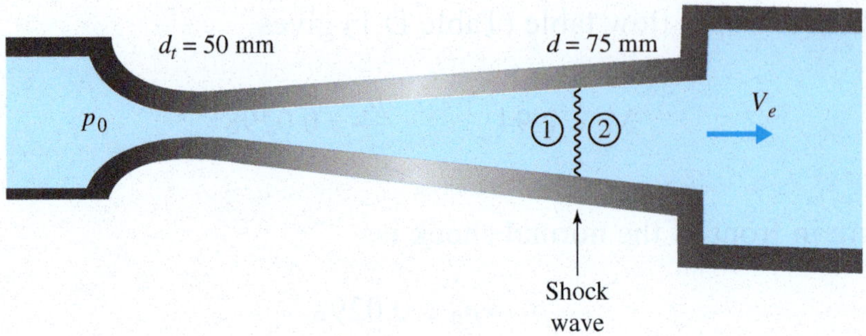

Figure E7.8

Solution

We will use the gas tables for this flow represented by the curve established by $p_r/p_0 = a$ in Figure 7.12. The throat is a critical area since for a supersonic flow $M_t = 1$. The area ratio is

$$\frac{A_1}{A^*} = \frac{7.5^2}{5^2} = 2.25$$

For this area ratio we find, from Table D.1, that

$$M_1 = 2.33$$

Then from Table D.2, for this Mach number, we obtain

$$M_2 = 0.531 \qquad \frac{p_{02}}{p_{01}} = 0.570$$

The reservoir pressure $p_0 = p_{01}$. Thus

$$p_{02} = 0.570 \times 200 = 114 \text{ kPa}$$

Isentropic flow occurs from state 2 immediately after the normal shock wave to the exit. Hence, for $M_2 = 0.531$, we find from Table D.1 that

$$\frac{A_2}{A^*} = 1.285$$

If A_e is the exit area, then

$$\frac{A_e}{A^*} = \frac{A_2}{A^*} \times \frac{A_e}{A_2} = 1.285 \times \frac{10^2}{7.5^2} = 2.284$$

The Mach number and pressure ratio corresponding to this area ratio are

$$M_e = 0.265 \qquad \frac{p_e}{p_{0e}} = 0.952$$

For isentropic flow between the shock and the exit, we know that $p_{02} = p_{0e}$; thus

$$p_e = p_{02} \times 0.952 = 114 \times 0.952 = \underline{109\ \text{kPa}}$$

Note the usefulness of the critical area ratio in obtaining the desired results.

Example 7.9

A pitot probe, the device used to measure the stagnation pressure in a flow, is inserted into an airstream and measures 300 kPa absolute, as shown in Figure E7.9. The pressure in the flow is measured to be 75 kPa absolute. If the temperature at the stagnation point of the probe is measured as 150°C, determine the free-stream velocity V.

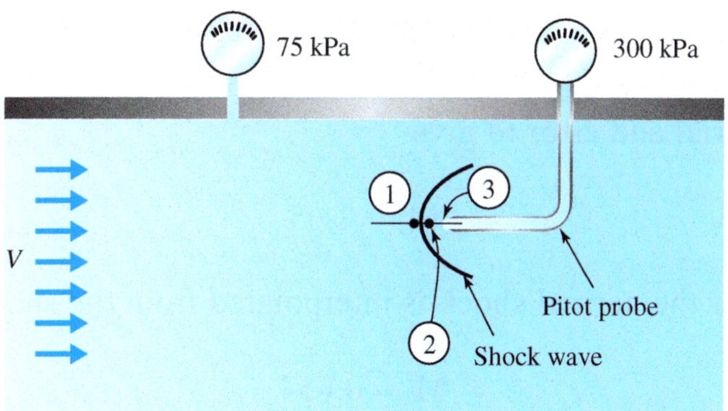

Figure E7.9

Solution

When a blunt object is placed in a supersonic flow a detached shock wave forms around the object, as it does around the front of the pitot probe shown. The flow that meets the front of the pitot probe at the stagnation point passes through a normal shock wave from state 1 to state 2; the subsonic flow at state 2 then decelerates isentropically to state 3, the stagnation point.

For the isentropic flow from state 2 to state 3 we can use Eq. 7.3.13,

$$\frac{p_3}{p_2} = \left(1 + \frac{k-1}{2}\text{M}_2^2\right)^{k/(k-1)}$$

Across the normal shock we know that (see Eq. 7.4.12)

$$\frac{p_2}{p_1} = \frac{2k}{k+1}\text{M}_1^2 - \frac{k-1}{k+1}$$

Also, the Mach numbers are related by Eq. 7.4.11,

$$\text{M}_2^2 = \frac{(k-1)M_1^2 + 2}{2kM_1^2 - k + 1}$$

The three equations above can be combined, with some algebraic manipulation, to yield the **Rayleigh-pitot-tube formula** for supersonic flows, namely,

$$\frac{p_3}{p_1}=\frac{\left(\dfrac{k+1}{2}\mathrm{M}_1^2\right)^{k/(k-1)}}{\left(\dfrac{2k\mathrm{M}_1^2}{k+1}-\dfrac{k-1}{k+1}\right)^{1/(k-1)}}$$

Substituting $k = 1.4$, $p_3 = 300$ kPa, and $p_1 = 75$ kPa, we have

$$\frac{300}{75}=\frac{\left(1.2\mathrm{M}_1^2\right)^{3.5}}{\left(1.167\mathrm{M}_1^2-0.1667\right)^{2.5}}$$

This can be solved by trial and error to give

$$\mathrm{M}_1 = 1.65$$

The Mach number, after the normal shock is interpolated from the shock table to be

$$\mathrm{M}_2 = 0.654$$

Using the isentropic flow table with this Mach number, we interpolate the temperature at state 2 to be

$$T_2 = T_3 \times 0.921$$
$$= 423 \times 0.921 = 389.6 \text{ K}$$

The temperature in front of the normal shock is found by using the normal shock table as follows:

$$\frac{T_2}{T_1}=1.423$$

$$\therefore T_1=\frac{T_2}{1.423}=274\,\text{K}$$

Finally, the velocity before the normal shock is given by

$$V_1 = \mathrm{M}_1 c_1 = \mathrm{M}_1\sqrt{kRT_1}$$
$$= 1.65\sqrt{1.4\times 287\times 274} = 547 \text{ m/s}$$

7.6 VAPOR FLOW THROUGH A NOZZLE

The flow of vapor through a nozzle forms a very important engineering problem. High-pressure steam flows through the nozzles of turbines in electrical generating plants; in this section we present the technique for analyzing such a problem. We recall, however, that vapor that is not substantially superheated does not behave very well as an ideal gas; the vapor tables must be consulted since c_p and c_v cannot be assumed to be constant. Consider the problem of a superheated vapor entering the nozzle of Figure 7.12. The flow is isentropic unless a shock wave is encountered, and the expansion is as sketched on the *T-s* diagram of Figure 7.13. Assume that the flow initiates in a reservoir with stagnation conditions, flows through a throat indicated by state *t*, and exits out a diverging section to the exit at state *e*.

Note that the exit state could very likely be in the quality region with the possibility of condensation of liquid droplets; however, for sufficiently high flow velocities there may not be sufficient time for the formation of the droplets and the associated heat transfer process. This produces a situation called **supersaturation** and a condition of **metastable equilibrium** exists; that is, the nonequilibrium state *e* is reached rather than the equilibrium state *e'*. To estimate the temperature of the metastable state *e*, we assume that the steam behaves as an ideal gas, so that $Tp^{k/(k-1)} = \text{const}$. If the exit state is sufficiently far into the quality region, a **condensation shock** will be experienced, a phenomenon not considered in this book.

Because of supersaturation it is possible to model the isentropic flow of a vapor through a nozzle with acceptable accuracy by considering the ratio of specific heats to be constant. For steam $k = 1.3$ gives acceptable results over a considerable range of temperatures. The critical pressure ratio given by Eq.7.3.14 for steam becomes

$$\frac{p^*}{p_0} = \left(\frac{2}{k+1}\right)^{k/(k-1)} = 0.546 \tag{7.6.1}$$

If saturated steam enters the nozzle, some liquid droplets will be formed and entrained by the vapor; provided that a condensation shock does not exist, a good approximation to the critical pressure ratio is 0.577 corresponding to $k = 1.14$.

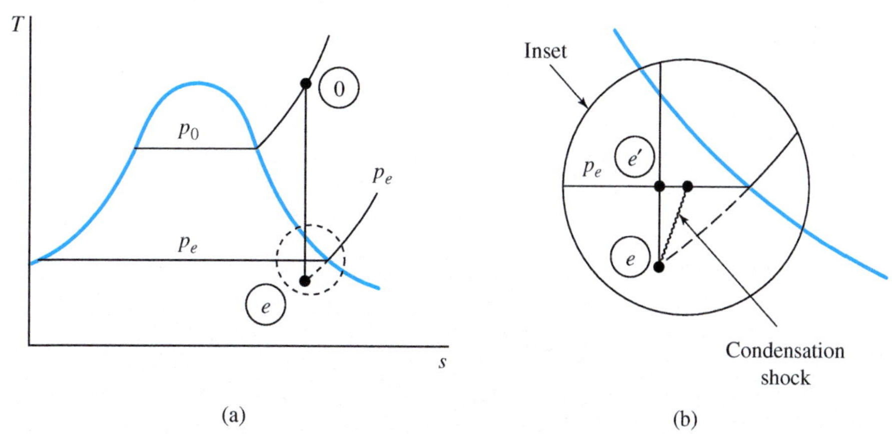

Figure 7.13 Isentropic expansion of a vapor.

Example 7.10

Steam is to be expanded isentropically from reservoir conditions of 300°C and 800 kPa absolute to an exit condition of 100 kPa absolute. If supersonic flow is desired, calculate the necessary throat and exit diameters if a mass flux of 2 kg/s is demanded.

Solution

From the steam tables (found in any thermodynamics textbook) we find that

$$s_0 = s_e = 7.2336 \text{ kJ/kg·K}$$
$$h_0 = 3056.4 \text{ kJ/kg}$$

To estimate the temperature of the metastable exit state, we use

$$T_e = T_0\left(\frac{p_e}{p_0}\right)^{(k-1)/k} = 593\left(\frac{100}{800}\right)^{0.3/1.3} = 367\,\text{K} \quad \text{or} \quad 94°\text{C}$$

Using the steam tables at this temperature, we interpolate, using the exit quality x_e, to find that

$$7.2336 = 1.239 + 6.90x_e$$
$$\therefore x_e = 0.968$$

Thus we have the enthalpy and specific volume at the exit:

$$h_e = 394 + 0.968 \times 2273 = 2594 \text{ kJ/kg}$$
$$v_e = 0.001 + 0.968 \times (2.06 - 0.001) = 1.99 \text{ m}^3/\text{kg}$$

Using the energy equation, the exit velocity is estimated as follows:

$$\cancelto{0}{\frac{V_0^2}{2}} + h_0 = \frac{V_e^2}{2} + h_e$$

$$\therefore V_e = \sqrt{2(h_0 - h_e)} = \sqrt{2(3056 - 2594) \times 1000} = 961 \text{ m/s}$$

where the 1000 converts kJ to J. From the definition of mass flux we have

$$m_e = \rho_e A_e V_e$$
$$2 = \frac{1}{1.99} \times \frac{\pi d_e^2}{4} \times 961$$
$$\therefore d_e = 0.0726 \text{ m} \quad \text{or} \quad \underline{7.26 \text{ cm}}$$

To determine the diameter of the throat, we recognize the throat to be the critical area; thus Eq. 7.6.1 gives

$$p^* = 0.546\, p_0 = 437 \text{ kPa}$$

Using this pressure and $s^* = s_0 = 7.2336$ kJ/kg·K we could use the steam tables to find h^* and v^*; the energy equation would then allow us to find V^* and thus d_t. However, a simpler, approximate technique, assuming constant specific heats, is to use Eq. 7.3.18 with $k = 1.3$ and obtain the following:

$$\dot{m} = p_0 A^* \sqrt{\frac{1.3}{RT_0}} \left(\frac{2.3}{2}\right)^{2.3/-0.6}$$

$$2 = 800\,000 \frac{\pi d_t^2}{4} \sqrt{\frac{13}{462 \times 573}} \times 0.585$$

$$\therefore d_1 = 0.049 \text{ m} \quad \text{or} \quad \underline{4.9 \text{ cm}}$$

This is reasonable since we have already assumed constant specific heats in predicting T_e. Obviously, the above is approximate; using $k = 1.3$ does, however, give reasonable predictions.

7.7 OBLIQUE SHOCK WAVE

In this section we investigate the oblique shock wave, a finite-amplitude wave that is not normal to the incoming flow. The flow approaching the oblique shock wave will be assumed to be in the x-direction. After the oblique shock the velocity vector will have a component normal to the flow direction. We will continue to assume that the flow before and after the oblique shock is uniform and steady.

Oblique shock waves form on the leading edge of a supersonic airfoil or in an abrupt corner, as sketched in Figure 7.14. Oblique shock waves may also be found on axisymmetric bodies such as a nose cone or a bullet traveling at supersonic speeds. In this book, we consider only plane flows.

The function of the oblique shock wave is to turn the flow so that the velocity vector $\mathbf{V}_2$ is parallel to the plane wall. The angle between the two velocity vectors introduces another variable into our analysis. The problem remains solvable, however, with the additional tangential momentum equation.

To analyze the oblique shock wave, consider a control volume enclosing a portion of the shock as shown in Figure 7.15. The velocity vector upstream is assumed to be in the x-direction only; the oblique shock wave makes an angle β with the upstream velocity vector and turns the flow through the **deflection angle** or **wedge angle** θ so that V_2 is parallel to the wall. The components of the velocity vectors are shown normal and tangential to the oblique

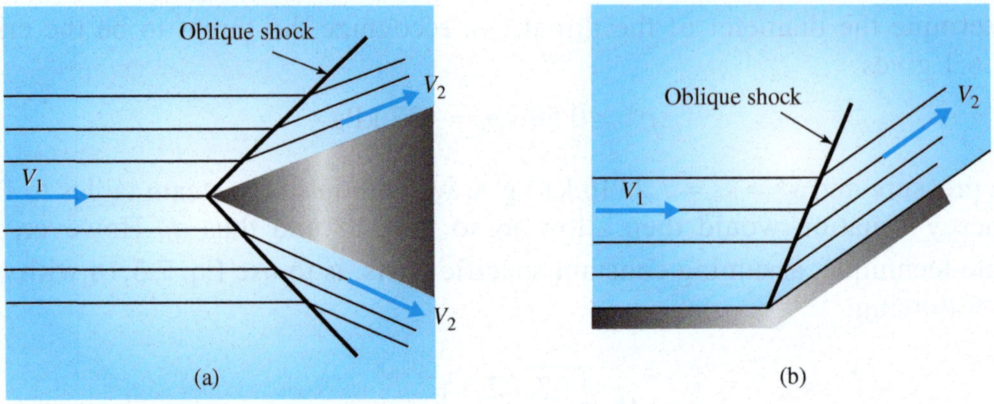

Figure 7.14 Oblique shock waves in a supersonic flow: (a) flow over a symmetrical wedge; (b) flow in a corner.

shock wave. The tangential components do not cause fluid to flow through the shock; hence the continuity equation, with $A_1 = A_2$, gives

$$\rho_1 V_{1n} = \rho_2 V_{2n} \tag{7.7.1}$$

where the normal components V_{1n} and V_{2n} are displayed in Figure 7.15. The pressure forces act normal to the oblique shock and produce no tangential components. Thus the momentum equation expressed in the tangential direction requires that the tangential momentum into the control volume equal the tangential momentum leaving the control volume; that is,

$$\dot{m}_1 V_{1t} = \dot{m}_2 V_{2t} \tag{7.7.2}$$

or, using $\dot{m}_1 = \dot{m}_2$, we demand that

$$V_{1t} = V_{2t} \tag{7.7.3}$$

The normal momentum equation takes the form

$$p_1 - p_2 = \rho_2 V_{2n}^2 - \rho_1 V_{1n}^2 \tag{7.7.4}$$

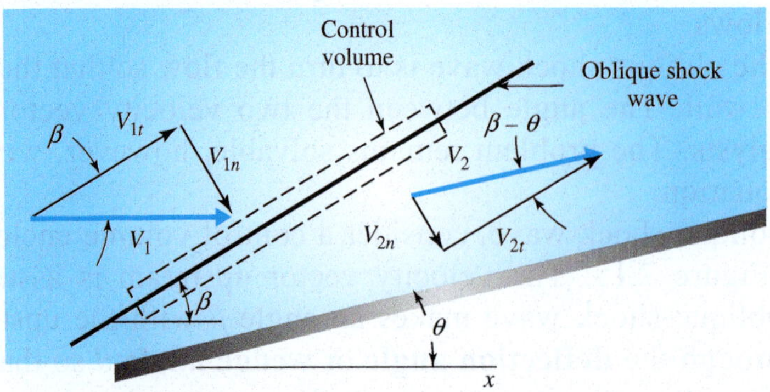

Figure 7.15 Control volume enclosing a small portion of an oblique shock wave.

The energy equation, using $V^2 = V_n^2 + V_t^2$, may be written as

$$\frac{V_{1n}^2}{2} + \frac{k}{k-1}\frac{p_1}{\rho_1} = \frac{V_{2n}^2}{2} + \frac{k}{k-1}\frac{p_2}{\rho_2} \tag{7.7.5}$$

where the tangential component terms have been canceled from both sides. Note that the tangential components of the two velocity vectors do not enter the continuity, normal momentum, or energy equations, the three equations used in the solution of the normal shock wave. Hence we may substitute V_{1n} and V_{2n} for V_1 and V_2, respectively, in the normal shock wave equations and obtain a solution. Either the normal shock wave equations or the normal shock wave table (Table D.2) may be used. We also replace M_1 and M_2 with M_{1n} and M_{2n}, respectively.

It is useful to relate the oblique shock angle β to the deflection angle θ. Using the continuity equation (7.7.1), with reference to Figure 7.15, there results

$$\frac{\rho_2}{\rho_1} = \frac{V_{1n}}{V_{2n}} = \frac{V_{1t}\tan\beta}{V_{2t}\tan(\beta-\theta)} = \frac{\tan\beta}{\tan(\beta-\theta)} \tag{7.7.6}$$

From the normal shock wave equations (7.4.12) and (7.4.13) we can find the density ratio to be

$$\frac{\rho_2}{\rho_1} = \frac{p_2T_1}{p_1T_2} = \frac{(k+1)M_{1n}^2}{(k-1)M_{1n}^2 + 2} \tag{7.7.7}$$

Substituting this into Eq. 7.7.6 gives

$$\tan(\beta-\theta) = \frac{\tan\beta}{k+1}\left[k-1+\frac{2}{M_1^2\sin^2\beta}\right] \tag{7.7.8}$$

For a flow with a given M_1, this equation relates the oblique shock angle β to the wedge or corner angle θ. The three variables β, θ, and M_1 in the equation above are often plotted as in Figure 7.16. We can observe several phenomena by studying the figure.

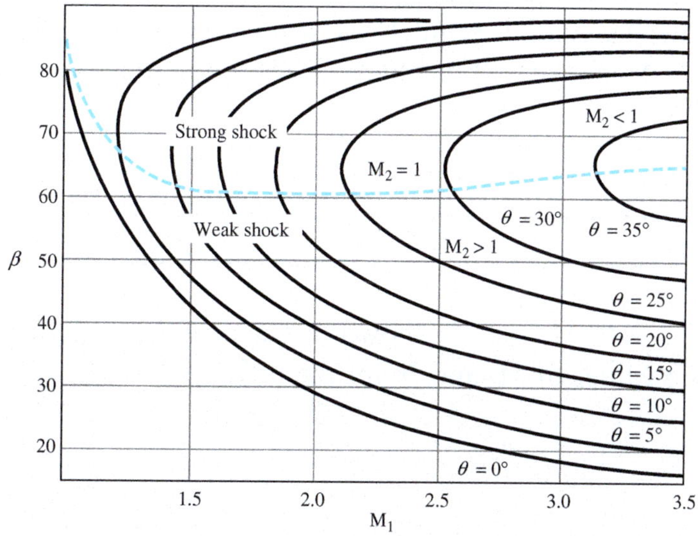

Figure 7.16 Oblique shock wave relationships for $k = 1.4$.

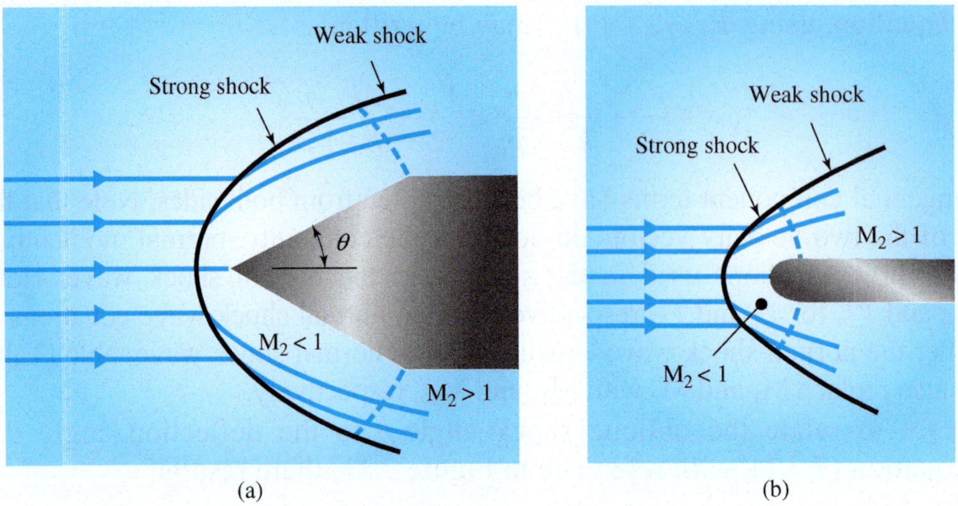

Figure 7.17 Detached shock waves: (a) flow around a wedge; (b) flow around a blunt object.

- For a specified upstream Mach number M_1 and a given wedge angle θ there are two possible oblique shock angles β, the larger one corresponding to a "strong" shock and the smaller corresponding to a "weak" shock.
- For a given wedge angle θ there is a minimum Mach number for which there is only one oblique shock angle β.
- For a given wedge angle θ, if M_1 is less than the minimum for that particular curve, no oblique shock wave exists and the shock wave becomes detached, as shown in Figure 7.17. Also, for a given M_1 there is a sufficiently large θ that will result in a detached shock wave.

The pressure rise across the oblique shock wave determines whether a weak shock or a strong shock occurs. For a relatively small pressure rise a weak shock will occur with $M_2 > 1$. If the pressure rise is relatively large a strong shock occurs with $M_2 < 1$. Note that for the detached shocks around bodies, a normal shock exists for the stagnation streamline; this is followed away from the stagnation point by a strong oblique shock, then a weak oblique shock, and eventually a Mach wave. For blunt bodies moving at supersonic speeds the shock wave is always detached.

Example 7.11

Air flows over a wedge with $M_1 = 3$, as shown in Figure E7.11. A weak shock reflects from the wall. Determine the values of M_3 and β_3 for the reflected wave.

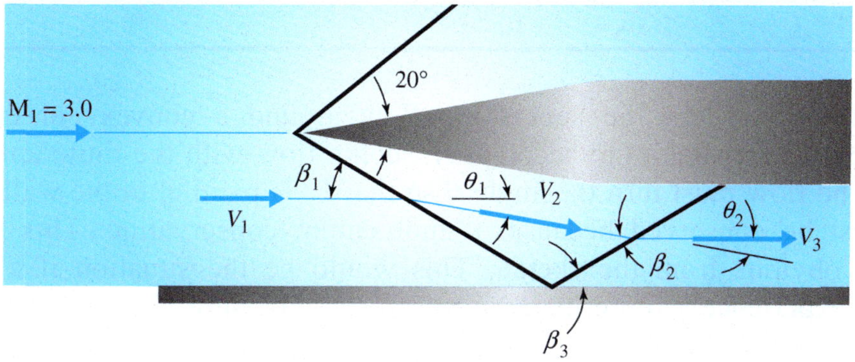

Figure E7.11

Solution

From Figure 7.16 with $\theta_1 = 10°$ and $M_1 = 3.0$, we find for the weak shock that $\beta_1 = 27.5°$. This yields

$$M_{1n} = 3 \sin 27.5° = 1.39$$

From the shock table we interpolate to find

$$M_{2n} = 0.744 = M_2 \sin (27.5° - 10°)$$
$$\therefore M_2 = 2.48$$

The reflected shock must again turn the flow through an angle of 10°, that is, $\theta_2 = 10°$. For this wedge angle and $M_2 = 2.48$ from Figure 7.16 for a weak shock, we see that $\beta_2 = 33°$. This results in

$$M_{2n} = 2.48 \sin 33° = 1.35$$

From the shock table

$$M_{3n} = 0.762 = M_3 \sin 23°$$
$$\therefore M_3 = \underline{1.95}$$

The desired angle is calculated to be

$$\beta_3 = \beta_2 - 10° = \underline{23}°$$

Note that Figure 7.16 does not allow precise calculations. Equation 7.7.8 could be used, by trial and error, to improve the accuracy of the β's and hence the quantities that follow.

7.8 ISENTROPIC EXPANSION WAVES

In this section we consider the supersonic flow around a convex corner, as shown in Figure 7.18. Let us first attempt to accomplish such a flow with the finite amplitude wave of Figure 7.18a. The flow must turn the angle θ so that V_2 is parallel to the wall. The tangential component must be conserved because of momentum conservation. This would result in $V_2 > V_1$, as is obvious from the sketch. This would be the situation if a subsonic flow, $M_{1n} < 1$, could experience a finite increase to a supersonic flow, $M_{2n} > 1$. This, of course, is impossible because of the second law, as was noted in the discussion associated with Figure 7.10. Consequently, we consider the turning of a flow around a convex corner using a finite wave as an impossibility.

Consider a second possible mechanism that would allow the flow to turn the corner, a fan composed of an infinite number of Mach waves, emanating from the corner, as shown in Figure 7.18b. The second law would not be violated with such a mechanism since each Mach wave is an isentropic wave. We will determine the effect of a single Mach wave on the flow and then integrate to obtain the total effect. Figure 7.19 shows the infinitesimal velocity change due to a single Mach wave. For the control volume enclosing the Mach wave, we know that the tangential momentum is conserved; thus the tangential velocity component remains unchanged as shown, as in the oblique shock wave. From the triangles in the figure we can write

$$V_t = V\cos\mu = (V + dV)\cos(\mu + d\theta) \tag{7.8.1}$$

Since $d\theta$ is small, using $\cos(\mu + d\theta) = \cos\mu - d\theta\sin\mu$, this becomes,[1]

$$V\sin\mu\, d\theta = \cos\mu\, dV \tag{7.8.2}$$

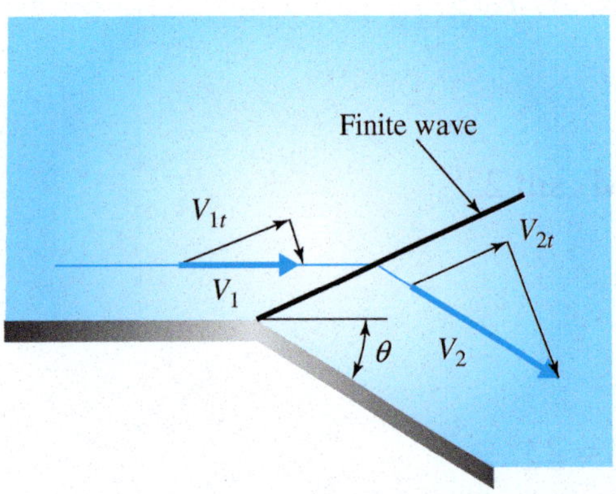

(a) Single finite wave

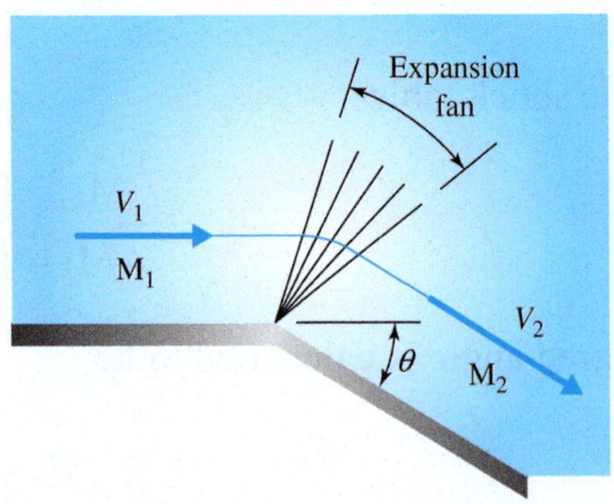

(b) Infinite number of Mach waves

Figure 7.18 Supersonic flow around a convex corner.

[1]Recall the trigonometric identity, $\cos(\alpha + \beta) \cos\alpha\cos\beta - \sin\alpha\sin\beta$. Then using $\cos d\theta = 1$ and $\sin d\theta = d\theta$, we have $\cos(\mu + d\theta) = \cos\mu - d\theta\sin\mu$.

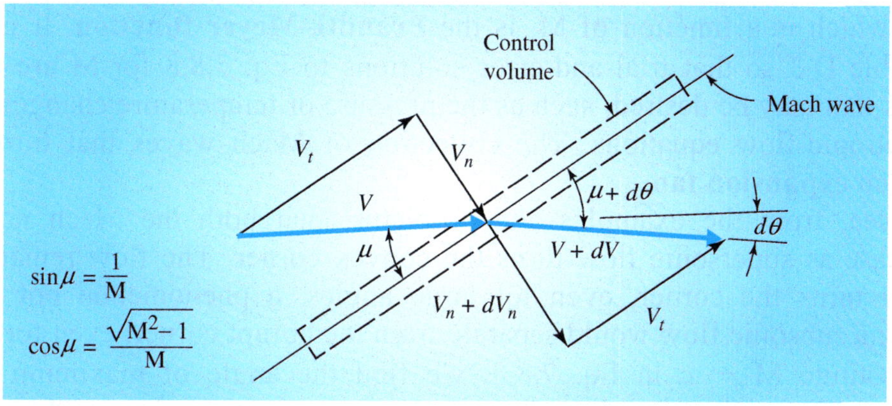

Figure 7.19 A single Mach wave.

Substituting $\sin\mu = 1/\mathrm{M}$ (see Eq. 7.2.13) and $\cos\mu = (\sqrt{\mathrm{M}^2-1})/\mathrm{M}$, we have

$$d\theta = \sqrt{\mathrm{M}^2-1}\frac{dV}{V} \tag{7.8.3}$$

The relationship $V = \mathrm{M}\sqrt{kRT}$ can be differentiated and rearranged to give

$$\frac{dV}{V} = \frac{d\mathrm{M}}{\mathrm{M}} + \frac{1}{2}\frac{dT}{T} \tag{7.8.4}$$

The energy equation (7.3.4), in the form $\frac{1}{2}V^2 + kRT/(k-1) = \text{const.}$ may also be differentiated to yield

$$\frac{dV}{V} + \frac{1}{(k-1)\mathrm{M}^2}\frac{dT}{T} = 0 \tag{7.8.5}$$

Eliminating dT/T by combining the two preceding equations, there results

$$\frac{dV}{V} = \frac{1}{2+(k-1)\mathrm{M}^2}\frac{dM}{M} \tag{7.8.6}$$

This can be substituted into Eq. 7.8.3, resulting in a relationship between θ and M. We have

$$d\theta = \frac{2\sqrt{\mathrm{M}^2-1}}{2+(k-1)\mathrm{M}^2}\frac{d\mathrm{M}}{\mathrm{M}} \tag{7.8.7}$$

This can be integrated, using $\theta = 0$ at M = 1, to provide a relationship between the resulting Mach number (M_2 in Figure 7.18b) and the angle, provided that the incoming Mach number is unity; the relationship is

$$\theta = \left(\frac{k+1}{k-1}\right)^{1/2} \tan^{-1}\left[\frac{k-1}{k+1}(\mathrm{M}^2-1)\right]^{1/2} - \tan^{-1}(\mathrm{M}^2-1)^{1/2} \tag{7.8.8}$$

The angle θ, which is a function of M, is the **Prandtl–Meyer function**. It is tabulated for $k = 1.4$ in Table D.3 so that trial-and-error solutions to Eq. 7.8.8 for M are not necessary. Other changes that may be desired, such as the pressure or temperature changes, can be found from the isentropic-flow equations. The collection of Mach waves that turn the flow are referred to as an **expansion fan**.

We will see, from the examples and problems, that both the Mach number and the velocity increase as supersonic flow turns the convex corner. The flow remains attached to the wall as it turns the corner, even for large angles, a phenomenon not observed in a subsonic flow; a subsonic flow would separate from the abrupt corner, even for small angles.

If we substitute $M = \infty$ in Eq. 7.8.8, we find the angle of maximum turning to be $\theta = 130.5°$. This would mean that both the temperature and pressure would be absolute zero; obviously, the gas would become a liquid before this would be possible. The angle of 130.5° is, however, an upper limit. The point is, rather large turning angles are possible in supersonic flows, angles that may exceed 90°, a rather surprising result. This introduces a design constraint on the exhaust nozzle of rocket engines that exhaust into the vacuum of space; the exhaust gases may turn through an angle so that impingement on the spacecraft body would occur if not designed properly.

Example 7.12

Air at a Mach number of 2.0 and a temperature and pressure of 500°C and 200 kPa absolute, respectively, flows around a corner with a convex angle of 20° (Figure E7.12). Find M_2, p_2, T_2, and V_2.

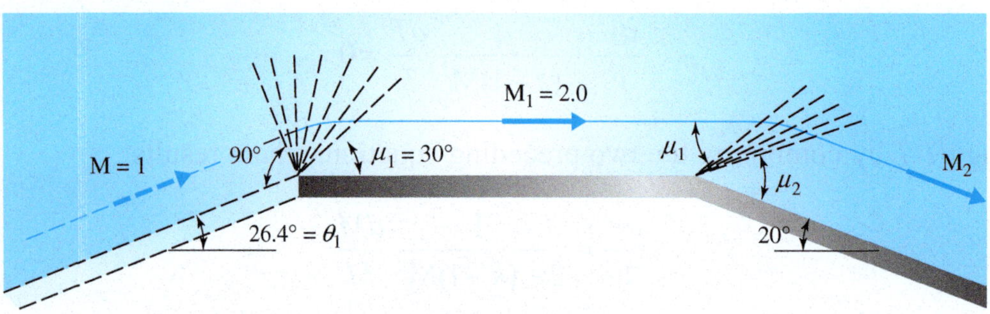

Figure E7.12

Solution

Table D.3 uses M = 1 as a reference condition; thus we visualize the flow as originating from a flow with M = 1 and turning through the angle θ_1 to $M_1 = 2$, as shown in the sketch. From the table, adding an additional 20° to the deflection angle, we find that $\theta_2 = 46.4°$. This would be equivalent to the flow at M = 1 turning a convex corner with $\theta = 46.4°$. Since the flow is isentropic, we can simply superimpose in this manner. Now, for an angle of $\theta = 46.4°$ from the table, we find that

$$M_2 = \underline{2.83}$$

From the isentropic flow table (Table D.1) we find that

$$p_2 = p_1 \frac{p_0}{p_1}\frac{p_2}{p_0}$$
$$= 200 \times \frac{1}{0.1278} \times 0.0352 = \underline{55.1\ \text{kPa}}$$
$$T_2 = T_1 \frac{T_0}{T_1}\frac{T_2}{T_0}$$
$$= 773 \times \frac{1}{0.5556} \times 0.3844 = 534.8\,\text{K} \quad \text{or} \quad \underline{261.8°\text{C}}$$

The velocity V_2 is found to be

$$V_2 = \text{M}_2\sqrt{kRT_2}$$
$$= 2.83\sqrt{1.4 \times 287 \times 534.8} = \underline{1312\ \text{m/s}}$$

7.9 SUMMARY

The zone of silence of a supersonic object that produces only Mach waves exists outside a cone with included angle ∞ found from

$$\sin\alpha = \frac{1}{\text{M}} \tag{7.9.1}$$

where the Mach number is $\text{M} = V/c$ and $c = \sqrt{kRT}$.

The relationship

$$\frac{dV}{V}(\text{M}^2 - 1) = \frac{dA}{A} \tag{7.9.2}$$

allows us to predict how subsonic and supersonic flows behave in converging and diverging nozzles. The mass flux through a nozzle with throat area A^* where $\text{M}^* = 1$ is given by

$$\dot{m} = p_0 A^* \sqrt{\frac{k}{RT_0}}\left(\frac{k+1}{2}\right)^{(k+1)/2(1-k)} \tag{7.9.3}$$

where p_0 and T_0 are the reservoir conditions. Temperature, pressure, and velocity in an isentropic flow are found from

$$\frac{T_0}{T}=1+\frac{k-1}{2}\mathrm{M}^2 \qquad \frac{p_0}{p}=\left(1+\frac{k-1}{2}\mathrm{M}^2\right)^{k/(k-1)} \qquad \frac{V^2}{2}=c_p(T_0-T) \tag{7.9.4}$$

The flow variables across a normal shock are found from continuity, momentum, and energy:

$$\rho_1 V_1=\rho_2 V_2, \quad p_1-p_2=\rho_1 V_1(V_2-V_1),$$
$$\frac{V_2^2-V_1^2}{2}+\frac{k}{k-1}\left(\frac{p_2}{\rho_2}-\frac{p_1}{\rho_1}\right)=0 \tag{7.9.5}$$

Rather than solve the above equations, we often use the normal-shock flow Table D.2.

Across an oblique shock the tangential velocity component does not change. The normal velocity component V_n simply replaces the velocity V in the normal-shock wave equations above, and Table D.2 may be used with M_n replacing M. The wedge angle θ through which the flow is turned is related to the oblique shock angle β by

$$\tan(\beta-\theta)=\frac{\tan\beta}{k+1}\left[k-1+\frac{2}{\mathrm{M}_1^2\sin^2\beta}\right] \tag{7.9.6}$$

or Figure 7.16 may be used to avoid a trial-and-error solution if β is desired.

In an expansion wave, the angle θ through which a flow with M = 1 can be turned is the Prandtl–Meyer function:

$$\theta=\left(\frac{k+1}{k-1}\right)^{1/2}\tan^{-1}\left(\frac{k-1}{k+1}(\mathrm{M}^2-1)\right)^{1/2}-\tan^{-1}(\mathrm{M}^2-1)^{1/2} \tag{7.9.7}$$

If we know θ and desire M, a trial-and-error solution is required. This is avoided by using Table D.3.

PROBLEMS

7.1 In thermodynamics c_p for air is often used as 0.24 Btu/lbm-°R. From Table B.4 c_p is 6012 ft-lb/slug-°R. Show these two values to be equal. Also, calculate c_v in both sets of units. (**Note:** The value of c_p used in thermodynamics and in fluid mechanics is the same when using the SI system of units.)

7.2 Show that $c_p = Rk/(k-1)$.

7.3 Show that Eqs.7.1.10 follow from Eqs. 7.1.9 and 7.1.7.

7.4. Verify the various forms of the energy equation in Eqs. 7.1.3, 7.1.4, and 7.1.5.

Speed of Sound

7.5 Show that Eq. 7.2.8 follows from Eq. 7.2.7.

7.6 Show that the velocity of sound in a high-frequency wave (i.e., a dog whistle) travels with speed $c = \sqrt{RT}$ if the process is assumed to be isothermal.

7.7 Show that for a small adiabatic disturbance in a steady flow, the energy equation takes the form $\Delta h = -c\Delta V$.

7.8 Verify that the speed of propagation of a small wave through water is about 1450 m/s.

7.9 Two rocks are slammed together on one side of a lake. An observer on the other side with his head under water "hears" the disturbance 0.6 s later. How far is it across the lake?

7.10 You and a friend are standing 10 m apart in waist-deep water. You slam two rocks together under water. How long after the rocks are slammed together does your friend hear the interaction if your friend's head is underwater?

7.11 Calculate the Mach number for an aircraft if it is flying at:
(a) Sea level at 200 m/s
(b) 15,000 ft at 600 fps
(c) 10 000 m at 200 m/s
(d) 60,000 ft at 600 fps
(c) 35 000 m at 200 m/s

7.12 A wood chopper is chopping wood some distance away. You carefully observe, using your numerical wrist timer, that it takes 1.21 s for the sound of the axe to reach your ears. Calculate the distance you are from the chopper if the temperature is −10°C.

7.13 You see a bolt of lightning and 2 s later you hear the sound of thunder. How far away did the lightning bolt strike? Use:
(a) SI units **(b)** English units

7.14 A needle-nosed projectile passes over you on a military testing field at a speed of 1000 m/s (Figure P7.14). You know that it has an elevation of 1000 m where $T =$ −10°C. How long after it passes overhead will you hear its sound? How far away will it be? Calculate its Mach number.

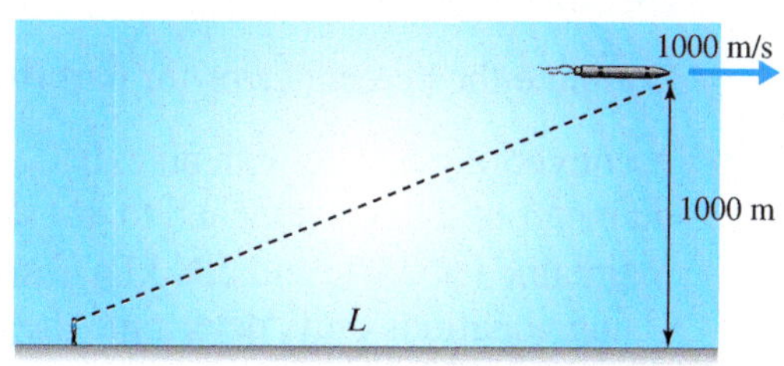

Figure P7.14

7.15 A special camera is capable of showing the Mach angle of a sharp-nosed bullet passing through the test section of a wind tunnel. If the Mach angle is measured to be 22°, calculate the speed of the bullet in:
(a) m/s
(b) ft/s
Assume standard conditions.

7.16 A small-amplitude wave passes through the standard atmosphere at sea level with a pressure rise of 0.3 psf. Estimate the associated induced velocity and temperature rise.

Isentropic Flow

7.17 Provide all the steps needed to proceed from:
(a) Eq.7.3.2 to Eq. 7.3.3
(b) Eq. 7.3.5 to Eq. 7.3.10
(c) Eq. 7.3.11 to Eq.7.3.12
(d) Eq. 7.3.16 to Eq. 7.3.18
(e) Eq. 7.3.17 to Eq. 7.3.19

7.18 A pitot probe, an instrument that measures stagnation pressure, is used to determine the velocity of an airplane. Such a probe, attached to an aircraft, measures 10 kPa. Determine the speed of the aircraft if it is flying at an altitude of:
(a) 3000 m
(b) 10 000 m
Assume an isentropic process from the free stream to the stagnation point.

7.19 A pitot probe, used to measure the stagnation pressure, indicates a pressure of 4 kPa at the nose of a surface vehicle traveling in 15°C-atmospheric air. Calculate its speed assuming:
(a) Isentropic flow
(b) Incompressible flow

Compute the percent error for part (b).

7.20 A converging nozzle with an exit diameter of 2 cm is attached to a reservoir maintained at 25°C and 200 kPa absolute. Using equations only, determine the mass flux of air if the receiver pressure is:
(a) 100 kPa absolute
(b) 130 kPa absolute

7.21 The converging nozzle of Figure P7.21 is attached to a reservoir maintained at 70°F and 30 psia. Using equations only, determine the mass flux of air if the receiver pressure is:
(a) 15 psia
(b) 20 psia

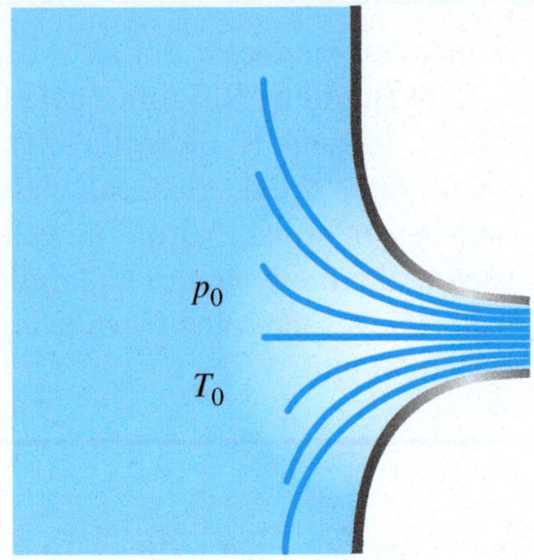

Figure P7.21

7.22 Using the isentropic flow table, rework:
(a) Problem 7.20a
(b) Problem 7.20b

7.23 Using the isentropic flow table, rework:
(a) Problem 7.21a
(b) Problem 7.21b

7.24 Air flows from a reservoir ($T_0 = 30$°C, $p_0 = 400$ kPa absolute) out a converging nozzle with a 10-cm-diameter exit. What exit pressure would just result in $M_e = 1$? Determine the mass flux for this condition. Use the isentropic flow table.

7.25 Air flows from a converging nozzle attached to a reservoir with $T_0 = 10$°C. What reservoir pressure is necessary to just cause $M_e = 1$ if the 6-cm-diameter nozzle exits to atmospheric pressure? Calculate the mass flux for this condition.

7.26 Air flows from a converging nozzle attached to a reservoir with $T_0 = 40$°F. If the 2.5-in.-diameter nozzle exits to atmospheric pressure, what reservoir pressure is necessary to just cause $M_e = 1$? Calculate the mass flux for this condition. Now double the reservoir pressure and determine the increased mass flux.

7.27 A 25-cm-diameter air line is pressurized to 500 kPa absolute and suddenly bursts (Figure P7.27). The exit area is later measured to be 30 cm^2. If 6 min elapsed before the 10°C air was turned off, estimate the cubic meters of air that was lost.

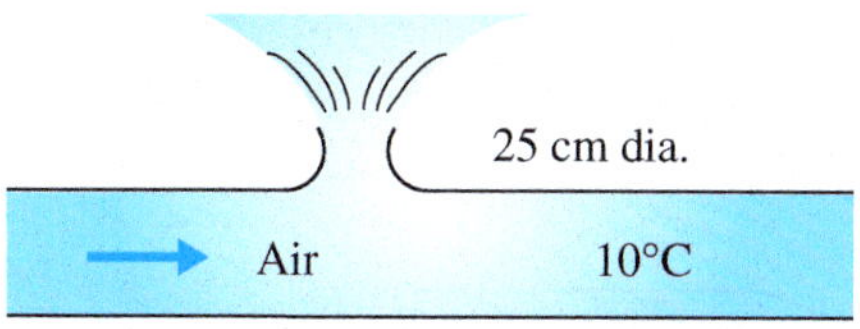

Figure P7.27

7.28 A converging nozzle is attached to a reservoir containing helium with $T_0 = 27°C$ and $p_0 = 200$ kPa absolute. Determine the receiver pressure that will just give $M_e = 1$. Now attach a diverging section with a 15-cm-diameter exit to the 6-cm-diameter throat. What is the highest receiver pressure that will give $M_t = 1$?

7.29 A venturi tube is used to measure the mass flux of air in a pipe by reducing the diameter from 10 cm to 5 cm and then back to 10 cm. The pressure at the inlet is 300 kPa and at the minimum diameter section it is 240 kPa. If the upstream temperature is 20°C, determine the mass flux.

7.30 Air enters a converting-diverging nozzle from a reservoir at 200 kPa (gage) and 22°C. The area at the throat is 9.7 cm^2, and at the exit it is 13 cm^2. Considering isentropic flow throughout with M = 1 at the throat, determine the velocity of air at the nozzle exit.

7.31 Consider the nozzle flow of Problem 7.30, but in this case the Mach number at the throat is 0.72. Determine the velocity at the nozzle exit for the given condition.

7.32 The mass flux of air flowing through the pipe of Figure P7.32 is desired. The pipe is reduced in diameter from 4 in. to 2 in. and then back to 4 in. The pressure at the inlet is 45 psi and at the minimum diameter section it is 36 psi. If the upstream temperature is 60°F, determine the mass flux.

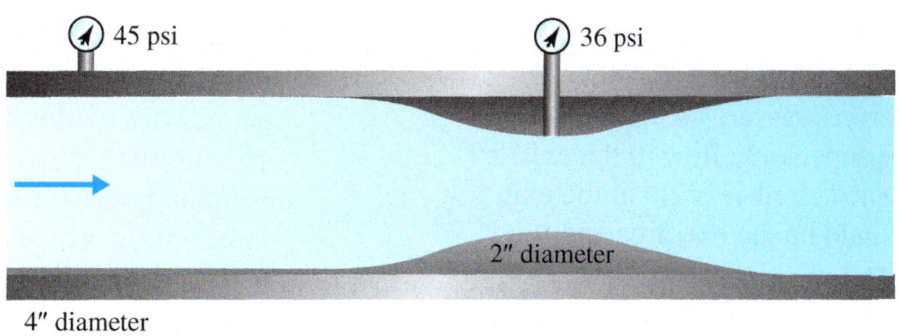

Figure P7.32

7.33 Air flows from a reservoir maintained at 30°C and 200 kPa absolute through a converging-diverging nozzle that has a 10-cm-diameter throat. Determine the diameter where M = 3. Use equations only.

7.34 Air flows in a converging-diverging nozzle from a reservoir maintained at 20°C and 500 kPa absolute. The throat and exit diameters are 5 cm and 15 cm, respectively. What two receiver pressures will result in M = 1 at the throat if isentropic flow occurs throughout? Use equations only.

7.35 Rework Problem 7.34 using the isentropic flow table.

7.36 Air flows from a nozzle with a mass flux of 1.0 slug/s. If $T_0 = 607°F$, $p_0 = 120$ psia, and $p_e = 15$ psia. Calculate the throat and exit diameters for an isentropic flow. Also, determine the exit velocity.

7.37 Air flows from a reservoir, maintained at 20°C and 2 MPa absolute, and exits from a nozzle with $M_e = 4$. The receiver pressure is then raised until the flow is just subsonic throughout the entire nozzle. Calculate this receiver pressure.

7.38 A small wind tunnel uses compressed air from a reservoir at 3.5 MPa and 320 K. The air flows isentropically from the reservoir through a nozzle, as shown in Figure P7.38. If the Mach number in the test section is 2.8, calculate the pressure and velocity in the test section.

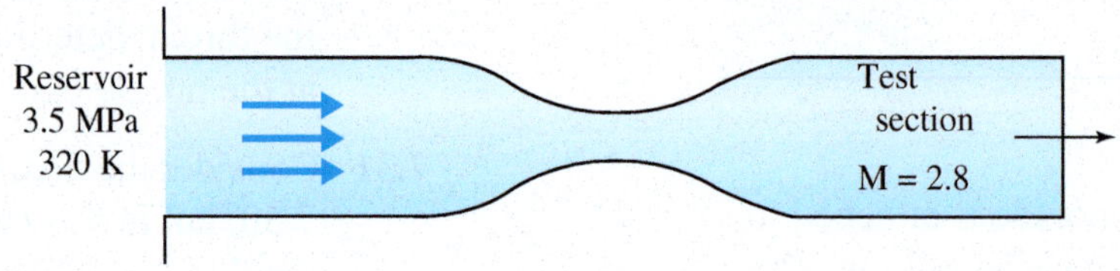

Figure P7.38

7.39 Consider isentropic air flow from a reservoir (where $p_0 = 600$ kPa, $T_0 = 30°C$) through a converging-diverging nozzle. At a section in the converging part of the nozzle before the throat, the Mach number is 0.50, and the cross-sectional area is 12.4 cm^2.

(a) If the area at the throat is 10 cm^2, calculate the pressure, temperature, and velocity at the throat.

(b) If M = 1 at the throat, what should the receiver pressure be to just produce supersonic flow at the exit?

(c) If the Mach number is 2.0 at the exit, what should be the exit area and the mass flux?

7.40 Air at 30°C flows in a 10-cm-diameter pipe at a velocity of 150 m/s. A venturi tube is used to measure the flow rate. What should be the minimum diameter of the tube so that supersonic flow does not occur?

7.41 For a nozzle efficiency of 96%, rework Problem 7.33.

7.42 Nitrogen enters a diffuser at 100 kPa absolute and 100°C with a Mach number of 3.0. The mass flux is 10 kg/s and the exit velocity is small. Sketch the diffuser, then determine the throat area and the exit pressure and temperature, assuming isentropic flow.

7.43 Nitrogen enters a diffuser at 15 psia absolute and 200°F with a Mach number of 3.0. The mass flux is 0.2 slug/s and the exit velocity is small. Sketch the diffuser, then determine the throat area and the exit pressure and temperature, assuming isentropic flow.

7.44 A rocket has a mass of 80 000 kg and is to be lifted vertically from a platform with six nozzles exiting exhaust gases with $T_e = 1000°C$. What should the exit velocity be from each 50-cm-diameter nozzle if the exhaust gases are assumed to be carbon dioxide?

7.45 A 100-kg man straps a small air-breathing jet engine on his back and just lifts off the ground vertically (Figure P7.45). The engine has an exit area of 200 cm^2. With what velocity must the 600°C exhaust gases exit the engine?

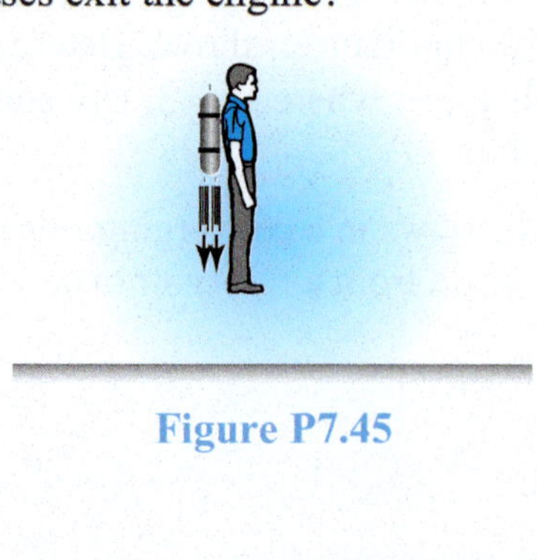

Figure P7.45

7.46 A converging-diverging nozzle is bolted into a reservoir at a diameter of 40 cm. The throat and exit diameters are 5 cm and 10 cm, respectively. If $T_0 = 27°C$ and $p_e = 100$ kPa absolute and isentropic flow of air occurs throughout the supersonic nozzle, calculate the force necessary to hold the nozzle onto the reservoir.

7.47 What is the maximum speed in **(a)** m/s and **(b)** mph that an aircraft can have during take-off or landing if the airflow around the aircraft is to be modeled as an incompressible flow? Allow a 3% error in the pressure from free stream to the stagnation point. Assume standard conditions.

Normal Shock

7.48 The pressure, temperature, and velocity before a normal shock wave are 80 kPa absolute, 10°C, and 1000 m/s, respectively. Calculate M_1, M_2, p_2, T_2, and ρ_2 for air. Use:
(a) Basic equations
(b) The normal shock table

7.49 The pressure, temperature, and velocity before a normal shock wave are 12 psia, 40°C, and 3000 ft/s, respectively. Calculate M_1, M_2, p_2, T_2, and ρ_2 for air. Use:
(a) Basic equations
(b) The normal shock table

7.50 Derive the **Rankine–Hugoniot relationship**,

$$\frac{\rho_2}{\rho_1} = \frac{(k+1)p_2/p_1 + k - 1}{(k-1)p_2/p_1 + k + 1}$$

which relates the density ratio to the pressure ratio across a normal shock wave. Find the limiting density ratio for air across a strong shock for which $p_2/p_1 >> 1$.

7.51 An explosion occurs just above the surface of the Earth, producing a shock wave that travels radially outward. At a given location it has a Mach number of 2.0. Determine the pressure just behind the shock and the induced velocity.

7.52 Air at 200 kPa absolute and 20°C passes through a normal shock wave with strength so that $M_2 = 0.5$. Calculate V_1, p_2, and ρ_2.

7.53 Air at 30 psia and 60°F passes through a normal shock wave with strength so that $M_2 = 0.5$. Calculate V_1, p_2, and ρ_2.

7.54 A blunt object travels at 1000 m/s at an elevation of 10 000 m. The flow approaching the stagnation point passes through a normal shock wave and then decelerates isentropically to the stagnation point. Calculate p_0 and T_0 at the stagnation point.

7.55 A pitot probe is inserted in an airflow in a pipe in which $p = 800$ kPa absolute, $T = 40°C$, and $M = 3.0$ (Figure P7.55). What pressure does it measure?

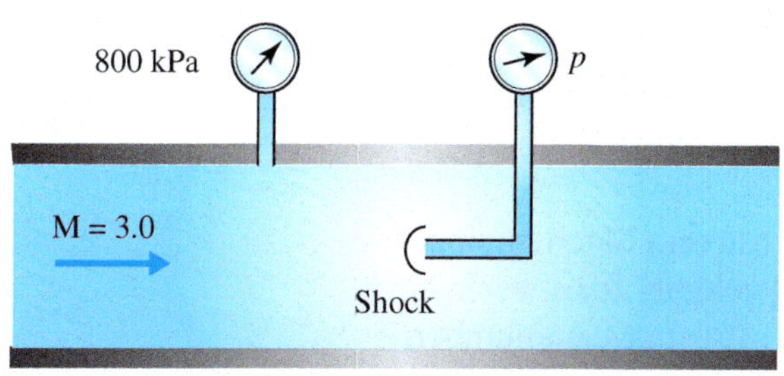

Figure P7.55

7.56 Air flows from a 25°C reservoir to the atmosphere through a nozzle with a 5-cm-diameter throat and a 10-cm-diameter exit. What reservoir pressure will just result in M = 1 at the throat? Also, calculate the mass flux. Maintaining this reservoir pressure, reduce the throat diameter to 4 cm and determine the resulting mass flux. Sketch the pressure distribution as in Figure 7.12.

7.57 Air flows from a 20°C reservoir to the atmosphere through a nozzle with a 5-cm-diameter throat and a 10-cm-diameter exit. What reservoir pressure is necessary to locate a normal shock wave at the exit? Also, calculate the velocity and pressure at the throat, before the shock, and after the shock.

7.58 Air flows from a 60°F reservoir to the atmosphere through a nozzle with a 2-in.-diameter throat and a 4-in.-diameter exit. What reservoir pressure is necessary to locate a normal shock wave at the exit? Also, calculate the velocity and pressure at the throat, before the shock, and after the shock.

7.59 Air flows from a reservoir maintained at 25°C and 500 kPa absolute. It flows through a nozzle with throat and exit diameters of 5 cm and 10 cm, respectively. What receiver pressure is necessary to locate a normal shock wave at a location where the diameter is 8 cm? Also, determine the velocity before the shock and at the exit.

Vapor Flow

7.60 Steam flows at a rate of 4 kg/s from reservoir conditions of 400°C and 1.2 MPa absolute through a converging–diverging nozzle to the atmosphere. Determine the throat and exit diameters if supersonic, isentropic flow exists throughout the diverging section.

7.61 Steam flows from reservoir conditions of 350°C and 1000 kPa absolute to the atmosphere at the rate of 15 kg/s. Estimate the converging nozzle exit diameter.

7.62 Steam flows from reservoir conditions of 700°F and 150 psia to the atmosphere at the rate of 0.25 slug/s. Estimate the converging nozzle exit diameter.

7.63 A header supplies steam at 400°C and 1.2 MPa absolute to a set of nozzles with throat diameters of 1.5 cm. The nozzles exhaust to a pressure of 120 kPa absolute. If the flow is approximately isentropic, calculate the mass flux and the exit temperature.

Oblique Shock Wave

7.64 An airflow with velocity, temperature, and pressure of 800 m/s, 30°C, and 40 kPa absolute, respectively, is turned with an oblique shock wave emanating from the wall, which makes an abrupt 20° corner.

(a) Find the downstream Mach number, pressure, and velocity for a weak shock.

(b) Find the downstream Mach number, pressure, and velocity for a strong shock.

(c) If the concave corner angle were 35°, sketch the corner flow situation.

7.65 Two oblique shocks intersect as shown in Figure P7.65. Determine the angle of the reflected shocks if the airflow must leave parallel to its original direction. Also find M_3.

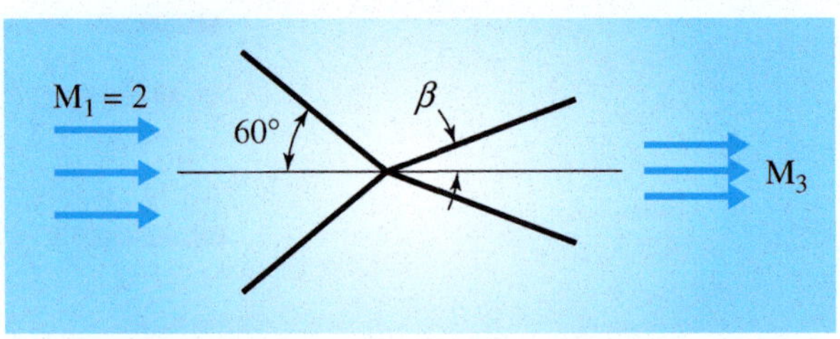

Figure P7.65

7.66 An oblique shock wave at a 35° angle is reflected from a plane wall. The upstream Mach number M_1 is 3.5 and $T_1 = 0°C$. Find V_3 after the reflected oblique shock wave for the airflow.

7.67 An oblique shock wave at a 35° angle is reflected from a plane wall. The upstream Mach number M_1 is 3.5 and $T_1 = 30°F$. Find V_3 after the reflected oblique shock wave for the airflow.

7.68 A supersonic inlet can be designed to have a normal shock wave oriented at the inlet, or a wedge can be used to provide a weak oblique shock wave, as shown in Figure P7.68. Compare the pressure p_3 from the flow shown to the pressure that would exist behind a normal shock wave with no oblique shock.

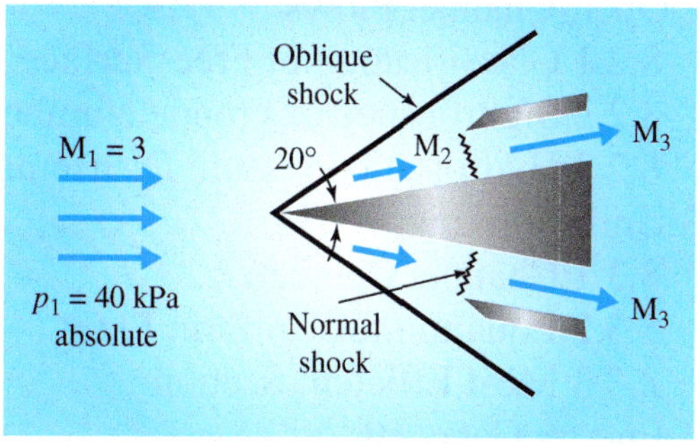

Figure P7.68

Expansion Waves

7.69 A supersonic airflow with $M_1 = 3$, $T_1 = -20°C$, and $p_1 = 20$ kPa absolute turns a convex corner of 25°. Calculate M_2, p_2, T_2, and V_2 after the expansion fan. Also calculate the included angle of the fan.

7.70 A supersonic airflow with $M_1 = 2$, $T_1 = 0°C$, and $p_1 = 20$ kPa absolute turns a convex corner. If $M_2 = 4$, what angle θ should the corner have? Also calculate T_2 and V_2.

7.71 A supersonic airflow with $M_1 = 2$, $T_1 = 30°F$, and $p_1 = 5$ psia turns a convex corner. If $M_2 = 4$, what angle θ should the corner have? Also calculate T_2 and V_2.

7.72 The flat plate shown in Figure P7.72 is used as an airfoil at an angle of attack of 5°. Oblique shock waves and expansion fans allow the air to remain attached to the plate with the flow behind the airfoil parallel to the original direction. Calculate

(a) The pressures on the upper and lower sides of the plate

(b) The downstream Mach numbers M_{2u} and M_{2l}

(c) The lift coefficient defined by $C_L =$ lift$/\left(\frac{1}{2}\rho_1 V_1^2 A\right)$. Noted that $\rho_1 V_1^2 = k M_1^2 p_1$

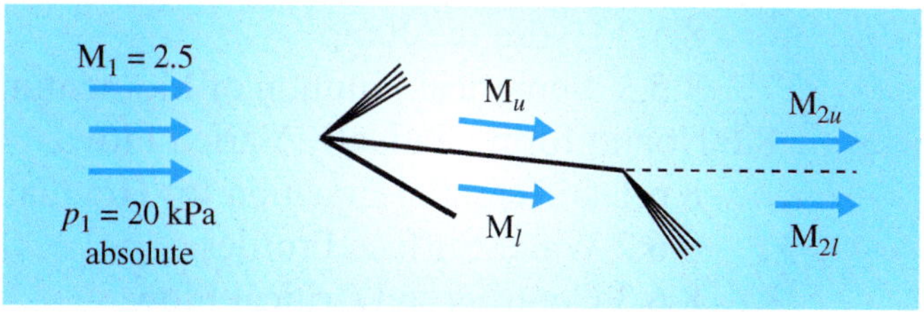

Figure P7.72

7.73 The supersonic airfoil shown in Figure P7.73 is to fly at zero angle of attack. Calculate the drag coefficient $C_D =$ drag$/\left(\frac{1}{2}\rho_1 V_1^2 A\right)$. Noted that $\rho_1 V_1^2 = k M_1^2 p_1$.

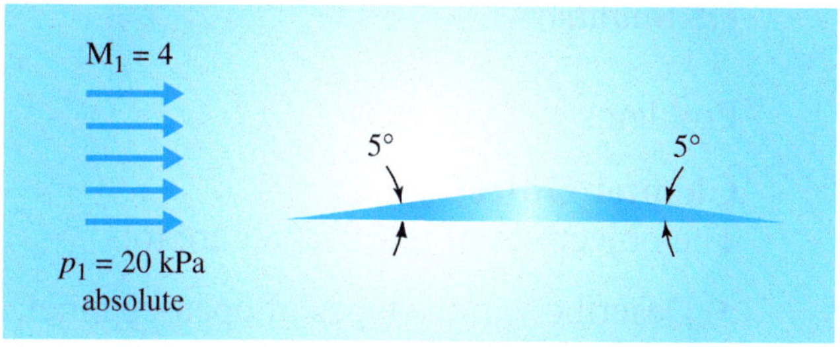

Figure P7.73

7.74 The airfoil in Problem 7.73 flies at an angle of attack of 5°. Determine the lift and drag coefficients, C_L and C_D. See Problems 7.72 and 7.73 for definitions of C_L and C_D.

8 Flow in Open Channels

Outline

Chapter Objectives

The objectives of this chapter are to:

- Describe various types of open-channel, free-surface flows
- Apply the Chezy–Manning equation with various geometric cross-sections for uniform flow applications
- Derive energy and momentum principles for rapidly varied flow situations
- Derive the differential equation for nonuniform gradually varied flow
- Present the method of profile synthesis—the qualitative description of open-channel flow, including establishment of controls and classification of water surface profiles
- Develop and present methods to numerically compute gradually varied flow
- Present numerous examples of uniform, rapidly varied, and gradually varied flow

M. C. Potter and D. C. Wiggert, *Applied Fluid Mechanics*,
https://doi.org/10.1007/978-3-031-65276-9_8

◄ Detail several examples of profile synthesis and numerical analysis of complex open-channel flow with emphasis on design application

8.1 INTRODUCTION

Free-surface flow is probably the most commonly occurring flow phenomenon that we encounter on the surface of the earth. Ocean waves, river currents, and overland flow of rainfall are examples that occur in nature. Man-made situations include flows in canals and culverts, drainage over impervious materials such as roofs and parking lots, and wave motion in harbors.

In all of these situations, the flow is characterized by an interface between the air and the upper layer of water, which is termed the **free surface**. At the free surface, the pressure is constant, and for nearly all situations, it is atmospheric. In such a case, the hydraulic grade line and the liquid free surface coincide. In engineering practice, most open channels convey water as the fluid. However, the principles developed and implemented in this chapter are also applicable for other liquids that flow with a free surface.

Generally, the elevation of the free surface does not remain constant; it can vary along with the fluid velocities. Another complexity is that the flow is often three-dimensional. Fortunately, there are many instances in which two-dimensional and even one-dimensional simplifications can be made. The flow patterns in estuaries[1] and lakes under certain circumstances can be treated as two-dimensional in the horizontal plane, averaged vertically over the depth. River and channel flows are usually treated as one-dimensional with respect to the position coordinate along the streambed. In this chapter we restrict our consideration to one-dimensional flows.

Figure 8.1a shows a representative centerline velocity distribution in a channel. The velocity profile is three-dimensional at a given cross section (Figure 8.1b). The boundary shear stress is nonuniform; at the free surface the shear stress is negligible, yet the shear stress varies

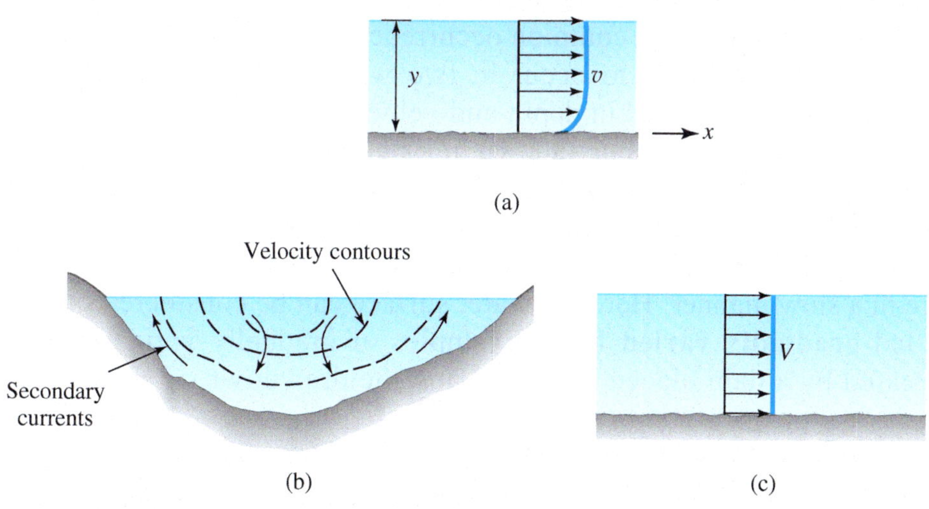

Figure 8.1 Free-surface flow: (a) centerline velocity distribution; (b) cross section; (c) one-dimensional model.

[1]An estuary is the lower course of a river that is influenced by the ocean tides.

around the wetted perimeter. In some circumstances the presence of secondary currents will force the maximum velocity to occur slightly below the free surface. By convention, y is defined as the depth from the deepest location to the free surface; note that y is not a coordinate. The mean velocity is given by the relation

$$V = \frac{1}{A}\int_A v dA \tag{8.1.1}$$

In the one-dimensional model, we assume the velocity to be equal to V everywhere at a given cross section. This model provides excellent results and is used widely. Flows in open channels are most likely turbulent, and the velocity profile can be assumed to be approximately constant, as in Figure 8.1c, without significant error. Hence the one-dimensional model is employed.

8.2 OPEN-CHANNEL FLOWS

8.2.1 Classification of Free-Surface Flows

Flow in a channel is characterized by the mean velocity, even though a velocity profile exists at a given section, as shown in Figure 8.1a. The flow is classified as a combination of steady or unsteady, and uniform or nonuniform. Steady flow signifies that the mean velocity V, as well as the depth y, is independent of time, whereas unsteady flow necessitates that time be considered as an independent variable. Uniform flow implies that V and y are independent of the position coordinate in the direction of flow; nonuniform flow signifies that V and y vary in magnitude along that coordinate. The possible combinations are shown in Table 8.1; the position coordinate is designated as x.

Steady, uniform flow is the situation where terminal velocity has been reached in a channel of constant cross section; not only is the mean velocity constant, but the depth is invariant as well. Steady, nonuniform flow is a common occurrence in rivers and in manmade channels. In those situations, it will be found to occur in two ways. In relatively short reaches, called **transitions**, there is a rapid change in depth and velocity; such flow is termed **rapidly varied flow**. Examples are the hydraulic jump (shown in Example 2.12), flow entering a steep channel from a lake or reservoir, flow downstream of a free out-fall from a channel, and flow in the vicinity of an obstruction such as a bridge pier or a sluice gate.

Along more extensive reaches of channel the velocity and depth may not vary rapidly, but rather, change in a slow manner. Here the water surface can be considered continuous, and the regime is called **gradually varied flow**. Examples of gradually varied steady flow are the backwater created by a dam placed in a river and the drawdown of a water surface as flow

Table 8.1 Combinations of One-Dimensional Free-Surface Flows

Type of flow	*Average velocity*	*Depth*
Steady, uniform	V = const.	y = const.
Steady, nonuniform	$V = V(x)$	$y = y(x)$
Unsteady, uniform	$V = V(t)$	$y = y(t)$
Unsteady, nonuniform	$V = V(x, t)$	$y = y(x, t)$

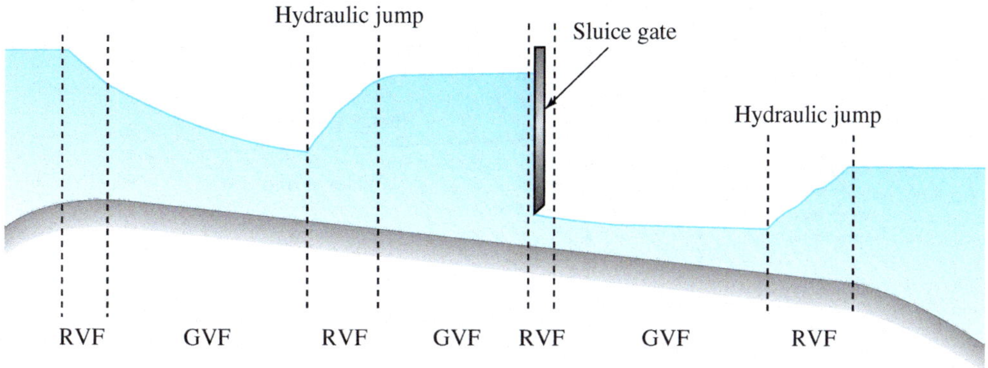

Figure 8.2 Steady nonuniform flow in a channel.

approaches a falls. Figure 8.2 illustrates how both rapidly varied flow (RVF) and gradually varied flow (GVF) can occur simultaneously in a reach of channel. Note that the vertical scale is larger than the horizontal scale; such scale distortion is common when representing open-channel flow situations.

Unsteady uniform flow rarely occurs, but unsteady nonuniform flow is common. Flood waves in rivers, hydraulic bores, and regulated flow in canals are all examples of the latter category. In many situations, these flows may be considered to behave sufficiently like steady uniform flow or steady nonuniform flow to justify treating them as such. Unsteady flows are beyond the scope of a fundamental treatment and will not be presented in this book; the one exception is the hydraulic bore—a moving hydraulic jump—which can be analyzed in a quasi steady fashion.

8.2.2 Significance of Froude Number

The primary mechanism for sustaining flow in an open channel is the gravitational force. For example, the elevation difference between two reservoirs will cause water to flow through a canal that connects them. The parameter that represents this gravitational effect is the **Froude number**,

$$\mathrm{Fr} = \frac{V}{\sqrt{gL}} \tag{8.2.1}$$

which has been observed in Chapter 4 to be the ratio of inertial force to gravity force. In the context of open-channel flow, V is the mean cross-sectional velocity, and L is a representative length. For a channel of rectangular cross section, L is the depth y of the flow.

The Froude number plays the dominant role in open-channel flow analysis. It appears in a number of relations that will be developed in this chapter. Furthermore, by knowing its magnitude, one can ascertain significant characteristics regarding the flow regime. For example, if $\mathrm{Fr} > 1$, the flow possesses a relatively high velocity and shallow depth; on the other hand, when $\mathrm{Fr} < 1$, the velocity is relatively low and the depth is relatively deep. Except in the vicinity of rapids, cascades, and waterfalls, most rivers experience a Froude number less than unity. Constructed channels may be designed for Froude numbers to be greater or less than unity, or to vary from greater than unity to less than unity along the length of the channel.

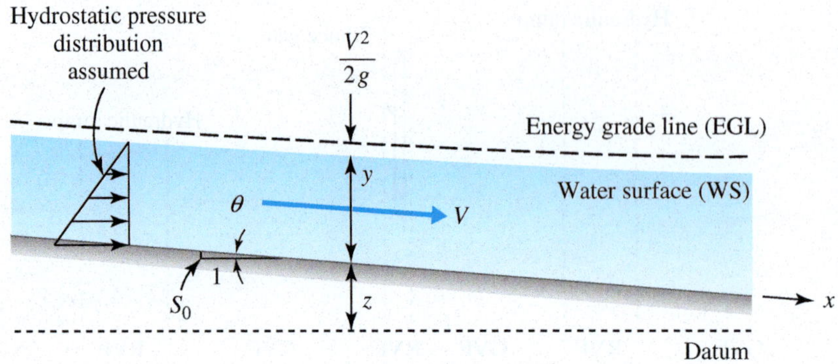

Figure 8.3 Reach of open-channel flow.

8.2.3 Hydrostatic Pressure Distribution

Consider a channel in which the flow is nearly horizontal, as shown in Figure 8.3. In this case, there is little or no vertical acceleration of fluid within the reach, and the streamlines remain nearly parallel. Such a condition is common in many open-channel flows, and indeed, if slight variations exist, the streamlines are assumed to behave as if they are parallel. Since the vertical accelerations are nearly zero, one can conclude that in the vertical direction the pressure distribution is hydrostatic. As a result, the sum $(p + \gamma z)$ remains a constant at any depth, and the hydraulic grade line coincides with the water surface. It is customary in open-channel flow to designate z as the elevation of the channel bottom and y as the depth of flow. Since at the channel bottom $p/\gamma = y$, the hydraulic grade line is given by the sum $(y + z)$. The concepts in this chapter are developed assuming a hydrostatic pressure distribution.

8.3 UNIFORM FLOW

Before we study nonuniform flow in Section 8.6, let us focus our attention on the simpler condition of steady, uniform flow or simply uniform flow. Such a flow is rare, but if it were to occur in a channel, the depth and velocity would not vary along its length, or in other words, terminal conditions would have been reached. In addition to uniform flow, this section covers the cross-sectional geometry of open channels; the formulations will apply to both uniform and nonuniform flow. Some of that material was covered in Section 5.7; it is reviewed here for completeness.

8.3.1 Channel Geometry

Channel cross sections can be considered to be either regular or irregular. A **regular section** is one whose shape does not vary along the length of the channel, whereas an irregular section will have changes in its geometry. In this chapter we consider mostly regular channel shapes; three common geometries are shown in Figure 8.4.

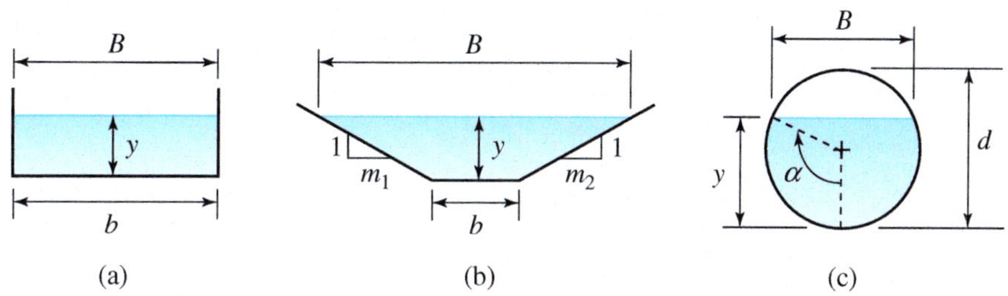

Figure 8.4 Representative regular cross sections: (a) rectangular; (b) trapezoidal; (c) circular.

The simplest channel shape is a rectangular section. The cross-sectional area is given by

$$A = by \tag{8.3.1}$$

The quantity b denotes the width of a channel bottom (see Figures 8.4a and 8.4b). Additional parameters of importance for open-channel flow are the wetted perimeter, the hydraulic radius, and the width of the free surface. The **wetted perimeter** P is the length of the line of contact between the liquid and the channel; for a rectangular channel it is

$$P = b + 2y \tag{8.3.2}$$

The **hydraulic radius** R is the area divided by the wetted perimeter, that is,

$$R = \frac{A}{P} = \frac{by}{b+2y} \tag{8.3.3}$$

The free-surface width B is equal to the bottom width b for a rectangular section.

A trapezoidal section (Figure 8.4b) has the added feature of sloped side walls. If m_1 is the ratio of the horizontal to vertical change of the wall on one side, and m_2 is the corresponding quantity on the other wall, the area, wetted perimeter, and free-surface width are given as

$$A = by + \frac{1}{2}y^2(m_1 + m_2) \tag{8.3.4}$$

$$P = b + y\left(\sqrt{1+m_1^2} + \sqrt{1+m_2^2}\right) \tag{8.3.5}$$

$$B = b + y(m_1 + m_2) \tag{8.3.6}$$

Note that the rectangular section is embodied in the trapezoidal definition, since for vertical side walls m_1 and m_2 are zero, and the relations for A, P, and B become identical to those for the rectangular channel. In addition, if b is set equal to zero, Eqs. 8.3.4 to 8.3.6 describe the geometry of a triangular-shaped channel.

The circular cross section is an important one to consider since many free-surface flows in drainage and sewer systems are conveyed in circular conduits. If d is the conduit diameter, the area, wetted perimeter, and free-surface width are given by

$$A = \frac{d^2}{4}(\alpha - \sin\alpha\cos\alpha) \tag{8.3.7}$$

$$P = \alpha d \tag{8.3.8}$$

$$B = d\sin\alpha \tag{8.3.9}$$

where

$$\alpha = \cos^{-1}\left(1 - 2\frac{y}{d}\right) \tag{8.3.10}$$

The angle α is displayed in Figure 8.4c.

A generalized cross-sectional geometry can be expressed in functional form as $A(y)$, $p(y)$, $R(y)$, and $B(y)$. The functional representations encompass all of the analytical forms given above, and they also may be used to describe an irregular channel. For example, at a river section, one can describe the area and wetted perimeter in tabular form and utilize techniques such as curve fitting or interpolation to extract the numerical information as functions of the depth. Such procedures are useful for computer-based analyses.

A composite section is one made up of several subsections; usually these subsections are of analytic form. The example shown in Figure 8.5a consists of a main channel and a floodplain. The main channel is approximated by a trapezoid and the floodplain by a rectangle, Figure 8.5b. One could derive analytical expressions for such a composite section; however, it may be more useful to consider the functional forms for the geometric parameters. Note that the functions will be discontinuous at depths where the two sections are matched.

Most of the theoretical developments in this chapter focus on cross sections that are rectangular. Such an assumption allows one to simplify the mathematics associated with open-channel flow analysis. Even though the equations will be simplified relative to more complicated geometries, the physical understanding of the phenomena and conclusions reached will apply to most generalized prismatic cross sections. A clear distinction will be made between rectangular and other types of geometry when various developments and concepts are presented.

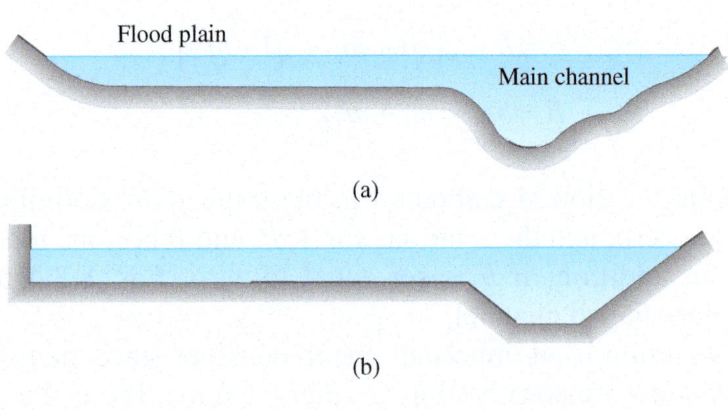

Figure 8.5 Generalized section representation: (a) actual cross section; (b) composite cross section.

8.3.2 Equation for Uniform Flow

Uniform flow occurs in a channel when the depth and velocity do not vary along its length, that is, when terminal conditions have been reached in the channel. Under such conditions, the energy grade line, water surface, and channel bottom are all parallel. Uniform flow can be predicted by an equation of the form

$$V = C\sqrt{RS_0} \tag{8.3.11}$$

in which S_0 is the slope of the channel bottom and C is the **Chezy coefficient**, which is independent of the Reynolds number since the flow is considered completely turbulent. It has become common engineering practice to relate C to the channel roughness and the hydraulic radius by use of the **Manning relation**

$$C = \frac{c_1}{n} R^{1/6} \tag{8.3.12}$$

where $c_1 = 1$ for SI units and $c_1 = 1.49$ for English units. Combining Eqs. 8.3.11 and 8.3.12 with the definition of discharge results in the **Chezy–Manning equation**

$$Q = \frac{c_1}{n} AR^{2/3}\sqrt{S_0} \tag{8.3.13}$$

Values of the Manning coefficient n are given in Table 5.3.

The depth associated with uniform flow is designated y_0; it is called either **uniform depth** or **normal depth**. Uniform flow rarely occurs in rivers because of the irregularity of the geometry. In man-made channels it is not always present, since the presence of controls such as sluice gates, weirs, or outfalls will cause the flow to become gradually varied. It is, however, necessary to determine y_0 when analyzing gradually varied flow conditions, since it provides a basis for evaluating the type of water surface that may exist in the channel. The design of gravity flow sewer networks is often based on assuming uniform flow and the use of Eq. 8.3.13, even though much of the time the flow in such systems may be nonuniform.

An examination of Eq. 8.3.13 reveals that it can be solved explicitly for Q, n, or S_0. Examples 5.19 and 5.20 provide illustrations. Use of a trial-and-error solution or an equation solver is necessary when it is required to find y_0 with the remaining parameters given.

Example 8.1

Water is flowing at a rate of 4.5 m^3/s in a trapezoidal channel (Figure 8.4b) whose bottom width is 2.4 m and side slopes are 1 vertical to 2 horizontal. Compute y_0 if $n = 0.012$ and $S_0 = 0.0001$.

Solution

Given geometrical data are $b = 2.4$ m and $m_1 = m_2 = 2$. Rearrange Eq. 8.3.13, noting that $R = A/P$ and $c_1 = 1$:

$$\frac{A^{5/3}}{P^{2/3}} = \frac{nQ}{\sqrt{S_0}}$$

Substituting in the known data and trapezoidal geometry, one has

$$\frac{\left[2.4y_0+\frac{1}{2}y_0^2(2+2)\right]^{5/3}}{\left[2.4+y_0(2\sqrt{1+2^2})\right]^{2/3}}=\frac{0.012\times 4.5}{\sqrt{0.0001}}$$

Solving for y_0, either by trial-and-error or by use of computational software, yields y_0 = 1.28 m.

Example 8.2

Uniform flow occasionally occurs in a 5-m-diameter circular concrete conduit (Figure 8.4c), but the depth of flow can vary. The Manning coefficient is $n = 0.013$, and the channel slope is $S_0 = 0.0005$. (a) Calculate the discharge for $y_0 = 3$ m, (b) Plot the discharge-depth curve.

Solution

(a) First, use Eq. 8.3.10 to find the angle α:

$$\alpha=\cos^{-1}(1-2y_0/d)$$

$$=\cos^{-1}(1-2\times 3/5)=101.54° \quad \text{or} \quad 101.54\times\pi/180=1.772 \text{ rad}$$

Using Eqs. 8.3.7 and 8.3.8, the area and wetted perimeter are

$$A=\frac{d^2}{4}(\alpha-\sin\alpha\cos\alpha)=\frac{5^2}{4}(1.772 \;-\; \sin 101.54°\cos 101.54°)=12.3 \text{ m}^2$$

$$P=\alpha d=1.772\times 5=8.86\text{m}$$

The hydraulic radius is then

$$R=\frac{A}{P}=\frac{12.3}{8.86}=1.388 \text{ m}$$

Finally, the discharge, when $y = 3$ m, is

$$Q=\frac{1}{n}AR^{2/3}S^{1/2}=\frac{1}{0.013}\times 12.3\times 1.388^{2/3}\times 0.0005^{0.5}=\underline{26.3\text{m}^3/\text{s}}$$

(b) Mathcad is employed to generate the curve shown in Fig. E8.2. Note that the solution is generalized, so that any diameter, Manning coefficient, or channel slope can be entered into the algorithm. Equations 8.3.7 and 8.3.8 are used to define the area and wetted perimeter, respectively. Either SI or English units can be employed by properly defining the parameter c_1. In this problem a value of 1.0 is used.

Input diameter, Manning coefficient, and channel slope:

$d := 5 \qquad n := 0.013 \qquad S_0 := 0.0005 \qquad c_1 := 1.0$

Define geometric functions:

$$\alpha(y) := a\cos\left(1-2\cdot\frac{y}{d}\right)$$

$$A(y) := \frac{d^2}{4}\cdot(\alpha(y)-\sin(\alpha(y))\cdot\cos(\alpha(y)))$$

$$P(y) := \alpha(y)\cdot d$$

$$R(y) := \frac{A(y)}{P(y)}$$

Define discharge function, (i.e., Manning's equation):

$$Q(y) := \frac{C_1}{n}\cdot A(y)\cdot R(y)^{\frac{2}{3}}\cdot\sqrt{S_0}$$

Plot depth versus discharge:

$$y := 0, 0.01..d$$

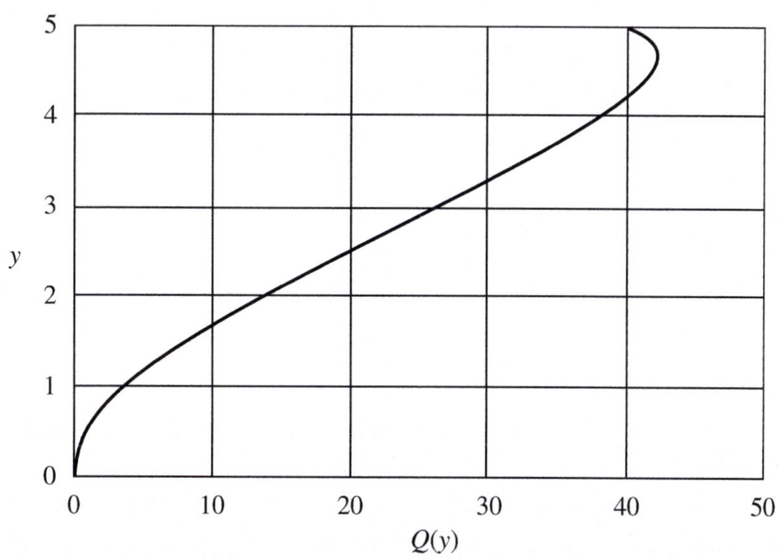

Figure E8.2

8.3.3 Most Efficient Section

Design of a channel to convey uniform flow typically consists of selecting or specifying the appropriate geometrical cross section provided that Q, n, and S_0 are known. Once chosen, an optimum sizing of the cross section can be based on the criteria of minimum resistance to flow. The flow resistance per unit length equals the average wall shear stress times the wetted perimeter. Using a control volume for uniform flow, assume a small slope S_0 so that $\sin\theta = S_0$; then, by summing forces there results

$$\tau_0 P = \gamma A S_0 \tag{8.3.14}$$

Hence, a least resistance criterion is equivalent to requiring a minimum cross-sectional area with respect to the parameters defining the area. Furthermore, we have to satisfy the Chezy–Manning equation. Since Q, n, and S_0 are given and $R = A/P$, Eq. 8.3.13 can be written as

$$P = cA^{5/2} \tag{8.3.15}$$

in which c is a constant. As an example, consider a rectangular channel with width b and depth y. The best hydraulic cross section is obtained by rewriting Eq. 8.3.15 such that A becomes a function of b only. Therefore, we express P in terms of A and b, using $A = by_0$ and $P = b + 2y_0$:

$$P = b + \frac{2A}{b} \tag{8.3.16}$$

Substitution of this equation into Eq. 8.3.15 yields

$$b + \frac{2A}{b} = cA^{5/2} \tag{8.3.17}$$

Now, differentiate Eq. 8.3.17 with respect to b, remembering that A is dependent on b:

$$1 + \frac{2}{b}\frac{dA}{db} - \frac{2A}{b^2} = \frac{5}{2}cA^{3/2}\frac{dA}{db} \tag{8.3.18}$$

Set $dA/db = 0$, since the objective is to find the value of b that minimizes A. The result is

$$\frac{2A}{b^2} = 1 \tag{8.3.20}$$

or, using $A = by$,

$$b = 2y_0 \tag{8.3.20}$$

Thus, if the width of a rectangular channel is twice the depth of the flowing water, the water will flow most efficiently.

For a trapezoidal cross section, it is simpler to begin with Eq. 8.3.15, eliminate b, and express P as a function of A, y_0, and m, where $m = m_1 = m_2$. Then by considering A as a function of m and y, the minimum of A is found by setting the gradient vector of A to zero. The result is $m = \sqrt{3}/3$ or a side slope angle of 60° with the horizontal. The resulting hexagonal shape is a trapezoid that best approximates a semicircle. The evaluation is left as an exercise for the reader.

The optimum criterion based on flow resistance cannot always be used, and in fact may be less significant than other design factors. Additional aspects to be considered are the type of excavation, and if the channel is unlined, the slope stability of the side walls and the possibility of erosion of the bed.

8.4 ENERGY CONCEPTS

The energy head at any position along the channel is the sum of the vertical distance measured from a horizontal datum z, the depth of flow y, and the kinetic energy head $V^2/2g$. That sum defines the energy grade line and is termed the **total energy** H:

$$H = z + y + \frac{V^2}{2g} \tag{8.4.1}$$

The kinetic-energy-correction factor associated with the $V^2/2g$ term is assumed to be unity (see Section 2.5.4); this is common practice for most prismatic channels of simple geometry since the velocity profiles are nearly uniform for the turbulent flows involved. The fundamental energy equation, developed in Section 2.5, states that losses will occur for a real fluid between any two sections of the channel, and hence the total energy will not remain constant. The energy balance is given simply by the relation

$$H_1 = H_2 + h_L \tag{8.4.2}$$

in which h_L is the head loss. The only manner in which energy can be added to an open-channel flow system is for mechanical pumping or lifting of the liquid to take place. Equation 8.4.2 is applicable for rapidly varied as well as gradually varied flow situations; it will be used in conjunction with the momentum and continuity equations in a number of applications.

8.4.1 Specific Energy

It is convenient in open-channel flow to measure the energy relative to the bottom of the channel; it provides a useful means to analyze complex flow situations. Such a measure is termed **specific energy** and is designated as E:

$$E = y + \frac{V^2}{2g} \tag{8.4.3}$$

Specific energy then is the sum of the flow depth y and kinetic energy head $V^2/2g$.

Rectangular Sections. For a rectangular section, the specific energy can be expressed as a function of the depth y. The **specific discharge** q is defined as the total discharge divided by the channel width, that is,

$$q = \frac{Q}{b} = Vy \tag{8.4.4}$$

The specific energy for a rectangular channel can thus be put in the form

$$E = y + \frac{q^2}{2gy^2} \tag{8.4.5}$$

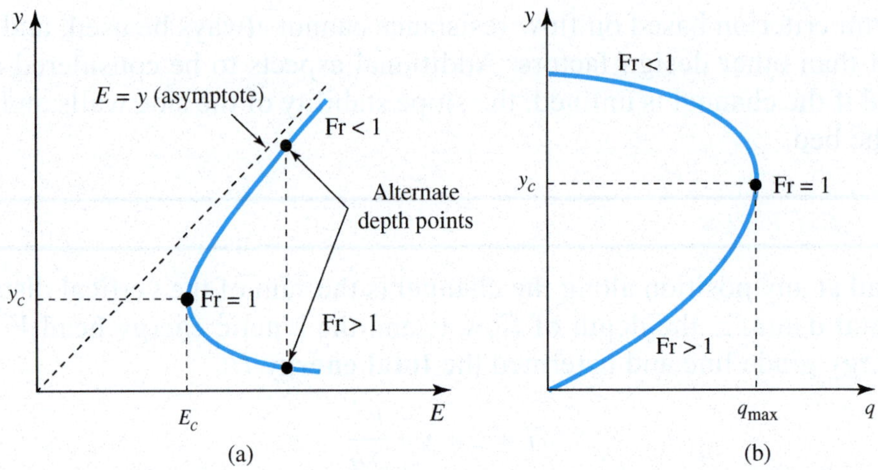

Figure 8.6 Variation of specific energy and specific discharge with depth: (a) E versus y for constant q; (b) q versus y for constant E.

This E-y relation is shown in Figure 8.6a. We may observe that a specific discharge requires at least a minimum energy; this minimum energy is referred to as **critical energy**, E_c. The corresponding depth y_c is called the **critical depth**. If the specific energy is greater than E_c, two depths are possible; those depths are referred to as **alternate depths**. For constant q, Eq. 8.4.5 is a cubic equation in y for a given value of E which is greater than E_c. The two positive solutions of y are the alternate depths.[2] Another way to express Eq. 8.4.5 is to consider E constant and vary q. Equation 8.4.5 can be solved for q as

$$q = \sqrt{2gy^2(E - y)} \tag{8.4.6}$$

This relation is shown in Figure 8.6b. This form of the energy relation is useful for analyzing flows in which the specific energy remains nearly constant throughout the transition region; examples are a change in the width of a channel and the variation of depth with discharge at the entrance to a channel. Note that maximum unit discharge, q_{max}, occurs at critical depth.

The critical depth y_c can be evaluated from Eq. 8.4.5 by setting the derivative of E with respect to y equal to zero:

$$\frac{dE}{dy} = 1 - \frac{q^2}{gy^3} = 0 \tag{8.4.7}$$

Since $q = Vy$, the condition of minimum E is

$$1 - \frac{V^2}{gy} = 1 - \mathrm{Fr}^2 = 0 \tag{8.4.8}$$

where the Froude number in a rectangular channel is

$$\mathrm{Fr} = \frac{V}{\sqrt{gy}} = \frac{q}{\sqrt{gy^3}} \tag{8.4.9}$$

[2]The E-y curve falls between the two asymptotes $E = y$ and $y = 0$. Another curve defined by the relation exists for negative y; it is not considered since it has no physical meaning for open-channel flow.

Thus, from Eq. 8.4.8 the Froude number is equal to unity for minimum energy. Solving for the depth in Eq. 8.4.7 in terms of q, we have

$$y = y_c = \left(\frac{q^2}{g}\right)^{1/3} \tag{8.4.10}$$

This relation gives the critical depth in terms of the specific discharge. Note that at critical flow conditions, E_c can conveniently be expressed by combining Eqs. 8.4.5 and 8.4.10 to eliminate q, resulting in

$$E_c = \frac{3}{2} y_c \tag{8.4.11}$$

On the E-y curve, for any depth greater than y_c, the flow is relatively slow or tranquil, and $\mathrm{Fr} < 1$; such a state is termed **subcritical flow**. Conversely, for a depth less than critical, the flow is relatively rapid or shooting, with $\mathrm{Fr} > 1$, and the regime is one of **supercritical flow**. The E-y diagram is a representation of the change in specific energy as the depth is varied, given a constant specific discharge. It is possible for q to vary, as when the width of a rectangular section changes in a transition region. As q increases, the E-y curve shifts to the right in Figure 8.6a. It is left as an exercise for the reader to show that for a given specific energy, maximization of q of Eq. 8.4.6 will produce critical conditions at maximum discharge (see Figure 8.6b).

Generalized Cross Section. For a generalized section, the specific energy is written in terms of the total discharge Q and the cross-sectional area A as

$$E = y + \frac{Q^2}{2gA^2} \tag{8.4.12}$$

The minimum-energy condition is obtained by differentiating Eq. 8.4.12 with respect y to obtain

$$\frac{dE}{dy} = 1 - \frac{Q^2}{gA^3}\frac{dA}{dy} \tag{8.4.13}$$

For incremental changes in depth, the corresponding change in area is $dA = B\,dy$. Thus setting Eq. 8.4.13 to zero, the minimum-energy condition becomes

$$1 - \frac{Q^2 B}{gA^3} = 0 \tag{8.4.14}$$

By analogy to Eq. 8.4.8, the second term in Eq. 8.4.14 is the square of the Froude number; hence

$$\mathrm{Fr} = \sqrt{\frac{Q^2 B}{gA^3}} = \frac{Q/A}{\sqrt{gA/B}} = \frac{V}{\sqrt{gA/B}} \tag{8.4.15}$$

The ratio A/B is termed the **hydraulic depth**, and its use allows the definition of the Froude number to be generalized. The reader can verify that A/B is equal to y for a rectangular channel.

The specific energy diagram of Figure 8.6 provides a useful means of visualizing the solution to a transition problem. Even though one probably would solve the problem numerically, an assessment of the graphical solution may provide useful physical insight as well as prevent the incorrect root from being selected.

Example 8.3

Water is flowing in a triangular channel with $m_1 = m_2 = 1.0$ at a discharge of $Q = 3\ \text{m}^3/\text{s}$. If the water depth is 2.5 m, determine the specific energy, Froude number, hydraulic depth, and alternate depth.

Solution

Recognizing that $b = 0$, the flow area and top width are computed from Eqs. 8.3.4 and 8.3.6 as follows:

$$A = \frac{1}{2} y^2 (m_1 + m_2)$$

$$= \frac{1}{2} \times 2.5^2 \times (1+1) = 6.25\ \text{m}^2$$

$$B = (m_1 + m_2) y$$

$$= (1+1) \times 2.5 = 5.0\ \text{m}$$

Using Eqs. 8.4.12 and 8.4.15, E and Fr are found to be

$$E = y + \frac{Q^2}{2gA^2}$$

$$= 2.5 + \frac{3^2}{2 \times 9.81 \times 6.25^2} = \underline{2.51\ \text{m}}$$

$$\text{Fr} = \sqrt{\frac{Q^2 B}{gA^3}}$$

$$= \sqrt{\frac{3^2 \times 5}{9.81 \times 6.25^3}} = \underline{0.137}$$

The hydraulic depth is

$$\frac{A}{B} = \frac{6.25}{5.0} = \underline{1.25\ \text{m}}$$

The alternate depth is calculated using the energy equation. Recognizing that $A = y^2$, we have

$$2.51 = y + \frac{3^2}{2 \times 9.81 \times (y^2)^2}$$

$$= y + \frac{0.459}{y^4}$$

A trial-and-error solution provides $y = \underline{0.71\ \text{m}}$.

8.4.2 Use of the Energy Equation in Transitions

As mentioned in Section 8.2.1, a transition is a relatively short reach of channel where the depth and velocity change, creating a nonuniform, rapidly varied flow. The mechanism for such changes in the flow is usually an alteration of one or more geometric parameters of the channel.[3] Within such regions, the energy equation can be used effectively to analyze the flow in a transition or to aid in the design of a transition. Two applications are given in this section to demonstrate the methodology; other types of transitions can be treated similarly.

Channel Constriction. Consider a rectangular channel whose bottom is raised by a distance h over a short region (Figure 8.7a). The change in depth in the transition can be analyzed by use of the energy equation, and as a first approximation, one can neglect losses. Assume that the specific energy upstream of the transition is known. Recognizing that $H = E + z$, Eq. 8.4.2 is applied from location 1 to the end of the transition region, location 2:

$$E_1 = E_2 + h \tag{8.4.16}$$

The depth y_2 at the end of the transition can be visualized by inspection of the E-y diagram (Figure 8.7b). If the flow at location 1 is subcritical, y_1 is located on the upper leg as shown in the specific energy diagram. The magnitude of h is selected to be relatively small such that $y_2 > y_c$, hence

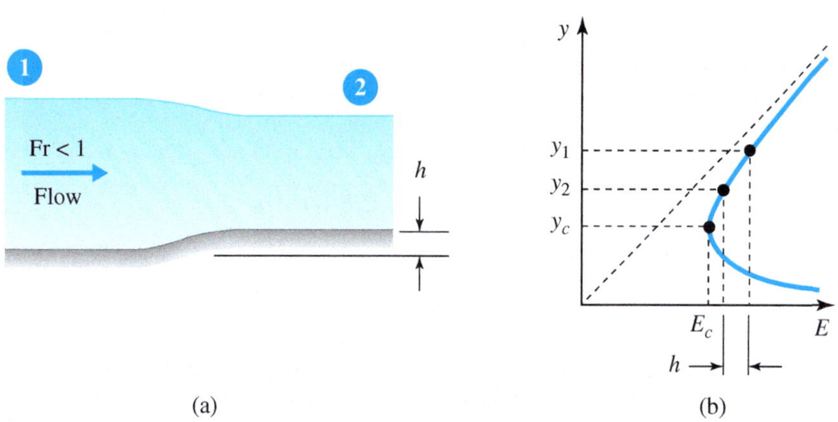

Figure 8.7 Channel constriction: (a) raised channel bottom; (b) specific energy diagram.

[3]One notable exception to this is the hydraulic jump, which requires use of the momentum principle and will be considered in Section 8.5.2.

the flow at location 2 is similarly subcritical; however, as h is increased still further, a state of minimum energy is ultimately reached in the transition. The condition of minimum energy is sometimes referred to as a **choking condition** or as **choked flow**. Once choked flow occurs, as h increases, the variations in depth and velocity are no longer localized in the vicinity of the transition. Influences may be observed for significant distances both upstream and downstream of the transition.

The reader should verify that in a transition region, a narrowing of the channel width will create a situation similar to an elevation of the channel bottom. The most general transition region is one that possesses a change in both width and in bottom elevation.

Example 8.4

A rectangular channel 3 m wide is conveying water at a depth $y_1 = 1.55$ m and velocity $V_1 = 1.83$ m/s. The flow enters a transition region as shown in Figure E8.4a, in which the bottom elevation is raised by $h = 0.20$ m. Determine the depth and velocity in the transition, and the value of h for choking to occur.

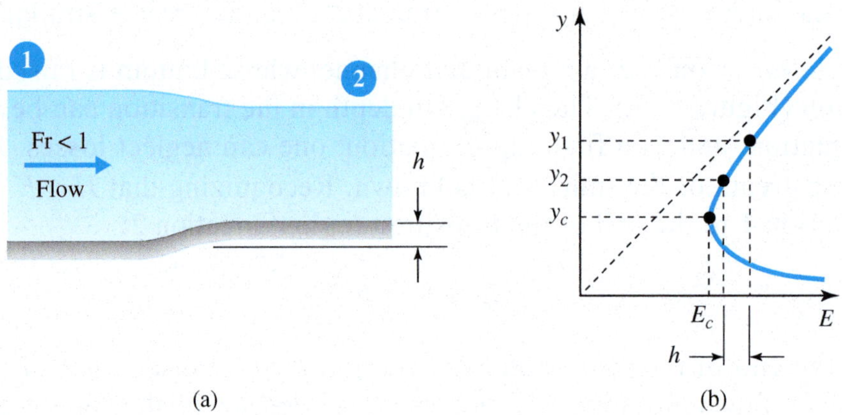

Figure E8.4

Solution

Use Eq. 8.4.4 to find the specific discharge to be

$$q = V_1 y_1$$
$$= 1.83 \times 1.55 = 2.84 \text{ m}^2/\text{s}$$

The Froude number at location 1 is

$$\text{Fr} = \frac{V_1}{\sqrt{g y_1}}$$
$$= \frac{1.83}{\sqrt{9.81 \times 1.55}} = 0.47$$

which is less than unity. Hence, the flow at location 1 is subcritical. The specific energy at location 1 is found, using Eq. 8.4.3, to be

$$E_1 = y_1 + \frac{V_1^2}{2g}$$
$$= 1.55 + \frac{1.83^2}{2 \times 9.81} = 1.72 \text{ m}$$

The specific energy at location 2 is found, using Eq. 8.4.16, to be

$$E_2 = E_1 - h$$
$$= 1.72 - 0.20 = 1.52 \text{ m}$$

If $E_2 > E_c$, it is possible to find the depth y_2. Therefore, E_c is calculated first. From Eqs. 8.4.10 and 8.4.11 the critical conditions are

$$y_c = \left(\frac{q^2}{g}\right)^{1/3} = \left(\frac{2.84^2}{9.81}\right)^{1/3} = 0.94 \text{ m}$$
$$E_c = \frac{3y_c}{2} = 3 \times \frac{0.94}{2} = 1.41 \text{ m}$$

Hence, since $E_2 > E_c$, we can proceed with calculating y_2. The depth y_2 can be evaluated by substituting known values into Eq. 8.4.5:

$$1.52 = y_2 + \frac{2.84^2}{2 \times 9.81 \times y_2^2}$$

The solution is

$$y_2 = \underline{1.26 \text{ m}}$$
$$\therefore V_2 = \frac{q}{y_2}$$
$$= \frac{2.84}{1.26} = 2.25 \text{ m/s}$$

The flow at location 2 is subcritical since there is no way in which the flow can become supercritical in the transition with the given geometry.

The value of h for critical flow to appear at location 2 is determined by setting $E_2 = E_c$ in Eq. 8.4.16.

$$h = E_1 - E_c = 1.72 - 1.40 = \underline{0.31 \text{ m}}$$

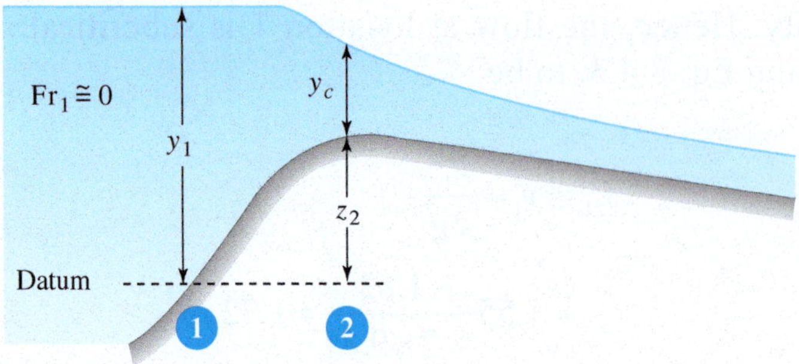

Figure 8.8 Outflow from a reservoir with critical flow at the channel entrance.

Channel Entrance with Critical Flow. Consider flow entering a channel from a lake or reservoir over a short rounded crest, as sketched in Figure 8.8. If the channel slope is steep, the flow will discharge freely into the channel and supercritical flow will occur downstream of the entrance region. Upstream of the crest the flow may be considered as subcritical. Since the flow regime at the crest changes from subcritical to supercritical, the flow at the crest must be critical. To determine the discharge, we assume that rapidly varied flow takes place over the crest in conjunction with the critical flow condition at the crest. An example illustrates the procedure.

Example 8.5

Water flows freely from a reservoir into a trapezoidal channel with bottom width $b = 5.0$ m and side slope parameters $m_1 = m_2 = 2.0$. The elevation of the water surface in the reservoir is 2.3 m above the entrance crest. Assuming negligible losses in the transition and a negligible velocity in the reservoir upstream of the entrance, find the critical depth at the transition and the discharge into the channel.

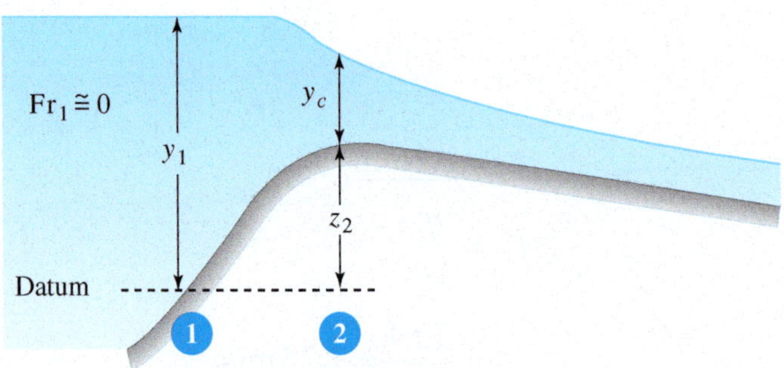

Figure E8.5

Solution

The total energy at location 1 in Figure E8.5 is y_1 since the kinetic energy in the reservoir is negligible ($V_1 \approx 0$). Equating the total energies at locations 1 and 2 gives

$$y_1 = E_2 + z_2$$

Since critical conditions occur at location 2, Eqs. 8.4.12 and 8.4.14 can be combined to eliminate the discharge, with the result

$$E_2 = y_c + \frac{A}{2B}$$

Elimination of E_2 in the two equations yields the expression

$$y_1 - z_2 = y_c + \frac{A}{2B} = y_c + \frac{by_c + \frac{1}{2}(m_1 + m_2)y_c^2}{2[b + (m_1 + m_2)y_c]}$$

or, with the given data, the expression becomes

$$2.3 = y_c + \frac{5y_c + \frac{1}{2}(2+2)y_c^2}{2[5 + (2+2)y_c]}$$

The relation above is a quadratic in y_c. The positive root is chosen, which is

$$y_c = \underline{1.70 \text{ m}}$$

Subsequently, one can find that A = 14.28 m^2 and B = 11.80 m. Use Eq. 8.4.14 to find the discharge to be

$$Q = \sqrt{\frac{gA^3}{B}}$$

$$= \sqrt{\frac{9.8 \times 14.3^3}{11.8}} = \underline{49.3 \text{ m}^3/\text{s}}$$

Energy Losses. Energy losses in expansions and contractions are known to be relatively small when the flow is subcritical; however, it may be necessary under certain circumstances that the losses be considered. Equation 8.4.2 includes a loss term that can account for transitional losses. The following experimentally derived formulas can be employed. For a channel expansion use

$$h_L = K_e\left(\frac{V_1^2}{2g} - \frac{V_2^2}{2g}\right) \tag{8.4.17}$$

and for a channel contraction use

$$h_L = K_c\left(\frac{V_2^2}{2g} - \frac{V_1^2}{2g}\right) \tag{8.4.18}$$

In Eq. 8.4.17, K_e is an expansion coefficient; it has been suggested (King and Brater, 1963) to use K_e = 1.0 for sudden or abrupt expansions, and K_e = 0.2 for well-designed or rounded expansions. For the contraction coefficient K_c in Eq. 8.4.18, use K_c = 0.5 for sudden contractions, and K_c = 0.10 for well-designed contractions. When the flow is supercritical, significant standing-wave patterns can be generated throughout and downstream of the transition region; for such flows, proper design requires that wave mechanics be considered (Chow, 1959; Henderson, 1966).

8.4.3 Flow Measurement

The most common means of metering discharge in an open channel is to use a weir. Basically, a **weir** is a device placed in the channel that forces the flow through an opening or aperture designed to measure the discharge. Specialized weirs have been designed for specific needs; in this section two fundamental types—broad-crested and sharp-crested—will be presented.

A properly designed weir will exhibit subcritical flow upstream of the structure, and the flow will converge and accelerate to a critical condition near the top or **crest** of the weir. As a result, a correlation between discharge and a depth upstream of the weir can be made. The downstream overflow is termed the **nappe**, which usually discharges freely into the atmosphere. There are a number of factors that affect the performance of a weir; significant among them are the three-dimensional flow pattern, the effects of turbulence, and frictional resistance, surface tension, and the amount of ventilation beneath the nappe. The simplified derivations presented here are based on the Bernoulli equation; the other effects can be accounted for by modifying the ideal discharge with a **discharge coefficient**, C_d. The actual discharge is the ideal discharge multiplied by the discharge coefficient. When possible, it is advantageous to calibrate a particular weir in place in order to obtain the desired accuracy.

Broad-Crested Weir. A broad-crested weir is shown in Figure 8.9. It has sufficient elevation above the channel bottom to choke the flow, and it is long enough so that the overflow streamlines become parallel, resulting in a hydrostatic pressure distribution. Then, at some position, say location 2, a critical flow condition exists.

Assume a rectangular horizontal channel and let the height of the weir be h. Location 1 is a point upstream of the weir where the flow is relatively undisturbed, and Y is the vertical distance from the top of the weir to the free surface at that point. Applying Bernoulli's equation from location 1 to location 2 at the free surface and neglecting the kinetic energy bead at location $1(V_1 \approx 0)$, one has the result

$$h + Y = h + y_c + \frac{V_c^2}{2g} \tag{8.4.19}$$

Solving for V_c,

$$V_c = \sqrt{2g(Y - y_c)} \tag{8.4.20}$$

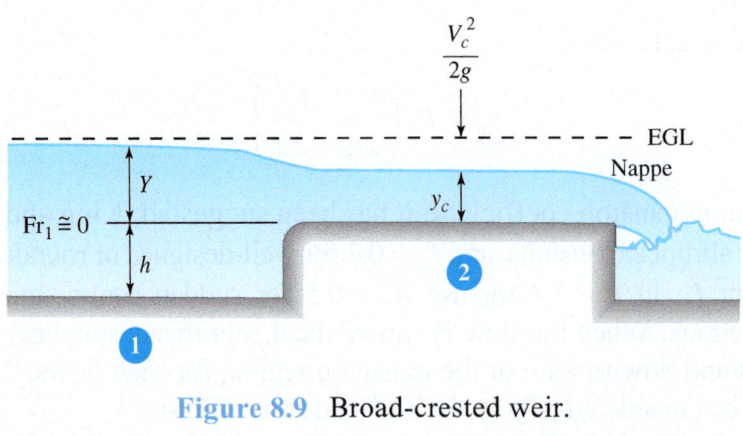

Figure 8.9 Broad-crested weir.

For a weir whose breadth normal to the flow is b, the ideal discharge is

$$Q = by_cV_c = by_c\sqrt{2g(Y - y_c)} \tag{8.4.21}$$

Recognizing that $Y = E_c$, Eq. 8.4.11 is used to relate y_c to Y, and when substituted into Eq. 8.4.21, the result is

$$Q = \frac{2}{3}\sqrt{\frac{2}{3}g}bY^{3/2} \tag{8.4.22}$$

For a properly rounded upstream edge on the weir, Eq. 8.4.22 is accurate to within several percent of the actual flow; hence, a discharge coefficient is not applied.

Sharp-Crested Weir. A sharp-crested weir is a vertical plate placed normal to the flow containing a sharp-edge crest so that the nappe will behave as a free jet. Figure 8.10 shows a rectangular weir with a horizontal crest extending across the entire width of the channel. Because of the presence of the side walls, lateral contractions are not present.

Let us define an idealized flow situation: The flow in the vertical plane does not contract as it passes over the crest so that the streamlines are parallel, atmospheric pressure is present in the nappe, and uniform flow exists at location 1 with negligible kinetic energy ($v_1 \approx 0$). The Bernoulli equation is applied along a representative streamline (Figure 8.10a) and solved for v_2, the local velocity in the nappe:

$$v_2 = \sqrt{2g\eta} \tag{8.4.23}$$

If b is the width of the crest normal to the flow, the ideal discharge is given as

$$Q = b\int_0^Y v_2 d\eta = b\int_0^Y \sqrt{2g\eta}\, d\eta = b\frac{2}{3}\sqrt{2g}Y^{3/2} \tag{8.4.24}$$

Experiments have shown that the magnitude of the exponent is nearly correct but that a discharge coefficient C_d must be applied to accurately predict the real flow, shown in Figure 8.10b:

$$Q = C_d\frac{2}{3}\sqrt{2g}bY^{3/2} \tag{8.4.25}$$

The discharge coefficient accounts for the effect of contraction, velocity of approach, viscosity, and surface tension. An experimentally (Chow, 1959) obtained formula for C_d has been given as

$$C_d = 0.61 + 0.08\frac{Y}{h} \tag{8.4.26}$$

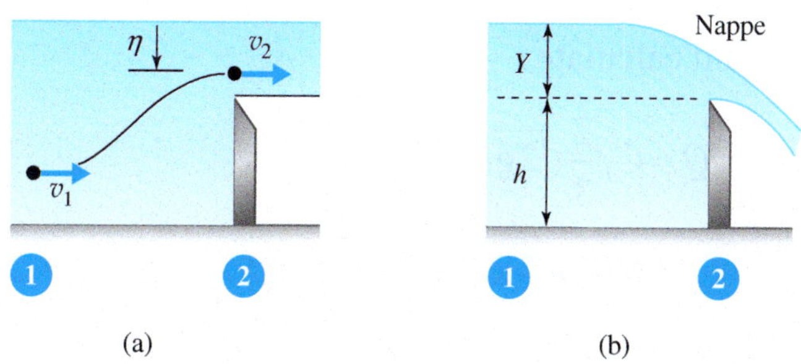

Figure 8.10 Rectangular sharp-crested weir: (a) ideal flow; (b) actual flow.

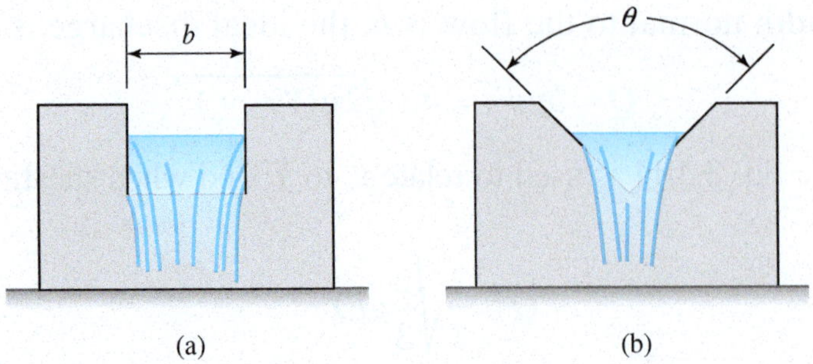

Figure 8.11 Contracted (a) rectangular and (b) V-notch weirs.

Normally, for small Y/h ratio, $C_d = 0.61$. If the weir crest does not extend to the side walls but allows for end contractions to appear, as in Figure 8.11a, the effective width of the weir can be approximated by $(b - 0.2Y)$.

The V-notch weir (Figure 8.11b) is more accurate than the rectangular weir for measurement of low discharge. In a manner similar to the development of the rectangular weir relation, the idealized discharge is found by integrating the local velocity throughout the nappe above the crest. A discharge coefficient is applied to yield

$$Q = C_d \frac{8}{15}\sqrt{2g}\left(\tan\frac{\theta}{2}\right)Y^{5/2} \tag{8.4.27}$$

For use with water, and for θ varying from 22.5° to 120°, experiments (King and Brater, 1963) have shown that a value of $C_d = 0.58$ is acceptable for engineering calculations. Derivation of Eq. 8.4.27 is left as an exercise.

The above equations can, as usual, be used with either set of units.

Example 8.6

Determine the discharge of water over a rectangular sharp-crested weir, $b = 1.25$ m, $Y = 0.35$ m, $h = 1.47$ m, with side walls and with end contractions. If a 90° V-notch weir were to replace the rectangular weir, what would be the required Y for a similar discharge?

Solution

For the rectangular weir, using Eq. 8.4.26, the discharge coefficient is

$$C_d = 0.61 + 0.08\frac{Y}{h} = 0.61 + 0.08 \times \frac{0.35}{1.47} = 0.63$$

Substitute into Eq. 8.4.25 and calculate

$$\begin{aligned} Q &= C_d \frac{2}{3}\sqrt{2g}\,bY^{3/2} \\ &= 0.63 \times \frac{2}{3} \times \sqrt{2 \times 9.81} \times 1.25 \times 0.35^{3/2} \\ &= 0.48 \text{ m}^3/\text{s} \end{aligned}$$

With end contractions the effective width of the weir is reduced by $0.2Y$, resulting in

$$Q = C_d \frac{2}{3}\sqrt{2g}(b - 0.2Y)Y^{3/2}$$

$$= 0.63 \times \frac{2}{3} \times \sqrt{2 \times 9.81} \times (1.25 - 0.2 \times 0.35) \times 0.35^{3/2}$$

$$= \underline{0.45 \text{ m}^3/\text{s}}$$

With a discharge of $Q = 0.48 \text{ m}^3/\text{s}$, use Eq. 8.4.27 to find Y for the 90° V-notch weir:

$$Y = \left[\frac{Q}{C_a \times \frac{8}{1.5} \times \sqrt{2g}\tan(\theta/2)}\right]^{2/5}$$

$$= \left[\frac{0.482}{0.58 \times \frac{8}{15} \times \sqrt{2 \times 9.81} \times \tan 45°}\right]^{2/5} = \underline{0.66 \text{ m}}$$

Additional Methods of Flow Measurement. Other types of weirs include those whose faces are inclined in the upstream and downstream direction (triangular, trapezoidal, irregular). In addition, the spillway section of a dam can be considered as a weir with a rounded crest. King and Brater (1963) give details on the selection and use of these types.

A special type of open flume is one in which the geometry of the throat is constricted in such a manner as to choke the flow, thereby creating critical flow followed by a hydraulic jump. When manufactured using a particular standardized section, the flume is called a **Parshall flume**, shown in Figure 8.12. Extensive calibrations have established reliable empirical formulas to predict the discharge. For throat widths from 1 to 8 ft (approximately 0.3 to 2.4 m), the discharge is given by the formula

$$Q = 4BH^{1.522B^{0.026}} \tag{8.4.28}$$

in which H is the depth measured at the upstream location shown in Figure 8.12. Note that H and B are measured in feet, and Q is in cubic feet per second. The other dimensions are used to develop Eq. 8.4.28.

In a natural section of a river it may be impractical to place a weir; in that case stream gaging can be used to measure the discharge. A control location is established upstream of the gaging site, and for a given depth, or **stage**, of the river, the two-dimensional velocity profile is measured using current meters. Subsequently, the profile is numerically integrated to yield the discharge. A series of such measurements will produce a stage-discharge curve, which can then be employed to estimate the discharge by measuring the river stage.

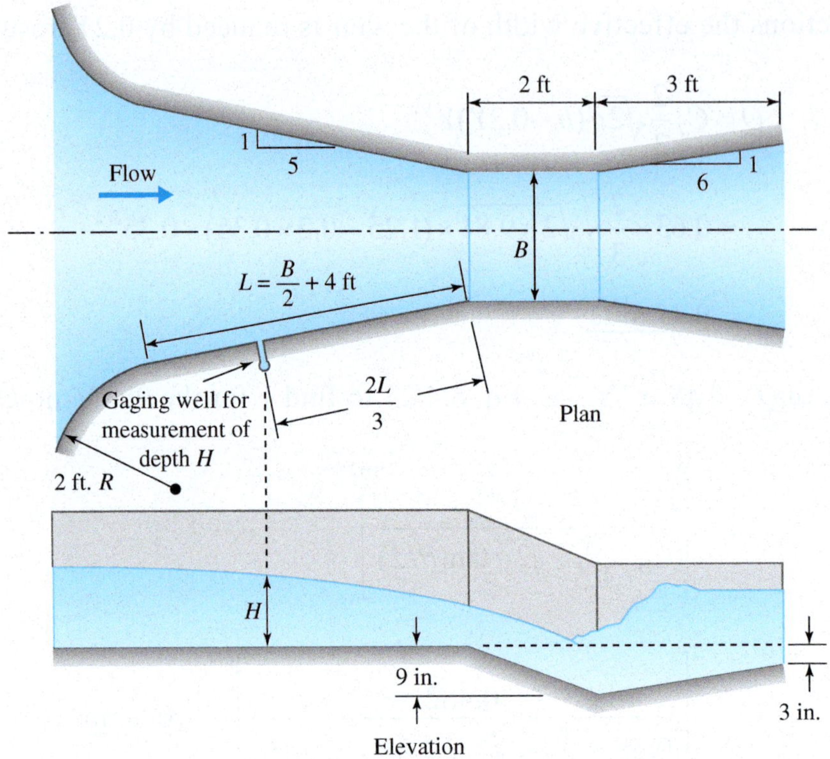

Figure 8.12 Parshall flume. (Based on Henderson, *Open Channel Flow*, 1st, 1966, Pearson Education, Inc, Upper Saddle River, New Jersey.)

8.5 MOMENTUM CONCEPTS

In the preceding section we have seen how the energy equation is applied to rapidly varied flow situations, and in particular, how it is utilized to analyze flows in transition regions. The momentum equation is also employed to study certain phenomena in those situations. When used in conjunction with the energy and continuity relations, the momentum equation gives the user a concise means of analyzing nearly all significant transition problems, including problems that involve a hydraulic jump.

8.5.1 Momentum Equation

Consider the open-channel reach with supercritical flow upstream of a submerged obstacle, as depicted in Figure 8.13a. It represents a rapidly varied flow with an abrupt change in depth but no change in width. In general, such a change may result from an obstacle in the flow or from a hydraulic jump. The upstream flow is supercritical and the downstream flow is subcritical. Other flow regimes may also be considered; for example, the conditions could be subcritical throughout the entire control volume. Each situation should be approached as a unique formulation.

The generalized flow situation may be used to develop the equation of motion for transition regions. The control volume corresponding to Figure 8.13a is shown in Figure 8.13b. The pressure distribution is assumed hydrostatic, and the resultant hydrostatic forces are given by $\gamma A\bar{y}$,

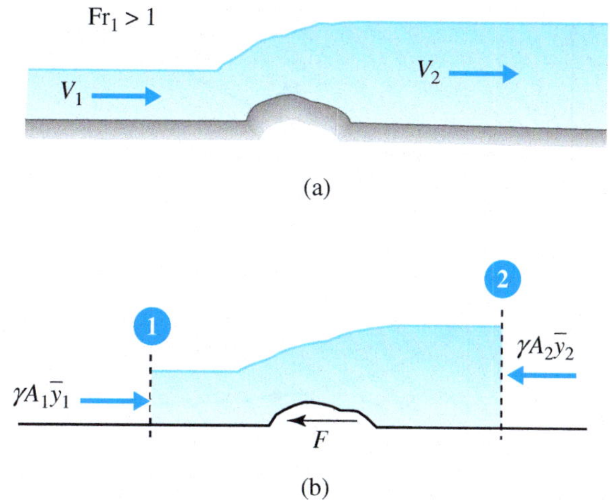

Figure 8.13 Channel flow over an obstacle: (a) idealized flow; (b) control volume.

where the distance $\bar{y}$ to the centroid of the cross-sectional area is measured from the free surface. The submerged obstacle (e.g., a rock or a two-dimensional obstacle) imparts a force F on the control volume with a direction opposite to the direction of flow. The linear momentum equation, presented in Section 2.6 is applied to the control volume in the x-direction to give

$$\gamma A_1\bar{y}_1 - \gamma A_2\bar{y}_2 - F = \rho Q(V_2 - V_1) \tag{8.5.1}$$

Note that frictional forces have not been included in Eq. 8.5.1; they are usually quite small relative to the other terms, so they can be ignored. Similarly, gravitational forces in the flow direction are negligible for the small channel slopes considered. Equation 8.5.1 can be rearranged in the form

$$M_1 - M_2 = \frac{F}{\gamma} \tag{8.5.2}$$

in which M_1 and M_2 are terms that contain the hydrostatic force and the momentum flux at locations 1 and 2, respectively. The quantity M is called the **momentum function**, and for a general prismatic section it is given by

$$M = A\bar{y} + \frac{Q^2}{gA} \tag{8.5.3}$$

For a rectangular section, $A\bar{y} = by^2/2$, and the momentum function is

$$M = \frac{by^2}{2} + \frac{bq^2}{gy} = b\left(\frac{y^2}{2} + \frac{q^2}{gy}\right) \tag{8.5.4}$$

Equation 8.5.4 is sketched in Figure 8.14. Two positive roots y exist for a given M and q; they are termed either **conjugate** or **sequent** depths. The upper leg of the curve ($y > y_c$) applies to subcritical flow and the lower leg ($y < y_c$) to supercritical flow. The values of M_1 and M_2 for the example of Figure 8.13a are shown, indicating supercritical flow upstream and subcritical flow

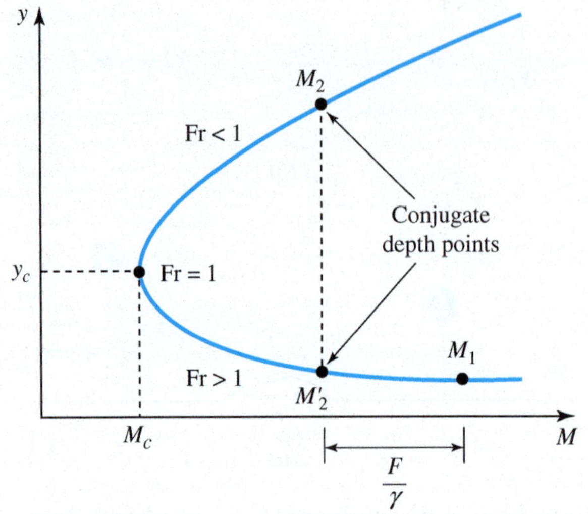

Figure 8.14 Variation of the momentum function with depth.

downstream of the submerged obstacle. The horizontal distance between M_1 and M_2 is equal to F/γ. The downstream flow would be supercritical if no hydraulic jump occurred; the corresponding value of the momentum function is indicated by M_2'.

The depth associated with a minimum M is found by differentiating M with respect to y in Eq. 8.5.3:

$$\frac{dM}{dy} = A - \frac{BQ^2}{gA^2} = 0 \tag{8.5.5}$$

Note[4] that $d(A\bar{y})/dy = A$. The condition for minimum M is thus,

$$Q^2 B = gA^3 \tag{8.5.6}$$

This result is identical to Eq. 8.4.14. Hence, the condition of minimum M is equivalent to that of minimum energy: critical flow with a Froude number equal to unity.

[4]This can be observed from the definition $\bar{y} = (1/A)\int (y-\eta)\, dA$. Differentiation of $(\bar{y}A)$ using Leibnitz's rule from calculus yields

$$\frac{d}{dy}(\bar{y}A) = \frac{d}{dy}\int_0^y (y-\eta)w(\eta)d\eta$$

$$= (y-\eta)w(\eta)\big|_{\eta=y} + \int_0^y w(\eta)d\eta$$

$$= 0 + A = A$$

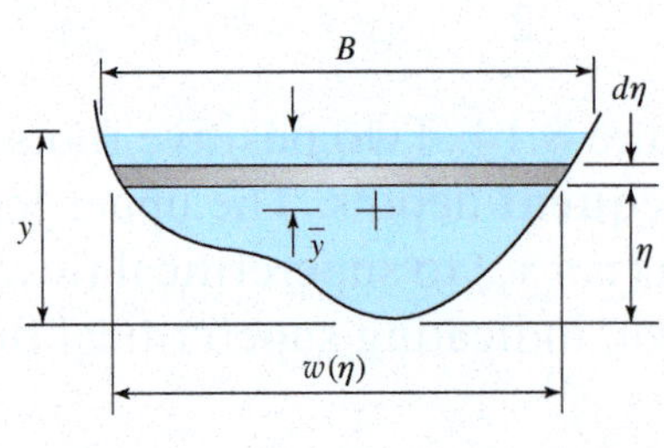

Momentum Equation Applied to a Transition Region. Quite often the momentum equation is applied in situations where it is desired to determine the resultant force acting at a specific location, or to find the change in depth or velocity when there is a significant undefined loss across the transition region. It is important to remember that the energy and continuity equations are also at our disposal, and we must determine which relations are necessary. In some instances, along with the continuity equation, both energy and momentum equations must be applied. An example is given below to illustrate the technique.

Example 8.7

In a rectangular 5-m-wide channel, water is discharging at 14.0 m^3/s (Figure E8.7). Find the force exerted on the sluice gate when $y_1 = 2$ m and $y_2 = 0.5$ m.

Figure E8.7

Solution

Using Eq. 8.5.3, the momentum functions at 1 and 2 are

$$M_1 = A_1\bar{y}_1 + \frac{Q^2}{gA_1}$$

$$= 5\times 2\times 1 + \frac{(14)^2}{9.81\times 5\times 2} = 12.0\ \text{m}^3$$

$$M_2 = A_2\bar{y}_2 + \frac{Q^2}{gA_2}$$

$$= 5\times 0.5\times 0.25 + \frac{(14)^2}{9.81\times 5\times 0.5} = 8.62\ \text{m}^3$$

The resultant force acting on the fluid control volume is determined, using Eq. 8.5.2, to be

$$F = \gamma(M_1 - M_2)$$

$$= 9800\times(12.0 - 8.62) = \underline{33\ 100\,\text{N}}$$

Hence, the force on the gate acts in the downstream sense with a magnitude of 33.1 kN.

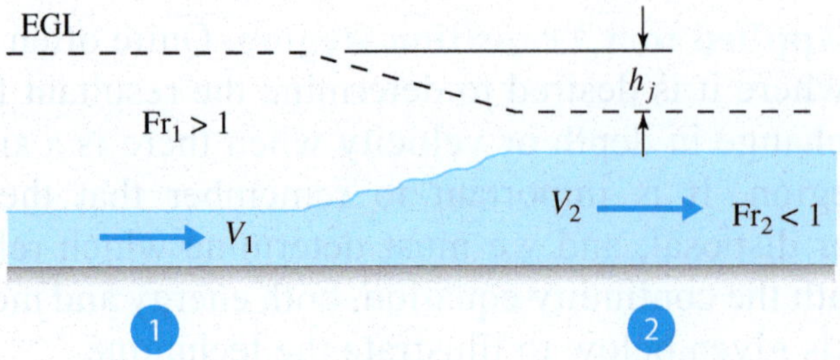

Figure 8.15 Idealized hydraulic jump.

8.5.2 Hydraulic Jump

A **hydraulic jump** is a phenomenon in which fluid flowing at a supercritical state will undergo a transition to a subcritical state. Boundary conditions upstream and downstream of the jump will dictate its strength as well as its location. An idealized hydraulic jump is shown in Figure 8.15. The strength of the jump varies widely, as shown in Table 8.2, with relatively mild disturbances occurring at one extreme, and significant separation and eddy formation taking place at the other. As a result, the energy loss associated with the jump is considered to be unknown, so the energy equation is not used in the initial analysis. Assuming no friction along the bottom and no submerged obstacle, that is, setting the friction force F equal to zero, Eq. 8.5.2 shows that $M_1 = M_2$. For a rectangular section, Eq. 8.5.4 can be substituted into the relation, allowing one to obtain

$$\frac{q^2}{g}\left(\frac{1}{y_1}-\frac{1}{y_2}\right)=\frac{1}{2}\left(y_2^2-y_1^2\right) \tag{8.5.7}$$

Table 8.2 Hydraulic Jumps in Horizontal Rectangular Channels

Upstream Fr	*Type*	*Description*	
1.0–1.7	Undular	Ruffled or undular water surface; surface rollers form near Fr = 1.7	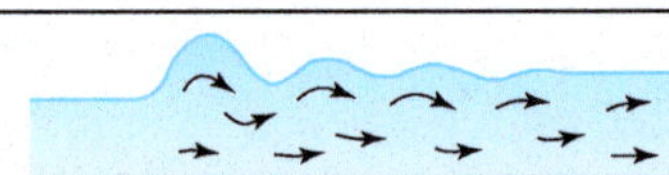
1.7–2.5	Weak	Prevailing smooth flow; low energy loss	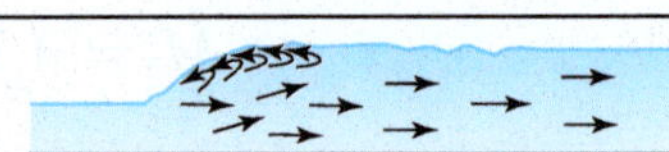
2.5–4.5	Oscillating	Intermittent jets from bottom to surface, causing persistent downstream waves	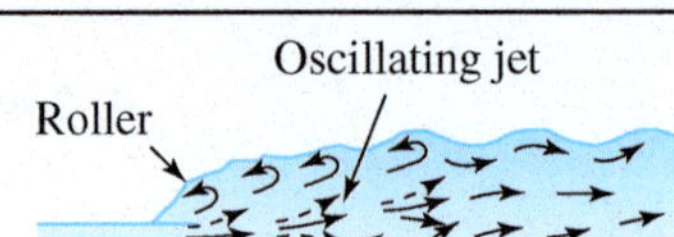
4.5–9.0	Steady	Stable and well-balanced; energy dissipation contained in main body of jump	
>9.0	Strong	Effective, but with rough, wavy surface downstream	

Source: Adapted from Chow, 1959.

Rearranging, factoring, and noting that $\mathrm{Fr}_1^2 = q^2/gy_1^3$ there results

$$\mathrm{Fr}_1^2 = \frac{1}{2}\frac{y_2}{y_1}\left(\frac{y_2}{y_1}+1\right) \tag{8.5.8}$$

This equation is dimensionless and it relates the Froude number upstream of the jump with the ratio of the downstream to upstream depths. It can be seen that Eq. 8.5.8 is quadratic with respect to y_2/y_1 provided that Fr_1 is known. Solving for y_2/y_1, one obtains

$$\frac{y_2}{y_1} = \frac{1}{2}\left(\sqrt{1+8\mathrm{Fr}_1^2}-1\right) \tag{8.5.9}$$

The positive sign in front of the radical has been chosen to yield a physically meaningful solution. It is worth noting that Eq. 8.5.9 is also valid if the subscripts on the depths and Froude number are reversed:

$$\frac{y_1}{y_2} = \frac{1}{2}\left(\sqrt{1+8\mathrm{Fr}_2^2}-1\right) \tag{8.5.10}$$

The theoretical energy loss associated with a hydraulic jump in a rectangular channel can be determined once the depths and flows at locations 1 and 2 are known. The energy equation is applied from 1 to 2, including the head loss h_j across the jump, as displayed in Figure 8.15. Combining the energy equation with Eq. 8.5.7 and the continuity equation, one can show after some algebra that

$$h_j = \frac{(y_2 - y_1)^3}{4y_1y_2} \tag{8.5.11}$$

Equations 8.5.8 through 8.5.11 are useful forms for solving most hydraulic jump problems in a rectangular channel.

Table 8.2 shows the various forms that a hydraulic jump may assume relative to the upstream Froude number. A steady, well-established jump, with $4.5 < \mathrm{Fr} < 9.0$, is often used as an energy dissipator downstream of a dam or spillway. It is characterized by the existence of breaking waves and rollers accompanied by a submerged jet with significant turbulence and dissipation of energy in the main body of the jump; downstream, the water surface is relatively smooth. For Froude numbers outside the range 4.5 to 9.0, less desirable jumps exist and may create undesirable downstream surface waves. The length of a jump is the distance from the front face to just downstream where smooth water exists; a steady jump has an approximate length of six times the upstream depth.

Example 8.8

A hydraulic jump is situated in a 4-m-wide rectangular channel. The discharge in the channel is 7.5 m^3/s, and the depth upstream of the jump is 0.20 m. Determine the depth downstream of the jump, the upstream and downstream Froude numbers, and the rate of energy dissipated by the jump.

Solution

Find the unit discharge and upstream Froude number:

$$q = \frac{Q}{b}$$

$$= \frac{7.5}{4} = 1.88 \text{ m}^2/\text{s}$$

$$\text{Fr}_1 = \frac{q}{\sqrt{gy_1^3}}$$

$$= \frac{1.88}{\sqrt{9.81 \times 0.20^3}} = \underline{6.71}$$

The downstream depth is computed, using Eq. 8.5.9, to be

$$y_2 = \frac{y_1}{2}\left(\sqrt{1+8\text{Fr}_1^2} - 1\right)$$

$$= \frac{0.20}{2}\left(\sqrt{1+8\times 6.71^3} - 1\right) = \underline{1.80 \text{ m}}$$

The downstream Froude number is

$$\text{Fr}_2 = \frac{q}{\sqrt{gy_2^3}}$$

$$= \frac{1.88}{\sqrt{9.81\times 1.80^3}} = \underline{0.25}$$

The head loss in the jump is given by Eq. 8.5.11:

$$h_j = \frac{(y_2 - y_1)^3}{4y_1y_2}$$

$$= \frac{(1.80 - 0.20)^3}{4\times 0.20\times 1.80} = 2.84 \text{ m}$$

Hence, the rate of energy dissipation in the jump is (refer to Eq. 2.5.25)

$$\gamma Q h_j = 9800 \times 7.5 \times 2.84 = 2.09 \times 10^5 \text{ W} \quad \text{or} \quad \underline{209 \text{ kW}}$$

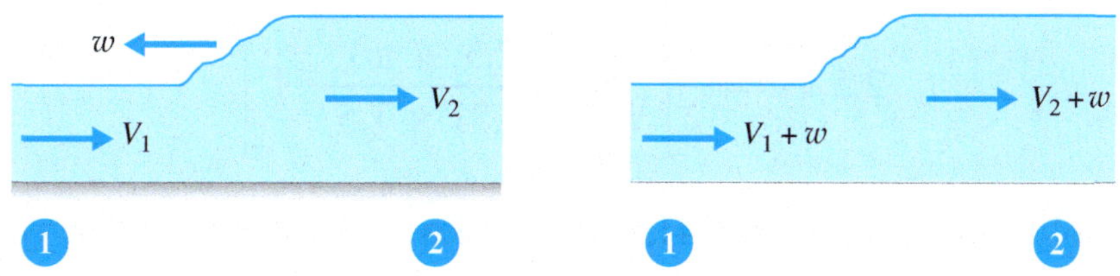

Figure 8.16 Translating hydraulic jump: (a) front moving upstream; (b) front appears stationary by superposition.

Translating Hydraulic Jump. A translating hydraulic jump, alternately termed a **surge** or a **hydraulic bore**, is shown in Figure 8.16a; it is termed a positive surge wave in the sense that it maintains a stable front as it propagates at speed w into an undisturbed region. Such a wave can be generated by abruptly lowering a downstream gate or by rapidly releasing water at an upstream location into a channel. This unsteady rapidly varied flow situation can be conveniently analyzed as a steady-state problem by superposing the bore speed w in the opposite sense on the control volume (Figure 8.16b). The front appears stationary, and the relative velocities at locations 1 and 2 are equal to $V_1 + w$ and $V_2 + w$, respectively.

Assume a bore translating in a rectangular, frictionless, horizontal channel. Equation 8.5.9 can be applied by substituting $V_1 + w$, for V_1 in the definition of Fr_1 to obtain

$$\frac{y_2}{y_1} = \frac{1}{2}\left[\sqrt{1 + 8\frac{(V_1 + w)^2}{gy_1}} - 1\right] \tag{8.5.12}$$

The continuity relation applied to the control volume of Figure 8.16b is

$$y_1(V_1 + w) = y_2(V_2 + w) \tag{8.5.13}$$

Equations 8.5.12 and 8.5.13 contain five parameters: y_1, y_2, V_1, V_2, and w. Three of them must be known to solve for the remaining two. Depending on which variables are unknown, the solution of Eqs. 8.5.12 and 8.5.13 will be either explicit or require a trial-and-error procedure.

Example 8.9

Water flows in a rectangular channel with a velocity of 2.5 m/s and a depth of 1.5 m. A gate is suddenly completely closed, forming a surge that travels upstream. Find the speed of the surge and the depth behind the surge.

Solution

Since the gate is closed, the downstream velocity is $V_2 = 0$. Equations 8.5.12 and 8.5.13 contain the two unknowns w and y_2. Combining them to eliminate y_2 and substituting $V_1 = 2.5$ and $y_1 = 1.5$ results in the relation

$$2.5+w=\frac{w}{2}\left[\sqrt{1+8\frac{(2.5\ +\ w)^2}{9.81\ \times\ 1.5}}-1\right]$$

A trial-and-error solution, yields w = 3.41 m/s. Use Eq. 8.5.13 to compute y_2:

$$y_2=y_1\frac{V_1+w}{V_2+w}$$

$$=1.5\times\frac{2.5+3.41}{3.41}=\underline{2.60\text{ m}}$$

Drag on Submerged Objects. If an object is submerged in the flow it is possible to describe the drag force in the manner

$$F=C_D A\,\rho\frac{V^2}{2} \tag{8.5.14}$$

in which C_D is the drag coefficient and A is the projected area normal to the flow. A discussion of drag coefficients for various immersed objects is given in Section 6.3.1. In open channel flow, since a free surface is present, the drag coefficient must be modified to account for wave drag as well as drag due to friction and separation. Examples of submerged objects in open channel flow include pipelines and bridge piers; in these situations, a hydraulic jump may not be present.

A design application is illustrated in Figure 8.17. Baffle blocks are devices placed in a reach of channel known as a **stilling basin** to stabilize the location of a hydraulic jump and to aid in the dissipation of flow energy. Typically they are used for Fr_1 greater than 4.5. Example 8.10 demonstrates how baffle blocks reduce the magnitude of the momentum function downstream of a hydraulic jump, resulting in reduced downstream depth and increased energy dissipation. In practice, Eq. 8.5.14 is seldom employed to analyze or design a stilling basin, since other factors such as additional appurtenances, approach velocity, scour, and cavitation must also be considered. Instead, design standards have been established based on observations of existing basins and laboratory model studies (U.S. Department of Interior,1974; Roberson et al., 1988).

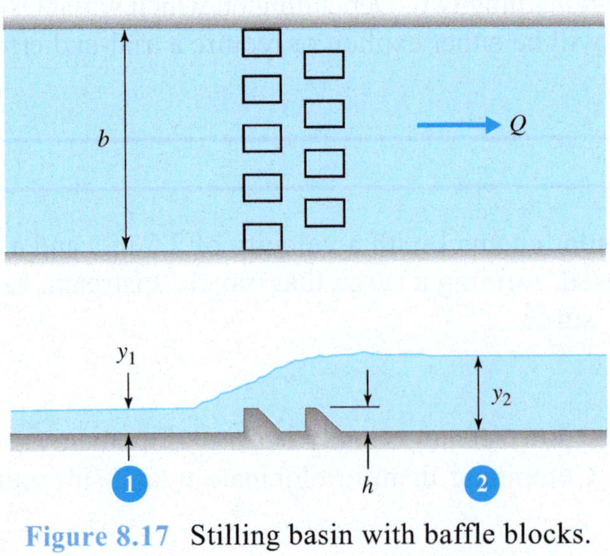

Figure 8.17 Stilling basin with baffle blocks.

Example 8.10

In the flow situation presented in Example 8.8, a series of baffle blocks is placed in the channel as shown in Figure 8.17. Laboratory experimentation has shown that the arrangement has an effective drag coefficient of 0.25, provided that the blocks are submerged in the flow. If the blocks are 0.15 m high, and if the discharge, upstream depth and width remain the same as in Example 8.8, determine the depth downstream of the jump and the rate of energy dissipated by the jump. The total width of the blocks is the channel width.

Solution

It is necessary to use Eq. 8.5.2 since obstacles (i.e., the baffle blocks) are placed within the control volume. The upstream velocity is

$$V_1 = \frac{Q}{A_1}$$

$$= \frac{7.5}{4 \times 0.2} = 9.83 \text{ m/s}$$

The force F due to the presence of the baffle blocks is computed using Eq. 8.5.14:

$$F = C_D A \rho \frac{V_1^2}{2}$$

$$= 0.25 \times (4 \times 0.15) \times 1000 \times \frac{9.38^2}{2} = 6600 \text{ N}$$

Note that the frontal area is the width of the channel multiplied by the height of the blocks. Substituting known conditions into Eq. 8.5.2, making use of Eq. 8.5.4 which defines M for a rectangular channel, and noting that $q = 7.5/4 = 1.88 \text{ m}^2/\text{s}$, we find

$$b\left(\frac{y_1^2}{2} + \frac{q^2}{g y_1}\right) - b\left(\frac{y_2^2}{2} + \frac{q^2}{g y_2}\right) = \frac{F}{\gamma}$$

$$4\left(\frac{0.2^2}{2} + \frac{1.88^2}{9.81 \times 0.2}\right) - 4\left(\frac{y_2^2}{2} + \frac{1.88^2}{9.81 y_2}\right) = \frac{6600}{9800}$$

The relation reduces to

$$y_2^2 + \frac{0.721}{y_2} = 3.31$$

The trial-and-error solution for y_2 is <u>1.70 m</u>. The change in specific energy between locations 1 and 2 is

$$E_1 - E_2 = y_1 + \frac{q^2}{2gy_1^2} - \left(y_2 + \frac{q^2}{2gy_2^2}\right)$$

$$= 0.2 + \frac{1.88^2}{2\times 9.81\times 0.2^2} - \left(1.70 + \frac{1.88^2}{2\times 9.81\times 1.70^2}\right)$$

$$= 2.94 \text{ m}$$

The rate of energy dissipation, therefore, is

$$\gamma Q(E_1 - E_2) = 9800 \times 7.5 \times 2.94$$

$$= 2.16 \times 10^5 \text{ W} \quad \text{or} \quad \underline{216 \text{ kW}}$$

8.5.3 Numerical Solution of the Momentum Equation

For nonrectangular channels, the momentum relation can be used directly to analyze the hydraulic jump or other problems requiring the momentum equation; the technique is demonstrated as follows. Consider a trapezoidal channel with conditions known at location 1 upstream of the jump. Consequently, M_1 and F are evaluated as constants and Eq. 8.5.2 can be written in the form

$$M_2 - M_1 + \frac{F}{\gamma} = 0 \tag{8.5.15}$$

in which M_2 is a function of y_2. Introducing the trapezoidal geometry at location 2, with $m_1 = m_2 = m$, the relation above can be written as

$$\frac{y_2^2}{6}(2my_2 + 3b) + \frac{Q^2}{g\left(by_2 + my_2^2\right)} - M_1 + \frac{F}{\gamma} = 0 \tag{8.5.16}$$

This can be solved for y_2 by a suitable numerical technique such as interval halving, falseposition, or Newton's method, which are discussed in any book on numerical methods (see, e.g., Chapra and Canale, 1998). Note that by setting $F = 0$, it becomes the relation for finding the conditions downstream of a hydraulic jump, and that by additionally letting $m = 0$, a rectangular hydraulic jump problem can be solved.

Example 8.11

A hydraulic jump occurs in a triangular channel with $m_1 = m_2 = 2.5$. The discharge is 20 m^3/s and $y_c = 1.67$ m. Upstream of the jump the following parameters are given: $y_1 = 0.75$ m, $\text{Fr}_1 = 7.42$, and $M_1 = 29.35$ m^3. Determine the conjugate depth y_2 downstream of the jump.

Solution

Use Eq. 8.5.16 with $F = 0$:

$$\frac{y_2^2}{6} \times 2 \times 2.5y_2 + \frac{20^2}{9.81 \times 2.5y_2^2} - 29.35 = 0$$

The relation reduces to

$$f(y_2) = y_2^3 + \frac{19.58}{y_2^2} - 35.23 = 0$$

The false position method is chosen to find y_2. The first step is to set the upper and lower limits, termed y_u and y_l. Since $Fr_1 > 1$, and $y_2 > y_c$, an appropriate lower limit is $y_l = y_c = 1.67$ m. The upper limit is assumed to be 5 m.

Iteration	y_u	y_l	$f(y_u)$	$f(y_l)$	y_r	$y(y_r)$	*Sign of* $f(y_l) \times (y_r)$	ε
1	5	1.67	90.55	−23.55	2.357	−18.61	–	
2	5	2.357	90.55	−18.61	2.808	−10.61	–	
3	5	2.808	90.55	−10.61	3.038	−5.076	–	
4	5	3.038	90.55	−5.067	3.142	−2.231	–	
5	5	3.142	90.55	−2.231	3.187	−0.944	–	
6	5	3.187	90.55	−0.944	3.205	−0.393	–	
7	5	3.205	90.55	−0.393	3.213	−0.162	–	2.5×10^{-3}
8	5	3.213	90.55	−0.162	3.216	−0.0672	–	9.5×10^{-4}
9	5	3.216	90.55	−0.0672	3.218	−0.0277	–	4.9×10^{-4}

The solution is tabulated above. In each iteration a new estimate y_r of the root is made:

$$y_r = \frac{y_u f(y_l) - y_l f(y_u)}{f(y_l) - f(y_u)}$$

The product $f(y_l) \times f(y_r)$ is formed to determine on which subinterval the root will be found. If $f(y_l) \times f(y_r) < 0$, then $y_u = y_r$; otherwise, $y_l = y_r$. It is required that initially $f(y_u)$ and $f(y_l)$ be of opposite sign. Iterations continue until a relative error ε, defined by

$$\varepsilon = \left| \frac{y_r^{\text{new}} - y_r^{\text{old}}}{y_r^{\text{old}}} \right|$$

is less than a specified value, which in the example is 0.0005. The result after nine iterations is $y_2 = 3.22$ m, rounded off to three significant figures.

Using Mathcad or MATLAB, one finds the solution to be less time consuming.

8.6 NONUNIFORM GRADUALLY VARIED FLOW

The evaluation of many open-channel flow situations must include accurate analyses of relatively long reaches where the depth and velocity may vary but do not exhibit rapid or sudden changes. In the preceding two sections we emphasized nonuniform, rapidly varied flow phenomena that occur over relatively short reaches, or transitions, in open channels. Attention is now focused on nonuniform, gradually varied flow, where the water surface is continuously smooth. One significant difference between the two is that for rapidly varied flow, losses will be neglected without severe consequences, whereas for gradually varied flow, it is necessary to include losses due to shear stress distributed along the channel length. The shear stress is the primary mechanism that resists the flow.

Gradually varied flow is a type of steady, nonuniform flow in which y and V do not exhibit sudden or rapid changes, but instead vary so gradually that the water surface can be considered continuous. As a result, it is possible to develop a differential equation that describes the incremental variation of y with respect to x, the distance along the channel.An analysis of this relation will enable one to predict the various trends that the water surface profile may assume based on the channel geometry, magnitude of the discharge, and the known boundary conditions. Numerical evaluation of the same equation will provide engineering design criteria.

8.6.1 Differential Equation for Gradually Varied Flow

A representative nonuniform gradually varied flow is shown in Figure 8.18. Over the incremental distance Δx, the depth and velocity are known to change slowly. The slope of the energy grade line is designated as S. In contrast to uniform flow, the slopes of the energy grade line, water surface, and channel bottom are no longer parallel. Since the changes in y and V are gradual, the energy loss over the incremental length Δx can be represented by the Chezy–Manning equation. This means that Eq. 8.3.13, which is valid for uniform flow, can also be used to evaluate S for a gradually varied flow situation, and that the roughness coefficients presented in Table 5.3 (page 327) are applicable. Additional assumptions include a regular cross section, small channel slope, hydrostatic pressure distribution, and one-dimensional flow, that is, $V = V(x)$ and $y = y(x)$.

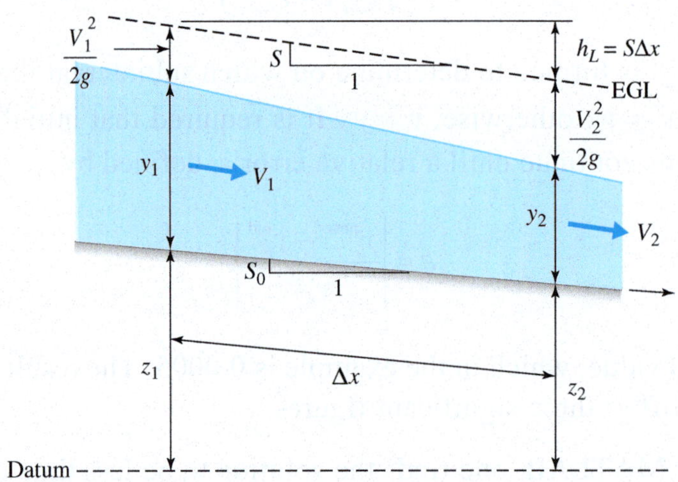

Figure 8.18 Nonuniform gradually varied flow.

The energy equation is applied from location 1 to location 2, with the loss term h_L given by $S\Delta x$. If the total energy at location 2 is expressed as the energy at location 1 plus the incremental change in energy over the distance Δx, Eq. 8.4.2 becomes

$$H_1 = H_2 + S\,\Delta x = H_1 + \frac{dH}{dx}\Delta x + S\,\Delta x \tag{8.6.1}$$

Substitute $H = y + z + V^2/2g$ and $dz/dx = -S_0$ into this relation and rearrange to find

$$S - S_0 = -\frac{d}{dx}\left(y + \frac{V^2}{2g}\right) \tag{8.6.2}$$

The right-hand term is $-dE/dx$, and it is transformed to

$$\frac{dE}{dx} = \frac{dE}{dy}\frac{dy}{dx} = (1 - \mathrm{Fr}^2)\frac{dy}{dx} \tag{8.6.3}$$

(Recall from Eqs. 8.4.13 and 8.4.15 that $dE/dy = 1 - Q^2B/gA^3 = 1 - \mathrm{Fr}^2$.) Finally, upon substitution into the energy relation and solving for the slope dy/dx of the water surface one finds that

$$\frac{dy}{dx} = \frac{S_0 - S}{1 - \mathrm{Fr}^2} \tag{8.6.4}$$

where S and Fr depend on x. This is the differential equation for gradually varied flow and is valid for any regular channel shape.

Example 8.12

Using an appropriate control volume for gradually varied flow, show that the slope S of the energy grade line is equivalent to $\tau_0/\gamma R$.

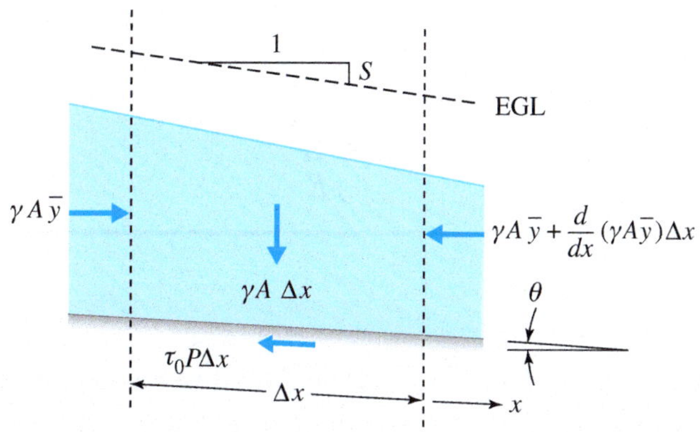

Figure E8.12

Solution

The control volume is shown in Figure E8.12. The resultant force acting on the control volume is due to the incremental change in hydrostatic pressure $[\gamma d(A\bar{y})/dx]\Delta x$, the component of weight in the x-direction $\gamma A \sin\theta\,\Delta x$, and the resistance term $\tau_0 P\Delta x$. Using the momentum equation

$$\Sigma F_x = \dot{m}\,(V_{2x} - V_{1x})$$

with $V_{2x} - V_{1x} = (dV/dx)\Delta x$ results in

$$-\gamma\frac{d}{dx}(A\bar{y})\Delta x + \gamma A\,\sin\theta\Delta x - \tau_0 P\Delta x = \rho VA\frac{dV}{dx}\Delta x$$

This relation can be simplified by noting that

$$\frac{d(A\bar{y})}{dx} = \frac{d(A\bar{y})}{dy}\frac{dy}{dx} = A\frac{dy}{dx}$$

and $P = A/R$. Substitute and divide the equation by $\gamma A\Delta x$, the weight of the control volume, and find that

$$-\frac{dy}{dx} + \sin\theta - \frac{\tau_0}{\gamma R} = \frac{V}{g}\frac{dV}{dx}$$

Since $\sin\theta \simeq S_0$ for small θ, the equation above can be rearranged in the form

$$\frac{\tau_0}{\gamma R} - S_0 = -\frac{dy}{dx} - \frac{V}{g}\frac{dV}{dx} = -\frac{d}{dx}\left(y + \frac{V^2}{2g}\right)$$

Upon comparison with Eq. 8.6.2, it is seen that the right-hand side is equivalent to $S - S_0$, and consequently,

$$\frac{\tau_0}{\gamma R} - S_0 = S - S_0$$

or

$$S = \frac{\tau_0}{\gamma R}$$

8.6.2 Water Surface Profiles

It is possible to identify a series of water surface profiles based on an evaluation of Eq. 8.6.4 (Bakhmeteff, 1932). Essential to the development is the determination of normal and critical depths. Note that y_0 and y_c are uniquely determined once the channel properties and discharge are established. Table 8.3 shows the classification of the water surface profiles. Associated

Table 8.3 Classification of Surface Profiles

Channel Slope	*Profile type*	*Depth range*	*Fr*	$\frac{dy}{dx}$	$\frac{dE}{dx}$	
Mild $S_0 < S_c$ $y_0 > y_c$	M_1	$y > y_0 > y_c$	< 1	> 0	> 0	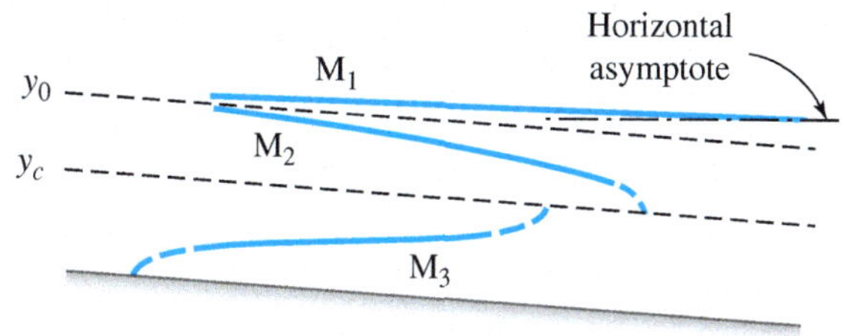
	M_2	$y_0 > y > y_c$	< 1	< 0	< 0	
	M_3	$y_0 > y_c > y$	> 1	> 0	< 0	
Steep $S_0 < S_c$ $y_0 > y_c$	S_1	$y > y_c > y_0$	< 1	> 0	> 0	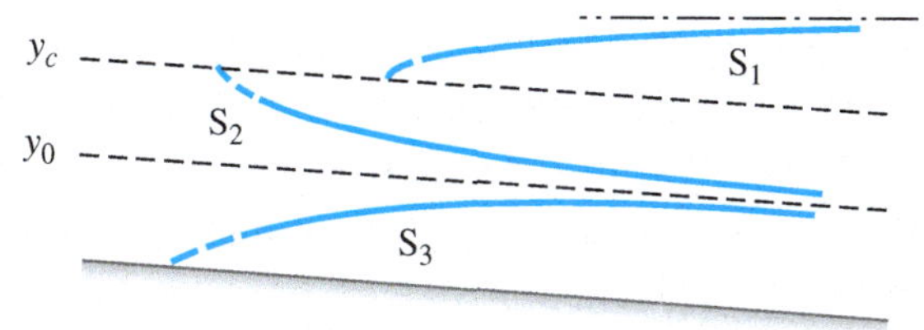
	S_2	$y_c > y > y_0$	> 1	< 0	> 0	
	S_3	$y_c > y_0 > y$	> 1	> 0	< 0	
Critical $S_0 = S_c$ $y_0 = y_c$	C_1	$y > y_c$ or y_0	< 1	> 0	> 0	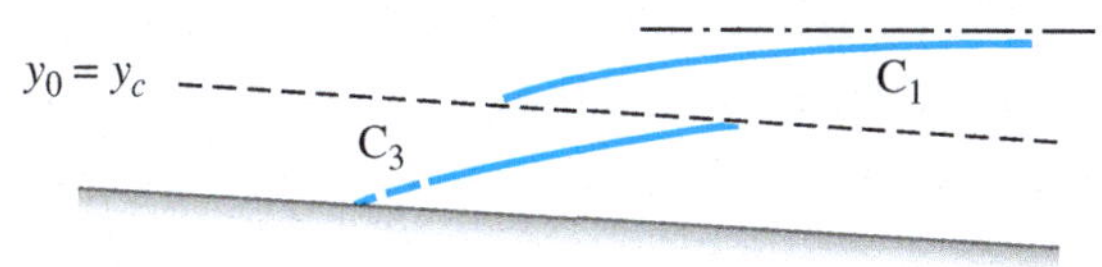
	C_3	y_c or $y_0 > y$	> 1	> 0	< 0	
Horizontal $S_0 = 0$ $y_0 \rightarrow \infty$	H_2	$y > y_c$	< 1	< 0	< 0	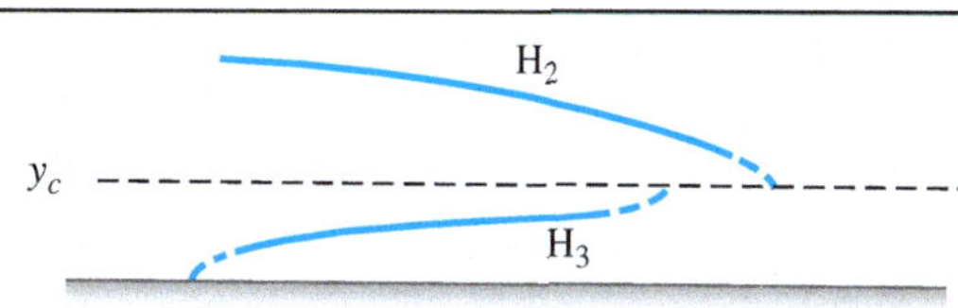
	H_3	$y_c > y$	> 1	> 0	< 0	
Adverse $S_0 < 0$ y_0 undefined	A_2	$y > y_c$	< 1	< 0	< 0	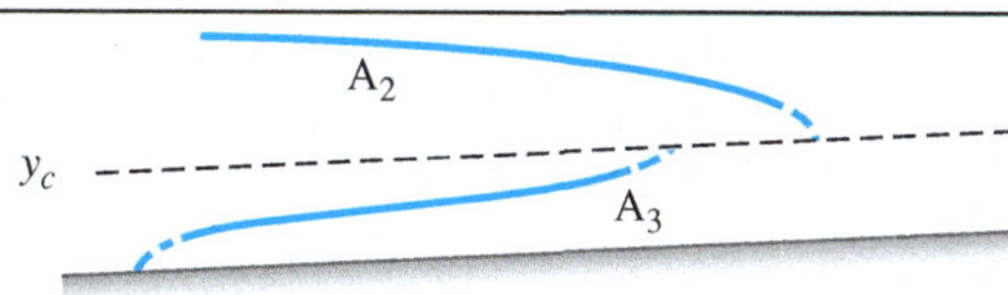
	A_3	$y_c > y$	> 1	> 0	< 0	

with y_c is a critical slope S_c, which is found by substituting y_c into the Chezy–Manning equation and solving for the slope. The channel slope can be designated as mild, steep, or critical, depending on whether S_0 is less than, greater than, or equal to S_c, respectively. A horizontal slope exists when $S_0 = 0$ and an adverse slope when $S_0 < 0$. Inspection of Table 8.3 shows that there are 12 possible profiles. Each profile is classified by a letter/number combination. The letter refers to the channel slope: M for mild, S for steep, C for critical, H for horizontal, and A for adverse. The numerical subscript designates the range of y relative to y_0 and y_c. Flow can occur at depths above or below y_c and at depths above or below y_0.

The variation of y with respect to x for each profile in Table 8.3 can now be found. For a given Q, n, S_0, and channel geometry, the analysis reduces to the determination of how S and Fr vary with y. An inspection of the Chezy–Manning equation will reveal that S decreases with increasing y; similarly, the Froude number decreases as y increases. The numerator in Eq. 8.6.4 will assume the following inequalities: $(S_0 - S) > 0$ for $y > y_0$, and $(S_0 - S) < 0$ for $y < y_0$. In addition, the denominator varies in the manner $(1 - \mathrm{Fr}^2) > 0$ for $y > y_c$ and $(1 - \mathrm{Fr}^2) < 0$ for $y < y_c$. With these criteria, the sign of dy/dx can be evaluated. In addition,with the use of Eq. 8.6.2, the sign of dE/dx is revealed.

The boundaries of the profiles can be established as follows[5]:

1. As y tends to y_0, S tends to S_0. Hence, dy/dx approaches zero; in other words, the water surface approaches y_0 asymptotically. This applies to curves M_1, M_2, S_2, and S_3.
2. As y becomes large, the velocity becomes small and S and Fr tend to zero, so that dy/dx approaches S_0. Hence the surface approaches a horizontal asymptote; curves M_1, S_1, and C_1 are of this type.
3. As y approaches y_c, dy/dx becomes infinite, a limit that is never reached. For supercritical flow, as the M_3, H_3, and A_3 curves approach y_c, a hydraulic jump will form and create a discontinuity in the water surface; at the beginning of the curve, rapid acceleration occurs with nonparallel streamlines. When the flow is subcritical (M_2, H_2, A_2), rapid drawdown takes place close to y_c, and the streamlines are no longer parallel. For all of these situations, Eq. 8.6.4 is invalid, since the flow is no longer one-dimensional.

For the profiles shown in Table 8.3, there is no physical significance to the theoretical limit of y approaching zero, since a finite depth is necessary for the existence of flow. It is left as an exercise to show that for the critical slope condition, $dy/dx > S_c$ as y approaches y_c from either a C_1 or C_2 profile, and that dy/dx approaches a horizontal asymptote as y becomes very large.

Example 8.13

By assuming a wide rectangular channel, develop the right-hand side of Eq. 8.6.4 to show how dy/dx varies with y.

Solution

For a wide rectangular channel, assume that $b >> y$, so that the wetted perimeter is approximated by $P \simeq b$. The hydraulic radius then becomes $R = A/P - (by)/b = y$. Noting that $Q = qb$, the Chezy–Manning equation, used to evaluate S, simplifies to

$$S = \frac{(qbn)^2}{(by^{5/3})^2} = \frac{(qn)^2}{y^{10/3}}$$

It is assumed that in the Chezy–Manning equation $c_1 = 1$. For a rectangular section the square of the Froude number (see Eq. 8.4.9) is

$$\mathrm{Fr}^2 = \frac{q^2}{gy^3}$$

Substituting into Eq. 8.6.4 gives

$$\frac{dy}{dx} = \frac{S_0 - (qn)^2/y^{10/3}}{1 - q^2/(gy)^3}$$

[5]Note that x is measured parallel to the channel bottom, and that y is measured vertically from the channel bottom; hence dy/dx and dE/dx are evaluated relative to the channel bottom and not to a horizontal datum.

Since $(qn)^2 y_0^{-10/3} = S_0$ and $\mathrm{Fr}_c^2 = q^2/(gy_c^3) = 1$, the relation can be written as

$$\frac{dy}{dx} = S_0 \frac{1-(y_0/y)^{10/3}}{1-(y_c/y)^3}$$

This equation can be used as an alternative to Eq. 8.6.4 to evaluate the water surface profiles shown in Table 8.3.

8.6.3 Controls and Critical Flow

The existence of the various profiles shown in Table 8.3 depend on the boundary conditions that are specified at given locations in the channel. Quite often, a control will define the boundary condition. A **control** exists when a depth-discharge relationship can be established at a section. The manner in which the control section affects the water surface away from its specific location can be studied by examining flow near the critical state in a rectangular channel.

Equation 8.5.12 is the expression for a surge of finite magnitude translating in the upstream direction in a rectangular channel. As y_2 approaches y_1, the magnitude of the surge becomes infinitesimal; these waves may be generated by the presence of control structures and other transition sections that tend to disturb the flow. In Eq. 8.5.12, one can replace y_1 and y_2 by y, and V_1 by V. In addition, let the wave travel in the downstream direction, so that the equation becomes

$$1 = \frac{1}{2}\left[\sqrt{1+8\frac{(V-w)^2}{gy}}-1\right] \tag{8.6.5}$$

Solving for w provides

$$w = V \pm \sqrt{gy} = V \pm c \tag{8.6.6}$$

in which $c = \sqrt{gy}$, termed the **celerity**. The celerity is the speed at which an infinitesimal wave will travel into an undisturbed region, that is, a region with zero velocity. For a non-rectangular channel, $c = \sqrt{gA/B}$. Thus if a disturbance is created at a midstream location in a channel, two infinitesimal waves are generated. One wave front will tend to propagate upstream at the speed $V - c$, and the second will move downstream at the speed $V + c$. These observations are made relative to an observer in a fixed position, that is, to one who is standing on the shore observing the wave motion.

In a rectangular channel with critical flow conditions, $\mathrm{Fr} = V/\sqrt{gy} = 1$, or $V = \sqrt{gy} = c$. Hence, at critical flow, the first wave front generated by the disturbance would not move upstream but would appear stationary and become a so-called **standing wave**. The opposite front would be swept downstream at the speed $2c$. If $Fr < 1$, the first wave would travel upstream, and the opposite wave would travel downstream at a speed less than $2c$. For $\mathrm{Fr} > 1$ in the channel, since $V > c$, both waves are swept downstream. Thus, since it contains amechanism that disturbs the flow, a control will influence upstream conditions only when the flow is

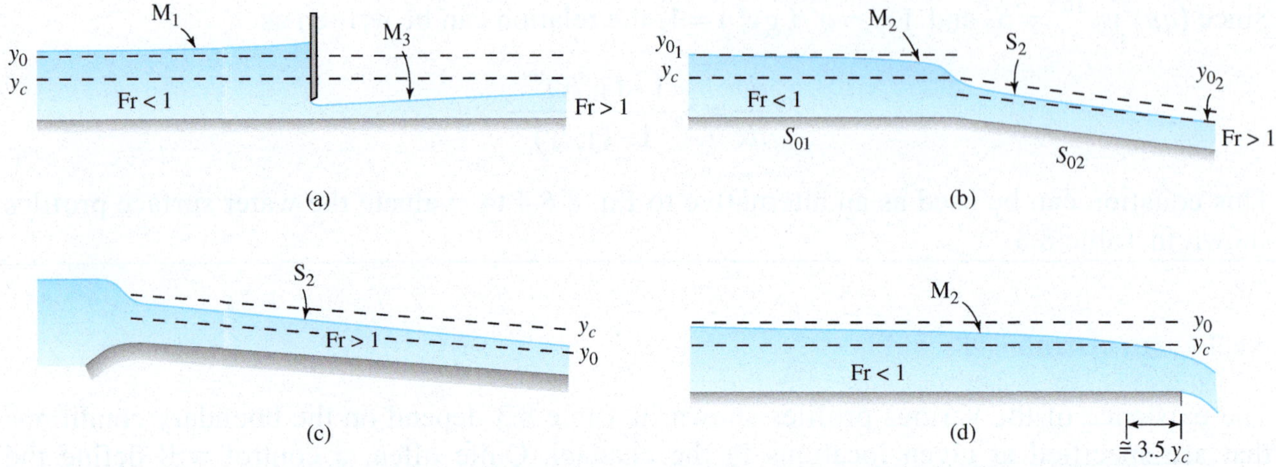

Figure 8.19 Representative controls: (a) sluice gate; (b) change in slope from mild (S_{01}) to steep (S_{02}); (c) entrance to a steep channel; (d) free outfall.

subcritical (Fr < 1). Similarly, when the flow is supercritical (Fr > 1), the control can influence only the downstream flow conditions. This is illustrated in Figure 8.19a, where a sluice gate is positioned in a channel with subcritical flow upstream and supercritical flow downstream. An M_1 profile is generated above the gate, and an M_3 profile exists below it.Any movement of the gate would influence the nature of the two profiles; lowering the gate would lengthen their range, and raising it would produce the opposite effect.

The profiles shown in Table 8.3 are all influenced by the presence of controls. In Figure 8.19, several controls are shown that create a variety of profiles. Critical depth often is associated with an effective control. Examples include sluice gates, weirs, dams, and flumes, all of which force critical flow to occur somewhere in the transition region. In addition, a control can be located at a break in channel slope from mild upstream to steep downstream (Figure 8.19b). The entrance to a steep channel (Figure 8.19c) is an example of an upstream control, with critical flow occurring at the crest. Critical flow will occur a short distance upstream from a free outfall on a mild slope (Figure 8.19d). Other controls, not shown in Figure 8.19, are a channel constriction acting as a choke, and for a mild slope, the existence of uniform flow at some location.

8.6.4 Profile Synthesis

Identification of controls and their interaction with the possible profiles is a requirement for proper understanding and correct design and analysis of open-channel flow. Since controls are essentially transition sections, the rapidly varied flow principles presented in Sections 8.4 and 8.5 can be used to determine the necessary depth-discharge relations. Once the controls have been identified, the profiles can be selected and the range of influence of the controls established.

As an example, consider the situation shown in Figure 8.20. Flow enters the steep channel from a reservoir, so that critical flow exists at the channel entrance. Both the magnitude of the discharge and the S_2 profile are influenced by the depth at the entrance; therefore, the depth acts as a control. At the downstream end of the channel, the lower reservoir acts as a control to establish an S_1 profile that projects upstream. At some interior location, a hydraulic jump occurs to allow the flow to pass from a supercritical to a subcritical state. The location of the jump can be found by plotting a curve of the depth conjugate to the S_2 profile and finding its point of

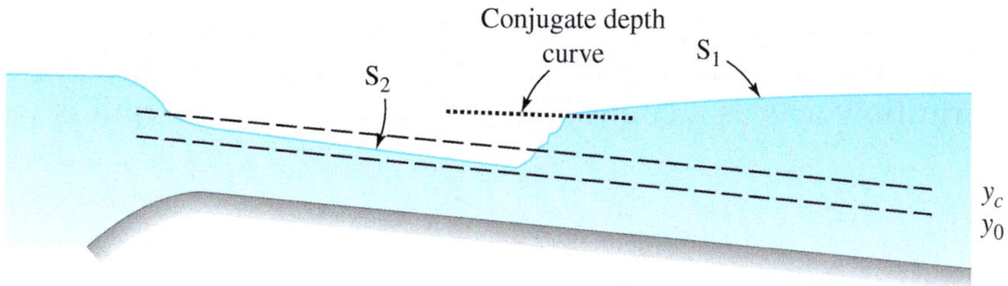

Figure 8.20 Example of profile synthesis.

intersection with the S_1 curve. A lowering of the lower reservoir elevation would cause the jump to move downstream and ultimately be swept out of the channel. Increasing the lower reservoir elevation would move the jump upstream; if it were to move into the entrance region, the upstream control would no longer exist. Additional examples of profile synthesis are given in Section 8.7.

Example 8.14

In a rectangular channel, $b = 3$ m, $n = 0.015$, $S_0 = 0.0005$, and $Q = 5$ m^3/s. At the entrance to the channel, flow issues from a sluice gate at a depth of 0.15 m. The channel is sufficiently long that uniform flow conditions are established away from the entrance region, Figure E8.14a. Find the nature of the water surface profile in the vicinity of the entrance and the depth before the hydraulic jump.

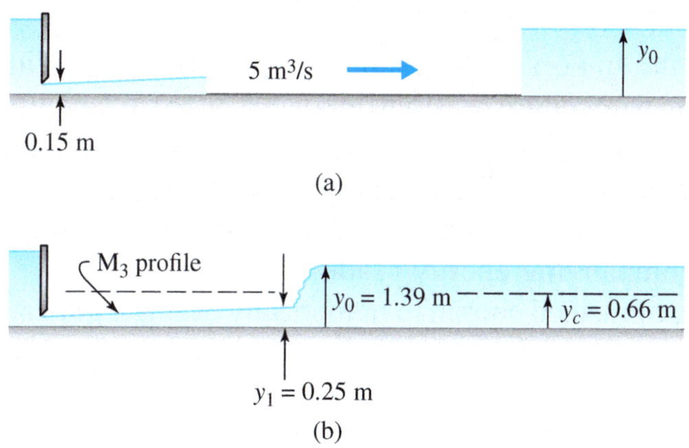

Figure E8.14

Solution

First find y_0 and y_c to determine the type of channel. To find y_0, follow the method shown in Example 8.1. Substitute known data into the Chezy–Manning equation:

$$\frac{3(y_0)^{5/3}}{(3+2y_0)^{2/3}} = \frac{0.015 \times 5}{\sqrt{0.0005}} = 3.354$$

Solving gives $y_0 = 1.39$ m. Next, the critical depth is computed to be

$$y_c = \left(\frac{q^2}{g}\right)^{1/3} = \left[\frac{(5/3)^2}{9.81}\right]^{1/3} = 0.66 \text{ m}$$

Since $y_0 > y_c$, a mild slope condition exists. The gate is a control and there will be an M_3 profile beginning at the entrance, terminated by a hydraulic jump. Downstream of the jump, the condition of uniform flow acts as a control, so at that location the depth is y_0, and the Froude number is

$$\mathrm{Fr}_0 = \frac{q}{\sqrt{g y_0^3}}$$

$$= \frac{5/3}{\sqrt{9.81 \times 1.39^3}} = 0.325$$

Using Eq. 8.5.10, the depth before the jump is

$$y_1 = \frac{y_0}{2}\left(\sqrt{1 + 8Fr_0^2} - 1\right)$$

$$= \frac{1.39}{2}\left(\sqrt{1 + 8 \times 0.325^2} - 1\right) = \underline{0.25 \text{ m}}$$

The depths y_c and y_1 have been calculated to two significant figures, since the Manning coefficient is known to only two significant figures. Figure 8.14b shows the profile.

8.7 NUMERICAL ANALYSIS OF WATER SURFACE PROFILES

Section 8.6 dealt with the understanding and interpretation of various aspects of gradually varied flow. An important procedure was outlined: A water surface profile can be synthesized, or predicted, by making use of relevant information about the channel geometry,roughness, and flow, and by determining or assuming the appropriate controls. Once a satisfactory profile prediction has been conducted, we are in a position to numerically calculate the desired water surface profiles and accompanying energy gradelines.

Regardless of the type of method chosen to numerically evaluate gradually varied flow, analysis of a water surface profile on a reach of channel with constant slope usually follows these steps:

1. The channel geometry, channel slope S_0, roughness coefficient n, and discharge Q are given or assumed.
2. Determine normal depth y_0 and critical depth y_c.
3. Establish the controls (i.e., the depth of flow) at the upstream and downstream ends of the channel reach.
4. Integrate Eq. 8.6.4 to find y and subsequently E as functions of x, allowing for the possibility of a hydraulic jump to occur within the reach.

For a prismatic channel, y_0 can be evaluated by applying the Chezy–Manning equation and finding the root of the function

$$\frac{Qn}{c_1 A R^{2/3} \sqrt{S_0}} - 1 = 0 \tag{8.7.1}$$

Similarly, y_c is found from application of the Froude number in the form

$$\frac{Q^2 B}{g A^3} - 1 = 0 \tag{8.7.2}$$

Numerical methods, such as the false position or interval halving (Chapra and Canale, 1988),or computational software can be used to solve Eqs. 8.7.1 and 8.7.2.

The Chezy–Manning equation, given here in the form of Eq. 8.7.1, is used to represent the depth–discharge variation for nonuniform as well as for uniform flow. Thus one can replace S_0 by the slope S of the energy grade line in Eq. 8.7.1 and solve for S in the manner

$$S(y) = \frac{Q^2 n^2}{c_1^2 [A(y)]^2 [R(y)]^{4/3}} \tag{8.7.3}$$

Since Q and n are either given or assumed, the right-hand side of Eq. 8.7.3 is a function of y alone.

Two numerical methods are presented in this section to compute water surface profiles along with normal and critical depths. The first, termed the **standard step method**, is the most widely used, and the second employs an accurate numerical integration scheme. Solutions will be illustrated using Excel, Mathcad, and MATLAB software.

8.7.1 Standard Step Method

To develop a numerical procedure for solving gradually varied flow problems, we use Eq. 8.6.2; it is applied over the reach of channel shown in Figure 8.21, resulting in

$$\frac{dE}{dx} = \frac{d}{dx}\left(y + \frac{V^2}{2g}\right) = S_0 - S(y) \tag{8.7.4}$$

For small changes in y and V between x_i and x_{i+1}, Eq. 8.7.4 can be approximated by

$$E_{i+1} - E_i = \int_{x_i}^{x_{i+1}} [S_0 - S(y)]\,dx$$

$$\simeq (x_{i+1} - x_i)\,[S_0 - S(y_m)] \tag{8.7.5}$$

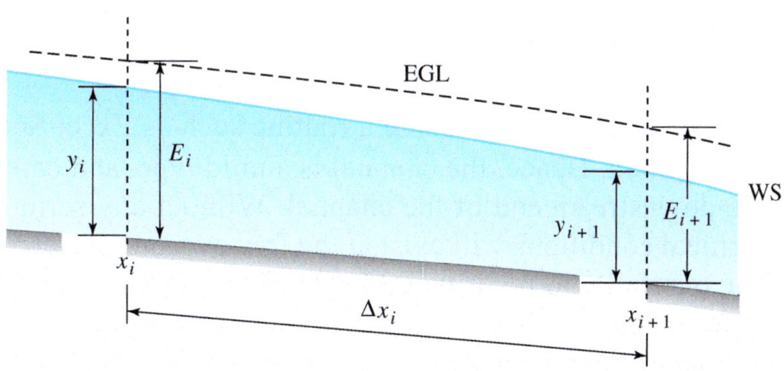

Figure 8.21 Notation for computing gradually varied flow.

in which $y_m = (y_{i+1} + y_i)/2$. Equation 8.7.5 is solved for x_{i+1} to yield

$$x_{i+1} = x_i + \frac{E_{i+1} - E_i}{S_0 - S(y_m)} \tag{8.7.6}$$

The calculations take place stepwise, beginning at a control point or other location where the depth is known. Assume that it is desired to generate a water surface profile along the channel, and compute values of x_i, y_i, and E_i, for $i = 1,\ldots, k$, where k is the location at the opposite end of the reach. Other than x_k, it is not necessary that location x_i be fixed, so itcan take on any value. Beginning at location i, the evaluation of conditions at location $i + 1$ proceeds as follows:

1. Choose y_{i+1}.
2. Compute E_{i+1} from Eq. 8.4.12, y_m knowing y_i and y_{i+1}, and $S(y_m)$ from Eq. 8.7.3.
3. Compute x_{i+1} from Eq. 8.7.6.
4. At location k, trial values of y_k are assumed until Eq. 8.7.6 is satisfied for the known value of x_k.

The standard step method can be carried out using a spreadsheet analysis.

Example 8.15

Water is flowing at $Q = 22\ \mathrm{m^3/s}$ in a long trapezoidal channel, $b = 7.5$ m, $m_1 = m_2 = 2.5$. A free overfall is located at the downstream end of the channel, where $x = 2000$ m. For $n = 0.015$, $S_0 = 0.0006$, find the water surface profile and energy grade line for a distance of approximately 800 m upstream from the free outfall.

Solution

Equations 8.7.1 and 8.7.2 are used to evaluate y_0 and y_c by substituting in known data:

$$\frac{22 \times 0.015 \times \left[7.5 + 2y_0\sqrt{1+(2.5)}\right]^{2/3}}{\left[7.5y_0 + \frac{1}{2}y_0^2(2.5+2.5)\right]^{5/3}\sqrt{0.006}} - 1 = 0$$

$$\frac{(22)^2 \times \left[7.5 + y_c(2.5+2.5)\right]}{9.8 \times \left[7.5y_c + \frac{1}{2}y_c^2(2.5+2.5)\right]^3} - 1 = 0$$

The roots of these equations can be found using a routine such as Excel Solver®; the solutions are $y_0 = 1.29$ m and $y_c = 0.86$ m. Hence, the channel is a mild type, and control will be close to the free overfall at the downstream end of the channel. Without any serious loss of accuracy, one can assume that critical conditions will exist at the free overfall. Referring to Table 8.3, the profile upstream of the overfall will be of type M_2. An Excel spreadsheet solution is shown in Figure E4(a). The upper part shows the values of critical and normal depths found by using Solver. In the residual column are very small numbers that should be close to zero; see the two above equations. The lower part of the table shows the step method solution. Calculations proceed from station 1 to station 5 in a straightforward manner, with arbitrary values of depth selected and placed in the y column. The beginning value of x (2000 m) is placed in the first cell of the x column, and the remaining distances are computed as explained previously. At station 6, different values of y are chosen until the distance is close to the desired value of 1200 m; a depth of 1.27 m results in a distance of 1230 m, which is acceptable. Then spreadsheet equations for computing normal and critical depths, and for the step method, follow in Figure E4(b).

	A	B	C	D	E	F	G	H	I
1									
2			Depth [m]	Residual					
3		Critical	0.865	1.08665E–06					
4									
5		Normal	1.292	1.81248E–06					
6									
7									
8	Station	y [m]	A [m^2]	V [m/s]	E [m]	y_m [m]	S(y_m)	Δx [m]	x [m]
9	1	0.865	8.358	2.632	1.218				2000
10	2	0.950	9.381	2.345	1.230	0.908	2.165E–03	–8	1992
11	3	1.050	10.631	2.069	1.268	1.000	1.527E–03	–41	1951
12	4	1.150	11.931	1.844	1.323	1.100	1.081E–03	–114	1837
13	5	1.250	13.281	1.656	1.390	1.200	7.866E–04	–357	1480
14	6	1.270	13.557	1.623	1.404	1.260	6.574E–04	–250	1230
15									

(a) Solution

	B	C	D
2		Depth [m]	Residual
3	Critical	0.86479527789122	=22^2*(7.5+2*2.5*C3)/(9.81*(7.5*C3+2.5*C3^2)^3)–1
4			
5	Normal	1.29156940192658	=22*0.015*(7.5+2*C5*SQRT(1+2.5^2))^0.6667/((7.5*C5+2.5*C5^2)^1.6667*SQRT(0.0006))–1

	A	B	C	D	E	F
8	Station	y [m]	A [m^2]	V [m/s]	E [m]	y_m [m]
9	1	0.865	=7.5*B9+B9^2*2.5	=22/C9	=B9+D9^2/(2*9.81)	
10	2	0.95	=7.5*B10+B10^2*2.5	=22/C10	=B10+D10^2/(2*9.81)	=0.5*(B9+B10)
11	3	1.05	=7.5*B11+B11^2*2.5	=22/C11	=B11+D11^2/(2*9.81)	=0.5*(B10+B11)
12	4	1.15	=7.5*B12+B12^2*2.5	=22/C12	=B12+D12^2/(2*9.81)	=0.5*(B11+B12)
13	5	1.25	=7.5*B13+B13^2*2.5	=22/C13	=B13+D13^2/(2*9.81)	=0.5*(B12+B13)
14	6	1.27	=7.5*B14+B14^2*2.5	=22/C14	=B14+D14^2/(2*9.81)	=0.5*(B13+B14)

	G	H	I
8	S(y_m)	Δx [m]	x [m]
9			2000
10	=22^2*0.015^2*(7.5*F10+F10^2*2.5)^–3.3333*((7.5+2*F10*SQRT(1+2.5^2))^1.3333)	=(E10–E9)/(0.0006–G10)	=I9+H10
11	=22^2*0.015^2*(7.5*F11+F11^2*2.5)^–3.3333*((7.5+2*F11*SQRT(1+2.5^2))^1.3333)	=(E11–E10)/(0.0006–G11)	=I10+H11
12	=22^2*0.015^2*(7.5*F12+F12^2*2.5)^–3.3333*((7.5+2*F12*SQRT(1+2.5^2))^1.3333)	=(E12–E11)/(0.0006–G12)	=I11+H12
13	=22^2*0.015^2*(7.5*F13+F13^2*2.5)^–3.3333*((7.5+2*F13*SQRT(1+2.5^2))^1.3333)	=(E13–E12)/(0.0006–G13)	=I12+H13
14	=22^2*0.015^2*(7.5*F14+F14^2*2.5)^–3.3333*((7.5+2*F14*SQRT(1+2.5^2))^1.3333)	=(E14–E13)/(0.0006–G14)	=I13+H14

(b) Spreadsheet formulas

Figure E4 Example 8.15 using Excel.

Example 8.16

A trapezoidal channel 300 m long conveys water flowing at $Q = 25\ m^3/s$. The inverse side slopes of the cross section are $m_1 = m_2 = 2.5$, the bottom width is 5 m, the bottom slope is $S_0 = 0.005$, and the Manning coefficient is $n = 0.013$. Compute the water surface profile and energy grade line if the upstream depth $y_u = 0.55$ m and the downstream depth $y_d = 2.15$ m.

Solution

An Excel spreadsheet solution using the step method is shown in Table E8.16. First,input data is entered, including the gravitational constant and the corresponding constant c_1 in the Manning equation. Since the two inverse side slopes are equal, a single value m is entered. Then critical and normal depths are evaluated using Excel Solver. Since $y_c > y_0$, the slope is steep. One control point is at the upstream end, and an S_3 profile exists in the upper reach of the channel. The profile is computed from station 1 beginning with y_u to station 7 where the depth is 0.85 m and the distance is 288 m. Then an S_1 profile is calculated from station 9 at the end of the channel beginning with y_d, and terminating when y reaches critical depth. A hydraulic jump is located approximately 190 m downstream from the channel entrance. It is located by plotting the curve of depth conjugate to the S_3 profile and finding its intersection with the S_1 profile as shown in Figure E8.16. The conjugate depths y_{cj} are computed using Excel Solver.

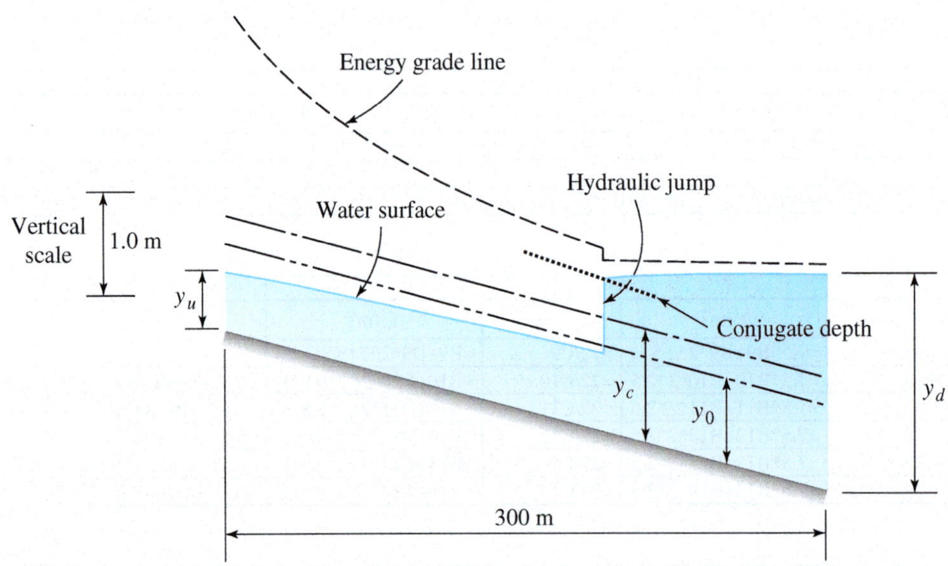

Figure E8.16

Table E8.16

$Q = 25$	$n = 0.013$	$S_0 = 0.005$	$L = 300$	$g = 9.81$	
$b = 5$	$m = 2.5$	$y_u = 0.55$	$y_d = 2.15$	$c_1 = 1.00$	

	Depth	Residual
$y_c =$	1.124	−2.26E-07
$y_0 =$	0.864	4.363E-07

Station	y	A	V	E	y_m	$S(y_m)$	Δx	x	y_{cj}	Residual
1	0.550	3.506	7.130	3.141						
2	0.600	3.900	6.410	2.694	0.575	2.189E-02	26	26	1.901	8.81E-07
3	0.650	4.306	5.806	2.368	0.625	1.624E-02	29	56	1.802	5.19E-07
4	0.700	4.725	5.291	2.127	0.675	1.231E-02	33	88	1.711	1.68E-07
5	0.750	5.156	4.848	1.948	0.725	9.500E-03	40	128	1.628	9.30E-07
6	0.800	5.600	4.464	1.816	0.775	7.449E-03	54	182	1.550	3.85-07
7	0.850	6.056	4.128	1.719	0.825	5.923E-03	105	288	1.477	5.07E-07
9	2.150	22.306	1.121	2.214				300		
10	2.050	20.756	1.204	2.124	2.100	1.575E-04	−19	281		
11	1.950	19.256	1.298	2.036	2.000	1.924E-04	−18	263		
12	1.850	17.806	1.404	1.950	1.900	2.371E-04	−18	245		
13	1.750	16.406	1.524	1.868	1.800	2.951E-04	−17	228		
14	1.650	15.056	1.660	1.791	1.700	3.713E-04	−17	211		
15	1.550	13.756	1.817	1.718	1.600	4.728E-04	−16	195		
16	1.450	12.506	1.999	1.654	1.500	6.102E-04	−15	180		
17	1.350	11.306	2.211	1.599	1.400	7.998E-04	−13	167		
18	1.250	10.156	2.462	1.559	1.300	1.067E-04	−10	157		
19	1.124	8.778	2.848	1.537	1.187	1.514E-03	−6	151		

8.7.2 Numerical Integration Method

There are a number of numerical methods available to solve Eq. 8.6.4 for channels with regular cross sections, any of which would provide a solution with sufficient accuracy for design and analysis purposes. Worthy of mention are integration techniques using trapezoidal or Simpson's rules, and solution of the differential form using a Runge–Kutta procedure. The underlying theory of these methods and algorithms for their implementation can be found in texts on numerical methods. McBean and Perkins (1975) discuss convergence problems associated with use of the trapezoidal rule applied to gradually varied flow.

A useful integration scheme is the two-point Gauss–Legendre quadrature (Chapra and Canale, 1998). It is particularly well suited for computer use and is usually more accurate than either the trapezoidal or Simpson's rule methods. Equation 8.6.4 is written in the integral form

$$x_{i+1} = x_i + \int_{y_i}^{y_{i+1}} \frac{1-\mathrm{Fr}^2}{S_0 - S}\,dy$$
$$= x_i + \int_{y_i}^{y_{i+1}} G(y)\,dy \qquad (8.7.7)$$

The last integral is approximated by the Gauss–Legendre formula, giving the result

$$\int_{y_i}^{y_{i+1}} G(y)dy = \frac{y_{i+1}-y_i}{2}\left[G\left(\frac{y_{i+1}+y_i-\sqrt{3}/3(y_{i+1}-y_i)}{2}\right)\right.$$
$$\left.+G\left(\frac{y_{i+1}+y_i+\sqrt{3}/3(y_{i+1}-y_i)}{2}\right)\right] \quad (8.7.8)$$

The following example illustrates use of the algorithm.

Example 8.17

Rework Example 8.15 using Gauss–Legendre quadrature.

Solution

A Mathcad solution is shown below. After the known data is entered and functions are defined, normal and critical depths are computed. Subsequently, the depth is iterated to find corresponding values of distance using Eqs.8.7.7 and 8.7.8.

Input data:

$$Q := 22 \qquad M := 2.5 \qquad b := 7.5 \qquad L := 2000$$
$$S_0 := 0.0006 \qquad n := 0.015 \qquad g := 9.81$$

Define functions:

$$A(y) := b\cdot y + M\cdot y^2 \qquad B(y) := b + 2\cdot M\cdot y \qquad P(y) := b + 2\cdot y\cdot\sqrt{1+M^2}$$

$$S(y) := \frac{Q^2\cdot n^2}{A(y)^{3.3333}\cdot P(y)^{-1.3333}} \qquad Fr(y) := \sqrt{\frac{Q^2\cdot B(y)}{g\cdot A(y)^3}} \qquad G(y) := \frac{1-Fr(y)^2}{S_0 - S(y)}$$

$$E(y) := y + \frac{Q^2}{2\cdot g\cdot A(y)^2}$$

Compute normal and critical depths:

$$y_n := \text{root}(S(y)-S_0, y, 0.01, 5) \qquad y_n = 1.292$$
$$y_c := \text{root}(Fr(y)-1, y, 0.01, 5) \qquad y_c = 0.865$$

Compute S2 profile from upstream location:

$$X_0 := 2000 \qquad y_0 := y_c \qquad \Delta y := 0.10 \qquad N := 4$$

$$i := 1, 2..N$$

$$y_i := y_{i-1} + \Delta y$$

$$x_i := x_{i-1} + \frac{\Delta y}{2} \cdot \left(G\left(\frac{y_i + y_{i-1} - \frac{\sqrt{3}}{3} \cdot \Delta y}{2} \right) + G\left(\frac{y_i + y_{i-1} + \frac{\sqrt{3}}{3} \cdot \Delta y}{2} \right) \right)$$

Computed depth, distance, and specific energy:

$$y = \begin{pmatrix} 0.865 \\ 0.965 \\ 1.065 \\ 1.165 \\ 1.265 \end{pmatrix} \quad x = \begin{pmatrix} 2 \times 10^3 \\ 1.988 \times 10^3 \\ 1.938 \times 10^3 \\ 1.798 \times 10^3 \\ 1.257 \times 10^3 \end{pmatrix} \quad E(y) = \begin{pmatrix} 1.218 \\ 1.235 \\ 1.275 \\ 1.333 \\ 1.4 \end{pmatrix}$$

8.7.3 Irregular Channels

The method demonstrated in Example 8.15 works well with regular channels. When calculating the profile for a reach of river channel in which the cross-sectional data are irregular and usually given at fixed locations, a trial solution is necessary at each location. Henderson (1966) outlines a method of computation that includes corrections for composite sections and eddy losses that occur at bends and expansions. For a composite section, the slope of the energy grade line is given by

$$S = \frac{Q^2}{(\Sigma K_i)^2} \tag{8.7.9}$$

in which K_i is termed the **conveyance** of subsection i:

$$K_i = \left(\frac{c_1 A R^{2/3}}{n} \right)_i \tag{8.7.10}$$

A kinetic-energy coefficient must be applied to the kinetic-energy term; for a composite section use

$$\alpha = \frac{(\Sigma A_i)^2}{(\Sigma K_i)^3} \sum \left(\frac{K_i^3}{A_i^2} \right) \tag{8.7.11}$$

With such complexity, it is most convenient to use the step method procedure in computer-coded form. The United States Army Corps of Engineers has developed the algorithm "HEC-RAS River Analysis System" for use in natural channels. The solution is based on the standard step method and is available for use on a personal computer. The program and documentation can be downloaded from the website *www.hec.usace.army.mil/software/hec-ras/*.

8.8 SUMMARY

As opposed to pipe flows, open channel flow is complicated by the presence of a free surface. In addition to velocity, the depth of flow is variable, so that even though it may be steady, the flow in a channel is generally nonuniform. In Section 8.3, as well in Chapter 5, we have introduced uniform flow as a limiting case–one that does not often occur but nonetheless is significant. Friction losses in open channel flow are described mathematically by the Chezy–Manning equation, which relates velocity or discharge to the geometric and hydraulic properties of the channel cross section. Energy and momentum principles were introduced in Sections 8.4 and 8.5 and applied to rapidly varied nonuniform flow that occurs at channel transitions and at a hydraulic jump. Critical and conjugate depths have been derived along with the definition of the Froude number. We have seen that specific energy is a useful concept for dealing with transition analysis. Much of the development in this chapter has used a rectangular cross section; this was done primarily for mathematical expediency. The guidelines that were developed and conclusions made using the rectangular section apply to nonrectangular sections as well. Table 8.4 summarizes several of the formulas developed in the chapter.

Gradually varied nonuniform flow occurs where the depth and velocity vary continuously with distance along a channel. In Section 8.6, application of the energy equation to gradually varied flow resulted in a mathematical description of the water surface profile. We have presented a classification of the various types of flow regimes that occur along a channel reach. In addition we have shown how the understanding of rapidly varied and gradually varied flow enable one to predict the manner in which water surfaces will behave, a task that is required prior to numerical computation of water surface profiles. Two numerical methods were introduced in Section 8.7 to integrate the gradually varied flow equation, and examples of solutions were provided using spreadsheets and or computational software. Perhaps the most useful means of solution is the spreadsheet, in which a generalized algorithm can be readily adapted for use in different problems.

Table 8.4 Formulas for Rectangular and General Section

Section	Fr	y_c	E	M
Rectangular	$\dfrac{q}{\sqrt{gy^3}}$	$\left(\dfrac{q^2}{g}\right)^{1/3}$	$y+\dfrac{q^2}{2gy^2}$	$\dfrac{by^2}{2}+\dfrac{q^2}{gy}$
General	$\dfrac{Q/A}{\sqrt{gA/B}}$	$\dfrac{Q^2B}{gA^3}=1$	$y+\dfrac{Q^2}{2gA^2}$	$A\overline{y}+\dfrac{Q^2}{gA}$

PROBLEMS

8.1 Give an example of each of the following types of flow for both closed-conduit and free-surface conditions. Include a sketch with each representation:
(a) Steady, uniform
(b) Unsteady, nonuniform
(c) Steady, nonuniform
(d) Unsteady, uniform

8.2 Water is flowing with a free surface at 4 m^3/s in a circular conduit. Find the diameter d such that the critical depth $y_c = 0.3d$. Under critical flow conditions, $Q^2B/(gA^3) = 1$.

8.3 Classify the following flows as steady or unsteady, and uniform or nonuniform:
(a) Water flowing through a logjam in a river with the observer standing on the shore.
(b) Water flowing through a river rapids with the observer on a raft traveling with the current.
(c) A flood wave traveling in a river which is generated by a gentle rainfall/runoff event, with the observer standing on the bank.
(d) A moving hydraulic jump in a prismatic channel, with the observer running alongside the bank at the same speed as the jump.

Uniform Flow

8.4 For each regular channel section, construct the curves R versus y and $AR^{2/3}$ versus y:
(a) Circular, $d = 2$ m
(b) Trapezoidal, $b = 3$ m, $m_1 = m_2 = 2.5$
(c) Composite (see Figure P8.4)

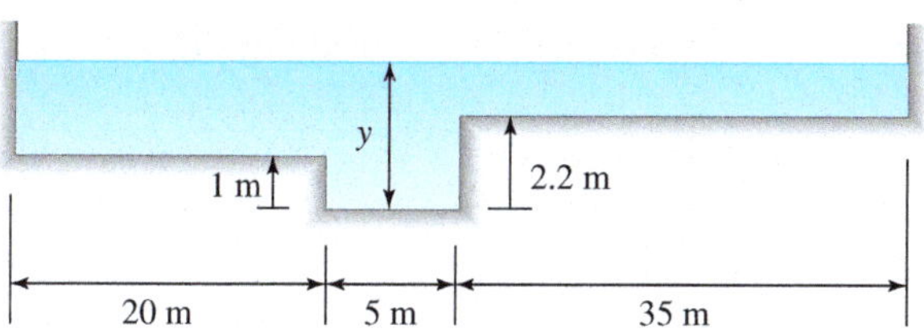

Figure P8.4

8.5 Determine the most efficient section based on flow resistance for a trapezoidal channel. Assume equal side slopes, that is, $m_1 = m_2$.

8.6 Determine the uniform depth y_0, area A,and wetted perimeter P for the following conditions:
(a) Trapezoidal channel, $m_1 = 2.5$, $m_2 = 3.5$, $b = 4.5$ m, $Q = 35$ m^3/s, $n = 0.015$, $S_0 = 0.00035$
(b) Circular channel, $d = 2.1$ m, $Q = 4.25$ m^3/s, $n = 0.012$, $S_0 = 0.001$
(c) Circular channel, $d = 6$ ft, $Q = 120$ ft^3/s, $n = 0.012$, $S_0 = 0.001$
(d) Trapezoidal channel, $Q = 15$ m^3/s, $n = 0.011$, $S_0 = 0.0013$, "most efficient" cross section
(e) Rectangular channel, $b = 25$ ft, $Q = 1200$ ft^3/s, $n = 0.020$, $S_0 = 0.0004$

8.7 Uniform flow occurs in a trapezoidal channel with $m_1 = m_2 = 1.75$. The channel is to convey a discharge of 4.0 m^3/s, with an average velocity of 1.4 m/s. If $n = 0.015$ and $S_0 = 0.001$, find the bottom width and the uniform depth.

8.8 A channel cross section, commonly called a **gutter**, forms at the side of a street next to a curb during rainfall conditions (Figure P8.8). The slope along the roadway is $S_0 = 0.0005$, and the Manning roughness coefficient is $n = 0.015$. Assuming that uniform flow conditions occur:
(a) Determine the discharge if the depth of flow is $y_0 = 12$ cm
(b) If $Q = 80$ L/s, what is the flow depth y_0?

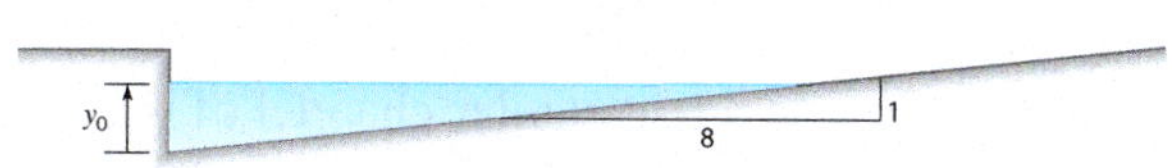

Figure P8.8

8.9 A sewer pipe of circular cross section is made of concrete ($n = 0.014$). It is to convey water at uniform flow conditions such that the pipe flows half full, that is, $y = d/2$, where d = diameter of the pipe. Evaluate the following conditions:

(a) Given $S_0 = 0.0003$ and $Q = 1.75$ m^3/s, find d
(b) Given $S_0 = 0.00005$ and $d = 1.3$ m, find Q
(c) Given $d = 0.75$ m and $Q = 0.45$ m^3/s, find S_0

Energy Concepts

8.10 For a given specific energy in a rectangular channel, show that critical flow exists when the specific discharge is at its maximum value q_{max}.

8.11 The specific energy relation for a rectangular section, Eq. 8.4.5, can be made dimensionless by normalizing it with respect to critical depth y_c. Prepare such a plot with y/y_c as the ordinate and E/y_c as the abscissa. Locate the minimum point mathematically.

8.12 Water is flowing in a rectangular channel at a velocity of 3 m/s and a depth of 2.5 m. Determine the changes in water surface elevation for the following alterations in the channel bottom:
(a) An increase (upward step) of 20 cm, neglecting losses.
(b) A "well-designed" increase of 15 cm.
(c) The maximum increase allowable for the specified upstream flow conditions to remain unchanged, neglecting losses.
(d) A "well-designed" decrease (downward step) of 20 cm.

8.13 Water is flowing in a rectangular channel whose width is 5 m. The depth of flow is 2 m and the discharge is 25 m^3/s. Determine the changes in depth for the following alterations in the channel width:
(a) An increase of 50 cm, neglecting losses
(b) A decrease of 25 cm, assuming a "well-designed" transition

8.14 Flow in a rectangular channel of width b takes place with a known depth y and velocity V. Downstream of this location, there is an upward step of height h. Neglecting losses, determine the change in width that must take place simultaneously for critical flow to occur within the transition.
(a) $b = 3$ m, $y = 3$ m, $V = 3$ m/s, $h = 70$ cm
(b) $b = 10$ ft, $y = 10$ ft, $V = 10$ ft/s, $h = 2.3$ ft

8.15 Water is flowing in a rectangular channel 2.0 m wide. At a transition section the channel bottom is lowered by $h = 0.1$ m for a short distance, and then is raised back to the original elevation (Figure P8.15). If $y = 1.22$ m and $Q = 4.8$ m^3/s, then, with losses neglected:
(a) Find the change in channel width necessary to maintain a horizontal water surface through the transition.
(b) What change in width would cause critical flow to occur in the transition?

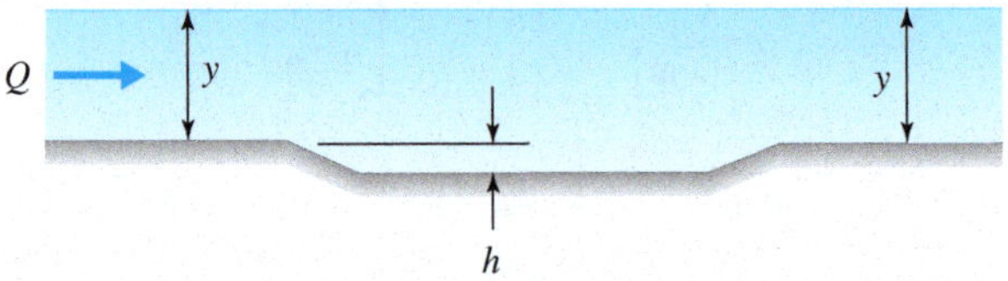

Figure P8.15

8.16 Water flows at a depth of 2.15 m and a unit discharge of 5.5 m^2/s in a rectangular channel. Energy losses can be neglected.
(a) What is the maximum height h of a raised bottom that will permit the flow to pass over it without increasing the upstream depth?
(b) Show the solution on an E-y diagram.
(c) Sketch the water surface and energy grade line.
(d) If the channel bottom is raised greater than h, discuss a type of change that may take place upstream of the transition.

8.17 A lake discharges into a steep channel. At the channel entrance the lake level is 2.5 m above the channel bottom. Neglecting losses, find the discharge for the following geometries:

(a) Rectangular section, $b = 4$ m
(b) Trapezoidal section, $b = 3$ m, $m_1 = m_2 = 2.5$
(c) Circular section, $d = 3.5$ m

8.18 Steep flow conditions exist at the outlet from a reservoir into an open canal. The canal width is not yet fixed, but the depth y at the entrance and the discharge Q are known. Determine the necessary widths for the following geometries:

(a) Rectangular section, $y = 1$ m, $Q = 18$ m^3/s
(b) Trapezoidal section, $m_1 = m_2 = 3$, $y = 1$ m, $Q = 18$ m^3/s
(c) Rectangular section, $y = 3$ ft, $Q = 635$ ft^3/s
(d) Trapezoidal section, $m_1 = m_2 = 3$, $y = 3$ ft, $Q = 635$ ft^3/s

8.19 Rapidly varied flow occurs over a rectangular sill (Figure P8.19), with free outfall conditions at location C. Throughout the reach losses can be neglected.

(a) Compute the discharge.
(b) Evaluate the depths at locations A, B, and C.

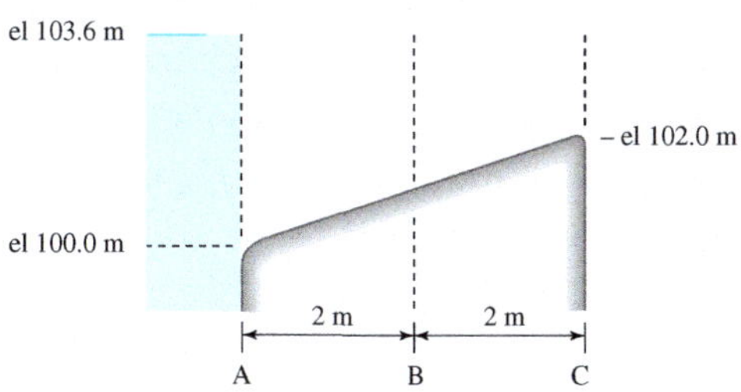

Figure P8.19

8.20 Prepare a computer algorithm that evaluates the critical depth of a regular channel using the false position technique or a method of your choice. Verify the program with the following data:

(a) Circular section, $Q = 5$ m^3/s, $d = 2.5$ m
(b) Rectangular section, $Q = 125$ cfs, $b = 12$ ft
(c) Trapezoidal section, $Q = 120$ m^3/s, $b = 10$ m, $m_1 = m_2 = 5$
(d) Triangular section, $Q = 250$ cfs, $m_1 = 2.5$, $m_2 = 3.0$

8.21 For the channel cross section shown in Figure P8.21, determine the critical depth if the discharge is 16.5 m^3/s.

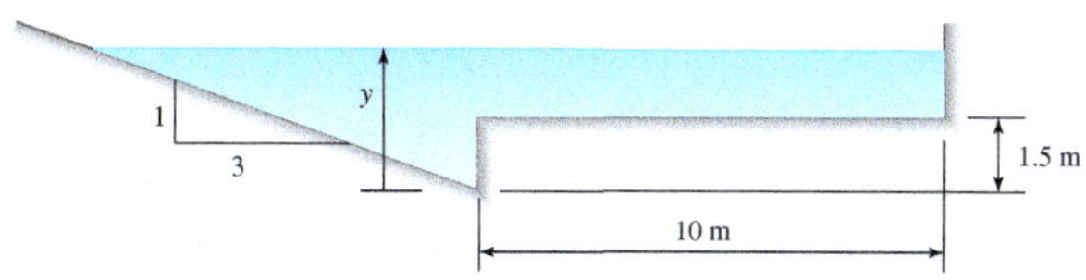

Figure P8.21

8.22 For the composite channel section shown in Figure P8.22, find the critical depth y_c if:

(a) $Q = 55$ m^3/s **(b)** $Q = 3.5$ m^3/s

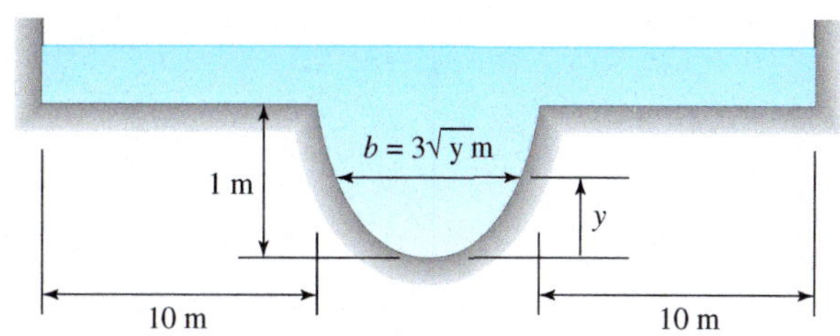

Figure P8.22

8.23 A parabolic channel (Figure P8.23) is described by the function $\eta = 0.1x^2$. If $Q = 25$ m^3/s and $y = 2.00$ m, find the alternate depth.

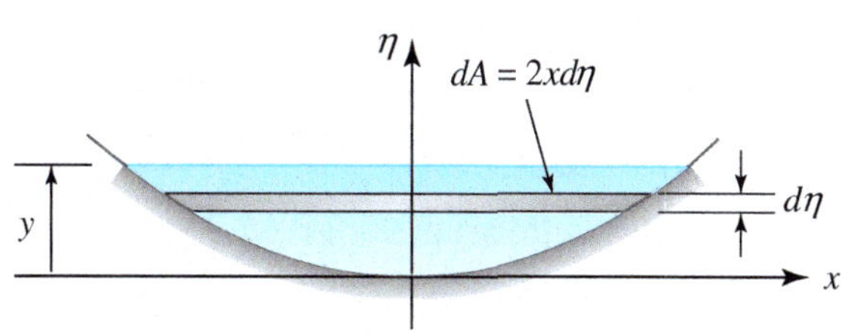

Figure P8.23

8.24 Prepare a computer algorithm that determines the alternate depth, given the geometry of a prismatic channel, the discharge, and the depth of flow. Use the interval-halving technique or a solution of your choice.

8.25 Derive the relation for a V-notch weir that expresses Q as a function of Y, Eq. 8.4.27.

8.26 Debris blocks the entire width of a river as shown in Figure P8.26. The cross section of the river is approximately rectangular, where $b = 15$ m, $Q = 22.5\ \text{m}^3/\text{s}$, $y_0 = 1.2$ m, and $h = 0.2$ m.

(a) Neglecting losses, analyze the rapidly varied water surface in the vicinity of the debris.

(b) Sketch the energy grade line and water surface.

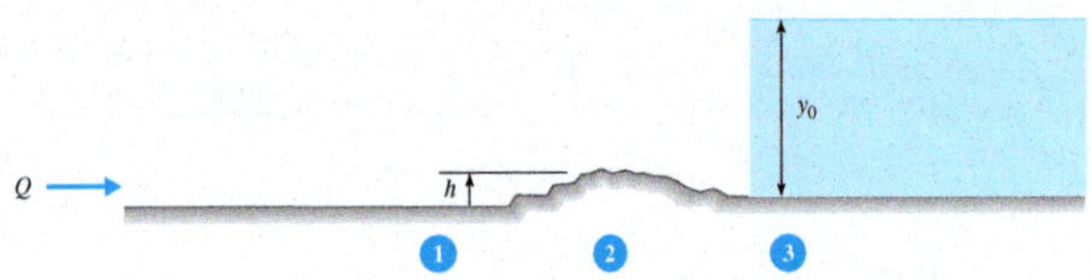

Figure P8.26

8.27 Develop the expression for the theoretical discharge for the weir shown in Figure P8.27.

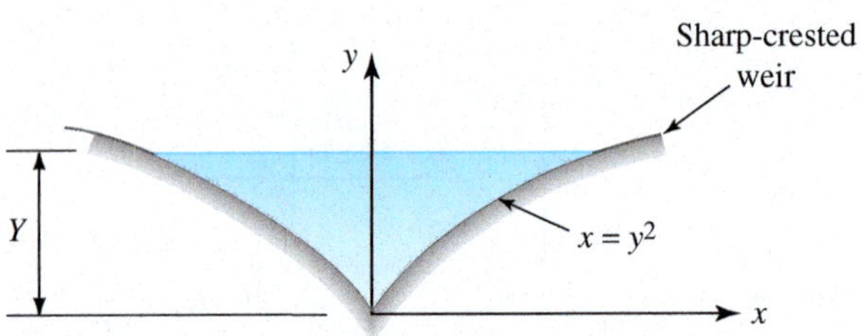

Figure P8.27

8.28 A Parshall flume is placed in a small stream to measure the discharge. It has a width of 3 ft at its throat.

(a) Compute the discharge if the measured upstream depth is 1.2 ft.

(b) Prepare a plot of Q versus H where H ranges from 0.5 to 1.5 ft.

8.29 Water is flowing at a given depth and discharge in a 3.5-m-wide rectangular channel. A 60-cm-high submerged raised section is placed in the channel as shown in Figure P8.29. The raised section spans the entire width of the channel, and is designed such that losses can be neglected. Calculate the water surface and energy grade line:

(a) When $y = 1.5$ m and $Q = 4\ \text{m}^3/\text{s}$

(b) When $y = 1.0$ m and $Q = 3\ \text{m}^3/\text{s}$

(c) Under what given conditions, (a) or (b) above, would the raised section act similar to a broad crested weir (explain your answer in words only)?

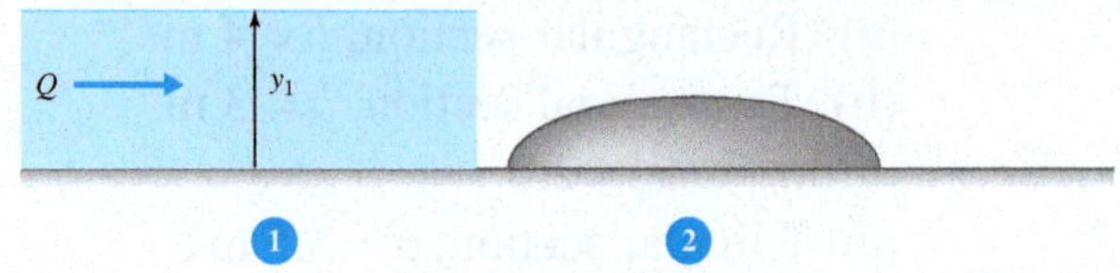

Figure P8.29

8.30 Design a broad-crested weir to convey a river discharge that varies between Q_1 and Q_2. The maximum water depth upstream of the weir is not to exceed y_2 and the minimum depth is not to be less than y_1.

(a) $Q_1 = 0.15\ \text{m}^3/\text{s}$, $Q_2 = 30\ \text{m}^3/\text{s}$, $y_1 = 1.05$ m, $y_2 = 1.75$ m

(b) $Q_1 = 5\ \text{ft}^3/\text{s}$, $Q_2 = 1000\ \text{ft}^3/\text{s}$, $y_1 = 3.45$ ft, $y_2 = 5.75$ ft

8.31 Design a canal to deliver water between the two reservoirs shown in Figure P8.31. The horizontal distance between the reservoirs is L. It is required that uniform flow exist throughout the reach. The water surface in reservoir A is to be no more than the distance h above the bottom of the canal at the entrance. The canal is rectangular in cross section, and made of concrete. Neglect entrance and exit losses, and assume uniform, subcritical flow conditions at the channel entrance.

(a) $\text{El}_A = 501.8$ m, $\text{El}_B = 500.2$ m, $L = 1500$ m, $h = 2$ m, $b = 2.5$ m

(b) $\text{El}_A = 1646$ ft, $\text{El}_B = 1641$ ft, $L = 4920$ ft, $h = 6.5$ ft, $b = 8$ ft

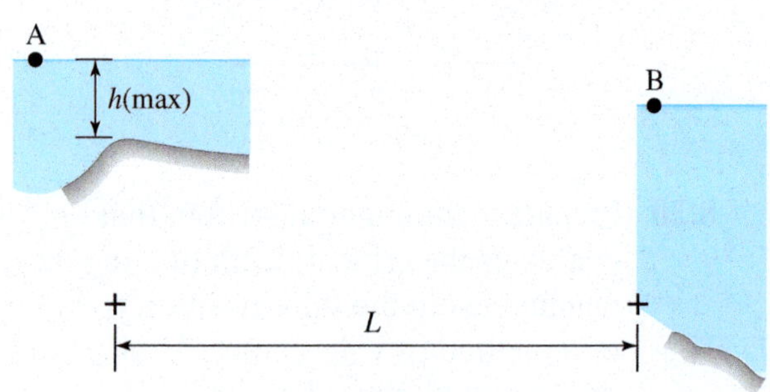

Figure P8.31

Momentum Concepts

8.32 The momentum function for a rectangular section, Eq. 8.5.4, can be made dimensionless by normalizing it with respect to $b(y_c)^2$. Prepare such a plot using y/y_c as ordinate and $M/(by_c)^2$ as abscissa.

8.33 Rework Problem 8.24 to evaluate the conjugate depth in place of the alternate depth.

8.34 Flow occurs in a wide rectangular channel at a depth y. A construction project over this channel requires that cofferdams spaced at distance w on center lines be placed in the channel (Figure P8.34). Assuming that the cofferdams create no energy losses between locations 1 and 2, determine the maximum permissible diameter of the cofferdams without creating upstream backwater effects (i.e., without increasing the upstream depth), and the resultant drag force on each cofferdam (assume that $C_D = 0.15$).

(a) $q = 1.5\ \text{m}^2/\text{s}$, $y = 1.8$ m, w = 6 m

(b) $q = 16\ \text{ft}^2/\text{s}$, $y = 6$ ft, w = 20 ft

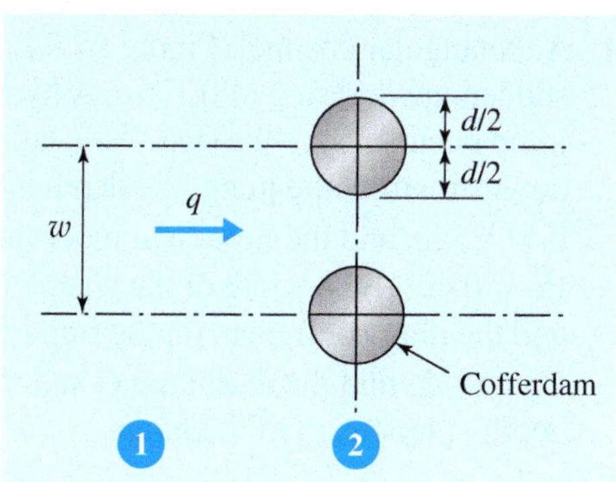

Figure P8.34

8.35 A temporary pipeline (diameter = 200 mm $C_D = 0.30$) is to be placed on a river bed normal to the direction of flow. The river is approximately rectangular in section, and is 100 m wide. The depth of flow downstream of the pipe is 2.5 m and the mean velocity is 3 m/s. Neglecting bed slope and resistance, determine the following:

(a) Flow conditions (i.e., depth and velocity) upstream of the pipe once it is in place

(b) Resultant drag force on the pipe

8.36 A hydraulic jump occurs over a sill located in a triangular channel with water flowing as shown in Figure P8.36. The inverse side slopes are $m_1 = m_2 = 3$, the drag coefficient $C_D = 0.40$, and the height of the sill $h = 0.3$ m. Determine the discharge Q if $y_1 = 0.50$ m and $y_2 = 1.8$ m.

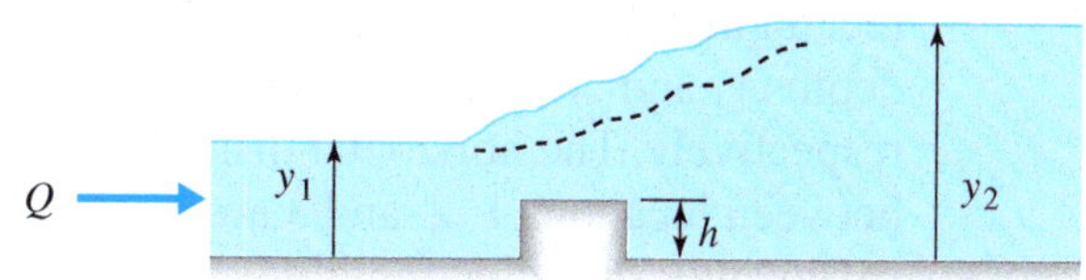

Figure P8.36

8.37 Water is flowing in a rectangular channel at a depth of 1.6 m and a velocity of 0.85 m/s. At a downstream location the discharge is suddenly reduced to zero, causing a surge to propagate upstream. Find the depth and velocity behind the surge, and the speed of the surge wave.

8.38 In a rectangular channel, water is flowing at a depth y and a velocity V. At a downstream location, the discharge is suddenly reduced by 60% (i.e., to 40% of the original value), causing a surge to propagate upstream. Determine the depth and velocity behind the surge, and the speed of the surge wave.

(a) $y = 1.5$ m, $V = 1$ m/s

(b) $y = 5$ ft, $V = 3$ ft/s

8.39 Water enters a reach of rectangular channel where $y_1 = 0.5$ m, $b = 7.5$ m, and $Q = 20\ \text{m}^3/\text{s}$ (Figure P8.39). It is desired that a hydraulic jump occur upstream (location 2) of the sill and on the sill critical conditions exist (location 3). Other than across the jump, losses can be neglected. Determine the following:

(a) Depths at locations 2 and 3
(b) Required height of the sill, h
(c) Resultant force acting on the sill
(d) Sketch the water surface and energy grade line.
(e) Describe the nature and character of the jump.

Figure P8.39

8.40 Water is flowing as shown in Figure P8.40 under the sluice gate in a horizontal rectangular channel that is 5 m wide. The depths y_1 and y_2 are 2.5 m and 10 cm, respectively. The horizontal distances between locations 1, 2, and 3 are sufficiently short that rapidly varied flow conditions can be assumed to occur. Determine the following:
(a) The discharge
(b) The depth downstream of the jump at location 3
(c) The power lost in the hydraulic jump

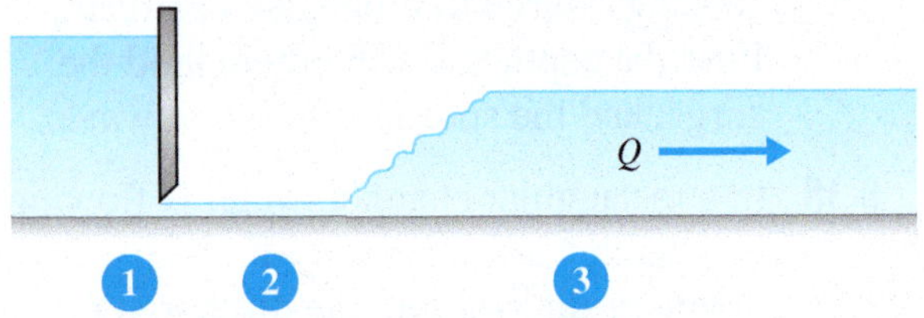

Figure P8.40

8.41 A transition section is located at the entrance to a rectangular channel as shown in Figure P8.41. At location 1, the depth is sufficiently large so that the velocity is negligible. If $b = 3$ m, $Fr_3 = 0.75$, and $Q = 5.55$ m^3/s, determine the following:
(a) The water surface elevation relative to the datum at locations 1, 2, and 3. Sketch the water surface and energy grade lines between locations 1 and 3.
(b) The resultant horizontal force acting on the channel bottom between locations 2 and 3

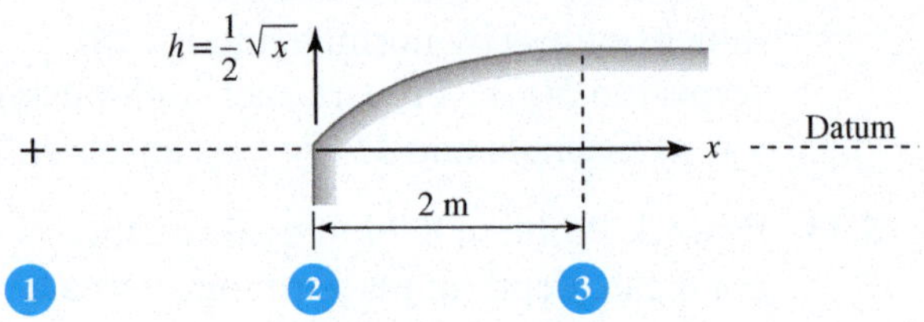

Figure P8.41

8.42 Water is flowing in a trapezoidal channel, $b = 5$ m, $m_1 = m_2 = 3$. A stationary hydraulic jump occurs, with the upstream depth $y_1 = 1.1$ m and the discharge $Q = 60$ m^3/s. Find the downstream depth y_2 and the power dissipated by the jump.

8.43 A 4-m-wide rectangular channel contains an upward rectangular step whose width is the same as the channel. The given flow conditions are $y_1 = 0.5$ m, $V_1 = 8$ m/s, and $y_2 = 2$ m.
(a) What is the height h of the step if $C_D = 1.2$?
(b) What would be y_2 if the step were not present?

8.44 A rectangular channel (Figure P8.44) has a sudden upward step of 0.17 m. A hydraulic jump occurs above the step. At location 2, downstream of the jump, the depth is $y_2 = 1.5$ m, and the Froude number there is $Fr_2 = 0.40$. If the width of the channel is 5 m and the drag coefficient on the step is $C_D = 0.35$, find the discharge Q and the depth y_1 upstream of the jump.

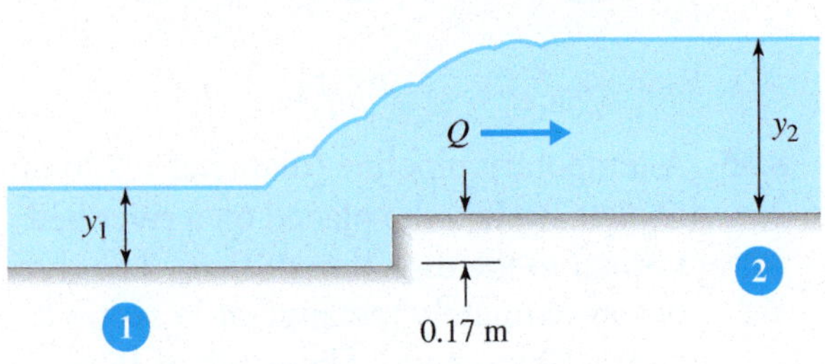

Figure P8.44

Nonuniform Gradually Varied Flow

8.45 The channel whose bottom profile is shown in Figure P8.45 has a rectangular cross section, with $b = 8$ m, $n = 0.014$, and $S_0 = 0.004$. Determine the following:

(a) The discharge in the channel

(b) Sketch the water surface and energy grade line.

(c) The possible existence of a hydraulic jump

(*Hint:* Assume critical flow conditions at the channel entrance.)

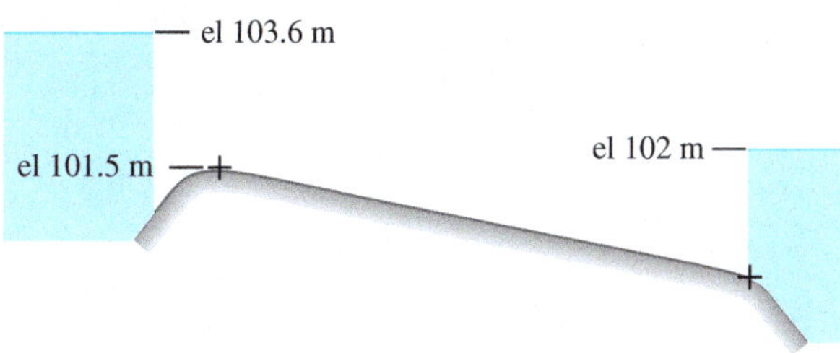

Figure P8.45

8.46 Uniform flow occurs in a long rectangular channel of width b, $S_0 = 0.0163$ and $n = 0.012$. The channel is altered by raising the channel bottom by Δz for a short distance. The alteration is intended to act as a "smooth transition region" so that losses can be considered negligible. Determine what changes take place due to the presence of the transition. In your analysis include all relevant rapidly varied flow calculations, identify any gradually varied flow profiles, and sketch the water surface and energy grade line.

(a) $Q = 0.35$ m^3/s, $b = 1.80$ m, $\Delta z = 100$ mm

(b) $Q = 12.5$ cfs, $b = 6$ ft, $\Delta z = 4$ in

8.47 Rework Problem 8.46 if $S_0 = 0.0013$.

8.48 Flow of water takes place in a rectangular channel with $b = 4$ m, $n = 0.012$, $L = 500$ m, $S_0 = 0.00087$. The discharge is 33 m^3/s. At the entrance to the channel, the depth is 0.68 m, and at the downstream end a free-outfall condition exists. Perform a profile synthesis to determine the nature of the water surface and energy grade line.

8.49 A flow of 8.5 m^3/s occurs in a long rectangular channel 3 m wide with $y_0 = 1.54$ m. There is a smooth constriction in the channel to 1.8 m width.

(a) Determine the depths to be expected in and just upstream of the constriction, neglecting losses. Show the solution on an *E-y* diagram.

(b) Classify the gradually varied flow profile upstream of the constriction.

8.50 Water is discharging at 20 m^3/s a triangular channel with $m_1 = 3.5$, $m_2 = 2.5$, $S_0 = 0.001$, $n = 0.014$. At the entrance to the channel the depth is 0.50 m, and at a distance 300 m downstream the depth is 2.5 m. Determine the nature of the water surface over the 300-m reach.

8.51 A long channel has a change in bottom slope 500 m from its downstream end; upstream of the transition the slope is $S_{01} = 0.0003$, and downstream of it the slope is $S_{02} = 0.0005$. The downstream reach of channel terminates at a reservoir whose elevation is 3 m above the channel bottom at that location. The upper reach is quite long, and very far upstream from the transition normal flow conditions exist. The channel section is trapezoidal, with a bottom width of 3 m, $m_1 = m_2 = 1.8$, $n = 0.012$, and a discharge of 17.5 m^3/s. Determine the nature of the water surface and energy grade line from a location far upstream of the transition to the end of the channel.

8.52 A rectangular channel with $b = 4$ m has a change in slope such that the normal depths are $y_{01} = 0.93$ m and $y_{02} = 1.42$ m. At locations far upstream and far downstream of the transition, flow occurs at the respective uniform depths that serve as controls. The discharge is 15 m^3/s. Find the variation of the water surface and energy grade line throughout the region.

8.53 The partial water profile shown in Figure P8.53 is for a rectangular channel of width $b = 3$ m, in which water is flowing at a discharge of $Q = 5$ m^3/s.

(a) Does a hydraulic jump occur in the channel? If so, is it located upstream or downstream of location A?

(b) Sketch the water surface and energy grade line, and identify any known water surface profiles.

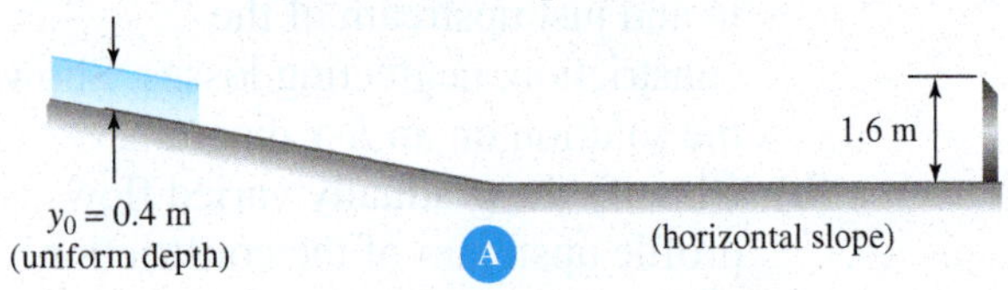

Figure P8.53

8.54 A very long rectangular channel (Figure P8.54) has normal flow conditions at locations A and C, and a change in the bottom slope occurs at location B.

(a) Somewhere between locations A and C a hydraulic jump will take place. Explain why this statement is true.

(b) Classify and sketch the two possible water surface profiles that exist between locations A and C.

(c) By appropriate reasoning and calculation, determine whether the jump will occur upstream of location B, or downstream of location B.

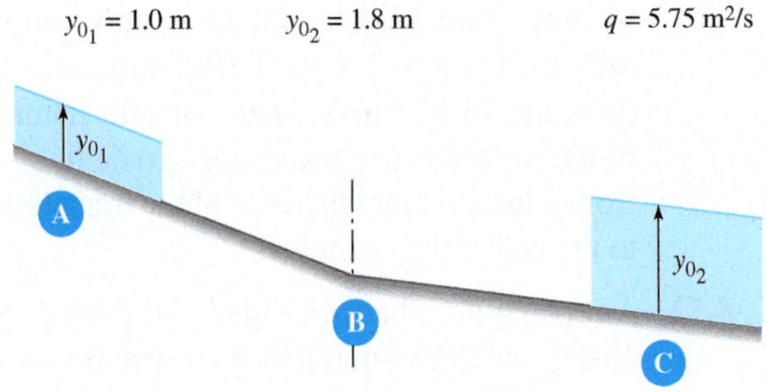

Figure P8.54

8.55 A long rectangular channel (Figure P8.55) has normal flow conditions at locations A and D, respectively, far upstream and downstream of location C. At C a change in slope occurs as shown.

(a) Classify the water surface, and sketch the water surface and the energy grade line between A and D.

(b) Determine the height h of a transition such that normal flow conditions occur between A and B, which is a short distance upstream of A.

(c) Explain what type of profile now exists downstream of B.

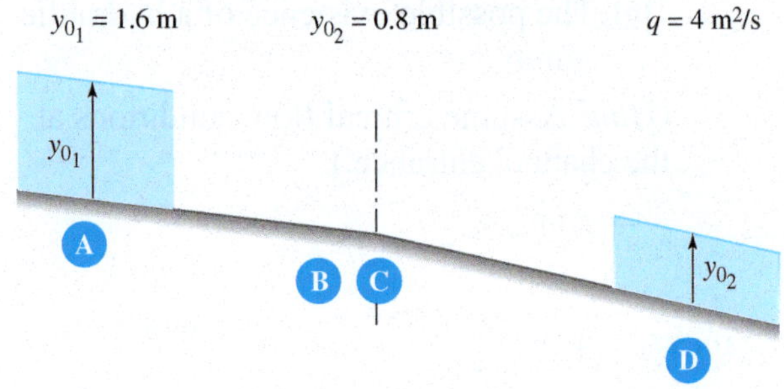

Figure P8.55

8.56 Water is flowing at a discharge Q in a circular conduit of diameter d and length L,with $S_0 = 0.001$, and $n = 0.015$. At the inlet the water depth is y_0 and at the outlet a free overfall exists. Determine the nature of the water surface in the conduit.

(a) $Q = 2.5$ m^3/s, $L = 500$ m, $d = 2.5$ m, $y_0 = 0.4$ m

(b) $Q = 88$ ft^3/s, $L = 1600$ ft, $d = 8$ ft, $y_0 = 1.3$ ft

8.57 With the results from Problem 8.51, compute the water surface profile and energy grade line using a numerical method of your choice.

8.58 A rectangular channel ($b = 5$ m, $S_0 = -0.001$) conveys flow at a rate of $Q = 3$ m^3/s. At a given location the depth is $y_1 = 2$ m and downstream of that location the depth is $y_2 = 1.8$ m. The Manning coefficient is $n = 0.02$.

(a) Determine the distance between locations 1 and 2.

(b) What type of water surface profile exists in the reach?

8.59 A rectangular channel (Figure P8.59) has the following properties: $b = 1$ m, $S_0 = 0.0002$, $n = 0.017$. A 120° V-notch weir is placed in the channel to measure the discharge, with the bottom notch of the weir located 0.5 m above the channel floor. Upstream of the weir the depth Y is measured to be 0.37 m.

(a) What is the discharge in the channel?

(b) If uniform flow conditions exist very far upstream of the weir, classify the nature of the water surface.

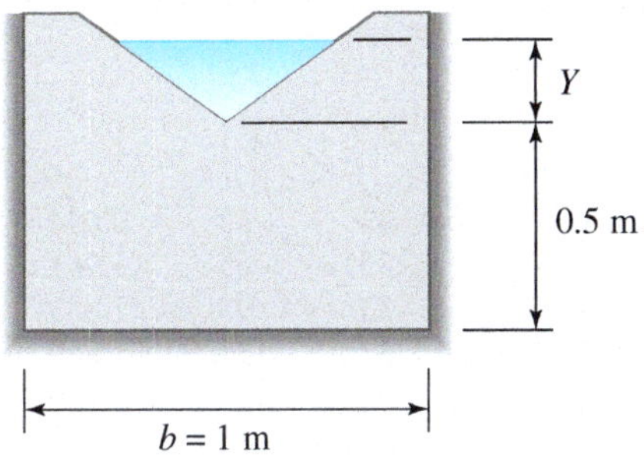

Figure P8.59

8.60 Gradually varied flow occurs over a reach of rectangular channel with $n = 0.013$, $S_0 = 0.005$, and $b = 2.5$ m. At location 1 the depth is 1.05 m and at location 2 the depth is 1.2 m. If the two locations are 50 m apart, find the discharge and identify the profile type.

8.61 Over a portion of a very long rectangular channel of width b (Figure P8.61), uniform flow occurs with discharge Q and depth y_0. At the end of the channel the flow is terminated by a free outfall. In addition, at the outlet the channel width is reduced to b_1 m for a very short distance. Determine and describe as completely as possible the changes that take place in the water surface between locations A and B. Use an E-y diagram for illustration.

(a) $Q = 5.5$ m^3/s, $y_0 = 0.5$ m, $b = 3$ m, $b_1 = 1.5$ m

(b) $Q = 200$ cfs, $y_0 = 1.65$ ft, $b = 10$ ft, $b_1 = 5$ ft

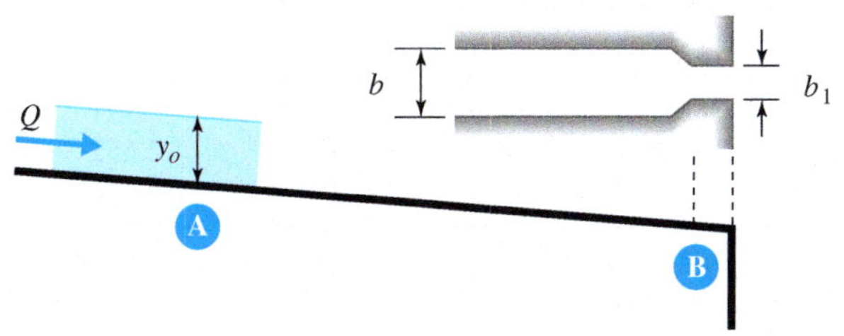

Figure P8.61

8.62 A sluice gate is placed in a long rectangular channel, $b = 4$ m, $n = 0.014$, and $S_0 = 0.0008$. The depth upstream of the gate is $y_1 = 1.85$ m and downstream of the gate the depth is $y_2 = 0.35$ m. Neglecting losses, identify and compute the water surface profiles on either side of the gate.

8.63 With the following SI data,[6] identify the profiles and plot the water surface and energy grade line to scale.

$Q = 3.5$	$n = 0.011$	$S_0 = 0.001$	$L = 200$	$g = 9.81$
$b = 5$	$m = 2.5$	$y_u = 0.80$	$y_d = 2.00$	$c_1 = 1.00$

	Depth	Residual
$y_c =$	1.356	– 3.44E-07
$y_0 =$	1.442	– 4.23E-07

Station	y	A	V	E	y_m	$S(y_m)$	Δx	x	y_{ci}	Residual
1	0.800	5.600	6.250	2.791						
2	0.850	6.056	5.779	2.552	0.825	8.311E-03	33	33	2.044	−5.98E-07
3	0.900	6.525	5.364	2.366	0.875	6.689E-03	33	65	1.961	−3.09E-07
4	0.950	7.006	4.996	2.222	0.925	5.443E-03	33	98	1.883	8.24E-07
5	1.000	7.500	4.667	2.110	0.975	4.472E-03	32	130	1.809	−7.46E-07
6	1.050	8.006	4.372	2.024	1.025	3.706E-03	32	162	1.740	−9.44E-08
7	1.100	8.525	4.106	1.959	1.075	3.097E-03	31	193	1.674	2.14E-07
8	1.150	9.056	3.865	1.911	1.125	2.606E-03	30	223	1.611	4.94E-07
7	2.000	20.000	1.750	2.156				200		
8	1.975	19.627	1.783	2.137	1.988	2.770E-04	−26	174		
9	1.950	19.256	1.818	2.118	1.963	2.916E-04	−26	147		
10	1.925	18.889	1.853	2.100	1.938	3.073E-04	−27	121		
11	1.900	18.525	1.889	2.082	1.913	3.240E-04	−27	94		
12	1.875	18.164	1.927	2.064	1.888	3.417E-04	−27	67		
13	1.850	17.806	1.966	2.047	1.863	3.607E-04	−27	40		
14	1.825	17.452	2.006	2.030	1.838	3.810E-04	−27	13		

8.64 With the following[7] SI data in (a) and English data in (b), identify the profiles and plot the water surface and energy grade line to scale.

[6] The data was generated by Excel spreadsheet.

[7] The data was generated by Excel spreadsheet

.

(a)

$Q = 3.5$ $n = 0.013$ $S_0 = 0.005$ $L = 100$ $g = 9.81$
$b = 2.5$ $m = 0$ $y_0 = 0.30$ $y_d = 0.95$ $c_1 = 1.00$

	Depth Residual
$y_c =$	0.585-8.58E-07
$y_0 =$	0.508-1.42E-07

Station	y	A	V	E	y_m	$S(y_m)$	Δx	x	y_{cj}	Residual
1	0.300	0.750	4.667	1.410						
2	0.325	0.813	4.308	1.271	0.313	2.154E-02	8	8	0.985	6.45E-07
3	0.350	0.875	4.000	1.165	0.338	1.702E-02	9	17	0.932	−4.60E-07
4	0.375	0.938	3.733	1.085	0.363	1.369E-02	9	26	0.884	6.34E-07
5	0.400	1.000	3.500	1.024	0.388	1.119E-02	10	36	0.840	3.55E-07
6	0.425	1.063	3.294	0.978	0.413	9.272E-03	11	47	0.799	2.00E-07
7	0.450	1.125	3.111	0.943	0.438	7.774E-03	13	60	0.762	4.40E-07
8	0.475	1.188	2.947	0.918	0.463	6.587E-03	16	76	0.727	9.05E-07
9	0.500	1.250	2.800	0.900	0.488	5.635E-03	29	104	0.694	2.22E-0
10	0.950	7.006	0.500	0.963				100		
11	0.850	6.056	0.578	0.867	0.900	9.698E-04	−24	76		
12	0.800	5.600	0.625	0.820	0.825	1.236E-03	−13	64		
13	0.750	5.156	0.679	0.773	0.775	1.474E-03	−13	51		
14	0.700	4.725	0.741	0.728	0.725	1.781E-03	−14	36		
15	0.650	4.306	0.813	0.684	0.675	2.183E-03	−16	21		
16	0.600	3.900	0.897	0.641	0.625	2.725E-03	−19	2		

(b)

$Q = 125$ $n = 0.013$ $S_0 = 0.005$ $L = 300$ $g = 32.2$
$b = 8$ $m = 0$ $y_u = 1.00$ $y_d = 3.00$ $c_1 = 1.49$

	Depth	Residual
$y_c =$	1.965	−1.04E-07
$y_0 =$	1.710	−4.30E-04

Station	y	A	V	E	y_m	$S(y_m)$	Δx	x	y_{cj}	Residual
1	1.000	8.000	15.625	4.791						
2	1.100	8.800	14.205	4.233	1.050	2.155E-02	34	34	3.311	−3.50E-07
3	1.200	9.600	13.021	3.833	1.150	1.634E-02	35	69	3.102	8.20E-07
4	1.300	10.400	12.019	3.543	1.250	1.269E-02	38	107	2.914	5.79E-07
5	1.400	11.200	11.161	3.334	1.350	1.007E-02	41	148	2.744	3.35E-07
6	1.500	12.000	10.417	3.185	1.450	8.135E-03	48	195	2.589	2.31E-07
7	1.600	12.800	9.766	3.081	1.550	6.674E-03	62	258	2.447	3.00E-07
8	1.700	13.600	9.191	3.012	1.650	5.549E-03	126	384	2.317	2.58E-07
9	3.000	37.500	3.333	3.173				300		
10	2.800	33.600	3.720	3.015	2.900	1.105E-03	−40	260		
11	2.600	29.900	4.181	2.871	2.700	1.349E-03	−39	220		
12	2.400	26.400	4.735	2.748	2.500	1.674E-03	−37	183		
13	2.200	23.100	5.411	2.655	2.300	2.121E-03	−32	151		
14	2.000	20.000	6.250	2.607	2.100	2.751E-03	−21	129		
15	1.965	19.478	6.417	2.605	1.983	3.247E-03	−1	128		

8.65 Consider a backwater curve situated upstream of a dam in a rectangular reservoir (Figure P8.65). Given data are: normal depth $y_0 = 3.92$ m, $Q = 125$ m^3/s, $S_0 = 0.0004$, $n = 0.025$, $y_A = 11$ m, $y_B = 12$ m

(a) Identify the type of profile that exists in the reservoir.

(b) Compute the distance between A and B.

(c) Sketch the water surface and energy grade line.

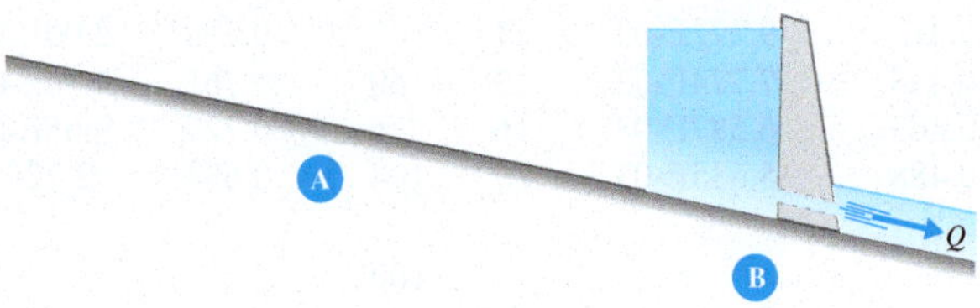

Figure P8.65

8.66 A rectangular channel has a change in slope as shown in Figure P8.66. The channel is 3.66 m wide, with $n = 0.017$, $S_{02} = 0.00228$, and $Q = 15.38$ m^3/s.

(a) Determine the depth that must exist in the downstream channel for a hydraulic jump to terminate at uniform flow conditions.

(b) If $y_{01} = 0.6$ m, calculate the length to the jump, L_j, using several increments of depth in a step calculation.

(c) Sketch the water surface and energy grade line, and identify all gradually varied flow profiles.

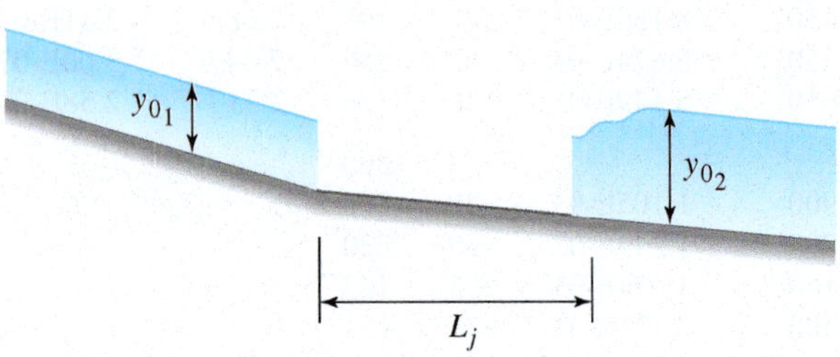

Figure P8.66

8.67 In the three channels shown in Figure P8.67, normal depths and critical depths are indicated by dashed (---) and dotted (· · ·) lines, respectively. Sketch one possible composite profile for each system, and identify all gradually varied surfaces.

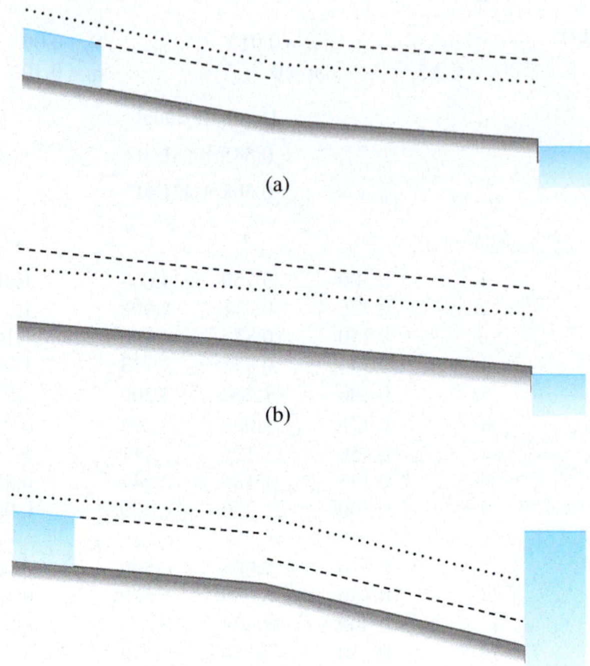

Figure P8.67

8.68 A design flow of $Q = 19$ m^3/s occurs in a river that can be approximated as a wide rectangular channel with the following properties: $b = 20$ m, $S_0 = 0.0005$, $n = 0.016$. A small dam is to be placed in the channel as part of a flood control plan (Figure P8.68). The height of the dam is $h = 6$ m, and the depth of flow at the toe of the dam is $y_{\text{toe}} = 0.10$ m. A concrete spillway apron ($n = 0.014$) is to be situated downstream of the dam. The purpose of the apron is to keep any hydraulic jump that may occur downstream of the dam off the river bed for the given design conditions to prevent erosion of the river bed. The depth behind the jump at the end of the apron is $y_2 = 0.75$ m. Determine the following:

(a) Power in kilowatts dissipated by the jump

(b) Design length of the apron

(c) The gradually varied-flow profiles upstream of the dam, and between the toe and the hydraulic jump

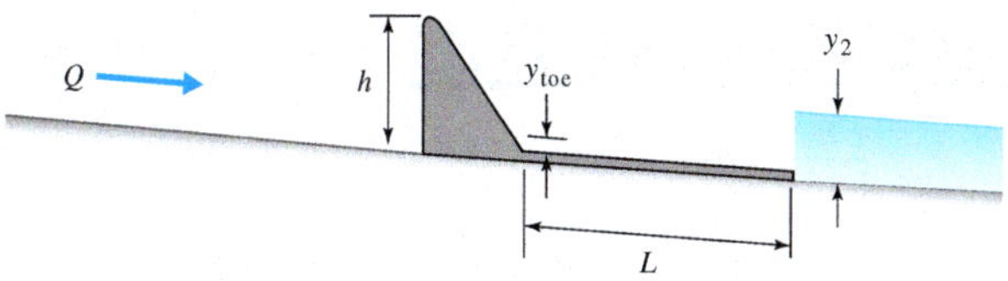

Figure P8.68

8.69 A very wide rectangular channel has the following properties: $S_0 = 0.0005$, $n = 0.017$, $q = 1.2\ \text{m}^2/\text{s}$. At some location, say location A, the depth is 0.65 m, and at location B, the depth is 0.90 m

(a) Is location B upstream or downstream of location A?

(b) Using the step method determine the distance between locations A and B.

8.70 An estuary to a river empties to the ocean. The estuary can be approximated as a very wide channel with a depth of 7 m at its outlet, and with $S_0 = 0.0001$, $n = 0.015$. If the discharge into the ocean from the estuary is $q = 1.5\ \text{m}^3/\text{s}$ per meter width, determine the distances upstream from the ocean where the depth is equal to 6, 5, 4, and 2 m.

8.71 A river cross section can be approximated by two rectangles as shown in Figure P8.71. The channel properties and dimensions are: $b_1 = 150$ m, $z_1 = 4$ m, $n_1 = 0.03$, $b_2 = 5$ m, $z_2 = 0$ m, and $n_2 = 0.02$. If the slope of the energy grade line is $S = 0.0005$ and the depth $y = 5$ m, find:

(a) The energy coefficient for the composite section

(b) The discharge

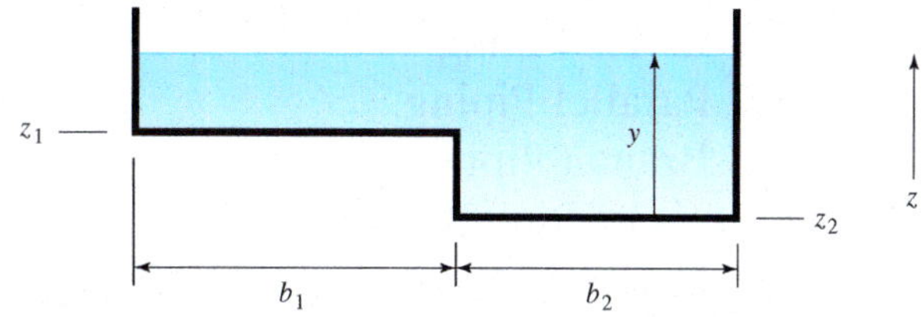

Figure P8.71

8.72 Consider the same river cross section of Problem 8.71. In the accompanying table, data are provided for two locations x along the river: At the downstream location $x = 400$ m, the depth $y = 3.0$ m when the discharge $Q = 280\ \text{m}^3/\text{s}$. Using a trial-and-error procedure, determine the depth y at the upstream location $x = 0$.

x (m)	b_1 (m)	b_2 (m)	z_1 (m)	z_2 (m)	n_1	n_2
0	115	107	16.7	15.0	0.05	0.03
400	149	91	17.7	15.1	0.05	0.03

9 Flows in Piping Systems

Outline

Chapter Objectives

The objectives of this chapter are to:

- Compare empirical frictional loss equations in piping
- Introduce the concept of hydraulic grade line as a variable for piping analysis
- Describe ad hoc methods of computing discharges and pressures in simple pipe systems
- Present linearized solutions (i.e., Hardy Cross method) for pipe networks using spreadsheet analysis
- Show use of computational software for pipe network analysis
- Introduce simplified analysis for unsteady flows in simple pipe systems

9.1 INTRODUCTION

Internal flows in pipelines and ducts are commonly encountered in all parts of our industrialized society. From delivering potable water to transporting chemicals and other industrial liquids, engineers have designed and constructed untold kilometers of relatively large-scale piping systems. Smaller piping units are also in abundance: in hydraulic controls, in heating and air conditioning systems, and in cardiovascular and pulmonary flow systems, to name only a few. These flows can be either steady or unsteady, uniform or nonuniform. The fluid can be either incompressible or compressible, and the piping material can be elastic, inelastic, or perhaps, viscoelastic. In this chapter we concentrate primarily on incompressible, steady flows in rigid piping. The piping system may be relatively simple,

M. C. Potter and D. C. Wiggert, *Applied Fluid Mechanics*,
https://doi.org/10.1007/978-3-031-65276-9_9

such that the variables can be solved rather easily using a calculator, or it may be sufficiently complicated that it is more convenient to use computational software.

Piping systems are considered to be composed of elements and components. Basically, pipe **elements** are reaches of constant-diameter piping, and the **components** consist of valves, tees, bends, reducers, or any other device that may create a loss to the system. In addition to components and elements, pumps add energy to the system and turbines extract energy. The elements and components are linked at junctions. Figure 9.1 illustrates several types of piping systems.

Following a discussion of piping losses, several piping systems, including series, branch, and parallel configurations, are analyzed. Attention is then directed to comprehensive network systems, where several methods of solution are presented. Most of the piping problems analyzed are those where the discharge is the unknown variable; this type of problem is classified as Category 2 in Section 5.6.3.

Finally, a brief introduction to unsteady flow in pipelines is presented. This topic is one that has been important for many years. It is taking on increasing importance as the demand for the construction of more cost-effective piping continues and the manner in which piping, valves, pumps, and other components interact and become more sophisticated. We address only the fundamental aspects of unsteady flows, focusing on two assumptions in a single pipe of constant diameter: an incompressible flow in an inelastic pipe, and a compressible liquid flow in an elastic pipe.

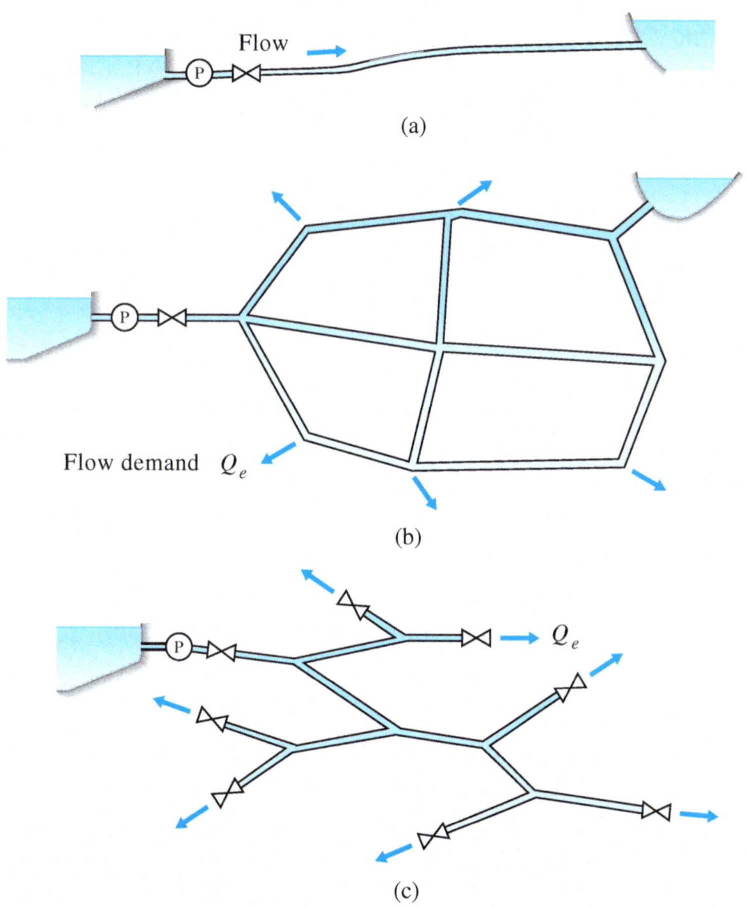

Figure 9.1 Pipe systems: (a) single pipe; (b) distribution network; (c) tree network.

9.2 LOSSES IN PIPING SYSTEMS

Losses can be divided into two categories: (a) those due to wall shear in pipe elements and (b) those due to piping components. The former are distributed along the length of pipe sections. The latter are treated as discrete discontinuities in the hydraulic and energy grade lines and are commonly referred to as minor losses; they are due primarily to separated or secondary flows.

The fundamental mechanics of wall shear and development of the empirical relations relating to pipe losses are treated in Chapter 5. Minor losses are covered in detail in Section 5.6.5 and are not discussed further here. The following material focuses on the treatment of losses in the analysis of piping systems.

9.2.1 Frictional Losses in Pipe Elements

It is convenient to express the pipe element frictional loss in the exponential form

$$h_L = RQ^{\beta} \tag{9.2.1}$$

in which h_L is the head loss over length L of pipe. R is the **resistance coefficient**, Q is the discharge in the pipe, and β is an exponent. Depending on the formulation chosen, the resistance coefficient may be a function of pipe roughness, Reynolds number, or length and diameter of the pipe element. In particular, the Darcy–Weisbach relation, Eq. 5.6.23, can be substituted into Eq. 9.2.1. Then with $\beta = 2$, the resulting expression for R is

$$\begin{aligned} R &= \frac{fL}{2gDA^2} \\ &= \frac{8fL}{g\pi^2 D^5} \end{aligned} \tag{9.2.2}$$

where f is the friction factor. The characteristics of f for commercial pipe flows are developed in Section 5.6.3. Specifically, the Moody diagram, Figure 5.13, presents a comprehensive picture of just how the friction factor varies over a wide range of Reynolds numbers and for a range of relative roughnesses. For pipe network analysis, it is convenient to express the behavior of f using approximate, equivalent empirical formulas in which the friction factor can be obtained directly in terms of the Reynolds number and relative roughness. A number of relations have been developed and shown to be reasonably accurate for engineering calculations (Benedict, 1980). In particular, the formulas of Swamee and Jain (1976) are presented in Section 5.6.3 and are shown to accurately represent the Colebrook relation, Eq. 5.6.28. A friction factor formula developed by Swamee and Jain is

$$f = 1.325\left\{\ln\left[0.27\left(\frac{e}{D}\right) + 5.74\left(\frac{1}{\mathrm{Re}}\right)^{0.9}\right]\right\}^{-2} \tag{9.2.3}$$

Combining Eqs. 9.2.2 and 9.2.3, one finds the resistance coefficient to be

$$R = 1.07\left(\frac{L}{gD^5}\right)\left\{\ln\left[0.27\left(\frac{e}{D}\right) + 5.74\left(\frac{1}{\text{Re}}\right)^{0.9}\right]\right\}^{-2} \tag{9.2.4}$$

Equations 9.2.3 and 9.2.4 are valid over the ranges $0.01 > e/D > 10^{-8}$, and $10^8 > \text{Re} > 5000$. The completely turbulent regime of Figure 5.13, where Re has a negligible effect on f, begins at a Reynolds number given by

$$\text{Re} = \frac{200D}{e\sqrt{f}} \tag{9.2.5}$$

For values of Re greater than this, the friction factor is a function only of e/D, and is given by

$$f = 1.325\left\{\ln\left[0.27\left(\frac{e}{D}\right)\right]\right\}^{-2} \tag{9.2.6}$$

Two additional expressions for pipe frictional losses, which have found wide use, are the Hazen–Williams and Chezy–Manning formulas. For water flow, the value of R in Eq. 9.2.1 for the Hazen–Williams relation is

$$R = \frac{K_1 L}{C^\beta D^m} \qquad K_1 = \begin{cases} 10.59 & \text{SI units} \\ 4.72 & \text{English units} \end{cases} \tag{9.2.7}$$

in which the exponents are $\beta = 1.85$ and $m = 4.87$, and C is the Hazen–Williams coefficient dependent only on the roughness. Values of the Hazen–Williams roughness coefficient C are given in Table 9.1.

The Chezy–Manning equation is more commonly associated with open-channel flow. However, in sewage and drainage systems in particular, it has been applied to conduits flowing under **surcharge**, that is, under pressurized conditions. The Chezy–Manning equation was introduced in Section 5.7. For a circular pipe flowing full, Eq. 5.7.6 can be substituted into Eq. 9.2.1 and solved for R:

$$R = \frac{10.29 n^2 L}{K_2 D^{5.33}} \qquad K_2 = \begin{cases} 1 & \text{SI units} \\ 2.22 & \text{English units} \end{cases} \tag{9.2.8}$$

in which n is the Manning roughness coefficient. In Eq. 9.2.1, the exponent $\beta = 2$.

Table 9.1 Nominal Values of the Hazen–Williams Coefficient C

Type of pipe	C
Extremely smooth; asbestos-cement	140
New or smooth cast iron; concrete	130
Wood stave; newly welded steel	120
Average cast iron; newly riveted steel; vitrified clay	110
Cast iron or riveted steel after some years of use	95–100
Deteriorated old pipes	60–80

An advantage to using Eq. 9.2.7 or Eq. 9.2.8 as opposed to Eq. 9.2.4 is that in the first two, C and n are dependent on roughness only, while in the last, f depends on the Reynolds number as well as the relative roughness. However, Eq. 9.2.4 is recommended since it provides a more precise representation of pipe frictional losses. Note that the Hazen–Williams and Chezy–Manning relations are dimensionally inhomogeneous,[1] whereas the Swamee–Jain equation is dimensionally homogeneous and contains the two parameters e/D and Re, which appropriately influence the losses.

The limitations of the Hazen–Williams and Chezy–Manning formulas are demonstrated as follows. Beginning with Eq. 9.2.1, the head loss h_L based on the Darcy–Weisbach resistance coefficient, Eq. 9.2.2, with the exponent $\beta = 2$ can be equated to h_L based on the Hazen–Williams coefficient, Eq. 9.2.7, with $\beta = 1.85$. Introducing the Reynolds number to eliminate Q and solving for the friction factor f results in

$$f = \frac{1.28 g K_1}{C^{1.85} D^{0.02} (\mathrm{Re}\, \nu)^{0.15}} \tag{9.2.9}$$

With SI units and water at 20°C, Eq. 9.2.9 reduces to

$$f = \frac{1056}{C^{1.85} D^{0.02} \mathrm{Re}^{0.15}} \tag{9.2.10}$$

Note that f is weakly dependent upon D in this equation. In a similar fashion, Eq. 9.2.8 can be substituted into Eq. 9.2.1 with $\beta = 2$ and equated to h_L based on the Darcy–Weisbach formulation to yield

$$f = \frac{124.5 n^2}{D^{0.33}} \tag{9.2.11}$$

Comparisons between the original Colebrook formula (Eq. 5.6.28), the formula of Swamee and Jain (Eq. 9.2.3), and the equivalent expressions for the Hazen–Williams and Manning coefficients (Eqs. 9.2.10 and 9.2.11) are shown in Figure 9.2 for a 1-m-diameter concrete pipe flowing full with water ($C = 130$, $n = 0.012$, $e/D = 0.00015$). It is evident that the Hazen–Williams and Chezy–Manning relations are valid over a limited range of Re, and that Eq. 9.2.3 provides a more versatile and accurate estimate of pipe losses. In the remainder of this chapter most of the frictional losses will be treated using the Darcy–Weisbach formulation, Eqs. 9.2.2 to 9.2.6. Any exceptions will be noted.

[1] In a dimensionally inhomogeneous equation, the constants in the equation have units assigned to them. In a dimensionally homogeneous equation the constants are dimensionless.

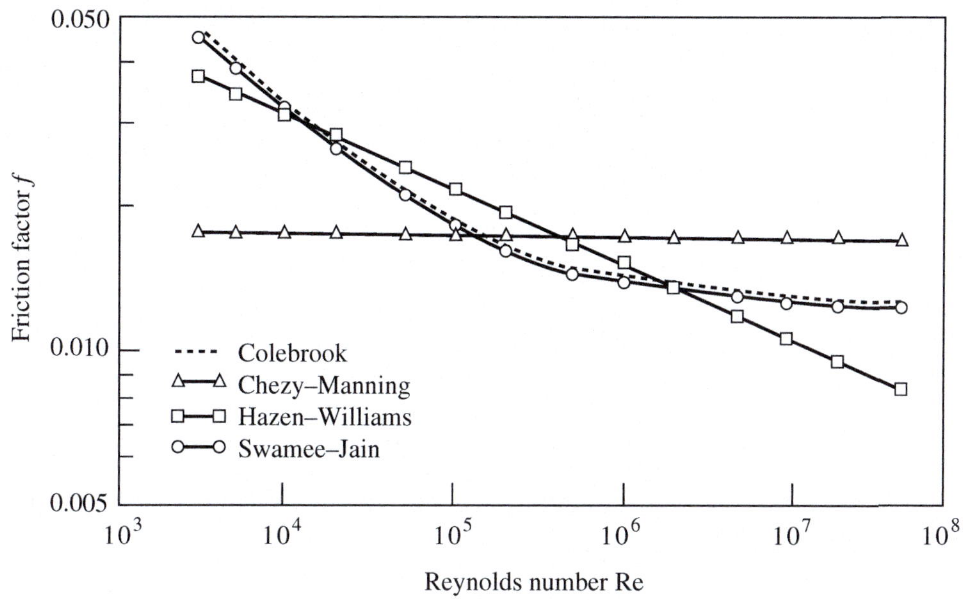

Figure 9.2 Comparison of several approximate formulas with the original Colebrook formula.

Example 9.1

A pipeline is conveying 0.05 m^3/s of water, at 30°C. The length of the line is 300 m, and the diameter is 0.25 m. Estimate the head loss due to friction using the three formulas (a) Darcy–Weisbach ($e = 0.5$ mm), (b) Hazen–Williams ($C = 110$), and (c) Chezy–Manning ($n = 0.012$).

Solution

The kinematic viscosity of water at 30°C is $\nu = 0.804 \times 10^{-6}$ m^2/s. The relative roughness, velocity, and Reynolds number are computed to be

$$\frac{e}{D} = 0.002 \qquad V = 1.02 \text{ m/s} \qquad \text{Re} = 3.17 \times 10^5$$

(a) Substitute values into Eq. 9.2.4:

$$R = 1.07 \frac{300}{9.81 \times (0.25)^5} \times \left\{ \ln \left[0.27 \times 0.002 + 5.74 (3.17 \times 10^5)^{-0.9} \right] \right\}^{-2} = 610$$

Then, with Eq. 9.2.1, the friction loss based on the Darcy–Weisbach formulation is

$$\begin{aligned} h_L &= RQ^2 \\ &= 610\,(0.05)^2 = \underline{1.52 \text{ m}} \end{aligned}$$

(b) For the Hazen–Williams formulation, use Eq. 9.2.7:

$$R = \frac{10.59 \times 300}{(110)^{1.85}(0.25)^{4.87}} = 454$$

$$h_L = 454(0.05)^{1.85} = \underline{1.78 \text{ m}}$$

(c) Use Eq. 9.2.8 for the Chezy–Manning formulation:

$$R = \frac{10.29(0.012)^2(300)}{1 \times (0.25)^{5.33}} = 719$$

$$h_L = 719(0.05)^2 = \underline{1.80 \text{ m}}$$

Part (a) with $h_L = 1.52$ m is the most accurate. The Hazen–Williams result is 17% too high, and the Chezy–Manning result is 18% too high.

9.3 SIMPLE PIPE SYSTEMS

The analysis of a single pipeline is presented in Section 5.6.3; it would be appropriate here for the reader to review the three categories of pipe problems in that section. For more complex piping systems, the methodology is similar. With relatively simple networks, such as series, parallel, and branching systems, ad hoc solutions can be developed that are suitable for use with calculators, spreadsheet algorithms, or computational software. Such approaches are relevant since they make use of the solver's ingenuity and require an understanding of the nature of the flow and piezometric head distributions for the particular piping arrangement. For systems of greater complexity, an alternative means of solving such problems is the Hardy Cross method, presented in Section 9.4.

The fundamental principle in the ad hoc approach is to identify all of the unknowns and write an equivalent number of independent equations to be solved. Subsequently, the system is simplified by eliminating as many unknowns as possible and reducing the problem to a series of single pipe problems—either Category 1 or Category 2 problems in Section 5.6.3.

9.3.1 Series Piping

Consider the series system shown in Figure 9.3. It consists of N pipe elements with a specified number of minor-loss components ΣK associated with each ith pipe element. A single minor loss is described in Section 5.6.5 to be equal to $h_L = KV^2/2g$. It is convenient here to express the minor loss in terms of the discharge rather than the velocity, so that $h_L = KQ^2/2gA^2$. For many flow situations, it is common practice to neglect the kinetic-energy terms at the inlet and outlet;

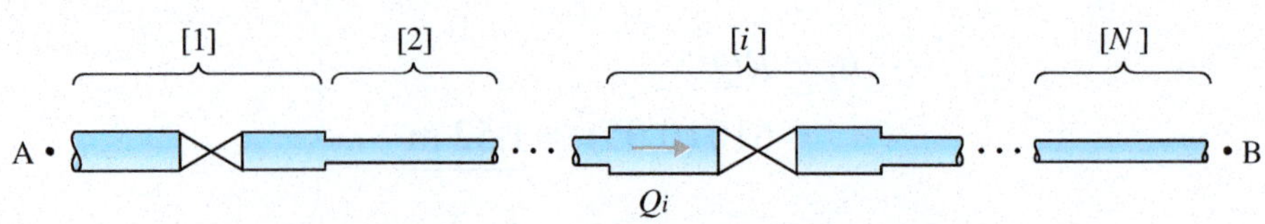

Figure 9.3 Series piping system.

they would be significant only if the velocities were relatively high. Assuming that the exponent in Eq. 9.2.1 is $\beta = 2$, the energy equation applied from location A to location B of Figure 9.3 is

$$\left(\frac{p}{\gamma}+z\right)_A-\left(\frac{p}{\gamma}+z\right)_B=\left(R_1+\frac{\Sigma K}{2gA_1^2}\right)Q_1^2+\left(R_2+\frac{\Sigma K}{2gA_2^2}\right)Q_2^2$$

$$+\cdots+\left(R_N+\frac{\Sigma K}{2gA_N^2}\right)Q_N^2$$

$$=\sum_{i=1}^{N}\left(R_i+\frac{\Sigma K}{2gA_i^2}\right)Q_i^2 \tag{9.3.1}$$

in which R_i is the resistance coefficient for pipe i.

The statement of continuity for the series system is that the discharge in every element is identical, or

$$Q_1=Q_2=\cdots=Q_i=\cdots=Q_N=Q \tag{9.3.2}$$

Replacing Q_i with Q, Eq. 9.3.1 becomes

$$\left(\frac{p}{\gamma}+z\right)_A-\left(\frac{p}{\gamma}+z\right)_B=\left[\sum_{i=1}^{N}\left(R_i+\frac{\Sigma K}{2gA_i^2}\right)\right]Q^2 \tag{9.3.3}$$

Equation 9.2.4 can be substituted for R_i, or alternatively, one may use Eq. 9.2.3 or the Moody diagram, Figure 5.13, to obtain values of the friction factor. The reader should recognize that in a series system, the discharge remains constant from one pipe element to another, and the losses are accumulative; that is, they are the sum of the minor loss components and the pipe frictional losses.

For a Category 1 problem, the right-hand side of Eq. 9.3.3 is known and the solution is straightforward. For a Category 2 problem, in which Q is unknown, a trial-and-error solution is required, since the Reynolds number, in terms of the unknown discharge ($\text{Re} = 4Q/\pi \nu D$), is present in the friction factor relation. Note that if flow in the fully turbulent regime is assumed, f is independent of Q and Eq. 9.3.3 reduces to a quadratic in Q. A Category 3 problem is not encountered in this type of analysis. A series piping system with minor losses and constant-diameter piping is applied to a Category 3 problem in Example 5.13. The solution for a somewhat more complex series system is illustrated in Example 9.2.

Example 9.2

For the system shown in Figure E9.2, find the required power to pump 100 L/s of liquid ($S = 0.85$, $\nu = 10^{-5}$ m^2/s). The pump is operating at an efficiency $\eta = 0.75$. Pertinent data are given in the figure. K_v and K_e are the valve and elbow coefficients, respectively.

$$\text{Line 1:}\quad L=10\text{ m},\quad D=0.20\text{ m},\quad e=0.05\text{ mm},\quad K_1=0.5,\quad K_v=2$$

$$\text{Line 2:}\quad L=500\text{ m},\quad D=0.25\text{ m},\quad e=0.05\text{ mm},\quad K_e=0.25,\quad K_2=1$$

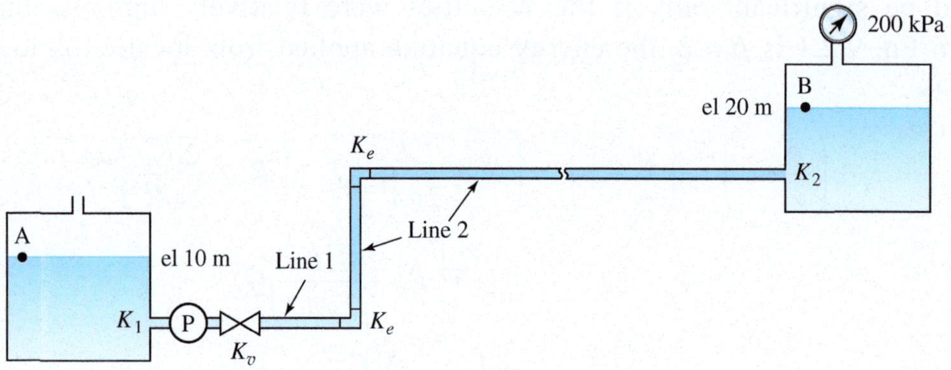

Figure E9.2

Solution

This is a Category 1 problem. The energy relation, Eq. 9.3.3, for the system is

$$\left(\frac{p}{\gamma}+z\right)_A + H_P = \left(\frac{p}{\gamma}+z\right)_B + \left[R_1 + \frac{K_1+K_v}{2gA_1^2} + R_2 + \frac{2K_e+K_2}{2gA_2^2}\right]Q^2$$

The resistance coefficients, R_1 and R_2, are calculated with Eq. 9.2.4 after first evaluating Re and e/D:

$$\text{Re}_1 = \frac{4Q}{\pi D_1 \nu} = \frac{4\times 0.10}{\pi \times 0.20 \times 10^{-5}} = 6.37\times 10^4 \qquad \left(\frac{e}{D}\right)_1 = \frac{0.05}{200} = 0.00025$$

$$\text{Re}_2 = \frac{4Q}{\pi D_2 \nu} = \frac{4\times 0.10}{\pi \times 0.25 \times 10^{-5}} = 5.09\times 10^4 \qquad \left(\frac{e}{D}\right)_2 = \frac{0.05}{250} = 0.0002$$

$$\text{R}_1 = 1.07\left(\frac{10}{9.81\times 0.20^5}\right) \times \left\{\ln\left[0.27\times 0.00025 + 5.74\times (6.37\times 10^4)^{-0.9}\right]\right\}^{-2} = 53.4$$

$$\text{R}_2 = 1.07\left(\frac{500}{9.81\times 0.25^5}\right) \times \left\{\ln\left[0.27\times 0.0002 + 5.74\times (5.09\times 10^4)^{-0.9}\right]\right\}^{-2} = 904$$

The minor loss coefficient terms are calculated to be

$$\frac{K_1+K_v}{2gA_1^2} = \frac{0.5+2}{2\times 9.81\times[\pi/4\times 0.20^2]^2} = 129.1$$

$$\frac{2K_e+K_2}{2gA_2^2} = \frac{2\times 0.25+1}{2\times 9.81\times[\pi/4\times 0.25^2]^2} = 31.7$$

Substitute these values into the energy equation and obtain

$$0 + 10 + H_P = \frac{200 \times 10^3}{0.85 \times 9800} + 20 + (53.4 + 129.1 + 904 + 31.7)0.1^2$$

This relation reduces to

$$10 + H_P = 24 + 20 + 11.2$$

Solving for the head across the pump gives $H_p = 45.2$ m. The required input power is

$$\begin{aligned}\dot{W}_P &= \frac{\gamma Q H_P}{\eta} \\ &= \frac{(9800 \times 0.85) \times 0.10 \times 45.2}{0.75} \\ &= 5.0 \times 10^4 \text{ W} \quad \text{or} \quad \underline{50 \text{ kW}}\end{aligned}$$

9.3.2 Parallel Piping

A parallel piping arrangement is shown in Figure 9.4; it is essentially an arrangement of N pipe elements joined at A and B with ΣK minor loss components associated with each pipe element i. The continuity equation applied at either location A or B is given by

$$Q = \sum_{i=1}^{N} Q_i \tag{9.3.4}$$

The algebraic sum of the energy grade line around any defined loop must be zero. As in the case of series piping, it is customary to assume that $V^2/2g << (p/\gamma + z)$. Hence, for any pipe element i, the energy equation from location A to B is

$$\left(\frac{p}{\gamma} + z\right)_A - \left(\frac{p}{\gamma} + z\right)_B = \left(R_i + \frac{\Sigma K}{2gA_i^2}\right)Q_i^2 \qquad i = 1, \ldots, N \tag{9.3.5}$$

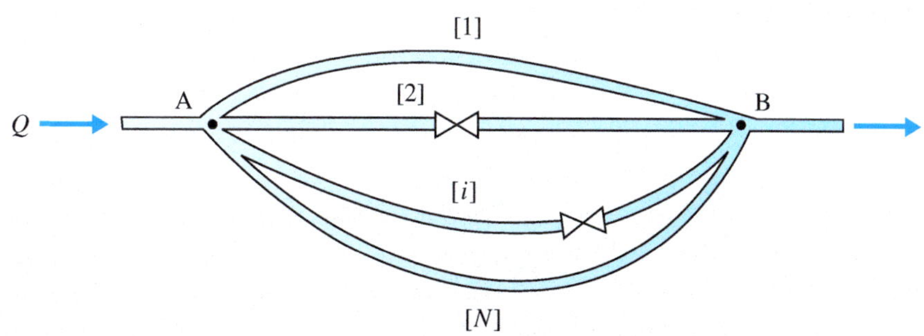

Figure 9.4 Parallel piping system.

The unknowns in Eqs. 9.3.4 and 9.3.5 are the discharges Q_i and the difference in piezometric head between A and B; the discharge Q into the system is assumed known. It is possible to convert the minor loss terms using an equivalent length as defined in Section 5.6.5. For each pipe element i the equivalent length L_e for ΣK minor loss components is

$$(L_e)_i = \frac{D_i}{f_i}\Sigma K \tag{9.3.6}$$

Thus, Eq. 9.3.5 simplifies to the form

$$\left(\frac{P}{\gamma}+z\right)_{\mathrm{A}} - \left(\frac{P}{\gamma}+z\right)_{\mathrm{B}} = \bar{R}_i Q_i^2 \tag{9.3.7}$$

in which the modified pipe resistance coefficient $\bar{R}_i$ is given by

$$\bar{R}_i = \frac{8f_i[L_i + (L_e)_i]}{g\pi^2 D_i^5} \tag{9.3.8}$$

Note that the right-hand side of Eq. 9.3.8 is equivalent to the term $R_i + \Sigma K/(2gA_i^2)$, where Eq. 9.2.2 expresses R_i.

A solution employing the method of successive substitution is developed in the following manner. Define the variable W to be the change in the hydraulic grade line between A and B; that is, $W = (p/\gamma + z)_{\mathrm{A}} - (p/\gamma + z)_{\mathrm{B}}$. Then Eq. 9.3.7 can be solved for Q_i in terms of W as

$$Q_i = \sqrt{\frac{W}{\bar{R}_i}} \tag{9.3.9}$$

Equations 9.3.4 and 9.3.9 are combined to eliminate the unknown discharges Q_i, resulting in

$$Q = \sum_{i=1}^{N}\sqrt{\frac{W}{\bar{R}_i}} = \sqrt{W}\sum_{i=1}^{N}\frac{1}{\sqrt{\bar{R}_i}} \tag{9.3.10}$$

The remaining unknown W is taken out of the summation sign since it is the same in all pipes. Solving for W in Eq. 9.3.10 results in

$$W = \left(\frac{Q}{\displaystyle\sum_{i=1}^{N}\frac{1}{\sqrt{\bar{R}_i}}}\right)^2 \tag{9.3.11}$$

An iterative procedure can be formulated to solve for W and the discharges Q_i, as follows:

1. Assume flows in each line to be in the completely turbulent regime, and compute an initial estimate of the friction factors in each line using Eq. 9.2.6.
2. Compute $\overline{R}_i$ for each pipe and evaluate W with Eq. 9.3.11.
3. Compute Q_i in each pipe with Eq. 9.3.9.
4. Update the estimates of the friction factors in each line using the current values of Q_i and Eq. 9.2.3.
5. Repeat steps 2 to 4 until the unknowns W and Q_i do not vary within a desired tolerance.

Note that if friction factors are in the completely turbulent regime so that they are independent of the discharge and therefore constant, steps 4 and 5 are unnecessary and a solution results on the first iteration. The technique is illustrated in Example 9.3.

Example 9.3

Find the distribution of flow and the drop in hydraulic grade line for the three-parallel-pipe arrangement shown in Figure E9.3. Use variable friction factors with $\nu = 10^{-6}$ m^2/s. The total water discharge is $Q = 0.020$ m^2/s.

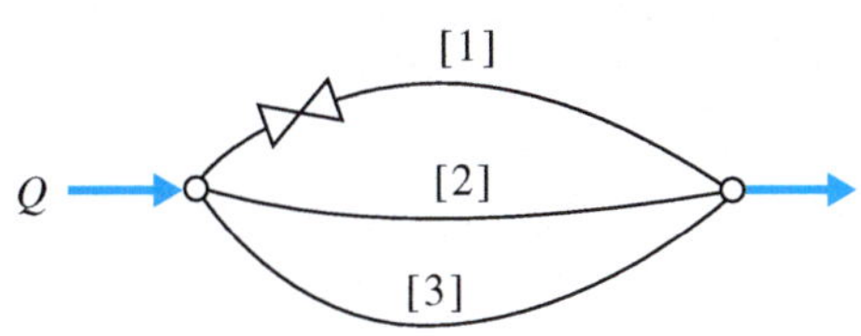

Figure E9.3

Pipe	L (m)	D (m)	e (mm)	ΣK
1	100	0.05	0.1	10
2	150	0.075	0.2	3
3	200	0.085	0.1	2

Solution

The initial estimates of f are based on Eq. 9.2.6. Preliminary calculations provide the following:

Pipe	e/D	f (Eq. 9.2.6)	L_e (Eq. 9.3.6)	$\overline{R}$ (Eq. 9.3.8)
1	0.002	0.023	21.7	7.40×10^5
2	0.0027	0.025	9.0	1.38×10^5
3	0.0012	0.021	8.1	8.14×10^4

Apply Eq. 9.3.11, and the first estimate of the drop W in the hydraulic grade line is

$$W = \left[\frac{0.020}{(7.40 \times 10^5)^{-1/2} + (1.38 \times 10^5)^{-1/2} + (8.14 \times 10^4)^{-1/2}} \right]^2$$
$$= 7.39 \text{ m}$$

Then, with Eq. 9.3.9, estimates of Q_i, are made:

$$Q_1 = \left(\frac{7.39}{7.40 \times 10^5} \right)^{1/2} = 0.00316\ \text{m}^3/\text{s}$$

$$Q_2 = \left(\frac{7.39}{1.38 \times 10^5} \right)^{1/2} = 0.00732\ \text{m}^3/\text{s}$$

$$Q_3 = \left(\frac{7.39}{8.14 \times 10^4} \right)^{1/2} = 0.00953\ \text{m}^3/\text{s}$$

A continuity check is made using Eq. 9.3.4:

$$\sum_{i=1}^{3} Q_i = 0.00316 + 0.00732 + 0.00953 = 0.0200\ \text{m}^3/\text{s}$$

Even though the sum satisfies the solution, another iteration will be carried out to study the convergence of the solution technique. First, the $\bar{R}$-values are updated using Eq. 9.2.3 to evaluate the friction factors.

Pipe	$\text{Re} = \dfrac{4Q}{\pi D \nu}$	f (Eq. 9.2.3)	$\bar{R}$ (Eq. 9.3.8)
1	8.05×10^4	0.026	8.20×10^5
2	1.24×10^5	0.027	1.49×10^5
3	1.43×10^5	0.022	8.51×10^4

Evaluating W using Eq. 9.3.11 yields $W = \underline{7.88\ \text{m}}$, and application of Eq. 9.3.9 gives the new estimates of Q_i:

$$Q_1 = \underline{0.00310\ \text{m}^3/\text{s}} \qquad Q_2 = \underline{0.00727\ \text{m}^3/\text{s}} \qquad Q_3 = \underline{0.00962\ \text{m}^3/\text{s}}$$

A continuity check shows that

$$\sum_{i=1}^{3} Q_i = 0.0200\ \text{m}^3/\text{s}$$

which is the same as in the first iteration.

9.3.3 Branch Piping

The branching network, illustrated in Figure 9.5a, is made up of three elements connected to a single junction. In contrast to the parallel system shown in Figure 9.4, no closed loops exist. In the analysis, one assumes the direction of flow in each element; then the energy equation for each element is written using an equivalent length to account for minor losses:

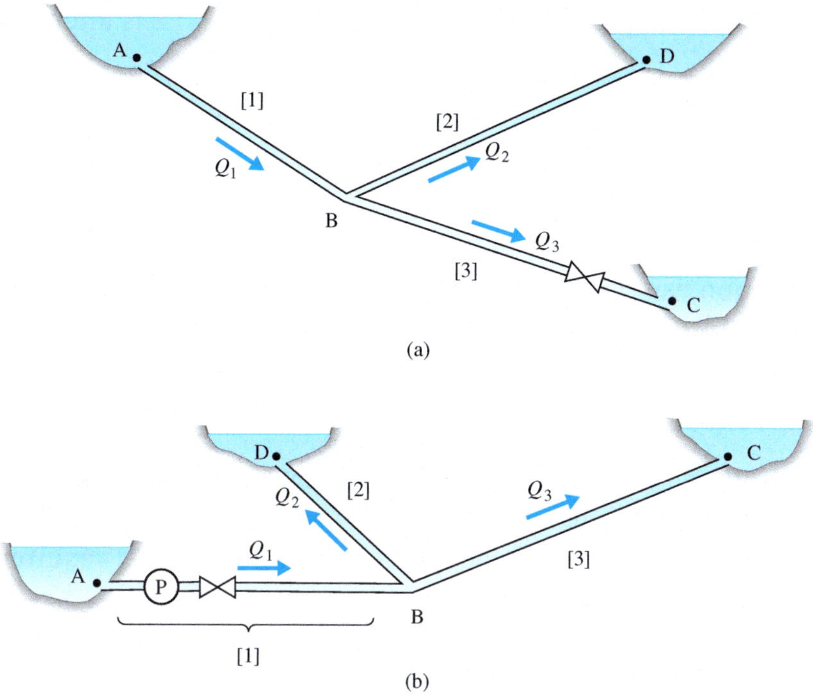

Figure 9.5 Branch piping systems: (a) gravity flow; (b) pump-driven flow.

$$\left(\frac{p}{\gamma}+z\right)_{\mathrm{A}}-\left(\frac{p}{\gamma}+z\right)_{\mathrm{B}}=\bar{R}_1Q_1^2 \tag{9.3.12}$$

$$\left(\frac{p}{\gamma}+z\right)_{\mathrm{B}}-\left(\frac{p}{\gamma}+z\right)_{\mathrm{C}}=\bar{R}_2Q_2^2 \tag{9.3.13}$$

$$\left(\frac{p}{\gamma}+z\right)_{\mathrm{B}}-\left(\frac{p}{\gamma}+z\right)_{\mathrm{D}}=\bar{R}_3Q_3^2 \tag{9.3.14}$$

The piezometric heads at locations A, C, and D are considered known. The unknowns are the piezometric head at B and the discharges Q_1, Q_2, and Q_3. The additional relation is the continuity balance at location B, which is

$$Q_1-Q_2-Q_3=0 \tag{9.3.15}$$

Thus, there are four equations with four unknowns. One convenient ad hoc method of solution is now outlined and illustrated in Example 9.4:

1. Assume a discharge Q_1 in element 1 (with or without a pump). Establish the piezometric head H at the junction by solving Eq. 9.3.12 (a Category 1 problem).
2. Compute the discharge Q_i in the remaining branches using Eqs. 9.3.13 and 9.3.14 (Category 2 problems).
3. Substitute the Q_i into Eq. 9.3.15 to check for continuity balance. Generally, the flow imbalance at the junction ΔQ will be nonzero. In Eq. 9.3.15, $\Delta Q = Q_1 - Q_2 - Q_3$.
4. Adjust the flow Q_1 in element 1 and repeat steps 2 and 3 until ΔQ is within desired limits.

If a pump exists in pipe 1 (see Figure 9.5b), Eq. 9.3.12 is altered in the manner

$$\left(\frac{p}{\gamma}+z\right)_A-\left(\frac{p}{\gamma}+z\right)_B+H_P=\bar{R}_1Q_1^2 \tag{9.3.16}$$

An additional unknown, namely, the pump head H_P, is introduced. The additional necessary relationship is the head-discharge curve for the pump; see, for example, Figure 5.18. The solution can proceed in a manner similar to that described in Example 9.5. It may be convenient to follow the solution graphically by plotting the assumed discharge in line 1 versus either the piezometric head at B or the imbalance of flow at B. Such a step is helpful to determine in what manner the discharge in line 1 should be altered for the next iteration.

An alternative method of solution for a single branching system is to eliminate all of the variables except the piezometric head H ($= p/\gamma + z$) at the junction, location B in Figure 9.5. Then a numerical solving technique can be employed. An additional requirement is to assume the direction of flow in each pipe. It may become necessary to correct the sign in one or more of the equations if during the solution, H moves from below one of the reservoir elevations to above it, or vice versa. Example 9.4 illustrates the procedure.

Examples 9.4 and 9.5 represent a level of complexity that one may not want to exceed when employing a calculator-based solution. To include additional pumps, reservoirs or pipes would make either ad hoc or simplified numerical approaches too cumbersome. For such systems, the more generalized network analysis described in Section 9.4 is recommended.

Example 9.4

For the three-branch piping system shown in Figure E9.4 we have the following data:

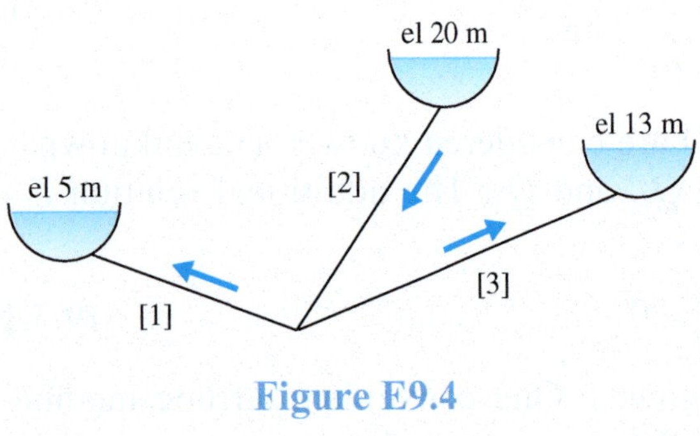

Figure E9.4

Pipe	L (m)	D (m)	f	ΣK
1	500	0.10	0.025	3
2	750	0.15	0.020	2
3	1000	0.13	0.018	7

Determine the flow rates Q_i and the piezometric head H at the junction. Assume constant friction factors.

Solution

The equivalent lengths and resistance coefficients are

$$(L_e)_1=\frac{0.10}{0.025}\times 3=12\text{ m}\qquad \bar{R}_1=\frac{8\times 0.025\times 512}{9.81\times\pi^2\times(0.10)^5}=1.06\times 10^5\text{ s}^2/\text{m}^5$$

$$(L_e)_2 = \frac{0.15}{0.020} \times 2 = 15 \text{ m} \qquad \bar{R}_2 = \frac{8 \times 0.020 \times 765}{9.81 \times \pi^2 \times (0.15)^5} = 1.66 \times 10^4 \text{ s}^2/\text{m}^5$$

$$(L_e)_3 = \frac{0.13}{0.018} \times 7 = 51 \text{ m} \qquad \bar{R}_3 = \frac{8 \times 0.018 \times 1051}{9.81 \times \pi^2 \times (0.13)^5} = 4.21 \times 10^4 \text{ s}^2/\text{m}^5$$

With the flow directions assumed as shown, the energy equation is written for each pipe and solved for the unknown discharge:

$$Q_1 = \left(\frac{H-5}{\bar{R}_1}\right)^{1/2} \qquad Q_2 = \left(\frac{20-H}{\bar{R}_2}\right)^{1/2} \qquad Q_3 = \left(\frac{H-13}{\bar{R}_3}\right)^{1/2}$$

The continuity equation is $-Q_1 + Q_2 - Q_3 = 0$. Eliminating Q_1, Q_2, and Q_3 with the energy relations results in an algebraic equation in terms of H:

$$w(H) = -\left(\frac{H-5}{1.06 \times 10^5}\right)^{1/2} + \left(\frac{20-H}{1.66 \times 10^4}\right)^{1/2} - \left(\frac{H-13}{4.21 \times 10^4}\right)^{1/2} = 0$$

Even though this can be solved as a quadratic, the method of false position is chosen to compute H, which would be required if the friction factors varied. The procedure was presented in Example 8.11. In the present example the recurrence formula is

$$H_r = \frac{H_l w(H_u) - H_u w(H_l)}{w(H_u) - w(H_l)}$$

The solution is shown in the table below. Note that with the initial guesses of H_l and H_u, the sign conventions in w require that $20 > H > 13$. Iteration continues until the convergence criterion shown in the last column becomes less than the arbitrary value of 0.005.

Iteration	H_u	H_l	$w(H_u)$	$w(H_l)$	H_r	$w(H_r)$	*Sign of* $w(H_l) \times w(H_r)$	$\varepsilon = \left\lvert \frac{H_r^{\text{new}} - H_r^{\text{old}}}{H_r^{\text{old}}} \right\rvert$
1	18	13	–0.01100	0.01185	15.59	–0.00154	+	
2	15.59	13	–0.00154	0.01185	15.29	–0.000384	+	0.019
3	15.29	13	–0.000384	0.01185	15.22	–0.000112	+	0.0046

Hence, $H \simeq \underline{15.2 \text{ m}}$. The discharges are now computed:

$$Q_1 = \left(\frac{15.2-5}{1.06 \times 10^5}\right)^{1/2} = \underline{0.0098 \text{ m}^3/\text{s}}$$

$$Q_2 = \left(\frac{20-15.2}{1.66 \times 10^4}\right)^{1/2} = \underline{0.0170 \text{ m}^3/\text{s}}$$

$$Q_3 = \left(\frac{15.2-13}{4.21 \times 10^4}\right)^{1/2} = \underline{0.0072 \text{ m}^3/\text{s}}$$

Note that continuity is satisfied.

Example 9.5

For the system shown in Figure E9.5, determine the flow distribution Q_i of water and the piezometric head H at the junction. The fluid power input by the pump is constant, equal to $\gamma Q H_P = 20$ kW. Assume constant friction factors.

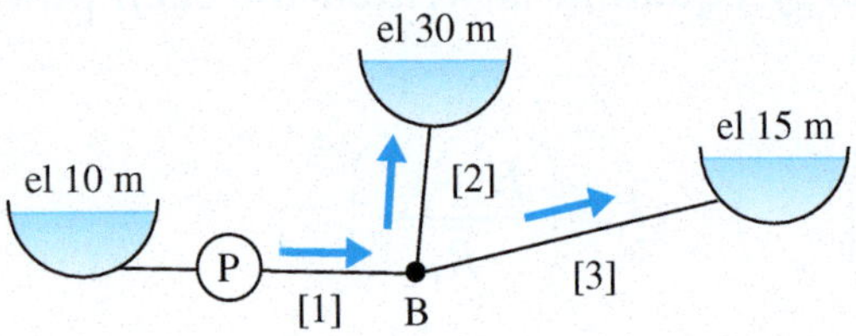

Figure E9.5

Pipe	L (m)	D (m)	f	ΣK
1	50	0.15	0.02	2
2	100	0.10	0.015	1
3	300	0.10	0.025	1

Solution

The equivalent lengths and resistance coefficients are computed from Eqs. 9.3.6 and 9.3.8 to be

$$(L_e)_1 = \frac{0.15}{0.02} \times 2 = 15 \text{ m} \qquad \bar{R}_1 = \frac{8 \times 0.02 \times 65}{9.81 \times \pi^2 \times (0.15)^5} = 1.42 \times 10^3 \text{ s}^2/\text{m}^5$$

$$(L_e)_2 = \frac{0.10}{0.015} \times 1 = 6.7 \text{ m} \qquad \bar{R}_2 = \frac{8 \times 0.015 \times 106.7}{9.81 \times \pi^2 \times (0.10)^5} = 1.32 \times 10^4 \text{ s}^2/\text{m}^5$$

$$(L_e)_3 = \frac{0.10}{0.025} \times 1 = 4 \text{ m} \qquad \bar{R}_3 = \frac{8 \times 0.025 \times 304}{9.81 \times \pi^2 \times (0.10)^5} = 6.28 \times 10^4 \text{ s}^2/\text{m}^5$$

Assume flow directions as shown. The energy equation for pipe 1 from the reservoir to the junction B is

$$z_1 + H_P = H + \bar{R}_1 Q_1^2$$

in which H is the piezometric head at B. Substituting in known parameters and solving for H results in

$$H = 10 + \frac{20 \times 10^3}{9800 Q_1} - 1.42 \times 10^3 Q_1^2$$

$$= 10 + \frac{2.04}{Q_1} - 1420 Q_1^2$$

An iterative solution is shown in the accompanying table. For each iteration, a value of Q_1 is estimated. Then H is calculated and Q_2 and Q_3 are evaluated from the relations

$$Q_2 = \left(\frac{H - z_2}{\bar{R}_2}\right)^{1/2} = \left(\frac{H - 30}{1.32 \times 10^4}\right)^{1/2}$$

$$Q_3 = \left(\frac{H - z_3}{\bar{R}_3}\right)^{1/2} = \left(\frac{H - 15}{6.28 \times 10^4}\right)^{1/2}$$

In the last column of the table, a continuity balance is employed to check the accuracy of the estimate of Q_1. The third estimate of Q_1 is based on a linear interpolation by setting $\Sigma Q = 0$ and using values of Q_1 and ΔQ from the first two iterations.

Iteration	Q_1	H	Q_2	Q_3	$\Delta Q = Q_1 - Q_2 - Q_3$
1	0.050	47.25	0.0362	0.0227	–0.0089
2	0.055	42.80	0.0311	0.0210	+0.0029
3	0.054	43.64	0.0322	0.0214	+0.0004

The approximate solution is $H = \underline{43.6 \text{ m}}$, $Q_1 = \underline{54 \text{ L/s}}$, $Q_2 = \underline{32 \text{ L/s}}$, and $Q_3 = \underline{21 \text{ L/s}}$. If greater precision is desired, a solution should be employed similar to that shown in Example 9.4.

9.4 ANALYSIS OF PIPE NETWORKS

Piping systems more complicated than those considered in Section 9.3 are best analyzed by formulating the solution for a network. Before considering a generalized set of network equations, it is worthwhile to examine a specific example of a piping network to observe the degree of complexity involved.

Figure 9.6a shows a relatively simple network consisting of seven pipes, two reservoirs, and one pump. The hydraulic grade lines at A and F are assumed known; these locations are termed **fixed-grade nodes**. Outflow demands are present at nodes C and D. Nodes C and D, along with nodes B and E are called **interior nodes** or **junctions**. Flow directions, even though not initially known, are assumed to be in the directions shown. The system equations are given as follows:

1. Energy balance for each pipe (seven equations):

$$\begin{aligned} H_A - H_B + H_P(Q_1) &= \bar{R}_1 Q_1^2 & \qquad H_C - H_E &= \bar{R}_5 Q_5^2 \\ H_B - H_D &= \bar{R}_2 Q_2^2 & H_E - H_D &= \bar{R}_6 Q_6^2 \\ H_C - H_D &= \bar{R}_3 Q_3^2 & H_F - H_E &= \bar{R}_7 Q_7^2 \\ H_B - H_C &= \bar{R}_4 Q_4^2 \end{aligned} \tag{9.4.1}$$

2. Continuity balance for each interior node (four equations):

$$\begin{aligned} Q_1 - Q_2 - Q_4 &= 0 \\ Q_2 + Q_3 + Q_6 &= Q_D \\ Q_4 - Q_3 - Q_5 &= Q_C \\ Q_5 - Q_6 + Q_7 &= 0 \end{aligned} \tag{9.4.2}$$

3. Approximation of pump curve (one equation):

$$H_P(Q_1) = a_0 + a_1 Q_1 + a_2 Q_1^2 \tag{9.4.3}$$

in which a_0, a_1, and a_2 are known constants.

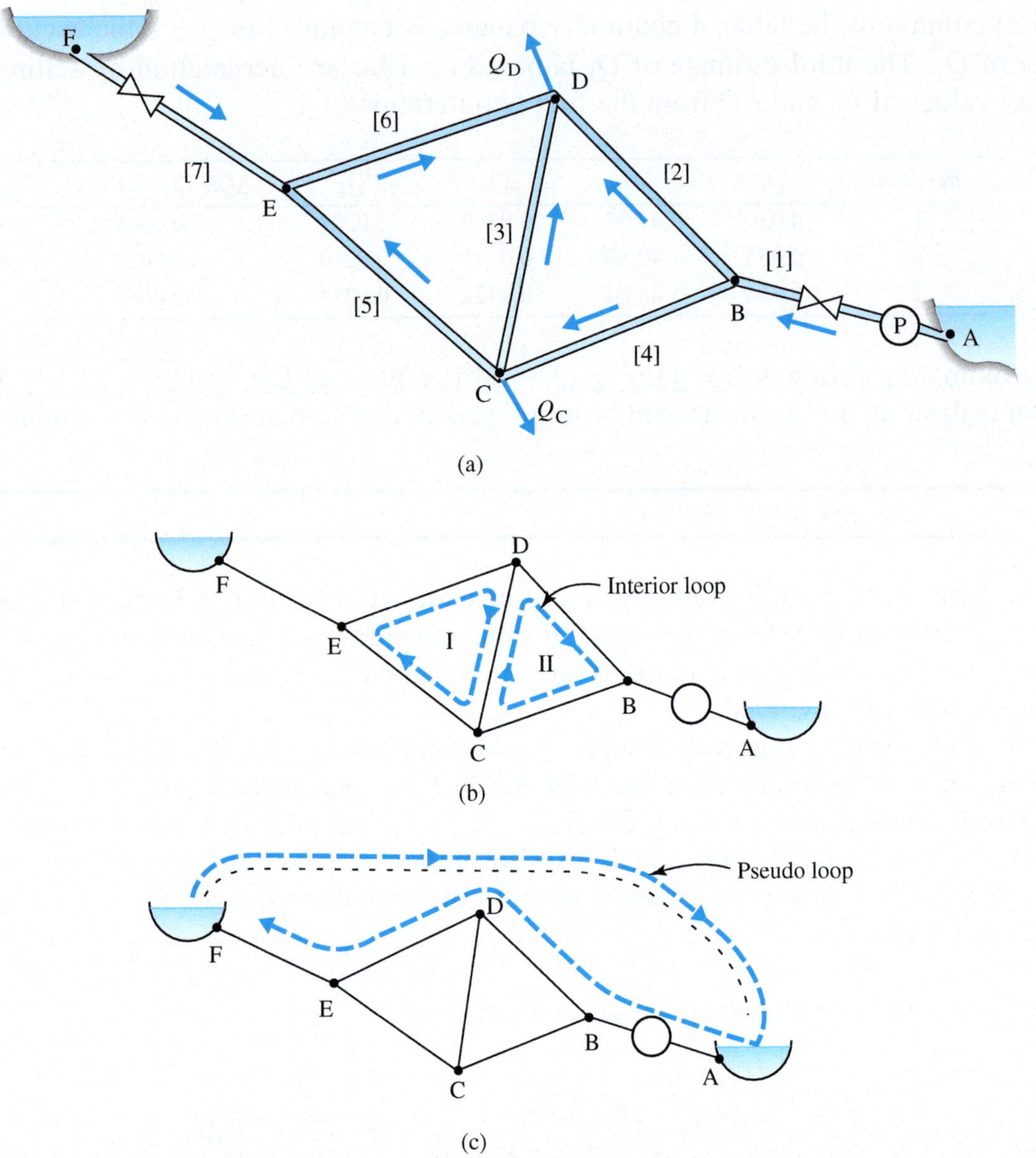

Figure 9.6 Representative piping network: (a) assumed flow directions and numbering scheme; (b) designated interior loops; (c) path between two fixed-grade nodes.

The unknowns are $Q_1, \ldots, Q_7$, H_B, H_C, H_D, H_E, and H_P. Thus there are 12 unknowns and 12 equations to solve simultaneously. Since the energy equations and the pump equation are nonlinear, it is necessary to resort to some type of approximate numerical solution. The 12 equations can be reduced in number by combining the energy equations along special paths. Let the drop in hydraulic grade line for any pipe element i be designated as W_i. Then

$$W_i = \overline{R}_i Q_1^2 \tag{9.4.4}$$

For the system under consideration, two closed paths, or **interior loops**, can be identified (see Figure 9.6b). Flow is considered positive in a clockwise sense around each loop. Energy balances, written around loops I and II, are

$$\begin{aligned} W_6 - W_3 + W_5 &= 0 \\ W_3 - W_2 + W_4 &= 0 \end{aligned} \tag{9.4.5}$$

To account for the flow in pipes 1 and 7, a path can be defined along nodes A, B, D, E, and F in Figure 9.6c. Then with the addition of the pump head, the energy balance from A to F is

$$H_A + H_P - W_1 - W_2 + W_6 + W_7 = H_F \tag{9.4.6}$$

Note that the path energy equation connects two fixed-grade nodes. Such a path is sometimes termed a **pseudoloop**, since an imaginary pipe with infinite resistance, or no flow, can be considered to connect the two reservoirs. The imaginary pipe is shown by a dotted line in Figure 9.6c. Substituting the pump equation and the friction equation into the energy relations above results in the following reduced set of equations:

$$\begin{aligned} -\bar{R}_3 Q_3^2 + \bar{R}_5 Q_5^2 + \bar{R}_6 Q_6^2 &= 0 \\ -\bar{R}_2 Q_2^2 + \bar{R}_3 Q_3^2 + \bar{R}_4 Q_4^2 &= 0 \\ -\bar{R}_1 Q_1^2 + \left(a_0 + a_1 Q_1 + a_2 Q_1^2\right) - \bar{R}_2 Q_2^2 + \bar{R}_6 Q_6^2 + \bar{R}_7 Q_7^2 + H_A - H_F &= 0 \\ Q_1 - Q_2 - Q_4 &= 0 \\ Q_2 + Q_3 + Q_6 &= Q_D \\ Q_4 - Q_3 - Q_5 &= Q_C \\ Q_5 - Q_6 + Q_7 &= 0 \end{aligned} \tag{9.4.7}$$

There are now seven unknowns $(Q_1, \ldots, Q_7)$ and seven equations to solve. The energy relations are nonlinear since the loss terms and the pump head are represented as polynomials with respect to the discharges.

9.4.1 Generalized Network Equations

Networks of piping, such as those shown in Figure 9.6, can be represented by the following equations.

1. Continuity at the j th interior node:

$$\Sigma(\pm)_j Q_j - Q_e = 0 \tag{9.4.8}$$

   in which the subscript j refers to the pipes connected to a node, and Q_e is the external demand. The algebraic plus or minus sign convention pertains to the assumed flow direction: Use the positive sign for flow into the junction, and the negative sign for flow out of the junction.

2. Energy balance around an interior loop:

$$\Sigma(\pm)_i W_i = 0 \tag{9.4.9}$$

   in which the subscript i pertains to the pipes that make up the loop. There will be a relation for each of the loops. Here it is assumed that there are no pumps located in the interior of the network. The plus sign is used if the flow in the element is positive in the clockwise sense; otherwise, the minus sign is employed.

3. Energy balance along a unique path or pseudoloop connecting two fixed-grade nodes:

$$\Sigma(\pm)_i[W_i - (H_P)_i] + \Delta H = 0 \tag{9.4.10}$$

where ΔH is the difference in magnitude of the two fixed-grade nodes in the path ordered in a clockwise fashion across the imaginary pipe in the pseudoloop. The term $(H_P)_i$ is the head across a pump that could exist in the ith pipe element. If F is the number of fixed-grade nodes, there will be $(F - 1)$ unique path equations. The plus and minus signs in Eq. 9.4.10 follow the same argument given for Eq. 9.4.9.

Let P be the number of pipe elements in the network, J the number of interior nodes, and L the number of interior loops. Then the following relation will hold if the network is properly defined:

$$P = J + L + F - 1 \tag{9.4.11}$$

In Figure 9.6 of the introductory example, $J = 4$, $L = 2$, $F = 2$, so that $P = 4 + 2 + 2 - 1 = 7$.

An additional necessary formulation is the relation between discharge and loss in each pipe; it is

$$W = RQ^\beta + \frac{\Sigma K}{2gA^2}Q^2 \tag{9.4.12}$$

If the minor losses can be defined in terms of an equivalent length, Eq. 9.4.12 can be replaced by[2]

$$W = \overline{R}Q^\beta \tag{9.4.13}$$

An approximate pump head-discharge representation is given by the polynomial

$$H_P(Q) = a_0 + a_1Q + a_2Q^2 \tag{9.4.14}$$

The coefficients a_0, a_1, and a_2 are assumed known; typically, they can be found by substituting three known data points from a specified pump curve and solving the three resulting equations simultaneously. In place of defining the pump head-discharge curve, an alternative means of including a pump in a line is to specify the useful power the pump puts into the system. The useful, or actual, power $\dot{W}_f$ is assumed to be constant and allows H_P to be represented in the manner

$$H_p(Q) = \frac{\dot{W}_f}{\gamma Q} \tag{9.4.15}$$

---

[2]In Eq. 9.4.13, the exponent β is used, since the Hazen–Williams formula is commonly employed in pipe network analysis. The equivalent length for use in the Hazen–Williams formulation is given by

$$L_e = 0.8106\frac{\Sigma K}{K_1}D^{0.87}C^{1.85}Q^{0.15}$$

Note that it is necessary to estimate the discharge when using this equation. The Darcy–Weisbach equation with $\beta = 2$ is recommended. An alternative to using the equivalent-length concept is to account for pipe friction and minor losses separately in each line, making use of Eq. 9.4.12.

This equation is particularly useful when the specific operating characteristics of a pump are unknown.

9.4.2 Linearization of System Energy Equations

Equation 9.4.10 is a general relation that can be applied to any path or closed loop in a network. If it is applied to a closed loop, ΔH is set equal to zero, and if no pump exists in the path or loop, $(H_P)_i$ is equal to zero. Note that Eq. 9.4.9 can be considered to be a subset of Eq. 9.4.10. In the following development, Eq. 9.4.10 will be used to represent any loop or path in the network.

Define the function $\phi(Q)$ to contain the nonlinear terms $W(Q)$ and $H_P(Q)$ in the form

$$\begin{aligned}\phi(Q) &= W(Q) - H_P(Q)\\ &= RQ^{\beta} - H_P(Q)\end{aligned} \tag{9.4.16}$$

Equation 9.4.16 can be expanded in a Taylor series as

$$\phi(Q) = \phi(Q_0) + \left.\frac{d\phi}{dQ}\right|_{Q_0}(Q - Q_0) + \left.\frac{d^2\phi}{dQ^2}\right|_{Q_0}\frac{(Q - Q_0)^2}{2} + \cdots \tag{9.4.17}$$

in which Q_0 is an estimate of Q. To approximate $\phi(Q)$ accurately, Q_0 should be chosen so that the difference $(Q - Q_0)$ is numerically small. Retaining the first two terms on the right-hand side of Eq. 9.4.17, and using Eq. 9.4.16, yields

$$\phi(Q) \simeq \bar{R}Q_0^{\beta} - H_P(Q_0) + \left[\beta\bar{R}Q_0^{\beta-1} - \left.\frac{dH_P}{dQ}\right|_{Q_0}\right](Q - Q_0) \tag{9.4.18}$$

Note that the approximation to $\phi(Q)$ is now linear with respect to Q.

The parameter G is introduced as

$$G = \beta\bar{R}Q_0^{\beta-1} - \left.\frac{dH_P}{dQ}\right|_{Q_0} \tag{9.4.19}$$

Using Eq. 9.4.14 to represent the pump head. Eq. 9.4.19 becomes

$$G = \beta\bar{R}Q_0^{\beta-1} - (a_1 + 2a_2Q_0) \tag{9.4.20}$$

Alternatively, with Eq. 9.4.15 substituted into Eq. 9.4.19, one has

$$G = \beta\bar{R}Q_0^{\beta-1} + \frac{\dot{W}_f}{\gamma Q_0^2} \tag{9.4.21}$$

Substituting Eq. 9.4.19 into Eq. 9.4.18 gives

$$\begin{aligned}\phi(Q) &= \overline{R}Q_0^{\beta} - H_P(Q_0) + (Q - Q_0)G \\ &= W_0 - H_{P0} + (Q - Q_0)G\end{aligned} \tag{9.4.22}$$

in which $W_0 = W(Q_0)$ and $H_{P0} = H_P(Q_0)$. Finally, Eq. 9.4.22 is substituted into Eq. 9.4.10 to produce the linearized loop or path energy equation

$$\Sigma(\pm)_i[(W_0)_i - (H_{P0})_i] + \Sigma[Q_i - (Q_0)_i]G_i + \Delta H = 0 \tag{9.4.23}$$

The second term does not contain the plus or minus sign since G_i is a monotonically increasing function of the flow correction. Since Q in the relations above can assume positive or negative values, Q_0^{β} and $Q_0^{\beta-1}$ are often replaced by $Q_0|Q_0|^{\beta-1}$ and $|Q_0|^{\beta-1}$, respectively, in solution algorithms. This is done, for example, in Eq. 9.4.23, which forms the basis for the Hardy Cross solution outlined below.

9.4.3 Hardy Cross Method

In Eq. 9.4.23, let $(Q_0)_i$ be the estimates of discharge from the previous iteration, and let Q_i be the new estimates of the discharge. Define a flow adjustment ΔQ for each loop to be

$$\Delta Q = Q_i - (Q_0)_i \tag{9.4.24}$$

The adjustment is applied independently to all pipes in a given loop. Hence, Eq. 9.4.23 can be written as

$$\Sigma(\pm)_i[(W_0)_i - (H_{P0})_i] + \Delta Q \Sigma G_i + \Delta H = 0 \tag{9.4.25}$$

Solving for ΔQ, one has

$$\Delta Q = \frac{-\Sigma(\pm)_i[(W_0)_i - (H_{P0})_i] - \Delta H}{\Sigma G_i} \tag{9.4.26}$$

It is necessary for the algebraic sign of Q to be positive in the direction of normal pump operation; otherwise, the pump curve will not be represented properly and Eq. 9.4.26 will be invalid. Furthermore, it is important that the discharge Q through the pump remain within the limits of the data used to generate the curve.

For a closed loop in which no pumps or fixed-grade nodes are present, Eq. 9.4.26 reduces simply to the form

$$\Delta Q = \frac{-\Sigma(\pm)_i(W_0)_i}{\Sigma G_i} \tag{9.4.27}$$

The Hardy Cross iterative solution is outlined in the following steps:

1. Assume an initial estimate of the flow distribution in the network that satisfies continuity, Eq. 9.4.8. The closer the initial estimates are to the correct values, the fewer will be the iterations required for convergence. One guideline to use is that in a pipe element, as $\overline{R}$ increases, Q decreases.
2. For each loop or path, evaluate ΔQ with Eq. 9.4.26 or 9.4.27. The numerators should approach zero as the loops or paths become balanced.
3. Update the flows in each pipe in all loops and paths, that is, from Eq. 9.4.24,

$$Q_i = (Q_0)_i + \Sigma\Delta Q \tag{9.4.28}$$

The term $\Sigma\Delta Q$ is used, since a given pipe may belong to more than one loop; hence the correction will be the sum of corrections from all loops to which the pipe is common.
4. Repeat steps 2 and 3 until a desired accuracy is attained. One possible criterion to use is

$$\frac{\Sigma\left|Q_i - (Q_0)_i\right|}{\Sigma\left|Q_i\right|} \le \varepsilon \tag{9.4.29}$$

in which ε is an arbitrarily small number. Typically, $0.001 < \varepsilon < 0.005$.

The Hardy Cross method of analysis is a simplified version of the method of successive approximations applied to a set of linearized equations. It does not require the inversion of a matrix; hence it can be used to solve relatively small networks using either a calculator or a spreadsheet algorithm on a personal computer. The continuity relations, Eq. 9.4.8, are in a sense "decoupled" from the solution of the energy relations, Eq. 9.4.26. Continuity is satisfied initially with assumed flows and remains satisfied throughout the solution process. Essentially, one computes a correction ΔQ to the flows Q_i in each loop separately, and then applies ΔQ to the entire network to bring flow through the loops into closer balance. In effect this is a superposition type of solution. Since the flow corrections ΔQ are applied to each loop independently, convergence may not be rapid.

Example 9.6

For the piping system shown in Figure E9.6a, determine the flow distribution and piezometric heads at the junctions using the Hardy Cross method of solution. Assume that losses are proportional to Q^2.

Solution

There are five junctions ($J = 5$), eight pipes ($P = 8$), and two fixed-grade nodes ($F = 2$). Hence, the number of closed loops is $L = 8 - 5 - 2 + 1 = 2$, plus one pseudoloop. The three loops and assumed flow directions are shown in Figure E9.6b. Equation 9.4.26 is applied to loop I:

$$\Delta Q_1 = \frac{-(\pm W_4 \pm W_3 \pm W_2 \pm W_1) - (z_A - z_B)}{G_4 + G_3 + G_2 + G_1}$$

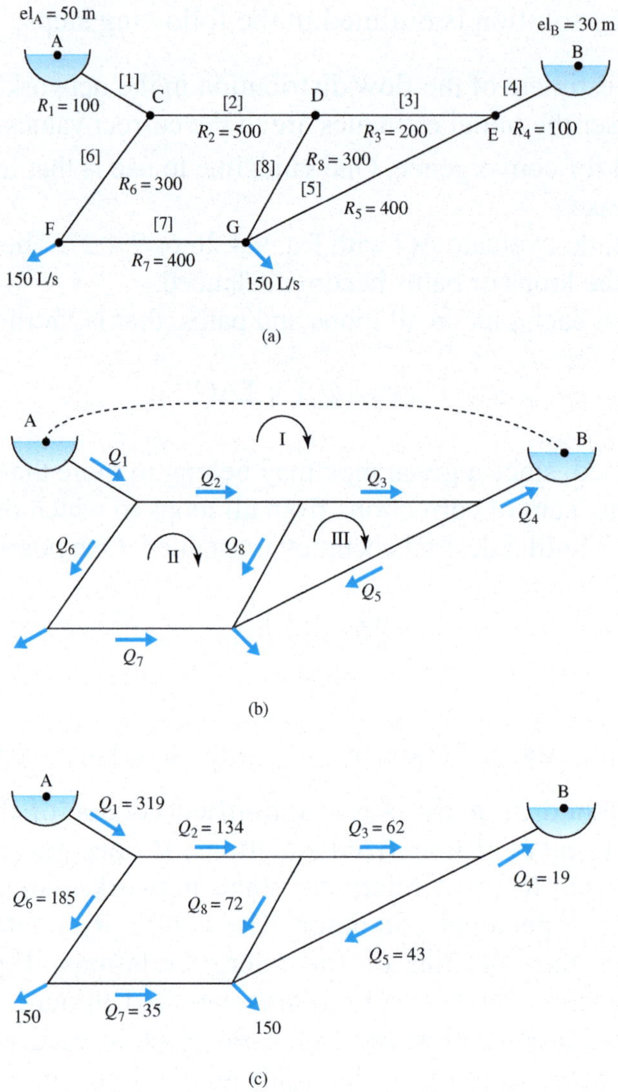

Figure E9.6

Apply Eq. 9.4.27 to loops II and III:

$$\Delta Q_{\text{II}} = \frac{-(\pm W_2 \pm W_8 \pm W_7 \pm W_6)}{G_2 + G_8 + G_7 + G_6}$$

$$\Delta Q_{\text{III}} = \frac{-(\pm W_3 + W_5 \pm W_8)}{G_3 + G_5 + G_8}$$

The problem is solved using an Excel spreadsheet, and the solution is shown in Table E9.6. In the loop equations note that W and G have the correct sign attributed automatically using the relations $\pm W = \overline{R}Q|Q|$ and $G = 2\overline{R}|Q|$. In addition, the initial values of each Q take on a positive or negative sign depending on the assumed flow direction relative to the positive clockwise direction for each loop. The initial flow estimates, shown under the heading "Iteration 1,"

Table E9.6 The Excel® Solution for Example 9.6

				Iteration 1			Iteration 2			Iteration 3			Iteration 4		
		R	Q	RQ|Q|	2R|Q|	Q	RQ|Q|	2R|Q|	Q	RQ|Q|	2R|Q|	Q	RQ|Q|	2R|Q|	Q
	ΔH			20.000			20.000			20.000			20.000		
	Pipe4	100	–0.020	–0.040	4.000	–0.022	–0.051	4.495	–0.019	–0.038	3.899	–0.020	–0.039	3.968	–0.019
Loop 1	Pipe3	200	–0.060	–0.720	24.000	–0.064	–0.809	25.440	–0.062	–0.770	24.817	–0.063	–0.787	25.098	–0.062
	Pipe2	500	–0.130	–8.450	130.000	–0.137	–9.433	137.352	–0.133	-8.907	133.466	–0.135	-9.150	135.276	–0.134
	Pipe1	100	–0.320	–10.240	64.000	–0.322	–10.399	64.495	–0.319	–10.208	63.899	–0.320	–10.230	63.968	–0.319
				0.550	222.000		–0.691	231.783		0.078	226.081		–0.206	228.309	
				ΔQ = –2.48E-03			ΔQ = 2.98E-03			ΔQ = –3.44E-04			ΔQ = 9.03E-04		
	Pipe2	500	0.130	8.450	130.000	0.137	9.433	137.352	0.113	8.907	133.466	0.135	9.150	135.276	0.134
Loop 2	Pipe8	300	0.070	1.470	42.000	0.074	1.632	44.251	0.071	1.530	42.853	0.073	1.578	43.519	0.072
	Pipe7	400	–0.040	–0.640	32.000	–0.035	–0.494	28.101	–0.036	–0.519	28.823	–0.035	–0.478	27.650	–0.035
	Pipe6	300	–0.190	–10.830	114.000	–0.185	–10.281	111.075	–0.186	–10.382	111.617	–0.185	–10.219	110.738	–0.185
				–1.550	318.000		0.290	320.779		–0.464	316.759		0.031	317.183	
				ΔQ = 4.87E-03			ΔQ = –9.03E-04			ΔQ = 1.47E-03			ΔQ = –9.83E-05		
	Pipe3	200	0.060	0.720	24.000	0.064	0.809	25.440	0.062	0.770	24.817	0.063	0.787	25.098	0.062
Loop 3	Pipe5	400	0.040	0.640	32.000	0.041	0.676	32.898	0.043	0.724	34.039	0.043	0.736	34.325	0.043
	Pipe8	300	–0.070	–1.470	42.000	–0.074	–1.632	44.251	–0.071	–1.530	42.853	–0.073	–1.578	43.519	–0.072
				–0.110	98.000		–0.146	102.589		–0.036	101.710		–0.054	102.941	
				ΔQ = 1.12E-03			ΔQ = 1.43E-03			ΔQ = 3.57E-04			ΔQ = 52.29E-04		

are chosen to satisfy continuity. Note that flow adjustments to each pipe element are made after all of the ΔQ's have been computed; for example, relative to loop I, $Q_2 = (Q_0)_2 + \Delta Q_{\text{I}} - \Delta Q_{\text{II}}$. After four iterations the absolute magnitude of the ΔQ's are all less than 0.001, and the left-hand side of Eq. 9.4.29 is 0.0051.

The values of Q_i after four iterations are shown in Figure E9.6a, along with the final flow directions. Flow rates are in liters per second. The piezometric heads are evaluated by computing the energy drop along designated paths, beginning at a known fixed-grade node, in this case, node A:

$$H_{\text{C}} = H_{\text{A}} - R_1 Q_1^2 = 50 - 100(0.319)^2 = 39.8 \text{ m}$$
$$H_{\text{D}} = H_{\text{C}} - R_2 Q_2^2 = 39.8 - 500(0.134)^2 = 30.8 \text{ m}$$
$$H_{\text{E}} = H_{\text{D}} - R_3 Q_3^2 = 30.8 - 200(0.062)^2 = 30.0 \text{ m}$$
$$H_{\text{F}} = H_{\text{C}} - R_6 Q_6^2 = 39.8 - 300(0.185)^2 = 29.5 \text{ m}$$
$$H_{\text{G}} = H_{\text{D}} - R_8 Q_8^2 = 30.8 - 300(0.072)^2 = 29.2 \text{ m}$$

Note that there is negligible loss in element four.

9.4.4 Network Analysis Using Generalized Computer Software

Spreadsheets and computational software such as Mathcad or MATLAB can be used to solve for the flow distribution in small networks, as illustrated in Example 9.6. However, for larger networks that contain more than several loops, the task of programming becomes cumbersome, especially since a unique description is required for each system accompanied by a large number of equations to solve. An advantage of using these methods of solution is that the user is applying the fundamental relations directly and can easily discern convergence of the solution. On the other hand, generalized computer codes have robust solution algorithms that allow a wide variety of network systems to be analyzed, and provide user friendly input/output schemes. They have become an indispensable tool for many engineers and practitioners in the water-supply industry, as well as finding use in other industrial applications. However, the user must become familiar with the code so that the correctness of the solution is assured.

With the advancement of software development and the increase of personal computer memory and computational speed, it is now possible to analyze large, highly complex networks with relative ease. In addition to the Hardy Cross method outlined in Section 9.4.3, other linearized methods have been used in computer codes. Herein is described briefly the use of the program EPANET developed by the United States Environmental Protection Agency.

EPANET is a comprehensive program that simulates both hydraulic flow and water quality in pressurized piping networks. For the hydraulic analysis, it uses a hybrid node-loop algorithm termed the **gradient method** to solve for the unknown piezometric heads and discharges. The network can contain pipes, pipe junctions, pumps, and various types of valves, reservoirs, and storage water tanks. Friction head loss is computed using the Darcy–Weisbach, Hazen–Williams, or Chezy–Manning formulas. Either pump curve data or useful power can be accommodated. Further capabilities of the program, a comprehensive description, and details of running the code are provided in the EPANET Users Manual (Rossman, 2000). The source code and users manual is available from the EPA Web site *www.epa.gov*.

Example 9.7

Rework Example 9.6 using EPANET.

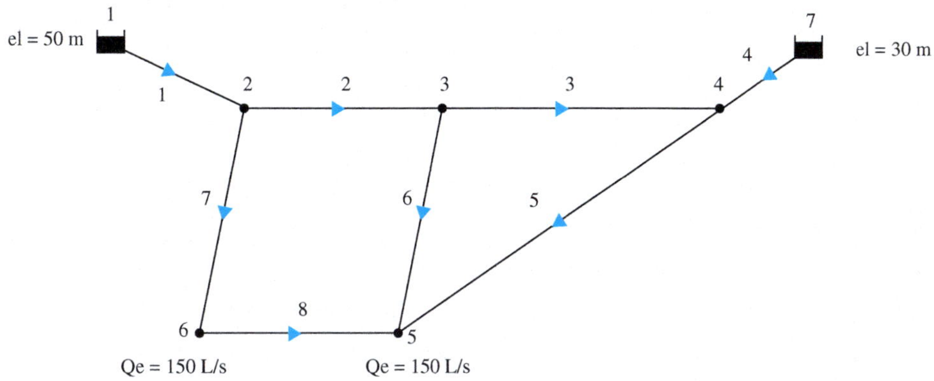

Figure E9.7

Solution

The user-defined network map is shown in Figure E9.7, and a partial listing of the computed results are provided in Table E9.7. For coding details refer to the EPANET Users Manual.

Table E9.7

Link – Node Table:

Link ID	Start Node	End Node	Length m	Diameter mm
1	1	2	66	250
2	2	3	330	250
3	3	4	130	250
4	4	7	66	250
5	4	5	260	250
6	3	5	200	250
7	2	6	200	250
8	6	5	260	250

Node Results:

Node ID	Demand LPS	Head m	Pressure m	Quality
2	0.00	40.45	40.45	0.00
3	0.00	30.96	30.96	0.00
4	0.00	30.05	30.05	0.00
5	150.00	29.16	29.16	0.00
6	150.00	29.82	29.82	0.00
1	–319.72	50.00	0.00	0.00 Reservoir
7	19.72	30.00	0.00	0.00 Reservoir

Link Results:

Link ID	Flow LPS	Velocity m/s	Headloss m/km	Status
1	319.72	6.52	144.71	Open
2	133.59	2.72	28.75	Open
3	62.19	1.27	6.98	Open
4	19.72	0.40	0.83	Open
5	42.47	0.87	3.44	Open
6	71.40	1.46	9.01	Open
7	186.12	3.79	53.13	Open
8	36.12	0.74	2.55	Open

In addition to Example 9.7, solutions of the networks shown in Figures 9.7a and 9.7b are provided in Appendix E, Figures E.1 and E.2. For the branching or tree-like system shown in Figure 9.7a, the Hazen–Williams friction formula is used with the coefficient $C = 130$.

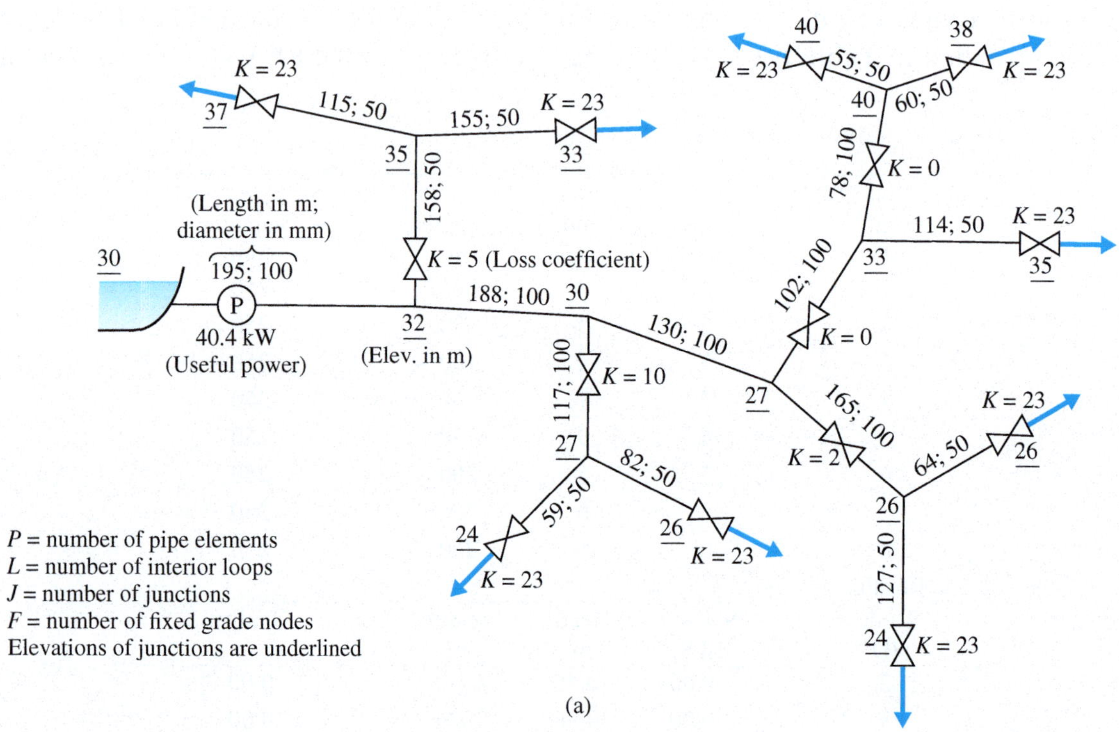

(a)

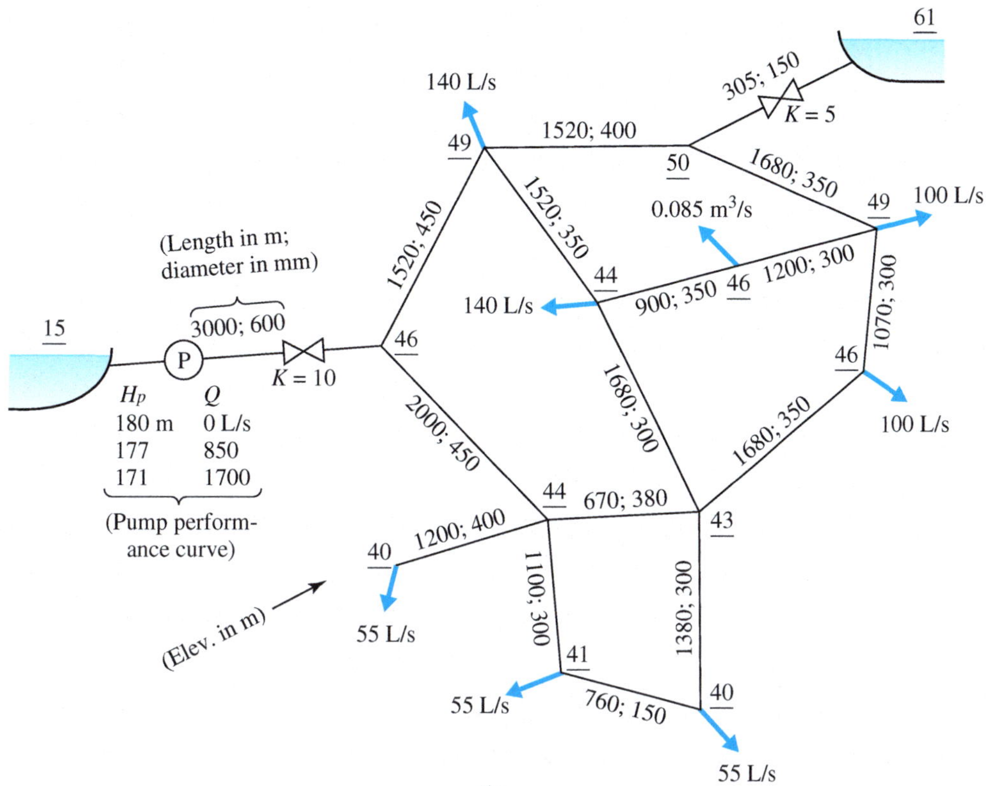

Figure 9.7 Two piping networks: (a) tree watering system: P = 17, L = 0, J = 8, F = 10; (b) water distribution system: P = 17, L = 4, J = 12, F = 2. (Wood, D. J. Algorithms for Pipe Network Analysis and Their Reliability, Research Report 127, Water Resources Research Institute, University of Kentucky, Lexington, KY., 1981)

The useful input power of the pump is 40.4 kW. For the looping network shown in Figure 9.7b, friction losses are based on the Darcy–Weisbach relation with an absolute roughness of 0.15 mm for all pipes and a kinematic viscosity of 10^{-6} m²/s. The pump performance curve is defined by three settings of pump head and discharge.

9.5 UNSTEADY FLOW IN PIPELINES

Unsteady, or transient flows in pipelines traditionally have been associated with hydropower piping and with long water and oil pipeline delivery systems. However, the application has broadened in recent years to include hydraulic control system operation, events that take place in power plant piping networks, fluid-structure interaction in liquid-filled piping, and pulsatile blood flow. For the unsteadiness to occur, some type of excitation to the system is necessary. Representative excitations are valve openings or closures, pump or turbine operations, pipe ruptures or breaks, and cavitation events. In this section we examine only the most fundamental aspects of unsteady flow in piping. First we study unsteady flow in a single constant-diameter pipe assuming inelastic and incompressible conditions; that is followed by the analysis of a system in which elasticity and compressibility play a significant role in the response of the pressure and velocity to an excitation. The first assumption results in the phenomenon called **surging**, while the second one leads to the phenomenon called **water hammer**.

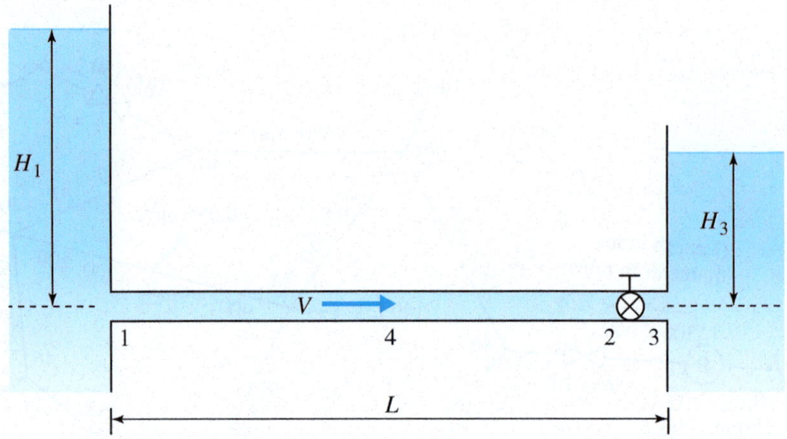

Figure 9.8 Horizontal pipe with a valve at the downstream end.

9.5.1 Incompressible Flow in an Inelastic Pipe

Consider a single horizontal pipe of length L and diameter D (Figure 9.8). The upstream end of the pipe is connected to a reservoir and a valve is situated at the downstream end. The upstream piezometric head is H_1, and downstream of the valve the piezometric head is H_3. Note that both H_1 and H_3 are constant time-independent reservoir elevations. The friction factor f is assumed constant, and the loss coefficient for the valve is K. There are a number of possible excitations that could be considered, but we will only look at the situation where initially there is a steady velocity V_0, the valve is then instantaneously opened to a new position, and subsequently the flow accelerates, increasing to a new steady-state velocity V_{ss}. For situations in which the valve is closed, either partially or completely, one should consider the possibility of water hammer taking place; this is treated in the next section. Here, we assume that the velocity does not vary with position, only with time.

In Figure 9.8 define a control volume for the liquid in the pipe between locations 1 and 2, whose mass is ρAL. Assume a long reach L so that the pipe entrance flow is of no concern. Note that location 2 is upstream of the valve. At the two ends of the control volume the pressures are p_1 and p_2, respectively, and on the pipe surface the wall shear stress is τ_0. The momentum equation for that liquid volume is given by

$$A(p_1 - p_2) - \tau_0 \pi DL = \rho AL \frac{dV}{dt} \tag{9.5.1}$$

Without serious consequences we can assume steady-flow conditions across the valve from location 2 to 3, and utilize the energy equation to provide

$$p_2 = p_3 + K\frac{\rho V^2}{2} \tag{9.5.2}$$

It is reasonable to assume that the Darcy–Weisbach friction factor based on steady-state flow can be employed without undue error. Then, from Section 5.3.3, Eq. 5.3.19, we have the relation

$$\tau_0 = \frac{\rho f V^2}{8} \tag{9.5.3}$$

Substituting Eqs. 9.5.2 and 9.5.3 into Eq. 9.5.1, dividing by the mass of the liquid column, and recognizing that $p_1 - p_3 = \rho g(H_1 - H_3)$, since the velocity heads are assumed to be negligible, there results

$$\frac{dV}{dt} + \left(\frac{f}{D} + \frac{K}{L}\right)\frac{V^2}{2} - g\frac{H_1 - H_3}{L} = 0 \tag{9.5.4}$$

Equation 9.5.4 is the relation that represents incompressible unsteady flow in the pipe. The initial condition at $t = 0$ is a given velocity $V = V_0$. When the final steady-state condition is attained, $dV/dt = 0$, and that steady-state velocity, designated as V_{ss}, can be obtained by setting the derivative to zero in Eq. 9.5.4:

$$V_{ss} = \sqrt{\frac{2g(H_1 - H_3)}{fL/D + K}} \tag{9.5.5}$$

Substituting Eq. 9.5.5 into Eq. 9.5.4, separating variables, and expressing the result in integral form, one has

$$\int_0^t dt = \frac{V_{ss}^2 L}{g(H_1 - H_3)} \int_{V_0}^{V} \frac{dV}{V_{ss}^2 - V^2} \tag{9.5.6}$$

Upon integration, the resulting relation defines the velocity V relative to time t after the valve excitation:

$$t = \frac{V_{ss} L}{2g(H_1 - H_3)} \ln \frac{(V_{ss} + V)(V_{ss} - V_0)}{(V_{ss} - V)(V_{ss} + V_0)} \tag{9.5.7}$$

There are some significant features to the solution. First, by studying Eq. 9.5.7 one concludes that infinite time is required to reach the steady-state velocity V_{ss}. In reality, V_{ss} will be reached some time sooner, because the losses have not been completely accounted for. However, it is possible to determine the time when a percentage, for example 99%, of V_{ss} has been reached that would be adequate for engineering purposes. In addition, it is possible for the fluid to be initially at rest, that is, $V_0 = 0$. Finally, remember that the solution is based on the assumptions that both liquid compressibility and pipe elasticity are ignored; the next section addresses the situation when those assumptions are invalid.

Example 9.8

A horizontal pipe 1000 m in length, with a diameter of 500 mm, and a steady velocity of 0.5 m/s, is suddenly subjected to a new piezometric head differential of 20 m when the downstream valve suddenly opens and its coefficient changes to $K = 0.2$. Assuming a friction factor of $f = 0.02$, determine the final steady-state velocity, and the time when the actual velocity is 75% of the final value.

Solution

Given are L = 1000 m, D = 0.5 m, V_0 = 0.5 m/s, f = 0.02, and K = 0.2. Setting $H_1 - H_3$ = 20 m, the final steady-state velocity V_{ss} is found by direct substitution into Eq. 9.5.5:

$$V_{ss} = \sqrt{\frac{2\times 9.81\times 20}{0.02\times 1000/0.5+0.2}} = \underline{3.12 \text{ m/s}}$$

The velocity that is 75% of V_{ss} is $V = 0.75\times 3.12 = 2.34$ m/s. Then, using Eq. 9.5.7, the time corresponding to that velocity is

$$t = \frac{3.12\times 1000}{2\times 9.81\times 20}\ln\frac{(3.12+2.34)\times(3.12-0.5)}{(3.12-2.34)\times(3.12+0.5)} = \underline{12.9 \text{ s}}$$

Hence, the final steady-state velocity is 3.12 m/s, and the time when 75% of that velocity is attained is approximately 13 s.

9.5.2 Compressible Flow in an Elastic Pipe

In contrast to the developments of Section 9.5.1, there are situations where the liquid is not incompressible and the piping is not rigid. Rather, the interaction between changes in momentum and applied forces causes the liquid to slightly compress and the pipe material to experience very small deformations. When this occurs, significant pressure changes can take place, and the phenomenon is termed **water hammer**. Water hammer is accompanied by pressure and velocity perturbations traveling at very high velocities, close to the speed of sound in the liquid. The resulting wave action occurs at relatively high frequencies. We will consider the situation in which a valve at the downstream end of a pipe closes or opens suddenly, either partially or completely, to initiate a water hammer response.

First let us develop the fundamental equations. Consider again the horizontal pipe shown in Figure 9.8, where now the valve will close so rapidly that elastic effects cause water hammer to occur. The movement of the valve will cause an acoustic, or pressure, wave with speed a to propagate upstream. A control volume of an incremental section of liquid contained in the pipe is shown in Figure 9.9a, where the pressure wave at a given instant is located. The presence of the wave implies that an unsteady flow is taking place within the control volume; at the entrance the velocity is V, and at the exit it is $V + \Delta V$. Steady-state conservation laws can be applied if the wavefront is made to appear stationary by an observer moving with the wavespeed (Figure 9.9b). (Refer to Section 2.6.3 for the development of inertial reference frames that move with a constant velocity.) In contrast to Figure 9.9a, the entrance velocity is now $V + a$, and at the exit it is $V + \Delta V + a$. The pressure, pipe area, and density at the entrance are p, A, and ρ, respectively. Due to the passage of the pressure wave, at the exit the pressure, pipe area, and liquid density are altered to $p + \Delta p$, $A + \Delta A$, and $\rho + \Delta\rho$, where Δp, ΔA, and $\Delta\rho$ are the respective changes in pressure, area, and density.

Applying the conservation of mass across the control volume in Figure 9.9b, we find that

$$0 = (\rho+\Delta\rho)(V+\Delta V+a)(A+\Delta A) - \rho(V+a)A \qquad (9.5.8)$$

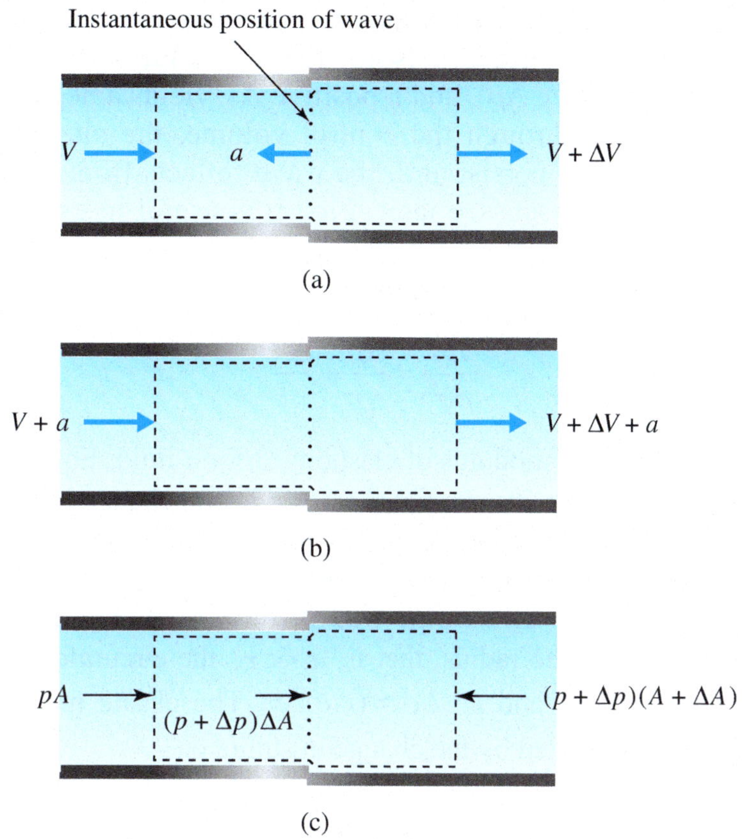

Figure 9.9 Pressure wave moving through a segment of horizontal pipe: (a) pressure wave moving to the left at speed a; (b) pressure wave appears stationary using the principle of superposition; (c) pressure forces acting on the control volume.

Neglecting frictional and gravitational forces, only pressure forces act on the control volume in the direction of flow, as shown in Figure 9.9c. The momentum equation across the control volume provides

$$pA + (p + \Delta p)\Delta A - (p + \Delta p)(A + \Delta A) = \rho A(V + a)[V + \Delta V + a - (V + a)] \tag{9.5.9}$$

Equations 9.5.8 and 9.5.9 are expanded, and the terms containing factors of Δ^2 and Δ^3 are dropped, since they are much smaller in magnitude than the remaining ones. Then Eqs. 9.5.8 and 9.5.9 become

$$\rho A \Delta V + (V + a)(A \Delta \rho + \rho \Delta A) = 0 \tag{9.5.10}$$

and

$$-A\Delta p = \rho A(V + a)\Delta V \tag{9.5.11}$$

In nearly all situations $V \ll a$, so that Eq. 9.5.11 becomes

$$\Delta P = -\rho a \Delta V \tag{9.5.12}$$

Equation 9.5.12, called the **Joukowsky equation**, relates the pressure change to the pressure wave speed and the change in velocity. Note that a velocity reduction (a negative ΔV) produces a pressure rise (a positive Δp), and a positive ΔV yields a negative Δp.

Once the wave has passed through the control volume, the altered conditions $p + \Delta p$, $V + \Delta V$, $A + \Delta A$, and $\rho + \Delta\rho$ will persist until the wave reflects from the upstream boundary; this return wave action will be discussed later. First it is useful to examine the nature of the pressure pulse wave speed a. Equations 9.5.10 and 9.5.11 are combined, again recognizing that $V \ll a$, to eliminate ΔV, with the result that

$$\frac{\Delta p}{\rho a^2} = \frac{\Delta\rho}{\rho} + \frac{\Delta A}{A} \tag{9.5.13}$$

From the definition of the bulk modulus of elasticity B for a fluid, Eq. 1.5.11, we can relate the change in density to the change in pressure as $\Delta\rho/\rho = \Delta p/B$. The change in pipe area can be related to the change in pressure by considering an instantaneous, elastic response of the pipe wall to pressure changes. Assuming a circular pipe cross section of radius r, we have $\Delta A/A = 2\Delta r/r$, and the change in circumferential strain ε in the pipe wall is $\Delta\varepsilon = \Delta r/r$. For a thin-walled pipe whose thickness e is much smaller than the radius, that is, $e \ll r$, the circumferential stress is given by $\sigma = pr/e$. For small changes in r and e, $\Delta\sigma \approx (r/e)\Delta p$. The elastic modulus for the pipe wall material is the change in stress divided by the change in strain, or

$$E = \frac{\Delta\sigma}{\Delta\varepsilon} \approx \frac{(r/e)\Delta p}{\Delta r/r} = \frac{(2r/e)\Delta p}{\Delta A/A} \tag{9.5.14}$$

Solving for $\Delta A/A$, and substituting the result into Eq. 9.5.13, along with the relative change in density related to the change in pressure and bulk modulus, there results

$$\frac{\Delta\rho}{\rho a^2} = \frac{\Delta p}{B} + \frac{2r\Delta p}{E} \tag{9.5.15}$$

The parameter Δp can be eliminated in Eq. 9.5.15, the diameter substituted for the radius, and the relation solved for a:

$$a = \sqrt{\frac{B/\rho}{1 + (D/e)(B/E)}} \tag{9.5.16}$$

Hence, the pressure pulse wave speed is shown to be related to the properties of the liquid (ρ and B) and those of the pipe wall (D, e, and E). If the pipe is very rigid, or stiff, then the term $DB/eE \ll 1$, and Eq. 9.5.16 becomes $a = \sqrt{B/\rho}$, which is the speed of sound in an unbounded liquid (see Eq. 9.5.12). Note that the effect of pipe elasticity is to reduce the speed of the pressure wave.

The use of Eqs. 9.5.12 and 9.5.16 will provide only the magnitude of the water hammer excitation. In addition, it is necessary to understand the nature of the pressure pulse wave as it travels throughout the pipe and is reflected from the boundaries. Consistent with the development of Eqs. 9.5.12 and 9.5.16, we neglect friction to simplify the analysis. Consider the situation in which the valve is suddenly closed and the pipe is rigid. Let us study in detail the sequence of events throughout one cycle of motion, as illustrated in Figure 9.10.

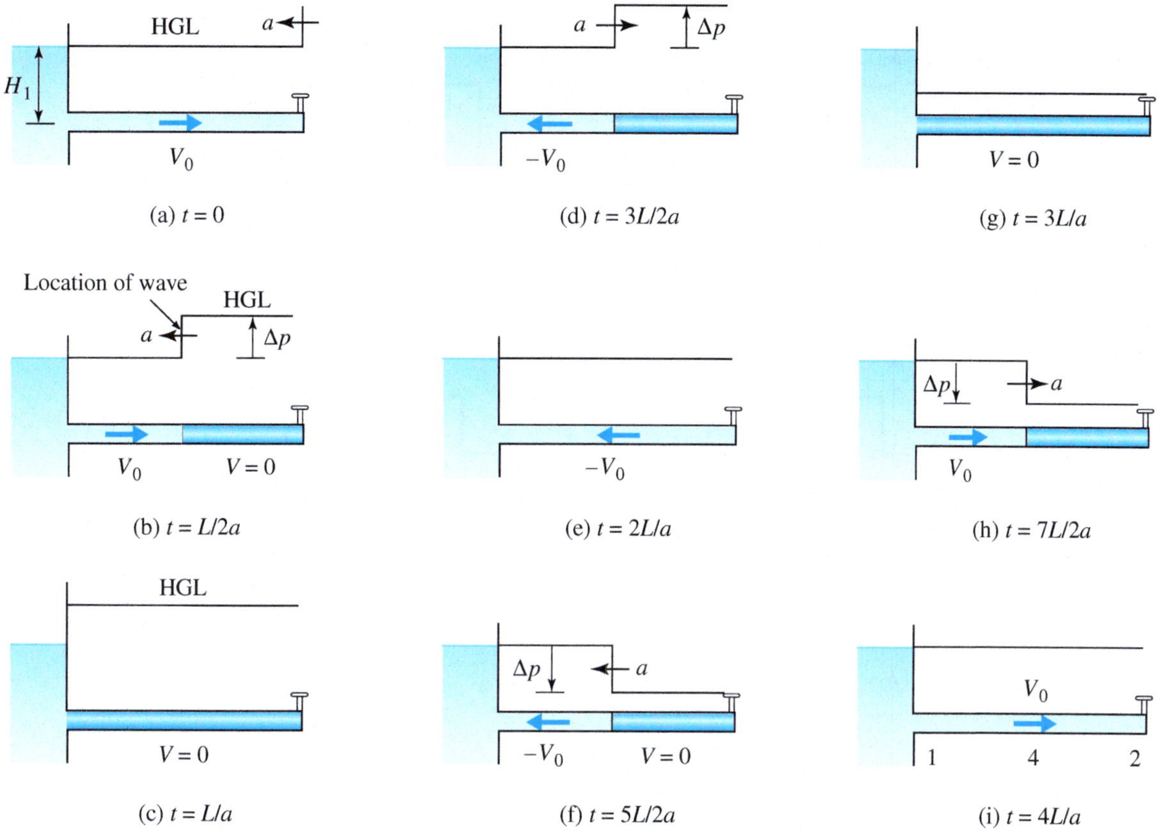

Figure 9.10 One cycle of wave motion in a pipe due to a sudden valve closure.

In Figure 9.10a an initial steady condition exists, the velocity is V_0, and the valve is suddenly closed at $t = 0$. Note that when friction is neglected, the hydraulic grade line appears horizontal. Subsequent to closure of the valve, the wave travels upstream (Figure 9.10b). Behind the wave, the velocity is reduced to zero (the value is closed), the pressure rises by the amount Δp, the liquid has been compressed, and the pipe slightly expand. At time L/a the wave reaches the reservoir, and an unbalanced force occurs at the pipe inlet (Figure 9.10c). In the pipe the pressure reduces to the reservoir pressure, and the velocity suddenly reverses direction; this process begins at the upstream end and propagates downstream at the wave speed a (Figure 9.10d).

When the wave reaches the valve at time $2L/a$, the velocity has magnitude $-V_0$ throughout the pipe (Figure 9.10e). Adjacent to the valve, which is now closed, the velocity reduces to zero, and the pressure reduces by the amount Δp (Figure 9.10f). The low-pressure wave travels upstream at speed a, and behind the wave the liquid is expanded and the pipe wall is contracted. (Note that if the pressure behind the wave reduces to vapor pressure, cavitation will occur, and some of the liquid will vaporize.) When the pressure wave reaches the reservoir at time $3L/a$, an unbalanced condition occurs again, opposite in magnitude to that at time L/a (Figure 9.10g). Equilibrium of forces will now cause a wave to travel downstream, with the pressure increased by the amount Δp and the velocity equal to $+V_0$ behind the front (Figure 9.10h). When the wave reaches the valve at time $4L/a$, initial steady-state conditions prevail once again throughout the pipe (Figure 9.10i).

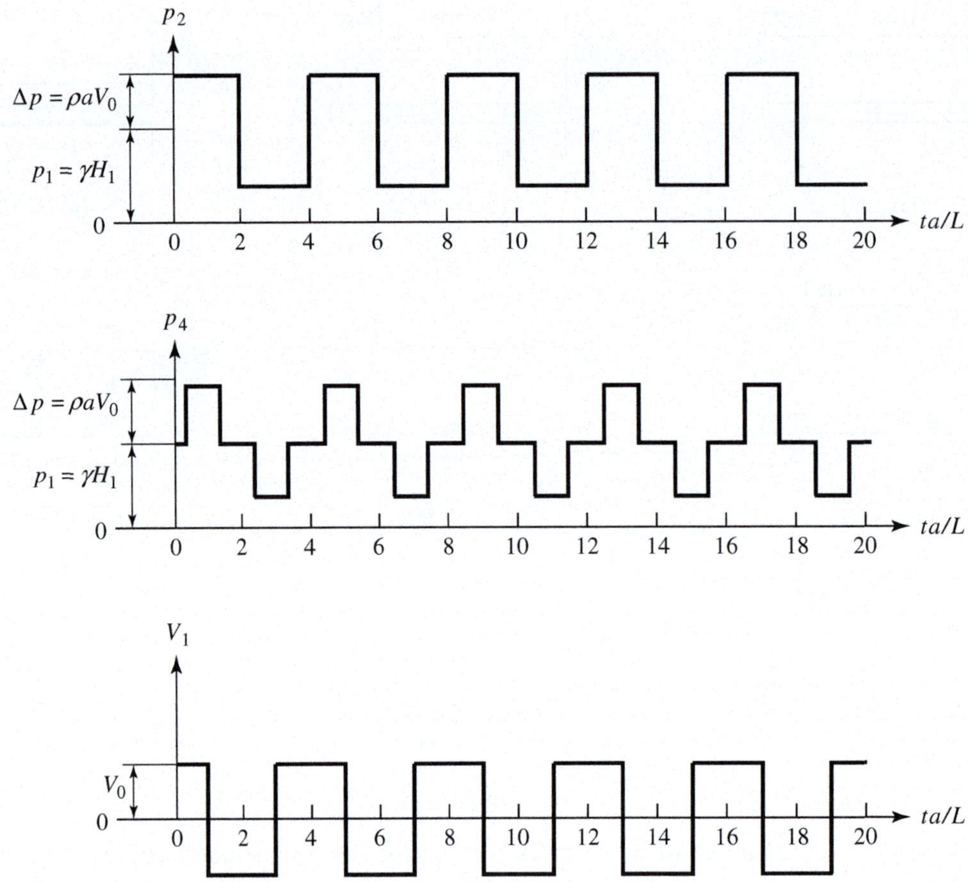

Figure 9.11 Pressure waveforms at the valve (p_2), pipe midpoint (p_4), and velocity waveform at the entrance to the pipe (V_1).

The process repeats itself every $4L/a$ seconds. For the ideal, frictionless situation shown here, the motion will perpetuate. The pressure waveform at the valve and at the midpoint of the pipe and the velocity at the pipe entrance are shown in Figure 9.11. For a real pipe system, friction, pipe motion, and inelastic behavior of the pipe material will eventually cause the oscillation to die down and cease (see Figure 9.12).

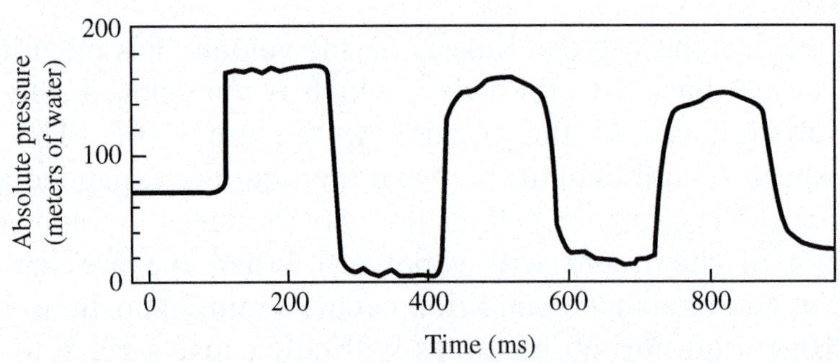

Figure 9.12 Pressure waveform at the valve for an actual pipe system following rapid valve closure. (Martin, C. D., Experimental Investigation of Column Separation with Rapid Valve Closure, Proceedings, 4th International Conference on Pressure Surges, BHRA Fluid Engineering, Cranfield, England, 1983, pp. 77–88. Reproduced with permission.)

The pressure rise Δp predicted by Eq. 9.5.13 is based on the assumption that the valve closes instantaneously. Actually, it can be used to predict the maximum pressure rise when the valve closes with any time less than $2L/a$, the time it takes for the pressure wave to travel from the valve to the reservoir and back again. For valve closure times greater than $2L/a$, a more comprehensive analysis is required (Wylie and Streeter, 1993). Keep in mind that the previous discussion pertains only to a single horizontal pipe with a reservoir at the upstream end and a valve closing instantaneously at the downstream end and one that contains a frictionless liquid. Most of the water hammer analysis performed by engineers today makes use of computer-based numerical methods for complex piping systems, incorporating a variety of excitation mechanisms such as pumps, surge suppressers, and various types of valves (Wylie and Streeter, 1993).

Example 9.9

A steel pipe (E = 207 GPa, L = 1500 m, D = 300 mm, e = 10 mm) conveys water at 20°C. The initial velocity is V_0 = 1 m/s. A valve at the downstream end is closed so rapidly that the motion is considered to be instantaneous, reducing the velocity to zero. Determine the pressure pulse wave speed in the pipe, the speed of sound in an unbounded water medium, the pressure rise at the valve, the time it takes for the wave to travel from the valve to the reservoir at the upstream end, and the period of oscillation.

Solution

The density and bulk modulus of water at 20°C are found in Table B.1: ρ = 998 kg/m^3 and $B = 220 \times 10^7$ Pa. The pressure wave speed, a, is given by Eq. 9.5.16:

$$a = \sqrt{\frac{220 \times 10^7/998}{1 + \frac{0.3}{0.01} \times \frac{220 \times 10^7}{207 \times 10^9}}} = \underline{1290 \text{ m/s}}$$

The speed of sound in an unbounded water medium is found (refer to the discussion after Eq. 9.5.16) to be

$$a = \sqrt{\frac{220 \times 10^7}{998}} = \underline{1485 \text{ m/s}}$$

Note that the sound speed is about 15% larger than the pressure wave speed. To compute the pressure rise at the valve, we recognize that the reduction in velocity is $-V_0 = -1$ m/s. Using Eq. 9.5.12, the increase in pressure upstream of the valve is

$$\Delta p = -998 \times 1290 \times (-1) = 1.29 \times 10^6 \text{ Pa} \quad \text{or} \quad \underline{1290 \text{ kPa}}$$

The wave travel time from the valve to the reservoir is L/a = 1500/1290 = $\underline{1.16 \text{ s}}$, and the period of oscillation is $4L/a = 4 \times 1.16 = \underline{4.65 \text{ s}}$.

9.6 Summary

A methodology for dealing with losses in piping has been provided in Section 9.2 using the resistance coefficient as a generalized term that can include minor losses along with a chosen empirical friction loss. For accurate representation of losses, the Darcy–Weisbach friction loss is recommended, with losses proportional to the square of the velocity (or discharge). Equation 9.2.3 is useful for estimating the Darcy–Weisbach friction factor in an iterative, or trial, solution. In Section 9.3 simple pipe systems have been defined as those that contain one to several pipes, arranged singly, in series, parallel, or branched. Typically, a problem consists of finding the flow distribution in the piping; along with discharges the piezometric head often is an unknown. Calculator-based solutions or computational software are useful tools to analyze these systems.

For more complex piping networks, a systematic solution approach has been presented in Section 9.4. First the network equations were linearized, and then the Hardy Cross method was applied to solve for the network flows and piezometric heads. A spreadsheet solution was applied to a network using the Hardy Cross solution; it allows one to correctly apply the energy and continuity equations as well as track the convergence of the solution. Note that even the simple pipe systems presented in Section 9.3 can be solved using the Hardy Cross method. The computer program EPANET was also introduced in Section 9.4. This software is useful for large systems in which spreadsheet programming becomes cumbersome.

Finally, analysis of unsteady flows in pipes has been introduced in Section 9.5. We have focused on flow that behaves either in an incompressible or compressible manner; the latter behavior is commonly called "water hammer." Water hammer is a result of acoustic waves propagating in a pipe. The waves travel at the speed of sound in the contained liquid and produce a periodic response of both pressures and velocities throughout the pipe. The Joukowsky relation, Eq. 9.5.12, gives the magnitude of the pressure wave.

PROBLEMS

Steady Flows

9.1 Using the results from Example 9.2, tabulate and plot to correct vertical scale the hydraulic and energy gradelines. Assume a 10-m length between the two elbows.

9.2 A pump is situated between two sections in a horizontal pipeline. The diameter D_1 and pressure p_1 are given at the upstream section, and D_2 and p_2 are given at the downstream section. Determine the required fluid power of the pump for the following conditions:

(a) $D_1 = 50$ mm, $p_1 = 350$ kPa, $D_2 = 80$ mm, $p_2 = 760$ kPa, $Q = 95$ L/min, $h_L = 6.6$ m, water is flowing at 20°C.

(b) $D_1 = 2$ in, $p_1 = 50$ lb/in^2, $D_2 = 3$ in, $p_2 = 110$ lb/in^2, $Q = 25$ gal/min, $h_L = 20$ ft, water is flowing at 70°F.

9.3 An oil ($S = 0.82$) is pumped between two storage tanks in a pipe with the following characteristics: $L = 2440$ m, $D = 200$ mm, $f = 0.02$, $\Sigma K = 12.5$. The upper tank is 32 m higher than the lower one. Using the pump data provided, determine: the oil discharge in the pipe and the power requirement for the pump.

Q (L/s)	0	15	30	45	60	75	100
H_P (m)	55	54	53	52	49	44	35
η	0	0.4	0.6	0.7	0.75	0.7	0.5

9.4 Consider the simple pumping system shown in Figure P9.4. Assume the following parameters to be known:

Reservoir elevations, z_1, z_2
Pipe length, L
Pipe roughness, e
Sum of minor loss coefficients, ΣK
Kinematic viscosity, v
Discharge, Q

Develop a numerical solution to find the required diameter. Use an equivalent length to represent the minor losses.

Determine the diameter for the following data:

(a) $z_2 - z_1 = 120$ ft, $L = 1500$ ft, $e = 0.003$ ft, $\Sigma K = 2.5$, water at 50°F, $Q = 15{,}000$ gal/min, pump characteristic curve:

H_P (ft)	490	486	473	453	423	406
Q (gal/min)	0	4000	8000	12,000	16,000	18,000

(b) $z_2 - z_1 = 40$ m, $L = 500$ m, $e = 1$ mm, $\Sigma K = 2.5$, water at 20°C, $Q = 1.1 \text{m}^3/\text{s}$, pump characteristic curve:

H_P (m)	160	158	154	148	138
Q (L/s)	0	400	800	1200	1600

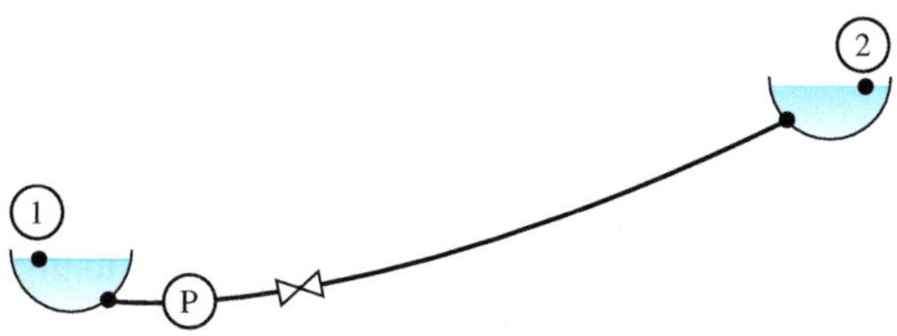

Figure P9.4

9.5 Gasoline is being pumped at 400 L/s in a pipeline from location A to B, as shown in Figure P9.5. The pipe follows the topography as shown, with the highest elevation shown at location C. The only contributions to minor losses are the two valves located at the ends of the pipe. If $S = 0.81$, $v = 4.26 \times 10^{-7}$ m^2/s, $p_v = 55.2$ kPa absolute, and $p_{atm} = 100$ kPa,

(a) Determine the necessary power to be delivered to the system to meet the flow requirement.

(b) What is the maximum elevation possible at location C without causing vapor pressure conditions to exist?

(c) Sketch the hydraulic grade line.

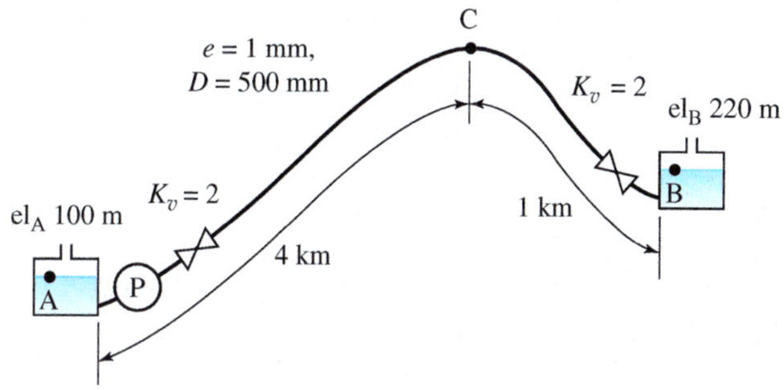

Figure P9.5

9.6 For the three pipes in series shown in Figure P9.6, minor losses are proportional to the discharge squared, and the Hazen–Williams formula is used to account for friction losses. With the data given, use Newton's method to determine the discharge. Note that minor losses can be neglected in order to initially estimate Q.

(a) $(p/\gamma + z)_A = 250$ m and $(p/\gamma + z)_B = 107$ m.

Pipe	L (m)	D (mm)	ΣK	C (*Hazen–Williams*)
1	200	200	2	100
2	150	250	3	120
3	300	300	0	90

(b) $(p/\gamma + z)_A = 820$ ft and $(p/\gamma + z)_B = 351$ ft.

Pipe	L (ft)	D (in.)	ΣK	C (*Hazen–Williams*)
1	600	8	2	100
2	300	10	3	120
3	900	12	0	90

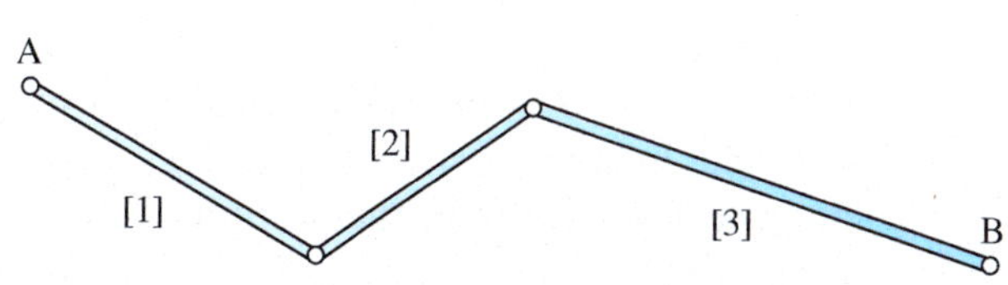

Figure P9.6

9.7 A long oil pipeline consists of the three segments shown in Figure P9.7. Each segment has a booster pump that is used primarily to overcome pipe friction. The two reservoirs are at the same elevation.

(a) Derive a single equation to determine the discharge in the system if the resistance factors $\overline{R}_i$ for each pipe and the useful power $\dot{W}_{fi}$ for each pump are known.

(b) Determine the discharge for the data given in the figure. The unit weight of oil is $\gamma = 8830$ N/m^3.

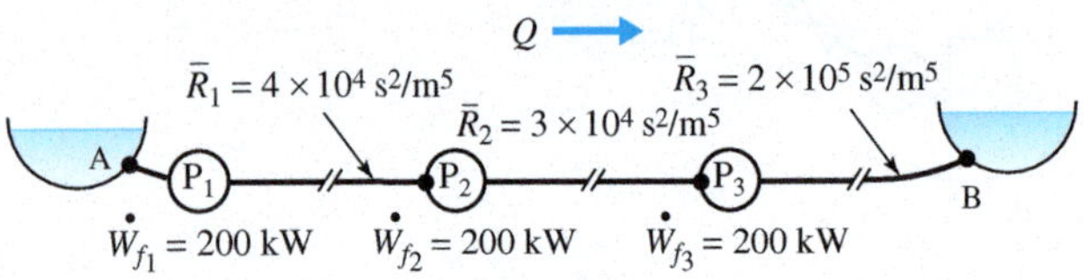

Figure P9.7

9.8 A liquid with a specific gravity of 0.68 is pumped from a storage tank to a free jet discharge through a pipe of length L and diameter D (Figure P9.8). The pump provides a known amount of fluid power $\dot{W}_f$ to the liquid. Assuming a constant friction factor of 0.015, determine the discharge for the following conditions:

(a) $z_1 = 24$ m, $p_1 = 110$ kPa, $z_2 = 18$ m, $L = 450$ m, $D = 300$ mm, $\dot{W}_f = 10$ kW.

(b) $z_1 = 75$ ft, $p_1 = 15$ lb/in^2, $z_2 = 60$ ft, $L = 1500$ ft, $D = 8$ in., $\dot{W}_f = 15$ hp.

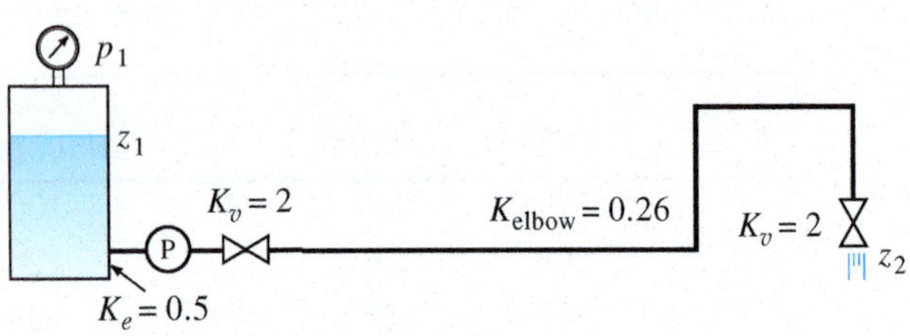

Figure P9.8

9.9 Water at 20°C is being pumped through the three pipes in series as shown in Figure P9.9. The power delivered to the pump is 1920 kW, and the pump efficiency is 0.82. Compute the discharge.

Pipe	L (m)	D (mm)	e (mm)	ΣK
1	200	1500	1	0
2	300	1000	1	2
3	120	1200	1	10

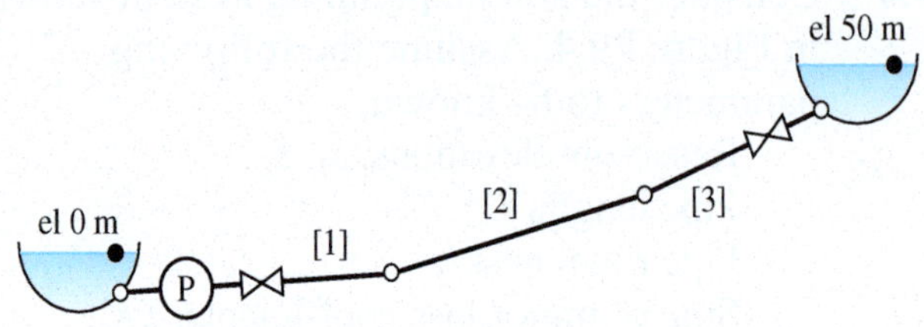

Figure P9.9

9.10 Find the water flow distribution in the parallel system shown in Figure P9.10, and the required pumping power if the discharge through the pump is $Q_1 = 3$ m^3/s. The pump efficiency is 0.75. Assume constant friction factors.

Pipe	L (m)	D (mm)	f	ΣK
1	100	1200	0.015	2
2	1000	1000	0.020	3
3	1500	500	0.018	2
4	800	750	0.021	4

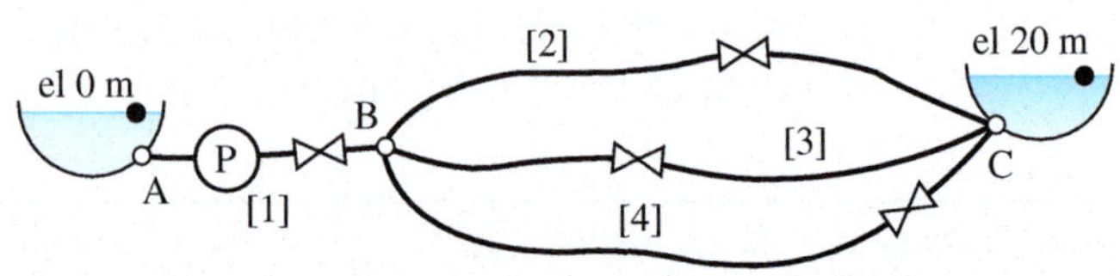

Figure P9.10

9.11 For the system shown in Figure P9.11, determine the water flow distribution and the piezometric head at the junction using an ad hoc approach. Assume constant friction factors. The pump characteristic curve is $H_P = a - bQ^2$.

(a) $a = 20$ m, $b = 30$ s^2/m^5, $z_1 = 10$ m, $z_2 = 20$ m, $z_3 = 18$ m.

Pipe	L (m)	D (cm)	f	K
1	30	24	0.020	2
2	60	20	0.015	2
3	90	16	0.025	0

(b) $a = 55$ ft, $b = 0.1$ s^2/ft^5, $z_1 = 20$ ft, $z_2 = 50$ ft, $z_3 = 45$ ft.

Pipe	L (ft)	D (in.)	f	K
1	100	10	0.020	2
2	200	8	0.015	2
3	300	6	0.025	0

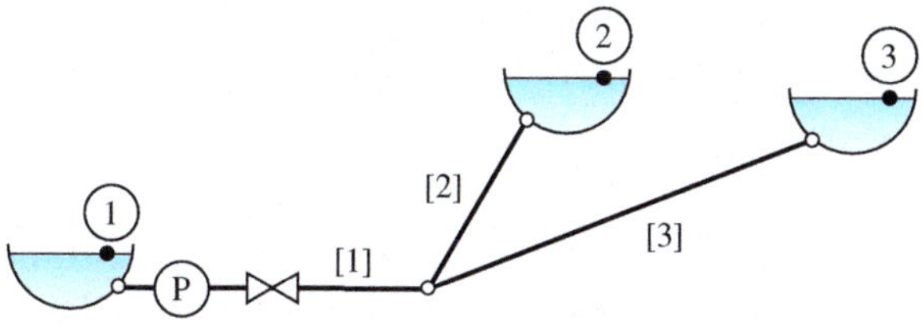

Figure P9.11

9.12 Work Problem 9.11 using the numerical False Position method.

9.13 Find the flow distribution in the parallel network shown in Figure P9.13. Assume constant friction factors. The change in hydraulic grade line between A and B is $(p/\gamma + z)_A - (p/\gamma + z)_B = 50$ m.

Pipe	L (m)	D (mm)	e (mm)	ΣK
1	600	1000	0.1	2
2	1000	1200	0.15	0
3	550	850	0.2	4
4	800	1000	0.1	1

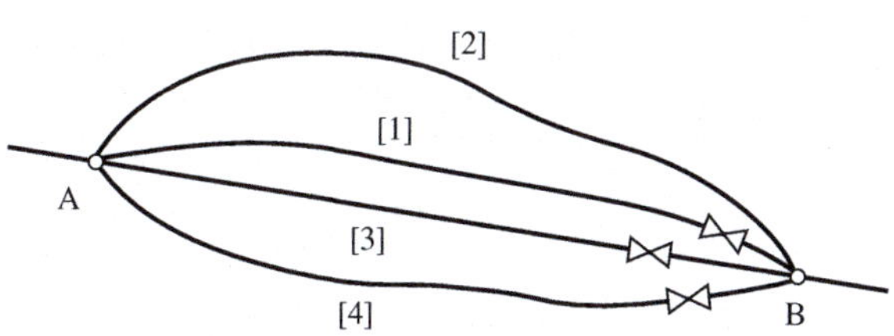

Figure P9.13

9.14 The water sprinkling system shown in Figure P9.14 is applied from a large diameter pipe with constant internal pressure $p_0 = 300$ kPa. The system is positioned in a horizontal plane. Determine the flow distribution Q_1, Q_2, Q_3, Q_4 for the given data. Valve losses are included in the $\bar{R}$, values. (*Hint:* If you are clever, no trial-and-error solution is necessary!)

Pipe	$\bar{R}$ (s^2/m^5)
1	1.6×10^4
2	5.3×10^5
3	1.0×10^6
4	1.8×10^6

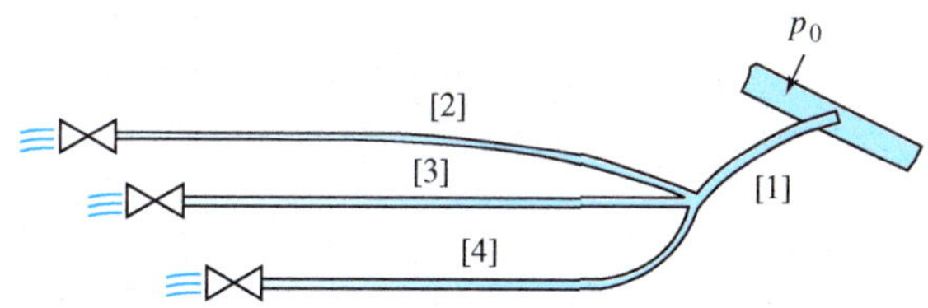

Figure P9.14

9.15 For the system shown in Figure P9.15, a fluid flows from A to B. Determine in which pipe there is the greatest velocity.

Pipe	L (m)	D (mm)	f
1	2000	450	0.012
2	650	150	0.020
3	1650	300	0.015

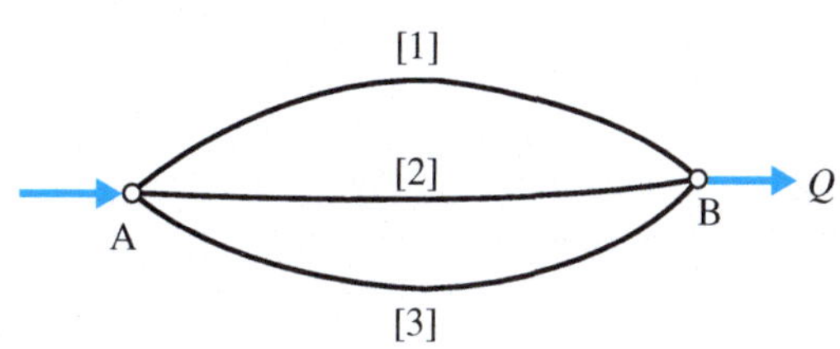

Figure P9.15

9.16 A proposed water irrigation system consists of one main pipe with a pump and three pipe branches (Figure P9.16). Each branch is terminated by an orifice, and each orifice has the same elevation. It is apparent that the flow distribution can be solved by treating the piping arrangement as a branched system. However, the piping can also be treated as a parallel system in order to determine the flows.

(a) Identify the equations and unknowns to satisfy the solution for a parallel system. Why is it possible to treat the irrigation system as a parallel piping problem?

(b) Why would the solution to a parallel system be preferred to that of a branching system?

(c) Determine the flow distribution and sketch the hydraulic grade line.

(d) What part of the piping would one change to approximately double the discharge, assuming that the individual lengths and pump curve could not be altered?

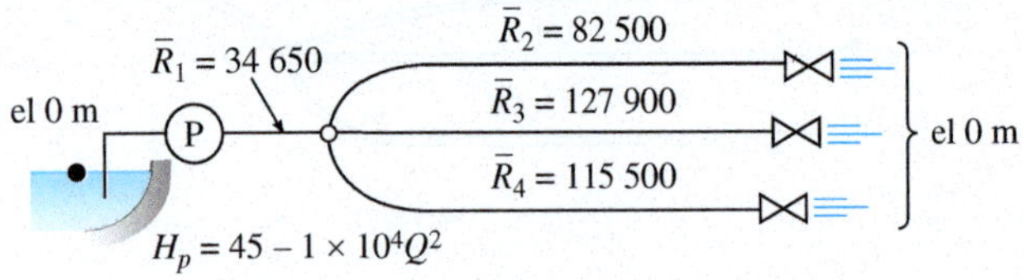

(H_p in m; Q in m³/s; $\bar{R}$ in s²/m⁵, orifice loss included in $\bar{R}$)

Figure P9.16

9.17 Determine the required head (H_P) and discharge (Q_1) of water to be handled by the pump for the system shown in Figure P9.17. The discharge in pipe 2 is $Q_2 = 35$ L/s in the direction shown.

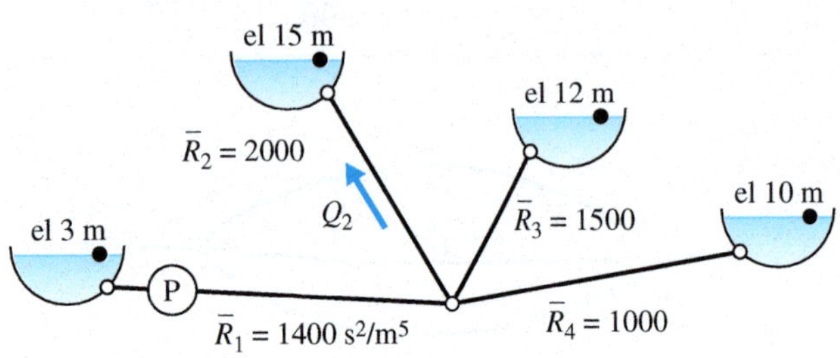

Figure P9.17

9.18 Determine the flow distribution of water in the parallel piping system shown in Figure P9.18.

(a) $Q_{in} = 600$ L/min

Pipe	L (m)	D (mm)	f	ΣK
1	30	50	0.020	3
2	40	75	0.025	5
3	60	60	0.022	1

(b) $Q_{in} = 0.35$ ft³/s

Pipe	L (ft)	D (in.)	f	ΣK
1	90	2	0.020	3
2	120	3	0.025	5
3	180	2.5	0.022	1

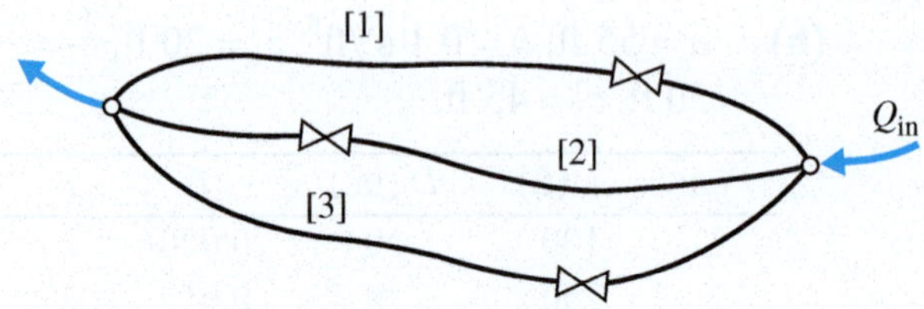

Figure P9.18

9.19 Water is being pumped in the piping system shown in Figure P9.19. The pump curve is approximated by the relation $H_P = 150 - 5Q_1^2$, with H_P in meters and Q_1 in m³/s. The pump efficiency is $\eta = 0.75$. Compute the flow distribution and find the required pump power.

Pipe	$\bar{R}$ (s²/m⁵)
1	400
2	1000
3	1500

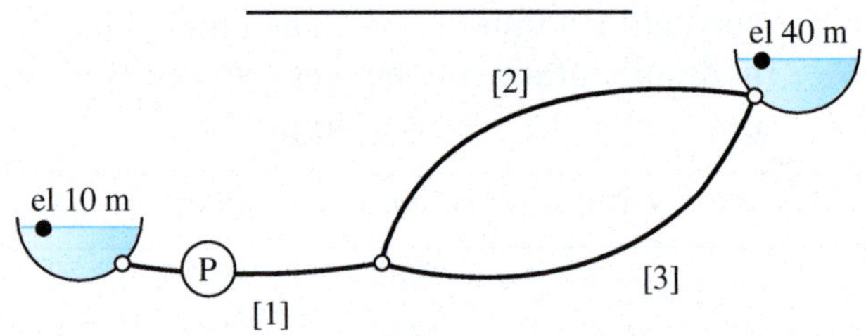

Figure P9.19

9.20 A pipeline consists of two pipe segments in series (Figure P9.20). The specific gravity of the fluid is 0.81. If pump A has a constant power input of 1 MW, find the discharge, the pressure head in pumps A and B, and the required power for pump B. The minimum allowable pressure on the suction side of pump B is 150 kPa, and both pumps have an efficiency of 0.76.

Pipe	L (m)	D (mm)	ΣK	f
1	5000	750	2	0.023
2	7500	750	10	0.023

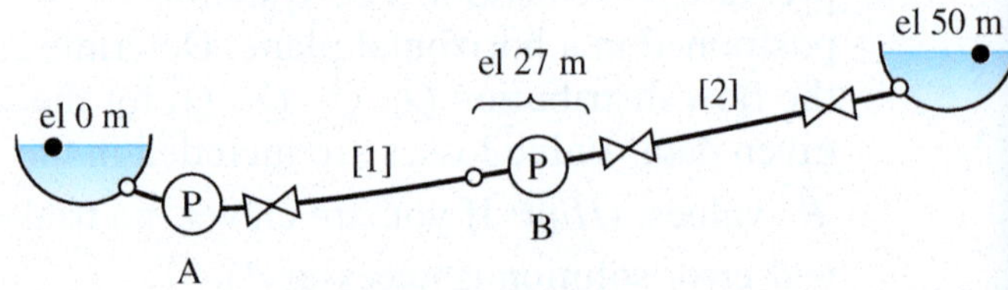

Figure P9.20

9.21 Determine the flow distribution of water in the system shown in Figure P9.21. The equivalent roughness for all elements is 0.1 mm.

Pipe	*L* (m)	*D* (mm)	Σ*K*
1	1000	200	3
2	200	25	0
3	250	25	0
4	340	30	2
5	420	40	0
6	500	175	5

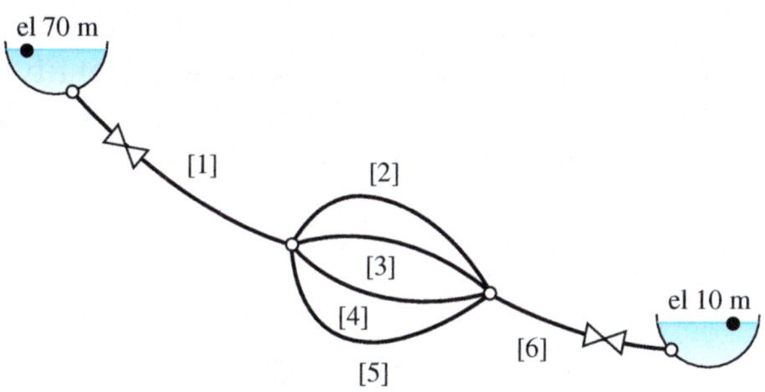

Figure P9.21

9.22 Determine the flow distribution of water in the branching pipe system shown in Figure P9.22:

(a) Without a pump in line 1.

(b) Include a pump located in line 1 adjacent to the lower reservoir. The pump characteristic curve is given by $H_P = 250 - 0.4Q = 0.1Q^2$, with H_P in meters and Q in m^3/s.

The Hazen–Williams coefficient $C = 130$ for all pipes.

Pipe	*L* (m)	*D* (mm)
1	200	500
2	600	300
3	1500	300
4	1500	400

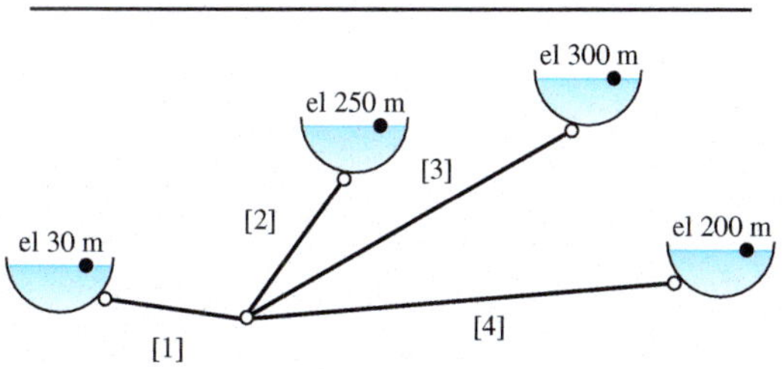

Figure P9.22

9.23 Determine the flow distribution of water in the system shown in Figure P9.23. Assume constant friction factors, with $f = 0.02$. The head-discharge relation for the pump is $H_P = 60 - 10Q^2$, where H_P is in meters and the discharge is in cubic meters per second.

Pipe	*L* (m)	*D* (mm)	Σ*K*
1	100	350	2
2	750	200	0
3	850	200	0
4	500	200	2
5	350	250	2

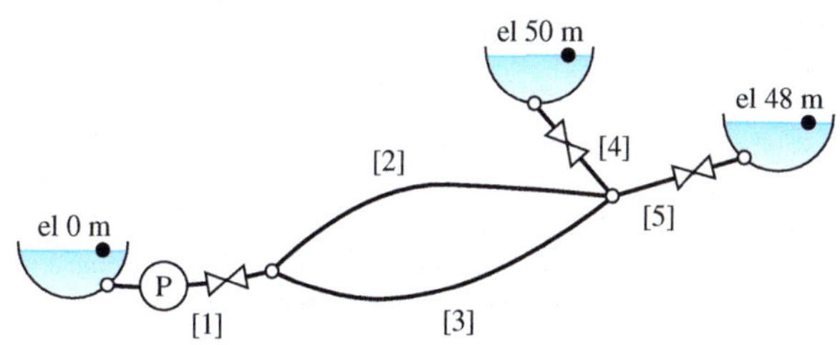

Figure P9.23

9.24 A pump, whose performance and efficiency curves are given in Figure P9.24a has been selected to deliver water in a piping system. The piping consists of four pipes arranged as shown in Figure P9.24b. Water at 60°F is being pumped from reservoir A and exits at either reservoir B or at location D, depending on whether the valves at those locations are open or closed. The pipe characteristics are shown in the accompanying table. All of the pipe diameters are 4 in., and the friction factor in each pipe is assumed to be $f = 0.04$.

(a) If the discharge through the pump is 5,000 gal/hr, what is the head loss across pipe 2?

(a) Compute the discharge in the system, assuming that the valve at location D is closed.

(c) If the valve at location D is open and the discharge through the pump is 11,000 gal/hr, determine the discharge in pipe 4.

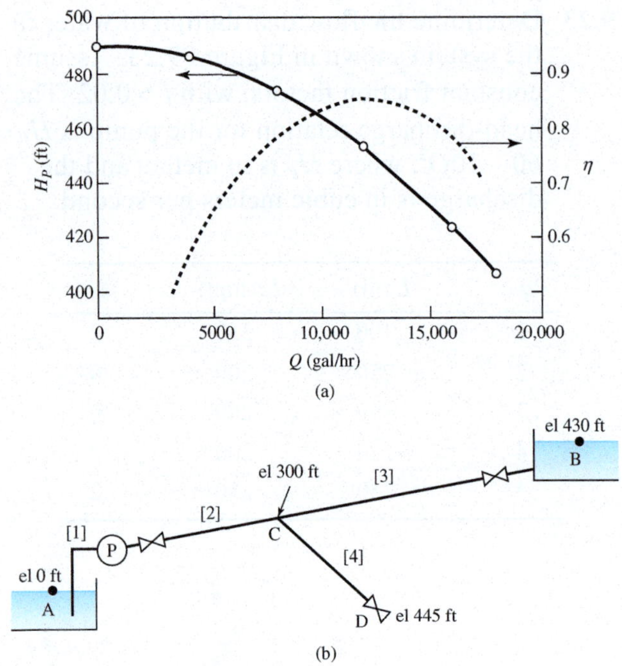

Figure P9.24

Pipe	L (ft)	ΣK
1	10	1
2	500	2
3	2000	2
4	750	4

9.25 Refer to the system and pump curves associated with Problem 9.24. Assume that the discharge through the pump is 12,000 gal/hr.

(a) What is the required power for the pump?

(b) Determine the pressure at location C.

(c) If the valve at location D is closed, what is the gage pressure at that location?

9.26 Refer to the system and pump curves associated with Problem 9.24.

(a) Assuming that the valve at location D is open and the head rise across the pump is 460 ft, determine the discharge in pipe 3.

(b) If the valve at location B was closed and the valve at location D was open, what is the head rise across the pump?

9.27 For the network shown in Figure P9.27, perform the following:

(a) Identify all of the unknown parameters and write the required system equations. Subsequently, reduce them to one equation in one unknown. Assume $\beta = 2$.

(b) Prepare the network for a Hardy Cross analysis by making a sketch, identifying necessary loops and paths, and writing the appropriate equations.

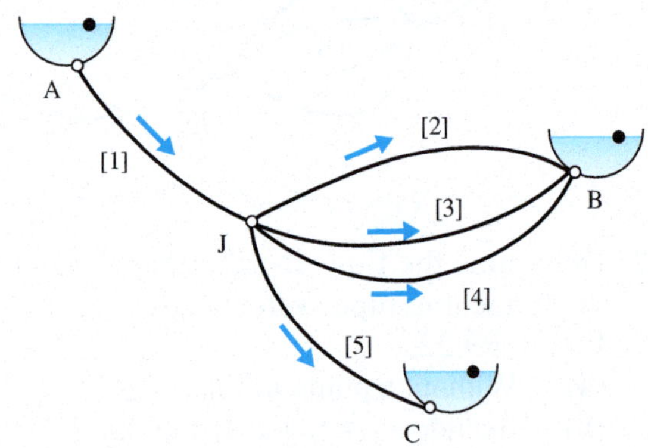

Figure P9.27

9.28 The branching system shown in Figure P9.28a is to be solved by using the Hardy Cross method.

(a) Write the equations for ΔQ_1, ΔQ_{11}, W, and G.

(b) Perform two iterations by filling in the blanks in the table shown in Figure P9.28b.

After two iterations, sketch the hydraulic grade line.

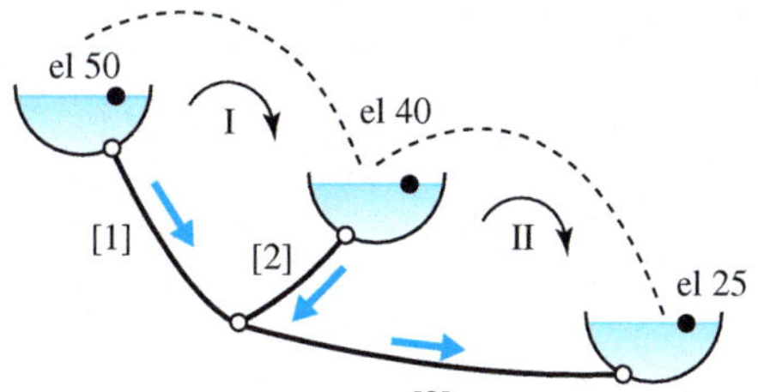

(a)

Loop	Pipe	$\bar{R}$	$2\bar{R}$	Q	W	G	Q	W	G	Q
I	1	2	4	–3.5						
	2	2	4	0						

$\Delta Q_{\mathrm{I}} =$ $\Delta Q_{\mathrm{I}} =$

Loop	Pipe	$\bar{R}$	$2\bar{R}$	Q	W	G	Q	W	G	Q
II	2	2	4	0						
	3	2	4	–3.5						

$\Delta Q_{\mathrm{II}} =$ $\Delta Q_{\mathrm{II}} =$

(b)

Figure P9.28

9.29 Determine the flow and the hydraulic grade line in the system shown in Figure P9.29 using the Hardy Cross analysis method. Use $\gamma = 9800\ \mathrm{N/m^3}$.

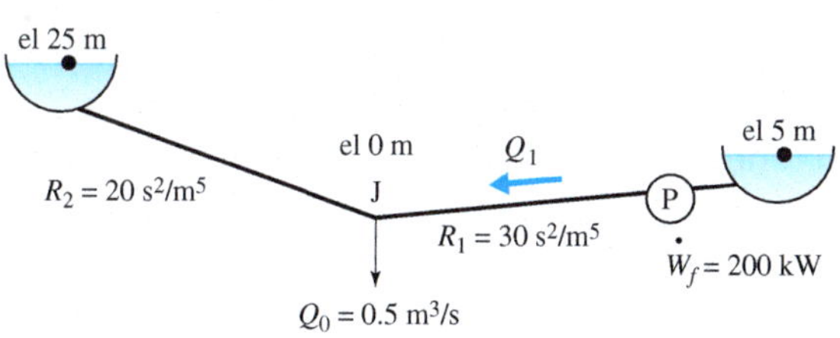

Figure P9.29

9.30 The solution of flow of water in a pipe network is shown in Figure P9.30. Compute:

(a) The hydraulic grade line throughout the system.

(b) The pressure at each node.

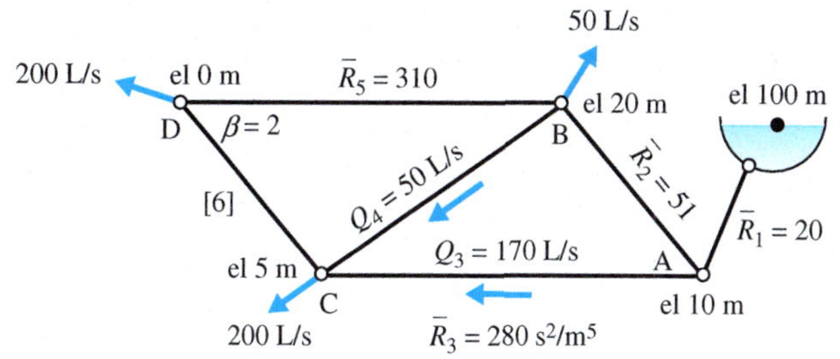

Figure P9.30

9.31 For the configuration shown in Problem 9.27, determine the water flow distribution and the hydraulic grade line at J. The system data are tabulated below.

Reservoir	*Elevation* (ft)	*Line*	L (ft)	D (in.)	f	ΣK
A	650	1	800	8	0.015	0
B	575	2	600	3	0.020	2
C	180	3	650	3	0.020	2
		4	425	3	0.025	3
		5	1000	4	0.015	4

9.32 Water is flowing in the piping system shown in Figure P9.32. Determine the flow distribution using the Hardy Cross method:

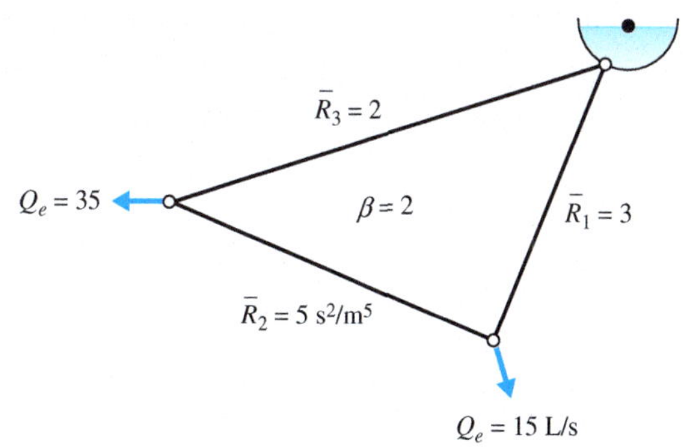

Figure P9.32

9.33 Determine the water flow distribution in the piping system shown in Figure P9.33 and the pressures at the internal nodes employing the Hardy Cross method of solution.

Pipe	L (m)	D (mm)	e (mm)	ΣK
1	500	300	0.15	0
2	600	250	0.15	0
3	50	150	0.15	10
4	200	250	0.15	2
5	200	300	0.15	2

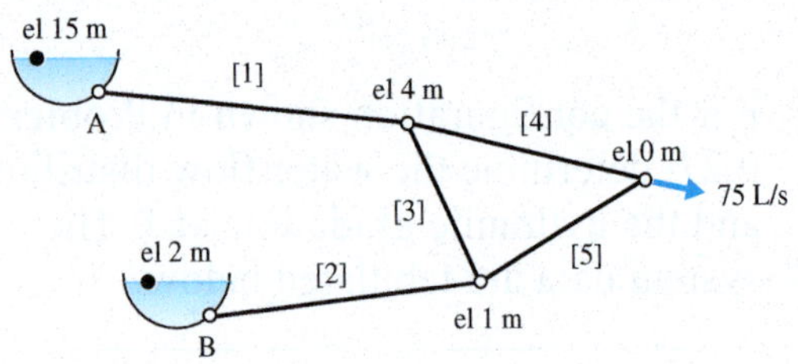

Figure P9.33

Line	L (m)	D (mm)	e (mm)	ΣK
1	200	100	0.1	2
2	150	50	0.1	30
3	500	100	0.1	0
4	35	50	0.1	35
5	120	100	0.1	2

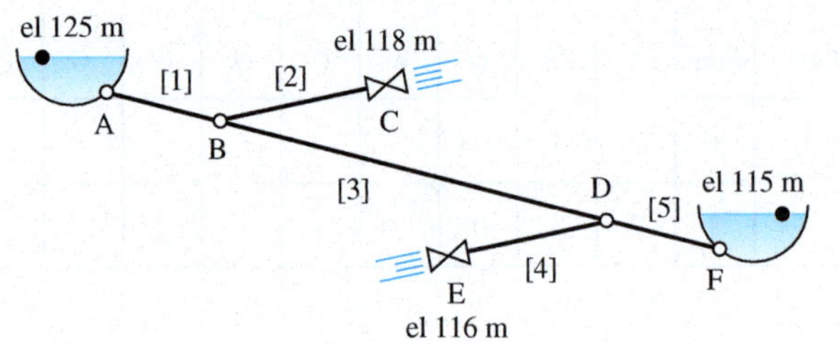

Figure P9.36

9.34 Solve Problem 9.9 by employing the Hardy Cross method. Use Eq. 9.2.6 to compute the friction factors.

9.35 Determine the discharge in the piping system shown in Figure P9.35:

(a) With an exact solution.

(b) Using the Hardy Cross method.

Water is the flowing liquid and the losses are proportional to the velocity squared ($\beta = 2$). The pump characteristic curve is $H_P = 100 - 826Q^2$ (H_p in meters and Q in m^3/s).

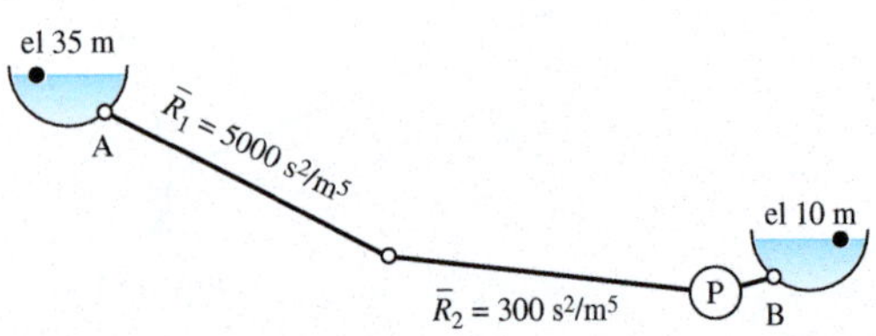

Figure P9.35

9.36 The piping system shown in Figure P9.36 delivers water to two side-branch sprinklers (C and E) and to a lower reservoir (F). The water is supplied by an upper reservoir (A). The sprinklers are represented hydraulically as orifices with relatively high K values. Determine the flow distribution using the Hardy Cross method.

9.37 In Problem 9.36, if a pump supplying 10 kW of useful power was inserted near A, what changes in flow and pressure would occur?

9.38 A cooling water condenser flow system for a thermal power plant is shown in Figure P9.38. Water is pumped from a reservoir at location A through a large-diameter pipe [1], passing through the condenser [2], and discharging through another large-diameter pipe [3] into the receiving pond, location B. The condenser consists of a large number N of elevated, small-diameter tubes placed in parallel; at either end of the condenser are large water-filled boxes called a headers, C and C′. Water surface elevations at A and B are known, as are the lengths and diameters of pipes 1 and 3. The condenser has N identical tubes, each with the same known diameter D_2 and length L_2. The friction factor for all pipes is the same constant value, and the pump curve is approximated by the relation $H_P = a_0 + a_1Q + a_2Q^2 + a_3Q^3$. Minor losses can be assumed negligible.

(a) Derive an equivalent single-pipe resistance coefficient R_2 for the condenser tubes.

(b) Write the energy equation for the entire system, using the required given variables and expressing flow resistance in terms of the resistance coefficients for pipes 1 and 3, and the equivalent resistance coefficient for pipe 2.

(c) Set up a Hardy Cross algorithm to determine the discharge through the system.

(d) If $z_A = 2$ m, $z_B = 0$ m, $L_1 = 100$ m, $D_1 = 2$ m, $L_2 = 15$ m, $D_2 = 0.025$ m, $N = 1000$, $L_3 = 200$ m, $D_3 = 2$ m, and $f = 0.02$, compute the discharge and hydraulic grade line. The coefficients for the pump curve are $a_0 = 30.4$, and $a_1 = -31.8$, $a_2 = 18.6$, and $a_3 = -4.0$, where Q is in cubic meters per second.

(e) If the top of the condenser is situated at an elevation of 6 m, what are the pressures at the top of the upstream and downstream headers?

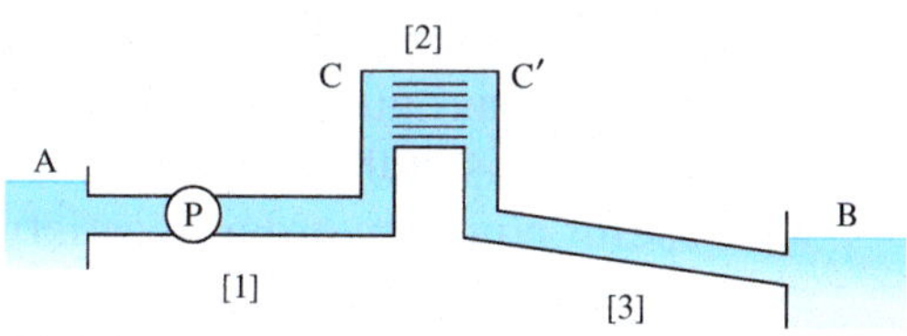

Figure P9.38

9.39 Determine the flow distribution in the branching or "tree-like" network shown in Figure P9.39. Water is the flowing liquid: $\gamma = 62.4$ lb/ft^3. The source of flow is a large pipeline (location A) maintained at a constant pressure of 60 psi. Each branch terminates at a location where the hydraulic grade line is known. (Courtesy of D. Wood)

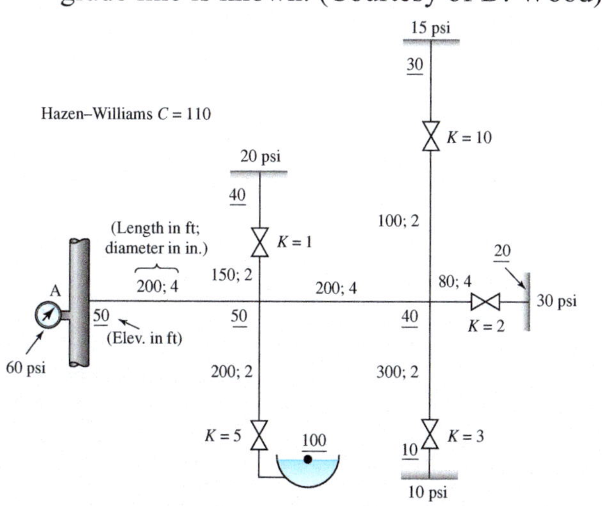

Figure P9.39

9.40 The 12-pipe system shown in Figure P9-40 represents a low-pressure region that is connected to a high-pressure system by pressure regulators set at 50 psi. Determine the pressure and water flow distribution for the demands shown. Assume Hazen–Williams $C = 120$ for all pipes. (Courtesy of D. Wood.)

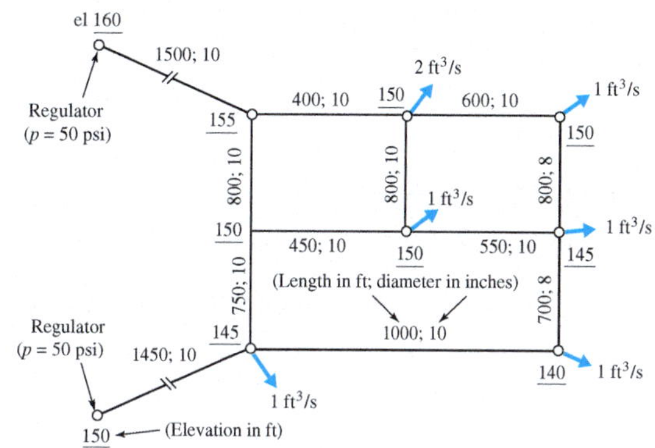

Figure P9.40

9.41 Determine the flow distribution for the 14-pipe water supply system shown in Figure P9.41. The characteristic curve for the pump is represented by the following data (courtesy of D. Wood):

Hp (m)	166	132	18
Q (L/s)	0	600	1000

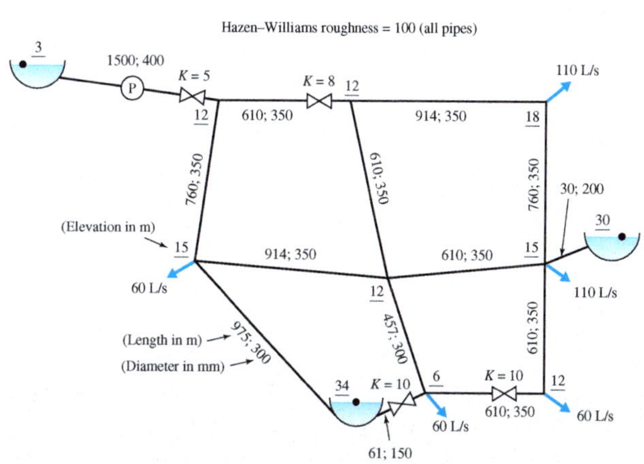

Figure P9.41

9.42 The layout of the required water demands for a proposed industrial complex are shown in Figure P9.42. Design an appropriate network and determine the appropriate useful power for a pump to meet the demand needs. The design criteria are as follows:

1. The lower reservoir supplies water to the system. The upper reservoir supplies water for emergency use only, such as fire, line breaks, and so on; under normal operation, there is no flow into or out of it.
2. Under normal operation, pressures in the network can range between 80 and 120 psi.
3. Two gate valves are to be placed in each line (one at each end) to isolate it in case of a break or for needed maintenance.
4. All demands must be met in case of a single line break with minimum allowable pressures of 20 psi.
5. Pipes are cast iron, with the following standard diameters available: 4, 6, 8, 12, 16, 20, 24, 30, 36, 42, and 48 in. Pipes can be placed anywhere within the region.

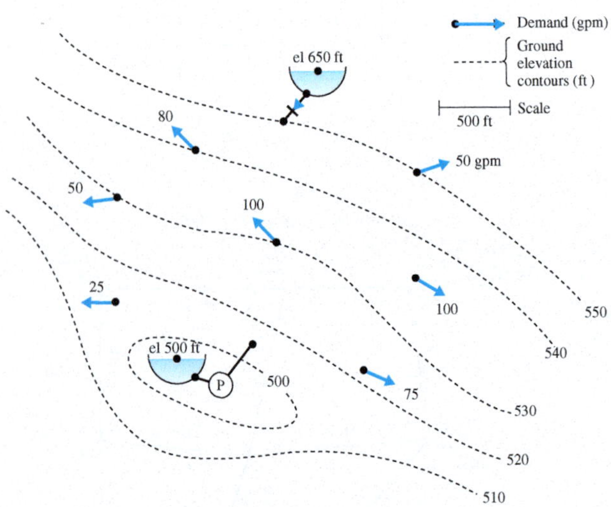

Figure P9.42

9.43 Design an irrigation system for part or all of the 18-hole golf course shown in Figure P9.43. The requirements are as follows:

1. Place the sprinkler couplings at approximately 100-ft intervals along the fairway centerline. The flow rate from each sprinkler is to be 40 gal/min. A maximum of two sprinklers on each fairway are to be in operation at any time.
2. Place four sprinklers around the periphery of each green. The flow rate from each sprinkler is to be 40 gal/min. A maximum of three sprinklers on each green are to be in operation at any time. As an alternative to specifying the sprinkler flow rate, specify the loss coefficient across each sprinkler (e.g., try $K_v = 35$), and with successive trials, adjust it until the desired flow rate is achieved.
3. Water is to be pumped from the well shown. The maximum possible flow from the well is 2000 gal/min. Work with useful power in order to satisfy the system demand.
4. System pressure is to be 80 ± 10 psi at all locations, and the design velocity in each pipe is not to exceed 5 ft/s. Use smooth plastic piping and fittings throughout.

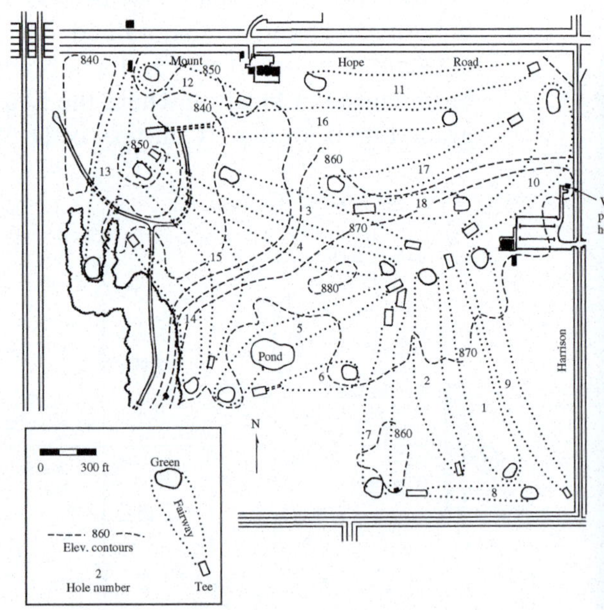

Figure P9.43

Unsteady Flow

9.44 An irrigation system has a nearly horizontal supply pipe that is connected to a tank at one end and a quick-opening valve at the other. The length of the pipe is 2500 m, and its diameter is 100 mm. If the elevation of water in the tank is 3 m, how long will it take for the flow to reach the 99% steadystate condition if the valve is suddenly opened from a closed position? Assume that the water is incompressible, the pipe is inelastic, $f = 0.025$, and $K = 0.15$ once the valve is opened.

9.45 In Problem 9.44 plot the velocity in the pipe as a function of time.

9.46 Gasoline is supplied by gravity without pumping from a storage tank through a 800-m-long 50-mm-diameter nearly horizontal pipe into a tanker truck. There is a quick-acting valve at the end of the pipe. The difference in elevations of gasoline between the reservoir and the truck tank is 8 m. Initially, the valve is partially closed so that $K = 275$. Then the operator decides to increase the discharge by opening the valve quickly to the position where $K = 5$. Assuming an incompressible fluid and an inelastic piping system, determine the new steady-state discharge and the time it takes to reach 95% of that value. Assume that $f = 0.015$.

9.47 Water is flowing through a pipe whose diameter is 200 mm and whose length is 800 m. The pipe is made of cast iron ($E = 150$ GPa) with a thickness of 12 mm. There is a reservoir at the upstream end of the pipe and a valve at the downstream end. Under steady-state conditions the discharge is $Q = 0.05$ m^3/s, when a valve at the end of the pipe is actuated very rapidly so that water hammer occurs.

(a) How long does it take for an acoustic wave to travel from the valve to the reservoir and back to the valve?

(b) Determine the change in pressure at the valve if the valve is opened such that the discharge is doubled.

(c) Determine the change in pressure at the valve if the valve is closed such that the discharge is halved.

9.48 When operating the system described in Problem 9.46, the operator decides to suddenly close the valve once steady-state flow has been reached. Assuming that the gasoline is slightly compressible and the piping is elastic, determine:

(a) The acoustic wave speed.

(b) The pressure rise at the valve once the valve is rapidly closed in a manner to cause water hammer.

(c) If the maximum allowable pressure in the pipe is 250 kPa, what can you conclude about the outcome of the water (gasoline) hammer activity?

The pipe is made of aluminum, $E = 70$ GPa, with 2.5-mm-thick walls, and the bulk modulus of elasticity of gasoline is $B = 1.05$ GPa.

9.49 Oil with a specific gravity of $S = 0.90$ is flowing at 20 ft^3/s through a 20-in.-diameter 13,000-ft-long pipe. The elastic modulus of the steel pipe is $E = 29 \times 10^6$ lb/in^2, its thickness is 0.40 in. and the bulk modulus of elasticity of the oil is $B = 217{,}000$ lb/in^2. A valve at the downstream end of the pipe is partially closed very rapidly so that a water hammer event is initiated and a pressure wave propagates upstream. If the magnitude of the wave is not to exceed 90 lb/in^2, determine:

(a) The percent decrease of flow rate tolerable during the valve closure.

(b) The time it takes the pressure wave to reach the upstream end of the pipe.

10 Turbomachinery

Outline

Chapter Objectives

The objectives of this chapter are to:

- Develop fundamental pump theory using the moment-of-momentum principle
- Describe radial-flow, mixed-flow, and axial-flow pumps, including presentation of prototype data and deviation from ideal conditions
- Introduce dimensionless pump coefficients and show how they are used in conjunction with similarity rules to develop data for turbomachinery design and analysis
- Demonstrate how pumps are incorporated into piping design and analysis
- Present introductory material on turbines: basic theory, types of turbines, and their selection and operation in conjunction with piping systems

10.1 INTRODUCTION

Turbomachines are the commonly employed devices that either supply or extract energy from a flowing fluid by means of rotating propellers or vanes. A turbopump, more commonly called a pump, adds energy to a system, with the result that the pressure is increased; it also causes flow to occur or it increases the rate of flow. Pumps are essential components of piping systems which are designed to convey liquids. Similarly, turbomachines are called blowers, fans, or compressors when performing work on air or other gases in ducts. A turbine extracts energy from a system and converts it to some other useful form, typically, to electric

M. C. Potter and D. C. Wiggert, *Applied Fluid Mechanics*,
https://doi.org/10.1007/978-3-031-65276-9_10

power. A hydroturbine, or simply a turbine, is a machine that generates power from high-pressure water; relatively large conduits or tunnels deliver fluid to closed turbines in order to generate power. Steam and air turbines are also of substantial engineering importance, but such devices are considered in other courses, such as thermodynamics.

The examples above are all those of turbomachines designed to facilitate or utilize internal flows. A wind turbine, on the other hand, makes use of the surrounding external flow to convert the energy contained in the natural movement of atmospheric air into useful electrical power. As shown in Section 2.6.4, a propeller performs work on the surrounding fluid to provide thrust and propel an object along a desired path, while a stationary fan performs work to circulate air. All turbomachines are characterized as being capable of either adding or subtracting energy from the fluid by means of rotating propellers or vanes. There is no contained volume of fluid being transported, as in a positive-displacement piston pump.

The bulk of this chapter is devoted to pumps, in particular, those that convey liquids such as water or gasoline. The centrifugal, or radial-flow pump is presented in detail, and to a lesser extent, mixed-flow and axial-flow pumps are studied. Dimensional analysis, an important tool for the selection and design of turbomachines, is presented next, followed by a discussion of the proper selection and implementation of pumps in piping systems. Finally, turbines are introduced by considering their fundamental characteristics.

10.2 TURBOPUMPS

A turbopump consists of two principal parts: an **impeller**, which imparts a rotary motion to the liquid, and the pump **housing**, or **casing**, which directs the liquid into the impeller region and transports it away under a higher pressure. Figure 10.1 shows a typical single-suction **radial-flow pump**. The impeller is mounted on a shaft and is often driven by an electric motor. The casing includes the suction and discharge nozzles and houses the impeller assembly. The portion of the casing surrounding the impeller is termed the **volute**. Liquid enters through the suction nozzle to the impeller eye and travels along the **shroud**, developing a rotary motion due to the impeller vanes. It leaves the volute casing peripherally at a higher pressure through the discharge nozzle. Some single-suction impellers are open, with the front shroud removed. Double-suction impellers have liquid entering from both sides.

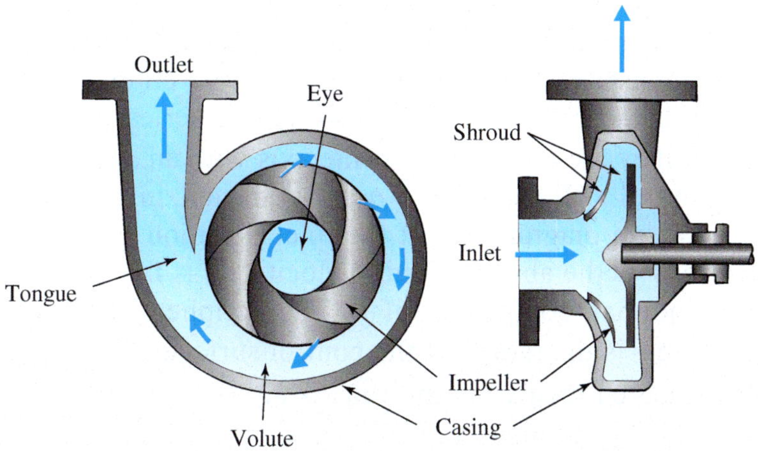

Figure 10.1 Single-suction pump.

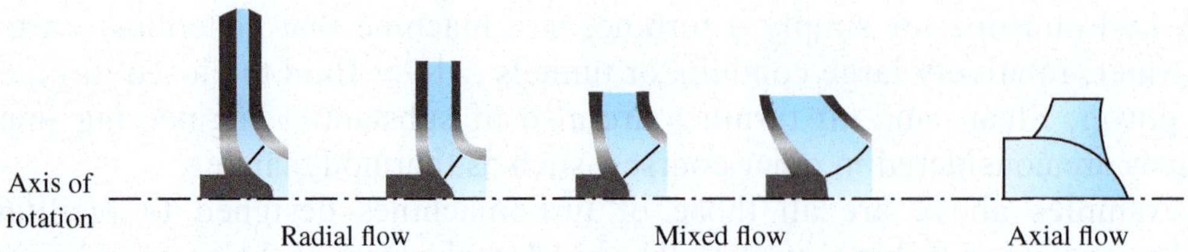

Figure 10.2 Various types of pump impellers.

In a radial-flow pump, the impeller vanes are commonly curved backward and the impeller is relatively narrow. As the impeller becomes wider, the vanes have a double curvature, becoming twisted at the suction end. Such pumps convey liquids with a lower-pressure rise than radial-flow pumps and are called **mixed-flow pumps**. At the opposite extreme from the radial-flow pump is the **axial-flow pump**; it is characterized by the flow entering and leaving the impeller region axially, parallel to the shaft axis. Typically, an axial-flow pump delivers liquid with a relatively low pressure rise. For axial-flow pumps and some mixed-flow pumps, the impellers are open; that is, there is no shroud surrounding them. Various types of impellers are shown in Figure 10.2.

10.2.1 Radial-Flow Pumps

A representative radial-flow pump is shown in Figure 10.1; this type of pump is commonly called a **centrifugal pump**, and it is the most commonly used pump in existence today. An elementary analysis will be performed on a radial-flow pump, which will provide a theoretical relationship between the discharge and the developed head rise. In addition, a better understanding of the manner in which momentum exchange takes place in such a turbomachine will be provided. Actual pumps operate at efficiencies less than unity; that is, they do not perform at the theoretical, idealized conditions. It is necessary, therefore, to conduct experiments so that the true operating characteristics of a turbopump can be determined.

Elementary Theory. The actual flow patterns in a turbopump are highly three-dimensional with significant viscous effects and separation patterns taking place. To construct a simplified theory for the radial-flow pump, it is necessary to neglect viscosity and to assume idealized two-dimensional flow throughout the impeller region. Figure 10.3a on the next page defines a control volume that encompasses the impeller region. Flow enters through the inlet control surface and exits through the outlet surface. Note that a series of vanes exists within the control volume, and that they are rotating about the axis with an angular speed ω.

A portion of the control volume is shown at an instant in time in Figure 10.3b. The idealized velocity vectors are diagrammed at the inlet, location 1, and the outlet, location 2. In the velocity diagrams, V is the absolute fluid velocity, V_t is the tangential component of V, and V_n is the radial, or normal, component of V. The peripheral or circumferential speed of the blade is $u = \omega r$, where r is the radius of the control surface. The angle between V and u is α. The fluid velocity measured relative to the vane is v.

The relative velocity is assumed to be always tangent to the vane; that is, perfect guidance of the fluid throughout the control volume takes place. The angle between v and u is designated as β; since perfect guidance along the vane is assumed, β designates the blade angle as well.

The moment-of-momentum relation, Eq. 2.7.3, can be written for steady flow in the form

$$\Sigma \mathbf{M} = \int_{\mathrm{c.s}} \rho \mathbf{r} \times \mathbf{V}(\mathbf{V} \cdot \hat{\mathbf{n}})\, dA \tag{10.2.1}$$

Applied to the control volume of Figure 10.3, this becomes

$$T = \rho Q(r_2 V_{t2} - r_1 V_{t1}) \tag{10.2.2}$$

in which T is the torque acting on the fluid in the control volume, and the right-hand side represents the flux of angular momentum through the control volume.

The power delivered to the fluid is the product of ω and T:

$$\omega T = \rho Q(u_2 V_{t2} - u_1 V_{t1}) \tag{10.2.3}$$

From the velocity vector diagrams in Figure 10.3b, $V_{t1} = V_1 \cos \alpha_1$, and $V_{t2} = V_2 \cos \alpha_2$ so that Eq. 10.2.3 can be written as

$$\omega T = \rho Q(u_2 V_2 \cos \alpha_2 - u_1 V_1 \cos \alpha_1) \tag{10.2.4}$$

For the idealized situation in which there are no losses, the delivered power must be equal to $\gamma Q H_t$, in which H_t is the theoretical pressure head rise across the pump (see Eq. 2.5.26). Then there results Euler's turbomachine relation,

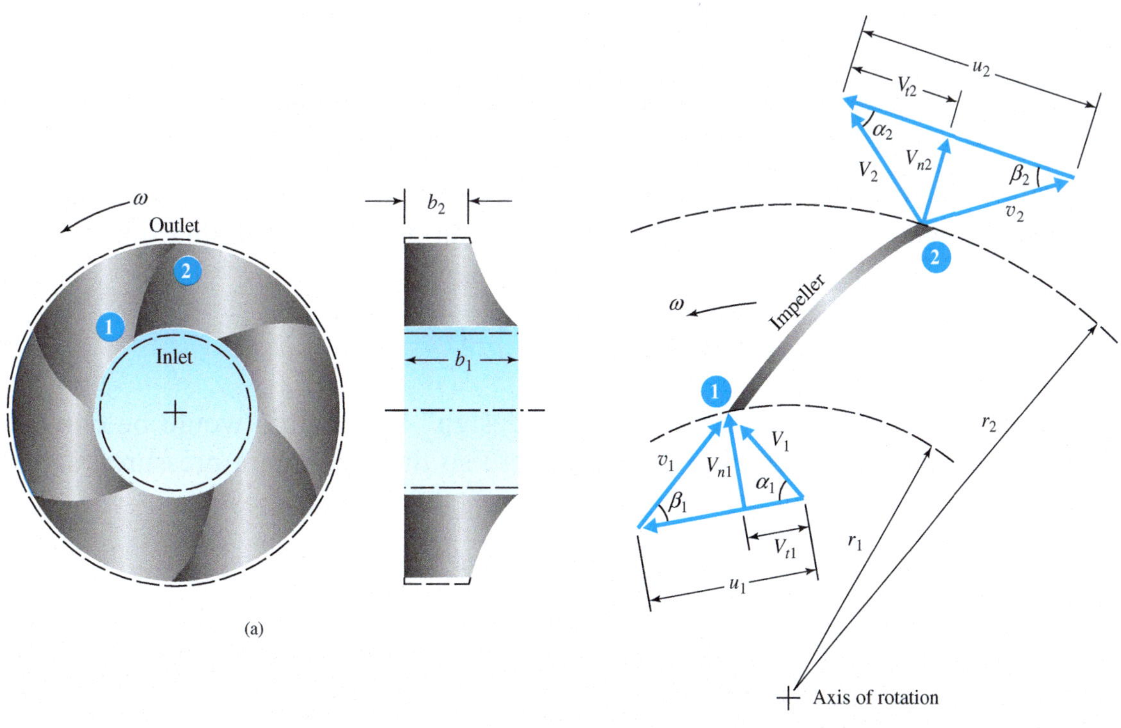

Figure 10.3 Idealized radial-flow impeller; (a) impeller control volume; (b) velocity diagrams at control surfaces.

$$H_t = \frac{\omega T}{\gamma Q}$$
$$= \frac{u_2 V_2 \cos\alpha_2 - u_1 V_1 \cos\alpha_1}{g} \tag{10.2.5}$$

Insight on the nature of flow through an impeller region can be obtained using Eq.10.2.5. From the law of cosines we can write

$$v_1^2 = u_1^2 + V_1^2 - 2u_1 V_1 \cos\alpha_1$$
$$v_2^2 = u_2^2 + V_2^2 - 2u_2 V_2 \cos\alpha_2$$

These can be substituted into Eq.10.2.5 to provide the relation

$$H_t = \frac{V_2^2 - V_1^2}{2g} + \frac{\left(u_2^2 - u_1^2\right) - \left(v_2^2 - v_1^2\right)}{2g} \tag{10.2.6}$$

The first term on the right-hand side represents the gain in kinetic energy as the fluid passes through the impeller; the second term accounts for the increase in pressure across the impeller. This can be seen by applying the energy equation across the impeller and solving for H_t:

$$H_t = \frac{p_2 - p_1}{\gamma} + z_2 - z_1 + \frac{V_2^2 - V_1^2}{2g} \tag{10.2.7}$$

Eliminating H_t, between Eqs. 10.2.6 and 10.2.7 yields the expression

$$\frac{p_1}{\gamma} + z_1 + \frac{v_1^2 - u_1^2}{2g} = \frac{p_2}{\gamma} + z_2 + \frac{v_2^2 - u_2^2}{2g} = \text{const} \tag{10.2.8}$$

This relation has historically been called the Bernoulli equation in rotating coordinates. Since $z_2 - z_1$ is usually much smaller than $(p_2 - p_1)/\gamma$, it can be eliminated, and thus the pressure difference is

$$p_2 - p_1 = \frac{\rho}{2}\left[\left(v_1^2 - u_1^2\right) - \left(v_2^2 - u_2^2\right)\right] \tag{10.2.9}$$

Returning to Eq. 10.2.5, we see that a "best design" for a pump would be one in which the angular momentum entering the impeller is zero so that maximum pressure rise can take place. Then in Figure 10.3b, $\alpha_1 = 90°$, $V_{n1} = V_1$, and Eq. 10.2.5 becomes

$$H_t = \frac{u_2 V_2 \cos\alpha_2}{g} \tag{10.2.10}$$

From the triangle geometry of Figure 10.3, $V_2 \cos\alpha_2 = u_2 - V_{n2} \cot\beta_2$, so that Eq. 10.2.10 takes the form

$$H_t = \frac{u_2^2}{g} - \frac{u_2 V_{n2} \cot\beta_2}{g} \tag{10.2.11}$$

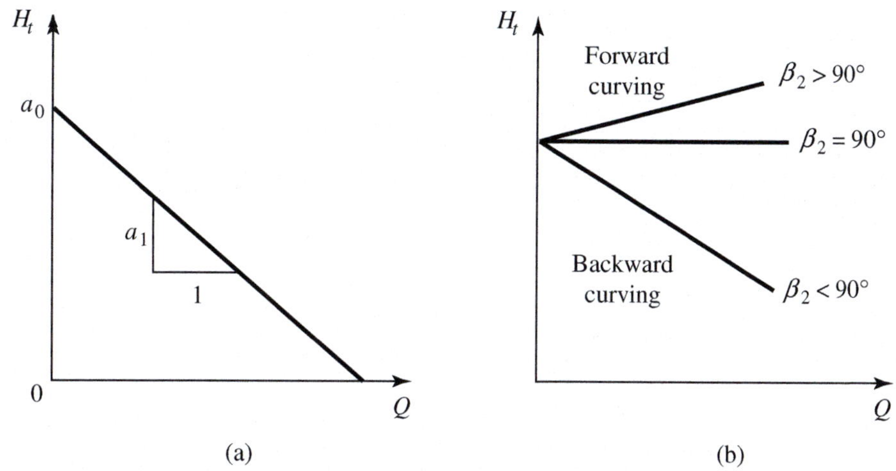

Figure 10.4 Ideal pump performance curves.

Applying the continuity principle at the outlet region to the control volume provides the relation

$$V_{n2} = \frac{Q}{2\pi r_2 b_2} \tag{10.2.12}$$

in which b_2 is the width of the impeller at location 2. Introducing Eq. 10.2.12 into Eq. 10.2.11, and recalling that $u_2 = \omega r_2$, one has the relation

$$H_t = \frac{\omega^2 r_2^2}{g} - \frac{\omega \cot \beta_2}{2\pi b_2 g} Q \tag{10.2.13}$$

For a pump running at constant speed, Eq. 10.2.13 takes the form

$$H_t = a_0 - a_1 Q \tag{10.2.14}$$

in which a_0 and a_1 are constants. Equation 10.2.13 is the **theoretical head curve** and is seen to be a straight line with a slope of $-a_1$, as shown in Figure 10.4a. The effect of the blade angle β_2 is shown in Figure 10.4b. A forward curving blade ($\beta_2 > 90°$) can be unstable and cause pump surge, where the pump oscillates in an attempt to establish an operating point. Backward-curving vanes ($\beta_2 < 90°$) are generally preferred.

Example 10.1

A radial-flow pump has the following dimensions;

$$\beta_1 = 44° \qquad r_1 = 21\text{ mm} \qquad b_1 = 11\text{ mm}$$
$$\beta_2 = 30° \qquad r_2 = 66\text{ mm} \qquad b_2 = 5\text{ mm}$$

For a rotational speed of 2500 rev/min, assuming ideal conditions (frictionless flow, negligible vane thickness, perfect guidance), with $\alpha_1 = 90°$ (no prerotation), determine (a) the discharge, theoretical head, required power, and pressure rise across the impeller, and (b) the theoretical head-discharge curve. Use water as the fluid.

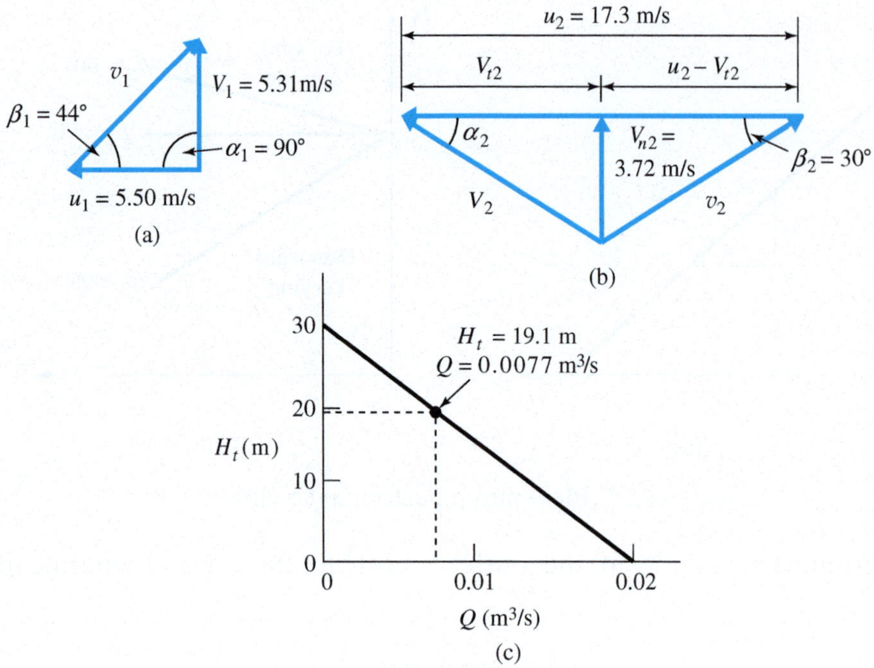

Figure E.10.1

Solution

(a) Construct the velocity diagram at location 1, as shown in Figure E10.1a. The rotational speed is converted to the appropriate units as

$$\omega = 2500\frac{2\pi}{60} = 261.8 \text{ rad/s}$$

The impeller speed at r_1 is then

$$u_1 = \omega r_1 = 261.8 \times 0.021 = 5.50 \text{ m/s}$$

From the velocity diagram we see that

$$V_1 = u_1 \tan\beta_1 = 5.50 \tan 44° = 5.31 \text{ m/s}$$

and since $\alpha_1 = 90°$, $V_1 = V_{n1}$, or $V_{n1} = 5.31$ m/s. The discharge is computed to be

$$\begin{aligned} Q &= 2\pi r_1 b_1 V_{n1} \\ &= 2\pi \times 0.021 \times 0.011 \times 5.31 = 7.71 \times 10^{-3} \text{ m}^3\text{/s} \quad \text{or} \quad \underline{7.71 \text{ L/s}} \end{aligned}$$

The normal component of velocity at location 2 is

$$V_{n2} = \frac{Q}{2\pi r_2 b_2} = \frac{7.71 \times 10^{-3}}{2\pi \times 0.066 \times 0.005} = 3.72 \text{ m/s}$$

and the impeller speed at the outlet is

$$u_2 = \omega r_2 = 261.8 \times 0.066 = 17.28 \text{ m/s}$$

The velocity diagram at location 2 is now sketched as shown in Figure E10.1b. From the velocity diagram we see that

$$u_2 - V_{t2} = \frac{V_{n2}}{\tan\beta_2} = \frac{3.72}{\tan 30°} = 6.44 \text{ m/s}$$

Therefore,

$$V_{t2} = u_2 - 6.44 = 17.28 - 6.44 = 10.84 \text{ m/s}$$

$$\alpha_2 = \tan^{-1}\frac{V_{n2}}{V_{t2}} = \tan^{-1}\frac{3.72}{10.84} = 18.9°$$

$$V_2 = \frac{V_{t2}}{\cos\alpha_2} = \frac{10.84}{\cos 18.9°} = 11.46 \text{ m/s}$$

The theoretical head is computed with Eq. 10.2.10:

$$H_t = \frac{u_2 V_2 \cos\alpha_2}{g} = \frac{17.28 \times 11.46 \times \cos 18.9°}{9.81} = \underline{19.1 \text{ m}}$$

Hence, the theoretical required power is

$$\dot{W}_p = \gamma Q H_t = 9810 \times (7.71 \times 10^{-3}) \times 19.1 = \underline{1440 \text{ W}}$$

The pressure rise is determined from the energy equation as follows;

$$\begin{aligned} p_2 - p_1 &= \left(H_t + \frac{V_1^2 - V_2^2}{2g} \right)\gamma \\ &= \left[19.1 + \frac{(5.31)^2 - (11.46)^2}{2 \times 9.81} \right] \times 9810 = 1.36 \times 10^5 \text{ Pa} \quad \text{or} \quad \underline{136 \text{ kPa}} \end{aligned}$$

(b) The theoretical head-discharge curve is Eq. 10.2.13. For the present example we have

$$\begin{aligned} H_t &= \frac{(\omega r_2)^2}{g} - \frac{\omega \cot\beta_2}{2\pi b_2 g} Q \\ &= \frac{(261.8 \times 0.066)^2}{9.81} - \frac{261.8 \cot 30°}{2\pi \times 0.005 \times 9.81} Q \\ &= 30.4 - 1471 Q \end{aligned}$$

The curve is shown in Figure E10.1c.

Head-Discharge Relations: Performance Curves. For real fluid flow, the theoretical head curve cannot be achieved in practice: it is necessary to resort to experimentation to determine the actual head-discharge curve. The energy equation written across a pump from the suction side (location 1, Figure 10.3) to the discharge side (location 2) is

$$\begin{aligned} H_p &= \left(\frac{p}{\gamma}+\frac{V^2}{2g}+z\right)_2 - \left(\frac{p}{\gamma}+\frac{V^2}{2g}+z\right)_1 \\ &= H_t - h_L \end{aligned} \tag{10.2.15}$$

in which H_P is the actual head across the pump, and h_L represents the losses through the pump. The actual power delivered to the fluid, designated as $\dot{W}_f$, is

$$\dot{W}_f = \gamma Q H_P \tag{10.2.16}$$

while the power delivered to the impeller is $\dot{W}_P$, often termed the **brake power**, and is given by

$$\dot{W}_P = \omega T \tag{10.2.17}$$

If there were no losses, $\dot{W}_f$ would be equal to $\dot{W}_P$. Since in actuality $\dot{W}_f < \dot{W}_P$, the **pump efficiency**[1] η_P is defined as

$$\eta_P = \frac{\dot{W}_f}{\dot{W}_P} = \frac{\gamma Q H_P}{\omega T} \tag{10.2.18}$$

One objective of pump design is to make the efficiency as high as possible. The theoretical head curve is compared with the actual head curve in Figure 10.5 on the next page. The difference between the two can be attributed to the following effects: (1) prerotation of the

---

[1]More precisely, η_P defined by Eq. 10.2.18 is termed the **overall efficiency**. It is possible to define three additional efficiencies for pumps as follows:

Hydraulic efficiency : $\eta_H = \dfrac{H_P}{H_t}$

Volumetric efficiency : $\eta_V = \dfrac{Q}{Q+Q_L}$, Q_L = leakage flow rate

Mechanic efficiency : $\eta_M = \dfrac{\gamma H_t (Q+Q_L)}{T\omega}$

These are related to the overall efficiency by $\eta_P = \eta_H\, \eta_V\, \eta_M$.

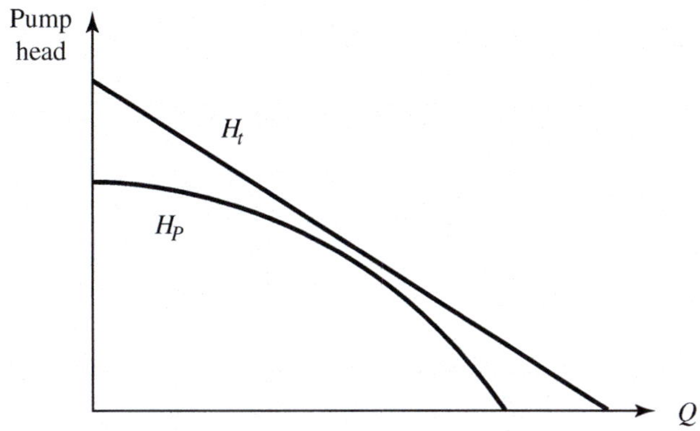

Figure 10.5 Comparison between theoretical and actual radial-flow pump performance curves.

fluid before entering the impeller region, (2) separation due to imperfect guidance of fluid as it enters the impeller region, and (3) separation due to expansion in the flow passages. Leakage and high shear rates created by the differences in velocity between fluid particles in the volute and impeller contribute further to the losses. Finally, impeller blades are designed to be most efficient at the so-called design discharge; at any other discharge—that is, "off-design"—the performance will deteriorate.

A representative pump and its performance curve are shown in Figure 10.6; this is the curve supplied by the pump manufacturer. Typically, more than one impeller diameter will be available; in the diagram, four different impellers are represented. In addition to the head-discharge performance curves, isoefficiency, power, and **net positive suction head** (*NPSH*) curves are usually provided. Net positive suction head is associated with concern for cavitation, which is discussed in Section 10.2.3.

10.2.2 Axial- and Mixed-Flow Pumps

The velocity diagram for an axial-flow pump is shown in Figure 10.7. In the axial-flow pump, there is no radial flow and the liquid particles leave the impeller at the same radius at which they enter, so that $u_1 = u_2 = u$. Furthermore, assuming a uniform flow, continuity considerations give $V_{n1} = V_{n2} = V_n$. Equation 10.2.5, which is valid for an axial-flow pump as well as a radial-flow pump, can be combined with the identities $V_2 \cos\alpha_2 = u - V_n \cot\beta_2$ and $V_1 \cos \alpha_1 = V_n \cot \alpha_1$ to produce

$$H_t = \frac{u^2}{g} - \frac{uV_n}{g}(\cot \alpha_1 + \cot \beta_2) \tag{10.2.19}$$

This form of the turbomachine relation is useful when the ideal absolute velocity entrance angle α_1 is established by a fixed vane, or stator. If there is no prerotation, $\alpha_1 = 90°$ and the theoretical head relation, Eq. 10.2.19, becomes

$$H_t = \frac{u^2}{g} - \frac{uV_n \cot \beta_2}{g} \tag{10.2.20}$$

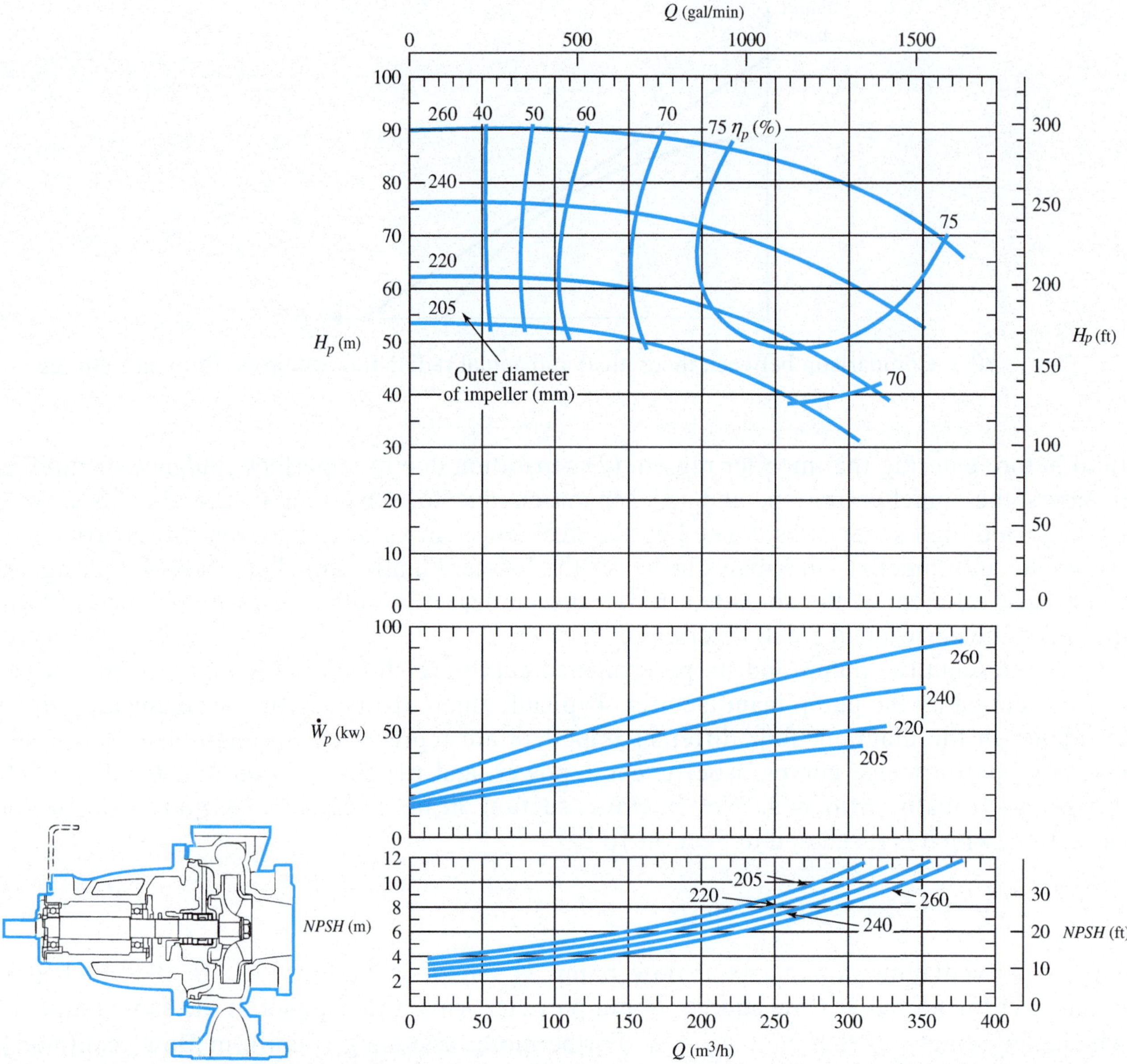

Figure 10.6 Radial-flow pump and performance curves for four different impellers with $N = 2900$ rpm ($\omega = 304$ rad/s). Water at 20°C is the pumped liquid.

This relation is identical to Eq. 10.2.11, which relates the theoretical head to the impeller outlet parameters for a radial flow pump. Note, however, that Eq. 10.2.11 is valid at the periphery of the impeller for the radial-flow pump, and that all fluid particles reach the maximum head at that location. By contrast, Eq. 10.2.20 pertains only at a specified radius for the axial-flow pump, since fluid particles enter and leave the control volume at their respective radii. In this case the head varies from a minimum at the axis to a maximum at the

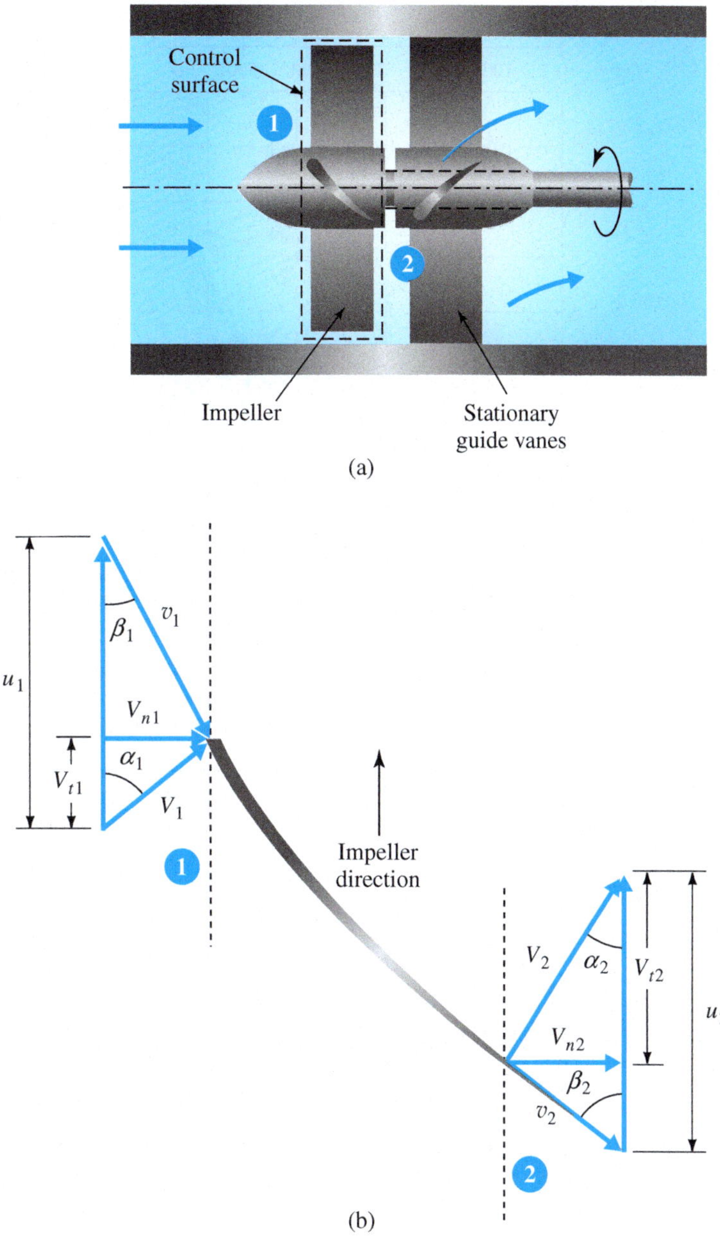

Figure 10.7 Idealized axial-flow impeller: (a) impeller control volume; (b) velocity diagrams at control surface.

periphery, and the total pump head is an integrated average. Axial-flow impellers are designed to maintain a constant axial velocity throughout the impeller region; this requires that the vane angles increase gradually from the periphery to the axis, as well as from the inlet to the outlet region. Figure 10.8 shows a representative axial-flow pump, and its performance curves.

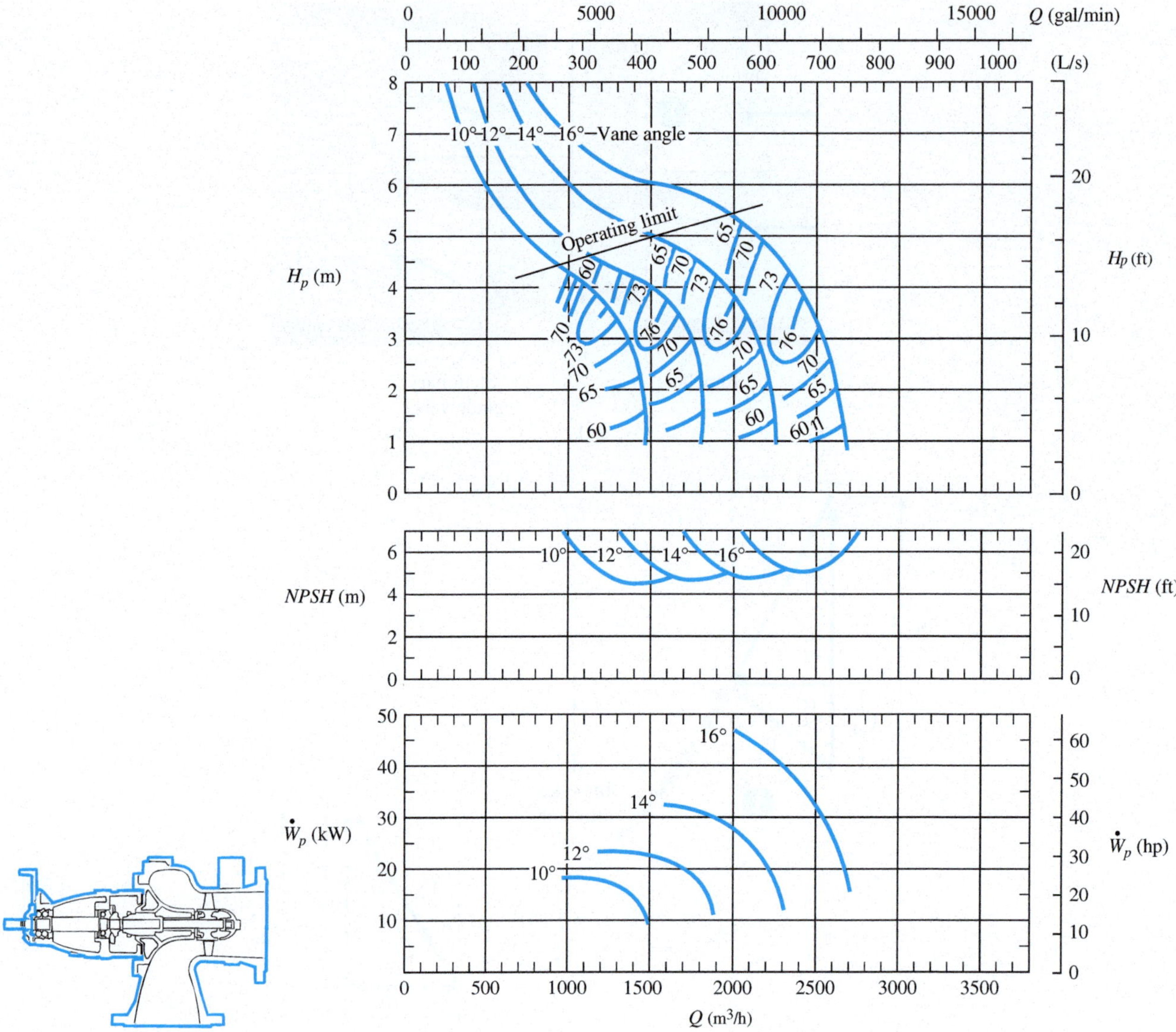

Figure 10.8 Axial-flow pump and performance curves for four different vane angles with $N = 880$ rpm ($\omega = 92.2$ rad/s). Propeller diameter is 500 mm. Water at 20°C is the pumped liquid.

Generally, flow in an impeller region consists of two components: the circular, vortex-like motion due to the action of the vanes, and the net flow through the impeller. The vortex motion is superposed on the radial outward flow in a radial-flow pump, and on the axial flow in an axial-flow pump. The action of the mixed-flow pump lies between these two types. The previous discussions relating to radial- and axial-flow pumps pertain to mixed-flow pumps as well. A mixed-flow pump and its characteristic curves are shown in Figure 10.9.

As in the case for a radial-flow pump, the performance characteristics of axial- and mixed-flow pumps deviate from the idealized considerations. Generally, radial-flow pumps

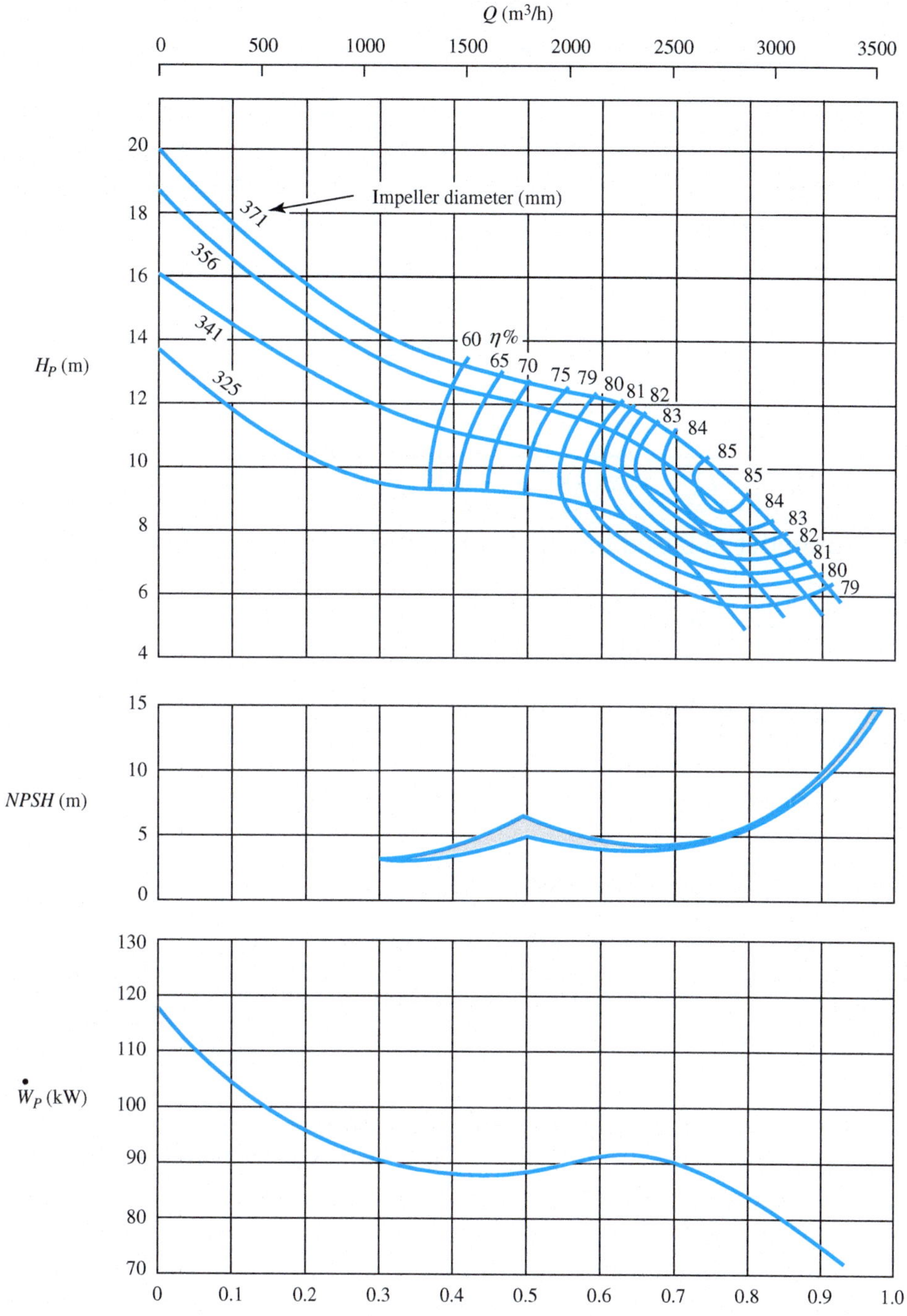

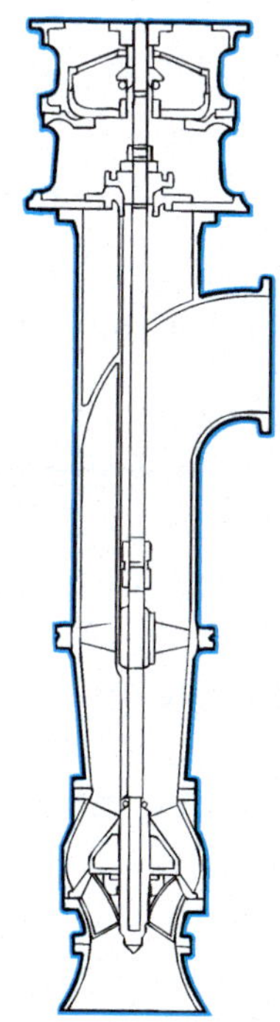

Figure 10.9 Mixed-flow pump and performance curves for four different impellers and vane angles with $N = 970$ rpm ($\omega = 102$ rad/s). Water at 20°C is the pumped liquid.

are designed to deliver relatively low discharges at high heads, and axial-flow pumps produce relatively large discharges at low heads. Mixed-flow pumps deliver heads and discharges between these two extremes. A method of selecting a pump appropriate for required design considerations is presented in Section 10.4.

Example 10.2

An axial-flow pump is designed with a fixed guide vane, or stator blade, located upstream of the impeller. The stator imparts an angle $\alpha_1 = 75°$ to the fluid as it enters the impeller region. The impeller has a rotational speed of 500 rpm with a blade exit angle of $\beta_2 = 70°$. The control volume has an outer diameter of $D_o = 300$ mm and an inner diameter of $D_i = 150$ mm. Determine the theoretical head rise and power required if 150 L/s of liquid ($S = 0.85$) is to be pumped.

Solution

First, the normal velocity component V_n is

$$V_n = \frac{Q}{A} = \frac{Q}{(\pi/4)(D_o^2 - D_i^2)} = \frac{0.15}{(\pi/4)(0.3^2 - 0.15^2)} = 2.83 \text{ m/s}$$

The peripheral speed u of the impeller is based on an average radius:

$$u \simeq \omega \frac{D_o + D_i}{4} = 500\left(\frac{2\pi}{60}\right)\frac{0.3 + 0.15}{4} = 5.89 \text{ m/s}$$

The theoretical head H_t is computed with Eq.10.2.19 to be

$$H_t = \frac{u}{g}[u - V_n(\cot\alpha_1 + \cot\beta_2)] = \frac{5.89}{9.81}[5.89 - 2.83(\cot 75° + \cot 70°)] = \underline{2.46 \text{ m}}$$

Finally, for the assumed ideal conditions, the required power is

$$\dot{W}_P = \gamma Q H_t = (9810 \times 0.85) \times 0.15 \times 2.46 = 3080 \text{ W or } \underline{3.1 \text{ kW}}$$

10.2.3 Cavitation in Turbomachines

Cavitation refers to conditions at certain locations within the turbomachine where the local pressure drops to the vapor pressure of the liquid, and as a result, vapor-filled cavities are formed. As the cavities are transported through the turbomachine into regions of greater pressure, they will collapse rapidly, generating extremely high localized pressures. Those bubbles that collapse close to solid boundaries can weaken the solid surface, and after repeated collapsing, pitting, erosion, and fatigue of the surface can occur. Signs of cavitation in turbopumps include noise, vibration, and lowering of the head-discharge and efficiency curves. Regions most susceptible to damage in a turbomachine are those slightly beyond the low-pressure zones on the back side of impellers (see Figure 10.10). In general, sudden changes in direction, sudden increases in area, and lack of streamlining are the culprits causing cavitation damage in turbopumps (Karassik, et al., 1986).

The proper design of turbopumps will minimize the possibility for cavitation to occur. Under adverse operating conditions, however, the pressures may decrease and cavitation may occur. Two parameters are used to designate the potential for cavitation: the cavitation number and the net positive suction head. Their interrelationship and use are now discussed.

Figure 10.10 Cavitation damage in an impeller. (Source: Roplant/shutterstock.com)

Consider a pump operating in the manner shown in Figure 10.11. Location 1 is on the liquid surface on the suction side, and location 2 is the point of minimum pressure within the pump. Write the energy equation from location 1 to location 2 and use an absolute pressure datum; there results

$$\frac{V_2^2}{2g} = \frac{p_{\text{atm}} - p_2}{\gamma} - \Delta z - h_L \tag{10.2.21}$$

in which h_L is the loss between location 1 and location 2, $\Delta z = z_2 - z_1$, and the kinetic energy at location 1 is assumed negligible. The minimum allowable pressure at location 2 is the vapor pressure p_v. Substituting this into the expression, one can say that the left-hand side of Eq. 10.2.21 represents the maximum kinetic energy head possible at location 2 when cavitation is imminent. Thus, the **net positive suction head** (*NPSH*) is defined as

$$NPSH = \frac{p_{\text{atm}} - p_v}{\gamma} - \Delta z - h_L \tag{10.2.22}$$

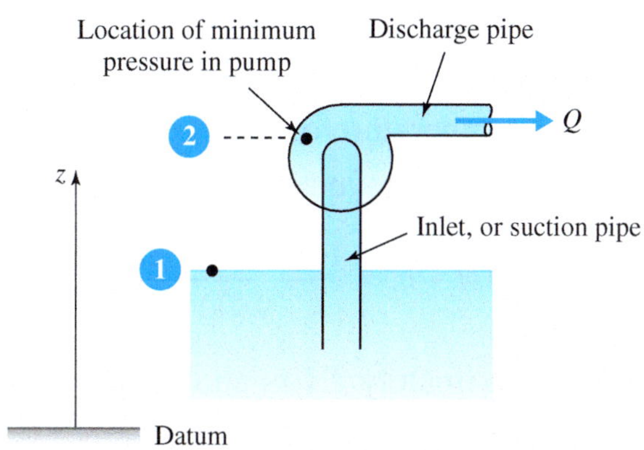

Figure 10.11 Cavitation setting for a pump.

The *NPSH is* also used for turbines; however, the sign of the h_L term in Eq. 10.2.22 changes, and location 1 refers to the liquid surface on the discharge side of the machine. The design requirement for a pump is thus established as follows:

$$NPSH \leq \frac{p_{\text{atm}} - p_v}{\gamma} - \Delta z - h_L \tag{10.2.23}$$

The performance data supplied by turbomachinery manufacturers usually includes *NPSH* curves; these are developed by testing a given family of machines in a laboratory environment. Figures 10.6, 10.8, and 10.9 show *NPSH* curves. The *NPSH* curve enables one to specify the required maximum value of Δz to be used for a given turbomachine; note that it is necessary to estimate h_L to obtain this.

The right-hand side of Eq. 10.2.22 can be divided by H_P, the total head across the pump to yield

$$\sigma = \frac{1}{H_P} \frac{(p_{\text{atm}} - p_v)}{\gamma} - \Delta z - h_L \tag{10.2.24}$$

in which σ is the **Thoma cavitation number**. This parameter is used as an alternative to the *NPSH* to establish design criteria for cavitation. A **critical cavitation number**, which signals that cavitation is imminent, is determined experimentally. Thus, for no cavitation to occur, σ must be greater than the critical cavitation number. The cavitation number is in dimensionless form, which is preferred to the dimensional form of the *NPSH*.

Example 10.3

Determine the elevation at which the 240-mm-diameter pump of Figure 10.6 can be situated above the water surface of the suction reservoir without experiencing cavitation. Water at 15°C is being pumped at 250 m^3/h. Neglect losses in the system. Use $p_{\text{atm}} = 101$ kPa.

Solution

From Figure 10.6, at a discharge of 250 m^3/h, the *NPSH* for the 240-mm-diameter impeller is approximately 7.4 m. For a water temperature of 15°C, $p_v = 1666$ Pa absolute, and $\gamma = 9800$ N/m^3. Equation 10.2.22 with $h_L = 0$ is employed to compute Δz to be

$$\Delta z = \frac{p_{\text{atm}} - p_v}{\gamma} - NPSH - h_L = \frac{101000 - 1666}{9800} - 7.4 - 0 = \underline{2.74 \text{ m}}$$

Thus, the pump can be placed approximately 2.7 m above the suction reservoir water surface.

10.3 DIMENSIONAL ANALYSIS AND SIMILITUDE FOR TURBOMACHINERY

The development and utilization of turbomachinery in engineering practice has benefited greatly from the application of dimensional analysis, probably to an extent more than any other area of applied fluid mechanics. It has enabled pump and turbine manufacturers to test and develop a relatively small number of turbomachines and subsequently produce a series of commercial units that cover a broad range of head and flow demands. The material developed in Chapter 4, which covers similitude and dimensional analysis, is applied to turbomachines in this section.

10.3.1 Dimensionless Coefficients

The following parameters can be considered significant ones for a turbomachine: power, rotational speed, outer diameter of the impeller, discharge, pressure change across the impeller, density of the fluid, and viscosity of the fluid. These are given in Table 10.1 along with the relevant dimensions. In functional form, the relation between these variables is given by

$$f(\dot{W},\ \omega,\ D,\ Q,\ \Delta p,\ \rho,\ \mu)=0 \tag{10.3.1}$$

Using ω, ρ, and D as repeating variables, a set of convenient dimensionless groupings can be derived. They are given as follows:

$$C_{\dot{W}}=\frac{\dot{W}}{\rho\omega^3 D^5} \quad \textbf{power coefficient} \tag{10.3.2}$$

$$C_P=\frac{\Delta p}{\rho\omega^2 D^2} \quad \textbf{pressure coefficient} \tag{10.3.3}$$

$$C_Q=\frac{Q}{\omega D^3} \quad \textbf{flow rate coefficient} \tag{10.3.4}$$

$$\mathrm{Re}=\frac{\omega D^2 \rho}{\mu} \quad \textbf{Reynolds number} \tag{10.3.5}$$

It is customary to equate Δp to γH, where H represents either the pressure head rise in the case of a pump or the pressure head drop for a turbine. Then Eq. 10.3.3 is replaced by

$$C_H=\frac{gH}{\omega^2 D^2} \quad \textbf{head coefficient} \tag{10.3.6}$$

Table 10.1 Turbomachinery Parameters

Parameter	*Symbol*	*Dimensions*
Power	$\dot{W}$	ML^2/T^3
Rotational speed	ω	T^{-1}
Outer diameter of impeller	D	L
Discharge	Q	L^3/T
Pressure change	Δp	M/LT^2
Fluid density	ρ	M/L^3
Fluid viscosity	μ	M/LT

Another dimensionless parameter for a pump can be found by grouping $C_Q, C_{\dot{W}}$, and C_H to form the ratio

$$\frac{C_Q C_H}{C_{\dot{W}}} = \frac{\gamma Q H}{\dot{W}} = \eta_P \quad \text{(pump)} \tag{10.3.7}$$

Hence, the efficiency is also a similitude parameter. For a turbine the efficiency grouping is given by

$$\frac{C_{\dot{W}}}{C_Q C_H} = \frac{\dot{W}}{\gamma Q H} = \eta_T \quad \text{(turbine)} \tag{10.3.8}$$

in which $\dot{W}$ is the output power.[2]

The interrelationships between the dimensionless parameters are expressed in the form of dimensionless performance curves, as shown in Figure 10.12[3]: these are for the same turbomachine as shown in Figure 10.6. Dimensionless performance curves for the axial-flow pump of Figure 10.8 and the mixed-flow pump of Figure 10.9 are shown on the next pages in Figures 10.13 and 10.14, respectively. The vertical dashed lines designate the operating conditions at maximum or so-called "best" efficiency.

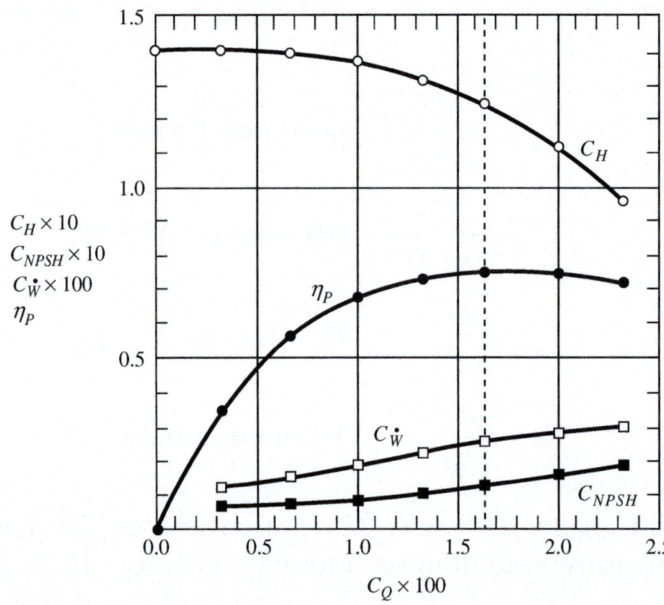

Figure 10.12 Dimensionless radial-flow pump performance curves for the pump presented in Figure 10.6; $D = 240$ mm; $\Omega_P = 0.61$.

---

[2]It is common industrial practice to use so-called *homologous units* in place of the dimensionless coefficients. The relationships between the two groupings are as follows:

Dimensionless coefficient	C_Q	C_H	$C_{\dot{W}}$
Homologous unit	$\frac{Q}{ND^3}$	$\frac{H}{N^2D^2}$	$\frac{\dot{W}}{N^3D^5}$

In the homologous units, density and gravitational acceleration have been eliminated, and the rotational speed is given in revolutions per minute, designated by *N*. Note that the homologous units are dimensional.

[3]In the caption, Ω_P is the specific speed, which will be defined in Section 10.3.3.

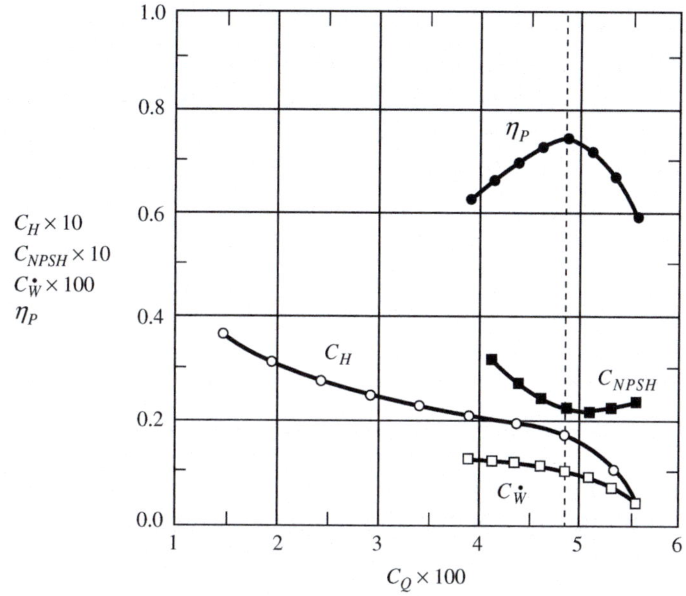

Figure 10.13 Dimensionless axial-flow pump performance curve for the pump presented in Figure 10.8; $D = 500$ mm; vane angle = 14°; $\Omega_P = 4.5$.

Note that the Reynolds number is not present as a parameter in Figures 10.12 to 10.14, even though viscous effects are significant in turbomachinery flow. In addition to the Reynolds number, one could include a relative roughness ratio. It is known that viscous and roughness effects will alter pump behavior; however, because of the difficulties involved, it is usual practice not to make a correlation between the Reynolds number, roughness, and pump losses. The flow regime in pump passages is usually very turbulent, and the roughness varies widely for different pumps. Furthermore, the friction losses may not be as significant as the losses due to eddy formation and separation caused by diffused flow occurring in the

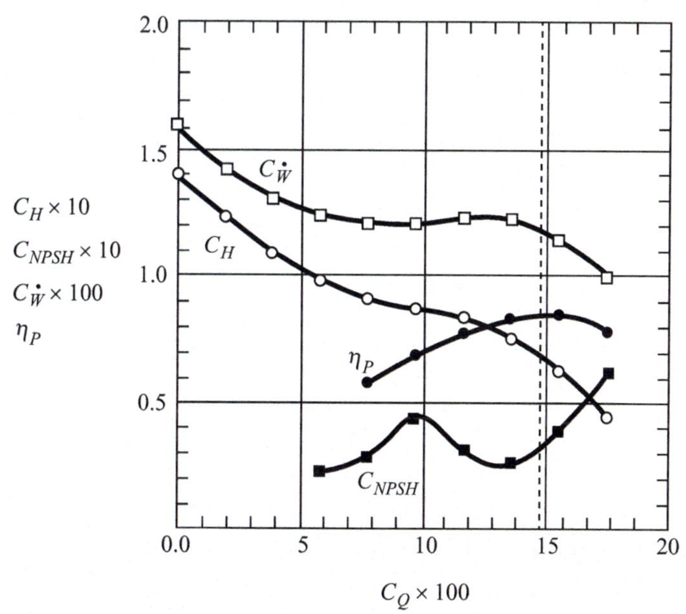

Figure 10.14 Dimensionless mixed-flow pump performance curve for the pump presented in Figure 10.9; $D = 371$ mm; $\Omega_P = 3.1$.

impeller and casing. As a result, the Re effect is usually not directly represented in similitude studies. Instead, the two following performance characteristics are recognized: (1) larger pumps exhibit relatively smaller losses and higher efficiencies than smaller pumps of the same family since they have smaller roughness and clearance ratios; and (2) higher viscosity liquids will cause reductions in pump head and efficiency and an increase in power requirement for a given discharge.

A dimensionless net positive suction head coefficient, C_{NPSH}, is also presented in Figures 10.12 to 10.14. It is formed by replacing H in Eq. 10.3.6 with the *NPSH* defined by Eq. 10.2.22.

Example 10.4

Determine the rotational speed, size, and required power for the axial-flow pump of Figure 10.13 to deliver 0.65 m^3/s of water at a head of 2.5 m.

Solution

The pump data are obtained from Figure 10.13. Reading from the figure we find that at the design, or maximum, efficiency of $\eta_P = 0.75$.

$$C_Q = 0.048 \qquad C_H = 0.018 \qquad C_{\dot{W}} = 0.0012$$

Equations 10.3.4 and 10.3.6 are employed to determine the two unknowns ω and D. Rearrange Eq. 10.3.6 to solve for ωD:

$$\omega D = \sqrt{\frac{gH_P}{C_H}} = \sqrt{\frac{9.81 \times 2.5}{0.018}} = 36.9 \text{ m/s}$$

Equation 10.3.4 is rearranged to include ωD as a known quantity to provide D:

$$D = \sqrt{\frac{Q}{C_Q \omega D}} = \sqrt{\frac{0.65}{0.048 \times 36.9}} = \underline{0.61 \text{ m}}$$

Substituting $D = 0.61$ m into the relation $\omega D = 36.9$, one finds that

$$\omega = \frac{36.9}{D} = \frac{36.9}{0.61} = 60.5 \text{ rad/s} \quad \text{or} \quad \underline{578 \text{ rpm}}$$

The density for water is ρ = 1000 kg/m^3. Then the pump power is determined by using Eq. 10.3.2:

$$\dot{W}_P = C_{\dot{W}} \rho \omega^3 D^5 = 0.0012 \times 1000 \times 60.5^3 \times 0.61^5 = 2.2 \times 10^4 \text{ W} \quad \text{or} \quad \underline{22 \text{ kW}}$$

Thus the speed, size, and required power are approximately 580 rpm, 610 mm, and 22 kW, respectively.

10.3.2 Similarity Rules

Following the principles outlined in Chapter 4 for similitude, similarity relationships between any two pumps from the same geometric family can be developed. They are:

$$(C_{\dot{W}})_1 = (C_{\dot{W}})_2 \quad \text{or} \quad \frac{\dot{W}_2}{\dot{W}_1} = \frac{\rho_2}{\rho_1}\left(\frac{\omega_2}{\omega_1}\right)^3\left(\frac{D_2}{D_1}\right)^5 \tag{10.3.9}$$

$$(C_H)_1 = (C_H)_2 \quad \text{or} \quad \frac{H_2}{H_1} = \frac{g_1}{g_2}\left(\frac{\omega_2}{\omega_1}\right)^2\left(\frac{D_2}{D_1}\right)^2 \tag{10.3.10}$$

$$(C_Q)_1 = (C_Q)_2 \quad \text{or} \quad \frac{Q_2}{Q_1} = \frac{\omega_2}{\omega_1}\left(\frac{D_2}{D_1}\right)^3 \tag{10.3.11}$$

These equations, called the **turbomachinery similarity rules**, are used to design or select a turbomachine from a family of geometrically similar units. Another use of them is to examine the effects of changing speed, fluid, or size on a given unit. They could also be used to design a pump to deliver flow on the moon or on a space station.

In light of Eq.10.3.7, one should also expect that $\eta_1 = \eta_2$, but as stated previously, larger pumps are more efficient than smaller ones of the same geometric family. An acceptable empirical correlation (Stepanoff, 1957) relating efficiencies to size is

$$\frac{1-(\eta_P)_2}{1-(\eta_P)_1} = \left(\frac{D_1}{D_2}\right)^{1/4} \tag{10.3.12}$$

This relation can also be used for turbines by replacing η_P with η_T.

Example 10.5

Determine the performance curve for the mixed-flow pump whose characteristic curves are shown in Figure 10.14. The required discharge is 2.5 m^3/s with the pump operating at a speed of 600 rpm. At design efficiency, what are the power and the *NPSH* requirements? Water is being pumped.

Solution

From Figure 10.14 the design efficiency is $\eta_P = 0.85$ and the corresponding design values for the coefficients are

$$C_Q = 0.15 \qquad C_H = 0.067 \qquad C_W = 0.0115 \qquad C_{NPSH} = 0.035$$

The rotational speed in radians per second is

$$\omega = 600 \times \frac{\pi}{30} = 62.8 \text{ rad/s}$$

The pump diameter is computed with Eq. 10.3.4:

$$D = \left(\frac{Q}{\omega C_Q}\right)^{1/3}$$
$$= \left(\frac{2.5}{62.8 \times 0.15}\right)^{1/3} = 0.64 \text{ m}$$

With Eqs.10.3.2 and 10.3.6, the required power and *NPSH* are computed:

$$\dot{W}_P = \rho\omega^3 D^5 C_{\dot{W}}$$
$$= 1000 \times 62.8^3 \times 0.64^5 \times 0.0115 = 3.1 \times 10^5 \text{ W} \quad \text{or} \quad \underline{310 \text{ kW}}$$

$$NPSH = \frac{\omega^2 D^2}{g} C_{NPSH}$$
$$= \frac{62.8^2 \times 0.64^2}{9.81} \times 0.035 = \underline{5.8 \text{ m}}$$

Hence, the required power is 310 kW and the required *NPSH* is 5.8 m.

The performance curve is constructed using Eqs. 10.3.10 and 10.3.11:

$$H_2 = \frac{\omega_2^2 D_2^2}{g}(C_H)_1$$
$$= \frac{62.8^2 \times 0.64^2}{9.81}(C_H)_1 = 165(C_H)_1$$

$$Q_2 = \omega_2 D_2^3 (C_Q)_1$$
$$= 62.8 \times 0.64^3 (C_Q)_1 = 16.5(C_Q)_1$$

Values of $(C_H)_1$ and $(C_Q)_1$ are read from Figure 10.14, along with η_P and placed in the table shown below. Columns 4 and 5 are Q_2 and H_2 computed from the similarity equations. At best efficiency, $H_2 = 10.4$ m and $Q_2 = 2.54$ m^3/s. The characteristic curve and efficiency curve are plotted in Figure E10.5.

$(C_Q)_1$	$(C_H)_1$	η_P	Q_2 (m^3/s)	H_2 (m)
0	0.138		0	22.8
0.019	0.123		0.31	20.3
0.039	0.110		0.64	18.2
0.058	0.099		0.96	16.3
0.077	0.091	0.59	1.27	15.0
0.096	0.088	0.70	1.58	14.5
0.116	0.085	0.78	1.91	14.0
0.135	0.076	0.84	2.23	12.5
0.154	0.063	0.85	2.54	10.4
0.174	0.046	0.79	2.87	7.6

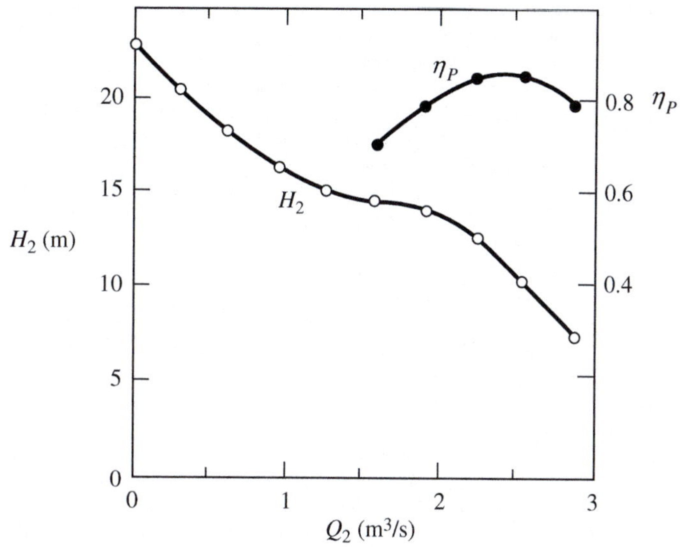

Figure E10.5

10.3.3 Specific Speed

It is possible to correlate a turbomachine of a given family to a dimensionless number that characterizes its operation at optimum conditions. Such a number is termed the **specific speed,** and it is determined in the following manner. The specific speed Ω_P of a pump is a dimensionless parameter associated with its operation at maximum efficiency, with known ω, Q, and H_P. It is obtained by eliminating D between Eqs. 10.3.4 and 10.3.6 and expressing the rotational speed ω to the first power:

$$\Omega_P = \frac{C_Q^{1/2}}{C_H^{3/4}} = \frac{\omega Q^{1/2}}{(gH_P)^{3/4}} \tag{10.3.13}$$

In Eq. 10.3.13, ω is usually based on motor requirements, and the values of Q and H_P are those at maximum efficiency.

A preliminary pump selection can be based on the specific speed. Figure 10.15 shows how the maximum efficiency and the discharge vary with Ω_P for radial-flow pumps. The specific speed can be correlated with the type of impeller shown in Figure 10.2; it is given as follows:

$$\begin{array}{ll} \Omega_P < 1 & \text{radial-flow pump} \\ 1 < \Omega_P < 4 & \text{mixed-flow pump} \\ \Omega_P > 4 & \text{axial-flow pump} \end{array} \tag{10.3.14}$$

For a turbine, the specific speed Ω_T is a dimensionless parameter associated with a given family of turbines operating at maximum efficiency, with known ω, H_T, and $\dot{W}_T$.

The diameter D is eliminated in Eqs. 10.3.2 and 10.3.6, and ω is expressed to the first power to obtain

$$\Omega_T = \frac{C_{\dot{W}}^{1/2}}{C_H^{5/4}} = \frac{\omega(\dot{W}_T/\rho)^{1/2}}{(gH_T)^{5/4}} \tag{10.3.15}$$

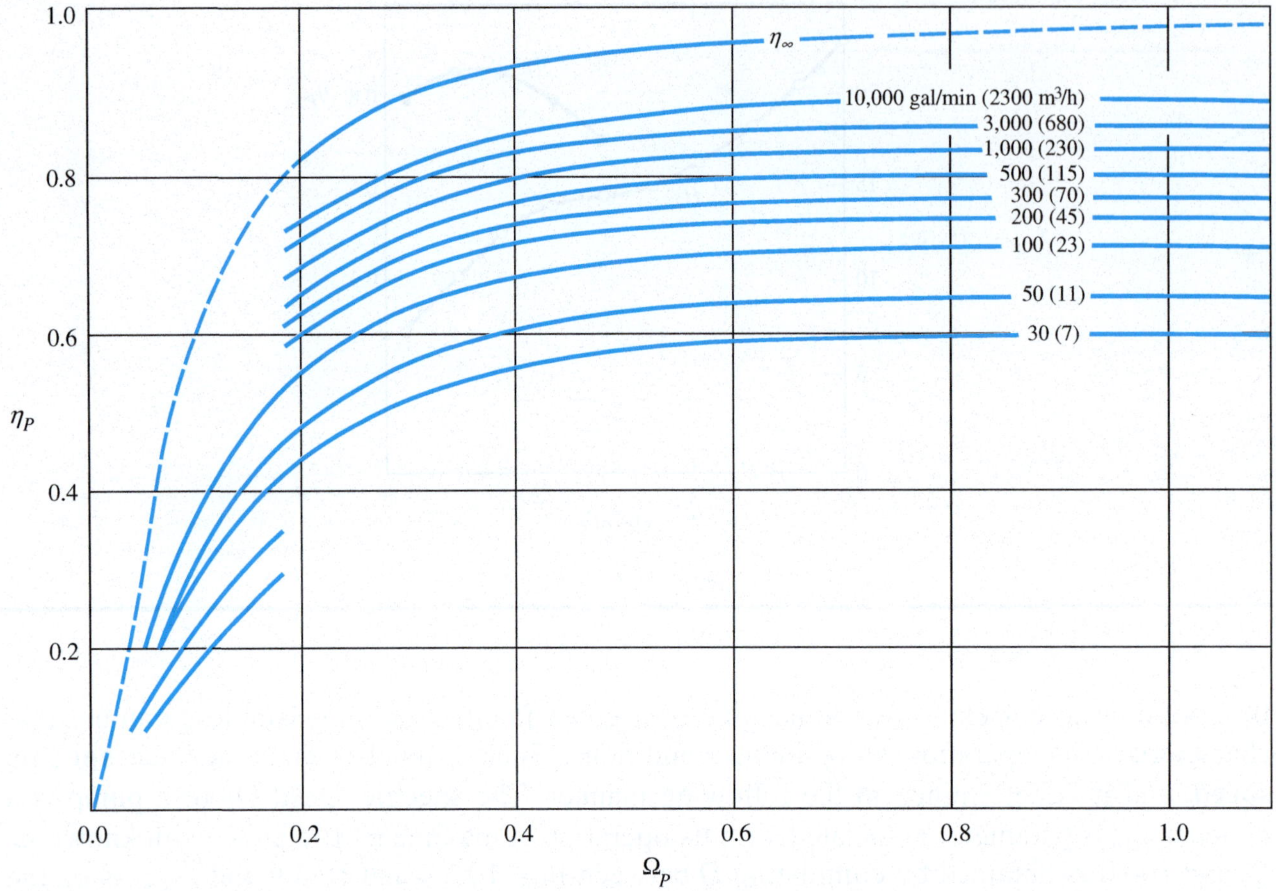

Figure 10.15 Maximum efficiency as a function of specific speed and discharge for radial-flow pumps. (Based on Karassik, *Pump Handbook*, 2nd edition, 1986, McGraw-Hill Companies, Inc.)

A comparison of Ω_T for different turbines will be made in Section 10.5.[4]

Another measure of cavitation is **suction specific speed**, S. Analogous to the formulation of specific speed of a pump, Eq. 10.3.13, suction specific speed for either a pump or turbine is given by

$$S = \frac{\omega Q^{1/2}}{(gNPSH)^{3/4}} \tag{10.3.16}$$

Here, two geometrically similar units will have the same S when operating at the same flow coefficient C_Q. Conversely, equal values of S indicate similar cavitation characteristics when the units are operating differently. Design values of S are determined by experiment; when cavitation is not present, Eq. 10.3.16 is no longer valid.

[4]In place of Eqs. 10.3.13 and 10.3.15, it has been common industrial practice to use specific speeds in dimensional form. Hence, for a pump, the dimensional, or *homologous specific speed* is $(N_s)_P = NQ^{1/2}/H^{3/4}$, and for a turbine the corresponding homologous specific speed is $(N_S)_T = N\dot{W}^{1/2}/H^{5/4}$. For these expressions, Q is in gal/min, ft³/s, or m³/s, H is in ft or m, $\dot{W}$ is in horsepower or kW, and N is in rpm.

Example 10.6

Select a pump to deliver 500 gal/min of water with a pressure rise of 65 psi. Assume a rotational speed not to exceed 3600 rpm.

Solution

To estimate the specific speed we need the following;

$$\omega = 3600 \times \frac{\pi}{30} = 377 \text{ rad/s}$$

$$H_P = \frac{\Delta p}{\rho g} = \frac{65 \times 144}{1.94 \times 32.2} = 150 \text{ ft}$$

$$Q = \frac{500}{7.48 \times 60} = 1.11 \text{ ft}^3/\text{s}$$

Use Eq. 10.3.13 to find the specific speed to be

$$\Omega_P = \frac{\omega\sqrt{Q}}{(gH_P)^{3/4}} = \frac{377\sqrt{1.11}}{(32.2 \times 150)^{3/4}} = 0.69$$

From Eq. 10.3.14, this would indicate a radial-flow pump. The pump of Figure 10.12 could be used, even though the specific speed (0.61) would result in a lower efficiency. With $\Omega_P \simeq 0.61$ (Figure 10.12), the speed is estimated with Eq. 10.3.13 as

$$\omega = \frac{\Omega_P (gH_P)^{3/4}}{\sqrt{Q}} = \frac{0.61 \times (32.2 \times 150)^{3/4}}{\sqrt{1.11}} = 335 \text{ rad/s}$$

Hence, the required speed is $335 \times 30/\pi = \underline{3200 \text{ rpm}}$, which does not exceed 3600 rpm. The required diameter is determined with use of Figure 10.12, where at maximum efficiency, $C_Q = 0.0165$. This is substituted into Eq. 10.3.4 to determine the diameter:

$$D = \left(\frac{Q}{C_Q \omega}\right)^{1/3} = \left(\frac{1.11}{0.0165 \times 335}\right)^{1/3} = \underline{0.59 \text{ ft}}$$

10.4 USE OF TURBOPUMPS IN PIPING SYSTEMS

The appropriate selection of one or more pumps to meet the flow demands of a piping system requires, in addition to a fundamental understanding of turbopumps, a hydraulic analysis of the pumps integrated into the piping system. A single pipe and pump arrangement was analyzed in Section 5.6.7. In Example 5.18 a system demand curve was obtained, and a trial solution to find the discharge and required pump head was shown. The technique was extended to simple piping systems in Section 9.3, wherein it was shown that one can employ either a pump performance curve or assume a constant power input to the pump. Section 9.4 dealt with piping networks. In those systems, the pump performance curve was represented by a polynomial relation (Eq. 9.4.14) and incorporated into the linearized energy equations, or alternatively, a constant pump power requirement was employed (Eq. 9.4.15).

In this section we look more closely at the manner in which pumps are selected. The procedure involves not only satisfying the required flow and pressure requirements for a piping system, but also making sure that cavitation is avoided and that the most efficient pump is selected.

10.4.1 Matching Pumps to System Demand

Consider a single pipeline that contains a pump to deliver fluid between two reservoirs. The **system demand curve** is defined as

$$H_P = (z_2 - z_1) + \left(f\frac{L}{D} + \Sigma K \right) \frac{Q^2}{2gA^2} \tag{10.4.1}$$

where atmospheric pressure is assumed at each reservoir, and the upstream and downstream reservoirs are at elevations z_1 and z_2, respectively. Note that z_2 need not be greater than z_1, and that the friction factor f can vary as the discharge (i.e., Reynolds number) varies. The term ΣK represents the minor losses in the pipe, see Section 5.6.4. Equation 10.4.1 is represented as curve (a) in Figure 10.16. The first term on the right-hand side is the static head and the second term is the head loss due to pipe friction and minor losses. The steepness of the demand curve is dependent on the sum of the loss coefficients in the system; as the loss coefficients increase, signified by curve (b) in Figure 10.16, the pumping head required for a given discharge is increased. Piping systems may experience short-term changes in the demand curve such as throttling of valves, and over the long term, aging of pipes may cause the loss coefficients to increase permanently. In either case the system demand curve could change from (a) to (b) as shown.

For a given design discharge, Eq. 10.3.14 allows the selection of a pump based on the specific speed. Once the type of pump is determined, an appropriate size is selected from a manufacturer's characteristic curve; a representative curve is shown in Figure 10.16. The intersection of the characteristic curve with the desired system demand curve will provide the **design head** H_D and **design discharge** Q_D. It is desirable to have the intersection occur at or close to the point of maximum efficiency of the pump, designated as the **best operating point**.

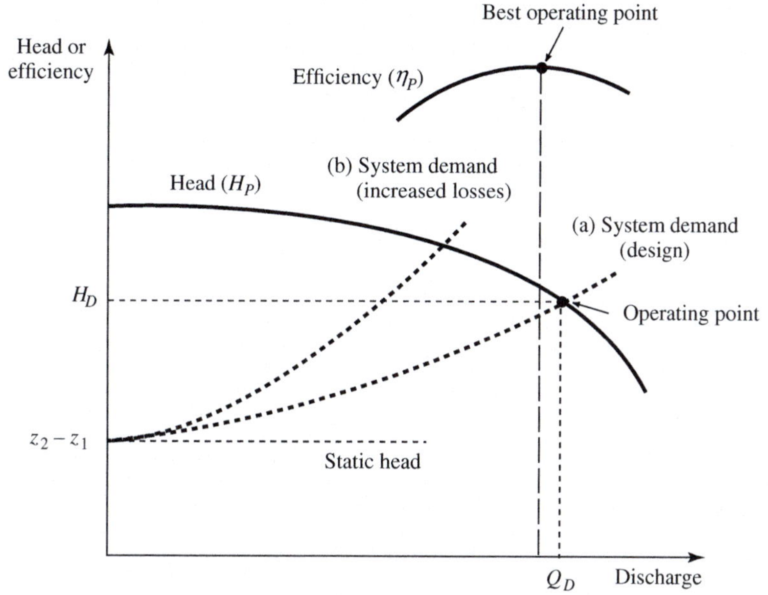

Figure 10.16 Pump characteristic curve and system demand curve.

10.4.2 Pumps in Parallel and in Series

In some instances, pumping installations may have a wide range of head or discharge requirements, so that a single pump may not meet the required range of demands. In these situations, pumps may be staged either in series or in parallel to provide operation in a more efficient manner. In this discussion, it is assumed that the pumps are placed at a single location with short lines connecting the separate units.

Where a large variation in flow demand is required, two or more pumps are placed in a parallel configuration (Figure 10.17). Pumps are turned on individually to meet the required flow demand; in this way operation at higher efficiency can be attained. It is not necessary to have identical pumps, but individual pumps, when running in parallel should not be operating in undesirable zones. For parallel pumping the combined characteristic curve is generated by recognizing that the head across each pump is identical, and the total discharge through the pumping system is ΣQ, the sum of the individual discharges through each pump.

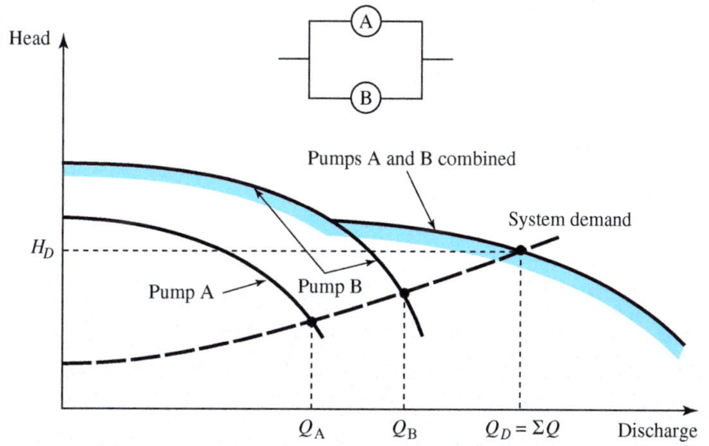

Figure 10.17 Characteristic curves for pumps operating in parallel.

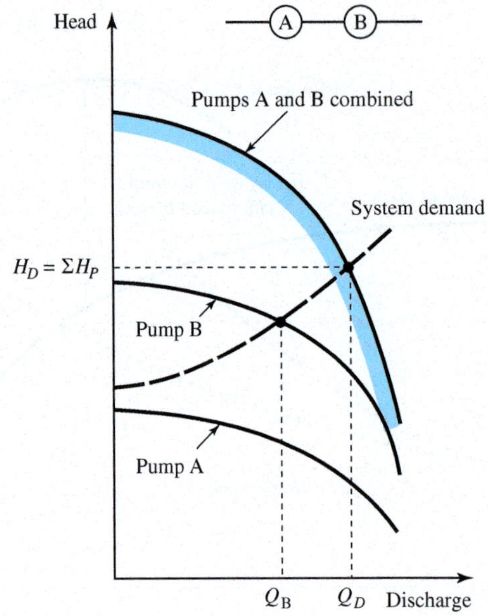

Figure 10.18 Characteristic curves for pumps operating in series.

Note the existence of three operating points in Figure 10.17, in which pump A or pump B is used separately, or in which pumps A and B are combined. Other design operating points could be obtained by throttling the flow or by changing the pump speeds. The overall efficiency of pumps in parallel is

$$\eta_P = \frac{\gamma H_D \Sigma Q}{\Sigma \dot{W}_P} \tag{10.4.2}$$

in which $\Sigma \dot{W}_P$ is the sum of the individual power required by each pump.

For high head demands, pumps placed in series will produce a head rise greater than those of the individual pumps (Figure 10.18). Since the discharge through each pump is identical, the characteristic curve is found by summing the head across each pump. Note that it is not necessary that the two pumps be identical. In Figure 10.18 the system demand curve is such that pump A operating alone cannot deliver any liquid because its shutoff head is lower than the static system head. There are two operating points, either with pump B alone or with pumps A and B combined. The overall **efficiency** is

$$\eta_P = \frac{\gamma (\Sigma H_P) Q_D}{\Sigma \dot{W}_P} \tag{10.4.3}$$

in which ΣH_P is the sum of the individual heads across each pump.

10.4.3 Multistage Pumps

Instead of placing several pumps in series, multistage pumps are available (Figure 10.19). Basically, the impellers are all housed in a single casing and the outlet from one impeller stage ejects into the eye of the next. Such pumps can provide extremely high heads. For the pump shown in Figure 10.19, the pressure head range is 1500 to 3400 m, and the discharge can vary from 4500 m^3/h down to 260 m^3/h. Up to eight stages can be selected and the maximum speed is about 8000 rpm.

Figure 10.19 Multistage centrifugal pump. (Source: Sergey Ryshov/shutterstock.com)

Example 10.7

Water is pumped between two reservoirs in a pipeline with the following characteristics: D = 300 mm, L = 70 m, f = 0.025, ΣK = 2.5. The radial-flow pump characteristic curve is approximated by the formula

$$H_P = 22.9 + 10.7Q - 111Q^2$$

where H_P is in meters and Q is in m^3/s.

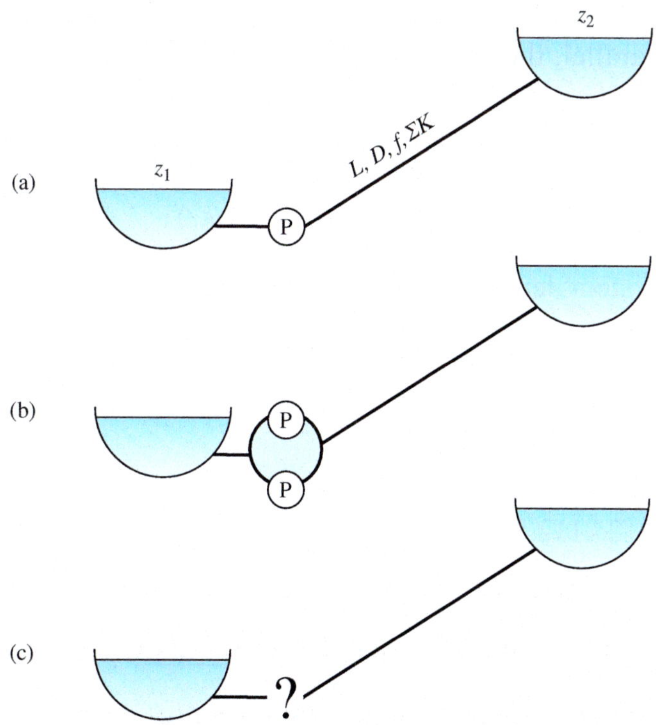

Figure E10.7

Determine the discharge Q_D and pump head H_D for the following situations (see Fig. E10.7): (a) $z_2 - z_1 = 15$ m, one pump placed in operation; (b) $z_2 - z_1 = 15$ m, with two identical pumps operating in parallel; and (c) the pump layout, discharge, and head for $z_2 - z_1 = 25$ m.

Solution

(a) The system demand curve (Eq. 10.4.1) is developed first:

$$H_P = (z_2 - z_1) + \left(f\frac{L}{D} + \Sigma K \right)\frac{Q^2}{2gA^2}$$

$$= 15 + \left(\frac{0.025 \times 70}{0.3} + 2.5 \right)\frac{Q^2}{2 \times 9.81(\pi/4 \times 0.3^2)^2} = 15 + 85Q^2$$

To find the operating point, equate the pump characteristic curve to the system demand curve,

$$15 + 85Q_D^2 = 22.9 + 10.7Q_D - 111Q_D^2$$

Reduce and solve for Q_D:

$$195Q_D^2 - 10.7Q_D - 7.9 = 0$$

$$\therefore Q_D = \frac{1}{2 \times 195}\left[10.7 + \sqrt{10.7^2 + 4 \times 195 \times 7.9} \right] = \underline{0.23 \text{ m}^3/\text{s}}$$

Using the system demand curve, H_D is computed as

$$H_D = 15 + 85 \times 0.23^2 = \underline{19.5 \text{ m}}$$

(b) For two pumps in parallel, the characteristic curve is

$$H_P = 22.9 + 10.7\left(\frac{Q}{2}\right) - 111\left(\frac{Q}{2}\right)^2$$

$$= 22.9 + 5.35Q - 27.75Q^2$$

Equate this to the system demand curve and solve for Q_D:

$$15 + 85Q_D^2 = 22.9 + 5.35Q_D - 27.75Q_D^2$$

$$112.8Q_D^2 - 5.35Q_D - 7.9 = 0$$

$$\therefore Q_D = \frac{1}{2 \times 112.8}\left[5.35 + \sqrt{5.35^2 + 4 \times 112.8 \times 7.9} \right] = \underline{0.29 \text{ m}^3/\text{s}}$$

The design head is calculated to be

$$H_D = 15 + 85 \times 0.29^2 = \underline{22.2 \text{ m}}$$

(c) Since $z_2 = z_1$ is greater than the single pump shutoff head (i.e., 25 > 22.9 m), it is necessary to operate with two pumps in series. The combined pump curve is

$$H_D = 2(22.9 + 10.7Q - 111Q^2)$$
$$= 45.8 + 21.4Q - 222Q^2$$

The system demand curve is changed since $z_2 - z_1 = 25$ m. It becomes

$$H_D = 25 + 85Q^2$$

Equating the two relations above and solving for Q_D and H_P results in

$$25 + 85Q_D^2 = 45.8 + 21.4Q_D - 222Q_D^2$$

or

$$307Q_D^2 - 21.4Q_D = 20.8 = 0$$

$$\therefore Q_D = \frac{1}{2\times 307}\left[21.4^2 + \sqrt{21.4^2 + 4\times 307\times 20.8}\right] = \underline{0.30\ \text{m}^3/\text{s}}$$

and

$$H_D = 25 + 85\times 0.30^2 = \underline{32.7\ \text{m}}$$

10.5 TURBINES

In many parts of the world, where sufficient head and large flow rates are possible, hydroturbines are used to produce electrical power. In contrast to pumps, turbines extract useful energy from the water flowing in a piping system. The moving component of a turbine is called a **runner**, which consists of vanes or buckets that are attached to a rotating shaft. The energy available in the liquid is transferred to the shaft by means of the rotating runner, and the resulting torque transferred by the rotating shaft can drive an electric generator. Hydroturbines vary widely in size and capacity, ranging from microunits generating 5 kW to those in large hydroelectric installations that produce over 400 MW.

There are two types of turbines. The **reaction turbine** utilizes both flow energy and kinetic energy of the liquid; energy conversion takes place in an enclosed space at pressures above atmospheric conditions. Reaction turbines can be subdivided further, according to the available head, as either Francis or propeller type. The **impulse turbine** requires that the flow energy in the liquid be converted into kinetic energy by means of a nozzle before the liquid impacts on the runner; the energy is in the form of a high-velocity jet at or near atmospheric pressure. Turbines can be classified according to turbine specific speed, as shown in Figure 10.20; the Pelton wheel is a particular type of impulse turbine (see Section 10.5.2).

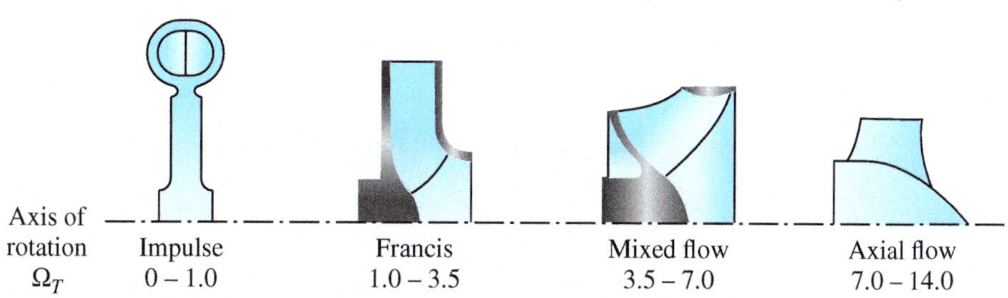

Figure 10.20 Various types of turbine runners.

10.5.1 Reaction Turbines

In reaction turbines, the flow is contained in a **volute** that channels the liquid into the runner (Figure 10.21). Adjustable **guide vanes** (also called **wicket gates**) are situated upstream of the runner; their function is to control the tangential component of velocity at the runner inlet. As a result, the fluid leaves the guide vane exit and enters the runner with an acquired angular momentum. As the fluid travels through the runner region, its angular momentum is reduced and it imparts a torque on the runner, which in turn drives the shaft to produce power. The flow exits from the runner into a diffuser, called the **draft tube**, which acts to convert the kinetic energy remaining in the liquid into flow energy.

Francis Turbine. In the Francis turbine, the incoming flow through the guide vanes is radial, with a significant tangential velocity component at the entrance to the runner

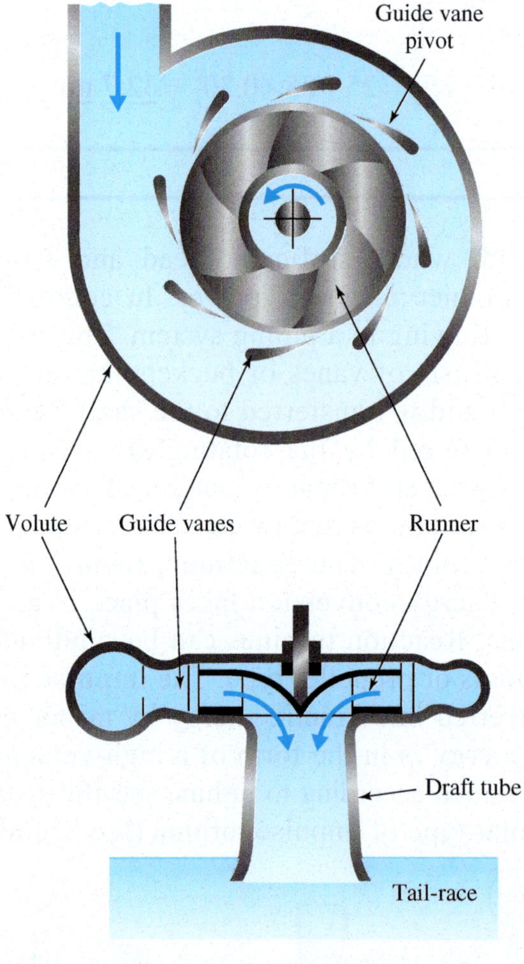

Figure 10.21 Francis type reaction turbine.

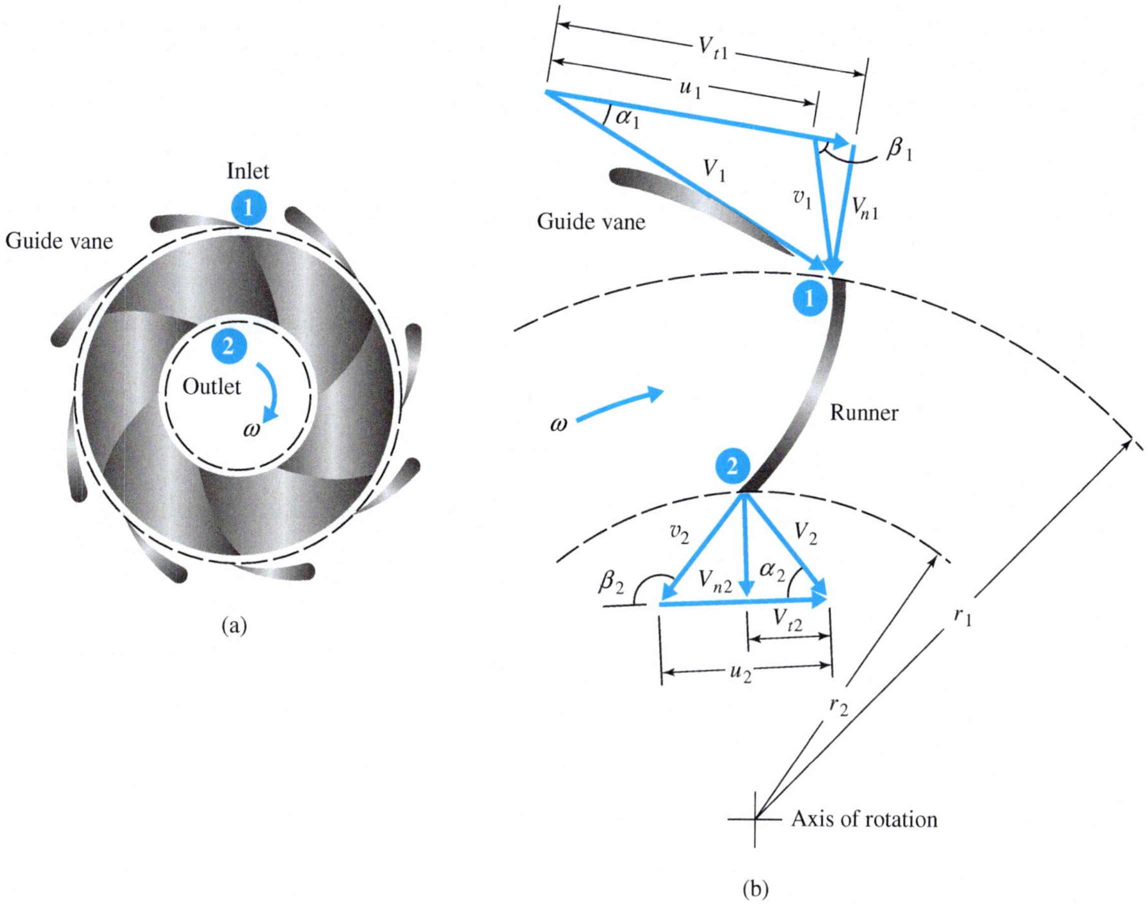

Figure 10.22 Idealized Francis turbine runner: (a) runner control volume; (b) velocity diagrams at control surfaces.

vanes (Figure 10.22). As the fluid traverses the runner the velocity takes on an axial component while the tangential component is reduced. As it leaves the runner, the fluid velocity is primarily axial with little or no tangential component. The pressure at the runner exit is below atmospheric.

The theoretical torque delivered to the runner is developed by applying Eq. 10.2.1 to the control volume shown in Figure 10.22; the same assumptions leading to the pump torque relation, Eq. 10.2.2, apply. The resulting relation is

$$T = \rho Q(r_1 V_{t1} - r_2 V_{t2}) \tag{10.5.1}$$

Multiplying the torque by the angular speed ω gives the power delivered to the shaft:

$$\begin{aligned} \dot{W}_T &= \omega T \\ &= \rho Q(u_1 V_1 \cos \alpha_1 - u_2 V_2 \cos \alpha_2) \end{aligned} \tag{10.5.2}$$

The fluid power input to the turbine is given by

$$\dot{W}_f = \gamma Q H_T \tag{10.5.3}$$

in which H_T is the actual head drop across the turbine. Thus, the overall efficiency is given by

$$\eta_T = \frac{\dot{W}_T}{\dot{W}_f} = \frac{\omega T}{\gamma Q H_T} \tag{10.5.4}$$

The action of the guide vanes can be described by considering the velocity vector diagrams in Figure 10.22b. Assume perfect guidance of the fluid along the guide vane; then the tangential velocity at the entrance to the runner is

$$V_{t1} = V_{n1} \cot \alpha_1 \tag{10.5.5}$$

From the velocity vector diagram the tangential velocity is also given by

$$V_{t1} = u_1 + V_{n1} \cot \beta_1 \tag{10.5.6}$$

The radial velocity component can be expressed in terms of the discharge Q and the width of the runner b_1:

$$V_{n1} = \frac{Q}{2\pi r_1 b_1} \tag{10.5.7}$$

Equations 10.5.5 to 10.5.7 are combined to eliminate V_{t1} and V_{n1}. There results

$$\cot \alpha_1 = \frac{2\pi r_1^2 b_1 \omega}{Q} + \cot \beta_1 \tag{10.5.8}$$

For a constant angular speed, in order to maintain the appropriate runner entry angle α_1, the guide vane angle is adjusted as Q changes.

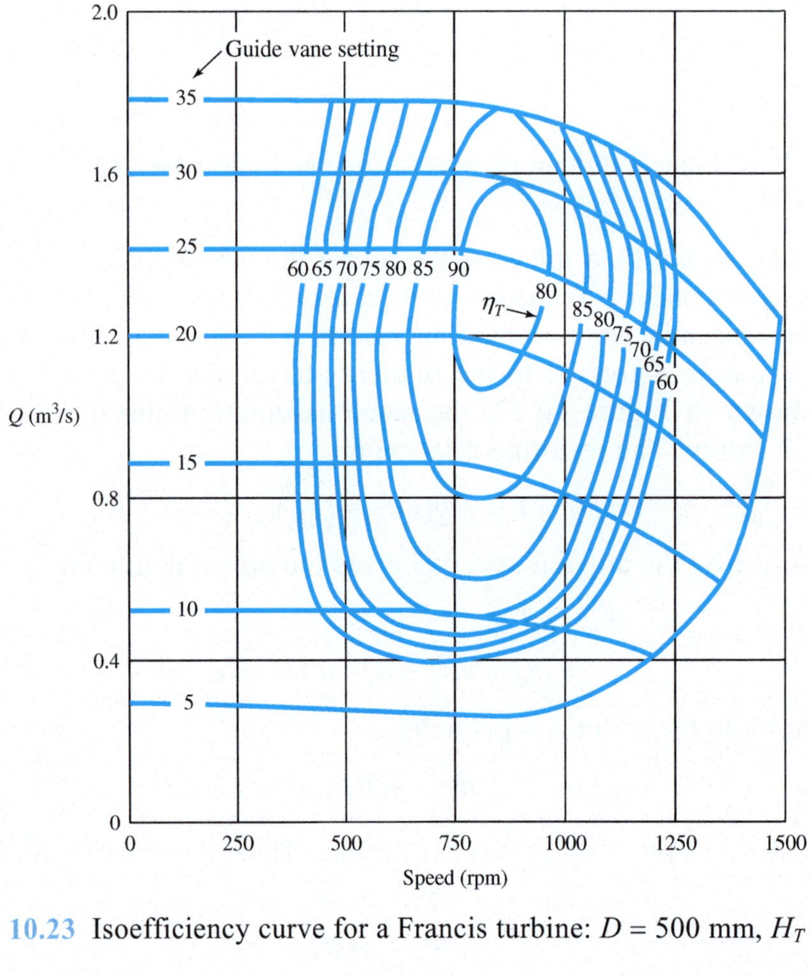

Figure 10.23 Isoefficiency curve for a Francis turbine: $D = 500$ mm, $H_T = 50$ m.

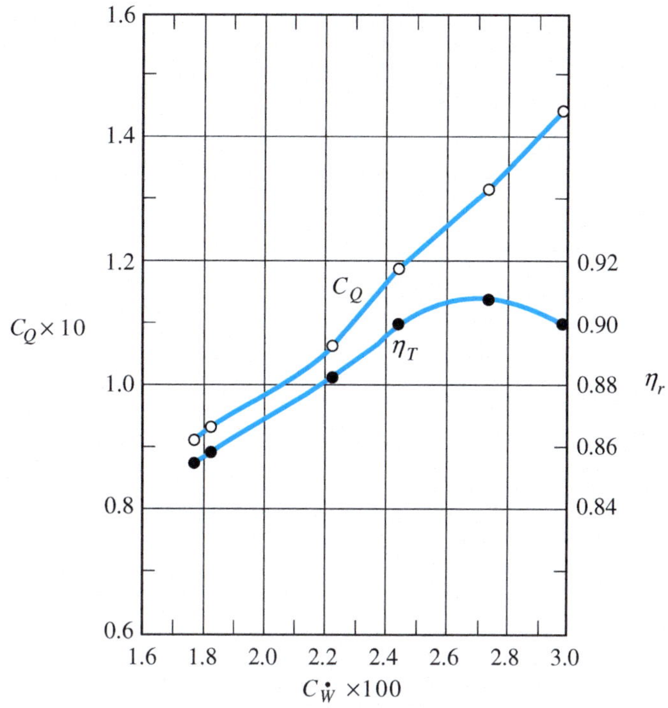

Figure 10.24 Performance curve for a prototype Francis turbine: D = 1000 mm, ω = 37.7 rad/s, $\Omega_T = 1.063$, $C_H = 0.23$.

Under normal operation, a turbine operates under a nearly constant head H_T so that the performance characteristics are viewed differently from those of a pump. For turbine operation under constant head, the important quantities are the variations of discharge, speed, and efficiency. The interrelationships among the three parameters are shown in the isoefficiency curve of Figure 10.23. Note the drop in discharge as the speed increases at a given guide vane setting. The efficiency is reduced by the following effects: (1) frictional head losses and draft tube head losses; (2) separation due to mismatch of flow entry angle with blade angle; (3) need to attain a certain turbine speed before useful power output is achieved; and (4) mechanical losses attributed to bearings, seals, and the like.

A representative dimensionless performance curve of a Francis turbine is shown in Figure 10.24. The speed and head are kept constant, and the guide vanes are automatically adjusted as the discharge varies in order to attain peak efficiency.

Example 10.8

A reaction turbine, whose runner radii are r_1 = 300 mm and r_2 = 150 mm, operates under the following conditions: $Q = 0.057\ \mathrm{m^3/s}$, ω = 25 rad/s, $\alpha_1 = 30°$, V_1 = 6 m/s, $\alpha_2 = 80°$, and V_2 = 3 m/s. Assuming ideal conditions, find the torque applied to the runner, the head on the turbine, and the fluid power. Use $\rho = 1000\ \mathrm{kg/m^3}$.

Solution

The applied torque is computed using Eq. 10.5.1:

$$\begin{aligned}T &= \rho Q(r_1 V_{t1} - r_2 V_{t2}) \\ &= \rho Q(r_1 V_1 \cos\alpha_1 - r_2 V_2 \cos\alpha_2) \\ &= 1000 \times 0.057(0.3 \times 6 \times \cos 30° - 0.15 \times 3 \times \cos 80°) \\ &= \underline{84.4\ \text{N}\cdot\text{m}}\end{aligned}$$

Under ideal conditions, the power delivered to the shaft is the same as the fluid power input to the turbine (i.e., $\eta_T = 1$). Thus,

$$\begin{aligned}\dot{W}_f &= \dot{W}_T \\ &= \omega T \\ &= 25 \times 84.4 = 2110\ \text{W} \quad \text{or} \quad \underline{2.11\ \text{kW}}\end{aligned}$$

The head on the turbine is found with the use of Eq. 10.5.3:

$$\begin{aligned}H_T &= \frac{\dot{W}_f}{\gamma Q} \\ &= \frac{2110}{9810 \times 0.057} = \underline{3.77\ \text{m}}\end{aligned}$$

Axial-FlowTurbine. In an axial-flow turbine the flow is parallel to the axis of rotation (Figure 10.25). Unlike the Francis turbine, the angular momentum of the liquid remains nearly constant and the tangential component of velocity is reduced across the blade. Both fixed-blade and pivoting-blade turbines are in use; the latter type, termed a **Kaplan turbine**,

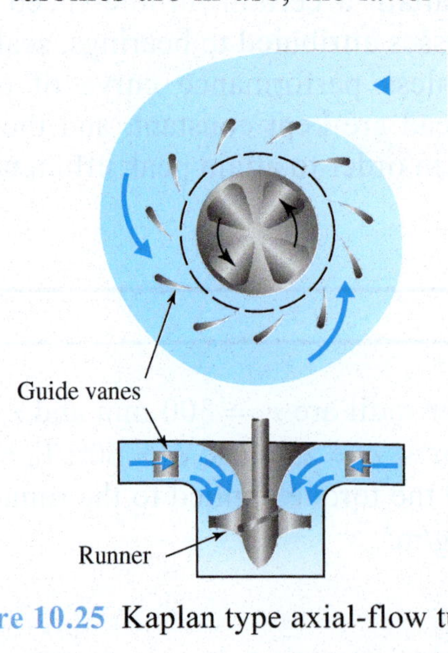

Figure 10.25 Kaplan type axial-flow turbine.

permits the blade angle to be adjusted to accommodate changes in head. Axial-flow turbines can be installed either vertically or horizontally. They are well suited for lowhead installations.

Cavitation Considerations. The net positive suction head (*NPSH*) and the cavitation number, defined in Section 10.2.3, are applicable for turbines. In Eqs. 10.2.21 to 10.2.23 the sign of the loss term becomes positive. The Thoma cavitation number, defined by Eq. 10.2.24 for a pump, is given in the following form for a turbine:

$$\sigma = \frac{1}{H_T}\left(\frac{p_{\text{atm}} - p_v}{\gamma} - \Delta z + h_L\right) \tag{10.5.9}$$

Typically, location 2 is defined at the outlet of the runner, and location 1 refers to the liquid surface at the draft tube outlet (Figure 10.26a).

The cavitation number is commonly used to characterize the cavitation behavior of a turbine. Figure 10.26b shows a representative plot of the cavitation number versus turbine efficiency, which is obtained experimentally by the turbine manufacturer. From such a plot operational procedures related to settings of the headwater and tailwater elevations can be established, and in addition, admissible levels of the cavitation number can be determined.

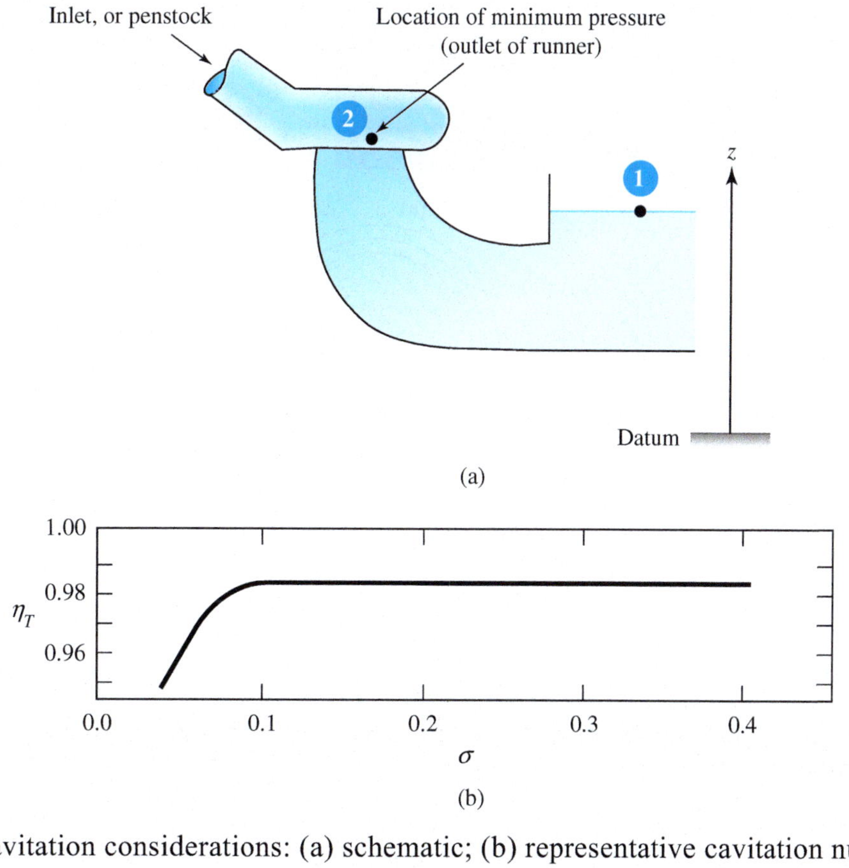

Figure 10.26 Cavitation considerations: (a) schematic; (b) representative cavitation number curve.

10.5.2 Impulse Turbines

The Pelton wheel (Figure 10.27) is an impulse turbine that consists of three basic components: one or more stationary inlet nozzles, a runner, and a casing. The runner consists of multiple buckets mounted on a rotating wheel. The pressure head upstream of the nozzle is converted into kinetic energy contained in the water jet leaving the nozzle. As the jet impacts the rotating buckets, the kinetic energy is converted into a rotating torque. The buckets are shaped in a manner to divide the flow in half and turn its relative velocity vector in the horizontal plane nearly 180°; the ideal angle of 180° is not attainable since the exiting liquid must stay free of the trailing buckets.

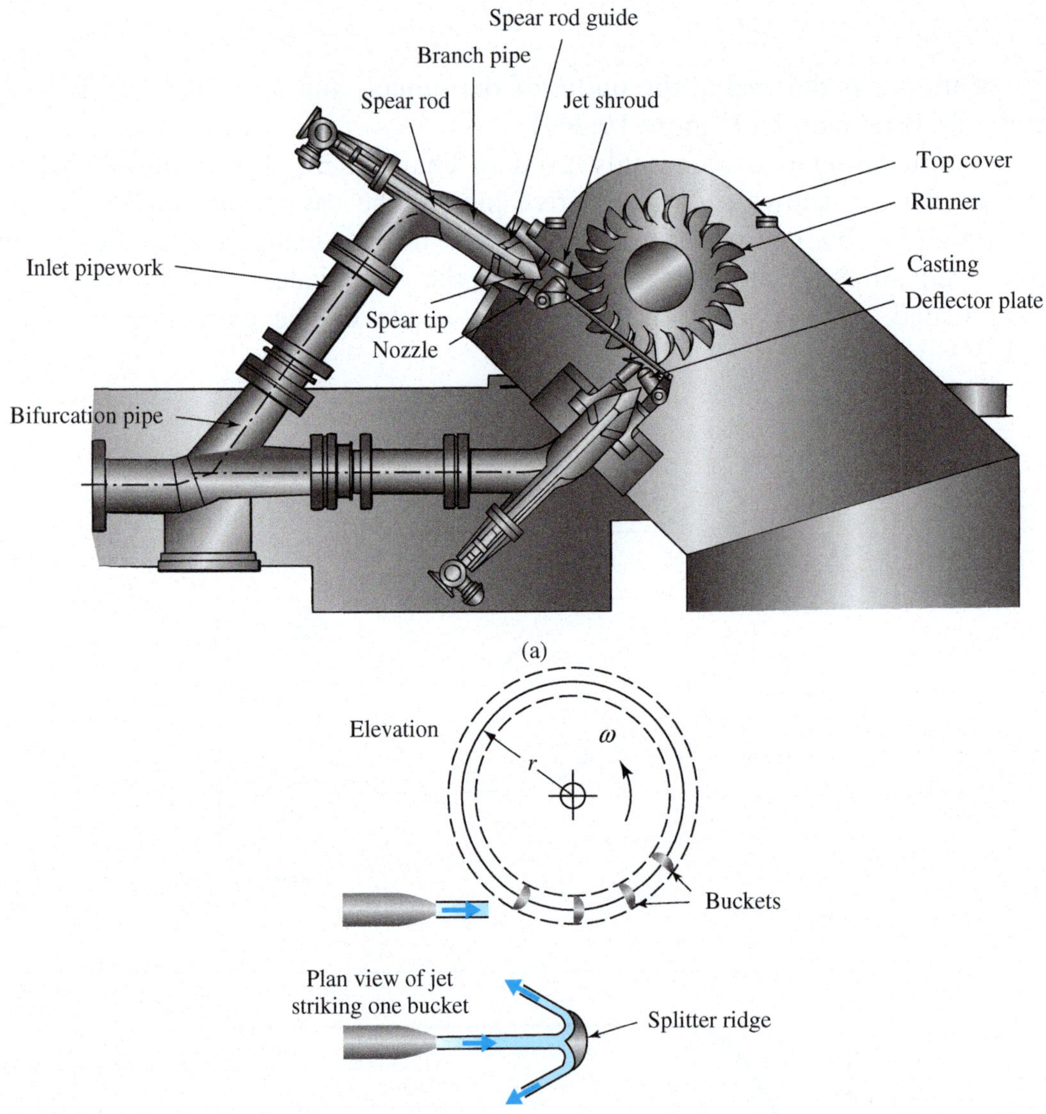

Figure 10.27 Impulse turbine (Pelton type): (a) typical arrangement of twin-jet machine; (b) jet striking a bucket.

The moment of momentum equation, illustrated in the preceding sections, can be applied to the control volume shown in Figure 10.28. Neglecting friction, the torque delivered to the wheel by the liquid jet is

$$T = \rho Q r(V_1 - u)(1 - \cos\beta_2) \tag{10.5.10}$$

in which Q is the discharge from all jets, and $u = r\omega$, r is the wheel radius as shown in Figure 10.27b. The power delivered by the fluid to the turbine runner is

$$\dot{W}_T = \rho Q u(V_1 - u)(1 - \cos\beta_2) \tag{10.5.11}$$

Usually, β_2 varies between 160 and 168°. Differentiation of Eq. 10.5.11 with respect to u and setting it equal to zero shows that maximum power occurs when $u = V_1/2$.

The jet velocity can be given in terms of the available head H_T:

$$V_1 = C_v\sqrt{2gH_T} \tag{10.5.12}$$

The **velocity coefficient** C_v accounts for the nozzle losses; typically, $0.92 \le C_v \le 0.98$.

The efficiency is

$$\eta_T = \frac{\dot{W}_T}{\gamma Q H_T} \tag{10.5.13}$$

Substituting Eqs. 10.5.11 and 10.5.12 into this relation and rearranging produces

$$\eta_T = 2\phi(C_v - \phi)(1 - \cos\beta_2) \tag{10.5.14}$$

in which the **speed factor** ϕ is defined as

$$\phi = \frac{r\omega}{\sqrt{2gH_T}} \tag{10.5.15}$$

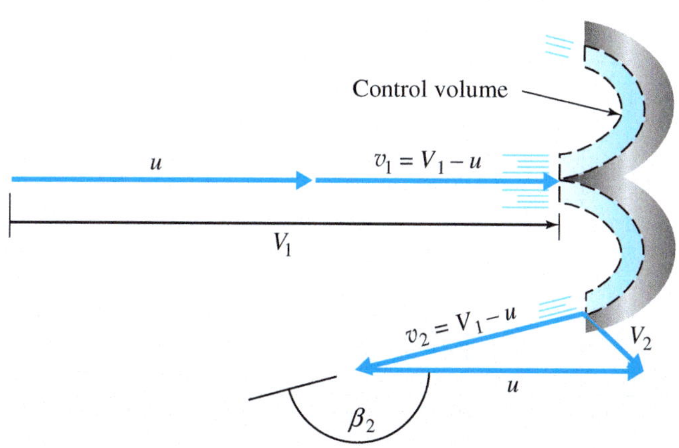

Figure 10.28 Velocity vector diagram for Pelton bucket.

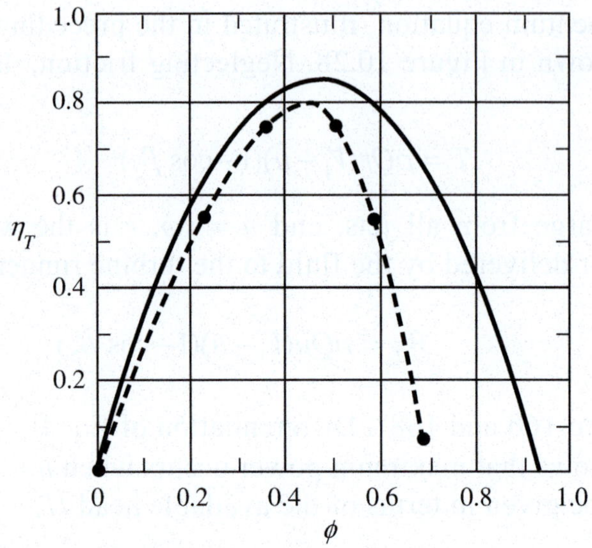

Figure 10.29 Speed factor versus efficiency for a laboratory-scale Pelton turbine: solid line, Eq. 10.5.14 ($C_v = 0.94$, $\beta_2 = 168°$); dashed line, experimental data.

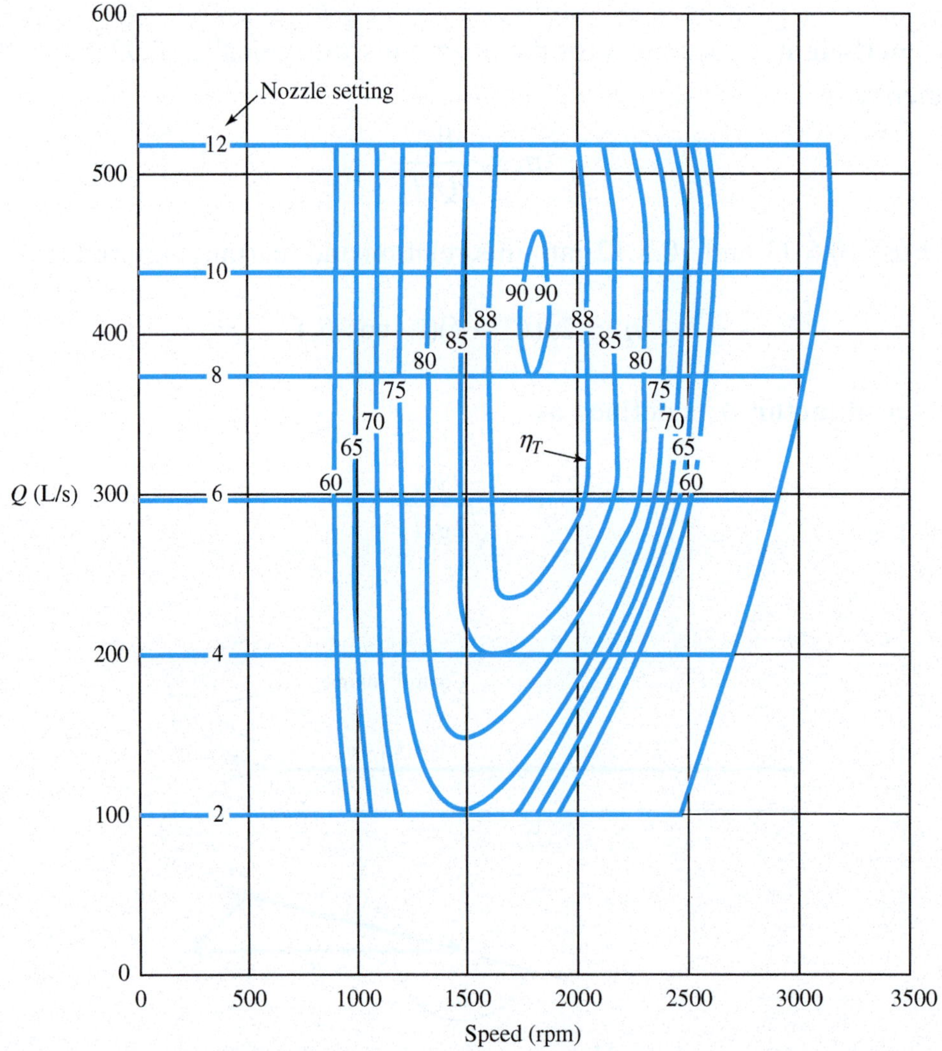

Figure 10.30 Isoefficiency curves for a Pelton turbine: D = 500 mm, H_T = 500 m.

Maximum efficiency occurs at $\phi = C_v/2$. Figure 10.29 shows Eq. 10.5.14 plotted for $C_v = 0.94$ and $\beta_2 = 168°$. In addition, data are plotted for a laboratory-scale Pelton wheel. The theoretical maximum efficiency is 0.874, while the experimental maximum efficiency is approximately 0.80. The efficiency is reduced because (1) the liquid jet does not strike the bucket with a uniform velocity, (2) losses occur when the jet strikes the bucket and splitter, (3) frictional loss is present due to the rotation of the bucket and splitter, and (4) there is frictional loss at the nozzle.

The interrelationships among Q, ω, and η_T are shown in the isoefficiency curves of Figure 10.30. For a constant head and nozzle setting, the discharge remains constant, since the nozzle outlet is at atmospheric pressure. A representative dimensionless performance curve for a prototype Pelton turbine is shown in Figure 10.31.

10.5.3 Selection and Operation of Turbines

A preliminary selection of the appropriate type of turbine for a given installation is based on the specific speed. Figure 10.20 shows how the turbine runner varies with Ω_T. Impulse turbines normally operate most economically at heads above 300 m, but small units can be

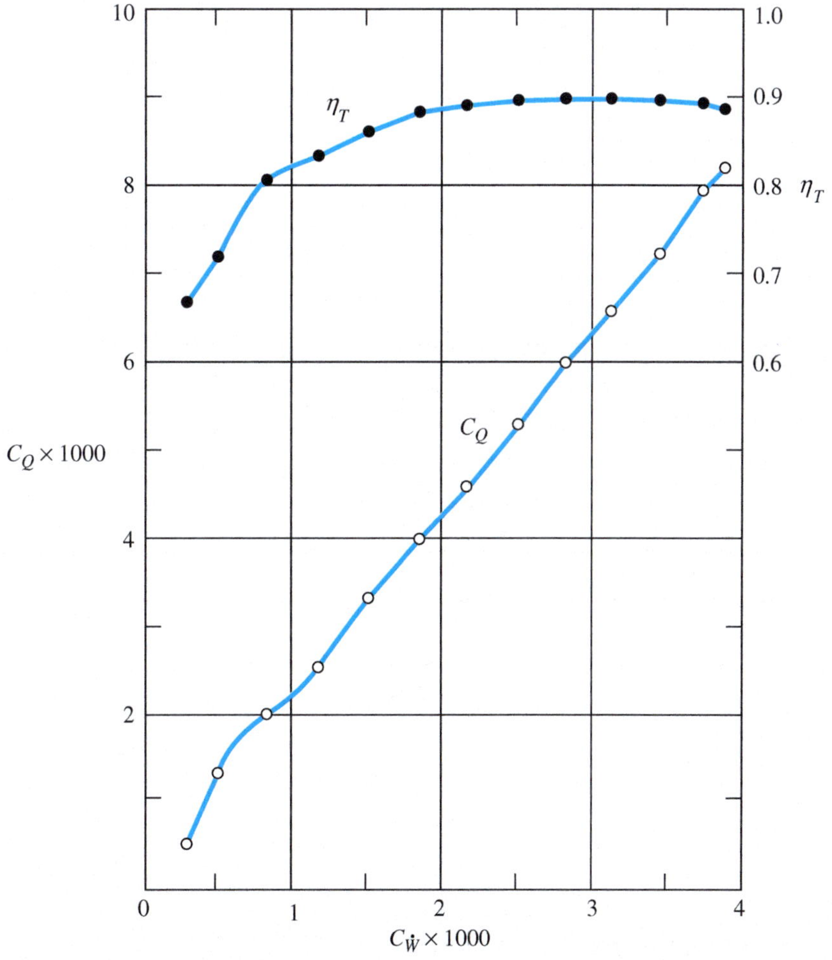

Figure 10.31 Performance and efficiency curves for a prototype Pelton turbine: D = 1000 mm, ω = 75.4 rad/s, Ω_T = 0.135, C_H = 0.52.

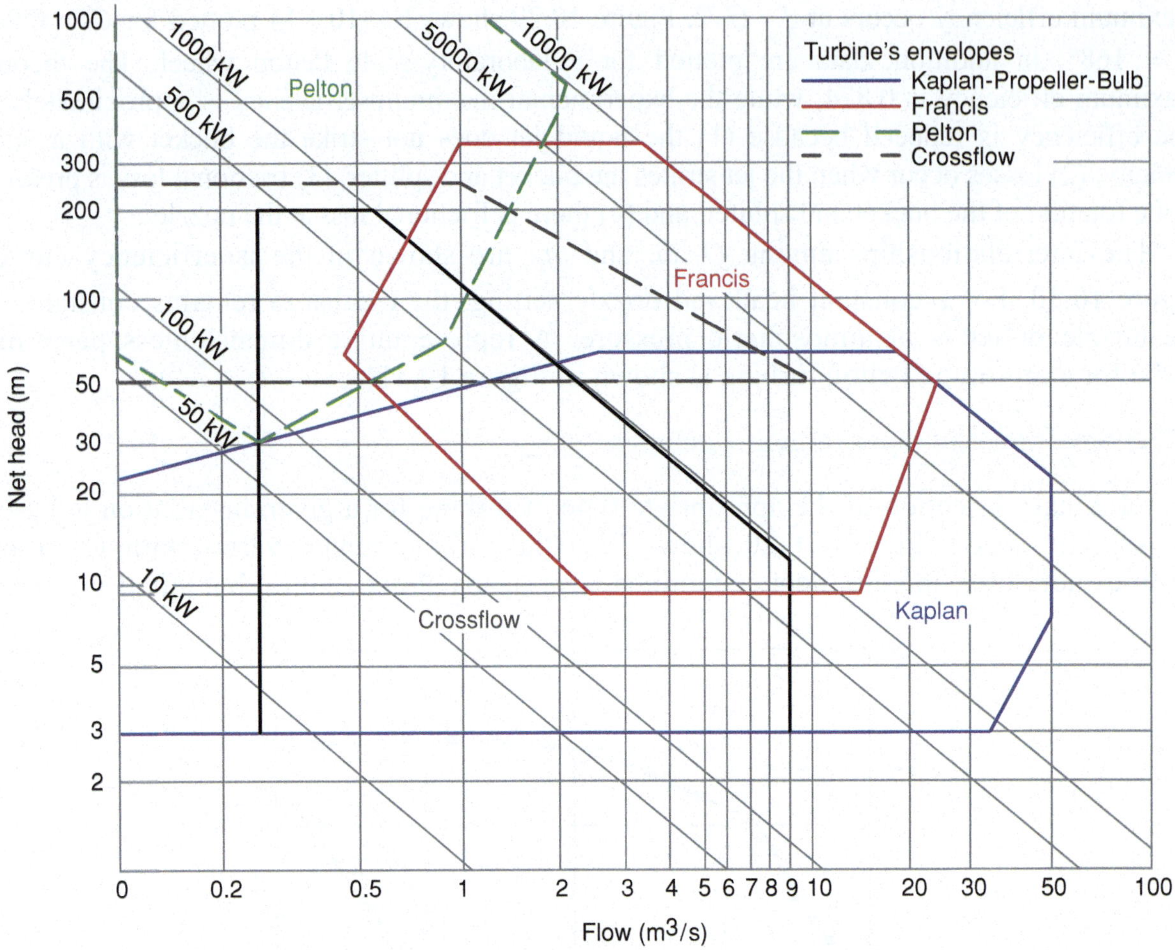

Figure 10.32 Application ranges for hydraulic turbines.

used for heads as low as 60 m. Heads up to 300 m are possible for Francis units, and propeller turbines are normally used for heads lower than 30 m. Figure 10.32 illustrates the ranges of application for a variety of hydraulic turbines.

Example 10.9

A Pelton turbine rotates at an angular speed of 400 rpm, developing 67.5 kW under a head of 60 m of water. The inlet pipe diameter at the base of the single nozzle is 200 mm. The operating conditions are $C_v = 0.97$, $\phi = 0.46$, and $\eta_T = 0.83$. Determine (a) the volumetric flow rate, (b) the diameter of the jet, (c) the wheel diameter, and (d) the pressure in the inlet pipe at the nozzle base.

Solution

(a) The discharge is computed from Eq. 10.5.13 to be

$$Q = \frac{\dot{W}_T}{\gamma H_T \eta_T} = \frac{67500}{9810 \times 60 \times 0.83} = \underline{0.138\ \text{m}^3/\text{s}}$$

(b) From Eq. 10.5.12, the velocity of the jet is

$$V_1 = C_v\sqrt{2gH_T} = 0.97\sqrt{2\times 9.81\times 60} = 33.3 \text{ m/s}$$

The area of the jet is the discharge divided by V_1, or

$$A_1 = \frac{Q}{V_1} = \frac{0.138}{33.3} = 4.14\times 10^{-3} \text{ m}^2$$

Hence, the jet diameter D_1 is

$$D_1 = \sqrt{\frac{4}{\pi}A_1} = \sqrt{\frac{4}{\pi}\times 4.14\times 10^{-3}} = 0.0726 \text{ m} \quad \text{or} \quad \underline{73 \text{ mm}}$$

(c) Use Eq. 10.5.15 to compute the wheel diameter D to be

$$\begin{aligned} D = 2r &= \frac{2\phi}{\omega}\sqrt{2gH_T} \\ &= \frac{2\times 0.46}{400\times \pi/30}\sqrt{2\times 9.81\times 60} = 0.754 \text{ m} \quad \text{or} \quad \underline{754 \text{ mm}} \end{aligned}$$

(d) The area of the inlet pipe is

$$A = \frac{\pi}{4}\times 0.20^2 = 0.0314 \text{ m}^2$$

The piezometric head just upstream of the nozzle is equal to H_T, so that pressure at that location is

$$\begin{aligned} p &= \gamma\left(H_T - \frac{Q^2}{2gA^2}\right) \\ &= 9810\left(60 - \frac{0.138^2}{2\times 9.81\times 0.0314^2}\right) \\ &= 5.79\times 10^5 \text{ Pa} \quad \text{or} \quad \text{approximately } \underline{580 \text{ kPa}} \end{aligned}$$

So-called minihydro and microhydro installations have recently been subject to renewed interest, not only in developing countries, but in industrially developed countries such as the United States, where the cost of energy has increased and incentive programs have been made available to promote their use. **Microhydro** has been classified as units that possess a capacity of less than 100 kW, and **minihydro** refers to system capacity from 100 to 1000 kW (Warnick, 1984). It is becoming common practice for manufacturers to develop a standardized turbine unit. In addition to specialized small Pelton, Francis, and propeller units, the **cross-flow turbine** is employed; basically, it is an impulse turbine that possesses a higher rotational speed than that of other impulse turbines.

Another alternative is to use a commercially available pump and operate it in the reverse direction; in the turbine mode, the best efficiency operating point requires a larger head and flow than when the pump operates at best efficiency in the pumping mode.

Reversible pump/turbine units are used in pumped/storage hydropower installations. Water is pumped from a lower reservoir to a higher storage reservoir during periods of low demand for conventional power plants. Subsequently, water is released from the upper reservoir, and the pumps are rotated in reverse to generate power during periods of high electric power demand. Such units require special design considerations to operate efficiently in either mode. The problem of converting the flow from the pump mode to the turbine mode, or vice versa, is a very complicated transient problem that is beyond the scope of this book. Water with enormous mass moving through the entire system must be reversed in direction; a special technique must be developed to accomplish this.

Example 10.10

A discharge of 2100 m^3/s and a head of 113 m are available for a proposed pumped-storage hydroelectric scheme. Reversible Francis pump/turbines are to be installed; in the turbine mode of operation, $\Omega_T = 2.19$, the rotational speed is 240 rpm, and the efficiency is 80%. Determine the power produced by each unit and the number of units required.

Solution

The power produced by each unit is found using the definition of specific speed, Eq. 10.3.15. Solving for the power, we have

$$\dot{W}_T = \rho\left[\frac{\Omega_T}{\omega}(gH_T)^{5/4}\right]^2$$

$$= 1000\left[\frac{2.19}{240\times\pi/30}(9.81\times113)^{5/4}\right]^2 = \underline{3.11\times10^8\ \text{W}}$$

From Eq. 10.5.13, the discharge in each unit is

$$Q = \frac{\dot{W}_T}{\gamma H_T \eta_T}$$

$$= \frac{3.11\times10^8}{9800\times113\times0.8} = 351\ \text{m}^3/\text{s}$$

The required number of units is equal to the available discharge divided by the discharge in each unit, or 2100/351 = 5.98. Hence, six units are required.

10.6 SUMMARY

This chapter focused on pumps or turbines that supply or extract energy by means of rotating impellers or vanes. First we emphasized the radial-flow, or centrifugal, pump and made use of the moment-of-momentum principle to derive an idealized bead-discharge relation. We then contrasted that with the actual head-discharge performance curve, one that is obtained experimentally and is required for use in pump analysis and design. Axial-flow and mixed-flow pumps were introduced, and the differences between these and the radial-flow pump were discussed. The significance of pump efficiency and net positive suction head (relating to cavitation) diagrams were shown for the proper design and selection of pumps.

Dimensional analysis and similitude were applied to both pumps and turbines; the three significant dimensionless numbers that were developed are the power coefficient, flow coefficient, and head coefficient. These coefficients were combined to yield pump or turbine efficiency, or to produce the specific speed for a pump or turbine. It was shown that a family of pumps or turbines could be represented in dimensionless form by a single curve. The grouping of pumps—either singly, in parallel, or in series—to meet appropriate head and discharge requirements for piping was illustrated.

Finally, a brief introduction to turbines was provided, by relating fundamental theory along with the application of reaction, axial-flow, and impulse turbines to various hydraulic design situations.

PROBLEMS

Elementary Theory

10.1 Determine the torque, power, and head input or output for each turbomachine shown in Figure P10.1. Is a pump or turbine implied? The following data are in common: outer radius 300 mm, inner radius 150 mm, $Q = 0.057\ m^3/s$, $\rho = 1000\ kg/m^3$, and $\omega = 25$ rad/s.

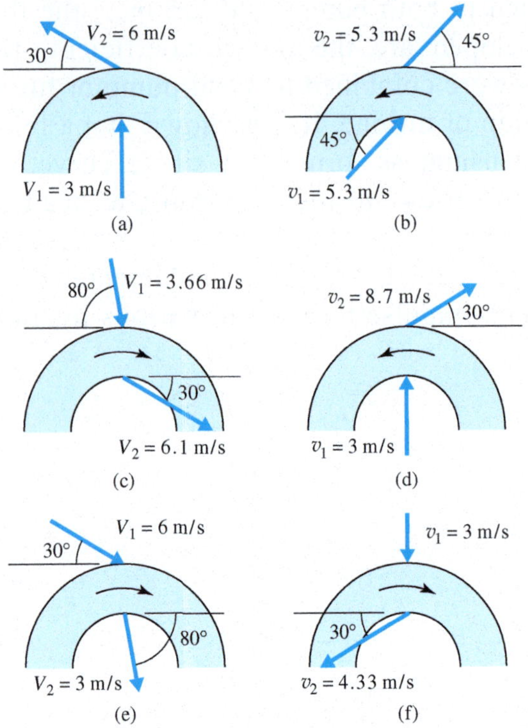

Figure P10.1

10.2 A centrifugal water pump rotates at 800 rpm. The impeller has uniform blade widths $b_1 = 50$ mm, $b_2 = 25$ mm, and radii $r_1 = 40$ mm, $r_2 = 125$ mm. The blade angles are $\beta_1 = 45°$, $\beta_2 = 30°$. Assuming no angular momentum of fluid at the blade entrance, determine the ideal flow rate, pressure head rise across the impeller, and the theoretical torque and power requirements.

10.3 The suction and discharge pipes of a pump are connected to a differential mercury manometer. The diameter of the suction pipe is 200 mm and the diameter of the discharge pipe is 150 mm. The specific gravity of the pumped liquid is $S = 0.81$, and the manometer reading is $H = 600$ mm. Determine the pressure head rise across the pump if the discharge is $Q = 115$ L/s. The centerlines of the suction and discharge pipes are at the same elevation.

10.4 A centrifugal pump with the dimensions $r_2 = 2.5$ in., $b_2 = 0.40$ in., $\beta_2 = 60°$, is pumping kerosene ($S = 0.80$) at a rate of 2 gal/s and rotating at 2000 rpm. For "best design" entry conditions, determine the theoretical pressure head rise and fluid power.

10.5 An axial-flow pump has a stator blade positioned upstream of the impeller, and it provides an angle $\alpha_1 = 60°$ to the flow. The radius of the impeller tip is 285 mm, and the hub radius is 135 mm. Determine the required average blade angle at the exit of the impeller if the pump is to deliver $0.57\ m^3/s$ of water with a theoretical head rise of 2.85 m. The rotational speed is 1500 rpm.

10.6 For the system shown in Figure P10.6, water flows through the pump at a rate of 50 L/s. The allowable *NPSH* provided by the manufacturer at that flow is 3 m. Determine the maximum height Δz above the water surface at which the pump can be located to operate without cavitating. Include all losses in the suction pipe.

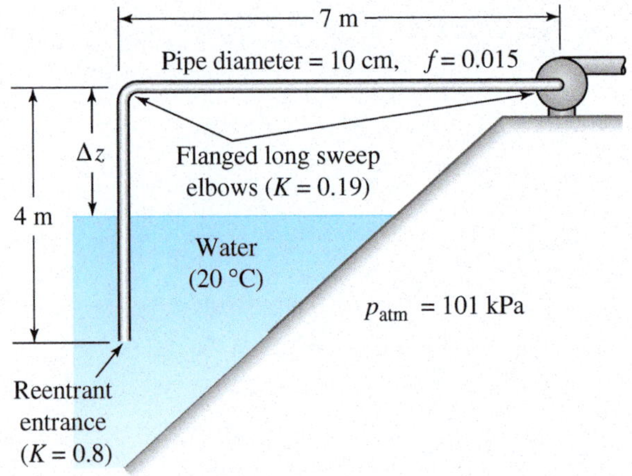

Figure P10.6

10.7 During a test on a water pump, no cavitation is detected when the gage pressure in the pipe at the pump inlet is −9.9 psi, and the water temperature is 80°F. The inlet pipe diameter is 4 in., the head across the pump is 115 ft, and the discharge is 13 gal/s.

(a) Compute the *NPSH* and the cavitation number if the atmospheric pressure is 14.7 psi.

(b) If the pump is to produce the same head and discharge at a location where the atmospheric pressure is 12 psi, what is the necessary change in elevation of the pump relative to the inlet reservoir to avoid cavitation?

Dimensional Analysis and Similitude

10.8 Consider the 371-mm-diameter pump curve of Figure 10.9. If the pump is run at 1200 rpm to deliver 0.8 m^3/s of water, find the resulting head rise and fluid power.

10.9 In Problem 10.8, how high above the suction reservoir can the pump running at 1200 rpm be located if the water temperature is 50°C and the atmospheric pressure is 101 kPa?

10.10 Prepare dimensionless performance curves for the 205-mm-diameter radial flow pump of Figure 10.6. Compare these with the curves shown in Figure 10.12. Can you explain why there may be differences between the two sets of curves?

10.11 If the 220-mm-diameter pump whose characteristics are shown Figure 10.6 is operating at 3300 rpm, what will be the head and the *NPSH* when the discharge is 200 m^3/h?

10.12 A pump is needed to transport 150 L/S of oil ($S = 0.86$) with an increase in head across the pump of 22 m when running at 1800 rpm. What type of pump is best suited for this operation?

10.13 A pump is required to deliver 0.17 m^3/s of water with a head increase of 104 m when operating at maximum efficiency at 2000 rpm. Determine the type of pump best suited for this duty.

10.14 Refer to Figure 10.13, the dimensionless performance curves for an axial-flow pump. It is desired to deliver water at a rate of 45 ft^3/s and a speed of 750 rpm. Determine:

(a) The available head, diameter, power, and *NPSH* requirements.

(b) The actual head-discharge and power-discharge curves.

10.15 A pump running at 400 rpm discharges water at a rate of 85 L/s. The total head rise across the pump is 7.6 m, and the efficiency is 0.7. A second pump, whose linear dimensions are two-thirds of the former, runs at 400 rpm under dynamically similar conditions; it is to be placed on a space satellite where simulated gravity produces a gravitational field which is 50% that of Earth's. What are the discharge, head rise, and power requirements for the second pump?

10.16 A pump is designed to operate under optimum conditions at 600 rpm when delivering water at 22.7 m^3/min against a head of 19.5 m. What type of pump is recommended?

10.17 A pump operating under the conditions stated in Problem 10.16 has a maximum efficiency of 0.70. If the same pump is now required to deliver water at a head of 30.5 m at maximum efficiency, determine the rotational speed, the discharge, and the required power.

10.18 A manufacturer desires to double both the discharge and head on a geometrically similar pump. Determine the change in rotational speed and diameter under the new conditions.

10.19 A pump is required to move water at 2900 gal/min against a head of 200 ft. The approximate rotational speed is 1400 rpm. Using the appropriate dimensionless performance curve in the text, determine the actual speed and size of the pump to operate at peak efficiency.

10.20 Select the type, size, and speed of pump to deliver a liquid ($\gamma = 8830\ \text{N/m}^3$) at $0.66\ \text{m}^3\text{/s}$ with fluid power requirements of 200 kW. (*Hint:* Assume that the pump speed can vary between 1000 and 2000 rpm.)

Use of Turbopumps

10.21 For the system shown in Figure P10.21, recommend the appropriate machine to pump $1\ \text{m}^3\text{/s}$ of water if the impeller speed is 600 rpm.

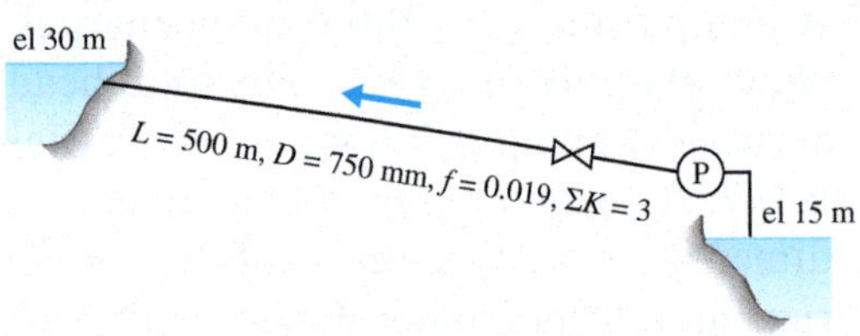

Figure P10.21

10.22 The following performance data were obtained from a test on a 216-mm double-entry centrifugal pump moving water at a constant speed of 1350 rev/min:

Q (m³/min)	H (m)	η_P
0	12.2	0
0.454	12.8	0.26
0.905	13.1	0.46
1.36	13.4	0.59
1.81	13.4	0.70
2.27	13.1	0.78
2.72	12.2	0.78
3.80	11.0	0.74

Plot H versus Q, η_P versus Q, and $\dot{W}_P$ versus Q. If the pump operates in a system whose demand curve is given by $5 + Q^2$, find the discharge, head, and power required. In the demand curve, Q is given in cubic meters per minute.

10.23 An oil ($S = 0.85$) is to be pumped through a pipe as shown in Figure P10.23. Using the characteristic curves of Figure 10.12 determine the required impeller diameter, speed, and input power to deliver the oil at a rate of 14.2 L/s. Would you expect the efficiency of the pump to be the value shown on the curve?

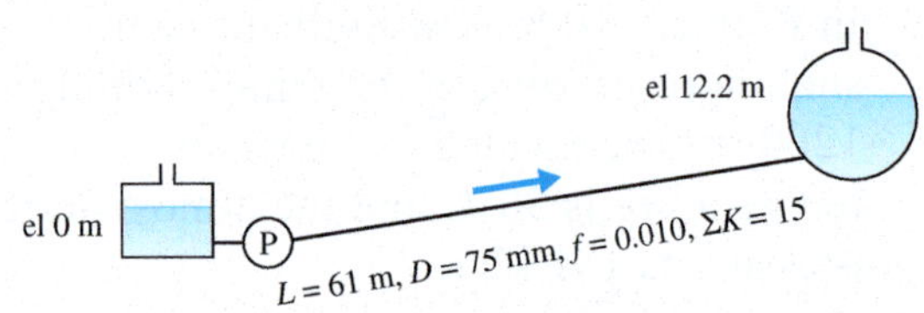

Figure P10.23

10.24 With reference to the pump data in Problem 10.22, if the pump is run at 1200 rpm, find the discharge, head, and required power. Is the operating efficiency under these conditions the same as the pump running at 1350 rpm?

10.25 For the pump data presented in Problem 10.22, plot the dimensionless pump curves C_H versus C_Q, $C_{\dot{W}}$ versus C_Q, and η_P versus C_Q. What is the specific speed of the pump? (**Note:** For double-entry pumps, use one-half the discharge when computing the specific speed.)

10.26 A closed-loop flow facility (Figure P10.26) is to be used to study water flow in a hydraulics laboratory. It is constructed of 300-mm-diameter hydraulically smooth pipe. The loss coefficient at each of the four bends is 0.1, and the total length of the loop is 14 m. The design velocity is $V = 3$ m/s. It is proposed that an axial flow pump running at about 300 rpm be used. Verify that an axial-flow pump is the correct type to employ.

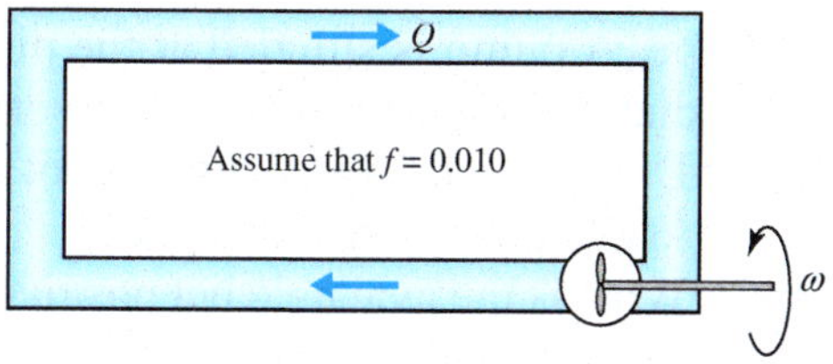

Figure P10.26

10.27 In Problem 10.26, assume that the only axial pump available is one from the family shown in Figure 10.13. If the diameter of the impeller were restricted to the pipe diameter, what would be the rotational speed and resulting head rise across the pump?

10.28 The 240-mm-diameter pump represented in Figure 10.6 is used to move water in a piping system whose demand curve is $62 + 270Q^2$, where Q is in cubic meters per second. Find the discharge, required input power, and the *NPSH* requirement for:

(a) One pump

(b) Two pumps in parallel

10.29 It is desired to pump 600 m^3/h of liquid ammonia ($\rho = 607$ kg/m^3) at a speed of 2500 rpm and a pressure rise of 1000 kPa.

(a) Determine the type of pump best suited for this duty.

(b) Using the appropriate dimensionless pump characteristic curve, find the diameter and number of stages required to meet the pumping requirements.

10.30 Crude oil ($S = 0.86$) is to be pumped through 3 miles of 18-in.-diameter cast iron pipe. The elevation rise from the upstream to the downstream end is 640 ft. If an available pump is the 240-mm-diameter radial-flow machine of Figure 10.6, how many pumps in series are required to provide the most efficient operation? Find the required power.

10.31 A water supply piping network designed for a community receives water from a 10-in.-diameter well (Figure P10.31). The pump is situated 150 ft below the 10-in.-diameter primary water main that delivers the flow to the network, and due to the available ground water, the hydraulic grade line on the suction side of the pump is 75 ft above the pump. The design discharge is 500 gal/min, and a pressure of 80 lb/in^2 is required in the primary water main. Making use of Figure 10.12, determine the size, speed, and number of stages of a pump to satisfy the design requirements, and the required horsepower. Assume that the pump impeller diameter is 8 in. and that the friction factor in the well delivery pipe is constant, $f = 0.02$.

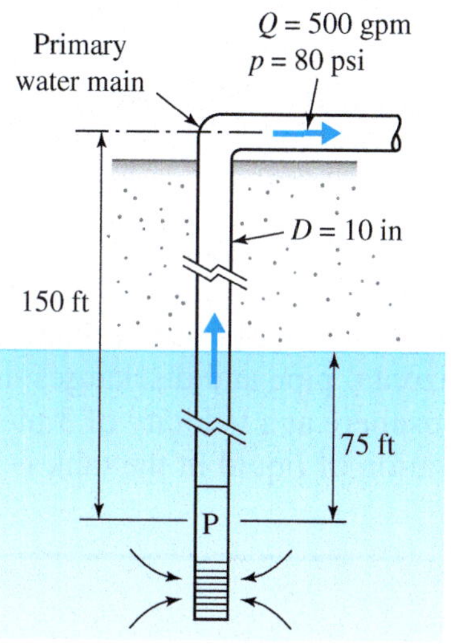

Figure P10.31

10.32 A model pump (Figure P10.32) delivers 80°C water at a speed of 2400 rpm. It begins to cavitate when the inlet pressure $p_i = 83$ kPa absolute and inlet velocity $V_i = 6$ m/s.

(a) Compute the *NPSH* of the model pump. Neglect losses and elevation changes between the inlet section where p_i is recorded and the cavitation region in the pump.

(b) What is the required *NPSH* of a prototype pump that is four times larger and runs at 1000 rpm?

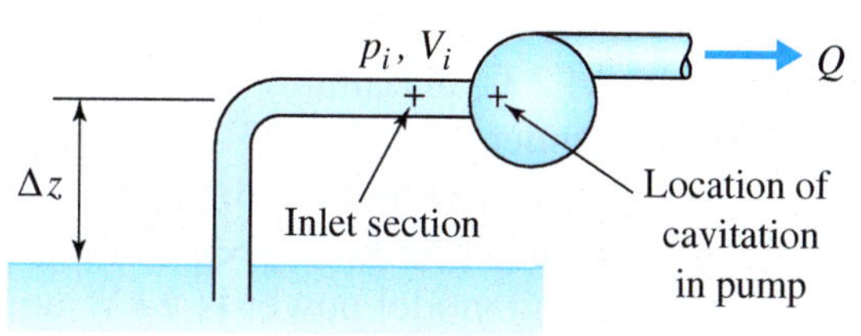

Figure P10.32

10.33 It is required to pump water at 1.25 m^3/s from a lower to an upper reservoir. The water surface in the lower reservoir is at an elevation of 20 m, and the upper reservoir elevation is 23 m. The length of the 750-mm-diameter pipe is 60 m. The friction factor can be assumed constant, $f = 0.02$, and the entrance and exit loss coefficients are 0.5 and 1.0, respectively.

(a) What type of pump is best suited for this system if the impeller speed is to be 600 rpm?

(b) Assuming a reasonable efficiency from an appropriate pump curve, estimate the power in kilowatts required to operate the pump.

10.34 Carbon tetrachloride at 20°C is pumped from a tank opened to the atmosphere through 200 m of 50-mm-diameter smooth horizontal pipe and discharges into the atmosphere at a velocity of 3 m/s. The elevation of liquid in the tank is 3 m above the pipe. A pump is situated at the pipe inlet immediately downstream of the reservoir.

(a) Determine the speed and size of a pump to meet the system demand.

(b) What is the available net positive suction head at the pump inlet? Neglect losses in the short suction pipe connecting the pump to the tank. Assume that the atmospheric pressure is 101 kPa.

(c) Sketch the hydraulic grade line for the system.

10.35 A wind tunnel with a test section 1 m × 2 m is powered by an axial fan 1.5 m in diameter. The velocity in the test section is 20 m/s, and the rotational speed of the fan is 2500 rpm. Assuming negligible losses in the tunnel and a performance curve for the fan defined by Figure 10.13:

(a) Calculate the power required to operate the fan.

(b) What is the pressure rise across the fan?

Turbines

10.36 A Francis turbine has the following dimensions: $r_1 = 4.5$ m, $r_2 = 2.5$ m, $b_1 = b_2 = 0.85$ m, $\beta_1 = 75°$, $\beta_2 = 100°$. The angular speed is 120 rpm, and the discharge is 150 m^3/s. For no separation at the entrance to the runner, determine the guide vane angle, theoretical torque, head, and power.

10.37 A model Francis turbine at 1/5 full scale develops 3 kW at 360 rpm under a head of 1.8 m. Find the speed and power of the full size turbine when operating under a head of 5.8 m. Assume that both units are operating at maximum efficiency.

10.38 A model is to be built for studying the performance of a turbine having a runner diameter of 3 ft, maximum output of 2200 kW under a head of 150 ft and a speed of 240 rpm. Determine the diameter of the model runner and its speed if the corresponding model power is 9 kW and the head is 25 ft.

10.39 A hydroelectric site has available $H_T = 80$ m and $Q = 3$ m^3/s. It is proposed to use a Francis turbine unit with the operating characteristics at optimum efficiency shown in Figure 10.24. Determine the required speed and diameter of the machine.

10.40 Determine the power output, the type of turbine, and the approximate speed for the installation shown in Figure P10.40. Neglect all minor losses except those existing at the valve.

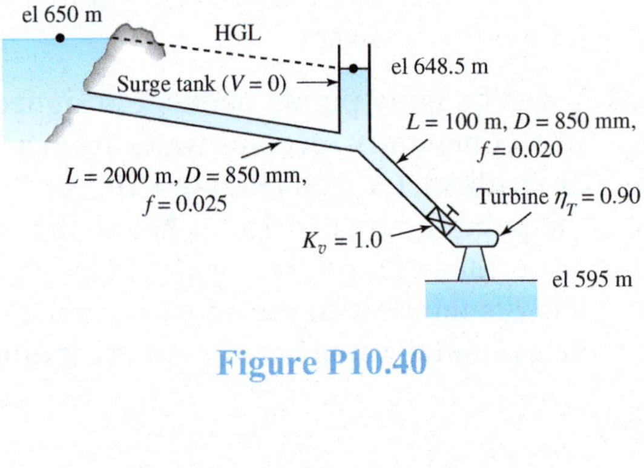

Figure P10.40

10.41 A water turbine is required to operate at 420 rpm under a net head of 3 m, a discharge of 0.312 m^3/s, and with an efficiency of 0.9. Tests on a 1/6-scale model running at 2000 rpm are to be carried out using water. For the model turbine, determine the head drop, flow rate, output power, and expected efficiency.

10.42 A Pelton wheel develops 4.5 MW under a head of 120 m at a speed of 200 rpm. The wheel diameter is eight times the jet diameter. Use the experimental data of Figure 10.29 at maximum efficiency to determine the required flow, wheel diameter, diameter of each jet, the number of jets required, and the specific speed.

10.43 A reaction turbine installation typically includes a draft tube (Figure 10.26a), whose function is to convert kinetic energy of the fluid exiting the runner into flow energy. Consider a turbine operating at $Q = 85$ m^3/s and $H_T = 31.8$ m. The radius of the draft tube at the outlet of the runner is 2.5 m, and the diameter at the end of the tube is 5.0 m. Losses in the draft tube can be neglected.

(a) For water at 20°C, what is the pressure at the runner outlet if in Figure P10.43, $z_2 - z_1 = 2.5$ m?

(b) What is the permissible $z_2 - z_1$ if the allowable cavitation number is $\sigma = 0.14$?

10.44 The Ludington pumped-storage system (Figure P10.44) on the eastern shore of Lake Michigan consists of six penstocks delivering a combined discharge of 73,530 ft^3/s in the power-producing mode of operation. Each turbine generates 427,300 hp at an efficiency of 0.85. What diameter penstock is required for the system to operate under design conditions? Classify the type of turbine employed.

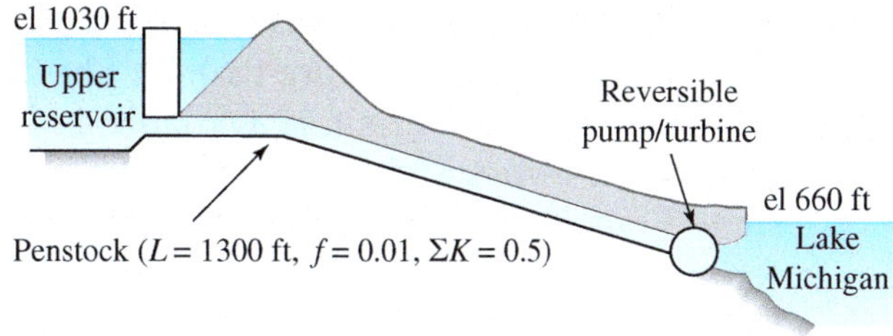

Figure P10.44

10.45 A Francis turbine of the same family as that represented in Figure 10.24 is running at a speed of 480 rpm under a head of 9.5 m. Determine the runner diameter, discharge, and developed power if the turbine is operating at maximum efficiency.

10.46 The available head from reservoir level to the four nozzles of a twin-runner Pelton wheel is 305 m. The length of the supply pipe, or penstock, is 3 km with a friction factor $f = 0.02$ and $\sum K = 2.0$. The turbine develops 10.4 MW with an efficiency of 0.85.

(a) If the head across the turbine is 95% of the available head, what is the diameter of the penstock?

(b) If $C_v = 0.98$ for each nozzle, calculate the diameter of each jet.

10.47 In a projected low-head hydroelectric scheme, 282 m^3/s of water is available under a head of 3.7 m. It is proposed to use Francis turbines with specific speed 2.42, for which the rotational speed is 50 rpm. Determine the number of units required and the power to be developed by each machine. Assume an efficiency of 0.9.

10.48 Repeat Problem 10.47 if propeller turbines are substituted for the Francis units. Assume a specific speed of 4.15 and an efficiency of 0.9.

10.49 A landowner has constructed an elevated reservoir as shown in Figure P10.49. She estimates that 1200 L/min is available as continuous flow and desires to install a small Francis turbine to generate electricity for her own use.

(a) What fluid power is expected to be delivered to the turbine?

(b) If a turbine similar to the one whose characteristics are shown in Figure 10.24 is to be installed, what are the speed and size required for this installation?

(c) Using an appropriate empirical equation, estimate the efficiency for this installation. What power is expected to be delivered to the generator?

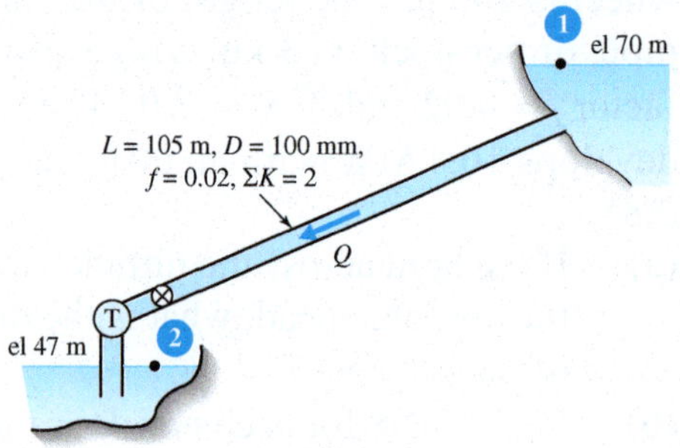

Figure P10.49

10.50 A proposed design for a hydroelectric project is based on a discharge of 0.25 m^3/s through the penstock and turbine (Figure P10.50). The friction factor can be assumed constant, $f = 0.015$, and the minor losses are negligible.

(a) Determine the power in kilowatts that can be expected from the facility, assuming that the turbine efficiency is 0.85.

(b) Show that the type of machine to be installed is a Francis turbine if the desired rotational speed is 1200 rpm.

(c) Determine the size, actual speed, and actual power output of the selected turbine.

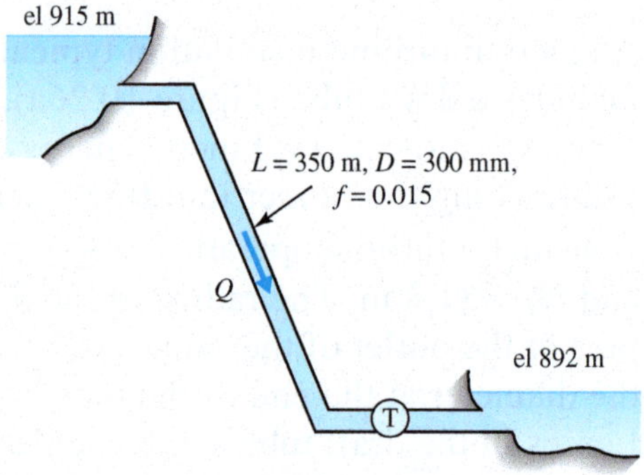

Figure P10.50

Appendix

A. UNITS AND CONVERSIONS AND MATHEMATICAL RELATIONSHIPS

Table A.1 U.S. Engineering Units, SI Units, and Their Conversion Factors

Quantity	*English units*	*International system[a] SI*	*Conversion factor*
Length	inch	millimeter	1 in. = 25.4 mm
	foot	meter	1 ft = 0.3048 m
	mile	kilometer	1 mile = 1.609 km
Area	square inch	square centimeter	1 in^2 = 6.452 cm^2
	square foot	square meter	1 ft^2 = 0.09290 m^2
Volume	cubic inch	cubic centimeter	1 in^3 = 16.39 cm^3
	cubic foot	cubic meter	1 ft^3 = 0.02832 m^3
	gallon		1 gal = 0.003789 m^3
Mass	pound-mass	kilogram	1 lb_m = 0.4536 kg
	slug		1 slug = 14.59 kg
Density	slug/cubic foot	kilogram/cubic meter	1 slug/ft^3 = 515.4 kg/m^3
Force	pound-force	newton	1 lb = 4.448 N
Work or torque	pound-foot	newton-meter	1 lb-ft = 1.356 N·m
Pressure	pound/square inch	newton/square meter (pascal)	1 psi = 6895 Pa
	pound/square foot		1 psf = 47.88 Pa
Temperature	degree Fahrenheit	degree Celsius	°F = $\frac{9}{5}$ °C + 32
	degree Rankine	kelvin	°R = $\frac{9}{5}$ K
Energy	British thermal unit	joule	1 Btu = 1055 J
	calorie		1 cal = 4.186 J
	pound-foot		1 lb-ft = 1.356 J
Power	horsepower	watt	1 hp = 745.7 W
	pound-foot/second		1 lb-ft/s = 1.356 W
Velocity	foot/second	meter/second	1 ft/s = 0.3048 m/s
Acceleration	foot/second squared	meter/second squared	1 ft/s^2 = 0.3048 m/s^2
Frequency	cycle/second	hertz	1 cps = 1.000 Hz
Viscosity	pound-second/square foot	newton-second/square meter	1 lb-s/ft^2 = 47.88 N·s/m^2

[a] The reversed initials in this abbreviation come from the French form of the name: Systéme International.

M. C. Potter and D. C. Wiggert, *Applied Fluid Mechanics*,
https://doi.org/10.1007/978-3-031-65276-9

Table A.2 Conversions

Length	*Force*	*Mass*	*Velocity*
1 cm = 0.3937 in.	1 1b = 0.4536 kg	1 oz = 28.35 g	1 mph = 1.467 ft/s
1 m = 3.281 ft	1 lb = 0.4448 × 10^6 dyn	1 1b = 0.4536 kg	1 mph = 0.8684 knot
1 km = 0.6214 mile	1 1b = 32.17 pdl	1 slug = 32.17 lb	1 ft/s = 0.3048 m/s
1 in. = 2.54 cm	1 kg = 2.205 lb	1 slug = 14.59 kg	1 m/s = 3.281 ft/s
1 = 0.3048 m	1 pdl = 0.03108 lb	1 kg = 2.205 lb	1 km/h = 0.278 m/s
1 mile = 1.609 km	1 dyn = 2.248 × 10^{-6} lb	1 kg = 0.06852 slug	
1 mile = 5280 ft	1 lb = 4.448 N		
1 mile = 1760 yd	1 N = 0.2248 lb		

Work, energy, and power	*Pressure*	*Volume*	*Flow rate*
1 Btu = 778.2 lb-ft	1 psi = 2.036 in. Hg	1 ft^3 = 28.32 L	1 ft^3/min = 4.719 × 10^{-4} m^3/s
1 J = 10^7 ergs 1 J = 0.7376 lb-ft	1 psi = 27.7 in. H_2O	1 ft^3 = 7.481 gal (U.S.)	1 ft^3/s = 0.02832 m^3/s
1 cal = 3.088 lb-ft	14.7 psi = 22.92 in. Hg	1 gal (U.S.) = 231 in^3	1 m^3/s = 35.31 ft^3/s
1 cal = 0.003968 Btu	14.7 psi = 33.93 ft H_2O	1 gal (Brit.) = 1.2 gal (U.S.)	1 gal/min = 0.002228 ft^3/s
1 kWh = 3413 Btu	14.7 psi = 1.0332 kg/cm^2	1 m^3 = 1000 L	1 ft^3/s = 448.9 gal/min
1 Btu = 1.055 kJ	14.7 psi = 1.0133 bar	1 ft^3 = 0.02832 m^3	
1 lb-ft = 1.356 J	1 kg/cm^2 = 14.22 psi	1 m^3 = 35.31 ft^3	*Viscosity*
1 hp = 550 lb-ft/s	1 in. Hg = 0.4912 psi		1 stoke = 10^{-4} m^2/s
1 hp = 0.7067 Btu/s	1 ft H_2O = 0.4331 psi		1 poise = 0.1 N·s/m^2
1 hp = 0.7455 kW	1 psi = 6895 Pa		1 lb-s/ft^2 = 47.88 N·s/m^2
1 W = 1 J/s	1 psf = 47.88 Pa		1 ft^2/s = 0.0929 m^2/s
1 W = 1.0 × 10^7 dyn-cm/s	10^5 Pa = 1 bar		
1 erg = 10^{-7} J	1 kPa = 0.145 psi		
1 quad = 10^{15} Btu			
1 therm = 10^{15} Btu			

Table A.3 Vector Relationships

$$\mathbf{A}\cdot\mathbf{B} = A_x B_x + A_y B_y + A_z B_z$$

$$\mathrm{A}\times\mathbf{B} = (A_y B_z - A_z B_y)\hat{\mathbf{i}} + (A_z B_x - A_x B_z)\hat{\mathbf{j}} + (A_x B_y - A_y B_x)\hat{\mathbf{k}}$$

If $\mathbf{A}\perp\mathbf{B},\ \mathbf{A}\cdot\mathbf{B}=0$

If $\mathbf{A}\parallel\mathbf{B},\ \mathbf{A}\times\mathbf{B}=0$

$$\text{gradient operator} = \nabla = \frac{\partial}{\partial x}\hat{\mathbf{i}} + \frac{\partial}{\partial y}\hat{\mathbf{j}} + \frac{\partial}{\partial z}\hat{\mathbf{k}}$$

$$\text{divergence of } \mathbf{V} = \nabla\cdot\mathbf{V} = \frac{\partial u}{\partial x} + \frac{\partial v}{\partial y} + \frac{\partial w}{\partial z}$$

$$\text{curl of } \mathbf{V} = \nabla\times\mathbf{V} = \left(\frac{\partial w}{\partial y} - \frac{\partial v}{\partial z}\right)\hat{\mathbf{i}} + \left(\frac{\partial u}{\partial z} - \frac{\partial w}{\partial x}\right)\hat{\mathbf{j}} + \left(\frac{\partial v}{\partial x} - \frac{\partial u}{\partial y}\right)\hat{\mathbf{k}}$$

Laplace's equation $= \nabla^2\phi = 0$

Irrotational (conservative) vector field: $\nabla\times\mathbf{V} = 0$

Stokes' theorem: $\oint_C \mathbf{V}\cdot d\mathrm{l} = \iint_A (\nabla\times\mathbf{V})\cdot\hat{\mathbf{n}}\, dA$

Gauss' theorem: $\oint\!\!\oint_A \mathbf{V}\cdot\hat{\mathbf{n}}\, dA = \iiint_{\mathcal{V}} \nabla\cdot\mathbf{V}\, d\mathcal{V}$

Table A.4 The Substantial Derivative, Vorticity, and Acceleration in Rectangular, Cylindrical, and Spherical Coordinates

Substantial Derivative

Rectangular

$$\frac{D}{Dt} = u\frac{\partial}{\partial x} + v\frac{\partial}{\partial y} + w\frac{\partial}{\partial z} + \frac{\partial}{\partial t}$$

Cylindrical

$$\frac{D}{Dt} = v_r\frac{\partial}{\partial r} + \frac{v_\theta}{r}\frac{\partial}{\partial\theta} + v_z\frac{\partial}{\partial z} + \frac{\partial}{\partial t}$$

Spherical

$$\frac{D}{Dt} = v_r\frac{\partial}{\partial r} + \frac{v_\theta}{r}\frac{\partial}{\partial\theta} + \frac{v_\phi}{r\sin\theta}\frac{\partial}{\partial\phi} + \frac{\partial}{\partial t}$$

Vorticity

Rectangular

$$\omega_x = \frac{\partial w}{\partial y} - \frac{\partial v}{\partial z}$$

$$\omega_y = \frac{\partial u}{\partial z} - \frac{\partial w}{\partial x}$$

$$\omega_z = \frac{\partial v}{\partial x} - \frac{\partial u}{\partial y}$$

Cylindrical

$$\omega_x = \frac{1}{r}\left(\frac{\partial v_Z}{\partial\theta}\right) - \frac{\partial v_\theta}{\partial z}$$

$$\omega_\theta = \frac{\partial v_r}{\partial z} - \frac{\partial v_Z}{\partial r}$$

$$\omega_z = \frac{1}{r}\left(\frac{\partial(rv_\theta)}{\partial r} - \frac{\partial v_r}{\partial\theta}\right)$$

Spherical

$$\omega_r = \frac{1}{r\sin\theta}\left[\frac{\partial}{\partial\theta}(v_\phi\sin\theta) - \frac{\partial v_\theta}{\partial\phi}\right]$$

$$\omega_\phi = \frac{1}{r}\left[\frac{\partial}{\partial r}(rv_\theta) - \frac{\partial v_r}{\partial\theta}\right]$$

$$\omega_\theta = \frac{1}{r}\left[\frac{1}{\sin\theta}\frac{\partial v_r}{\partial\phi} - \frac{\partial}{\partial r}(rv_\phi)\right]$$

Acceleration

Rectangular

$$a_x = \frac{\partial u}{\partial t} + u\frac{\partial u}{\partial x} + v\frac{\partial u}{\partial y} + w\frac{\partial u}{\partial z}$$

$$a_y = \frac{\partial v}{\partial t} + u\frac{\partial v}{\partial x} + v\frac{\partial v}{\partial y} + w\frac{\partial v}{\partial z}$$

$$a_z = \frac{\partial w}{\partial t} + u\frac{\partial w}{\partial x} + v\frac{\partial w}{\partial y} + w\frac{\partial w}{\partial z}$$

Cylindrical

$$a_r = \frac{\partial v_r}{\partial t} + v_r\frac{\partial v_r}{\partial r} + \frac{v_\theta}{r}\frac{\partial v_r}{\partial \theta} + v_z\frac{\partial v_r}{\partial z} - \frac{v_\theta^2}{r}$$

$$a_\theta = \frac{\partial v_\theta}{\partial t} + v_r\frac{\partial v_\theta}{\partial r} + \frac{v_\theta}{r}\frac{\partial v_\theta}{\partial \theta} + v_z\frac{\partial v_\theta}{\partial z} + \frac{v_r v_\theta}{r}$$

$$a_z = \frac{\partial v_z}{\partial t} + v_r\frac{\partial v_z}{\partial r} + \frac{v_\theta}{r}\frac{\partial v_z}{\partial \theta} + v_z\frac{\partial v_z}{\partial z}$$

Spherical

$$a_r = \frac{\partial v_r}{\partial t} + v_r\frac{\partial v_r}{\partial r} + \frac{v_\theta}{r}\frac{\partial v_r}{\partial \theta} + \frac{v_\phi}{r\sin\theta}\frac{\partial v_r}{\partial \phi} - \frac{v_\phi^2 + v_\theta^2}{r}$$

$$a_\theta = \frac{\partial v_\theta}{\partial t} + v_r\frac{\partial v_\theta}{\partial t} + \frac{v_\theta}{r}\frac{\partial v_\theta}{\partial \theta} + \frac{v_\phi}{r\sin\phi}\frac{\partial v_\theta}{\partial \phi} + \frac{v_r v_\theta - v_\phi^2\cot\theta}{r}$$

$$a_\phi = \frac{\partial v_\phi}{\partial t} + v_r\frac{\partial v_\phi}{\partial r} + \frac{v_\theta}{r}\frac{\partial v_\phi}{\partial \theta} + \frac{v_\phi}{r\sin\theta}\frac{\partial v_\phi}{\partial \phi} + \frac{v_r v_\phi + v_\theta v_\phi\cot\theta}{r}$$

Table A.5 Integral Forms of the Fundamental Laws

Continuity	*Energy*	*Momentum*
	General Form	
$0 = \frac{d}{dt}\int_{\text{c.v.}} \rho d\mathcal{V} + \int_{\text{c.v.}} \rho \mathbf{V} \cdot \hat{\mathbf{n}}\, dA$	$-\sum \dot{W} = \frac{d}{dt}\int_{\text{c.v.}} \left(\frac{V^2}{2} + gz\right)\rho\, d\mathcal{V}$ $+ \int_{\text{c.s.}} \left(\frac{V^2}{2} + \frac{p}{\rho} + gz\right)\rho \mathbf{V} \cdot \hat{\mathbf{n}}\, dA + \text{losses}$	$\sum \mathbf{F} = \frac{d}{dt}\int_{\text{c.v.}} \rho \mathbf{V}\, d\mathcal{V} + \int_{\text{c.s.}} \rho \mathbf{V}(\mathbf{V} \cdot \hat{\mathbf{n}})\, dA$
	Steady Flow	
$0 = \int_{\text{c.s.}} \rho \mathbf{V} \cdot \hat{\mathbf{n}}\, dA$	$-\sum \dot{W} = \int_{\text{c.s.}} \left(\frac{V^2}{2} + \frac{p}{\rho} + gz\right)\rho \mathbf{V} \cdot \hat{\mathbf{n}}\, dA + \text{losses}$	$\sum \mathbf{F} = \int_{\text{c.s.}} \rho \mathbf{V}(\mathbf{V} \cdot \hat{\mathbf{n}})\, dA$
	Steady Nouniform Flow[a]	
$\dot{m} = \rho_1 A_1 \overline{V}_1 = \rho_2 A_2 \overline{V}_2$	$\frac{-\sum \dot{W}}{\dot{m}g} = \alpha_2 \frac{\overline{V}_2^2}{2g} + \frac{p_2}{\gamma_2} + z_2 - \alpha_1 \frac{\overline{V}_1^2}{2g} - \frac{p_1}{\gamma_1} - z_1 + h_L$	$\sum F_x = \dot{m}(\beta_2 \overline{V}_{2x} - \beta_1 \overline{V}_{1x})$ $\sum F_y = \dot{m}(\beta_2 \overline{V}_{2y} - \beta_1 \overline{V}_{1y})$
	Steady Uniform Form[a]	
$\dot{m} = \rho_1 A_1 V_1 = \rho_2 A_2 V_2$	$\frac{-\sum \dot{W}}{\dot{m}g} = \frac{V_2^2}{2g} + \frac{p^2}{\gamma^2} + z_2 - \frac{V_1^2}{2g} - \frac{p_1}{\gamma} - z_1 + h_L$	$\sum \mathbf{F} = \dot{m}(\mathbf{V}_2 - \mathbf{V}_1)$
	Steady Uniform Incompressible Flow[a]	
$Q = A_1 V_1 = A_2 V_2$	$\frac{-\sum \dot{W}}{\dot{m}g} = \frac{V_2^2}{2g} + \frac{p_2}{\gamma} + z_2 - \frac{V_1^2}{2g} - \frac{p_1}{\gamma} - z_1 + h_L$ $H_P + \frac{V_1^2}{2g} + \frac{p_1}{\gamma} + z_1 = H_T + \frac{V_2^2}{2g} + \frac{p_2}{\gamma} + z_2 + h_L$	$\sum \mathbf{F} = \dot{m}(\mathbf{V}_2 - \mathbf{V}_1)$
$\dot{m}$ = mass flux Q = flow rate $\overline{V}$ = average velocity $= \frac{\int V dA}{A}$	α = kinetic energy correction factor $= \frac{\int V^3 dA}{\overline{V}^3 A}$ β = momentum correction factor $= \frac{\int V^2 dA}{\overline{V}^2 A}$	h_L = head loss $\sum \dot{W} = \dot{W}_S + \dot{W}_{\text{shear}} + \dot{W}_I$ H_P = pump head = $\dot{W}_p / \dot{m}g$ H_T = turbine head = $\dot{W}_T / \dot{m}g$

[a]The control volume has one entrance (section 1) and one exit (section 2).

Table A.6 Fundamental Laws for Incompressible Flows

Continuity

Rectangular

$$\frac{\partial u}{\partial x}+\frac{\partial v}{\partial y}+\frac{\partial w}{\partial z}=0$$

Cylindrical

$$\frac{1}{r}\frac{\partial}{\partial r}(rv_r)+\frac{1}{r}\frac{\partial v_\theta}{\partial \theta}+\frac{\partial v_z}{\partial z}=0$$

Spherical

$$\frac{1}{r^2}\frac{\partial}{\partial r}(r^2 v_r)+\frac{1}{r\sin\theta}\frac{\partial}{\partial\theta}(v_\theta\sin\theta)+\frac{1}{r\sin\theta}\frac{\partial v_\phi}{\partial\phi}=0$$

Momentum

Rectangular

$$\frac{Du}{Dt}=-\frac{1}{\rho}\frac{\partial p}{\partial x}+g_x+v\nabla^2 u$$

$$\frac{Dv}{Dt}=-\frac{1}{\rho}\frac{\partial p}{\partial y}+g_y+v\nabla^2 v$$

$$\frac{Dw}{Dt}=-\frac{1}{\rho}\frac{\partial p}{\partial z}+g_z+v\nabla^2 w$$

$$\frac{D}{Dt}=\frac{\partial}{\partial t}+u\frac{\partial}{\partial x}+v\frac{\partial}{\partial y}+w\frac{\partial}{\partial z}$$

$$\nabla^2=\frac{\partial^2}{\partial x^2}+\frac{\partial^2}{\partial y^2}+\frac{\partial^2}{\partial z^2}$$

Cylindrical

$$\frac{Dv_r}{Dt}-\frac{v_\theta^2}{r}=-\frac{1}{\rho}\frac{\partial p}{\partial r}+g_r+\nu\left(\nabla^2 v_r-\frac{v_r}{r^2}-\frac{2}{r^2}\frac{\partial v_\theta}{\partial\theta}\right)$$

$$\frac{Dv_\theta}{Dt}+\frac{v_r v_\theta}{r}=-\frac{1}{\rho r}\frac{\partial p}{\partial\theta}+g_\theta+\nu\left(\nabla^2 v_\theta+\frac{2}{r^2}\frac{\partial v_r}{\partial\theta}-\frac{v_\theta}{r^2}\right)$$

$$\frac{Dv_z}{Dt}=-\frac{1}{\rho}\frac{\partial p}{\partial z}+g_z+\nu\nabla^2 v_z$$

$$\frac{D}{Dt}=\frac{\partial}{\partial t}+v_r\frac{\partial}{\partial r}+\frac{v_\theta}{r}\frac{\partial}{\partial\theta}+v_z\frac{\partial}{\partial z}$$

$$\nabla^2=\frac{\partial^2}{\partial r^2}+\frac{1}{r}\frac{\partial}{\partial r}+\frac{1}{r^2}\frac{\partial^2}{\partial\theta^2}+\frac{\partial^2}{\partial z^2}$$

Spherical

$$\frac{Dv_r}{Dt}-\frac{v_\theta^2+v_\phi^2}{r}=-\frac{1}{\rho}\frac{\partial p}{\partial r}+g_r+\nu\left(\nabla^2 v_r-\frac{2v_r}{r^2}-\frac{2}{r^2}\frac{\partial v_\theta}{\partial\theta}-\frac{2v_\theta\cot\theta}{r^2}-\frac{2}{r^2\sin\theta}\frac{\partial v_\phi}{\partial\phi}\right)$$

$$\frac{Dv_\theta}{Dt}+\frac{v_r v_\theta-v_\phi^2\cot\theta}{\partial t}=-\frac{1}{\rho r}\frac{\partial p}{\partial\theta}+g_\theta+\nu\left(\nabla^2 v_\theta+\frac{2}{r^2}\frac{\partial v_r}{\partial\theta}-\frac{v_\theta}{r^2\sin^2\theta}-\frac{2\cos\theta}{r^2\sin^2\theta}\frac{\partial v_\phi}{\partial\phi}\right)$$

$$\frac{Dv_\phi}{Dt}+\frac{v_\phi v_r}{r}+\frac{v_\theta v_\phi\cot\theta}{r}=-\frac{1}{\rho r\sin\theta}\frac{\partial p}{\partial\phi}+g_\theta+\nu\left(\nabla^2 v_\theta-\frac{v\phi}{r^2\sin^2\theta}+\frac{2}{r^2\sin^2\theta}\frac{\partial v_r}{\partial\phi}+\frac{2\cos\theta\partial v_\theta}{r^2\sin^2\theta\partial\phi}\right)$$

$$\frac{D}{Dt}=\frac{\partial}{\partial t}+v_r\frac{\partial}{\partial r}+\frac{v_\theta}{r}\frac{\partial}{\partial\theta}+\frac{v_\phi}{r\sin\theta}\frac{\partial}{\partial\phi}$$

$$\nabla^2=\frac{1}{r^2}\frac{\partial}{\partial r}\left(r^2\frac{\partial}{\partial r}\right)+\frac{1}{r^2\sin\theta}\frac{\partial}{\partial\theta}\left(\sin\theta\frac{\partial}{\partial\theta}\right)+\frac{1}{r^2\sin^2\theta}\frac{\partial^2}{\partial\phi^2}$$

Energy

Rectangular

$$\rho\frac{Dh}{Dt}=K\nabla^2T+2\mu\left[\left(\frac{\partial u}{\partial x}\right)^2+\left(\frac{\partial v}{\partial y}\right)^2+\left(\frac{\partial w}{\partial z}\right)^2+\frac{1}{2}\left(\frac{\partial u}{\partial y}+\frac{\partial v}{\partial x}\right)^2+\frac{1}{2}\left(\frac{\partial v}{\partial z}+\frac{\partial w}{\partial y}\right)^2+\frac{1}{2}\left(\frac{\partial u}{\partial z}+\frac{\partial w}{\partial x}\right)\right]$$

Cylindrical

$$\rho\frac{Dh}{Dt}=K\nabla^2T+2\mu\left[\left(\frac{\partial v_r}{\partial r}\right)^2+\left(\frac{1}{r}\frac{\partial v_\theta}{\partial\theta}+\frac{v_r}{r}\right)^2+\left(\frac{\partial v_z}{\partial z}\right)^2+\frac{1}{2}\left(\frac{1}{r}\frac{\partial v_z}{\partial\theta}+\frac{\partial v_\theta}{\partial z}\right)^2+\frac{1}{2}\left(\frac{\partial v_r}{\partial z}+\frac{\partial v_z}{\partial r}\right)^2+\frac{1}{2}\left(\frac{1}{r}\frac{\partial v_r}{\partial\theta}+\frac{\partial v_\theta}{\partial r}-\frac{v_\theta}{r}\right)^2\right]$$

Spherical

$$\rho\frac{Dh}{Dt}=K\nabla^2T+2\mu\left[\left(\frac{\partial v_r}{\partial r}\right)^2+\left(\frac{1}{r}\frac{\partial v_\theta}{\partial\theta}+\frac{v_r}{r}\right)^2+\left(\frac{1}{r\sin\theta}\frac{\partial v_\phi}{\partial\phi}+\frac{v_r}{r}\frac{v_\theta\cot\theta}{r}\right)^2\right]+\mu\left[\left(\frac{1}{r\sin\theta}\frac{\partial v_\theta}{\partial\phi}+\frac{\sin\theta}{r}\frac{\partial}{\partial\theta}\left(\frac{v_\theta}{\sin\theta}\right)\right)^2+\left(\frac{1}{r\sin\theta}\frac{\partial v_r}{\partial\phi}+r\frac{\partial}{\partial r}\left(\frac{v_\phi}{r}\right)\right)^2+\left(r\frac{\partial}{\partial r}\left(\frac{v_\theta}{r}\right)+\frac{1}{r}\frac{\partial v_r}{\partial\theta}\right)^2\right]$$

Stresses

Rectangular

$$\sigma_{xx}=-p+2\mu\frac{\partial u}{\partial x}\qquad \tau_{xy}=\mu\left(\frac{\partial u}{\partial y}+\frac{\partial v}{\partial x}\right)$$

$$\sigma_{yy}=-p+2\mu\frac{\partial v}{\partial y}\qquad \tau_{yz}=\mu\left(\frac{\partial v}{\partial z}+\frac{\partial w}{\partial y}\right)$$

$$\sigma_{zz}=-p+2\mu\frac{\partial w}{\partial z}\qquad \tau_{xz}=\mu\left(\frac{\partial u}{\partial z}+\frac{\partial w}{\partial x}\right)$$

Cylindrical

$$\sigma_{rr}=-p+2\mu\frac{\partial v_r}{\partial r}\qquad \tau_{r\theta}=\mu\left[r\frac{\partial}{\partial r}\left(\frac{v_\theta}{r}\right)+\frac{1}{r}\frac{\partial v_r}{\partial\theta}\right]$$

$$\sigma_{\theta\theta}=-p+2\mu\left(\frac{1}{r}\frac{\partial v_\theta}{\partial\theta}+\frac{v_r}{r}\right)\qquad \tau_{\theta z}=\mu\left[\frac{\partial v_\theta}{\partial z}+\frac{1}{r}\frac{\partial v_z}{\partial\theta}\right]$$

$$\sigma_{zz}=-p+2\mu\frac{\partial v_z}{\partial z}\qquad \tau_{rz}=\mu\left[\frac{\partial v_r}{\partial z}+\frac{\partial v_z}{\partial r}\right]$$

Spherical

$$\sigma_{rr}=-p+2\mu\frac{\partial v_r}{\partial r}$$

$$\sigma_{\theta\theta}=-p+2\mu\left(\frac{1}{r}\frac{\partial v_\theta}{\partial\theta}+\frac{v_r}{r}\right)$$

$$\sigma_{\phi\phi}=-p+2\mu\left(\frac{1}{r\sin\theta}\frac{\partial v_\phi}{\partial\phi}+\frac{v_r}{r}+\frac{v_\theta\cot\theta}{r}\right)$$

$$\tau_{r\theta}=\mu\left[r\frac{\partial}{\partial r}\left(\frac{v_\theta}{r}\right)+\frac{1}{r}\frac{\partial v_r}{\partial\theta}\right]$$

$$\tau_{\theta\phi}=\mu\left[\frac{\sin\theta}{r}\frac{\partial}{\partial\theta}\left(\frac{v_\phi}{\sin\theta}\right)+\frac{1}{r\sin\theta}\frac{\partial v_\theta}{\partial\phi}\right]$$

$$\tau_{r\phi}=\mu\left[\frac{1}{r\sin\theta}\frac{\partial v_r}{\partial\phi}+r\frac{\partial}{\partial r}\left(\frac{v_\phi}{r}\right)\right]$$

B. FLUID PROPERTIES

Table B.1 Properties of Water

Temperature T (°C)	*Density* ρ (kg/m^3)	*Viscosity* μ (N· s/m^2)	*Kinematic viscosity* ν (m^2/s)	*Surface tension* σ (N/m)	*Vapor pressure* p_v (kPa)	*Bulk modulus* B (Pa)
0	999.9	1.792×10^{-3}	1.792×10^{-6}	0.0762	0.610	204×10^7
5	1000.0	1.519	1.519	0.0754	0.872	206
10	999.7	1.308	1.308	0.0748	1.13	211
15	999.1	1.140	1.141	0.0741	1.60	214
20	998.2	1.005	1.007	0.0736	2.34	220
30	995.7	0.801	0.804	0.0718	4.24	223
40	992.2	0.656	0.661	0.0701	3.38	227
50	988.1	0.549	0.556	0.0682	12.3	230
60	983.2	0.469	0.477	0.0668	19.9	228
70	977.8	0.406	0.415	0.0650	31.2	225
80	971.8	0.357	0.367	0.0630	47.3	221
90	965.3	0.317	0.328	0.0612	70.1	216
100	958.4	0.284×10^{-3}	0.296×10^{-6}	0.0594	101.3	207×10^7

Table B.1–English Properties of Water

Temperature (°F)	*Density* (slug/ft^3)	*Viscosity* (lb-s/ft^2)	*Kinematic viscosity* (ft^2/s)	*Surface tension* (lb/ft)	*Vapor pressure* (psi)	*Bulk modulus* (psi)
32	1.94	3.75×10^{-5}	1.93×10^{-5}	0.518×10^{-2}	0.089	293,000
40	1.94	3.23	1.66	0.514	0.122	294,000
50	1.94	2.74	1.41	0.509	0.178	305,000
60	1.94	2.36	1.22	0.504	0.256	311,000
70	1.94	2.05	1.06	0.500	0.340	320,000
80	1.93	1.80	0.93	0.492	0.507	322,000
90	1.93	1.60	0.83	0.486	0.698	323,000
100	1.93	1.42	0.74	0.480	0.949	327,000
120	1.92	1.17	0.61	0.465	1.69	333,000
140	1.91	0.98	0.51	0.454	2.89	330,000
160	1.90	0.84	0.44	0.441	4.74	326,000
180	1.88	0.73	0.39	0.426	7.51	318,000
200	1.87	0.64	0.34	0.412	11.53	308,000
212	1.86	0.59×10^{-5}	0.32×10^{-5}	0.404×10^{-2}	14.7	300,000

Table B.2 Properties of Air at Atmospheric Pressure

Temperature T (°C)	*Density* ρ (kg/m^3)	*Viscosity* μ (N·s/m^2)	*Kinematic viscosity* ν (m^2/s)	*Velocity of sound* c (m/s)
−50	1.582	1.46×10^{-5}	0.921×10^{-5}	299
−30	1.452	1.56	1.08×10^{-5}	312
−20	1.394	1.61	1.16	319
−10	1.342	1.67	1.24	325
0	1.292	1.72	1.33	331
10	1.247	1.76	1.42	337
20	1.204	1.81	1.51	343
30	1.164	1.86	1.60	349
40	1.127	1.91	1.69	355
50	1.092	1.95	1.79	360
60	1.060	2.00	1.89	366
70	1.030	2.05	1.99	371
80	1.000	2.09	2.09	377
90	0.973	2.13	2.19	382
100	0.946	2.17	2.30	387
200	0.746	2.57	3.45	436
300	0.616	2.93×10^{-5}	4.75×10^{-5}	480

Table B.2–English Properties of Air at Atmospheric Pressure

Temperature (°F)	*Density* (slug/ft^3)	*Viscosity* (lb-s/ft^2)	*Kinematic viscosity* (ft^2/s)	*Velocity of sound* (ft/s)
−20	0.00280	3.34×10^{-7}	11.9×10^{-5}	1028
0	0.00268	3.38	12.6	1051
20	0.00257	3.50	13.6	1074
40	0.00247	3.62	14.6	1096
60	0.00237	3.74	15.8	1117
68	0.00233	3.81	16.0	1125
80	0.00228	3.85	16.9	1138
100	0.00220	3.96	18.0	1159
120	0.00213	4.07	18.9	1180
160	1.00199	4.23	21.3	1220
200	0.00187	4.50	24.1	1258
300	0.00162	4.98	30.7	1348
400	0.00144	5.26	36.7	1431
1000	0.000844	7.87×10^{-7}	93.2×10^{-5}	1839

Table B.3 Properties of the Standard Atmosphere

Altitude (m)	*Temperature* (K)	*Pressure* (kPa)	*Density* (kg/m³)	*Velocity of sound* (m/s)
0	288.2	101.3	1.225	340
500	284.9	95.43	1.167	338
1 000	281.7	89.85	1.112	336
2 000	275.2	79.48	1.007	333
4 000	262.2	61.64	0.8194	325
6 000	249.2	47.21	0.6602	316
8 000	236.2	35.65	0.5258	308
10 000	223.3	26.49	0.4136	300
12 000	216.7	19.40	0.3119	295
14 000	216.7	14.17	0.2278	295
16 000	216.7	10.35	0.1665	295
18 000	216.7	7.563	0.1216	295
20 000	216.7	5.528	0.0889	295
30 000	226.5	1.196	0.0184	302
40 000	250.4	0.287	4.00×10^{-3}	317
50 000	270.7	0.0798	1.03×10^{-3}	330
60 000	255.8	0.0255	3.06×10^{-4}	321
70 000	219.7	0.00551	8.75×10^{-5}	297
80 000	180.7	0.00103	2.00×10^{-5}	269

Table B.3–English Properties of the Standard Atmosphere

Altitude (ft)	*Temperature* (°F)	*Pressure* (lb/ft²)	*Density* (slugs/ft³)	*Velocity of sound* (ft/s)
0	59.0	2116	0.00237	1117
1 000	55.4	2014	0.00231	1113
2 000	51.9	1968	0.00224	1109
5 000	41.2	1760	0.00205	1098
10 000	23.4	1455	0.00176	1078
15 000	5.54	1194	0.00150	1058
20 000	−12.3	973	0.00127	1037
25 000	−30.1	785	0.00107	1016
30 000	−48.0	628	0.000890	995
35 000	−65.8	498	0.000737	973
36 000	−67.6	475	0.000709	971
40 000	−67.6	392	0.000586	971
50 000	−67.6	242	0.000362	971
100 000	−51.4	23.2	3.31×10^{-5}	971

Table B.4 Properties of Ideal Gases at 300 K

Gas	*Chemical formula*	*Molar mass*	*R* (lb-ft/slug-°R)	*R* (kJ/kg·K)	c_p (lb-ft/slug-°R)	c_p (kJ/kg·K)	k
Air	—	28.97	1,716	0.287	6,012	1.004	1.40
Argon	Ar	39.94	1,244	0.2081	3,139	0.5203	1.667
Carbon dioxide	CO_2	44.01	1,129	0.1889	5,085	0.8418	1.287
Carbon monoxide	CO	28.01	1,775	0.2968	6,238	1.041	1.40
Ethane	C_2H_6	30.07	1,653	0.2765	10,700	1.766	1.184
Helium	He	4.003	12,420	2.077	31,310	5.193	1.667
Hydrogen	H_2	2.016	24,660	4.124	85,930	14.21	1.40
Methane	CH_4	16.04	3,100	0.5184	13,330	2.254	1.30
Nitrogen	N_2	28.02	1,774	0.2968	6,213	1.042	1.40
Oxygen	O_2	32.00	1,553	0.2598	5,486	0.9216	1.394
Propane	C_3H_8	44.10	1,127	0.1886	10,200	1.679	1.12
Steam	H_2O	18.02	2,759	0.4615	11,150	1.872	1.33

$c_v = c_p - R, k = c_p/c_v$

Table B.5 Properties of Common Liquids at Atmospheric Pressure and Approximately 60 to 70°F (16 to 21°C)

Liquid	*Specific weight γ* lb/ft³	*Specific weight γ* N/m³	*Density ρ* slugs/ft³	*Density ρ* kg/m³	*Surface tension*[a] *σ* lb/ft	*Surface tension*[a] *σ* N/m	*Vapor pressure* p_v psia	*Vapor pressure* p_v kPa
Alcohol, ethyl	49.3	7 744	1.53	789	0.0015	0.022	—	—
Benzene	56.2	8 828	1.75	902	0.0020	0.029	1.50	10.3
Carbon tetrachloride	99.5	15 629	3.09	1 593	0.0018	0.026	12.50	86.2
Gasoline	42.4	6 660	1.32	680	—	—	—	—
Glycerin	78.6	12 346	2.44	1 258	0.0043	0.063	2×10^{-6}	1.4×10^{-5}
Kerosene	50.5	7 933	1.57	809	0.0017	0.25	—	—
Mercury	845.5	132 800	26.29	13 550	0.032	0.467	2.31×10^{-5}	1.59×10
SAE 10 oil	57.4	9 016	1.78	917	0.0025	0.036	—	—
SAE 30 oil	57.4	9 016	1.78	917	0.0024	0.035	—	—
Turpentine	54.3	8 529	1.69	871	0.0018	0.026	7.7×10^{-3}	5.31×10^{-2}
Water	62.4	9 810	1.94	1000	0.0050	0.073	0.34	2.34

[a]In contact with air

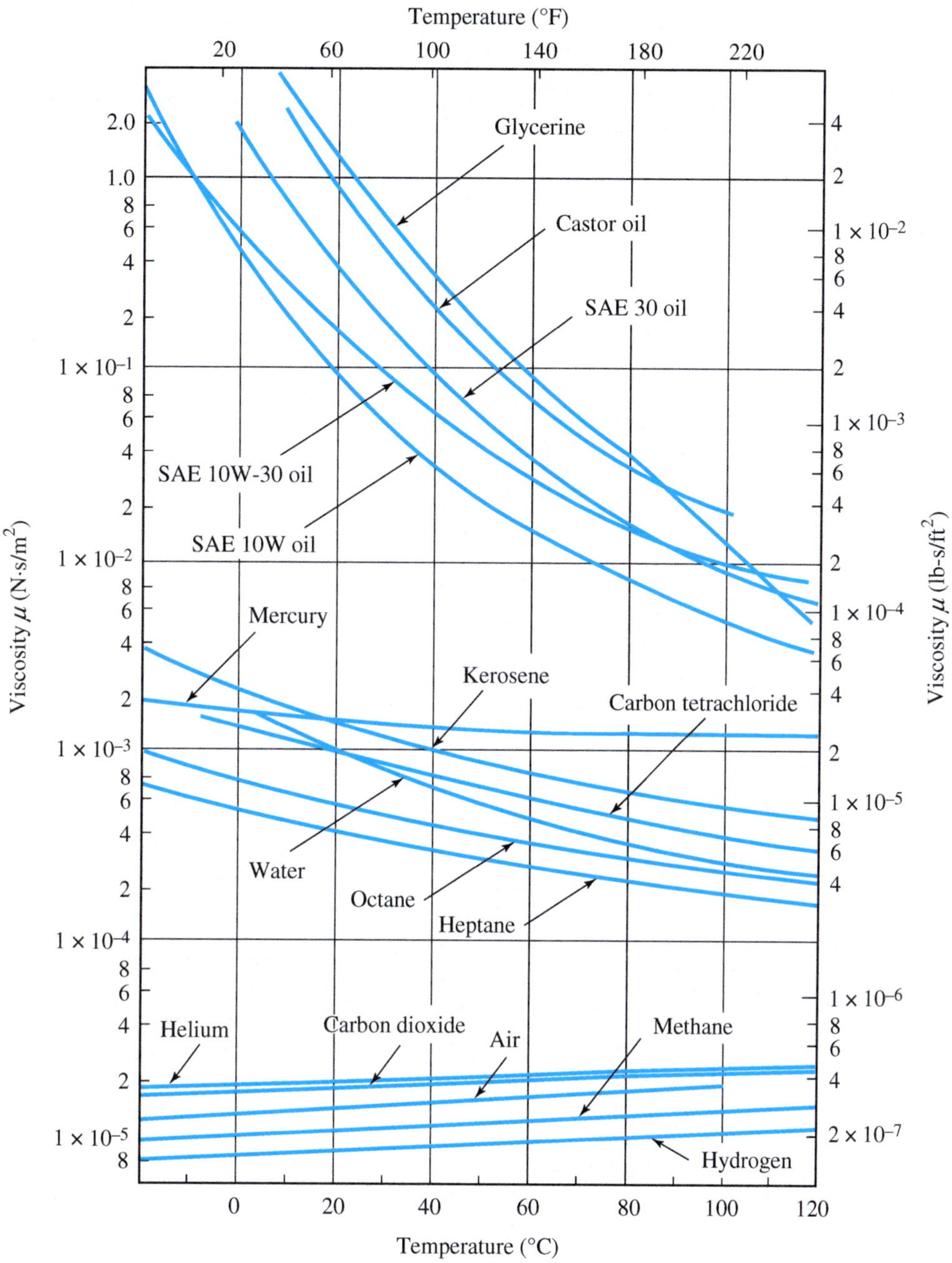

Figure B.1 Viscosity as a function of temperature. (From R. W. Fox and T. A. McDonald, *Introduction to Fluid Mechanics*, 2nd ed., John Wiley & Sons, Inc., New York, 1978.)

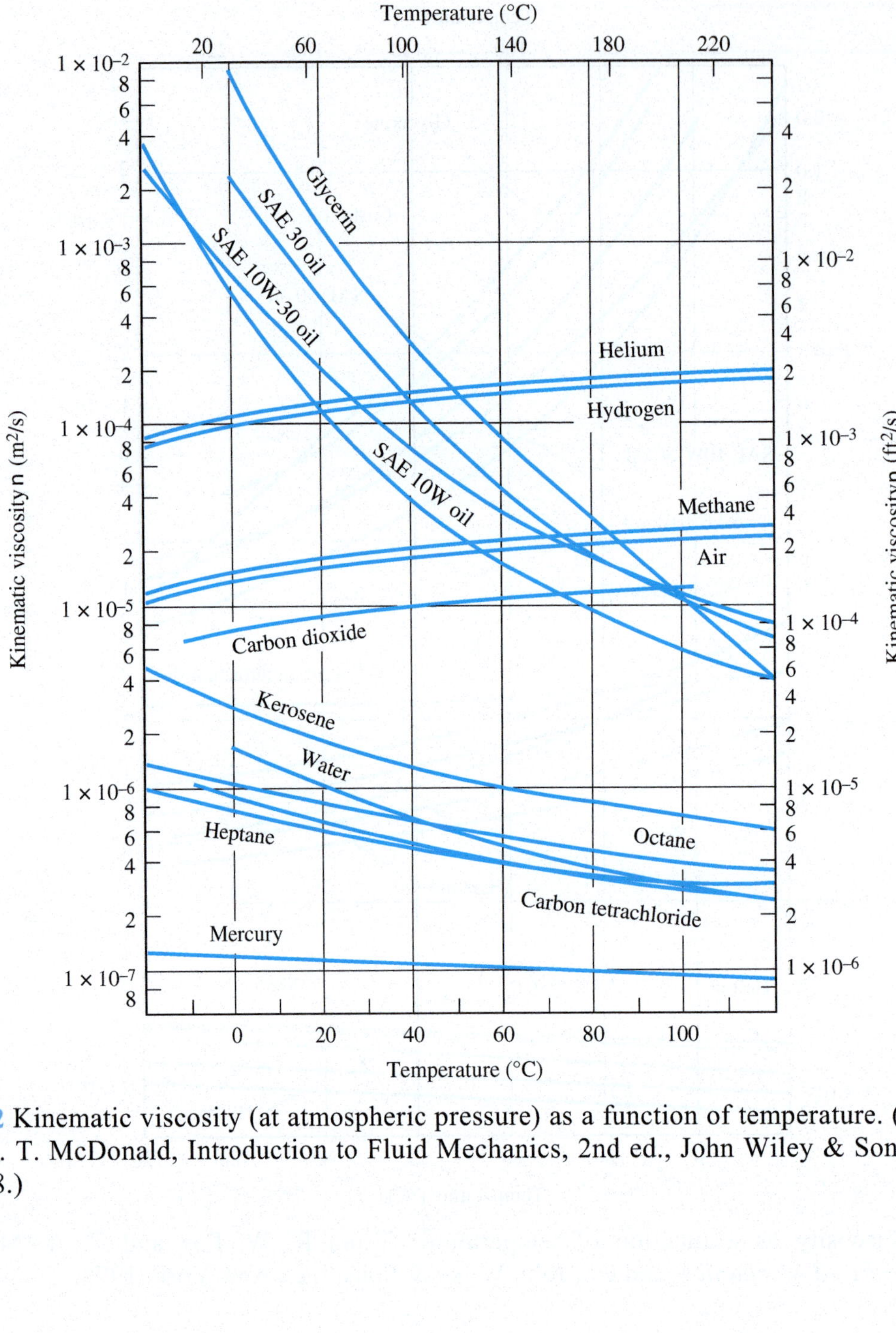

Figure B.2 Kinematic viscosity (at atmospheric pressure) as a function of temperature. (From R. W. Fox and A. T. McDonald, Introduction to Fluid Mechanics, 2nd ed., John Wiley & Sons, Inc., New York, 1978.)

C. PROPERTIES OF AREAS AND VOLUMES

Table C.1 Areas

	Sketch	*Area*	*Centroid*	*Second moment*
Rectangle	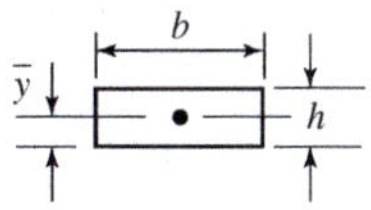	bh	$\bar{y} = h/2$	$\bar{I} = bh^3/12$ $\bar{I}_{xy} = 0$
Triangle	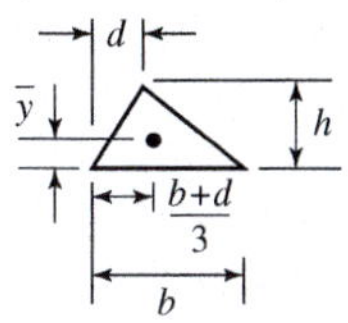	$bh/2$	$\bar{y} = h/3$	$\bar{I} = bh^3/36$ $\bar{I}_{xy} = (b-2d)bh^3/72$
Circle	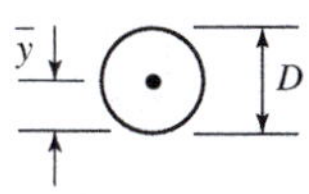	$\pi D^2/4$	$\bar{y} = r$	$\bar{I} = \pi D^4/64$
Semicircle	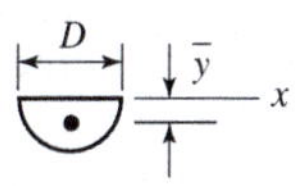	$\pi D^2/8$	$\bar{y} = 4r/3\pi$	$I_x = \pi D^4/128$
Ellipse	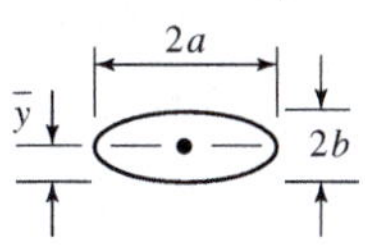	πab	$\bar{y} = b$	$\bar{I} = \pi ab^3/4$
Semiellipse	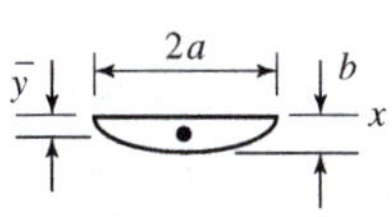	$\pi ab/2$	$\bar{y} = 4b/3\pi$	$I_x = \pi ab^3/8$

Table C.2 Volumes

	Sketch	*Surface area*	*Volume*	*Centroid*
Cylinder	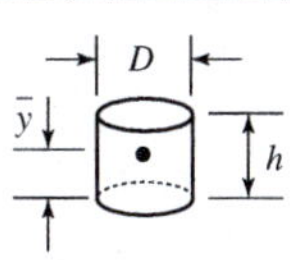	$\pi Dh + \pi D^2/2$	$\pi D^2 h/4$	$\bar{y} = h/2$
Sphere	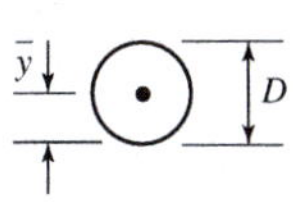	πD^2	$\pi D^3/6$	$\bar{y} = r$
Cone	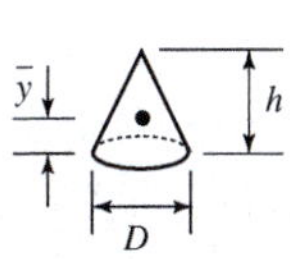	$\pi(r^2 + r\sqrt{r + h^2})$	$\pi D^2 h/12$	$\bar{y} = h/4$
Hemisphere	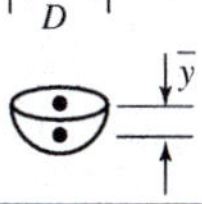	$3\pi D^2/4$	$\pi D^2/12$	$\bar{y} = 3r/8$

D. COMPRESSIBLE-FLOW TABLES FOR AIR

Table D.1 Isentropic Flow

M	p/p_0	T/T_0	A/A^*	M	p/p_0	T/T_0	A/A^*
0	1.0000	1.0000	∞	.74	.6951	.9013	1.0681
.02	.9997	.9999	28.9421	.76	.6821	.8964	1.0570
.04	.9989	.9997	14.4815	.78	.6691	.8915	1.0471
.06	.9975	.9993	9.6659	.80	.6560	.8865	1.0382
.08	.9955	.9987	7.2616	.82	.6430	.8815	1.0305
.10	.9930	.9980	5.8218	.84	.6300	.8763	1.0237
.12	.9900	.9971	4.8643	.86	.6170	.8711	1.0179
.14	.9864	.9961	4.1824	.88	.6041	.8659	1.0129
.16	.9823	.9949	3.6727	.90	.5913	.8606	1.0089
.18	.9776	.9936	3.2779	.92	.5785	.8552	1.0056
.20	.9725	.9921	2.9635	.94	.5658	.8498	1.0031
.22	.9668	.9904	2.7076	.96	.5532	.8444	1.0014
.24	.9607	.9886	2.4956	.98	.5407	.8389	1.0003
.26	.9541	.9867	2.3173	1.00	.5283	.8333	1.000
.28	.9470	.9846	2.1656	1.02	.5160	.8278	1.000
.30	.9395	.9823	2.0351	1.04	.5039	.8222	1.001
.32	.9315	.9799	1.9219	1.06	.4919	.8165	1.003
.34	.9231	.9774	1.8229	1.08	.4800	.8108	1.005
.36	.9143	.9747	1.7358	1.10	.4684	.8052	1.008
.38	.9052	.9719	1.6587	1.12	.4568	.7994	1.011
.40	.8956	.9690	1.5901	1.14	.4455	.7937	1.015
.42	.8857	.9659	1.5289	1.16	.4343	.7879	1.020
.44	.8755	.9627	1.4740	1.18	.4232	.7822	1.025
.46	.8650	.9594	1.4246	1.20	.4124	.7764	1.030
.48	.8541	.9560	1.3801	1.22	.4017	.7706	1.037
.50	.8430	.9524	1.3398	1.24	.3912	.7648	1.043
.52	.8317	.9487	1.3034	1.26	.3809	.7590	1.050
.54	.8201	.9449	1.2703	1.28	.3708	.7532	1.058
.56	.8082	.9410	1.2403	1.30	.3609	.7474	1.066
.58	.7962	.9370	1.2130	1.32	.3512	.7416	1.075
.60	.7840	.9328	1.1882	1.34	.3417	.7358	1.084
.62	.7716	.9286	1.1657	1.36	.3323	.7300	1.094
.64	.7591	.9243	1.1452	1.38	.3232	.7242	1.104
.66	.7465	.9199	1.1265	1.40	.3142	.7184	1.115
.68	.7338	.9153	1.1097	1.42	.3055	.7126	1.126
.70	.7209	.9107	1.0944	1.44	.2969	.7069	1.138
.72	.7080	.9061	1.0806	1.46	.2886	.7011	1.150

Table D.1 Isentropic Flow (*continued*)

M	p/p_0	T/T_0	A/A^*	M	p/p_0	T/T_0	A/A^*
1.48	.2804	.6954	1.163	2.30	$.7997^{-1}$	.4859	2.193
1.50	.2724	.6897	1.176	2.32	$.7751^{-1}$	.4816	2.233
1.52	.2646	.6840	1.190	2.34	$.7512^{-1}$	.4773	2.274
1.54	.2570	.6783	1.204	2.36	$.7281^{-1}$	.4731	2.316
1.56	.2496	.6726	1.219	2.38	$.7057^{-1}$	.4688	2.359
1.58	.2423	.6670	1.234	2.40	$.6840^{-1}$	.4647	2.403
1.60	.2353	.6614	1.250	2.42	$.6630^{-1}$	.4606	2.448
1.62	.2284	.6558	1.267	2.44	$.6426^{-1}$	.4565	2.494
1.64	.2217	.6502	1.284	2.46	$.6229^{-1}$	.4524	2.540
1.66	.2151	.6447	1.301	2.48	$.6038^{-1}$	.4484	2.588
1.68	.2088	.6392	1.319	2.50	$.5833^{-1}$	.4444	2.637
1.70	.2026	.6337	1.338	2.52	$.5674^{-1}$	.4405	2.686
1.72	.1966	.6283	1.357	2.54	$.5500^{-1}$	.4366	2.737
1.74	.1907	.6229	1.376	2.56	$.5332^{-1}$	.4328	2.789
1.76	.1850	.6175	1.397	2.58	$.5169^{-1}$	.4289	2.842
1.78	.1794	.6121	1.418	2.60	$.5012^{-1}$	.4252	2.896
1.80	.1740	.6068	1.439	2.62	$.4859^{-1}$	.4214	2.951
1.82	.1688	.6015	1.461	2.64	$.4711^{-1}$	.4177	3.007
1.84	.1637	.5963	1.484	2.66	$.4568^{-1}$	.4141	3.065
1.86	.1587	.5910	1.507	2.68	$.4429^{-1}$	.4104	3.123
1.88	.1539	.5859	1.531	2.70	$.4295^{-1}$	.4068	3.183
1.90	.1492	.5807	1.555	2.72	$.4165^{-1}$	.4033	3.244
1.92	.1447	.5756	1.580	2.74	$.4039^{-1}$	.3998	3.306
1.94	.1403	.5705	1.606	2.76	$.3917^{-1}$	.3963	3.370
1.96	.1360	.5655	1.633	2.78	$.3799^{-1}$	.3928	3.434
1.98	.1318	.5605	1.660	2.80	$.3685^{-1}$	.3894	3.500
2.00	.1278	.5556	1.688	2.82	$.3574^{-1}$	.3860	3.567
2.02	.1239	.5506	1.716	2.84	$.3467^{-1}$	.3827	3.636
2.04	.1201	.5458	1.745	2.86	$.3363^{-1}$	.3794	3.706
2.06	.1164	.5409	1.775	2.88	$.3263^{-1}$	.3761	3.777
2.08	.1128	.5361	1.806	2.90	$.3165^{-1}$	.3729	3.850
2.10	.1094	.5313	1.837	2.92	$.3071^{-1}$	.3696	3.924
2.12	.1060	.5266	1.869	2.94	$.2980^{-1}$	.3665	3.999
2.14	.1027	.5219	1.902	2.96	$.2891^{-1}$	.3633	4.076
2.16	$.9956^{-1}$	.5173	1.935	2.98	$.2805^{-1}$	.3602	4.155
2.18	$.9649^{-1}$	.5127	1.970	3.00	$.2722^{-1}$	.3571	4.235
2.20	$.9352^{-1}$	.5081	2.005	3.02	$.2642^{-1}$	.3541	4.316
2.22	$.9064^{-1}$	.5036	2.041	3.04	$.2564^{-1}$	.3511	4.399
2.24	$.8785^{-1}$	.4991	2.078	3.06	$.2489^{-1}$	.3481	4.483
2.26	$.8514^{-1}$	.4947	2.115	3.08	$.2416^{-1}$	.3452	4.570
2.28	$.8251^{-1}$	.4903	2.154	3.10	$.2345^{-1}$	.3422	4.657

Table D.1 Isentropic Flow (*continued*)

M	p/p_0	T/T_0	A/A^*	M	p/p_0	T/T_0	A/A^*
3.12	$.2276^{-1}$	.3393	4.747	3.92	$.7332^{-2}$	.2455	9.977
3.14	$.2210^{-1}$	.3365	4.838	3.94	$.7137^{-2}$	.2436	10.16
3.16	$.2146^{-1}$	.3337	4.930	3.96	$.6948^{-2}$	.2418	10.34
3.18	$.2083^{-1}$	.3309	5.025	3.98	$.6764^{-2}$	.2399	10.53
3.20	$.2023^{-1}$	.3281	5.121	4.00	$.6586^{-2}$	.2381	10.72
3.22	$.1964^{-1}$	.3253	5.219	4.02	$.6413^{-2}$	.2363	10.91
3.24	$.1908^{-1}$	.3226	5.319	4.04	$.6245^{-2}$	.2345	11.11
3.26	$.1853^{-1}$	.3199	5.420	4.06	$.6082^{-2}$	.2327	11.31
3.28	$.1799^{-1}$	.3173	5.523	4.08	$.5923^{-2}$	.2310	11.51
3.30	$.1748^{-1}$	.3147	5.629	4.10	$.5769^{-2}$	.2293	11.71
3.32	$.1698^{-1}$	.3121	5.736	4.12	$.5619^{-2}$	.2275	11.92
3.34	$.1649^{-1}$	.3095	5.845	4.14	$.5474^{-2}$	.2258	12.14
3.36	$.1602^{-1}$	.3069	5.956	4.16	$.5333^{-2}$	.2242	12.35
3.38	$.1557^{-1}$	.3044	6.069	4.18	$.5195^{-2}$	.2225	12.57
3.40	$.1512^{-1}$	.3019	6.184	4.20	$.5062^{-2}$	.2208	12.79
3.42	$.1470^{-1}$	.2995	6.301	4.22	$.4932^{-2}$	.2192	13.02
3.44	$.1428^{-1}$	.2970	6.420	4.24	$.4806^{-2}$	.2176	13.25
3.46	$.1388^{-1}$	.2946	6.541	4.26	$.4684^{-2}$	.2160	13.48
3.48	$.1349^{-1}$	.2922	6.664	4.28	$.4565^{-2}$	.2144	13.72
3.50	$.1311^{-1}$	.2899	6.790	4.30	$.4449^{-2}$	.2129	13.95
3.52	$.1274^{-1}$	.2875	6.917	4.32	$.4337^{-2}$	.2113	14.20
3.54	$.1239^{-1}$	.2852	7.047	4.34	$.4228^{-2}$	.2098	14.45
3.56	$.1204^{-1}$	.2829	7.179	4.36	$.4121^{-2}$	.2083	14.70
3.58	$.1171^{-1}$	.2806	7.313	4.38	$.4018^{-2}$	.2067	14.95
3.60	$.1138^{-1}$	.2784	7.450	4.40	$.3918^{-2}$	.2053	15.21
3.62	$.1107^{-1}$	.2762	7.589	4.42	$.3820^{-2}$	.2038	15.47
3.64	$.1076^{-1}$	.2740	7.730	4.44	$.3725^{-2}$	.2023	15.74
3.66	$.1047^{-1}$	.2718	7.874	4.46	$.3633^{-2}$	.2009	16.01
3.68	$.1018^{-1}$	.2697	8.020	4.48	$.3543^{-2}$	.1994	16.28
3.70	$.9903^{-2}$	.2675	8.169	4.50	$.3455^{-2}$	.1980	16.56
3.72	$.9633^{-2}$	.2654	8.320	4.52	$.3370^{-2}$	.1966	16.84
3.74	$.9370^{-2}$	.2633	8.474	4.54	$.3288^{-2}$	.1952	17.13
3.76	$.9116^{-2}$	.2613	8.630	4.56	$.3207^{-2}$	.1938	17.42
3.78	$.8869^{-2}$	.2592	8.789	4.58	$.3129^{-2}$	.1925	17.72
3.80	$.8629^{-2}$	.2572	8.951	4.60	$.3053^{-2}$	.1911	18.02
3.82	$.8396^{-2}$	.2552	9.115	4.62	$.2978^{-2}$	.1898	18.32
3.84	$.8171^{-2}$	.2532	9.282	4.64	$.2906^{-2}$	.1885	18.63
3.86	$.7951^{-2}$	.2513	9.451	4.66	$.2836^{-2}$	.1872	18.94
3.88	$.7739^{-2}$	.2493	9.624	4.68	$.2768^{-2}$	.1859	19.26
3.90	$.7532^{-2}$	.2474	9.799	4.70	$.2701^{-2}$	.1846	19.58

$.9956^{-1} = 0.09956$ $.9903^{-2} = 0.009903$

Table D.1 Isentropic Flow (*continued*)

M	p/p_0	T/T_0	A/A^*	M	p/p_0	T/T_0	A/A^*
4.72	$.2637^{-2}$	.1833	19.91	4.94	$.2028^{-2}$	.1700	23.82
4.74	$.2573^{-2}$	.1820	20.24	4.96	$.1981^{-2}$	.1689	24.21
4.76	$.2512^{-2}$	.1808	20.58	4.98	$.1935^{-2}$	.1678	24.60
4.78	$.2452^{-2}$	.1795	20.92	5.00	$.1890^{-2}$	.1667	25.00
4.80	$.2394^{-2}$	.1783	21.26	6.00	$.0633^{-2}$	.1219	53.19
4.82	$.2338^{-2}$	.1771	21.61	7.00	$.0242^{-2}$	.0926	104.14
4.84	$.2283^{-2}$	.1759	21.97	8.00	$.0102^{-2}$	.0725	109.11
4.86	$.2229^{-2}$	.1747	22.33	9.00	$.0474^{-3}$	.0582	327.19
4.88	$.2177^{-2}$	.1735	22.70	10.00	$.0236^{-3}$	.0476	535.94
4.90	$.2126^{-2}$	.1724	23.07	∞	0	0	∞
4.92	$.2076^{-2}$	.1712	23.44				

Table D.2 Normal Shock Flow

M_1	M_2	p_2/p_1	T_2/T_1	p_{02}/p_{01}
1.00	1.000	1.000	1.000	1.000
1.02	.9805	1.047	1.013	1.000
1.04	.9620	1.095	1.026	.9999
1.06	.9444	1.144	1.039	.9997
1.08	.9277	1.194	1.052	.9994
1.10	.9118	1.245	1.065	.9989
1.12	.8966	1.297	1.078	.9982
1.14	.8820	1.350	1.090	.9973
1.16	.8682	1.403	1.103	.9961
1.18	.8549	1.458	1.115	.9946
1.20	.8422	1.513	1.128	.9928
1.22	.8300	1.570	1.141	.9907
1.24	.8183	1.627	1.153	.9884
1.26	.8071	1.686	1.166	.9857
1.28	.7963	1.745	1.178	.9827
1.30	.7860	1.805	1.191	.9794
1.32	.7760	1.866	1.204	.9758
1.34	.7664	1.928	1.216	.9718
1.36	.7572	1.991	1.229	.9676
1.38	.7483	2.055	1.242	.9630
1.40	.7397	2.120	1.255	.9582
1.42	.7314	2.186	1.268	.9531
1.44	.7235	2.253	1.281	.9476
1.46	.7157	2.320	1.294	.9420
1.48	.7083	2.389	1.307	.9360
1.50	.7011	2.458	1.320	.9298
1.52	.6941	2.529	1.334	.9233

Table D.2 Normal Shock Flow (*continued*)

M_1	M_2	p_2/p_1	T_2/T_1	p_{02}/p_{01}
1.54	.6874	2.600	1.347	.9166
1.56	.6809	2.673	1.361	.9097
1.58	.6746	2.746	1.374	.9026
1.60	.6684	2.820	1.388	.8952
1.62	.6625	2.895	1.402	.8877
1.64	.6568	2.971	1.416	.8799
1.66	.6512	3.048	1.430	.8720
1.68	.6458	3.126	1.444	.8640
1.70	.6405	3.205	1.458	.8557
1.72	.6355	3.285	1.473	.8474
1.74	.6305	3.366	1.487	.8389
1.76	.6257	3.447	1.502	.8302
1.78	.6210	3.530	1.517	.8215
1.80	.6165	3.613	1.532	.8127
1.82	.6121	3.698	1.547	.8038
1.84	.6078	3.783	1.562	.7948
1.86	.6036	3.870	1.577	.7857
1.88	.5996	3.957	1.592	.7765
1.90	.5956	4.045	1.608	.7674
1.92	.5918	4.134	1.624	.7581
1.94	.5880	4.224	1.639	.7488
1.96	.5844	4.315	1.655	.7395
1.98	.5808	4.407	1.671	.7302
2.00	.5774	4.500	1.688	.7209
2.02	.5740	4.594	1.704	.7115
2.04	.5707	4.689	1.720	.7022
2.06	.5675	4.784	1.737	.6928
2.08	.5643	4.881	1.754	.6835
2.10	.5613	4.978	1.770	.6742
2.12	.5583	5.077	1.787	.6649
2.14	.5554	5.176	1.805	.6557
2.16	.5525	5.277	1.822	.6464
2.18	.5498	5.378	1.839	.6373
2.20	.5471	5.480	1.857	.6281
2.22	.5444	5.583	1.875	.6191
2.24	.5418	5.687	1.892	.6100
2.26	.5393	5.792	1.910	.6011
2.28	.5368	5.898	1.929	.5921
2.30	.5344	6.005	1.947	.5833
2.32	.5321	6.113	1.965	.5745
2.34	.5297	6.222	1.984	.5658

Table D.2 Normal Shock Flow (*continued*)

M_1	M_2	p_2/p_1	T_2/T_1	p_{02}/p_{01}
2.36	.5275	6.331	2.002	.5572
2.38	.5253	6.442	2.021	.5486
2.40	.5231	6.553	2.040	.5401
2.42	.5210	6.666	2.059	.5317
2.44	.5189	6.779	2.079	.5234
2.46	.5169	6.894	2.098	.5152
2.48	.5149	7.009	2.118	.5071
2.50	.5130	7.125	2.138	.4990
2.52	.5111	7.242	2.157	.4991
2.54	.5092	7.360	2.177	.4832
2.56	.5074	7.479	2.198	.4754
2.58	.5056	7.599	2.218	.4677
2.60	.5039	7.720	2.238	.4601
2.62	.5022	7.842	2.259	.4526
2.64	.5005	7.965	2.280	.4452
2.66	.4988	8.088	2.301	.4379
2.68	.4972	8.213	2.322	.4307
2.70	.4956	8.338	2.343	.4236
2.72	.4941	8.465	2.364	.4166
2.74	.4926	8.592	2.386	.4097
2.76	.4911	8.721	2.407	.4028
2.78	.4896	8.850	2.429	.3961
2.80	.4882	8.980	2.451	.3895
2.82	.4868	9.111	2.473	.3829
2.84	.4854	9.243	2.496	.3765
2.86	.4840	9.376	2.518	.3701
2.88	.4827	9.510	2.540	.3639
2.90	.4814	9.645	2.563	.3577
2.92	.4801	9.781	2.586	.3517
2.94	.4788	9.918	2.609	.3457
2.96	.4776	10.06	2.632	.3398
2.98	.4764	10.19	2.656	.3340
3.00	.4752	10.33	2.679	.3283
3.02	.4740	10.47	2.703	.3327
3.04	.4729	10.62	2.726	.3172
3.06	.4717	10.76	2.750	.3118
3.08	.4706	10.90	2.774	.3065
3.10	.4695	11.05	2.799	.3012
3.12	.4685	11.19	2.823	.2960
3.14	.4674	11.34	2.848	.2910
3.16	.4664	11.48	2.872	.2860

Table D.2 Normal Shock Flow (*continued*)

M_1	M_2	p_2/p_1	T_2/T_1	p_{02}/p_{01}
3.18	.4654	11.63	2.897	.2811
3.20	.4643	11.78	2.922	.2762
3.22	.4634	11.93	2.947	.2715
3.24	.4624	12.08	2.972	.2668
3.26	.4614	12.23	2.998	.2622
3.28	.4605	12.38	3.023	.2577
3.30	.4596	12.54	3.049	.2533
3.32	.4587	12.69	3.075	.2489
3.34	.4578	12.85	3.101	.2446
3.36	.4569	13.00	3.127	.2404
3.38	.4560	13.16	3.154	.2363
3.40	.4552	13.32	3.180	.2322
3.42	.4544	13.48	3.207	.2382
3.44	.4535	13.64	3.234	.2243
3.46	.4527	13.80	3.261	.2205
3.48	.4519	13.96	3.288	.2167
3.50	.4512	14.13	3.315	.2129
3.52	.4504	14.29	3.343	.2093
3.54	.4496	14.45	3.370	.2057
3.56	.4489	14.62	3.398	.2022
3.58	.4481	14.79	3.426	.1987
3.60	.4474	14.95	3.454	.1953
3.62	.4467	15.12	3.482	.1920
3.64	.4460	15.29	3.510	.1887
3.66	.4453	15.46	3.539	.1855
3.68	.4446	15.63	3.568	.1823
3.70	.4439	15.81	3.596	.1792
3.72	.4433	15.98	3.625	.1761
3.74	.4426	16.15	3.654	.1731
3.76	.4420	16.33	3.684	.1702
3.78	.4414	16.50	3.713	.1673
3.80	.4407	16.68	3.743	.1645
3.82	.4401	16.86	3.772	.1617
3.84	.4395	17.04	3.802	.1589
3.86	.4389	17.22	3.832	.1563
3.88	.4383	17.40	3.863	.1536
3.90	.4377	17.58	3.893	.1510
3.92	.4372	17.76	3.923	.1485
3.94	.4366	17.94	3.954	.1460
3.96	.4360	18.13	3.985	.1435
3.98	.4355	18.31	4.016	.1411

Table D.2 Normal Shock Flow (*continued*)

M_1	M_2	p_2/p_1	T_2/T_1	p_{02}/p_{01}
4.00	.4350	18.50	4.047	.1388
4.02	.4344	18.69	4.078	.1364
4.04	.4339	18.88	4.110	.1342
4.06	.4334	19.06	4.141	.1319
4.08	.4329	19.25	4.173	.1297
4.10	.4324	19.45	4.205	.1276
4.12	.4319	19.64	4.237	.1254
4.14	.4314	19.83	4.269	.1234
4.16	.4309	20.02	4.301	.1213
4.18	.4304	20.22	4.334	.1193
4.20	.4299	20.41	4.367	.1173
4.22	.4295	20.61	4.399	.1154
4.24	.4290	20.81	4.432	.1135
4.26	.4286	21.01	4.466	.1116
4.28	.4281	21.20	4.499	.1098
4.30	.4277	21.41	4.532	.1080
4.32	.4272	21.61	4.566	.1062
4.34	.4268	21.81	4.600	.1045
4.36	.4264	22.01	4.633	.1028
4.38	.4260	22.22	4.668	.1011
4.40	.4255	22.42	4.702	$.9948^{-1}$
4.42	.4251	22.63	4.736	$.9787^{-1}$
4.44	.4247	22.83	4.771	$.9628^{-1}$
4.46	.4243	23.04	4.805	$.9473^{-1}$
4.48	.4239	23.25	4.840	$.9320^{-1}$
4.50	.4236	23.46	4.875	$.9170^{-1}$
4.52	.4232	23.67	4.910	$.9022^{-1}$
4.54	.4228	23.88	4.946	$.8878^{-1}$
4.56	.4224	24.09	4.981	$.8735^{-1}$
4.58	.4220	24.31	5.017	$.8596^{-1}$
4.60	.4217	24.52	5.052	$.8459^{-1}$
4.62	.4213	24.74	5.088	$.8324^{-1}$
4.64	.4210	24.95	5.124	$.8192^{-1}$
4.66	.4206	25.17	5.160	$.8062^{-1}$
4.68	.4203	25.39	5.197	$.7934^{-1}$
4.70	.4199	25.61	5.233	$.7809^{-1}$
4.72	.4196	25.82	5.270	$.7685^{-1}$
4.74	.4192	26.05	5.307	$.7564^{-1}$
4.76	.4189	26.27	5.344	$.7445^{-1}$
4.78	.4186	26.49	5.381	$.7329^{-1}$
4.80	.4183	26.71	5.418	$.7214^{-1}$

Table D.2 Normal Shock Flow (*continued*)

M_1	M_2	p_2/p_1	T_2/T_1	p_{02}/p_{01}
4.82	.4179	26.94	5.456	$.7101^{-1}$
4.84	.4176	27.16	5.494	$.6991^{-1}$
4.86	.4173	27.39	5.531	$.6882^{-1}$
4.88	.4170	27.62	5.569	$.6775^{-1}$
4.90	.4167	27.85	5.607	$.6670^{-1}$
4.92	.4164	28.07	5.646	$.6567^{-1}$
4.94	.4161	28.30	5.684	$.6465^{-1}$
4.96	.4158	28.54	5.723	$.6366^{-1}$
4.98	.4155	28.77	5.761	$.6268^{-1}$
5.00	.4152	29.00	5.800	$.6172^{-1}$
6.00	.4042	41.83	7.941	$.2965^{-1}$
7.00	.3974	57.00	10.469	$.1535^{-1}$
8.00	.3929	74.50	13.387	$.0849^{-1}$
9.00	.3898	94.33	16.693	$.0496^{-1}$
10.00	.3875	116.50	20.388	$.0304^{-1}$
∞	.3780	∞	∞	0

Table D.3 Prandtl–Meyer Function

M	θ	μ	M	θ	μ
1.00	0	90.00	1.50	11.905	41.81
1.02	.1257	78.64	1.52	12.495	41.14
1.04	.3510	74.06	1.54	13.086	40.49
1.06	.6367	70.63	1.56	13.677	39.87
1.08	.9680	67.81	1.58	14.269	39.27
1.10	1.336	65.38	1.60	14.861	38.68
1.12	1.735	63.23	1.62	15.452	38.12
1.14	2.160	61.31	1.64	16.043	37.57
1.16	2.607	59.55	1.66	16.633	37.04
1.18	3.074	57.94	1.68	17.222	36.53
1.20	3.558	56.44	1.70	17.810	36.03
1.22	4.057	55.05	1.72	18.397	35.55
1.24	4.569	53.75	1.74	18.981	35.08
1.26	5.093	52.53	1.76	19.565	34.62
1.28	5.627	51.38	1.78	20.146	34.18
1.30	6.170	50.28	1.80	20.725	33.75
1.32	6.721	49.25	1.82	21.302	33.33
1.34	7.280	48.27	1.84	21.877	32.92
1.36	7.844	47.33	1.86	22.449	32.52
1.38	8.413	46.44	1.88	23.019	32.13
1.40	8.987	45.58	1.90	23.586	31.76
1.42	9.565	44.77	1.92	24.151	31.39
1.44	10.146	43.98	1.94	24.712	31.03
1.46	10.731	43.23	1.96	25.271	30.68
1.48	11.317	42.51	1.98	25.827	30.33

Table D.3 Prandtl–Meyer Function (*continued*)

M	θ	μ	M	θ	μ
2.00	26.380	30.00	2.94	48.586	19.89
2.02	26.929	29.67	2.96	48.980	19.75
2.04	27.476	29.35	2.98	49.370	19.61
2.06	28.020	29.04	3.00	49.757	19.47
2.08	28.560	28.74	3.02	50.142	19.34
2.10	29.097	28.44	3.04	50.523	19.20
2.12	29.631	28.14	3.06	50.902	19.07
2.14	30.161	27.86	3.08	51.277	18.95
2.16	30.689	27.58	3.10	51.560	18.82
2.18	31.212	27.30	3.12	52.020	18.69
2.20	31.732	27.04	3.14	52.386	18.57
2.22	32.250	26.77	3.16	52.751	18.45
2.24	32.763	26.51	3.18	53.112	18.33
2.26	33.273	26.26	3.20	53.470	18.21
2.28	33.780	26.01	3.22	53.826	18.09
2.30	34.283	25.77	3.24	54.179	17.98
2.32	34.783	25.53	3.26	54.529	17.86
2.34	35.279	25.30	3.28	54.877	17.75
2.36	35.771	25.07	3.30	55.222	17.64
2.38	36.261	24.85	3.32	55.564	17.53
2.40	36.746	24.62	3.34	55.904	17.42
2.42	37.229	24.41	3.36	56.241	17.31
2.44	37.708	24.19	3.38	56.576	17.21
2.46	38.183	23.99	3.40	56.907	17.10
2.48	38.655	23.78	3.42	57.237	17.00
2.50	39.124	23.58	3.44	57.564	16.90
2.52	39.589	23.38	3.46	57.888	16.80
2.54	40.050	23.18	3.48	58.210	16.70
2.56	40.509	22.99	3.50	58.530	16.60
2.58	40.963	22.81	3.52	58.847	16.51
2.60	41.415	22.62	3.54	59.162	16.41
2.62	41.863	22.44	3.56	59.474	16.31
2.64	42.307	22.26	3.58	59.784	16.22
2.66	42.749	22.08	3.60	60.091	16.13
2.68	43.187	21.91	3.62	60.397	16.04
2.70	43.621	21.74	3.64	60.700	15.95
2.72	44.053	21.57	3.66	61.000	15.86
2.74	44.481	21.41	3.68	61.299	15.77
2.76	44.906	21.24	3.70	61.595	15.68
2.78	45.327	21.08	3.72	61.899	15.59
2.80	45.746	20.92	3.74	62.181	15.51
2.82	46.161	20.77	3.76	62.471	15.42
2.84	46.573	20.62	3.78	62.758	15.34
2.86	46.982	20.47	3.80	63.044	15.26
2.88	47.388	20.32	3.82	63.327	15.18
2.90	47.790	20.17	3.84	63.608	15.10
2.92	48.190	20.03	3.86	63.887	15.02

Table D.3 Prandtl–Meyer Function (*continued*)

M	θ	μ	M	θ	μ
3.88	64.164	14.94	4.46	71.386	12.96
3.90	64.440	14.86	4.48	71.610	12.90
3.92	64.713	14.78	4.50	71.832	12.84
3.94	64.983	14.70	4.52	72.052	12.78
3.96	65.253	14.63	4.54	72.271	12.73
3.98	65.520	14.55	4.56	72.489	12.67
4.00	65.785	14.48	4.58	72.705	12.61
4.02	66.048	14.40	4.60	72.919	12.56
4.04	66.309	14.33	4.62	73.132	12.50
4.06	66.569	14.26	4.64	73.344	12.45
4.08	66.826	14.19	4.66	73.554	12.39
4.10	67.082	14.12	4.68	73.763	12.34
4.12	67.336	14.05	4.70	73.970	12.28
4.14	67.588	13.98	4.72	74.176	12.23
4.16	67.838	13.91	4.74	74.381	12.18
4.18	68.087	13.84	4.76	74.584	12.13
4.20	68.333	13.77	4.78	74.786	12.08
4.22	68.578	13.71	4.80	74.986	12.03
4.24	68.821	13.64	4.82	75.186	11.97
4.26	69.053	13.58	4.84	75.383	11.92
4.28	69.302	13.51	4.86	75.580	11.87
4.30	69.541	13.45	4.88	75.775	11.83
4.32	69.777	13.38	4.90	75.969	11.78
4.34	70.012	13.32	4.92	76.162	11.73
4.36	70.245	13.26	4.94	76.353	11.68
4.38	70.476	13.20	4.96	76.544	11.63
4.40	70.706	13.14	4.98	76.732	11.58
4.42	70.934	13.08	5.00	76.920	11.54
4.44	71.161	13.02			

E. Numerical

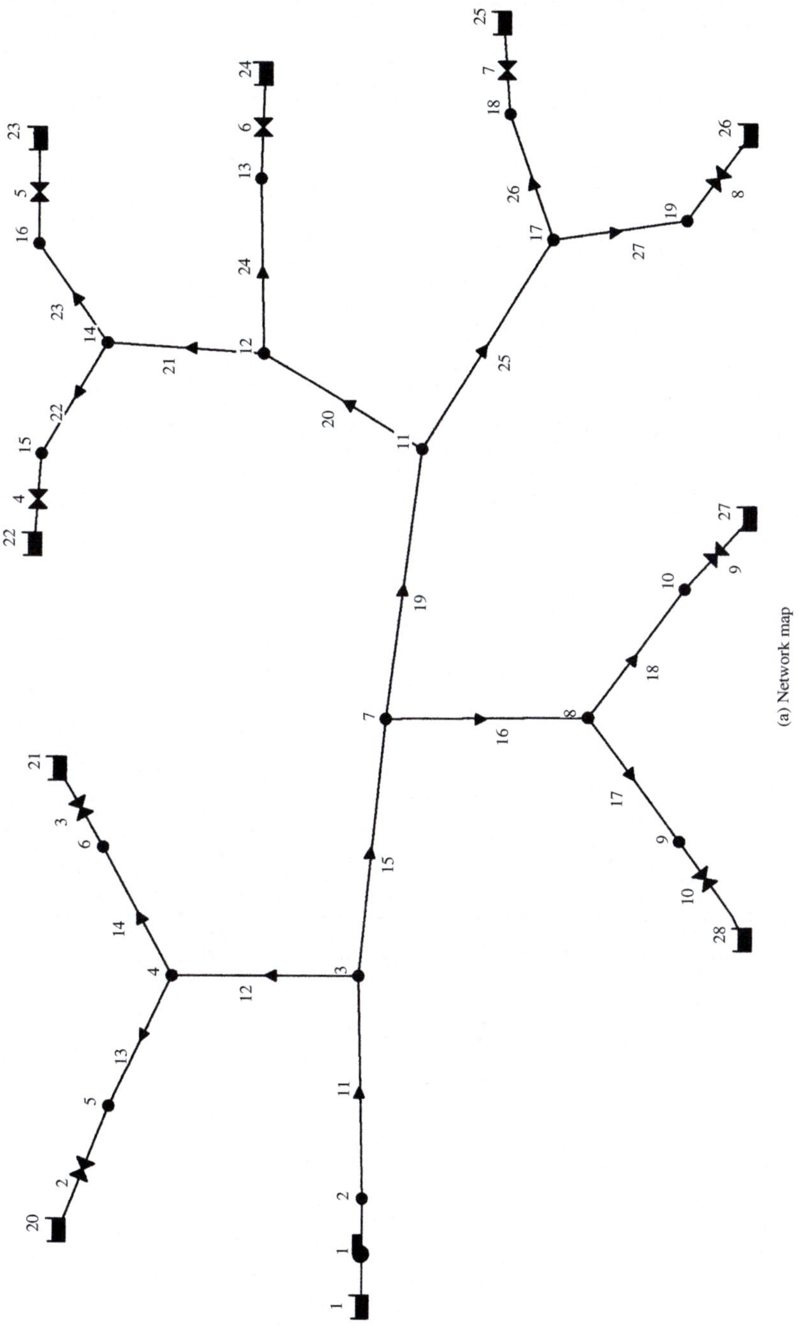

Figure E.1(a) Figure E9.7a EPANET 2 solution.

Link - Node Table:

Link ID	Start Node	End Node	Length m	Diameter mm
11	2	3	195	100
12	3	4	158	50
13	4	5	115	50
14	4	6	155	50
15	3	7	188	100
16	7	8	117	100
17	8	9	59	50
18	8	10	82	50
19	7	11	130	100
20	11	12	102	100
21	12	14	78	100
22	14	15	55	50
23	14	16	60	50
24	12	13	114	50
25	11	17	165	100
26	17	18	64	50
27	17	9	127	50
1	1	2	#N/A	#N/A Pump
2	5	20	#N/A	50 Valve
3	6	21	#N/A	50 Valve
4	15	22	#N/A	50 Valve
5	16	23	#N/A	50 Valve
6	13	24	#N/A	50 Valve
7	18	25	#N/A	50 Valve
8	19	26	#N/A	50 Valve
9	10	27	#N/A	50 Valve
10	9	28	#N/A	50 Valve

Figure E.1(b) Figure E9.7a EPANET 2 solution (Computed results.)

Node Results:

Node ID	Demand LPS	Head m	Pressure m	Quality
2	0.00	135.85	105.85	0.00
3	0.00	89.79	57.79	0.00
4	0.00	47.59	12.59	0.00
5	0.00	40.02	3.02	0.00
6	0.00	36.36	3.36	0.00
7	0.00	58.08	28.08	0.00
8	0.00	53.37	26.37	0.00
9	0.00	37.67	13.67	0.00
10	0.00	36.41	10.41	0.00
11	0.00	49.22	22.22	0.00
12	0.00	47.26	14.26	0.00
13	0.00	38.55	3.55	0.00
14	0.00	46.56	6.56	0.00
15	0.00	42.98	2.98	0.00
16	0.00	41.74	3.74	0.00
17	0.00	46.01	20.01	0.00
18	0.00	34.75	8.75	0.00
19	0.00	30.07	6.07	0.00
1	-38.91	30.00	0.00	0.00 Reservoir
20	3.15	37.00	0.00	0.00 Reservoir
21	3.32	33.00	0.00	0.00 Reservoir
22	3.13	40.00	0.00	0.00 Reservoir
23	3.51	38.00	0.00	0.00 Reservoir
24	3.42	35.00	0.00	0.00 Reservoir
25	5.36	26.00	0.00	0.00 Reservoir
26	4.47	24.00	0.00	0.00 Reservoir
27	5.85	26.00	0.00	0.00 Reservoir
28	6.70	24.00	0.00	0.00 Reservoir

Figure E.1(b) Computed results. (*continued*)

Link Results:

Link ID	Flow LPS	Velocity m/s	Headloss m/km	Status
11	38.91	4.96	236.24	Open
12	6.47	3.30	267.05	Open
13	3.15	1.61	65.81	Open
14	3.32	1.69	72.49	Open
15	32.44	4.13	168.65	Open
16	12.55	1.60	40.21	Open
17	6.70	3.42	266.15	Open
18	5.85	2.98	206.84	Open
19	19.89	2.53	68.13	Open
20	10.06	1.28	19.27	Open
21	6.64	0.85	8.94	Open
22	3.13	1.60	65.00	Open
23	3.51	1.79	80.23	Open
24	3.42	1.74	76.36	Open
25	9.83	1.25	19.45	Open
26	5.36	2.73	176.01	Open
27	4.47	2.28	125.52	Open
1	38.91	0.00	-105.85	Open Pump
2	3.15	1.61	3.02	Active Valve
3	3.32	1.69	3.36	Active Valve
4	3.13	1.60	2.98	Active Valve
5	3.51	1.79	3.74	Active Valve
6	3.42	1.74	3.55	Active Valve
7	5.36	2.73	8.75	Active Valve
8	4.47	2.28	6.07	Active Valve
9	5.85	2.98	10.41	Active Valve
10	6.70	3.42	13.67	Active Valve

Figure E.1(b) Computed results. (*continued*)

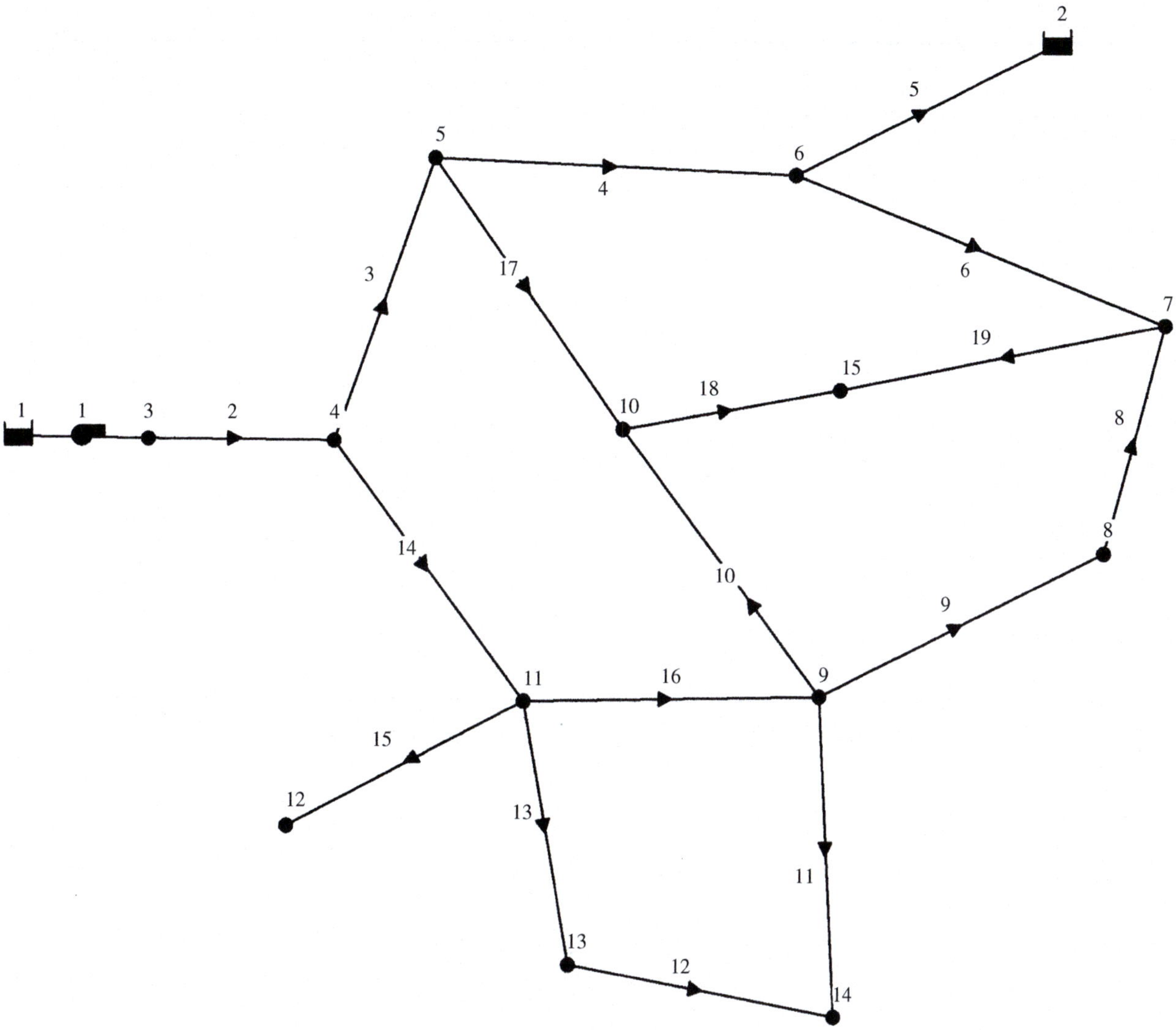

(a) Network map

Figure E.2(a) Figure E9.7b EPANET 2 solution.

Link - Node Table:

Link ID	Start Node	End Node	Length m	Diameter mm
2	3	4	3000	600
3	4	5	1520	450
4	5	6	1520	400
5	6	2	305	150
6	6	7	1680	350
8	7	8	1070	300
9	8	9	1680	350
10	9	10	1680	300
11	9	14	1380	300
12	14	13	760	150
13	13	11	1100	300
14	11	4	2000	450
15	11	12	1200	400
16	9	11	670	380
17	10	5	1520	350
18	10	15	900	350
19	15	7	1200	350
1	1	3	#N/A	#N/A Pump

Figure E.2(b) Figure E9.7b EPANET 2 solution (b) Computed results.

Node Results:

Node ID	Demand LPS	Head m	Pressure m	Quality
3	0.00	192.15	177.15	0.00
4	0.00	155.46	109.49	0.00
5	140.00	132.55	83.55	0.00
6	0.00	126.03	76.03	0.00
7	100.00	122.64	73.64	0.00
8	100.00	123.04	77.04	0.00
9	0.00	130.00	87.00	0.00
10	140.00	124.13	80.13	0.00
11	0.00	136.56	92.56	0.00
12	55.00	136.02	96.02	0.00
13	55.00	133.12	92.12	0.00
14	55.00	128.61	88.61	0.00
15	85.00	122.59	76.59	0.00
1	-823.11	15.00	0.00	0.00 Reservoir
2	93.11	61.00	0.00	0.00 Reservoir

Link Results:

Link ID	Flow LPS	Velocity m/s	Headloss m/km	Status
2	823.11	2.91	12.22	Open
3	460.26	2.90	15.10	Open
4	177.63	1.41	4.29	Open
5	93.11	5.27	213.20	Open
6	84.52	0.88	2.01	Open
8	-22.87	0.32	0.37	Open
9	-122.87	1.28	4.15	Open
10	74.98	1.06	3.50	Open
11	39.19	0.55	1.01	Open
12	-15.81	0.89	5.94	Open
13	-70.81	1.00	3.13	Open
14	-362.85	2.28	9.46	Open
15	55.00	0.44	0.45	Open
16	-237.04	2.09	9.80	Open
17	-142.63	1.48	5.54	Open
18	77.61	0.81	1.71	Open
9	-7.39	0.10	0.05	Open
1	823.11	0.00	-177.15	Open Pump

Figure E.2(b) Computed results. *(continued)*

Bibliography

REFERENCES

ADRIAN, R. J., "Multi-point Optical Measurements of Simultaneous Vectors in Unsteady Flow—A Review," *Int. J. Heat Fluid Flow*, Vol. 7, No. 2, June 1986, pp. 127–145.

BAKHMETEFF, B.A., *Hydraulics of Open Channels*, McGraw-Hill Book Company, New York, 1932.

BEAN, H. S., ed., *Fluid Meters: Their Theory and Application*, 6th ed.. American Society of Mechanical Engineers, New York, 1971.

BENEDICT, R. R, *Fundamentals of Pipe Flow*, John Wiley & Sons, Inc.. New York, 1980.

CHAPR A, S. C., and CANALE, R. P., *Numerical Methods for Engineers*, 3rd ed., McGraw-Hill Book Company, New York, 1998.

CHOW, V. T., *Open Channel Hydraulics*, McGraw-Hill Book Company. New York, 1959.

COLEMAN, HUGH W, and STEELE, W. GLENN, JR, *Experimentation and Uncertainty Analysis for Engineers*, 2nd ed. John Wiley & Sons, New York, 1998.

CROSS, H., "Analysis of Flow in Networks of Conduits or Conductors," *University of Illinois Bulletin 286*, November 1936.

GOLDSTEIN, R. J., ed., *Fluid Mechanics Measurements*, 2nd ed., Taylor and Francis, Washington, DC, 1996.

HEC-RAS River Analysis System, Hydrologic Engineering Center: U.S. Army Corps of Engineers, Davis, CA, July 1995.

HENDERSON, F. M., *Open Channel Flow*, Macmillan Publishing Company, New York, 1966.

KARASSIK, I. J., MESSINA, J. P., COOPER, P., HEALD, C. C., eds., *Pump Handbook*, 3rd ed., McGraw-Hill Professional Publishing, 2000.

KING, H. W, and BRATER, E. F., *Handbook of Hydraulics*, 7th ed., McGraw-Hill Book Company, New York, 1996.

KLINE, S. J., ed., Proceedings of the Symposium on Uncertainty Analysis. *J. Fluids Eng.*, Vol. 107, No. 2, June 1985, pp. 153–178.

MARTIN, C. S., "Experimental Investigation of Column Separation with Rapid Valve Closure," *Proceedings, 4th International Conference on Pressure Surges*, BHRA Fluid Engineering, Cranfield, England, 1983, pp. 77-88.

MCBEAN, E. A., and PERKINS, F. E., "Convergence Schemes in Water Profile Computation," *J. Hydraulics Div., ASCE*, Vol. 101, No. HY10, October 1975, pp. 1380–1384.

ORMSBEE, L. E., and WOOD, D. J., "Hydraulic Design Algorithms for Pipe Networks," *J. Hydraulic Eng., ASCE*, Vol. 112, No. 12, December 1986, pp. 1195–1207.

ROBERSON, J. A., CASSIDY, J. J., and CHAUDHRY, M. H., *Hydraulic Engineering*, Houghton Mifflin Company, Boston, MA, 1988.

ROBERSON, J. A., and CROWE, C.T., *Engineering Fluid Mechanics*, 6th ed., John Wiley & Sons, 1997.

ROSSMAN, J. A., *EPANETZ Users Manual*, United States Environmental Protection Agency, June 2000.

STEPANOFF, A. J., *Centrifugal and Axial Flow Pumps*, 2nd ed., John Wiley & Sons, Inc., New York, 1957.

SWAMEE, P. K., and JAIN, A. K., "Explicit Equations for Pipe-Flow Problems," *J. Hydraulics Div., ASCE*, Vol. 102, No. HY5, May 1976, pp. 657-664.

TAYLOR, G. I., "Dispersion of Soluble Matter in Solvent Flowing Slowly Through a Tube," *Proc. Royal Soc. London*, Vol. 219A, 1953, pp. 186–203.

M. C. Potter and D. C. Wiggert, *Applied Fluid Mechanics*,
https://doi.org/10.1007/978-3-031-65276-9

U. S. DEPARTMENT OF INTERIOR, BUREAU OF RECLAMATION, *Design of Small Dams*, U.S. Government Printing Office, Washington DC, 1974.
VAN DYKE, M., *An Album of Fluid Motion*, Parabolic Press, Stanford, CA, 1982.
WARNICK, C. C., *Hydropower Engineering*, Prentice-Hall, Inc.: Englewood Cliffs, NJ, 1984.
WOOD, D. J., "Algorithms for Pipe Network Analysis and Their Reliability," *Research Report 127*, Water Resources Research Institute, University of Kentucky: Lexington, KY, 1981.
WYLIE, E. B., and STREETER, V.L., *Fluid Transients in Systems*, Prentice Hall: Englewood Cliffs, N.J., 1993.

GENERAL INTEREST

DAILY, J.W, and HARLEMAN, D. R. E, Fluid Dynamics, Addison-Wesley Publishing Company, Inc., Reading, MA, 1968.
ESKINAZI, S., *Principles of Fluid Mechanics*, 2nd ed., Allyn and Bacon, Inc.: Needham Heights, MA, 1968.
FOX, R.W, and MCDONALD, A. T, *Introduction to Fluid Mechanics*, 4th ed., John Wiley & Sons, Inc., New York, 1992.
FUNG, Y. C., *Continuum Mechanics*, 2nd ed., Prentice-Hall, Inc., Englewood Cliffs, NJ, 1977.
HANSON, A. G., *Fluid Mechanics*, John Wiley & Sons, Inc., New York, 1967.
JOHN, E.A. J., *Gas Dynamics*, Allyn and Bacon, Inc., Needham Heights, MA, 1969.
MUNSON, B. R., YOUNG, D. F, and OKIISHI, T. H., *Fundamentals of Fluid Mechanics*, 2nd ed., John Wiley & Sons, Inc., New York, 1994.
OWCZAREK, J. A., *Introduction to Fluid Mechanics*, International Textbook Company, Scranton, PA, 1968.
POTTER, M. C., and Foss, J. F., *Fluid Mechanics*, Great Lakes Press,Okemos, MI, 1979.
ROBERSON, J. A., and CROWE, C. T., *Engineering Fluid Mechanics*, 6th ed., Houghton-Mifflin Company, Boston, MA, 1996.
SABERSKY, R. H., ACOSTA, A. J., and HAUPTMAN. E. G., *Fluid Flow*, 5th ed., Macmillan Publishing Company, New York, 2001.
SCHLICHTING, H., *Boundary Layer Theory*, 8th ed., McGraw-Hill Book Company, New York, 2000.
SHAMES., L, *Mechanics of Fluids*, 3rd ed., McGraw-Hill Book Company, New York, 1992.
SHAPIRO, A. H., *The Dynamics and Thermodynamics of Compressible Fluid Flow*, Ronald Press, New York, 1953.
STREETER, V. L., and WYLIE, E. B., *Fluid Mechanics*, 8th ed., McGraw-Hill Book Company, New York, 1985.
VAN WYLEN, G. J., and SONNTAG, R. E., *Fundamentals of Classical Thermodynamics*, 2nd ed., John Wiley & Sons, Inc., New York, 1976.
WHITE, F. M. *Fluid Mechanics*, 3rd ed., McGraw-Hill Book Company, New York, 1994.
YIH, C. S., *Fluid Mechanics*, West River Press, 1988.
YUAN, S.W, *Foundations of Fluid Mechanics*, Prentice-Hall, Inc., Englewood Cliffs, NJ, 1967.

Selected Answers

CHAPTER 1

1.5 **a)** $x^2 - 2xy + y^2 = 4y$. A parabola

1.7 Eulerian: Several college students would be positioned at each intersection and quantities would be recorded as a function of time.

1.9 **a)** $\hat{\mathbf{n}} = (2\hat{\mathbf{i}} - 3\hat{\mathbf{j}})/\sqrt{3}$ **c)** $\hat{\mathbf{n}} = (8\hat{\mathbf{i}} + 5\hat{\mathbf{j}})/\sqrt{89}$

1.11 **b)** $8\hat{\mathbf{i}} - 4\hat{\mathbf{j}}$ **d)** $2\hat{\mathbf{i}} - 114\hat{\mathbf{j}} + 15\hat{\mathbf{k}}$

1.13 **a)** $-40\hat{\mathbf{i}}$ **c)** $12\hat{\mathbf{i}} - 4\hat{\mathbf{k}}$

1.14 **a)** rate-of strain = $\begin{matrix} 0 & 20 & 0 \\ 20 & 0 & 0 \\ 0 & 0 & 0 \end{matrix}$ rad/s

1.15 **a)** (9.375, 0, 0) m/s^2

1.17 $(\partial u/\partial t)\mathbf{i}$

1.19 **b)** 1.875 m/s at $t = \infty$

1.21 -9.11×10^{-4} kg/m^3·s

1.23 0.04 kg/m^3·s

1.25 $\mathbf{a} = \partial\mathbf{V}/\partial t + (\mathbf{V}\cdot\nabla)\mathbf{V}$

1.27 $-52 \times 10^{-5}\,\hat{\mathbf{j}} + 0.0224\,\hat{\mathbf{i}}$ m/s^2

1.29 Steady: a, c, e, f, h. Unsteady: b, d, g

1.32 **a)** inviscid. **c)** inviscid.
e) viscous inside the boundary layers and separated regions.
g) viscous

1.37 11 400. ∴Turbulent

1.39 **a)** Re = 795. Laminar

1.41 x_T = 8.17 m. Laminar

1.43 $\partial(\rho)/\partial t + u\partial(\rho)/\partial x + v\partial(\rho)/\partial y = 0$

1.45 **b)** 351 fps

1.47 57 m/s

1.49 **b)** $r = r_c : p_T = \rho U_\infty^2/2$
d) $\theta = 90°$: $p_{90} = -3\rho U_\infty^2/2$

1.51 **a)** $-50\rho(2/x + 1/x^2)$ **c)** $-450\rho(2/x + 1/x^2)$

1.53 9.39 m/s

1.55 **a)** 3.14 m/s **c)** 11.62 fps

1.57 12.76 m/s

1.59 **b)** 3.516 m/s, 193.6 kPa

1.61 19.82 cm

1.63 **a)** –6195 Pa **c)** –5470 Pa

1.65 **a)** 37.42 m/s, 71.4 m **c)** 121.8 fps, 231 ft

1.67 **a)** 37.4 m/s **c)** 118.6 fps

1.69 **b)** –43.3 kPa **d)** –59 Pa

1.71 Negative pressure, so it will rise

1.73 High pressure

1.75 The higher pressure at B will force the fluid toward the lower pressure at A, thereby causing a secondary flow normal to the pipe's axis. This results in a relatively high loss for an elbow.

M. C. Potter and D. C. Wiggert, *Applied Fluid Mechanics*,
https://doi.org/10.1007/978-3-031-65276-9

CHAPTER 2

2.5	**b)** $Q - W = 0$
2.7	**b)** Conservation of mass
	d) Energy equation
2.9	$-0.707\,(\hat{\mathbf{i}} + \hat{\mathbf{j}})$, $0.866\hat{\mathbf{i}} - 0.5\hat{\mathbf{j}}$, $-\hat{\mathbf{j}}$, -7.07 fps, 8.66 fps, 0
2.11	1559 cm^3
2.17	$dm_V/dt = \dot{m} - pQ$
2.19	15 fps, 3.97 slug/s, 2.045 ft^3/s
2.21	0.1509 m^3/s
2.23	320 m/s, 20.9 kg/s, 4.712 m^3/s, 2.512 m^3/s
2.25	**b)** 4.478 m
2.27	**a)** 5 m/s, 320 kg/s, 0.32 m^3/s
	c) 7.5 m/s, 480 kg/s, 0.48 m^3/s
2.29	5.08 kg/s
2.31	0.82 kg/s
2.33	4.528 slug/s, 4.6 slug/s. $\therefore \rho\overline{V} \neq \overline{\rho}\overline{V}$
2.35	0.565 m
2.37	12.04 m/s
2.39	0.244 m/s
2.41	27.3 m/s
2.43	$\rho\pi[(d^2 - d_2^2)\dot{h}_2/4 + h_1^2\dot{h}_1 \tan^2\phi]$
2.45	3.99×10^{-4} kg/s
2.47	**b)** 8.84 mm/s
2.49	–14.5 kPa/s
2.51	1847 W
2.53	0.836°C
2.55	**b)** 524 kW
2.57	7.93 ft or 1.76 ft
2.59	9.94 m/s
2.61	60.44 psi
2.63	32.1 MPa
2.65	**a)** 5.01 m^3/s
2.67	**b)** 0.485 m^3/s
2.69	$12.35 d_1^2 d_2^2 (H/(d_1^4 - d_2^4))^{1/2}$
2.71	**a)** 0.663 m
2.73	0.4 m
2.75	**a)** 0.131 m
2.77	23.5 hp
2.79	1304 kW
2.81	**a)** 299°C
2.83	5390 kW
2.85	0.5 W
2.87	3820 W, 39.4 kPa, 416 kPa
2.89	1.85 m or 2.22 m
2.91	0.815 m
2.93	**a)** 2.00
2.95	$\dot{m}_f q_f = F_D V_\infty + \dot{m}\,[(V_2^2 - V_1^2)/2 + c_v(T_2 - T_1)]$
2.97	7.37×10^{-5} m^3/s
2.99	**a)** ≅ 0.32 m^3/s
2.101	**b)** 679 N **d)** 127 lb
2.103	**a)** 692 N **c)** 2275 N
2.105	**b)** 2280 N, 1071 N
2.107	49.6 N
2.109	618 kN
2.111	**a)** 3.2 m, 8.86 m/s
2.113	3.8 ft, 15.8 fps
2.115	4420 N
2.117	**a)** 11.46 kN **b)** 6.25 kN
2.119	**b)** 202 kg/s
2.121	147.3 kW
2.123	986 kW
2.125	**b)** 47°, 29.5°, 3.6 MW
2.127	**a)** 11.08 kN, 240 kg/s, 80 kg/s
2.129	13.33 m/s
2.131	647 hp
2.133	110 kN
2.135	1440 N, 69.9 m^3/s, 62.5%
2.137	291 hp, 186.2 slug/s
2.139	525 m^3/day
2.141	**a)** 20.2 kWh **b)** 12.0 kWh
	c) 733 kWh **d)** 8.41 kWh
2.143	47.1 N, –4.79 N
2.145	141 N
2.147	191 ft/s/ft
2.149	3780 N
2.151	**a)** 54.9 kW/m
2.155	46.9 rad/s
2.157	39.6 rad/s
2.159	$11.73(1 - e^{-31.8t})$ rad/s

CHAPTER 3

3.2 See Table 5.1

3.4 $\rho du/dx + ud\rho/dx = 0$

3.6 $u\partial\rho/dx + w\partial\rho/\partial z = 0, \partial u/\partial x + \partial w/\partial z = 0$

3.8 **a)** $v_r = C/r$

3.10 **a)** $3y/(0.15x + 0.5)^2$
b) $-9y/(0.15x + 0.5)^4$

3.12 $v = \text{const}$

3.14 $-Ay$

3.16 $(10 - 0.4/r^2) \sin\theta$

3.18 $(10 - 80/r^3) \cos\theta$

3.20 0.541 m/s

3.22 0.296 m/s

3.24 **a)** −0.88 m/s **b)** 2772 m/s²

3.26 $[100\rho/(x^2 + y^2)^2](x\hat{\mathbf{i}} + y\hat{\mathbf{j}})$

3.28 $(\partial p/\partial\theta)/r = -\rho v_r v_\theta/r - \rho v_r \partial v_\theta/\partial r - \rho(v_\theta/r)\partial v_\theta/\partial\theta$

3.30 $\rho(D\mathbf{V}/Dt + 2\boldsymbol{\Omega}\times\mathbf{V} + \boldsymbol{\Omega}\times(\boldsymbol{\Omega}\times\mathbf{r}) +$
$(d\boldsymbol{\Omega}/dt)\times\mathbf{r}) = -\nabla p - \rho\mathbf{g}$

3.32 $252y^2x^{-9/5} - 5270y^3x^{-13/5}$, $\tau_{xy} = 5.01 \times 10^{-5}$ Pa

3.34 $-\nabla p - \rho\mathbf{g} + \mu\nabla^2\mathbf{V} + \mu\nabla(\nabla\cdot\mathbf{V})/3$

3.36 $-[(V_1 + V_2)y/h] + V_1$

3.38 $\partial p/\partial z = \mu[\partial^2 v_z/\partial r^2 + \partial v_z/dr)/r]$

3.40 $\partial p/\partial\theta = \mu[\partial/\partial r(r^2\partial v_\theta/\partial r)]/r - \mu v_\theta/r \sin^2\theta$

3.44 $\rho\partial u/\partial t = \mu\partial^2 u/\partial y^2$

3.48 $0.3 + 10y$

3.50 $\rho D\tilde{u}/Dt = K\nabla^2 T$

3.52 $\rho Dh/Dt = K\nabla^2 T + Dp/Dt$

3.56 **a)** $\partial u/\partial t = v\partial^2 u/\partial y^2$,
$\rho c\partial T/\partial t = K\partial^2 T/\partial y^2 + \mu(\partial u/\partial y)^2$

CHAPTER 4

4.2 **a)** FT/L **c)** FT^2/L^4 **e)** FL

4.4 $V\ell\rho/\mu = \text{const}$

4.6 $V = \sqrt{gH/C}$

4.8 $F_D/\rho\ell^2V^2 = f_1(d/\ell, \mu/\rho\ell V)$

4.10 $h/d = f_1(\sigma/\gamma d^2, \beta)$

4.12 $\sigma = CMy/I$

4.14 $V = \text{const}\sqrt{gH}$

4.16 $\Delta p/\rho V^2 = f_1(v/Vd, L/d, e/d)$

4.18 $Q/\sqrt{gR^5} = f_1(A/R^2, e/R, s)$

4.20 $F_D/\rho V^2D^2 = f_1(\mu/\rho Vd, e/d, I)$

4.22 $T/\rho V^2D^2 = f[\text{Re}]$

4.24 $F_D/\rho V^2d^2 = f_1(\mu/\rho Vd, e/d, r/d, Cd^2)$

4.26 $F_L/\rho V^2\ell_c^2 = f_1(c/V, t/\ell_c, \alpha)$

4.28 $F_D/\rho V^2d^2 = f_1(Vd\rho/\mu, d/L, \rho/\rho_c, V/\omega d)$

4.30 $T/\rho\omega^2d^5 = g_1(f/\omega, H/d, \ell/d, h/d)$

4.32 $d/D = f_1(V/V_j, \sigma/\rho V_j^2D, \mu/\rho V_jD, \rho_u/\rho)$

4.34 $\mu/\rho VD = f_1(H/D, \ell/D, gD/V^2)$

4.36 $y_2/y_1 = f(gy_1/V_1^2)$

4.38 $Q_m/Q_p = V_m\ell_m^2/V_p\ell_p^2$,
$(F_p)_m/(F_p)_p = \rho_mV_m^2\ell_m^2/\rho_pV_p^2\ell_p^2$,
$T_m/T_p = \rho_mV_m^2\ell_m^3/\rho_pV_p^2\ell_p^3$

4.40 **a)** 160 kg/s, 15 MPa

4.42 Full-scale is recommended.

4.44 1184 N

4.46	0.0048 ft, 1.11		
4.48	$v_m = 6.1 \times 10^{-9}$ m^2/s. Impossible!		
4.50	**a)** 0.00632 m^3/s	**b)** 12 kN	
4.52	120 kN·m		
4.54	1633 kW		
4.56	4.16 N, 95 hp		
4.58	**b)** 186 m/s, 2080 N		
4.60	**a)** 6320 rpm	**b)** 2000 rpm	
4.62	5 motions/s		
4.64	$f\ell/V$		
4.66	$g\ell/U^2$		
4.68	**b)** $\frac{\partial u^*}{\partial t^*} = -\mathrm{Re}\frac{\partial p^*}{\partial x^*} + \frac{\partial^2 u^*}{\partial r^{*2}} + \frac{1}{r^*}\frac{\partial u^*}{\partial r^*}$		
4.70	$(K/\mu c_p)(\mu/\rho U\ell) = \mathrm{Pr}^{-1}\,\mathrm{Re}^{-1}$		

CHAPTER 5

5.2	**b)** 0.0015 m/s		
5.4	Re = 700 000. Turbulent		
5.6	**a)** 12.6 m	**c)** 25.0 m	
5.8	$L_E = 7.2$ m. Developed		
5.10	3.7 m vs 0.72 m		
5.12	**a)** 20.8 m, 5.2 m		
5.18	0.123°, 7.79×10^{-6} m^3/s		
5.20	**b)** 1.04×10^{-4} m^3/s. Laminar		
5.22	1.15×10^{-5} m^3/s, 1.2 m		
5.24	0.396 psf		
5.26	12.1 m/s, 3.5 Pa, 260 m		
5.28	3.85×10^{-6} m^3/s		
5.30	**a)** $0.707r_0$	**b)** $0.5r_0$	
5.32	0.0188 ft, 0.00163 psf, 217 ft		
5.34	0.462 m/s, 0.0054 Pa		
5.38	**a)** 0.0274 N·s/m^2		
5.40	12.1 m^3/s, 241 000, 60.4 m/s, 20.1 Pa		
5.42	0.028 m^3/s, 2030		
5.44	0.0556 cfs		
5.46	**a)** 18.5 m/s	**c)** –6.15 m/s	
5.48	32 kPa/m		
5.50	151 N·m		
5.52	7.9 N·m		
5.54	18.9 W		
5.56	0.040 N·m, 12.6 W, 1650		
5.58	$4\pi\mu r_1^2 r_2^2 L\omega_2/(r_2^2 - r_1^2)$		
5.64	$\overline{u'^2} = 51.2$ m^2/s^2, $\overline{u'v'} = 9.7$ m^2/s^2		
5.66	0.146 ft^2/s, –0.118, 0.342 in.		
5.68	**b)** Rough		
5.70	**b)** 0.262 m/s		
5.72	8.5, 21.7 fps		
5.74	–40.8 kPa/m		
5.76	24.2 m/s		
5.78	**a)** 1.12 m/s, 0.0127 m^3/s		
5.80	**a)** 0.0143	**b)** 0.0146	
5.82	**a)** 0.064	**c)** 0.034	
5.84	**a)** 62 ft	**c)** 42 ft	
5.86	**a)** 136 kPa	**c)** 20 kPa	
5.88	11,200 psf		
5.90	147 kPa		
5.92	**b)** 0.0033 m^3/s		
5.94	**b)** 0.069 kg/s		
5.96	**a)** 0.032 m	**c)** 0.031 m	
5.98	**a)** 0.121 m	**b)** 0.127 m	
5.100	0.000143 m^3/s		
5.102	**b)** 0.257 m^3/s		
5.106	**a)** 52.6 kPa		
5.108	11.7		
5.110	0.0044 m^3/s		
5.112	**b)** 0.011 m^3/s		
5.114	1.3 min		
5.116	46.1 m/s		
5.118	Oscillates between laminar and turbulent flow		
5.120	117 kW, 13.0 m		
5.122	187 kW, 13 m		
5.124	0.23 m^3/s, 190 kW		

5.126 a) 195 kW b) –81 kPa c) 625 kPa

5.128 $0.016\ m^3/s$, 593 W

5.130 0.170 psf

5.132 a) $1.64\ m^3/s$

5.134 a) $0.794\ m^3/s$ b) $0.86\ m^3/s$

5.136 3.91 m

5.138 2.04 ft

CHAPTER 6

6.2 3.78×10^{-5} m

6.6 b) 7.9×10^6. Separated

6.8 2659 N, 469 N, 2.63, 0.417

6.10 b) 4.58×10^{-5} N

6.12 $C_{D,\text{high}}/C_{D,\text{low}} = 2.5$

6.14 0.5, 1.46×10^5

6.16 a) 1380 N, 25.7 kN·m

6.18 36 kN

6.20 a) 27 fps

6.22 a) 10 hp

6.24 a) 93.6 N

6.26 $8800

6.28 1.24 hp

6.30 a) 1.58 m/s c) 2.86 m/s

6.32 $683

6.34 125 km/hr

6.36 8.13×10^{-5} ft, 0.24 in.

6.38 0.0191 m/s

6.40 3.5 hp, 0.12 hp

6.42 900 N, 60 N

6.44 3.3 N, 0.29 N

6.46 4800 N, no cavitation

6.48 52 kN

6.50 b) $1.24\ m/s^2$

6.52 0.587

6.54 13.8 hp

6.56 38 m/s

6.58 39.9 m/s

6.60 a) 7.77% increase

6.66 a) $5x^2 + 10y^2$ c) $\sqrt{x^2 + y^2}$

6.68 $100y - 50x$, $100x + 50y$

6.70 $-40 \tan^{-1} y/x$

6.72 b) $10y + 10 \tan^{-1} y/x$
c) $100 - 50(2/x + 1/x^2)$kPa
e) $-12.5\ m/s^2$

6.74 a) (–1″, 0) b) 0.119′
c) 1.257″ d) 27.5 fps

6.76 c) $(\sqrt{2},0), (\sqrt{2},0), 0, 0$

6.78 a) 0.314 m

6.80 b) $(0, -104)\ m/s^2$ c) 13.76 kPa

6.82 b) $-8 \sin\theta$ c) $58 - 32 \sin^2\theta$ kPa
d) 42.7 kN

6.84 100 rad/s, –3660 Pa

6.86 472, 000 lb

6.88 (0.729, 0.481) m/s

6.90 b) 5.04″

6.92 a) 3 cm c) 3 cm
e) 6 mm

6.94 $20 - 200 \sin^2(x/2)$ kPa, $20 \sin(x/2)$ m/s

6.96 $16/(2.67 - x)$, $256/(2.67 - x)^3$

6.100 a) $4.79\sqrt{\nu x/U_\infty}$ b) $0.328\mu U_\infty \sqrt{U_\infty/\nu x}$
c) $\int_0^\delta y \cos[(0.189\sqrt{U_\infty/\nu})y]\, dy$

6.102 $6.65\sqrt{\nu x/U_\infty}$, $0.451\rho U_\infty^2 \mathrm{Re}_x^{-1/2}$, 33%, 36%

6.104 21 cm, 0.001 m, 0.001 m

6.106 a) $1.74\sqrt{\nu x/U_\infty}$, $0.648\sqrt{\nu x/U_\infty}$, 1.2%, 0.62%

6.108 a) 0.0221 m b) 0.00498 Pa
c) 0.299 N d) 0.00391 m/s

6.110 a) 0.0949 m, 0.6 Pa

6.112 a) 0.31 lb

6.114 a) 235 m, 0.0618 Pa b) 0.151 Pa, 585 m

6.116 **a)** 0.00212 **b)** 0.229 psf **c)** 8.09×10^{-5} m **d)** 0.228 ft

6.118 163 kN, 0.89 m

6.124 **a)** 0.0124 Pa **b)** 0.0122 m **c)** 0.00527 m/s **d)** 0.04 m^3/s/m

6.126 **a)** 32.1 Pa **b)** 0.0248 m **c)** 0.109 m^3/s/m

6.128 3.85 mm, 0.00127 m/s, 0.011 Pa

6.132 negative, zero, positive, positive, negative

CHAPTER 7

7.10 0.0069 s

7.12 393 m

7.14 3.776 s

7.16 –0.113 fps, 0.021°F

7.18 **a)** 77.3 m/s

7.20 **a)** 0.1473 kg/s **b)** 0.1393 kg/s

7.22 **a)** 0.1472 kg/s **b)** 0.1423 kg/s

7.24 211.3 kPa abs, 7.29 kg/s

7.26 27.83 psia, 0.101 slug/s, 2.202 slug/s

7.28 97.45 kPa abs, 199.4 kPa abs

7.30 168 m/s, 476 m/s

7.32 0.1025 slug/s

7.34 494.2 kPa abs, 4.29 kPa abs

7.36 0.319 ft, 684 fps, 0.417 ft

7.38 129 kPa abs, 1004 m/s

7.40 8.16 cm

7.42 22.1 cm^2, 772°C, 3670 kPa abs

7.44 1260 m/s

7.46 412 kN

7.48 **a)** 3.774 kg/m^3, 808 kPa abs, 2.966, 473°C, 0.477

7.50 $(k + 1)/(k - 1)$

7.52 908 m/s, 1600 kPa abs, 8.33 kg/m^3

7.54 391 kPa abs, 448°C

7.56 102.5 kPa abs, 0.471 kg/s, 0.302 kg/s

7.58 49.7 psia, 1020 fps, 1989 fps, 523 fps

7.60 6 cm, 9.2 cm

7.62 2.39″

7.64 **a)** 1.49, 120.4 kPa abs, 620 m/s

7.66 780 m/s

7.68 555 kPa abs, 413 kPa abs

7.70 39.4°, 867 m/s, –156°C

7.72 **a)** 14.24 kPa abs, 26.4 kPa abs

7.74 0.0854, 0.010

CHAPTER 8

8.2 2.86 m

8.6 **a)** 2.15 m, 23.5 m^2, 18.6 m **c)** 4.15 ft, 20.9 ft^2, 11.8 ft

8.8 **a)** 0.0121 m^3/s **b)** 0.244 m

8.12 **a)** –0.17 m **c)** –0.43 m

8.14 **a)** 3.46 m **b)** 11.51 ft

8.16 **a)** 0.3 m

8.18 **a)** 5.75 m **c)** 21.5 ft

8.20 **a)** 1.00 m **c)** 1.81 m

8.22 **a)** 1.75 m **b)** 0.83 m

8.26 ≅ 0.95 m

8.28 **a)** 16 cfs

8.30 **a)** 1.03 m × 28.8 ft

8.32 $M/by_c^2 = y^2/2y_c^2 + y_c/y$

8.34 **a)** 3.87 m, 363 N

8.36 7.03 m^3/s

8.38 **a)** 1.77 m, 0.339 m/s, 3.33 m/s

8.40 **a)** 3.43 m^3/s **b)** 0.932 m **c)** 52 kW

8.42 2.55 m, 3.38 kW

8.44 11.51 m^3/s, 0.346 m

8.46 **a)** $y_0 = 0.094$ m, $y_1 = 0.316$ m, $y_3 = 0.088$ m, $y_{cj} = 0.243$ m

8.48 $y_0 = 2.98$ m, M_3 followed by hydraulic jump to M_2

8.50 M_3 followed by hydraulic jump to M_1

8.52 $y_c = 1.13$ m. $y_{cj} = 1.35$ m. Hydraulic jump occurs upstream, followed by an S_1 curve.

8.54 **a)** $y_c = 1.50$ m. A hydraulic jump between *A* and *C*.

8.56 **a)** $y_c = 0.7$ m. $y_0 = 0.99$ m. M_3 followed by a hydraulic jump to an M_2

8.58 **a)** 20.2 m **b)** A_2 profile

8.60 15.5 m^3/s, S_3

8.62 $Q = 7.72$ m^3/s, $y_c = 0.72$ m, $y_0 = 1.17$ m, $y_{cj} = 0.41$ m. M_1 upstream, M_3 downstream

8.64 S_3 followed by hydraulic jump to S_1

8.66 **a)** 0.87 m **b)** 48 m

8.68 **a)** 721 kW **b)** 59.3 m

8.70 At $y = 2$ m, $x \cong 55$ km

8.72 3.2 m

CHAPTER 9

9.2 **a)** 750 W

9.4 **a)** 1.29 ft

9.6 **a)** 0.265 m^3/s

9.8 **a)** 0.32 m^3/s

9.10 1.78 m^3/s, 0.285 m^3/s, 0.935 m^3/s, 1.07 MW

9.12 0.198 m^3/s, 0.138 m^3/s, 0.06 m^3/s

9.14 16 L/s, 7.1 L/s, 5.1 L/s, 3.8 L/s

9.16 **c)** 28.2 L/s, 10.7 L/s, 8.6 L/s, 9.1 L/s

9.18 **a)** 143 L/s, 290 L/s, 166 L/s

9.20 1.05 m^3/s, 90.8 m, 73.6 m, 810 kW

9.22 **a)** 2.27 m^3/s, 0.811 m^3/s, 0.580 m^3/s, 0.880 m^3/s

9.24 **a)** 2.28 ft **c)** 3600 gal/hr

9.26 **a)** 7400 gal/hr

9.28 2.30, 0.43, 2.73 flow units

9.30 **a)** 96, 92, 88, 82 m **b)** 843, 706, 813, 804 kPa

9.32 21.6 L/s, 6.6 L/s, 28.4 L/s

9.34 2.77 m^3/s

9.36 9.3, 1.6, 7.7, 0.6, 7.1 L/s

9.38 **d)** 2.06 m^3/s

9.44 69 s

9.46 1.57 L/s, 8.03 s

9.48 **a)** 1090 m/s **b)** 590 kPa

CHAPTER 10

10.2 42.1 L/s, 35.6 N, 2980 W, 7.22 m

10.4 54.1 ft, 1.31 hp

10.6 1.23 m

10.8 17.6 m, 172 kW

10.12 $\Omega_P = 1.30$, so mixed-flow pump

10.14 **a)** 2.27 ft, 17.8 ft, 22.7 ft, 113 hp

10.16 $\Omega_P = 0.751$, so radial-flow pump

10.18 Speed and diameter are 1.19 times greater

10.20 Mixed flow, 0.251 m, 282 rad/s

10.22 2.75 m^3/min, 12.6 m, 7.2 kW

10.24 2.44 m^3/ min, 10.0 m, 5.1 kW

10.26 $\Omega_P = 5.19$ so axial flow

10.28 280 m^3/h, 64 kW, 8.3 m

10.30 Three pumps, 153.6 kW

10.32 **a)** 5.67 m **b)** 15.8 m

10.34 **a)** radial flow, 90 mm, 4680 rpm **b)** 3.95 m

10.36 6.1°, 28.3 MN·m, 243 m, 360 MW

10.38 0.221 m, 399 rpm

10.40 189 kW, Francis, 2920 rpm

10.42 4.78 m^3/s, 1.95 m, 244 mm, 3 jets, 0.204

10.44 25.8 ft, Francis or pump/turbine

10.46 **a)** 1.45 m **b)** 13.6 cm

10.48 9.2 MW total, two units

10.50 **a)** 24.6 kW **b)** $\Omega_T = 1.65$, Francis
c) 0.3 m, 680 rpm, 26 kW

Index

A

B

M. C. Potter and D. C. Wiggert, *Applied Fluid Mechanics*,
https://doi.org/10.1007/978-3-031-65276-9

G

J

K

L

O

P

Q

R

S

T

W

Z

SPRINGER NATURE

GPSR Compliance

The European Union's (EU) General Product Safety Regulation (GPSR) is a set of rules that requires consumer products to be safe and our obligations to ensure this.

If you have any concerns about our products, you can contact us on ProductSafety@springernature.com

In case Publisher is established outside the EU, the EU authorized representative is:

Springer Nature Customer Service Center GmbH
Europaplatz 3
69115 Heidelberg, Germany

The manufacturer's authorised representative in the EU is Springer Nature Customer Service Centre GmbH, Europaplatz 3, 69115 Heidelberg, Germany. If you have any concerns regarding our products, please contact ProductSafety@springernature.com

Printed and bound by CPI Group (UK) Ltd, Croydon, CR0 4YY
30/09/2026
02225476-0009